Philosophy of Science

BLACKWELL PHILOSOPHY ANTHOLOGIES

Each volume in this outstanding series provides an authoritative and comprehensive collection of the essential primary readings from philosophy's main fields of study. Designed to complement the *Blackwell Companions to Philosophy* series, each volume represents an unparalleled resource in its own right, and will provide the ideal platform for course use.

1 Cottingham: *Western Philosophy: An Anthology* (second edition)
2 Cahoone: *From Modernism to Postmodernism: An Anthology* (expanded second edition)
3 LaFollette: *Ethics in Practice: An Anthology* (third edition)
4 Goodin and Pettit: *Contemporary Political Philosophy: An Anthology* (second edition)
5 Eze: *African Philosophy: An Anthology*
6 McNeill and Feldman: *Continental Philosophy: An Anthology*
7 Kim and Sosa: *Metaphysics: An Anthology*
8 Lycan and Prinz: *Mind and Cognition: An Anthology* (third edition)
9 Kuhse and Singer: *Bioethics: An Anthology* (second edition)
10 Cummins and Cummins: *Minds, Brains, and Computers – The Foundations of Cognitive Science: An Anthology*
11 Sosa, Kim, Fantl, and McGrath *Epistemology: An Anthology* (second edition)
12 Kearney and Rasmussen: *Continental Aesthetics – Romanticism to Postmodernism: An Anthology*
13 Martinich and Sosa: *Analytic Philosophy: An Anthology*
14 Jacquette: *Philosophy of Logic: An Anthology*
15 Jacquette: *Philosophy of Mathematics: An Anthology*
16 Harris, Pratt, and Waters: *American Philosophies: An Anthology*
17 Emmanuel and Goold: *Modern Philosophy – From Descartes to Nietzsche: An Anthology*
18 Scharff and Dusek: *Philosophy of Technology – The Technological Condition: An Anthology*
19 Light and Rolston: *Environmental Ethics: An Anthology*
20 Taliaferro and Griffiths: *Philosophy of Religion: An Anthology*
21 Lamarque and Olsen: *Aesthetics and the Philosophy of Art – The Analytic Tradition: An Anthology*
22 John and Lopes: *Philosophy of Literature – Contemporary and Classic Readings: An Anthology*
23 Cudd and Andreasen: *Feminist Theory: A Philosophical Anthology*
24 Carroll and Choi: *Philosophy of Film and Motion Pictures: An Anthology*
25 Lange: *Philosophy of Science: An Anthology*
26 Shafer-Landau and Cuneo: *Foundations of Ethics: An Anthology*
27 Curren: *Philosophy of Education: An Anthology*
28 Shafer-Landau: *Ethical Theory: An Anthology*
29 Cahn and Meskin: *Aesthetics: A Comprehensive Anthology*
30 McGrew, Alspector-Kelly and Allhoff: *Philosophy of Science: An Historical Anthology*

Forthcoming

31 May: *Philosophy of Law: Classic and Contemporary Readings*
32 Rosenberg and Arp: *Philosophy of Biology: An Anthology*

PHILOSOPHY OF SCIENCE

AN HISTORICAL ANTHOLOGY

Edited by

Timothy McGrew, Marc Alspector-Kelly

and

Fritz Allhoff

A John Wiley & Sons, Ltd., Publication

This edition first published 2009

Blackwell Publishing was acquired by John Wiley & Sons in February 2007. Blackwell's publishing program has been merged with Wiley's global Scientific, Technical, and Medical business to form Wiley-Blackwell.

Registered Office
John Wiley & Sons Ltd, The Atrium, Southern Gate, Chichester, West Sussex, PO19 8SQ, United Kingdom

Editorial Offices
350 Main Street, Malden, MA 02148-5020, USA
9600 Garsington Road, Oxford, OX4 2DQ, UK
The Atrium, Southern Gate, Chichester, West Sussex, PO19 8SQ, UK

For details of our global editorial offices, for customer services, and for information about how to apply for permission to reuse the copyright material in this book please see our website at www.wiley.com/wiley-blackwell.

Library of Congress Cataloging-in-Publication Data
Philosophy of science: an historical anthology / edited by Timothy McGrew, Marc Alspector-Kelly, and Fritz Allhoff.
p. cm. – (Blackwell philosophy anthologies)
Includes bibliographical references and index.
ISBN 978-1-4051-7543-2 (hardcover : alk. paper) – ISBN 978-1-4051-7542-5 (pbk. : alk. paper)
1. Science–Philosophy–History. I. McGrew, Timothy J., 1965– II. Alspector-Kelly, Marc. III. Allhoff, Fritz.
Q175.P51235 2009
501–dc22

2008036230

A catalogue record for this book is available from the British Library.

Set in 9.5/11.5 Minion by Graphicraft Limited, Hong Kong
Printed in Singapore by Ho Printing Pte Ltd

04 2013

εἰ δή τις ἐξ ἀρχῆς τὰ πράγματα φυόμενα βλέψειεν,
ὥσπερ ἐν τοῖς ἄλλοις, καὶ ἐν τούτοις κάλλιστ' ἄν οὕτω θεωπήσειεν.

On these subjects as with others, he would study things best who observed them growing from the beginning.
Aristotle, *Politics* I.1, 1252a24–26

To Jerry Gough, whose teaching first opened my eyes to the history of science, and to Greg Fisk, whose astonishing generosity opened the door for me
T.M.

To my father, Fredrick Miles Kelly, a physicist who brought science alive by smashing frozen balloons with mercury hammers
M.A.

To my parents, my first and most inspirational teachers
F.A.

Contents

List of Figures and Table

Figures

Table

Notes on Editors

Timothy McGrew is Professor and Chairman of the Department of Philosophy at Western Michigan University. His research interests include the history and philosophy of science, probability, and epistemology. He is the author of *The Foundations of Knowledge* (Littlefield Adams, 1995) and co-author with Lydia McGrew of *Internalism and Epistemology* (Routledge, 2007). He has published articles in *Mind, The Monist, Analysis, British Journal for the Philosophy of Science, Erkenntnis* and many other journals and contributed the article on "Physics in Islam" in Helaine Selin (ed.), *The Encyclopaedia of the History of Science, Technology and Medicine in Non-Western Cultures* (Kluwer, 1997).

Marc Alspector-Kelly is an Associate Professor of Philosophy at Western Michigan University. His research interests include the history of analytic philosophy, the realism debate in the philosophy of science, and epistemology. He has published articles in *Philosophy of Science, Synthese, Philosophical Studies, Philosophy and Phenomenological Research, Canadian Journal of Philosophy, Pacific Philosophical Quarterly, Philosophical Psychology,* and *Erkenntnis.*

Fritz Allhoff is an Assistant Professor of Philosophy, an Adjunct Assistant Professor in the Mallinson Institute for Science Education, and the Director of the History and Philosophy of Science Workgroup at Western Michigan University. He is also an Associate in the Center for Philosophy of Science at the University of Pittsburgh, where he has held a Visiting Fellowship. Fritz's research interests are in philosophy of biology, ethics of science and technology, and various other facets of ethics; he has published over 20 essays and edited over 10 volumes.

Personal Acknowledgments

In putting together this volume, we have received help from many people to whom we owe a large debt of gratitude. Much of the first part of the book was developed during Tim McGrew's graduate seminars in the History and Philosophy of Science, and he would like to thank his students for their interest in the material and their willingness to serve as guinea pigs through various refinements of the course. Much of the second part of the book has been covered, in various capacities, in Marc Alspector-Kelly's under-graduate courses and graduate seminars, and he would like to extend similar appreciation to his students. Marc would also like to thank his wife, Tammy Alspector-Kelly, for extensive comments and editing on the second part.

We also had substantial feedback from some experts on particular parts of the book. Jim Lennox was essential to our attempts to tame Aristotle; he gave both poignant and detailed feedback on our selections and helped us to convey Aristotle's philosophy of science in a reasonable number of words (from which there were too many to choose!). Jim also offered the translation of the book's epigraph and made terrific suggestions regarding the Herschel and Darwin selections; Jeff Schwarz offered important feedback on the Darwin selections as well. Gyula Klima provided invaluable feedback on the Ancient and Medieval periods. Johannes Person helped with the selections on Hume. Finally, Zvi Biener contributed to the bibliographies at the end of each of the unit commentaries.

A very special mention should go to Trin Turner, our graduate assistant, without whom we would have never made it this far. Trin, while assigned as a teaching assistant to Fritz Allhoff's philosophy of science course, was "exapted" into a serious production role which included retrieving many of the primary sources (both from our libraries and from others around the country), assembling the manuscript, and counting (and then re-counting) the words for each selection. This was a tremendous amount of work – mostly supererogatory – and we are grateful to him for his efforts.

Much of the editing for this book took place while Fritz was a Visiting Fellow in the Center for Philosophy of Science at the University of Pittsburgh. He would like to thank all of his friends at Pitt HPS – especially John Norton, the Center's Director – for providing such a fantastic place to think about the history and philosophy of science.

Wiley-Blackwell has, from the start, demonstrated exceptional enthusiasm for this project. In particular, we acknowledge Jeff Dean, Danielle Descoteaux, and Jamie Harlan. Jeff has been especially helpful, and we thank him for his constant feedback and entertainment of our seemingly never-ending emails. We also thank the production team at Wiley-Blackwell – especially Barbara Duke, Gail Ferguson, and Louise Spencely – for their formidable undertaking and fantastic work.

Finally, we thank each other: we are lucky enough to be good friends and all the more so for being able to work together on a project like this.

F.A., M.A., T.M.
Kalamazoo, MI
January 2009

Text Acknowledgments

Units 1.1 and 1.2 from *The Book of the Cosmos*, ed. Dennis Richard Danielson (Cambridge, MA: Perseus, 2000), pp. 25, 25–7. © 2000 by Dennis Richard Danielson. Reprinted by permission of Basic Books, a member of Perseus Books Group.

Units 1.3, 1.6, 1.7, 1.8 from *The Complete Works of Aristotle*, ed. Jonathan Barnes, Vol. 1 (Princeton, NJ: Princeton University Press, 1984), extracts from pp. 114–17, 120–2, 329–42, 393–4, 404–5, 439, 473–8, 474–6, 477–8, 480, 482–4, 486–9 ("Physics V & VIII"). © 1984 by The Jowett Copyright Trustees. Reprinted with permission from Princeton University Press.

Unit 1.4 from *Timaeus*, trans. by Donald Zeyl (Hackett, 2000), pp. 14–15, 17–18, 23–5, 26–7, 39–40, 42–3. © 1997 by Hackett Publishing Company, Inc. Reprinted with permission from Hackett Publishing Company, Inc. All rights reserved.

Unit 1.5 from *The Complete Works of Aristotle*, ed. Jonathan Barnes, vol. 2 (Princeton, NJ: Princeton University Press, 1984), pp. 1695–8. © 1984 by The Jowett Copyright Trustees. Reprinted with permission from Princeton University Press.

Unit 1.9 from *Aristotle On the Parts of Animals I–IV*, trans. James G. Lennox (Oxford: Clarendon Press, 2001), pp. 1–8. © 2001 by James G. Lennox. Reprinted by permission of Oxford University Press.

Unit 1.10 from Phillip Howard DeLacy and Estell Allen DeLacy, "Philodemus: On Methods of Inference. A Study in Ancient Empiricism," *The American Journal of Philology* 68: 3 (1947): 321–2 (extracts). © 1947 by The Johns Hopkins University Press. Reprinted with permission from The Johns Hopkins University Press.

Unit 1.11 from *On the Nature of the Universe*, trans. R. E. Latham, revised with an introduction by John Goodwin (Penguin, 1951, 1994), from book 1, pp. 11, 13–27. Translation copyright © R. E. Latham, 1951. Revisions, introduction and notes copyright © John Goodwin, 1994. Reproduced by permission of Penguin Books Ltd.

Unit 1.12 from *The Book of the Cosmos*, ed. Dennis Richard Danielson (Cambridge, MA: Perseus, 2000), extracts from pp. 69–74. Adapted from Claudius Ptolemy, *Almagest*, trans. G. J. Toomer (New York: Springer-Verlag, 1984). © 1998 by Princeton University Press, 1998 paperback edition. Reprinted by permission of Princeton University Press and Gerald Duckworth & Co. Ltd.

Unit 1.13 from Proclus, "Hypotyposis astronomicarum positionum," trans. A. Wasserstein, in *Physical Thought from the Presocratics to the Quantum Physicists*, ed. Shmuel Sambursky (London: Hutchinson, 1974). © 1974. Reprinted with permission from The Random House Group Ltd. and Basic Books, a member of Perseus Books Group.

Units 1.14 and 1.15 from Morris R. Cohen and I. E. Drabkin (eds.), *A Source Book in Greek Science* (Cambridge, MA: Harvard University Press, 1966), pp. 221–3, 217–21. © 1948, 1976 by the President and Fellows of Harvard College. Reprinted with permission from Harvard University Press.

Unit 1.16 from *The Guide of the Perplexed*, Vol. 1, trans. Shlomo Pines (Chicago: University of Chicago Press, 1963). Reprinted with permission from the University of Chicago Press.

Unit 1.17 from Marshall Clagett, *The Science of Mechanics in the Middle Ages* (Madison, WI: University of Wisconsin Press, 1959), pp. 532–8. © 1959. Reprinted with permission from The University of Wisconsin Press.

Unit 1.18 from Oresme, *Le Livre du Ciel et du Monde*, ed. Albert D. Menut and Alexander J. Denomy, trans. Albert D. Menut (Madison, WI: University of Wisconsin Press, 1968), selections from pp. 519–39. © 1968. Reprinted with permission from The University of Wisconsin Press.

Unit 2.1 from Edward Rosen (trans. and ed.), *Three Copernican Treatises*, 2nd edn. (New York: Dover, 1959), pp. 162, 164–8 (selections). © 1959. Reprinted with permission from Dover Publications.

Unit 2.2 from Osiander's preface to *On the Revolutions of the Heavenly Spheres*, 1939, Encyclopaedia Britannica, Inc., pp. 505–6. © 1952, 1990, Encyclopaedia Britannica, Inc. Reprinted with permission from Great Books of the Western World.

Unit 2.3 from Dennis Danielson (ed.), *The Book of the Cosmos: Imagining the Universe from Heraclitus to Hawking* (Cambridge, MA: Perseus, 2000), selections from pp. 104–17. © 2000 by Dennis Richard Danielson. Reprinted by permission of Basic Books, a member of Perseus Books Group.

Unit 2.4 from Brahe, *De Nova Stella*, "Opera Omnia, Tomus I,", ed. J. L. E. Dreyer, 1913, translated by John H. Walden, 1928. From Harlow Shapley and Helen E. Howarth (eds.), *A Source Book on Astronomy* (Cambridge, MA: Harvard University Press, 1969), selections from pp. 233–9. © 1969. Reprinted with permission from Harvard University Press.

Unit 2.5 from Gerald Holton, "Johannes Kepler's Universe: Its Physics and Metaphysics," *American Journal of Physics* Vol. 24, No. 5 (1956), selections from pp. 340–51. © 1956 by American Association of Physics Teachers. Reprinted by permission of the American Institute of Physics.

Unit 2.6 from William H. Donahue, *Selections from Kepler's* Astronomia Nova (Santa Fe, NM: Green Lion Press, 2004), pp. 3–28, selections. © 2004. Reprinted by permission of William H. Donahue and Green Lion Press.

Unit 2.7 from William H. Donahue and Owen Gingerich, trans. and ed., *Johannes Kepler's New Astronomy* (Cambridge University Press, 1993), ch. 19, selections. © 1993. Reprinted by permission of William H. Donahue and Green Lion Press.

Unit 2.8 from "The Assayer," in *Discoveries and Opinions of Galileo*, trans. with an introduction and notes by Stillman Drake (New York: Anchor Books, Doubleday, 1957), selections from pp. 237–41, 258, 271–2. © 1957 by Stillman Drake. Used by permission of Doubleday, a division of Random House, Inc.

Units 2.9, 2.10 and 2.11 from Galileo Galilei, *Dialogue Concerning the Two Chief World Systems: Ptolemaic and Copernican.* Trans. with notes by Stillman Drake (New York: The Modern Library, 1953), pp. 132–41,161–73, 388–400 © 1953, 1962, 1967 by The Regents of the University of California. Copyright renewed © 1981 by Stillman Drake. Reprinted by permission of University of California Press.

Unit 2.12 from M. A. Stewart (ed.), *Selected Philosophical Papers of Robert Boyle* (Indianapolis, IN: Hackett, 1991), selections from pp. 18–20, 23–30 (all footnotes cut).

Unit 2.13 from Huygens, *Treatise on Light*, trans. Sylvanus P. Thompson (University of Chicago Press, 1912).

Unit 2.14 from Newton, *Optiks*, 4th edn. "From Query 31" (New York: Dover Publications, 1952), pp. 404–5. © 1952. Reprinted with permission from Dover Publications. And from *The Correspondence of Isaac Newton*, ed. H. W.

Turnbull, J. F. Scott, A. Rupert Hall, and Laura Tilling, 7 vols. (Cambridge: Cambridge University Press, 1959–77), vol. 5, pp. 396–7, Vol. 1, pp. 96–7, Vol. 1, p. 169, Vol. 1, pp. 209–10. © 1959–77. Reprinted with permission from The Royal Society.

Units 2.15, 2.16 and 2.17 from Newton, *The* Principia: *Mathematical Principles of Natural Philosophy*, trans. I. Bernard Cohen and Anne Whitman (Berkeley: University of California Press, 1999), selections from pp. 403–15, 794–6, 939–40, 943. © 1999 by the Regents of the University of California. Reprinted with permission from the University of California Press.

Unit 2.18 from *Newton*, Norton Critical Edition, ed. I. Bernard Cohen and Richard Westfall (New York: W. W. Norton & Co., 1995), selections from pp. 259–66. © 1995 by W. W. Norton & Company Inc. Used by permission of W. W. Norton and Company, Inc.

Unit 3.1 from *Francis Bacon: The New Organon*, ed. Lisa Jardine and Michael Silverthorne (West Nyack, NY: Cambridge University Press, 2000), pp. 33, 34, 36–7, 39, 40–2, 45–6, 49–50, 83–4. © 2000 by Cambridge University Press. Reprinted with permission from the editors and publisher.

Units 3.2 and 3.3 from *The Philosophical Writings of Descartes*, Vol. 1. Trans. J. Cottingham, R. Stoothoff and D. Murdoch (Cambridge: Cambridge University Press, 1985), pp. 111, 118–20, 223–5, 230–8, 240–5, 247. © 1985 by Cambridge University Press. Reprinted with permission from the translators and publisher.

Unit 3.4 from Locke, *An Essay Concerning Human Understanding*, Vol. 2 (New York: Dover Publications, 1959), selections from pp. 212–18, 220–3.

Units 3.5 and 3.6 from Leibniz, *Philosophical Papers and Letters*, Vol. II, ed. and trans by L. E. Loemker (Chicago: University of Chicago Press, 1956), pp. 777–83, 785–8, 1104–9, 1145–8.

Units 3.7 and 3.8 from *The Philosophical Works of David Hume*, Vol. 4 (London: Little, Brown, and Co., 1854), pp. 30–46, selections from pp. 71–90.

Unit 3.9 from Kant, *Metaphysical Foundations of Natural Science*, trans. and ed. by Michael Friedman (Cambridge: Cambridge University Press, 2004), pp. 3–14. © 2004 by Cambridge University Press. Reprinted with permission from the translator/editor and publisher.

Unit 4.1 from Lavoisier, *Elements of Chemistry*, trans. Robert Kerr (New York: Dover, 1965), pp. xiii–xxxvii © 1965. Reprinted with permission from Dover Publications.

Unit 4.2 from Laplace, *Philosophical Essay on Probabilities*, trans. Andrew I. Dale (New York: Springer-Verlag, 1995), selections from pp. 2–6. © 1995 by Springer. Reprinted with permission from Springer Science and Business Media.

Unit 4.3 from Herschel, *Preliminary Discourse on the Study of Natural Philosophy* (London: Longman, Kees, Orme, Brown, & Green, 1830), pp. 191–3, 195–9, 201–2, 206–9.

Unit 4.4 from Mill, *A System of Logic, Ratiocinative and Inductive*, Vol. 2 (London: John W. Parker, 1843), selections from Book III, pp. 185–7, 191–4, 197–8, 200–3.

Units 4.5 and 4.6 from Whewell, *The Philosophy of the Inductive Sciences Founded Upon Their History* (London: John W. Parker, 1847), pp. 165–70, 471, 95–8.

Unit 4.7 from Cuvier, *Discourse on the Revolutionary Upheavals on the Surface of the Globe and on the Changes which they have Produced in the Animal Kingdom*, trans. Ian Johnston, pp. 1–9; http://www.mala.bc.ca/~johnstoi/cuvier/cuvier-e.htm

Unit 4.8 from Lyell, *The Principles of Geology* (London: John Murray, 1830), selections from pp. 1–4, 382–5.

Unit 4.9 from Darwin, *The Variation of Animals and Plants under Domestication*, 2nd edn., Vol. 1 (New York: D. Appleton and Co., 1876), selections from pp. 9–14.

Units 4.10 and 4.11 from *Collected Papers of Charles Sanders Peirce*, Vols. 5–6, ed. Charles Hartshorne and Paul Weiss (Cambridge, MA: The Belknap Press of Harvard University Press, 1960), 6.40–2, 5.189, 6.522–8. © 1934 by the President and Fellows of Harvard College. Reprinted by permission of Harvard University Press.

Unit 4.12 from Poincaré, *Science and Hypothesis* (Dover Publications, Inc., 1952), pp. 140–53. © 1952 by Dover Publications, Inc. Reprinted by permission of Dover Publications.

Unit 4.13 from Duhem, *The Aim and Structure of Physical Theory* (New York: Atheneum, 1962) (originally published by Princeton University Press), pp. 182–90, 208–12. © 1954 by Princeton University Press, 1982 renewed PUP. Reprinted with permission from Princeton University Press.

Unit 4.14 from Einstein, "On the Method of Theoretical Physics," *Philosophy of Science* (Chicago: University of Chicago Press), Vol. 1, No. 2 (Apr, 1934), pp. 163–9. © 1934, Philosophy of Science Association. Reprinted with permission from The University of Chicago Press.

Unit 5.1 from Carnap, *Philosophical Foundations of Physics: An Introduction to the Philosophy of Science*, Martin Gardner (ed.), (New York: Basic Books, 1966), pp. 3–8, 16–22, 32–9, 225–39.

Unit 5.2 from Hempel, *Aspects of Scientific Explanation and Other Essays in the Philosophy of Science* (New York: Free Press, 1965), pp. 335–8, 380–4, 394–403. © 1965 by The Free Press. Copyright renewed © 1997 by Carl G. Hempel. Reprinted with the permission of The Free Press, a Division of Simon & Schuster, Inc.

Unit 5.3 from Carnap, "Empiricism, Semantics and Ontology," *Revue Internationale de Philosophie* 4 (1950): 20–40. As reprinted in the *Supplement to Meaning and Necessity: A Study in Semantics and Modal Logic*, enlarged edn. (University of Chicago Press, 1956). © 1950. Reprinted by permission of Revue Internationale de Philosophie.

Unit 5.4 from Reichenbach, *Experience and Prediction* (Chicago: University of Chicago Press, 1961), pp. 339–42, 348–57. © 1961 by Hans Reichenbach. Reprinted with permission from Maria Reichenbach.

Unit 5.5 from Strawson, *Introduction to Logical Theory* (New York: John Wiley & Sons, 1952), pp. 248–52, 256–63. © 1952. Reproduced by permission of Taylor & Francis Books UK.

Unit 6.1 from Hempel, *Aspects of Scientific Explanation and Other Essays in the Philosophy of Science* (New York: Free Press, 1965), pp. 101–19. © 1965. This essay combines, with certain omissions and some changes, the contents of two articles: "Problems and Changes in the Empiricist Criterion of Meaning," *Revue Internationale de Philosophie* No. 11, pp. 41–63 (January, 1950); and "The Concept of Cognitive Significance: A Reconsideration," *Proceedings of the American Academy of Arts and Sciences* 80, No. 1, pp. 61–77 (1951). This material is reprinted with kind permission of Revue Internationale de Philosophie and of the American Academy of Arts and Sciences.

Unit 6.2 from Hempel, "Studies in the Logic of Confirmation (I.)" *Mind*, Vol. LIV, No. 213 (1945): 1–26. © 1945 by the Mind Association. Reprinted with permission from Oxford University Press.

Unit 6.3 from Quine, *From a Logical Point of View*, Harvard University Press, 1961. Original version appeared in *The Philosophical Review*, Vol. 69 (1951): 20–43 (excerpts). © 1951 by Duke University Press. All rights reserved. Used by permission of the publisher.

Unit 6.4 from Goodman's *Fact, Fiction and Forecast*, Fourth Edition (Cambridge, MA: Harvard University Press, 1983), pp. 72–81. © 1979, 1983 by Nelson Goodman. Reprinted with permission from Harvard University Press.

Unit 6.5 from Putnam's *Mathematics, Matter, and Method, Philosophical Papers Vol. 1* (2nd edn.), Cambridge: Cambridge University Press, 1979, pp. 215–20. First published in E. Nagel, P. Suppes, and A. Tarski (eds.), *Logic, Methodology and Philosophy of Science* (Stanford: Stanford University Press, 1962). © 1962 by the Board of Trustees of the Leland Stanford Jr. University. All rights reserved. Used with the permission of Stanford University Press, www.sup.org

Unit 6.6 from Hanson, *Patterns of Discovery: An Inquiry into the Conceptual Foundations of Science* (Cambridge: Cambridge University Press, 1965), pp. 4–25. © 1958 by Cambridge University Press. Reprinted with permission from the author and publisher.

Unit 6.7 from Maxwell, "The Ontological Status of Theoretical Entities," from *Minnesota Studies in the Philosophy of Science, Vol. 3: Scientific Explanation, Space, and Time*, ed. Herbert Feigl and Grover Maxwell (University of Minnesota Press, 1962), pp. 3–27. © 1962 by the University of Minnesota Press. Reprinted with permission of the publisher, the University of Minnesota Press.

Unit 7.1 from a lecture given at Peterhouse, Cambridge, summer 1953. Originally published by the British Council under the title "Philosophy of Science: a Personal Report," in *British Philosophy in Mid-Century*, ed. C. A. Mace (George Allen & Unwin, 1957).

Unit 7.2 Selections from Kuhn, *The Structure of Scientific Revolutions*, 3rd edn. (Chicago: University of Chicago Press, 1996). © 1996 by Thomas S. Kuhn. Reprinted with permission from The University of Chicago Press.

Unit 7.3 from Lakatos, *The Methodology of Scientific Research Programmes: Vol. 1: Philosophical Papers* (Cambridge: Cambridge University Press, 1977), pp. 1–7. © 1978 by Cambridge University Press. Reprinted with permission from the Imre Lakatos Memorial Fund and the publisher.

Unit 8.1 from Salmon, *Four Decades of Scientific Explanation* (Pittsburgh, PA: University of Pittsburgh Press, 2006). "2.3 Famous Counterexamples to the Deductive-Nomological Model" is from *Four Decades of Scientific Explanation*, by Wesley C. Salmon. © 2006. Reprinted with permission of the University of Pittsburgh Press.

Unit 8.2 from Salmon, "Statistical Explanation and Statistical Relevance," from *Nature and Function of Scientific Theories*, Robert G. Colodny (ed.) (Pittsburgh: University of Pittsburgh Press, 1970), pp. 173–231. Reprinted in Wesley C. Salmon, Richard Jeffrey, and James Greeno (eds.), *Statistical Explanation and Statistical Relevance* (Pittsburgh: University of Pittsburgh Press, 1971), extracts from pp. 29–87. © 1970. Reprinted by permission of the University of Pittsburgh Press.

Unit 8.3 from *The Proceedings and Addresses of the American Philosophical Association*, Vol. 51, No. 6 (August, 1978), extracts from pp. 688–96, 697, 699–700 plus notes 702–5. © 1978. Reprinted by permission of the American Philosophical Association.

Units 8.4 and 9.2 from *Philosophy of Science* 48 (1981), Kitcher extracts from pp. 509–10, 512–26, 529–31: Laudan extract from pp. 19–49. © 1981 by the Philosophy of Science Association. Reprinted with permission from The University of Chicago Press.

Unit 9.1 from Boyd, "The Current Status of the Issue of Scientific Realism," in *Erkenntnis*, Vol. 19, Nos. 1–3 (May 1983), selections from pp. 45–90. © 2006 by Springer. Reprinted with permission from the author and Springer Science and Business Media.

Unit 9.3 from van Fraassen, *The Scientific Image* (Oxford: Clarendon Press, 1980), pp. 1–21, 23–5, 38–40, 56–9, 64, 67–9, 80–3. © 1980 by Bas C. van Fraassen. Reprinted by permission of Oxford University Press.

Unit 9.4 from A. Fine, *The Shaky Game: Einstein, Realism, and the Quantum Theory* (Chicago: University of Chicago Press, 1986), pp. 112–35. © 1986 by The University of Chicago. Reprinted by permission of The University of Chicago Press.

Volume Introduction

This anthology provides a broadly chronological survey of the history and philosophy of science from the pre-Socratics through the twentieth century. The very attempt to cover so much ground, even in a volume of nearly half a million words, is of course wildly ambitious; we could easily have doubled or even tripled the length of the volume without running short of interesting and valuable material. Yet the project of putting together an anthology on this scale seemed to us not only worth the pain of compression but also pedagogically unavoidable.

Existing anthologies in the philosophy of science almost invariably present science as seen through the lens of twentieth-century developments. In our teaching of the philosophy of science at both the undergraduate and the graduate level, we found ourselves supplying a great deal of material from the history of science to provide a context for the philosophical discussions in these textbooks. Some gesture toward the history of science has, of course, become *de rigueur* in philosophical circles since Kuhn's critique in *The Structure of Scientific Revolutions* of ahistorical philosophy of science. But serious engagement with the history of science has not, until now, worked its way down to the level of anthologies designed for use in philosophy of science courses.

A serious commitment to incorporating the history of science into a philosophy of science course opens the door to a wider view of both science and philosophy. In earlier ages, the two disciplines were not as sharply distinguished as they are today; it is sometimes difficult to classify passages from Aristotle as clearly one or the other. And even where this is possible, scientific theories have their own history and their own philosophical context. Increasingly, we realized that contemporary philosophy of science could not be properly understood in isolation either from the history of science or from the history of philosophy itself. Debates prominent in late twentieth and early twenty-first century philosophy of science are variations, often enormously sophisticated variations, on themes that have resonated throughout a 2,500 year history. The nature of explanation, the goals of science, knowledge of the invisible, super-empirical virtues, questions of realism and anti-realism, the status of explanatory inference – all of these themes recur not only in the history of philosophy but in the work of scientists themselves, work that in turn provides the motivation for contemporary philosophizing.

The challenge, for us, has been to condense this material into a single volume and to make it accessible for students with little background in either science or philosophy. To this end, we made several conscious decisions. First, the readings in this anthology are frequently highly excerpted, particularly in Part I where the material is historical. This is not a volume for specialists who will want to examine every document in full; such specialists can take care of themselves. This is an anthology for students who need to get to the heart of the texts without slogging through hundreds of pages of less relevant material. We thought and talked quite a bit about our prospective readers, their background, their interests, and their needs. In the end, we decided that excerpting the selections would enable us to pack the greatest amount of essential material into the volume.

Second, approximately a sixth of the volume is devoted to a running commentary that provides a very substantial framework for understanding the context and significance of the readings. The commentary is what makes the book more than just a set of readings, and in writing it we have kept the needs of students very clearly in view. We have tried not only to draw the reader's attention to key issues in the readings but also to connect those readings to their historical context and, where appropriate, to connect them to each other. Anyone would profit from reading Herschel and Darwin, for example, but the selections offered here show the direct influence of Herschel on Darwin's work. Connections like this illustrate the fact that science and philosophy are not merely usefully compared but actually organically connected. The commentary, both in the historical and in the contemporary material, draws many of these connections.

Third, we chose to give very thorough coverage to what one might term the "main trunk" of philosophy of science in the twentieth century rather than trying to give scantier coverage of the various thematic branches that have sprung from it. In part this was a choice dictated by considerations of space; in part it was a decision based on the realization that different instructors would naturally wish to pursue different topics in more detail than we could reasonably hope to cover. But anyone using this book as a text who wants to focus on recent developments in areas like methodology, laws, or realism will find the foundation for such discussions firmly laid.

Because of the way we have constructed this anthology, it can be used to support a number of different sorts of courses. There is enough material here for a two-semester sequence in which Part I lays the groundwork in both history of philosophy and history of science and Part II explores the rise, fall, and aftermath of the "Received View." But it can also support courses built around any of a number of themes. One such course might cover the development of astronomy and dynamics from antiquity through the scientific revolution. Another might focus on the transition from the Greek notion of scientific understanding, with its emphasis on certainty and teleological explanations, to the modern concept based on probability and mechanical explanations. A third might trace the theme of realism in science and philosophy – its birth, growth, triumphs, changes, and contemporary fortunes.

Though the volume as a whole is a collaborative effort, we have each taken responsibility for different parts. Tim McGrew selected the pieces and wrote the commentary and reading introductions for Part I; Marc Alspector-Kelly did the same for Part II. Fritz Allhoff provided extensive feedback both on the selection of the readings and on drafts of every part of the volume and kept it from falling into asymptotic completion.

Part I

Introduction to Part I

1 The Unbroken Thread

The various scientific disciplines as we know them today have distinctive characteristics by which even the uninitiated can recognize them. Physics has intricate equations and imposing particle accelerators, astronomy boasts orbital telescopes and maps of supergalactic clusters, biology has a sweeping evolutionary narrative and describes proteins that fold into exact three-dimensional shapes to catalyze reactions, chemistry proudly displays its periodic table and produces numberless new compounds that find their way into the surgical room and the supermarket. And the disciplines are linked in important ways. Geology intersects with both evolutionary biology and physics. Physicists are as interested in the periodic table as chemists. Biochemistry straddles the disciplinary division between biology and chemistry. Astronomers make use of physics to describe exotic objects like black holes and to understand stellar evolution.

But as we work our way backwards in time, the view becomes less and less familiar. Much of the technology that is the cultural signature of physics and chemistry disappears by the time we cross the 1900 line. When we move back to 1850, biology has lost its grand unifying vision and chemistry has lost its table of the elements. Prior to 1800 chemistry looks even less familiar, with odd substances like phlogiston and caloric populating the textbooks. Biology is largely reduced to anatomy and conjectural taxonomy, and its connections to chemistry and geology are severed.

Even as far back as 1700, at least physics and astronomy look like threadbare versions of their modern selves. But when we retreat backwards across the 1600 line, the fabric of science unravels with shocking abruptness. Physics is shorn of its mathematical tools, including calculus and modern algebra, and the foundations of elementary dynamics are in disarray. Telescopes and microscopes have disappeared. What passes for chemistry is inseparable from alchemy, and astronomers eke out their incomes by casting horoscopes for the nobility. The majority opinion is that the earth does not orbit the sun or rotate on its axis, and fundamental biological facts such as the circulation of the blood are unknown.

We might stop there and declare that, prior to 1600, science simply did not exist. But on a closer look this drastic pronouncement appears too simplistic. Though the technology and even the fundamental principles of all the major sciences are absent, there remain recognizable similarities of aim and purpose. It is not just that we recognize, beneath the strange terminology of alchemy and surrounded by a good deal of nonsense, bits of true empirical lore; the more significant fact is that the alchemists are actively seeking a set of basic principles that govern the transformations of matter. In astronomy, the models of the solar system are complex geometrical nightmares; but the underlying aim of giving a rational account of the motions of the sun, moon, planets, and stars remains. In anatomy the leading thinkers are investigating the structure of the human body with great attention to detail, and in medicine the best minds are hunting for the causes of disease and the mechanisms of contagion. In the aims of these Renaissance disciplines, we can see a thread of continuity stretching from that era to our own.

Moving carefully backwards through the Middle Ages we find that this thread, though perilously thin, is never quite broken. We see it in Paris during the 1300s. It virtually disappears from the west prior to 1100 but remains visible in the work of Arab and Persian and Jewish scholars in Baghdad and Cordoba and Damascus and Fez. In the 500s, it surfaces in Alexandria. Moving back before the fall of Rome we find it in Italy, and other threads appear across the Roman Empire. Before the dawn of the Christian era these threads stretch back into Greece, where for a brief and glorious period they draw together into a vivid conceptual ribbon woven in rich texture with strange but explicit designs. Back through Archimedes and Euclid, Aristotle and Plato, Epicurus and Democritus and Empedocles and Pythagoras and Thales we can trace it, though the ribbon narrows to a mere thread again at the earliest parts of this period.

And there at last, somewhere around 600 BC, the thread disappears altogether.

This is why we must begin our study of the history of science with the Greeks. They were the first thinkers known to us who clearly articulated the idea that the world is a *cosmos*, an ordered whole governed by rationally discoverable principles. The earliest conjectures regarding the nature of those principles look wholly foreign to us, even absurd. Thales guessed that the fundamental stuff out of which all things are made is water, Anaximander suggested that it was something formless and boundless, Anaximenes thought it was air, and Xenophanes took it to be earth. But their varied and incompatible answers are far less interesting than the assumption lying behind the question they all seek to answer: the assumption that there *is* a fundamental kind of stuff and that the bewildering complexity of the visible world is in some way a function of a simpler underlying order.

2 Greek Science and Greek Philosophy

The early history of science is largely a history of philosophical ideas regarding the nature and structure both of the cosmos and of our knowledge of it. Philosophy and science flourished together in the golden age of Greece; indeed, the line between them was not as sharply drawn as it is today, and in many cases it is difficult to say where philosophy ends and science begins. Aristotle's voluminous writings provide a case in point. In founding the discipline of logic, for example, Aristotle does more than to codify some principles of correct reasoning and list some common fallacies; he fashions syllogistic logic with one eye on the metaphysics of essences and the other on the structure of scientific knowledge. Scientific knowledge, according to Aristotle, is *demonstrative*: it is arrived at by rigorous reasoning from premises that are themselves known and not merely conjectured. An infinite regress threatens, each item of knowledge requiring others to be known first. Aristotle sees and resolves this problem by developing a version of foundationalism, the view that all scientific knowledge rests in the end on primary certainties that are known immediately rather than by derivation from anything else.

The analogy here with mathematics is very strong, and this is no accident. Aristotle wrote about one generation before Euclid, and geometry, already a well-developed branch of learning, provided a compelling model of structured and certain knowledge. But in the case of geometry, the foundations are – most of them – intuitively compelling. It is natural to ask whether the analogy can be pressed at this point. Are there also intuitively compelling foundations for biology, astronomy, and the other sciences? If so, what are they? And by what means can we identify them?

Aristotle's answer is suggestive but frustratingly inexplicit: in experience we *recognize*, somehow, the universal implicit in a few particular instances. In one sense this is plausible enough. We see many cats that differ in size and color and shape of head, but they are all similar; later when we see another cat we can recognize that it is the same type of creature. But does this experience give us knowledge of the *essence* of felines – that is, of their true nature? And is that knowledge certain, so that the conclusions we deduce from it will also be certainties? These are high standards. Aristotle is well aware that mere enumeration of instances does not guarantee knowledge of essences; the mental move he calls *epagoge*, in which the mind rises from perception of particulars to knowledge of universals, cannot be flattened out into anything like induction in the modern sense. For the next two millennia, some of the best minds in the world wrestled with the problems posed by Aristotle's account of the structure and foundations of scientific knowledge.

A second strand of Greek thought about scientific knowledge is no less important: science aims to give us not merely *foresight* but *understanding*. Predictive astronomy is older than the attempt to understand *why* things happen in the heavens. Babylonian cuneiform tablets bearing dated records of astronomical events reach as far back as 747 BC, and star catalogues and records of planetary motions go back a millennium before that. From such data it is possible to extrapolate the motions of the sun, moon, and planets against the background of the fixed stars. This is no mean computational feat. Mars, for example, generally progresses through the constellations of the zodiac in a fixed direction; but at intervals it appears to reverse itself, backtracks, then doubles back again and proceeds in its original direction, losing about four months and creating a looped path against the background of stars. Something similar occurs in the motion of all of the planets. Astronomers call this detour "retrograde motion." The Babylonians knew of retrograde motion and could even predict it. But we have no record that they ever speculated as to *why* it occurs.

By contrast, the Greeks were almost obsessed with understanding why things happen. In the *Republic* Plato puts in the mouth of Socrates an allegory in which the cosmos is depicted as a sort of spinning top with eight concentric shells, corresponding to the stars, sun, moon, and the five known planets, rotating independently at different rates of speed. The model is crude, and it makes no provision for retrograde motion, but it is a first step toward trying to make the motions of the heavenly bodies understandable. In the second century AD, Claudius Ptolemy, relying heavily on the work of Hipparchus from three centuries earlier, produced a completely different model in which the planets ride on the circumferences of circles called epicycles, and each epicycle rides on the circumference of a larger circle, with the earth standing motionless near the center of the whole system. The complete construction, which sometimes involves epicycles upon epicycles and epicycles oriented vertically to account for latitudinal motions of the planets, is enormously complex. It yields, however, reasonably accurate predictions for the visible motions of all of the major heavenly bodies.

But was the resulting construct real? Was this the way the heavens were constituted? Once asked, the question could not be retracted. If the goal of science is to give knowledge not just of the fact but of the *reasoned* fact, as Aristotle puts it, then mere guesses at the hidden mechanism of the universe could never suffice. In the long history of commentary both on Aristotle and on Ptolemy there is much more than mere servile endorsement. Aristotle's views were widely contested in both space and time – in Alexandria in the sixth century, in Baghdad and Damascus and Fez in the tenth through the twelfth, in Paris in the thirteenth and fourteenth, in Saragossa in the fifteenth. The criticism was sharper and more widespread in active centers of learning than it was in places where, for whatever reason, intellectual activity had fallen off. But the Victorian notion that Aristotle's teachings went unchallenged during the Middle Ages is simply a myth.

Similarly, one thinker after another struggled to come to terms with Ptolemy's geometrically outlandish system. Is predictive accuracy evidence of truth? If not, how can we ever hope to discover the real causes of phenomena? Should we dismiss epicycles as useful fictions, mere calculating devices not to be taken literally? Where, then, is the reasoned fact? Are we doomed to agnosticism regarding the true explanations of planetary motions? If so, how much more profound must be our ignorance of the more complex phenomena of the natural world around us.

Despite fairly widespread skepticism about its parts, Aristotle's view of nature remained dominant through the end of the sixteenth century. If his views on projectile motion, for example, seemed far-fetched, the problem could be tabled pending further research or patched up by replacing faulty components piecemeal, as happened with the introduction of impetus into the larger Aristotelian framework. A grand synthesis, even one with known flaws, is not easily set aside; it generally requires a rival synthesis to challenge and replace it.

Although telescopes and microscopes come readily to mind when we think of the interplay of science and technology, arguably the single most significant technological contribution to the overthrow of the Aristotelian world picture was the invention of moveable type around 1450. From the early sixteenth century onward, information and arguments were disseminated at a rate unimaginable in medieval times, exponentially increasing the exposure given to new theories and ideas.

3 The Revolution from Copernicus to Newton

Ironically, the first printed book to create a major crack in the Aristotelian edifice was inspired by the desire to be faithful to Aristotle. Nicolas Copernicus, horrified by the liberties Ptolemy had taken with Aristotle's views on motion, inverted the structure of the Ptolemaic cosmos and set the earth in motion while holding the sun still. Such a view accorded better than the Ptolemaic one with Aristotle's prescription of uniform circular motion for heavenly bodies, and it revealed some truly beautiful harmonies among the motions of the earth and the planets. Yet Copernicus's new view faced observational difficulties and caused tensions with Aristotle's physics, tensions that Copernicus himself was unable to resolve. From the publication of Copernicus's work in 1543 it took almost a century and a half and the work of three brilliant and utterly different minds to overcome these obstacles and found a new physics.

The first was Galileo Galilei, the brilliant Italian astronomer, mathematician, and physicist, who used the newly invented telescope to overturn observational objections to the Copernican view and even to turn observations into arguments in its favor. That alone would have guaranteed Galileo immortality in the annals of science; but it was the lesser half of his achievement. Utterly persuaded of the truth of the Copernican view, Galileo set out to overthrow the principles of Aristotelian physics that were at variance with it. The discussion of physics in Galileo's *Dialogue Concerning the Two Chief Systems of the World* provides both a powerful critique of Aristotelian physics and the first recognizably modern, if still not quite correct, account of inertia.

The second was one of Galileo's contemporaries, the intense German mathematician and astronomer Johannes Kepler. Where Galileo's mind cut through the clutter of details to get to the heart of the matter, Kepler's mind reveled in details and found myriad patterns, some of them spurious, hidden in mountains of data. Galileo, moved by the need to present a simple and persuasive account of the Copernican theory, omits mention of Copernicus's numerous epicycles; Kepler sets out to find the true orbit of Mars and ends up dispensing with circular motion altogether in favor of an elliptical orbit. Yet the longing for celestial harmonies binds all three great astronomers together. In Kepler's case it issues in three elegant laws of planetary motion that bear his name.

The third intellectual giant, and the one who actually completed the revolution, was Isaac Newton. With unerring instinct Newton took the best from each of his great predecessors and left the dross. From Galileo he took the idea of terrestrial inertia, but in the corrected form advanced by Descartes. From Kepler he took the three laws of planetary motion, but he left aside Kepler's wilder speculations regarding the Platonic solids and the music of the spheres. Combining these materials with the accurate observations of John Flamsteed, the first Astronomer Royal, Newton created a synthesis of unparalleled scope that accounted, using one unified set of laws, for the fall of an apple, the orbit of the moon, and the ebb and flow of the tides. And Newton had the mathematical power to go beyond plausible speculation. In an intellectual *tour de force* he proved that a centrally directed force would result in Kepler's second law, proved that the mass of a spherical solid would exert a gravitational attraction on other bodies as if all of its mass were concentrated at the central point, and proved that an inverse square law of gravitational attraction would produce an elliptical orbit. It is only a slight overstatement to say that before Newton no one was quite sure whether one set of laws could account for all the phenomena of the heavens and the earth, while after Newton no one doubted it.

4 The Biological Side of the Scientific Revolution

While the revolution in astronomy and dynamics was flowering, a parallel revolution was unfolding in anatomy and biology. In 1543, the same year that Copernicus's *On the Revolutions of the Heavenly Spheres* appeared, the young physician Andreas Vesalius published *On the Fabric of the Human Body* with outstanding woodcuts by the Flemish artist Jan Kalkar. Vesalius challenged a number of assertions of the Greek physician Galen, notably regarding the supposed existence of pores in the septum of the heart. The ink on the pages of Vesalius's book had scarcely dried when the Italian physician Girolamo Fracastoro published *On Contagion* (1546), in which he advanced the thesis that various diseases were each caused by a specific agent that could be spread either by direct contact with someone already infected or by more indirect means such as contact with cloth or linens that could foster the "essential seeds" of the contagion. The crowning achievement of Renaissance anatomy was the discovery of the circulation of the blood, postulated by Michael Servetus in 1553 and announced, with compelling arguments, by William Harvey in 1616.

All of these advances took place before the development of even crude microscopes. But advances in optics in the seventeenth century opened new vistas for biology as well as for astronomy. Robert Hooke's *Micrographia* (1664), which contained large fold-out pictures of insects such as the flea and the louse as seen through the microscope, was wildly popular and established the reputation of the Royal Society. The Dutch scientist Antony van Leeuwenhoek (1632–1723) discovered an ingenious trick for creating tiny spherical lenses that could magnify up to 300 diameters, and using these lenses he discovered microscopic single-celled organisms.

5 The Rise of Modern Philosophy

As it had in the golden age of Greece, the sudden upsurge in scientific theorizing in the sixteenth and seventeenth centuries dovetailed with fresh work on the foundations of scientific knowledge. We can date the beginnings of self-conscious reflection on the philosophical implications of the new science and the scientific implications of the emerging philosophy from about 1600. In the early years of the seventeenth century, Francis Bacon (1561–1626), Lord Chancellor of England, wrote several works decrying the sterility of scholastic philosophy and outlining a new program for experimental philosophy and the organization of scientific societies. The French philosopher and mathematician René Descartes (1596–1650) strove to provide both better foundations for knowledge and better mechanics than Aristotle. John Locke (1632–1704), fellow of the Royal Society and friend of such luminaries as Robert Hooke (1635–1703) and Isaac Newton (1642–1727), wrote his monumental *Essay Concerning Human Understanding* in an attempt to explain the empiricist epistemological underpinnings of the new science.

The enormous success of Newton's mechanical picture of the universe and the emergence of the new optical technology triggered an avalanche of scientific and philosophical work in the eighteenth century. The mathematical elaboration of calculus, using Leibniz's notation rather than Newton's, enabled astronomers to put the finishing touches on celestial mechanics. But with the new science came new questions. What is light, a rain of particles or a vibration in an intervening medium between the luminous object and the eye? What is the nature of space? Would it still exist if all objects were to disappear? How is learning from experience possible at all? What do we really mean by "cause and effect," and how can we move from bare sensory impressions to justified belief in causal claims? How do we come to know the geometric structure of space?

One critical philosophical development accompanying the scientific revolution was the shift from the Aristotelian conception of science as absolutely certain knowledge derived from first principles to a more modest conception of science as a rational but fallible discipline. It proved extraordinarily difficult to find the right balance between optimism and pessimism. Locke laid out an empiricist approach that would suffice to let us know what we need for practical purposes, but he had doubts about our ability to discover the hidden mechanisms of nature. David Hume (1711–1776) refined Locke's modest empiricism and pushed it into skepticism. Immanuel Kant (1724–1804), provoked by Hume's arguments but dissatisfied with his skepticism, developed an elaborate form of idealism in which part of our knowledge of the world of appearances is guaranteed by the very structures of our understanding. The tension between empiricism and idealism continues to shape philosophical reflection to this day.

6 From the Scientific Revolution to the Twentieth Century

Throughout the eighteenth century and into the nineteenth, scientists and philosophers debated the issue of scientific methodology. Newton's disparaging remarks about "hypotheses" led some of his disciples to take a strong stand against the hypothetico-deductive approach advocated by Christiaan Huygens (1629–1695) and actively pursued on the continent. Advocates of the method of hypothesis saw science as a matter of trying to find plausible guesses that fit existing evidence, then testing those guesses by deriving further consequences to see whether they would continue to hold up; inductivists saw science as a bottom-up activity in which the data from careful experimental work preceded and suggested the theories. The contrast was sometimes drawn more sharply than the actual practice of the scientists would support; Newton himself frequently advanced hypotheses, and Huygens was a careful observer. But the polemical polarization of the two methodological schools had the positive effect of bringing philosophical questions into focus. It is easy, inductivists claimed, to frame a scientific hypothesis to account for all of the available data – so easy that the mere fact one has constructed an empirically adequate hypothesis is no evidence of its truth; but in that case it is a waste of time to construct hypotheses. Their opponents shot back that it is not so easy as all that to come up with even one good hypothesis that fits all of the data and that the mere hoarding up of observations does little to suggest an interesting theory.

The problem of induction, conceived as the problem of determining what, if anything, we are reasonably entitled to infer from an observed uniformity or statistical preponderance, provoked a great deal of thoughtful work but little consensus. Hume's skeptical arguments set the terms of the debate, and John Stuart Mill famously responded that our inductive reasoning depends on the uniformity of nature. The problem became more than academic when uniformity turned out to be the issue in question between two dominant views of geology: catastrophism, represented by Georges Cuvier, and uniformitarianism, championed by James Hutton and Charles Lyell. The problem, in a nutshell, was whether we are entitled to extrapolate backwards in time the sorts and rates of processes we see around us. If so, then various phenomena seem to point to a very old earth; but if we are to believe that catastrophic events of a sort unseen in our time shaped the earth as we find it, such extrapolations may be wide of the mark. Through the course of the nineteenth century, thanks largely to the tireless work of Lyell, the uniformitarians gradually gained the upper hand in this argument.

The first volume of Lyell's *Principles of Geology* appeared in 1830, just in time to be placed in the hands of a young naturalist named Charles Darwin as he set out on a survey expedition aboard HMS *Beagle*. Darwin began to look at rock formations through Lyell's eyes. Geologic "deep time," Darwin reasoned, provided nature with sufficient resources to account for the evolution of species through blind variation and selective retention. The geographic distribution of wildlife seemed inexplicable on a view of special creation. Gradually and cautiously Darwin assembled the evidence for his theory of evolution and finally published it in 1859. A tempestuous public debate ensued and lasted into the early twentieth century when the scientific community was persuaded by the confluence of evidence from geology, paleontology, biology, genetics, astronomy and physics. Biology had found its grand unifying narrative.

The great strength of Darwin's theory is its explanatory power. Around the turn of the twentieth century, philosophers and scientists took a fresh look at the nature of scientific explanation in light of the scientific advances of the preceding three centuries. The American logician and philosopher Charles Sanders Peirce, in particular, gave careful attention to what he called *abduction*, a mode of inference that leads from facts to theories that account for them. Some scientists, including Henri Poincaré, Pierre Duhem, and Albert Einstein, gave serious consideration to the philosophical underpinnings of physical theorizing. Yet they came to no agreement, and their very lack of unanimity posed a puzzle that cried out for resolution. How the logical positivists tried to resolve it, and what became of their solution, is the story told in Part II.

Unit 1
The Ancient and Medieval Periods

From its beginnings in conceptual puzzles regarding motion and the constitution of matter, Greek science developed into a sweeping cosmology and natural philosophy fashioned principally by Aristotle, with the astronomical part brought to maturity by Claudius Ptolemy. The history of science in the Middle Ages is in large part a series of reactions to Aristotle and Ptolemy, including some very critical reactions. Yet the Aristotelian and Ptolemaic view remained dominant through the Middle Ages until the dawn of the scientific revolution.

1 Zeno's Paradoxes

One of the earliest signs of the vitality of Greek thinking about nature was the emergence of puzzles about the fundamental notions of motion, space, and time. In reading 1.3, Zeno of Elea, an older contemporary of Socrates, provides us with some examples in his paradoxes of motion. The paradox of division is typical. In crossing a room, one comes to the halfway point; what remains is half of the distance to be covered. Continuing, one crosses half of that remainder, leaving a quarter of the original distance still to go. We can continue analyzing the situation like this, cutting the remaining distance in half again and again ad infinitum, but no matter how far we go in this analysis there will be some distance left over that is not yet covered. Does this show that one can never cross the room, or more generally that one can never complete the trip from point A to some other point B?

It is hard to say how seriously Zeno took the conclusion that motion is impossible. On the one hand he was a disciple of Parmenides, who held that change is unreal. But on the other hand the conclusion seems so obviously false that it is difficult to imagine anyone's taking it at face value. Whatever Zeno's beliefs, however, his argument requires some sort of answer from those who would maintain the commonsense position that motion is possible.

In the work of Arisotle we have both an account of Zeno's paradoxes and an attempt to resolve them. Aristotle's answer is that both distance (magnitude, or length) and time are potentially infinitely divisible, even though neither a distance nor a time can actually be divided an infinite number of times. For someone moving across the room at a steady rate of speed, the first half of the trip takes half of the time, the next quarter takes a quarter of the time, and so forth. This is true as far as it goes, but Zeno might have responded that Aristotle is still accumulating an infinite number of fractions for the time taken to cover each part of the path:

$$1/2,\ 1/4,\ 1/8, \ldots,\ 1/2^n,\ 1/2^{n+1}, \ldots$$

Each of these fractions is greater than zero, yet for the commonsense position to work the sum of the infinite series must be finite. Modern mathematics is up to the task: one of the classic proofs in undergraduate math shows that the sum of this series is 1. But it was not until the work of Cauchy in the nineteenth century that we possessed a rigorous justification for the summing of an infinite series. Aristotle's response would be rather different: that at any point in the process of dividing a time or a space we have, not an infinite number of these fractions, but a finite number of them.

2 The Atomists

The central insight of Greek science is that the universe is a comprehensible whole and the phenomena around us are governed by rationally discoverable principles. The search for underlying order led some of the Greeks to an astonishing hypothesis: everything we see around us is made of small indivisible components, which they called atoms, and the varied phenomena of nature can be explained in terms of atoms moving in the void, the properties of objects depending on the ordering of the atoms just as the meaning of words and sentences depends on the ordering of letters. Democritus, a contemporary of Socrates, is one of the first writers we know of to advance this position; reading 1.1, from an account preserved in the writings of Diogenes Laertius, gives a brief and tantalizing glimpse of his thought. In reading 1.2, a letter from the atomist philosopher Epicurus to a student named Herodotus, we find a fuller display of the explanatory power of atomism. Reading 1.11, from the Roman poet Lucretius's work *On the Nature of Things*, gives a later and more detailed defense of atomism.

With only atoms and the void, Democritus and his followers attempted to resolve a major conceptual issue that occupied some of the best Greek minds: to give an account of *change*, considered as a general phenomenon. Children are born, grow old, and die; a leaf flourishes and withers; wet clothes hung out on the line dry in the wind; wax left by the fire softens and melts into a puddle; a bird flickers across the space from one tree to another; a ring worn next to the finger grows thin over the course of many years; the sun rises and sets. All of these phenomena exhibit change, yet change with unity: it is in some sense the same leaf that withers, the same wax that melts, the same ring that is gradually worn. According to the atomists, all of the varieties of change could be reduced to one: the change of position of atoms with respect to one another. Like the letters of the alphabet, the atoms could be rearranged to produce different and distinctive entities, and the possibilities for ever more complex combinations of atoms were limitless.

From the outset, the atomist project was profoundly reductive. Democritus explicitly says that the soul itself is composed of atoms, and in Lucretius's hands atomism becomes an argument for atheism. If the effects commonly ascribed to the gods are in fact explicable in natural terms, then the gods are out of a job; we have no need to hypothesize their existence. But the very scope of the project was also a handicap. Atoms were imperceptible, and their reality had to be defended by plausible arguments rather than by demonstrative proofs. Lacking any compelling account of how atoms interact, the early atomists were unable to give a detailed account of how rearrangements of atoms underwrite the spectacular diversity of nature's effects. The existence of a void was also a theoretical postulate, though one supported by an ingenious thought experiment: if there were no void, motion would be impossible since all material things would be caught up in cosmic gridlock, the ultimate traffic jam.

Democritus viewed the cosmos of atoms and void as a deterministic system. Subsequent atomists were not always content or comfortable with this, and some critics found determinism unacceptable or absurd. The metaphysical simplicity of atomism proved to be both a strength and a weakness; a strength, insofar as simplicity is an attractive feature of any theory, but a weakness when it came to giving a detailed account of the wild variety of physical phenomena, particularly since Greek atomism offered no account of the cohesion of atoms. Atomism never died out, but it drifted to the margins of science. There it waited over the centuries until some of its ideas were revived and reformulated by scientists like Galileo, Boyle, and Newton during the scientific revolution.

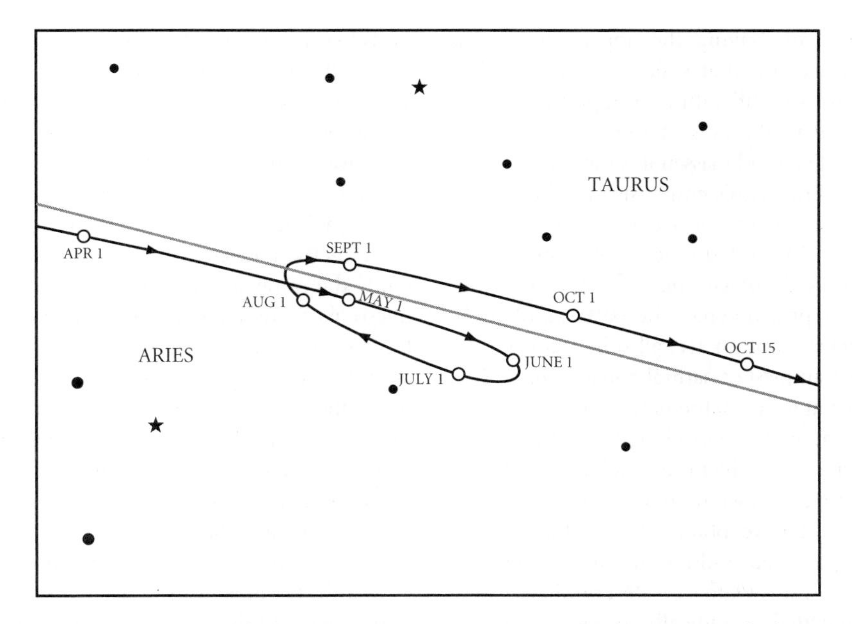

Figure 1 Retrograde motion. The diagram shows the apparent path of Mars through the constellations Aries and Taurus during a six-month period. Starting in mid June the planet stops, reverses its course, stops again, and finally heads back in the original direction

3 Plato's Cosmology

In reading 1.4 from the *Timaeus*, Plato offers a very different image of the world. In the form of a story he tells of the creation of the universe by a benevolent demiurge who strives, with mixed success, to represent in a physical medium the structure of the eternal Forms. Mathematical ideas dominate the story. There are four elements, earth, air, fire, and water, each with its distinctive properties. Why four? Plato conjectures – he makes it plain, through the mouth of the character Timaeus, that he is not pretending to deduce all of this with certainty but only to tell a "likely tale" – that the four elements may correspond to four of the regular convex solids: fire to the tetrahedron, earth to the cube, air to the octahedron, water to the icosahedron. (Another mathematical argument Plato offers for there being four elements is that between any two cubes, a^3 and b^3, there are two other terms that fill in a geometrical progression, a^2b and ab^2.) Each of these solids has sides that can be disassembled into triangles and then reassembled, permitting in principle the transformation of one element into another. There is in addition a fifth solid, the dodecahedron, which has 12 pentagonal faces. Plato suggests that this may correspond to the quintessence of which the heavens themselves are made. Celestial motions are dictated by the mathematical analysis of the musical scale. The universe is spherical, as are the individual stars, and the system taken together is "an intricately wrought whole."

Plato was well aware that there were details not yet accounted for by the sort of system he favored. In Book 10 of the *Republic* he makes a passing reference to the retrograde motion of Mars (see Figure 1), and he gave his pupils in the Academy the task of creating a model that would more accurately reflect the visible motions of the planets and bring their seemingly capricious reversals under the control of an overarching mathematical ideal. One of those students, Eudoxus, refined Plato's model by employing 27 spheres and allowing the axes of each sphere to be attached to the next at an angle. By means of this ingenious construction, Eudoxus was able to reproduce, at least qualitatively, the retrograde motions of the planets.

The goal of "saving the appearances" – of giving an account that squares with the visible phenomena – is still with us as a part of modern science, and in the eyes of some contemporary empiricists like van Fraassen it is the principal goal of science. But Plato's emphasis on mathematics came at the expense of a detailed study of the physical world, and this left him and his students at a disadvantage when it came to the study of complex phenomena. The patterns of stellar and planetary motion are, all things considered, among the simplest of natural phenomena and the most amenable to mathematical description. In contrast, the mathematical analysis of biological phenomena is a problem so much more difficult, and requiring so much more information than the Greeks could have obtained with the unaided eye, that Plato had nothing interesting to say on the subject. It was Plato's greatest pupil, Aristotle, who articulated a radically different view of nature that opened the way to intelligible explanations of more complicated phenomena.

4 Aristotle's Natural Philosophy

No other figure in history has ever dominated the intellectual landscape in as many fields as Aristotle. His views were almost always the point of departure for a discussion of any scientific topic; for the better part of two millennia, the accepted method of making a contribution to scientific thought was to write a commentary on one of Aristotle's scientific works. Many of his works were lost to the west in the early Middle Ages, but in the Arab world the tradition of commentary on Aristotle continued. During the eleventh century Aristotle's works returned to the west, first in translations from Arabic to Latin and later directly from Greek into Latin, and the most prominent scholars of the high Middle Ages discussed them vigorously. At the Council of Trent in the mid 1500s Aristotle was designated by the Catholic Church as the preeminent authority in matters of philosophy.

Aristotle's sweeping synthesis offered an impressively unified view of logic, metaphysics, epistemology, biology, physics, and astronomy. Aristotle did not build this system entirely from scratch, but his thorough review of prior theories and his ingenious and interconnected resolutions to existing problems gave his views substance and authority. When it was needed, as reading 1.6 shows, he would ask and then answer the simple but deep questions that provide the framework for scientific reasoning, questions about the fundamental constituents of nature, the sorts of causes, and the conceptual relations of necessity, spontaneity, and chance. In reading 1.9 he extends the same sort of care to the fundamental questions about living things, asking how they differ from abstract objects, what sorts of causes are appropriate to inquiry about the living world, and the method of grasping the real causes of things. The method of classification he proposes, based on looking at the similarities and differences of things, is the foundation of comparative morphology; it was the principle that undergirded the dominant taxonomic scheme (kingdom, phylum, class, order . . .) until the advent of molecular genetics.

But he could also engage in system building on the grandest scale. Noting the unchanging nature of the heavens and the mutable nature of the earth and its nearby atmosphere, Aristotle makes a fundamental cut between the celestial and terrestrial realms and constructs distinct sets of fundamental principles to govern each. He addresses the problem of change through a distinction between form and matter and an analysis of four senses in which we might say that one thing causes another. He answers the question of which sorts of changes require explanation and which do not by laying down the doctrine of essential natures and natural motions.

The notion of essential natures is foreign to our present view of science, but it is critical to the Aristotelian system. Each natural object, according to Aristotle, belongs to a kind, and objects of each kind have a nature that determines how they behave naturally, that is, when not acted upon by other objects. It is the nature of a seed to grow into a plant, but it is not the nature of a seed to move across the landscape; for that, it must be blown along by the wind or caught on the flank of a passing animal. By appealing to the essential natures of kinds of things, which is to say, their natural ends, Aristotle could widen the scope of science to include all natural phenomena and make a principled distinction between events or sorts of behavior that require explanation and events that do not. When things behave according to their natures, no explanation is required;

when they do not, we must seek an explanation. True scientific knowledge, Aristotle argues in reading 1.7, consists in demonstration from a knowledge of the essential natures of things.

Another way of expressing this same thought is that each natural object has a *telos* – a goal or an end which it has an intrinsic tendency to reach. An acorn's *telos* is an oak tree; or, to use the term a bit more freely, we might say that its *telos* is its tendency to become an oak tree. This teleological conception of nature shapes the explanations Aristotle offers: we can rest content with an explanation when we have traced the behavior of an object back to an understanding of its *telos*.

The appeal to natures is evident in Aristotle's account of free fall. All mundane physical objects are either heavy or light, depending on which elements preponderate in their makeup. A heavy object such as a stone will fall naturally downward, that is, toward the center of the universe, where the center of the spherical earth is located. The stone will continue to fall even through a medium thicker than air, such as water or honey, but its rate of fall is slowed considerably. Though he does not use a mathematical formula to express his views, Aristotle gives enough verbal description of the relationship between force and resistance to suggest that he accepted, within certain limits, the idea that velocity is directly proportional to force and inversely proportional to resistance, or $V \propto F/R$. This view of the relation of these variables has an interesting consequence: in the limit as resistance goes to zero, the velocity of an object acted upon by even the slightest force goes to infinity. Aristotle quite reasonably objected that infinite velocity is an absurdity; an object moving at an infinite velocity would literally be in two places at the same time. Therefore there must always be some resistance – and hence, contrary to the atomists, there cannot be a void.

Aristotle's cosmology exhibits some of the central features of his scientific thought. In reading 1.5, he takes up the question of the motions of the stars and the planets. The stars wheel around us every 24 hours, but the planets gradually work their way through the band of constellations that make up the ecliptic, and they do so in irregular ways. All of these motions require a cause, that is, something in contact with the thing moved that keeps it moving. Aristotle develops a famous argument that the regress of causes cannot go on to infinity: there must be something that moves without itself being moved. In reading 1.8, Aristotle gives us more details about the large-scale structure of the universe and the shape and size of the earth. Contrary to a nineteenth-century urban legend that the medievals believed in a flat earth, virtually every educated person from the time of Aristotle onward realized that the earth is approximately spherical.[1]

Although Aristotle was a plenist – that is, he denied that there is any void or vacuum anywhere in nature – he understood the appeal of the gridlock argument advanced by the atomists. He countered it with a circulatory theory of motion called *antiperistasis* according to which adjacent objects could move in a circle, or at least in a closed curve, like a long line of train cars on a circular track with the engine coming around to touch the caboose from behind. Formally, the solution was adequate: objects could move this way in a plenum without gridlock. Its merits as an explanation of motion were less obvious. Aristotle suggests with some hesitancy that the continued motion of a flung stone may be due to the rush of displaced air around behind it, pushing it onward.

Aristotle conceived of science as knowledge of reasoned facts deduced from evident first principles. The distinction between knowledge of facts and knowledge of reasoned facts is significant. In a passage from the *Posterior Analytics* I.13 illustrating the distinction, Aristotle considers the planets, which we know from observation do not twinkle as the stars do but rather shine with a steady light. Why is this? Aristotle's answer is that they are relatively near, and that which is near does not twinkle. Here we have not only a demonstration but a causal explanation for the observed phenomenon, a paradigm case of knowledge of the reasoned fact. But how do we know that the planets are close? We can construct another argument using the same basic materials: what does not twinkle is relatively near, and the planets do not twinkle; therefore, they are relatively near. This second argument, however, does not give us an *explanation* of the nearness of the planets. Here, in Aristotle's terms, we have knowledge of the fact but not of the reasoned fact.

This example, however, raises more problems than it resolves. The ultimate starting points for scientific demonstration must, according to Aristotle, be known with certainty. But how can we know with certainty the critical premise that what is near does not twinkle? Accidental correlation of nearness with non-twinkling will not do the job; non-twinkling must be an essential characteristic of nearby objects or an essential consequence of their nearness. For that matter, how can we know that the correlation invariably holds? These two problems – the problem of distinguishing generalizations that hold necessarily or essentially from accidentally true ones, and the problem of inferring universal generalizations, even accidental ones, from mere instances of correlation – cast a long shadow over the philosophy of science, resurfacing time and again up through the twentieth century.

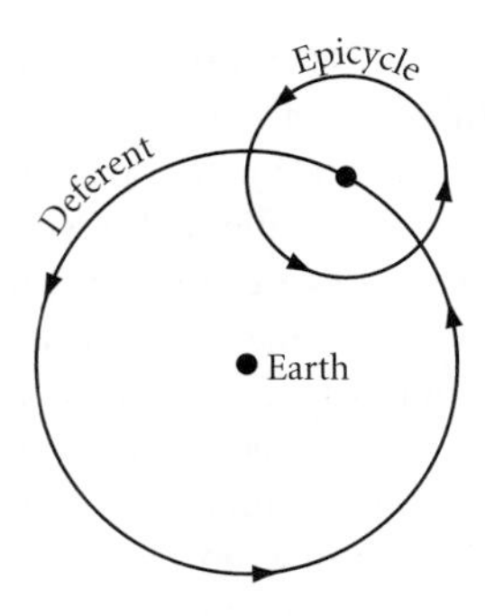

Figure 2 The epicycle. The planet travels on the rotating epicycle, which is itself carried around the deferent. By selecting the rotational speeds and the radius of the epicycle carefully, one can reproduce the retrograde motion of a planet as it is seen from earth

5 Ptolemaic Astronomy

In the second century AD, Claudius Ptolemy synthesized Aristotelian ideas regarding motion with a great mass of observational data and considerable geometric ingenuity to produce the definitive work on positional astronomy. Ptolemy's tome, best known by its Arabic title *Almagest*, is a curious and difficult book. His stated aim is to construct a positional astronomy that will yield predictions in good agreement with observation, to "save the phenomena." Three critical conclusions for which he argues in the first book (reading 1.12) are that the earth is spherical, that its size in comparison with the distance to the stars is negligible, and that it does not move.

Ptolemy's argument that the earth is motionless provides a good illustration of his use of Aristotle. The fundamental principle of forced (unnatural) motion, according to Aristotle, is that everything that moves unnaturally is moved by something else. Suppose, then, that the earth rotates on its axis once every 24 hours. An observer whose feet are planted firmly on the ground is moved by, and therefore moves with, the earth. But if he jumps, his feet lose contact with the earth; and therefore, according to Aristotle, he should no longer move with it. If we take Aristotle's dynamics seriously, we should expect anyone who jumps up into the air to be slammed into the west wall of the room at speeds up to 1000 miles per hour (depending on the latitude of the unfortunate jumper). Obviously this does not happen. Ptolemy, who takes Aristotle's dynamics very seriously, draws the conclusion that the earth is not rotating.

But Ptolemy takes liberties with Aristotle whenever doing so increases his predictive accuracy, and he shows a shocking lack of concern for the consistency of his geometric constructions with one another. His constructions, compounded of little circles (called *epicycles*) that ride on other circles (called *deferents*), with the heavenly bodies riding on the epicycles, provide a tolerably good guide to the apparent positions of the planets in the night sky (see Figure 2). In particular, by a careful combination of motions – the planet moving on the epicycle at one speed, the epicycle moving on the deferent at another – Ptolemy is able to reproduce the periodic retrograde motions of the planets. But the selection of radii for these circles are in many cases arbitrary – other sizes of radius might do equally well – and in some cases (e.g. with the moon) there is no three-dimensional path through space that corresponds to all of Ptolemy's constructions, since he uses one radius for the computation of the moon's visible size and a different one for its motion. One can build scale models of pieces of the Ptolemaic system, but one cannot build a single scale model that represents all of the incompatible pieces.

Is this a problem? The answer depends very much on one's view of the aims of science, and

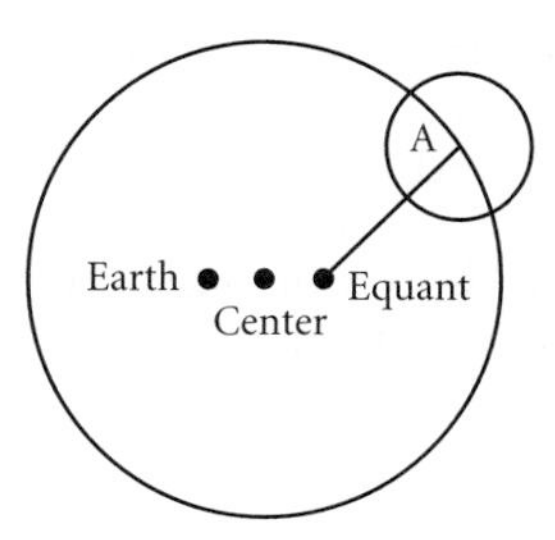

Figure 3 The equant. The planet travels on the epicycle with center A. The line from the equant to A rotates at a constant speed. But this means that the speed of A around the deferent, viewed either from earth or from the center of the circle, will not be constant

of astronomy in particular. If the geometric models of Ptolemaic astronomy are considered as mere calculating devices, they do their job well enough. His system as a whole is a reasonably accurate predictive instrument, and if prediction is all we ask from a theory, we have little ground for complaint. But if we want a believable picture of the paths of the planets through three-dimensional space or an account of the causes of their motions, Ptolemy's account in the *Almagest* does not offer what we seek. Even waiving the problem of incompatible constructions, his system invokes many strange and implausible devices. Planets ride on epicycles while the centers of those epicycles travel along deferents, yet the center of every circle involved is not a heavenly body – even the earth is not quite at the geometric center of Ptolemy's solar system – but an empty point in space. And the rotational speeds of these circles are not always uniform in the sense Aristotle intended; in multiple cases a point moving along the circumference of the deferent sweeps out equal angles in equal times only when viewed from the *equant*, a point other than the center of the deferent (see Figure 3).

6 The Critical Reaction to Aristotle and Ptolemy

Some casual observers of the history of science have mistaken the apparent obsession of medieval thinkers with Aristotle for blind, uncritical acceptance of his teachings. This misconception, like the myth that the medievals believed in a flat earth, was a Victorian invention, and it still works its way into textbooks and popular essays despite the best efforts of professional historians of science. Many of the commentaries on Aristotle are quite critical. Some offer experimental evidence for or against his views, particularly his views on motion, and against Ptolemy's astronomy.

We can get a glimpse of dissatisfaction with the Aristotelian conception of science in a fragment from the writings of Philodemus, an Epicurean philosopher writing in the first century BC (reading 1.10). Philodemus is concerned with the discovery of essential characteristics, and he stresses the importance of having a wide sample and varying the other characteristics. Though Philodemus's immediate target is the Stoic position, the methodological precautions he advocates are pointless on the Aristotelian view. If the abstraction of essences is the ineffable operation of *nous*, or the mind, there is no need for large and varied samples.

In the sixth century, the Christian philosopher John Philoponus offered more direct and vigorous criticisms of Aristotle's physical theories. Aristotle's suggested account of projectile motion, Philoponus argues in reading 1.14, is not credible; various everyday experiences and easily constructed experiments show it to be false. But if the air is not pushing the flung object, why does it continue to move? Philoponus advances a novel solution: the hand imparts to the object an *impetus*, an internal force or tendency to motion, and that impetus then carries the object along after it is released from the hand. From a modern perspective this is wrong, but it is in an important sense less wrong than Aristotle's original conception. It is not an accident that Galileo adopted the theory of impetus in his youth before he developed his more powerful and more nearly accurate concept of inertia – which he still calls "impetus" – in the *Dialogue Concerning the Two Chief Systems of the World* in 1632.

The Aristotelian position on free fall fares no better at Philoponus's hands (reading 1.15). Though he agrees with Aristotle that there is in fact no void, Philoponus thinks this is not a matter of necessity; since it follows from Aristotle's analysis of velocity in terms of force and resistance, that analysis must be wrong. Philoponus's position can, to a first approximation, be expressed in a modern notation as $V \propto F - R$. One consequence of this is that motion in the absence of resistance need not be infinitely swift. Again, although from a modern standpoint this analysis is still misguided, it represents a step forward in the analysis of motion.

Ptolemy also comes in for his share of criticism. In reading 1.13, the Neoplatonic philosopher Proclus, writing in the fifth century, expresses dismay at the "casual attitude" of the astronomers in expounding the hypothetical devices they use to account for planetary motions. Significantly, he stresses that mere successful prediction does not satisfy our yearning for complete understanding, for the real explanations behind the mathematical constructions.

These aspects of Ptolemy's system in the *Almagest* bothered subsequent thinkers. Toward the end of the twelfth century, the Jewish thinker Maimonides echoed Proclus's complaints (reading 1.16). Yet Maimonides is genuinely perplexed as to how to account for retrograde motion without the use of epicycles. By the fourteenth century, dissatisfaction with Ptolemy's astronomy was well entrenched in Paris. Both Jean Buridan (1.17) and his brilliant student Nicole Oresme (1.18) openly discussed the compatibility of the earth's rotation with astronomical data, even suggesting modifications to Aristotle's physics in order to reconcile the conjecture with observation.

All of these arguments were extant in the literature; many of them were available to subsequent generations of thinkers, at least those who lived near centers of learning. But in the absence of some better system of comparable scope, Aristotle's view of science and the Ptolemaic model of the heavens remained dominant. The questions and alternative proposals of the critics remained scattered across the intellectual landscape until they were drawn together in the scientific revolution.

Note

1 The chief perpetrator of this urban legend was Washington Irving, better known as the author of "The Legend of Sleepy Hollow." See Jeffrey Burton Russell, *Inventing the Flat Earth* (New York: Praeger, 1997).

Suggestions for Further Reading

The following brief list offers a sampling of the rich literature in the history of science pertaining to the material covered in Unit 1. Items marked with an asterisk assume little or no background in the subject; these would be a good place to begin further reading.

Barbour, Julian, 2001, *The Discovery of Dynamics*. Oxford: Oxford University Press, chapters 1–4.

*Clagett, Marshall, 1994, *Greek Science in Antiquity*. New York: Barnes and Noble.

Cohen, Morris R. and Drabkin, I. E. (eds.), 1966, *A Source Book in Greek Science*. Cambridge, MA: Harvard University Press.

*Crowe, Michael, 2001, *Theories of the World from Antiquity to the Copernican Revolution*, 2nd edn. New York: Dover, chapters 1–5.

Dijksterhuis, E. J., 1961, *The Mechanization of the World Picture*. New York: Oxford University Press, Parts I–III.

Ferejohn, Michael, 1991, *The Origins of Aristotelian Science*. New Haven: Yale University Press.

Franklin, Allan, 1976, *The Principle of Inertia in the Middle Ages*. Boulder, CO: Colorado Associated University Press.

Grant, Edward, 1977, *Physical Science in the Middle Ages*. Cambridge, UK: Cambridge University Press.

Grant, Edward, 1996, *The Foundations of Modern Science in the Middle Ages: Their Religious, Institutional and Intellectual Contexts*. Cambridge, UK: Cambridge University Press.

Huggett, Nick, 1999, *Space from Zeno to Einstein*. Cambridge, MA: MIT Press.

*Lindberg, David C., 1992, *The Beginnings of Western Science*. Chicago: University of Chicago Press.

McMullin, Ernan, 1992, *The Inference That Makes Science*. Milwaukee: Marquette University Press.

Palter, Robert M. (ed.), 1961, *Toward Modern Science*, 2 vols. New York: Noonday Press.

Pyle, Andrew, 1997, *Atomism and Its Critics: From Democritus to Newton*. Bristol: Thommes Press.

Sambursky, Samuel, 1962, *The Physical World of Late Antiquity*. New York: Basic Books.

Sambursky, Samuel, 1992, *The Physical World of the Greeks*. Princeton: Princeton University Press.

Sargent, Steven D., 1982 (trans. and ed.), *On the Threshold of Exact Science: Selected Writings of Anneliese Maier on Late Medieval Natural Philosophy*. Philadelphia: University of Pennsylvania Press.

*Toulmin, Stephen and Goodfield, June, 1961, *The Fabric of the Heavens*. Chicago: University of Chicago Press, chapters 1–5.

Whyte, Lancelot Law, 1961, *Essay on Atomism: From Democritus to 1960*. Middletown, CT: Wesleyan University Press.

1.1

Atoms and Empty Space

Diogenes Laertius

Democritus (c.460–c.370 BC), a student of Leucippus, gave the first clear and widely read expression of atomism. Though we no longer have any complete works by Democritus, the following summary of his teachings on atoms and the void is preserved in the ninth volume of Diogenes Laertius' *Lives of the Philosophers*, a work probably written in the third century AD. Diogenes reports a good deal of gossip about the philosophers he discusses, not all of it reliable. He attributes to Aristoxenus the claim that Plato wanted to burn all of the available copies of Democritus' numerous works.

The first principles of the universe are atoms and empty space. Everything else is merely thought to exist. The worlds are unlimited [or boundless]. They come into being and perish. Nothing can come into being from that which is not nor pass away into that which is not. Further, the atoms are unlimited in size and number, and they are borne along in the whole universe in a vortex, and thereby generate all composite things – fire, water, air, earth. For even these are conglomerations of given atoms. And it is because of their solidity that these atoms are impassive and unalterable. The sun and the moon have been composed of such smooth and spherical masses [i.e., atoms], and so also the soul, which is identical with reason.

In *The Book of the Cosmos*, ed. Dennis Richard Danielson (Cambridge, MA: Perseus, 2000), p. 25.

1.2

Letter to Herodotus

Epicurus

Epicurus (c.341–c.271 BC) was an articulate spokesman for atomism. We are indebted again to Diogenes Laertius for preserving, in the tenth book of his *Lives of the Philosophers*, a considerable amount of Epicurus' own work, including the letter to Herodotus (not to be confused with the historian by the same name) from which the following selection is taken. In this letter Epicurus summarizes his atomism and sketches the way that it can be employed to account for observable phenomena.

Epicurus to Herodotus, greeting.

For those who are unable to study carefully all my physical writings or to go into the longer treatises at all, I have myself prepared an epitome of the whole system, Herodotus, to preserve in the memory enough of the principal doctrines, to the end that on every occasion they may be able to aid themselves on the most important points, so far as they take up the study of physics. . . .

To begin with, nothing comes into being out of what is non-existent. For in that case anything would have arisen out of anything, standing as it would in no need of its proper germs. And if that which disappears had been destroyed and become nonexistent, everything would have perished, that into which the things were dissolved being nonexistent. Moreover, the sum total of things was always such as it is now, and such it will ever remain. For there is nothing into which it can change. For outside the sum of things there is nothing which could enter into it and bring about the change.

Further, the whole of being consists of bodies and space. For the existence of bodies is everywhere attested by sense itself, and it is upon sensation that reason must rely when it attempts to infer the unknown from the known. And if there were no space (which we call also void and place and intangible nature), bodies would have nothing in which to be and through which to move, as they are plainly seen to move. Beyond bodies and space there is nothing which by mental apprehension or on its analogy we can conceive to exist. When we speak of bodies and space, both are regarded as wholes or separate things, not as the properties or accidents of separate things.

Again, of bodies some are composite, others the elements of which these composite bodies are made. These elements are indivisible ["*atoma*"] and unchangeable, and necessarily so, if things are not all to be destroyed and pass into non-existence, but are to be strong enough to endure when the composite bodies are broken up, because they possess a solid nature and are incapable of being anywhere or anyhow dissolved. It follows that the first beginnings must be indivisible, corporeal entities.

Again, the sum of things is infinite [or boundless]. For what is finite has an extremity, and the extremity of anything is discerned only by comparison with something else. Now the sum of things is not discerned by comparison with anything else. Hence, since it has no extremity, it has no limit. And since it has no limit, it must be unlimited or infinite.

Moreover, the sum of things is unlimited both by reason of the multitude of the atoms and the extent of the void. For if the void were infinite and bodies finite, the bodies would not have stayed anywhere but would have been dispersed in their course through the infinite void, not having any supports or counter-checks to send them back on their upward rebound. Again, if the void were finite, the infinity of bodies would not have anywhere to be.

Furthermore, the atoms, which have no void in them – out of which composite bodies arise and into which they are dissolved – vary indefinitely in their shapes. For so many varieties of things as we see could never have arisen out of a recurrence of a definite number of the same shapes. The like atoms of each shape are absolutely infinite. But the variety of shapes, though indefinitely large, is not absolutely infinite.

The atoms are in continual motion through all eternity.

1.3

The Paradoxes of Motion

Zeno

Zeno of Elea (c.490–c.430 BC) was a native of southern Italy and a disciple of Parmenides. His fame rests entirely on four paradoxes recounted here by Aristotle in his *Physics*. The paradoxes seem silly if taken literally, as their conclusions contradict the evidence of the senses; they are better understood as shrewd challenges to the adequacy of existing concepts of space, time, and motion. Obviously something goes wrong in the reasoning presented in each case. But what is it? A satisfactory answer did not emerge until the development of the calculus in the seventeenth century.

[...]

Moreover, the current arguments make it plain that, if time is continuous, magnitude is continuous also, inasmuch as a thing passes over half a given magnitude in half the time, and in general over a less magnitude in less time; for the divisions of time and of magnitude will be the same. And if either is infinite, so is the other, and the one is so in the same way as the other; i.e. if time is infinite in respect of its extremities, length is also infinite in respect of its extremities; if time is infinite in respect of divisibility, length is also infinite in respect of divisibility; and if time is infinite in both respects, magnitude is also infinite in both respects.

Hence Zeno's argument makes a false assumption in asserting that it is impossible for a thing to pass over or severally to come in contact with infinite things in a finite time. For there are two ways in which length and time and generally anything continuous are called infinite: they are called so either in respect of divisibility or in respect of their extremities. So while a thing in a finite time cannot come in contact with things quantitatively infinite, it can come in contact with things infinite in respect of divisibility; for in this sense the time itself is also infinite: and so

From *The Complete Works of Aristotle*, ed. Jonathan Barnes, Vol. 1 (Princeton, NJ: Princeton University Press, 1984), extracts from pp. 393–4, 404–5, 439 ("Physics V and VIII").

we find that the time occupied by the passage over the infinite is not a finite but an infinite time, and the contact with the infinites is made by means of moments not finite but infinite in number.

[. . .]

Zeno's arguments about motion, which cause so much trouble to those who try to answer them, are four in number. The first asserts the non-existence of motion on the ground that that which is in locomotion must arrive at the half-way stage before it arrives at the goal. This we have discussed above.

The second is the so-called Achilles, and it amounts to this, that in a race the quickest runner can never overtake the slowest, since the pursuer must first reach the point whence the pursued started, so that the slower must always hold a lead. This argument is the same in principle as that which depends on bisection, though it differs from it in that the spaces with which we have successively to deal are not divided into halves. The result of the argument is that the slower is not overtaken; but it proceeds along the same lines as the bisection-argument (for in both a division of the space in a certain way leads to the result that the goal is not reached, though the Achilles goes further in that it affirms that even the runner most famed for his speed must fail in his pursuit of the slowest), so that the solution too must be the same. And the claim that that which holds a lead is never overtaken is false: it is not overtaken while it holds a lead; but it is overtaken nevertheless if it is granted that it traverses the finite distance. These then are two of his arguments.

The third is that already given above, to the effect that the flying arrow is at rest, which result follows from the assumption that time is composed of moments: if this assumption is not granted, the conclusion will not follow.

The fourth argument is that concerning equal bodies which move alongside equal bodies in the stadium from opposite directions – the ones from the end of the stadium, the others from the middle – at equal speeds, in which he thinks it follows that half the time is equal to its double. The fallacy consists in requiring that a body travelling at an equal speed travels for an equal time past a moving body and a body of the same size at rest. That is false. E.g. let the stationary equal bodies be AA; let BB be those starting from the middle of the A's (equal in number and in magnitude to them); and let CC be those starting from the end (equal in number and magnitude to them, and equal in speed to the B's). Now it follows that the first B and the first C are at the end at the same time, as they are moving past one another. And it follows that the C has passed all the A's and the B half; so that the time is half, for each of the two is alongside each for an equal time. And at the same time it follows that the first B has passed all the C's. For at the same time the first B and the first C will be at opposite ends, being an equal time alongside each of the B's as alongside each of the A's, as he says, because both are an equal time alongside the A's. That is the argument, and it rests on the stated falsity.

[. . .]

But, although this solution is adequate as a reply to the questioner (the question asked being whether it is impossible in a finite time to traverse or count an infinite number of units), nevertheless as an account of the fact and the truth it is inadequate. For suppose the distance to be left out of account and the question asked to be no longer whether it is possible in a finite time to traverse an infinite number of distances, and suppose that the inquiry is made to refer to the time itself (for the time contains an infinite number of divisions): then this solution will no longer be adequate, and we must apply the truth that we enunciated in our recent discussion.

1.4

Plato's Cosmology

Plato

Plato (427–347 BC) was a pupil of Socrates and the teacher of Aristotle. His dialogues exerted a tremendous influence on the development of most of the major areas of philosophy. This selection from the *Timaeus* presents a speculative creation myth that gives a sense of both the nature and the limitations of Plato's cosmology. His treatment of space and his description of the nature and interactions of the four elements provide a striking contrast to the notions of Democritus, and his account of the composition of bodies by triangles reveals his profoundly geometric, a priori turn of mind. Yet he seems conscious of the limitation of this method, since he leaves open the possibility that there may be "principles yet more ultimate than these."

Now as to the whole universe or world order [*kosmos*] – let's just call it by whatever name is most acceptable in a given context – there is a question we need to consider first. This is the sort of question one should begin with in inquiring into any subject. Has it always existed? Was there no origin from which it came to be? Or did it come to be and take its start from some origin? It has come to be. For it is both visible and tangible and it has a body – and all things of that kind are perceptible. And, as we have shown, perceptible things are grasped by opinion, which involves sense perception. As such, they are things that come to be, things that are begotten. Further, we maintain that, necessarily, that which comes to be must come to be by the agency of some cause. Now to find the maker and father of this universe [*to pan*] is hard enough, and even if I succeeded, to declare him to everyone is impossible. And so we must go back and raise this question about the universe: Which of the two models did the maker use when he fashioned it? Was it the one that does not change and stays the same, or the one that has come to be? Well, if this world of ours is beautiful and its craftsman good, then clearly he looked at the eternal model. But if

From *Timaeus*, trans. by Donald Zeyl (Hackett, 2000), pp. 14–15, 17–18, 23–5, 26–7, 39–40, 42–3.

what it's blasphemous to even say is the case, then he looked at one that has come to be. Now surely it's clear to all that it was the eternal model he looked at, for, of all the things that have come to be, our universe is the most beautiful, and of causes the craftsman is the most excellent. This, then, is how it has come to be: it is a work of craft, modeled after that which is changeless and is grasped by a rational account, that is, by wisdom.

Since these things are so, it follows by unquestionable necessity that this world is an image of something. Now in every subject it is of utmost importance to begin at the natural beginning, and so, on the subject of an image and its model, we must make the following specification: the accounts we give of things have the same character as the subjects they set forth. So accounts of what is stable and fixed and transparent to understanding are themselves stable and unshifting. We must do our very best to make these accounts as irrefutable and invincible as any account may be. On the other hand, accounts we give of that which has been formed to be like that reality, since they are accounts of what is a likeness, are themselves likely, and stand in proportion to the previous accounts, i.e., what being is to becoming, truth is to convincingness. Don't be surprised then, Socrates, if it turns out repeatedly that we won't be able to produce accounts on a great many subjects – on gods or the coming to be of the universe – that are completely and perfectly consistent and accurate. Instead, if we can come up with accounts no less likely than any, we ought to be content, keeping in mind that both I, the speaker, and you, the judges, are only human. So we should accept the likely tale on these matters. It behooves us not to look for anything beyond this.

[...]

Now that which comes to be must have bodily form, and be both visible and tangible, but nothing could ever become visible apart from fire, nor tangible without something solid, nor solid without earth. That is why, as he began to put the body of the universe together, the god came to make it out of fire and earth. But it isn't possible to combine two things well all by themselves, without a third; there has to be some bond between the two that unites them. Now the best bond is one that really and truly makes a unity of itself together with the things bonded by it, and this in the nature of things is best accomplished by proportion. For whenever of three numbers which are either solids or squares the middle term between any two of them is such that what the first term is to it, it is to the last, and, conversely, what the last term is to the middle, it is to the first, then, since the middle term turns out to be both first and last, and the last and the first likewise both turn out to be middle terms, they will all of necessity turn out to have the same relationship to each other, and, given this, will all be unified.

So if the body of the universe were to have come to be as a two dimensional plane, a single middle term would have sufficed to bind together its conjoining terms with itself. As it was, however, the universe was to be a solid, and solids are never joined together by just one middle term but always by two. Hence the god set water and air between fire and earth, and made them as proportionate to one another as was possible, so that what fire is to air, air is to water, and what air is to water, water is to earth. He then bound them together and thus he constructed the visible and tangible universe. This is the reason why these four particular constituents were used to beget the body of the world, making it a symphony of proportion. They bestowed friendship upon it, so that, having come together into a unity with itself, it could not be undone by anyone but the one who had bound it together.

Now each one of the four constituents was entirely used up in the process of building the world. The builder built it from all the fire, water, air and earth there was, and left no part or power of any of them out. His intentions in so doing were these: First, that as a living thing it should be as whole and complete as possible and made up of complete parts. Second, that it should be just one universe, in that nothing would be left over from which another one just like it could be made. Third, that it should not get old and diseased. He realized that when heat or cold or anything else that possesses strong powers surrounds a composite body from outside and attacks it, it destroys that body prematurely, brings disease and old age upon it and so causes it to waste away. That is why he concluded that he should fashion the world as a single whole, composed of all wholes, complete and free of old age and disease, and why he fashioned it that way. And he gave it a shape appropriate to the kind of thing it was. The appropriate shape for that living thing that is to contain within itself all the living things would be the one which embraces within itself all the shapes there are. Hence he gave it a round shape, the form of a sphere, with its center equidistant from its extremes in all directions. This of all shapes is the most complete and most like itself, which he gave to it because he believed that likeness is incalculably more excellent than unlikeness.

[. . .]

Now when the Father who had begotten the universe observed it set in motion and alive, a thing that had come to be as a shrine for the everlasting gods, he was well pleased, and in his delight he thought of making it more like its model still. So, as the model was itself an everlasting Living Thing, he set himself to bringing this universe to completion in such a way that it, too, would have that character to the extent that was possible. Now it was the Living Thing's nature to be eternal, but it isn't possible to bestow eternity fully upon anything that is begotten. And so he began to think of making a moving image of eternity: at the same time as he brought order to the universe, he would make an eternal image, moving according to number, of eternity remaining in unity. This number, of course, is what we now call "time."

For before the heavens came to be, there were no days or nights, no months or years. But now, at the same time as he framed the heavens, he devised their coming to be. These all are parts of time, and *was* and *will be* are forms of time that have come to be. Such notions we unthinkingly but incorrectly apply to everlasting being. For we say that it *was* and *is* and *will be*, but according to the true account only *is* is appropriately said of it. *Was* and *will be* are properly said about the becoming that passes in time, for these two are motions. But that which is always changeless and motionless cannot become either older or younger in the course of time – it neither ever became so, nor is it now such that it has become so, nor will it ever be so in the future. And all in all, none of the characteristics that becoming has bestowed upon the things that are borne about in the realm of perception are appropriate to it. These, rather, are forms of time that have come to be – time that imitates eternity and circles according to number. And what is more, we also say things like these: that what has come to be *is* what has come to be, that what is coming to be *is* what is coming to be, and also that what will come to be *is* what will come to be, and that what is not *is* what is not. None of these expressions of ours is accurate. But I don't suppose this is a good time right now to be too meticulous about these matters.

Time, then, came to be together with the universe so that just as they were begotten together, they might also be undone together, should there ever be an undoing of them. And it came to be after the model of that which is sempiternal, so that it might be as much like its model as possible. For the model is something that has being for all eternity, while it, on the other hand, has been, is, and shall be for all time, forevermore. Such was the reason, then, such the god's design for the coming to be of time, that he brought into being the Sun, the Moon and five other stars, for the begetting of time. These are called "wanderers," and they came to be in order to set limits to and stand guard over the numbers of time. When the god had finished making a body for each of them, he placed them into the orbits traced by the period of the Different – seven bodies in seven orbits. He set the Moon in the first circle, around the earth, and the Sun in the second, above it. The Dawnbearer (the Morning Star, or Venus) and the star said to be sacred to Hermes (Mercury) he set to run in circles that equal the Sun's in speed, though they received the power contrary to its power. As a result, the Sun, the star of Hermes and the Dawnbearer alike overtake and are overtaken by one another. As for the other bodies, if I were to spell out where he situated them, and all his reasons for doing so, my account, already a digression, would make more work than its purpose calls for. Perhaps later on we could at our leisure give this subject the exposition it deserves.

[. . .]

The gods he made mostly out of fire, to be the brightest and fairest to the eye. He made them well-rounded, to resemble the universe, and placed them in the wisdom of the dominant circle [i.e., of the Same], to follow the course of the universe. He spread the gods throughout the whole heaven to be a true adornment [*kosmos*] for it, an intricately wrought whole. And he bestowed two movements upon each of them. The first was rotation, an unvarying movement in the same place, by which the god would always think the same thoughts about the same things. The other was revolution, a forward motion under the dominance of the circular carrying movement of the Same and uniform. With respect to the other five motions, the gods are immobile and stationary, in order that each of them may come as close as possible to attaining perfection.

This, then, was the reason why all those everlasting and unwandering stars – divine living things which stay fixed by revolving without variation in the same place – came to be. Those that have turnings and thus wander in that sort of way came to be as previously described.

[. . .]

. . . Suppose you were molding gold into every shape there is, going on non-stop remolding one shape into the next. If someone then were to point at one of them and ask you, "What *is* it?," your safest answer by far, with respect to truth, would be to say, "gold," but never "triangle" or any of the other shapes that come to be in the gold, as though it *is* these, because they change even while you're making the statement. However, that answer, too, should be satisfactory, as long as the shapes are willing to accept "what is such" as someone's designation. This has a degree of safety.

Now the same account, in fact, holds also for that nature which receives all the bodies. We must always refer to it by the same term, for it does not depart from its own character in any way. Not only does it always receive all things, it has never in any way whatever taken on any characteristic similar to any of the things that enter it. Its nature is to be available for anything to make its impression upon, and it is modified, shaped and reshaped by the things that enter it. These are the things that make it appear different at different times. The things that enter and leave it are imitations of those things that always are, imprinted after their likeness in a marvellous way that is hard to describe. This is something we shall pursue at another time. For the moment, we need to keep in mind three types of things: that which comes to be, that in which it comes to be, and that after which the thing coming to be is modeled, and which is the source of its coming to be. It is in fact appropriate to compare the receiving thing to a mother, the source to a father, and the nature between them to their offspring. We also must understand that if the imprints are to be varied, with all the varieties there to see, this thing upon which the imprints are to be formed could not be well prepared for that role if it were not itself devoid of any of those characters that it is to receive from elsewhere. For if it resembled any of the things that enter it, it could not successfully copy their opposites or things of a totally different nature whenever it were to receive them. It would be showing its own face as well. This is why the thing that is to receive in itself all the elemental kinds must be totally devoid of any characteristics. Think of people who make fragrant ointments. They expend skill and ingenuity to come up with something just like this [i.e., a neutral base], to have on hand to start with. The liquids that are to receive the fragrances they make as odorless as possible. Or think of people who work at impressing shapes upon soft materials. They emphatically refuse to allow any such material to already have some definite shape. Instead, they'll even it out and make it as smooth as it can be. In the same way, then, if the thing that is to receive repeatedly throughout its whole self the likenesses of the intelligible objects, the things which always are – if it is to do so successfully, then it ought to be devoid of any inherent characteristics of its own. This, of course, is the reason why we shouldn't call the mother or receptacle of what has come to be, of what is visible or perceivable in every other way, either earth or air, fire or water, or any of their compounds or their constituents. But if we speak of it as an invisible and characterless sort of thing, one that receives all things and shares in a most perplexing way in what is intelligible, a thing extremely difficult to comprehend, we shall not be misled. And in so far as it is possible to arrive at its nature on the basis of what we've said so far, the most correct way to speak of it may well be this: the part of it that gets ignited appears on each occasion as fire, the dampened part as water, and parts as earth or air in so far as it receives the imitations of these.

[. . .]

The third type is space, which exists always and cannot be destroyed. It provides a fixed state for all things that come to be. It is itself apprehended by a kind of bastard reasoning that does not involve sense perception, and it is hardly even an object of conviction. We look at it as in a dream when we say that everything that exists must of necessity be somewhere, in some place and occupying some space, and that that which doesn't exist somewhere, whether on earth or in heaven, doesn't exist at all.

[. . .]

Now as the wetnurse of becoming turns watery and fiery and receives the character of earth and air, and as it acquires all the properties that come with these characters, it takes on a variety of visible aspects, but because it is filled with powers that are neither similar nor evenly balanced, no part of it is in balance. It sways irregularly in every direction as it is shaken by those things, and being set in motion it in turn shakes them. And as they are moved, they drift continually, some in one direction and others in others, separating from one another. They are winnowed out, as it were, like grain that is sifted by winnowing sieves or other such implements. They are carried off and settle down, the dense and heavy ones in one direction, and the rare and light ones to another place.

That is how at that time the four kinds were being shaken by the receiver, which was itself agitating like a shaking machine, separating the kinds most unlike each other furthest apart and pushing those most like each other closest together into the same region. This, of course, explains how these different kinds came to occupy different regions of space, even before the universe was set in order and constituted from them at its coming to be.

[. . .]

Everyone knows, I'm sure, that fire, earth, water and air are bodies. Now everything that has bodily form also has depth. Depth, moreover, is of necessity comprehended within surface, and any surface bounded by straight lines is composed of triangles. Every triangle, moreover, derives from two triangles, each of which has one right angle and two acute angles. Of these two triangles, one [the isosceles right-angled triangle] has at each of the other two vertices an equal part of a right angle, determined by its division by equal sides; while the other [the scalene right-angled triangle] has unequal parts of a right angle at its other two vertices, determined by the division of the right angle by unequal sides. This, then, we presume to be the originating principle of fire and of the other bodies, as we pursue our likely account in terms of Necessity. Principles yet more ultimate than these are known only to the god, and to any man he may hold dear.

1.5

The Structure and Motion of the Heavenly Spheres

Aristotle

Aristotle (384–322 BC), the most famous pupil of Plato and the personal tutor of Alexander the Great, was perhaps the most influential philosopher of all time. In this selection from the twelfth book of his *Metaphysics*, Aristotle makes the connection between his conception of substances and the details of his cosmology. The sphere of the stars and the spheres of individual planets must each be moved by something else, a substance unmovable in itself and eternal. The actual motions of the planets, however, are very troublesome; each planet exhibits multiple motions and requires a separate cause for each motion. To take the sun and moon into account as well requires a total of 55 spheres. Aristotle's discussion is noteworthy for the caution he displays, repeatedly referring to his conjectures as "reasonable" rather than as demonstrative and leaving the issue of whether the number of the spheres is a matter of necessity to "more powerful minds."

We must not ignore the question whether we have to suppose one such substance or more than one, and if the latter, how many; we must also mention, regarding the opinions expressed by others, that they have said nothing that can even be clearly stated about the number of the substances. For the theory of Ideas has no special discussion of the subject; for those who believe in Ideas say the Ideas are numbers, and they speak of numbers now as unlimited, now as limited by the number 10; but as for the reason why there should be just so many numbers, nothing is said with any demonstrative exactness.

We however must discuss the subject, starting from the presuppositions and distinctions we have mentioned. The first principle or primary being is not movable either in itself or accidentally, but produces the primary eternal and single movement. And since that which is moved must be moved by something, and the first

From *The Complete Works of Aristotle*, ed. Jonathan Barnes, Vol. 2 (Princeton, NJ: Princeton University Press, 1984), pp. 1695–8.

mover must be in itself unmovable, and eternal movement must be produced by something eternal and a single movement by a single thing, and since we see that besides the simple spatial movement of the universe, which we say the first and unmovable substance produces, there are other spatial movements – those of the planets – which are eternal (for the body which moves in a circle is eternal and unresting; we have proved these points in the *Physics*), each of *these* movements also must be caused by a substance unmovable in itself and eternal. For the nature of the stars is eternal, being a kind of substance, and the mover is eternal and prior to the moved, and that which is prior to a substance must be a substance. Evidently, then, there must be substances which are of the same number as the movements of the stars, and in their nature eternal, and in themselves unmovable, and without magnitude, for the reason before mentioned.

That the movers are substances, then, and that one of these is first and another second according to the same order as the movements of the stars, is evident. But in the number of movements we reach a problem which must be treated from the standpoint of that one of the mathematical sciences which is most akin to philosophy – viz. of astronomy; for this science speculates about substance which is perceptible but eternal, but the other mathematical sciences, i.e. arithmetic and geometry, treat of no substance. That the movements are more numerous than the bodies that are moved, is evident to those who have given even moderate attention to the matter; for each of the planets has more than one movement. But as to the actual number of these movements, we now – to give some notion of the subject – quote what some of the mathematicians say, that our thought may have some definite number to grasp; but, for the rest, we must partly investigate for ourselves, partly learn from other investigators, and if those who study this subject form an opinion contrary to what we have now stated, we must esteem both parties indeed, but follow the more accurate.

Eudoxus supposed that the motion of the sun or of the moon involves, in either case, three spheres, of which the first is the sphere of the fixed stars, and the second moves in the circle which runs along the middle of the zodiac, and the third in the circle which is inclined across the breadth of the zodiac; but the circle in which the moon moves is inclined at a greater angle than that in which the sun moves. And the motion of the planets involves, in each case, four spheres, and of these also the first and second are the same as the first two mentioned above (for the sphere of the fixed stars is that which moves all the other spheres, and that which is placed beneath this and has its movement in the circle which bisects the zodiac is common to all), but the *poles* of the third sphere of each planet are in the circle which bisects the zodiac, and the motion of the fourth sphere is in the circle which is inclined at an angle to the equator of the third sphere; and the poles of the third spheres are different for the other planets, but those of Venus and Mercury are the same.

Callippus made the position of the spheres the same as Eudoxus did, but while he assigned the same number as Eudoxus did to Jupiter and to Saturn, he thought two more spheres should be added to the sun and two to the moon, if we were to explain the phenomena, and one more to each of the other planets.

But it is necessary, if all the spheres combined are to explain the phenomena, that for each of the planets there should be other spheres (one fewer than those hitherto assigned) which counteract those already mentioned and bring back to the same position the first sphere of the star which in each case is situated below the star in question; for only thus can all the forces at work produce the motion of the planets. Since, then, the spheres by which the planets themselves are moved are eight and twenty-five, and of these only those by which the lowest-situated planet is moved need not be counteracted, the spheres which counteract those of the first two planets will be six in number, and the spheres which counteract those of the next four planets will be sixteen, and the number of all the spheres – those which move the planets and those which counteract these – will be fifty-five. And if one were not to add to the moon and to the sun the movements we mentioned, all the spheres will be forty-nine in number.

Let this then be taken as the number of the spheres, so that the unmovable substances and principles may reasonably be taken as just so many; the assertion of *necessity* must be left to more powerful thinkers.

If there can be no spatial movement which does not conduce to the moving of a star, and if further every being and every substance which is immune from change and in virtue of itself has attained to the best must be considered an end, there can be no other being apart from these we have named, but this must be the number of the substances. For if there are others, they will cause change as being an end of movement; but there *cannot* be other movements besides those mentioned. And it is reasonable to infer this from a consideration of the bodies that are moved; for if everything that moves is for the sake of that which is moved, and every movement belongs to something that is moved, no movement can be for the sake of itself or of another movement, but all movements must be for the sake of the stars. For if a movement is to be for the sake of a movement, this latter also will have to be for the sake of something else; so that since there cannot be an infinite regress, the end of every movement will be one of the divine bodies which move through the heaven.

Evidently there is but one heaven. For if there are many heavens as there are many men, the moving principles, of which each heaven will have one, will be one in form but in number many. But all things that are many in number have matter. (For one and the same formula applies to *many* things, e.g. the formula of man; but Socrates is *one.*) But the primary essence has not matter; for it is fulfillment. So the unmovable first mover is one both in formula and in number; therefore also that which is moved always and continuously is one alone; therefore there is one heaven alone.

Our forefathers in the most remote ages have handed down to us their posterity a tradition, in the form of a myth, that these substances are gods and that the divine encloses the whole of nature. The rest of the tradition has been added later in mythical form with a view to the persuasion of the multitude and to its legal and utilitarian expediency; they say these gods are in the form of men or like some of the other animals, and they say other things consequent on and similar to these which we have mentioned. But if we were to separate the first point from these additions and take it alone – that they thought the first substances to be gods – we must regard this as an inspired utterance, and reflect that, while probably each art and science has often been developed as far as possible and has again perished, these opinions have been preserved like relics until the present. Only thus far, then, is the opinion of our ancestors and our earliest predecessors clear to us.

1.6

Change, Natures, and Causes

Aristotle

It is easy for us to take for granted the organization of the various branches of science and the principal concepts that are used to frame scientific inquiry. But when science was first being born, none of these things was obvious. In this selection from the second book of his *Physics*, Aristotle methodically addresses a number of fundamental questions: How many elements, or simple bodies, are there? Which things have innate impulses to change, and which are changed only by external causes? What senses can be given to the notion of the natures of things, and which sense is primary for science? In how many distinct senses do we use the word "cause"? What are the relations among necessity, spontaneity, and chance?

1 Of things that exist, some exist by nature, some from other causes. By nature the animals and their parts exist, and the plants and the simple bodies (earth, fire, air, water) – for we say that these and the like exist by nature.

All the things mentioned plainly differ from things which are *not* constituted by nature. For each of them has within itself a principle of motion and of stationariness (in respect of place, or of growth and decrease, or by way of alteration). On the other hand, a bed and a coat and anything else of that sort, *qua* receiving these designations – i.e. in so far as they are products of art – have no innate impulse to change. But in so far as they happen to be composed of stone or of earth or of a mixture of the two, they *do* have such an impulse, and just to that extent – which seems to indicate that nature is a principle or cause of being moved and of being at rest in that to which it belongs primarily, in virtue of itself and not accidentally.

I say 'not accidentally', because (for instance) a man who is a doctor might himself be a cause of health to himself. Nevertheless it is not in so far as he is a patient that he possesses the art of medicine: it merely has happened that the same man is doctor and patient – and that is why these attributes are not always found together. So

From *The Complete Works of Aristotle*, ed. Jonathan Barnes, Vol. 1 (Princeton, NJ: Princeton University Press, 1984), pp. 329–42.

it is with all other artificial products. None of them has in itself the principle of its own production. But while in some cases (for instance houses and the other products of manual labour) that principle is in something else external to the thing, in others – those which may cause a change in themselves accidentally – it lies in the things themselves (but not in virtue of what they are).

Nature then is what has been stated. Things have a nature which have a principle of this kind. Each of them is a substance; for it is a subject, and nature is always in a subject.

The term 'according to nature' is applied to all these things and also to the attributes which belong to them in virtue of what they are, for instance the property of fire to be carried upwards – which is not a nature nor has a nature but is by nature or according to nature.

What nature is, then, and the meaning of the terms 'by nature' and 'according to nature', has been stated. *That* nature exists, it would be absurd to try to prove; for it is obvious that there are many things of this kind, and to prove what is obvious by what is not is the mark of a man who is unable to distinguish what is self-evident from what is not. (This state of mind is clearly possible. A man blind from birth might reason about colours.) Presumably therefore such persons must be talking about words without any thought to correspond.

Some identify the nature or substance of a natural object with that immediate constituent of it which taken by itself is without arrangement, e.g. the wood is the nature of the bed, and the bronze the nature of the statue.

As an indication of this Antiphon points out that if you planted a bed and the rotting wood acquired the power of sending up a shoot, it would not be a bed that would come up, but *wood* which shows that the arrangement in accordance with the rules of the art is merely an accidental attribute, whereas the substance is the other, which, further, persists continuously through the process.

But if the material of each of these objects has itself the same relation to something else, say bronze (or gold) to water, bones (or wood) to earth and so on, *that* (they say) would be their nature and substance. Consequently some assert earth, others fire or air or water or some or all of these, to be the nature of the things that are. For whatever any one of them supposed to have this character – whether one thing or more than one thing – this or these he declared to be the whole of substance, all else being its affections, states, or dispositions. Every such thing they held to be eternal (for it could not pass into anything else), but other things to come into being and cease to be times without number.

This then is one account of nature, namely that it is the primary underlying matter of things which have in themselves a principle of motion or change.

Another account is that nature is the shape or form which is specified in the definition of the thing.

For the word 'nature' is applied to what is according to nature and the natural in the same way as 'art' is applied to what is artistic or a work of art. We should not say in the latter case that there is anything artistic about a thing, if it is a bed only potentially, not yet having the form of a bed; nor should we call it a work of art. The same is true of natural compounds. What is potentially flesh or bone has not yet its own nature, and does not exist by nature, until it receives the form specified in the definition, which we name in defining what flesh or bone is. Thus on the second account of nature, it would be the shape or form (not separable except in statement) of things which have in themselves a principle of motion. (The combination of the two, e.g. man, is not nature but by nature.)

The form indeed is nature rather than the matter; for a thing is more properly said to be what it is when it exists in actuality than when it exists potentially. Again man is born from man but not bed from bed. That is why people say that the shape is not the nature of a bed, but the wood is – if the bed sprouted, not a bed but wood would come up. But even if the shape *is* art, then on the same principle the shape of man is his nature. For man is born from man.

Again, nature in the sense of a coming-to-be proceeds towards nature. For it is not like doctoring, which leads not to the art of doctoring but to health. Doctoring must start from the art, not lead to it. But it is not in this way that nature is related to nature. What grows *qua* growing grows from something into something. Into what then does it grow? Not into that from which it arose but into that to which it tends. The shape then is nature.

Shape and nature are used in two ways. For the privation too is in a way form. But whether in unqualified coming to be there is privation, i.e. a contrary, we must consider later.

2 We have distinguished, then, the different ways in which the term 'nature' is used.

The next point to consider is how the mathematician differs from the student of nature; for natural bodies contain surfaces and volumes, lines and points, and these are the subject-matter of mathematics.

Further, is astronomy different from natural science or a department of it? It seems absurd that the student of nature should be supposed to know the nature of sun or moon, but not to know any of their essential attributes, particularly as the writers on nature obviously do discuss their shape and whether the earth and the world are spherical or not.

Now the mathematician, though he too treats of these things, nevertheless does not treat of them as the limits of a natural body; nor does he consider the attributes indicated as the attributes of such bodies. That is why he separates them; for in thought they are separable from motion, and it makes no difference, nor does any falsity result, if they are separated. The holders of the theory of Forms do the same, though they are not aware of it; for they separate the objects of natural science, which are less separable than those of mathematics. This becomes plain if one tries to state in each of the two cases the definitions of the things and of their attributes. Odd and even, straight and curved, and likewise number, line, and figure, do not involve motion; not so flesh and bone and man – *these* are defined like snub nose, not like curved.

Similar evidence is supplied by the more natural of the branches of mathematics, such as optics, harmonics, and astronomy. These are in a way the converse of geometry. While geometry investigates natural lines but not *qua* natural, optics investigates mathematical lines, but *qua* natural, not *qua* mathematical.

Since two sorts of thing are called nature, the form and the matter, we must investigate its objects as we would the essence of snubness, that is neither independently of matter nor in terms of matter only. Here too indeed one might raise a difficulty. Since there are two natures, with which is the student of nature concerned? Or should he investigate the combination of the two? But if the combination of the two, then also each severally. Does it belong then to the same or to different sciences to know each severally?

If we look at the ancients, natural science would seem to be concerned with the *matter*. (It was only very slightly that Empedocles and Democritus touched on form and essence.)

But if on the other hand art imitates nature, and it is the part of the same discipline to know the form and the matter up to a point (e.g. the doctor has a knowledge of health and also of bile and phlegm, in which health is realized and the builder both of the form of the house and of the matter, namely that it is bricks and beams, and so forth): if this is so, it would be the part of natural science also to know nature in both its senses.

Again, that for the sake of which, or the end, belongs to the same department of knowledge as the means. But the nature is the end or that for the sake of which. For if a thing undergoes a continuous change toward some end, that last stage is actually that for the sake of which. (That is why the poet was carried away into making an absurd statement when he said 'he has the end for the sake of which he was born'. For not every stage that is last claims to be an end, but only that which is best.)

For the arts make their material (some simply make it, others make it serviceable), and we use everything as if it was there for our sake. (We also are in a sense an end. 'That for the sake of which' may be taken in two ways, as we said in our work *On Philosophy*.) The arts, therefore, which govern the matter and have knowledge are two, namely the art which uses the product and the art which directs the production of it. That is why the using art also is in a sense directive; but it differs in that it knows the form, whereas the art which is directive as being concerned with production knows the matter. For the helmsman knows and prescribes what sort of form a helm should have, the other from what wood it should be made and by means of what operations. In the products of art, however, we make the material with a view to the function, whereas in the products of nature the matter is there all along.

Again, matter is a relative thing – for different forms there is different matter.

How far then must the student of nature know the form or essence? Up to a point, perhaps, as the doctor must know sinew or the smith bronze (i.e. until he understands the purpose of each); and the student of nature is concerned only with things whose forms are separable indeed, but do not exist apart from matter. Man is begotten by man and by the sun as well. The mode of existence and essence of the separable it is the business of first philosophy to define.

3 Now that we have established these distinctions, we must proceed to consider causes, their character and number. Knowledge is the object of our inquiry, and men do not think they know a thing till they have grasped the 'why' of it (which is to grasp its primary cause). So clearly we too must do this as regards both coming to be and passing away and every kind of natural change, in order that, knowing their principles, we may try to refer to these principles each of our problems.

In one way, then, that out of which a thing comes to be and which persists, is called a cause, e.g. the bronze of the statue, the silver of the bowl, and the genera of which the bronze and the silver are species.

In another way, the form or the archetype, i.e. the definition of the essence, and its genera, are called causes (e.g. of the octave the relation of 2:1, and generally number), and the parts in the definition.

Again, the primary source of the change or rest; e.g. the man who deliberated is a cause, the father is cause of the child, and generally what makes of what is made and what changes of what is changed.

Again, in the sense of end or that for the sake of which a thing is done, e.g. health is the cause of walking about. ('Why is he walking about?' We say: 'To be healthy', and, having said that, we think we have assigned the cause.) The same is true also of all the intermediate steps which are brought about through the action of something else as means towards the end, e.g. reduction of flesh, purging, drugs, or surgical instruments are means towards health. All these things are for the sake of the end, though they differ from one another in that some are activities, others instruments.

This then perhaps exhausts the number of ways in which the term 'cause' is used.

As things are called causes in many ways, it follows that there are several causes of the same thing (not merely accidentally), e.g. both the art of the sculptor and the bronze are causes of the statue. These are causes of the statue qua statue, not in virtue of anything else that it may be – only not in the same way, the one being the material cause, the other the cause whence the motion comes. Some things cause each other reciprocally, e.g. hard work causes fitness and vice versa, but again not in the same way, but the one as end, the other as the principle of motion. Further the same thing is the cause of contrary results. For that which by its presence brings about one result is sometimes blamed for bringing about the contrary by its absence. Thus we ascribe the wreck of a ship to the absence of the pilot whose presence was the cause of its safety.

All the causes now mentioned fall into four familiar divisions. The letters are the causes of syllables, the material of artificial products, fire and the like of bodies, the parts of the whole, and the premisses of the conclusion, in the sense or 'that from which'. Of these pairs the one set are causes in the sense of what underlies, e.g. the parts, the other set in the sense of essence – the whole and the combination and the form. But the seed and the doctor and the deliberator, and generally the maker, are all sources whence the change or stationariness originates, which the others are causes in the sense of the end or the good of the rest; for that for the sake of which tends to be what is best and the end of the things that lead up to it. (Whether we call it good or apparently good makes no difference.)

Such then is the number and nature of the kinds of cause.

Now the modes of causation are many, though when brought under heads they too can be reduced in number. For things are called causes in many ways and even within the same kind one may be prior to another: e.g. the doctor and the expert are causes of health, the relation 2:1 and number of the octave, and always what is inclusive to what is particular. Another mode of causation is the accidental and its genera, e.g. in one way Polyclitus, in another a sculptor is the cause of a statue, because being Polyclitus and a sculptor are accidentally conjoined. Also the classes in which the accidental attribute is included; thus a man could be said to be the cause of a statue or, generally, a living creature. An accidental attribute too may be more or less remote, e.g. suppose that a pale man or a musical man were said to be the cause of the statue.

All causes, both proper and accidental, may be spoken of either as potential or as actual; e.g. the cause of a house being built is either a house-builder or a house-builder building.

Similar distinctions can be made in the things of which the causes are causes, e.g. of this statue or of a statue or of an image generally, of this bronze or of bronze or of material generally. So too with the accidental attributes. Again we may use a complex expression for either and say, e.g., neither 'Polyclitus' nor a 'sculptor' but 'Polyclitus, the sculptor'.

All these various uses, however, come to six in number, under each of which again the usage is twofold. It is either what is particular or a genus, or an accidental attribute or a genus of that, and these either as a complex or each by itself; and all either as actual or as potential. The difference is this much, that causes which are actually at work and particular exist and cease to exist simultaneously with their effect, e.g. this healing person with this being-healed person and that house-building man with that being-built house; but this is not always true of potential causes – the house and the housebuilder do not pass away simultaneously.

In investigating the cause of each thing it is always necessary to seek what is most precise (as also in other things): thus a man builds because he is a builder, and a builder builds in virtue of his art of building. This last cause then is prior; and so generally.

Further, generic effects should be assigned to generic causes, particular effects to particular causes, e.g. statue to sculptor, this statue to this sculptor; and powers are relative to possible effects, actually operating causes to things which are actually being effected.

This must suffice for our account of the number of causes and the modes of causation.

4 But chance and spontaneity are also reckoned among causes: many things are said both to be and to come to be as a result of chance and spontaneity. We must inquire therefore in what manner chance and spontaneity are present among the causes enumerated, and whether they are the same or different, and generally what chance and spontaneity are.

Some people even question whether there are such things or not. They say that nothing happens by chance, but that everything which we ascribe to chance or spontaneity has some definite cause, e.g. coming by chance into the market and finding there a man whom one wanted but did not expect to meet is due to one's wish to go and buy in the market. Similarly, in other so-called cases of chance it is always possible, they maintain, to find something which is the cause; but not chance, for if chance were real, it would seem strange indeed, and the question might be raised, why on earth none of the wise men of old in speaking of the causes of generation and decay took account of chance; whence it would seem that they too did not believe that anything is by chance. But there is a further circumstance that is surprising. Many things both come to be and are by chance and spontaneity, and although all know that each of them can be ascribed to some cause (as the old argument said which denied chance), nevertheless they all speak of some of these things as happening by chance and others not. For this reason they ought to have at least referred to the matter in some way or other.

Certainly the early physicists found no place for chance among the causes which they recognized – love, strife, mind, fire, or the like. This is strange, whether they supposed that there is no such thing as chance or whether they thought there is but omitted to mention it – and that too when they sometimes used it, as Empedocles does when he says that the air is not always separated into the highest region, but as it may chance. At any rate he says in his cosmogony that 'it happened to run that way at that time, but it often ran otherwise'. He tells us also that most of the parts of animals came to be by chance.

There are some who actually ascribe this heavenly sphere and all the worlds to spontaneity. They say that the vortex arose spontaneously, i.e. the motion that separated and arranged the universe in its present order. This statement might well cause surprise. For they are asserting that chance is not responsible for the existence or generation of animals and plants, nature or mind or something of the kind being the cause of them (for it is not any chance thing that comes from a given seed but an olive from one kind and a man from another); and yet at the same time they assert that the heavenly sphere and the divinest of visible things arose spontaneously, having no such cause as is assigned to animals and plants. Yet if this is so, it is a fact which deserves to be dwelt upon, and something might well have been said about it. For besides the other absurdities of the statement, it is the more absurd that people should make it when they see nothing coming to be spontaneously in the heavens, but much happening by chance among the things which as they say are not due to chance; whereas we should have expected exactly the opposite.

Others there are who believe that chance is a cause, but that it is inscrutable to human intelligence, as being a divine thing and full of mystery.

Thus we must inquire what chance and spontaneity are, whether they are the same or different, and how they fit into our division of causes.

5 First then we observe that some things always come to pass in the same way, and others for the most part. It is clearly of neither of these that chance, or the result of chance, is said to be the cause – neither of that which is by necessity and always, nor of that which is for the most part. But as there is a third class of events besides these two – events which all say are by chance – it is plain that there is such a thing as chance and spontaneity; for we know that things of this kind are due to chance and that things due to chance are of this kind.

Of things that come to be, some come to be for the sake of something, others not. Again, some of the former class are in accordance with intention, others not, but both are in the class of things which are for the sake of something. Hence it is clear that even among the things which are outside what is necessary and what is for the most part, there are some in connexion with which the phrase 'for the sake of something' is applicable. (Things that are for the sake of something include whatever may be done as a result of thought or of nature.) Things of this kind, then, when they come to pass accidentally are said to be by chance. For just as a thing is something either in virtue of itself or accidentally, so may it be a cause. For instance, the housebuilding faculty is in virtue of itself a cause of a house, whereas the pale or the musical is an accidental cause. That which is *per se* cause is determinate, but the accidental cause is indeterminable; for the possible attributes of an individual are innumerable. As we said, then, when a thing of this kind comes to pass among events which are for the sake of something, it is said to be spontaneous or by chance. (The distinction between the two must be made later – for the present it is sufficient if it is plain that both are in the sphere of things done for the sake of something.)

Example: A man is engaged in collecting subscriptions for a feast. He would have gone to such and such a place for the purpose of getting the money, if he had known. He actually went there for another purpose, and it was only accidentally that he got his money by going there; and this was not due to the fact that he went there as a rule or necessarily, nor is the end effected (getting the money) a cause present in himself – it belongs to the class of things that are objects of choice and the result of thought. It is when these conditions are satisfied that the man is said to have gone by chance. If he had chosen and gone for the sake of this – if he always or normally went there when he was collecting payments – he would not be said to have gone by chance.

It is clear then that chance is an accidental cause in the sphere of those actions for the sake of something which involve choice. Thought, then, and chance are in the same sphere, for choice implies thought.

It is necessary, no doubt, that the causes of what comes to pass by chance be indefinite; and that is why chance is supposed to belong to the class of the indefinite and to be inscrutable to man, and why it might be thought that, in a way, nothing occurs by chance. For all these statements are correct, as might be expected. Things *do*, in a way, occur by chance, for they occur accidentally and chance is an accidental cause. But it is not the cause without qualification of anything; for instance, a housebuilder is the cause of a house; accidentally, a fluteplayer may be so.

And the causes of the man's coming and getting the money (when he did not come for the sake of that) are innumerable. He may have wished to see somebody or been following somebody or avoiding somebody, or may have gone to see a spectacle. Thus to say that chance is unaccountable is correct. For an account is of what holds always or for the most part, whereas chance belongs to a third type of event. Hence, since causes of this kind are indefinite, chance too is indefinite. (Yet in some cases one might raise the question whether *any* chance fact might be the cause of the chance occurrence, e.g. of health the fresh air or the sun's heat may be the cause, but having had one's hair cut *cannot*; for some accidental causes are more relevant to the effect than others.)

Chance is called good when the result is good, evil when it is evil. The terms 'good fortune' and 'ill fortune' are used when either result is of considerable magnitude. Thus one who comes within an ace of some great evil or great good is said to be fortunate or unfortunate. The mind affirms the presence of the attribute, ignoring the hair's breadth of difference. Further, it is with reason that good fortune is regarded as unstable; for chance is unstable, as none of the things which result from it can hold always or for the most part.

Both are then, as I have said, accidental causes – both chance and spontaneity – in the sphere of things which are capable of coming to pass not simply, nor for the most part and with reference to such of these as might come to pass for the sake of something.

6 They differ in that spontaneity is the wider. Every result of chance is from what is spontaneous, but not everything that is from what is spontaneous is from chance.

Chance and what results from chance are appropriate to agents that are capable of good fortune and of action generally. Therefore necessarily chance is in the sphere of actions. This is indicated by the fact that good fortune is thought to be the same, or nearly the same, as happiness, and happiness to be a kind of action, since it is well-doing. Hence what is not capable of action cannot do anything by chance. Thus an inanimate thing or a beast or a child cannot do anything by chance, because it is incapable of choice; nor can good fortune or ill fortune be ascribed to them, except metaphorically, as Protarchus, for example, said that the stones of which altars are made are fortunate because they are held in honour, while their fellows are trodden under foot. Even these things, however, can in a way be affected by chance, when one who is dealing with them does something to them by chance, but not otherwise.

The spontaneous on the other hand is found both in the beasts and in many inanimate objects. We say, for example, that the horse came spontaneously, because, though his coming saved him, he did not come for the sake of safety. Again, the tripod fell spontaneously, because, though it stood on its feet so as to serve for a seat, it did not fall so as to serve for a seat.

Hence it is clear that events which belong to the general class of things that may come to pass for the sake of something, when they come to pass not for the sake of what actually results, and have an external cause, may be described by the phrase 'from spontaneity'. These spontaneous events are said to be from chance if they have the further characteristics of being the objects of choice and happening to agents capable of choice. This is indicated by the phrase 'in vain', which is used when one thing which is for the sake of another, does not result in it. For instance, taking a walk is for the sake of evacuation of the bowels; if this does not follow after walking, we say that we have walked in vain and that the walking was vain. This implies that what is naturally for the sake of an end is in vain, when it does not effect the end for the sake of which it was the natural means – for it would be absurd for a man to say that he had bathed in vain because the sun was not eclipsed, since the one was not done for the sake of the other. Thus the spontaneous is even according to its derivation the case in which the thing itself happens in vain. The stone that struck the man did not fall for the sake of striking him; therefore it fell spontaneously, because it might have fallen by the action of an agent and for the sake of striking. The difference between spontaneity and what results by chance is greatest in things that come to be by nature; for when anything comes to be contrary to nature, we do not say that it came to be by chance, but by spontaneity. Yet strictly this too is different from the spontaneous proper; for the cause of the latter is external, that of the former internal.

We have now explained what chance is and what spontaneity is, and in what they differ from each other. Both belong to the mode of causation 'source of change', for either some natural or some intelligent agent is always the cause; but in this sort of causation the number of possible causes is infinite.

Spontaneity and chance are causes of effects which, though they might result from intelligence or nature, have in fact been caused by something accidentally. Now since nothing which is accidental is prior to what is *per se*, it is clear that no accidental cause can be prior to a cause *per se*. Spontaneity and chance, therefore, are posterior to intelligence and nature. Hence, however true it may be that the heavens are due to spontaneity, it will still be true that intelligence and nature will be prior causes of this universe and of many things in it besides.

7 It is clear then that there are causes, and that the number of them is what we have stated. The number is the same as that of the things comprehended under the question 'why'. The 'why' is referred ultimately either, in things which do not involve motion, e.g. in mathematics, to the 'what' (to the definition of straight line or commensurable or the like); or to what initiated a motion, e.g. 'why did they go to war? – because there had been a raid'; or we are inquiring 'for the sake of what?' – 'that they may rule'; or in the case of things that come into being, we are looking for the matter. The causes, therefore, are these and so many in number.

Now, the causes being four, it is the business of the student of nature to know about them all, and if he refers his problems back to all of them, he will assign the 'why' in the way proper to his science – the matter, the form, the mover, that for the sake of which. The last three often coincide; for the what and that for the sake of which are one, while the primary source of motion is the same in species as these. For man generates man – and so too, in general, with all things which cause movement by being themselves moved; and such as are not of this kind are no longer inside the province of natural science, for they cause motion not by possessing motion or a source of motion in themselves, but being themselves incapable of motion. Hence there are three branches of study, one of things which are incapable of motion, the second of things in motion, but indestructible, the third of destructible things.

The question 'why', then, is answered by reference to the matter, to the form, and to the primary moving cause. For in respect of coming to be it is mostly in this last way that causes are investigated – 'what comes to be after what? what was the primary agent or patient?' and so at each step of the series.

Now the principles which cause motion in a natural way are two, of which one is not natural, as it has no principle of motion in itself. Of this kind is whatever causes movement, not being itself moved, such as that which is completely unchangeable, the primary reality, and the essence of a thing, i.e. the form; for this is the end or that for the sake of which. Hence since nature is for the sake of something, we must know this cause also. We must explain the 'why' in all the senses of the term, namely, that from this that will necessarily result ('from this' either without qualification or for the most part); that this must be so if that is to be so (as the conclusion presupposes the premisses); that this was the essence of the thing; and because it is better thus (not without qualification, but with reference to the substance in each case).

8 We must explain then first why nature belongs to the class of causes which act for the sake of something; and then about the necessary and its place in nature, for all writers ascribe things to this cause, arguing that since the hot and the cold and the like are of such and such a kind, therefore certain things *necessarily are* and come to be – and if they mention any other cause (one friendship and strife, another mind), it is only to touch on it, and then good-bye to it.

A difficulty presents itself: why should not nature work, not for the sake of something, nor because it is better so, but just as the sky rains, not in order to make the corn grow, but of necessity? (What is drawn up must cool, and what has been cooled must become water and descend, the result of this being that the corn grows.) Similarly if a man's crop is spoiled on the threshing-floor, the rain did not fall for the sake of this – in order that the crop might be spoiled – but that result just followed. Why then should it not be the same with the parts in nature, e.g. that our teeth should come up of necessity – the front teeth sharp, fitted for tearing, the molars broad and useful for grinding down the food – since they did not arise for this end, but it was merely a coincident result; and so with all other parts in which we suppose that there is purpose? Wherever then all the parts came about just what they would have been if they had come to be for an end, such things survived, being organized spontaneously in a fitting way; whereas those which grew otherwise perished and continue to perish, as Empedocles says his 'man-faced oxprogeny' did.

Such are the arguments (and others of the kind) which may cause difficulty on this point. Yet it is impossible that this should be the true view. For teeth and all other natural things either invariably or for the most part come about in a given way; but of not one of the results of chance or spontaneity is this true. We do not ascribe to chance or mere coincidence the frequency of rain in winter, but frequent rain in summer we do; nor heat in summer but only if we have it in winter. If then, it is agreed that things are either the result of coincidence or for the sake of something, and these cannot be the result of coincidence or spontaneity, it follows that they must be for the sake of something; and that such things are all due to nature even the champions of the theory which is before us would agree. Therefore action for an end is present in things which come to be and are by nature.

Further, where there is an end, all the preceding steps are for the sake of that. Now surely as in action, so in nature; and as in nature, so it is in each action, if nothing interferes. Now action is for the sake of an end; therefore the nature of things also is so. Thus if a house, e.g., had been a thing made by nature, it would have been made in the same way as it is now by art; and if things made by nature were made not only by nature but also by art, they would come to be in the same way as by nature. The one, then, is for the sake of the other; and generally art in some cases completes what nature cannot bring to a finish, and in others imitates nature. If, therefore, artificial products are for the sake of an end, so clearly also are natural products. The relation of the later to the earlier items is the same in both.

This is most obvious in the animals other than man: they make things neither by art nor after inquiry or deliberation. That is why people wonder whether it is by intelligence or by some other faculty that these creatures work, – spiders, ants, and the like. By gradual advance in this direction we come to see clearly that in plants too that is produced which is conducive to the end – leaves, e.g. grow to provide shade for the fruit. If then it is both by nature and for an end that the swallow makes its nest and the spider its web, and plants grow leaves for the sake of the fruit and send their roots down (not up) for the sake of nourishment, it is plain that this kind of cause is operative in things which come to be and are by nature. And since nature is twofold, the matter and the form, of which the latter is the end, and since all the rest is for the sake of the end, the form must be the cause in the sense of that for the sake of which.

Now mistakes occur even in the operations of art: the literate man makes a mistake in writing and the doctor pours out the wrong dose. Hence clearly mistakes are possible in the operations of nature also. If then in art there are cases in which what is rightly produced serves a purpose, and if where mistakes occur there was a purpose in what was attempted, only it was not attained, so must it be also in natural products, and monstrosities will be failures in the purposive effort. Thus in the original combinations the 'ox-progeny', if they failed to reach a determinate end must have arisen through the corruption of some principle, as happens now when the seed is defective.

Further, seed must have come into being first, and not straightway the animals: what was 'undifferentiated first' was seed.

Again, in plants too we find that for the sake of which, though the degree of organization is less. Were there then in plants also olive-headed vine-progeny, like the 'man-headed ox-progeny', or not? An absurd suggestion; yet there must have been, if there were such things among animals.

Moreover, among the seeds anything must come to be at random. But the person who asserts this entirely does away with nature and what exists by nature. For those things are natural which, by a continuous movement originated from an internal principle, arrive at some end: the same end is not reached from every principle; nor any chance end, but always the tendency in each is towards the same end, if there is no impediment.

The end and the means towards it may come about by chance. We say, for instance, that a stranger has come by chance, paid the ransom, and gone away, when he does so as if he had come for that purpose, though it was not for that that he came. This is accidental, for chance is an accidental cause, as I remarked before. But when an event takes place always or for the most part, it is not accidental or by chance. In natural products the sequence is invariable, if there is no impediment.

It is absurd to suppose that purpose is not present because we do not observe the agent deliberating. Art does not deliberate. If the ship-building art were in the wood, it would produce the same results by nature. If, therefore, purpose is present in art, it is present also in nature. The best illustration is a doctor doctoring himself: nature is like that.

It is plain then that nature is a cause, a cause that operates for a purpose.

9 As regards what is of necessity, we must ask whether the necessity is hypothetical, or simple as well. The current view places what is of necessity in the process of production, just as if one were to suppose that the wall of a house necessarily comes to be because what is heavy is naturally carried downwards and what is light to the top, so that the stones and foundations take the lowest place, with earth above because it is lighter, and wood at the top of all as being the lightest. Whereas, though the wall does not come to be *without* these, it is not *due* to these, except as its material cause: it comes to be for the sake of sheltering and guarding certain things. Similarly in all other things which involve that for the sake of which: the product cannot come to be without things which have a necessary nature, but it is not due to these (except as its material); it comes to be for an end. For instance, why is a saw such as it is? To effect so-and-so and for the sake of so-and-so. This end, however, cannot be realized unless the saw is made of iron. It is, therefore, necessary for it to be of iron, if we are to have a saw and perform the operation of sawing. What is necessary then, is necessary on a hypothesis, not as an end. Necessity is in the matter, while that for the sake of which is in the definition.

Necessity in mathematics is in a way similar to necessity in things which come to be through the operation of nature. Since a straight line is what it is, it is necessary that the angles of a triangle should equal two right angles. But not conversely; though if the angles are *not* equal to two right angles, then the straight line is not what it is either. But in things which come to be for an end, the reverse is true. If the end is to exist or does exist, that also which precedes it will exist or does exist; otherwise just as there, if the conclusion is not true, the principle will not be true, so here the end or that for the sake of which will not exist. For this too is itself a principle, but of the reasoning, not of the action. (In mathematics the principle is the principle of the reasoning only, as there is no action.) If then there is to be a house, such-and-such things must be made or be there already or exist, or generally the matter relative to the end, bricks and stones if it is a house. But the end is not due to these except as the matter, nor will it come to exist because of them. Yet if they do not exist at all, neither will the house, or the saw – the former in the absence of stones, the latter in the absence of iron – just as in the other case the principles will not be true, if the angles of the triangle are not equal to two right angles.

The necessary in nature, then, is plainly what we call by the name of matter, and the changes in it. Both causes must be stated by the student of nature, but especially the end; for that is the cause of the matter, not *vice versa*; and the end is that for the sake of which, and the principle starts from the definition or essence: as in artificial products, since a house is of such-and-such a kind, certain things must *necessarily* come to be or be there already, or since health is this, these things must necessarily come to be or be there already, so too if man is this, then these; if these, then those. Perhaps the necessary is present also in the definition. For if one defines the operation of sawing as being a certain kind of dividing, then this cannot come about unless the saw has teeth of a certain kind; and these cannot be unless it is of iron. For in the definition too there are some parts that stand as matter.

1.7

Scientific Inference and the Knowledge of Essential Natures

Aristotle

Aristotle's work broke new ground not only in science but also in logic. The two pursuits are related. For to do science well we must reason well, and this requires that we understand the structure of good reasoning and recognize the difference between cogent reasoning and sophistry. In this selection from the first book of the *Posterior Analytics*, Aristotle discusses the nature of demonstrative inference and then argues that the proper subject matter of scientific demonstration is not just what happens accidentally but rather what belongs to things in themselves, on account of their essences, or real natures. A knowledge of those natures is, therefore, required for scientific thought.

1 All teaching and all intellectual learning come about from already existing knowledge. This is evident if we consider it in every case; for the mathematical sciences are acquired in this fashion, and so is each of the other arts. And similarly too with arguments – both deductive and inductive arguments proceed in this way; for both produce their teaching through what we are already aware of, the former getting their premisses as from men who grasp them, the latter proving the universal through the particular's being clear. (And rhetorical arguments too persuade in the same way; for they do so either through examples, which is induction, or through enthymemes, which is deduction.)

It is necessary to be already aware of things in two ways: of some things it is necessary to believe already that they are, of some one must grasp what the thing said is, and of others both – e.g. of the fact that everything is either affirmed or denied truly, one must believe that it is; of the triangle, that it signifies *this*; and of the unit both (both what it signifies and that it is). For each of these is not equally clear to us.

But you can become familiar by being familiar earlier with some things but getting knowledge

From *The Complete Works of Aristotle*, ed. Jonathan Barnes, Vol. 1 (Princeton, NJ: Princeton University Press, 1984), extracts from pp. 114–7, 120–2.

of the others at the very same time – i.e. of whatever happens to be under the universal of which you have knowledge. For that every triangle has angles equal to two right angles was already known; but that there is a triangle in the semicircle here became familiar at the same time as the induction. (For in some cases learning occurs in this way, and the last term does not become familiar through the middle – in cases dealing with what are in fact particulars and not said of any underlying subject.)

Before the induction, or before getting a deduction, you should perhaps be said to understand in a way – but in another way not. For if you did not know if it is *simpliciter*, how did you know that it has two right angles *simpliciter*? But it is clear that you understand it in *this* sense – that you understand it universally – but you do not understand it *simpliciter*. (Otherwise the puzzle in the *Meno* will result; for you will learn either nothing or what you know.)

For one should not argue in the way in which some people attempt to solve it: Do you or don't you know of every pair that it is even? And when you said Yes, they brought forward some pair of which you did not think that it was, nor therefore that it was even. For they solve it by denying that people know of every pair that it is even, but only of anything of which they know that it is a pair. – Yet they know it of that which they have the demonstration about and which they got their premisses about; and they got them not about everything of which they know that it is a triangle or that it is a number, but of every number and triangle *simpliciter*. For no proposition of such a type is assumed (that *what you know to be a number* . . . or *what you know to be rectilineal* . . .), but they are assumed as holding of every case.

But nothing, I think, prevents one from in a sense understanding and in a sense being ignorant of what one is learning; for what is absurd is not that you should know in some sense what you are learning, but that you should know it in *this* sense, i.e. in the way and sense in which you are learning it.

2 We think we understand a thing *simpliciter* (and not in the sophistic fashion accidentally) whenever we think we are aware both that the explanation because of which the object is is its explanation, and that it is not possible for this to be otherwise. It is clear, then, that to understand is something of this sort; for both those who do not understand and those who do understand – the former think they are themselves in such a state, and those who do understand actually are. Hence that of which there is understanding *simpliciter* cannot be otherwise.

Now whether there is also another type of understanding we shall say later; but we say now that we do know through demonstration. By demonstration I mean a scientific deduction; and by scientific I mean one in virtue of which, by having it, we understand something.

If, then, understanding is as we posited, it is necessary for demonstrative understanding in particular to depend on things which are true and primitive and immediate and more familiar than and prior to and explanatory of the conclusion (for in this way the principles will also be appropriate to what is being proved). For there will be deduction even without these conditions, but there will not be demonstration; for it will not produce understanding.

Now they must be true because one cannot understand what is not the case – e.g. that the diagonal is commensurate. And they must depend on what is primitive and non-demonstrable because otherwise you will not understand if you do not have a demonstration of them; for to understand that of which there is a demonstration non-accidentally is to have a demonstration. They must be both explanatory and more familiar and prior – explanatory because we only understand when we know the explanation; and prior, if they are explanatory, and we are already aware of them not only in the sense of grasping them but also of knowing that they are.

Things are prior and more familiar in two ways; for it is not the same to be prior by nature and prior in relation to us, nor to be more familiar and more familiar to us. I call prior and more familiar in relation to us what is nearer to perception, prior and more familiar *simpliciter* what is further away. What is most universal is furthest away, and the particulars are nearest; and these are opposite to each other.

Depending on things that are primitive is depending on appropriate principles; for I call the same thing primitive and a principle. A principle of a demonstration is an immediate proposition, and an immediate proposition is one to which there is no other prior. A proposition is the one part of a contradiction, one thing said of one; it is dialectical if it assumes indifferently either part, demonstrative if it determinately assumes the one that is true. [A statement is either part of a contradiction.] A contradiction is an opposition of which of itself excludes any intermediate; and the part of a contradiction saying something *of* something is an affirmation, the one saying something *from* something is a denial.

An immediate deductive principle I call a posit if one cannot prove it but it is not necessary for anyone who is to learn anything to grasp it; and one which it is necessary for anyone who is going to learn anything whatever to grasp, I call an axiom (for there are some such things); for we are accustomed to use this name especially of such things. A posit which assumes either of the parts of a contradiction – i.e., I mean, that something is or that something is not – I call a supposition; one without this, a definition. For a definition is a posit (for the arithmetician posits that a unit is what is quantitatively indivisible) but not a supposition (for what a unit is and that a unit is are not the same).

Since one should both be convinced of and know the object by having a deduction of the sort we call a demonstration, and since this is the case when *these* things on which the deduction depends are the case, it is necessary not only to be already aware of the primitives (either all or some of them) but actually to be better aware of them. For a thing always belongs better to that thing because of which it belongs – e.g. that because of which we love is better loved. Hence if we know and are convinced because of the primitives, we both know and are convinced of them better, since it is because of them that we know and are convinced of what is posterior.

It is not possible to be better convinced than one is of what one knows, of what one in fact neither knows nor is more happily disposed toward than if one in fact knew. But this will result if someone who is convinced because of a demonstration is not already aware of the primitives, for it is necessary to be better convinced of the principles (either all or some of them) than of the conclusion.

Anyone who is going to have understanding through demonstration must not only be familiar with the principles and better convinced of them than of what is being proved, but also there must be no other thing more convincing to him or more familiar among the opposites of the principles on which a deduction of the contrary error may depend – if anyone who understands *simpliciter* must be unpersuadable.

3 Now some think that because one must understand the primitives there is no understanding at all; others that there is, but that there are demonstrations of everything. Neither of these views is either true or necessary.

For the one party, supposing that one cannot understand in another way, claim that we are led back *ad infinitum* on the grounds that we would not understand what is posterior because of what is prior if there are no primitives; and they argue correctly, for it is impossible to go through infinitely many things. And if it comes to a stop and there are principles, they say that these are unknowable since there is no *demonstration* of them, which alone they say is understanding; but if one cannot know the primitives, neither can what depends on them be understood *simpliciter* or properly, but only on the supposition that they are the case.

The other party agrees about understanding; for it, they say, occurs only through demonstration. But they argue that nothing prevents there being demonstration of everything; for it is possible for the demonstration to come about in a circle and reciprocally.

But *we* say that neither is all understanding demonstrative, but in the case of the immediates it is non-demonstrable – and that this is necessary is evident; for if it is necessary to understand the things which are prior and on which the demonstration depends, and it comes to a stop at some time, it is necessary for these immediates to be non-demonstrable. So as to that we argue thus; and we also say that there is not only understanding but also some principle of understanding by which we become familiar with the definitions.

And that it is impossible to demonstrate *simpliciter* in a circle is clear, if demonstration must depend on what is prior and more familiar; for it is impossible for the same things at the same time to be prior and posterior to the same things – unless one is so in another way (i.e. one in relation to us, the other *simpliciter*), which induction makes familiar. But if so, knowing *simpliciter* will not have been properly defined, but will be twofold. Or is the other demonstration not demonstration *simpliciter* in that it comes from about what is more familiar *to us*?

There results for those who say that demonstration is circular not only what has just been described, but also that they say nothing other than that this is the case if this is the case – and it is easy to prove everything in this way. It is clear that this results if we posit three terms. (For it makes no difference to say that it bends back through many terms or through few, or through few or two.) For whenever if *A* is the case, of necessity *B* is, and if this then *C*, then if *A* is the case *C* will be the case. Thus given that if *A* is the case it is necessary that *B* is, and if this is that *A* is (for that is what being circular is) – let *A* be *C*: so to say that if *B* is the case *A* is, is to say that *C* is, and this implies that if *A* is the case *C* is. But *C* is the same as *A*. Hence it results that those who assert that demonstration is circular say nothing but that if *A* is the case *A* is the case. And it is easy to prove everything in this way.

[. . .]

6 Now if demonstrative understanding depends on necessary principles (for what one understands cannot be otherwise), and what belongs to the objects in themselves is necessary (for in the one case it belongs in what they are; and in the other they belong in what they are to what is predicated of them, one of which opposites necessarily belongs), it is evident that demonstrative deduction will depend on things of this sort; for everything belongs either in this way or accidentally, and what is accidental is not necessary.

Thus we must either argue like this, or, positing as a principle that demonstration is necessary and that if something has been demonstrated it cannot be otherwise – the deduction, therefore, must depend on necessities. For from truths one can deduce *without* demonstrating, but from necessities one cannot deduce without demonstrating; for this is precisely the mark of demonstration.

There is evidence that demonstration depends on necessities in the fact that this is how we bring our objections against those who think they are demonstrating – saying that it is not necessary, if we think either that it is absolutely possible for it to be otherwise, or at least for the sake of argument.

From this it is clear too that those people are silly who think they get their principles correctly if the proposition is reputable and true (e.g. the sophists who assume that to understand is to have understanding). For it is not what is reputable or not that is a principle, but what is primitive in the genus about which the proof is; and not every truth is appropriate.

That the deduction must depend on necessities is evident from this too: if, when there is a demonstration, a man who has not got an account of the reason why does not have understanding, and if it might be that *A* belongs to *C* from necessity but that *B*, the middle term through which it was demonstrated, does not hold from necessity, then he does not know the reason why. For this is not so because of the middle term; for it is possible for that not to be the case, whereas the conclusion is necessary.

Again, if someone does not know now, though he has got the account and is preserved, and the object is preserved, and he has not forgotten, then he did not know earlier either. But the middle term might perish if it is not necessary; so that though, being himself preserved and the object preserved, he will have the account, yet he does not know. Therefore, he did not know earlier either. And if it has not perished but it is possible for it to perish, the result would be capable of occurring and possible; but it is impossible to know when in such a state.

Now when the conclusion is from necessity, nothing prevents the middle term through which it was proved from being non-necessary; for one can deduce a necessity from a non-necessity, just as one can deduce a truth from non-truths. But when the middle term is from necessity, the conclusion too is from necessity, just as from truths it is always true; for let *A* be said of *B* from necessity, and this of *C* – then that *A* belongs to *C* is also necessary. But when the conclusion is not necessary, the middle term cannot be necessary either; for let *A* belong to *C* not from necessity, but to *B* and this to *C* from necessity – therefore *A* will belong to *C* from necessity too; but it was supposed not to.

Since, then, if a man understands demonstratively, it must belong from necessity, it is clear that he must have his demonstration through a middle term that is necessary too; or else he will not understand either why or that it is necessary for that to be the case, but either he will think but not know it (if he believes to be necessary what is not necessary) or he will not even think it (equally whether he knows the fact through middle terms or the reason why actually through immediates).

Of accidentals which do not belong to things in themselves in the way in which things belonging in themselves were defined, there is no demonstrative understanding. For one cannot prove the conclusion from necessity; for it is possible for what is accidental not to belong – for that is the sort of accidental I am talking about. Yet one might perhaps puzzle about what aim we should have in asking these questions about them, if it is not necessary for the conclusion to be the case; for it makes no difference if one asks chance questions and then says the conclusion. But we must ask not as though the conclusion were necessary because of what was asked, but because it is necessary for anyone who says them to say it, and to say it truly if they truly hold.

Since in each kind what belongs to something in itself and as such belongs to it from necessity, it is evident that scientific demonstrations are about what belongs to things in themselves, and depend on such things. For what is accidental is not necessary, so that you do not necessarily know why the conclusion holds – not even if it should always be the case but not in itself (e.g. deductions through signs). For you will not understand in itself something that holds in itself; nor will you understand why it holds. (To understand why is to understand through the explanation.) Therefore the middle term must belong to the third, and the first to the middle, because of itself.

7 One cannot, therefore, prove anything by crossing from another genus – e.g. something geometrical by arithmetic. For there are three things in demonstrations: one, what is being demonstrated, the conclusion (this is what belongs to some genus in itself); one, the axioms (axioms are the things on which the demonstration depends); third, the underlying genus of which the demonstration makes clear the attributes and what is accidental to it in itself.

Now the things on which the demonstration depends may be the same; but of things whose genus is different – as arithmetic and geometry, one cannot apply arithmetical demonstrations to the accidentals of magnitudes, unless magnitudes are numbers. (How this is possible in some cases will be said later.)

Arithmetical demonstrations always include the genus about which the demonstration is, and so also do the others; hence it is necessary for the genus to be the same, either *simpliciter* or in some respect, if the demonstration is going to cross.

1.8

The Cosmos and the Shape and Size of the Earth

Aristotle

One of Aristotle's great contributions was his methodical exposition and critique of the theories of his predecessors to discover what was established. In this selection from the second book of *On the Heavens*, Aristotle argues that the universe is spherical, that its movement is regular, that the stars move around a motionless earth, and that the planets have an additional motion by which they work their way slowly backward against the wheeling background of the stars. As for the earth, Aristotle argues that it stands at the center of the universe, that it is motionless, that in comparison to the height of the stars it is of no great size, and that it is spherical. This last conclusion surprises some people who have been taught that Columbus discovered the sphericity of the earth; but this was in fact a commonplace among the educated from Aristotle's day through the Renaissance.

4 The shape of the heaven is of necessity spherical; for that is the shape most appropriate to its substance and also by nature primary.

First, let us consider generally which shape is primary among planes and solids alike. Every plane figure must be either rectilinear or curvilinear. Now the rectilinear is bounded by more than one line, the curvilinear by one only. But since in any kind the one is naturally prior to the many and the simple to the complex, the circle will be the first of plane figures.

[. . .]

Now the first figure belongs to the first body, and the first body is that at the farthest circumference. It follows that the body which revolves with a circular movement must be spherical. The

From *The Complete Works of Aristotle*, ed. Jonathan Barnes, Vol. 1 (Princeton, NJ: Princeton University Press, 1984), extracts from pp. 473, 474–6, 477–8, 480, 482–4, 486–9.

same then will be true of the body continuous with it; for that which is continuous with the spherical is spherical. The same again holds of the bodies between these and the centre. Bodies which are bounded by the spherical and in contact with it must be, as wholes, spherical; and the lower bodies are contiguous with the sphere above them. The sphere then will be spherical throughout; for every body within it is contiguous and continuous with spheres.

Again, since the whole seems – and has been assumed – to revolve in a circle, and since it has been shown that outside the farthest circumference there is neither void nor place, from these grounds also it will follow necessarily that the heaven is spherical. For if it is to be rectilinear in shape, it will follow that there is place and body and void without it. For a rectilinear figure as it revolves never continues in the same room, but where formerly was body, is now none, and where now is none, body will be in a moment because of the changing positions of the corners. Similarly, if the world had some other figure with unequal radii, if, for instance, it were lentiform, or oviform, in every case we should have to admit space and void outside the moving body, because the whole body would not always occupy the same room.

Again, if the motion of the heaven is the measure of all movements in virtue of being alone continuous and regular and eternal, and if, in each kind, the measure is the minimum, and the minimum movement is the swiftest, then the movement of the heaven must be the swiftest of all movements. Now of lines which return upon themselves the line which bounds the circle is the shortest; and that movement is the swiftest which follows the shortest line. Therefore, if the heaven moves in a circle and moves more swiftly than anything else, it must necessarily be spherical.

Corroborative evidence may be drawn from the bodies whose position is about the centre. If earth is enclosed by water, water by air, air by fire, and these similarly by the upper bodies – which while not continuous are yet contiguous with them – and if the surface of water is spherical, and that which is continuous with or embraces the spherical must itself be spherical, then on these grounds also it is clear that the heavens are spherical. But the surface of water is seen to be spherical if we take as our starting-point the fact that water naturally tends to collect in the more hollow places – and the more hollow are those nearer the centre. Draw from the centre the lines *AB*, *AC*, and let them be joined by the straight line *BC*. The line *AD*, drawn to the base of the triangle, will be shorter than either of the radii. Therefore the place in which it terminates will be more hollow. The water then will collect there until equality is established. But the line *AE* is equal to the radii. Thus water lies at the ends of the radii, and there will it rest; but the line which connects the extremities of the radii is circular: therefore the surface of the water *BEC* is spherical.

It is plain from the foregoing that the universe is spherical. It is plain, further, that it is so accurately turned that no manufactured thing nor anything else within the range of our observation can even approach it. For the matter of which these are composed does not admit of anything like the same regularity and finish as the substance of the enveloping body; since with each step away from earth the matter manifestly becomes finer in the same proportion as water is finer than earth.

5 Now there are two ways of moving along a circle, from *A* to *B* or from *A* to *C*, and we have already explained that these movements are not contrary to one another. But nothing which concerns the eternal can be a matter of chance or spontaneity, and the heaven and its circular motion are eternal. We must therefore ask why this motion takes one direction and not the other. Either this is itself a principle or there is a principle behind it. It may seem evidence of excessive folly or excessive zeal to try to provide an explanation of some things, or of everything, admitting no exception. The criticism, however, is not always just: one should first consider what reason there is for speaking, and also what kind of certainty is looked for, whether human merely or of a more cogent kind. When any one shall succeed in finding proofs of greater precision, gratitude will be due to him for the discovery, but at present we must be content with what seems to be the case. If nature always follows the best course possible, and, just as upward movement is the superior form of rectilinear movement, since the upper region is more divine than the lower, so forward movement is superior to backward, then front and back exhibits, like right and left, as we said before and as the difficulty just stated itself suggests, the distinction of prior and posterior, which provides a reason and so solves

our difficulty. Supposing that nature is ordered in the best way possible, this may stand as the reason of the fact mentioned. For it is best to move with a movement simple and unceasing, and, further, in the superior of two possible directions.

6 We have next to show that the movement of the heaven is regular and not irregular. This applies to the first heaven and the first movement; for the lower spheres exhibit a composition of several movements into one. If the movement is uneven, clearly there will be acceleration, maximum speed, and retardation, since these appear in all irregular motions. The maximum may occur either at the starting-point or at the goal or between the two; and we expect natural motion to reach its maximum at the goal, unnatural motion at the starting-point, and missiles midway between the two. But circular movement, having no beginning or limit or middle without qualification, has neither whence nor whither nor middle; for in time it is eternal, and in length it returns upon itself without a break. If then its movement has no maximum, it can have no irregularity, since irregularity is produced by retardation and acceleration. Further, since everything that is moved is moved by something, the cause of the irregularity of movement must lie either in the mover or in the moved or in both. For if the mover moved not always with the same force, or if the moved were altered and did not remain the same, or if both were to change, the result might well be an irregular movement in the moved. But none of these possibilities can occur in the case of the heavens. As to that which is moved, we have shown that it is primary and simple and ungenerated and indestructible and generally unchanging; and it is far more reasonable to ascribe those attributes to the mover. It is the primary that moves the primary, the simple the simple, the indestructible and ungenerated that which is indestructible and ungenerated. Since then that which is moved, being a body, is nevertheless unchanging, how should the mover, which is incorporeal, be changed?

For if irregularity occurs, there must be change either in the movement as a whole, from fast to slow and slow to fast, or in its parts. That there is no irregularity in the parts is obvious, since, if there were, some divergence of the stars would have taken place before now in the infinity of time, as one moved slower and another faster; but no alteration of their intervals is ever observed. Nor again is a change in the movement as a whole admissible. Retardation is always due to incapacity, and incapacity is unnatural. The incapacities of animals, age, decay, and the like, are all unnatural, due, it seems, to the fact that the whole animal complex is made up of materials which differ in respect of their proper places, and no single part occupies its own place. If therefore that which is primary contains nothing unnatural, being simple and unmixed and in its proper place and having no contrary, then it has no place for incapacity, nor, consequently, for retardation or (since acceleration involves retardation) for acceleration. Again, it is unreasonable that the mover should first show incapacity for an infinite time, and capacity afterwards for another infinity. For clearly nothing which, like incapacity, is unnatural ever continues for an infinity of time; nor does the unnatural endure as long as the natural, or any form of incapacity as long as the capacity. But if the movement is retarded it must necessarily be retarded for an infinite time. Equally impossible is perpetual acceleration or perpetual retardation. For such movement would be infinite and indefinite; but every movement, in our view, proceeds from one point to another and is definite in character. Again, suppose one assumes a minimum time in less than which the heaven could not complete its movement. For, as a given walk or a given exercise on the harp cannot take any and every time, but every performance has its definite minimum time which is unsurpassable, so, one might suppose, the movement of the heaven could not be completed in any and every time. But in that case perpetual acceleration is impossible (and, equally, perpetual retardation; for the argument holds of both and each), if we may take acceleration to proceed by identical or increasing additions of speed and for an infinite time. The remaining possibility is to say that the movement exhibits an alternation of slower and faster; but this is a mere fiction and quite unreasonable. Further, irregularity of this kind would be particularly unlikely to pass unobserved, since contrast makes observation easy.

That there is one heaven, then, only, and that it is ungenerated and eternal, and further that its movement is regular, has now been sufficiently explained.

[. . .]

8 Since changes evidently occur not only in the position of stars but also in that of the whole heaven, there are three possibilities: either both are at rest, or both are in motion, or the one is at rest and the other in motion.

That both should be at rest is impossible; for, if the earth is at rest, the hypothesis does not account for the phenomena; and we take it as granted that the earth is at rest. It remains either that both are moved, or that the one is moved and the other at rest.

On the view, first, that both are in motion, we have the absurdity that the stars and the circles move with the same speed, i.e. that the pace of every star is that of the circle in which it moves. For star and circle are seen to come back to the same place at the same moment; from which it follows that the star has traversed the circle and the circle has completed its own movement, i.e. traversed its own circumference, at one and the same moment. But it is unreasonable that the pace of each star should be exactly proportioned to the size of its circle. That the pace of each circle should be proportionate to its size is not absurd but inevitable; but that the same should be true of the movement of the stars contained in the circles is quite unreasonable. For if the star which moves on the greater circle is necessarily swifter, clearly if the stars shifted their position so as to exchange circles, the slower would become swifter and the swifter slower. But this would show that their movement was not their own, but due to the circles. If, on the other hand, the arrangement was a chance combination, the coincidence in every case of a greater circle with a swifter movement of the star contained in it is unreasonable. In one or two cases it might not inconceivably fall out so, but to imagine it in every case alike is a mere fiction. Besides, chance has no place in that which is natural, and what happens everywhere and in every case is no matter of chance.

The same absurdity is equally plain if it is supposed that the circles stand still and that it is the stars themselves which move. For it will follow that the outer stars are the swifter, and that the pace of the stars corresponds to the size of circles.

Since, then, we cannot reasonably suppose either that both are in motion or that the star alone moves, it remains that the circles should move, while the stars are at rest and move with the circles to which they are attached. Only on this supposition are we involved in no absurd consequence. For, in the first place, the quicker movement of the larger circle is reasonable when all the circles are attached to the same centre. Whenever bodies are moving with their proper motion, the larger moves quicker. It is the same here with the revolving bodies; for the arc intercepted by two radii will be larger in the larger circle, and hence it is reasonable that the revolution of the larger circle should take the same time as that of the smaller. And secondly, the fact that the heavens do not break in pieces follows not only from this but also from the proof already given of the continuity of the whole.

[. . .]

10 With their order – I mean the movement of each, as involving the priority of some and the posteriority of others, and their distances from each other – astronomy may be left to deal, since the astronomical discussion is adequate. This discussion shows that the movements of the several stars depend, as regards the varieties of speed which they exhibit, on their distances. It is established that the outermost revolution of the heavens is a simple movement and the swiftest of all, and that the movement of all other bodies is composite and relatively slow, for the reason that each is moving on its own circle with the reverse motion to that of the heavens. This at once makes it reasonable that the body which is nearest to that first simple revolution should take the longest time to complete its circle, and that which is farthest from it the shortest, the others taking a longer time the nearer they are and a shorter time the farther away they are. For it is the nearest body which is most strongly influenced, and the most remote, by reason of its distance, which is least affected, the influence on the intermediate bodies varying, as the mathematicians show, with their distance.

[. . .]

13 It remains to speak of the earth, of its position, of the question whether it is at rest or in motion, and of its shape.

As to its *position* there is some difference of opinion. Most people – all, in fact, who regard the whole heaven as finite – say it lies at the centre. But the Italian philosophers known as Pythagoreans take the contrary view. At the centre, they say, is fire, and the earth is one of the stars, creating night and day by its circular motion about the centre. They further construct another earth in opposition to ours to which they give the name counter-earth. In all this they are not seeking for theories and causes to account for the phenomena, but rather forcing the phenomena and trying to accommodate them to certain theories and opinions of their own. But there are many others who would agree that it is wrong to give the earth the central position, looking for confirmation rather to theory than to the phenomena. Their view is that the most precious place befits the most precious thing; but fire, they say, is more precious than earth, and the limit than the intermediate, and the circumference and the centre are limits. Reasoning on this basis they take the view that it is not earth that lies at the centre of the sphere, but rather fire. The Pythagoreans have a further reason. They hold that the most important part of the world, which is the centre, should be most strictly guarded, and name the fire which occupies that place the 'Guard-house of Zeus', as if the word 'centre' were quite unequivocal, and the centre of the mathematical figure were always the same with that of the thing or the natural centre. But it is better to conceive of the case of the whole heaven as analogous to that of animals, in which the centre of the animal and that of the body are different. For this reason they have no need to be so disturbed about the world, or to call in a guard for its centre: rather let them look for the centre in the other sense and tell us what it is like and where nature has set it. That centre will be something primary and precious; but to the mere position we should give the last place rather than the first. For the middle is what is defined, and what defines it is the limit, and that which contains or limits is more precious than that which is limited, seeing that the latter is the matter and the former the substance of the system.

As to the position of the earth, then, this is the view which some advance, and the views advanced concerning its *rest or motion* are similar. For here too there is no general agreement. All who deny that the earth lies at the centre think that it revolves about the centre, and not the earth only but, as we said before, the counter-earth as well. Some of them even consider it possible that there are several bodies so moving, which are invisible to us owing to the interposition of the earth. This, they say, accounts for the fact that eclipses of the moon are more frequent than eclipses of the sun; for in addition to the earth each of these moving bodies can obstruct it. Indeed, as in any case the earth is not actually a centre but distant from it a full hemisphere, there is no more difficulty, they think, in accounting for the phenomena on their view that we do not dwell at the centre, than on the view that the earth is in the middle. Even as it is, there is nothing to suggest that we are removed from the centre by half the diameter of the earth. Others, again, say that the earth, which lies at the centre, is rolled, and thus in motion, about the axis of the whole heaven. So it stands written in the *Timaeus*.

There are similar disputes about the *shape* of the earth. Some think it is spherical, others that it is flat and drum-shaped. For evidence they bring the fact that, as the sun rises and sets, the part concealed by the earth shows a straight and not a curved edge, whereas if the earth were spherical the line of section would have to be circular. In this they leave out of account the great distance of the sun from the earth and the great size of the circumference, which, seen from a distance on these apparently small circles appears straight. Such an appearance ought not to make them doubt the circular shape of the earth. But they have another argument. They say that because it is at rest, the earth must necessarily have this shape.

There are many different ways in which the movement or rest of the earth has been conceived. The difficulty must have occurred to every one. It would indeed be a complacent mind that felt no surprise that, while a little bit of earth, let loose in mid-air, moves and will not stay still, and the more there is of it the faster it moves, the whole earth, free in mid-air, should show no movement at all. Yet here is this great weight of earth, and it is at rest. And again, from beneath one of these moving fragments of earth, before it falls, take away the earth, and it will continue its downward movement with nothing to stop it. The difficulty then, has naturally passed into a commonplace of philosophy; and one may well wonder that the solutions offered are not seen to involve greater absurdities than the problem itself.

By these considerations some, like Xenophanes of Colophon, have been led to assert that the earth below us is infinite, [saying that it has 'pushed its roots to infinity'] in order to save the trouble of seeking for the cause. Hence the sharp rebuke of Empedocles, in the words 'if the deeps of the earth are endless and endless the ample ether – such is the vain tale told by many a tongue, poured from the mouths of those who have seen but little of the whole'. Others say the earth rests upon water. This, indeed, is the oldest theory that has been preserved, and is attributed to Thales of Miletus. It was supposed to stay still because it floated like wood and other similar substances, which are so constituted as to rest upon water but not upon air. As if the same account had not to be given of the water which carries the earth as of the earth itself! It is not the nature of water, any more than of earth, to stay in mid-air: it must have something to rest upon. Again, as air is lighter than water, so is water than earth: how then can they think that the naturally lighter substance lies below the heavier? Again, if the earth as a whole is capable of floating upon water, that must obviously be the case with any part of it. But observation shows that this is not the case. Any piece of earth goes to the bottom, the quicker the larger it is. These thinkers seem to push their inquiries some way into the problem, but not so far as they might. It is what we are all inclined to do, to direct our inquiry not to the matter itself, but to the views of our opponents; for even when inquiring on one's own one pushes the inquiry only to the point at which one can no longer offer any opposition. Hence a good inquirer will be one who is ready in bringing forward the objections proper to the genus, and that he will be when he has gained an understanding of all the differences.

[. . .]

[. . .] There are some, Anaximander, for instance, among the ancients, who say that the earth keeps its place because of its indifference. Motion upward and downward and sideways were all, they thought, equally inappropriate to that which is set at the centre and indifferently related to every extreme point; and to move in contrary directions at the same time was impossible: so it must needs remain still. This view is ingenious but not true. The argument would prove that everything which is put at the centre must stay there. Fire, then, will rest at the centre; for the proof turns on no peculiar property of earth. But in any case it is superfluous. The observed facts about earth are not only that it remains at the centre, but also that it moves to the centre. The place to which any fragment of earth moves must necessarily be the place to which the whole moves; and in the place to which a thing naturally moves, it will naturally rest. The reason then is not in the fact that the earth is indifferently related to every extreme point; for this would apply to any body, whereas movement to the centre is peculiar to earth. Again it is absurd to look for a reason why the earth remains at the centre and not for a reason why fire remains at the extremity. If the extremity is the natural place of fire, clearly earth must also have a natural place. But suppose that the centre is not its place, and that the reason of its remaining there is this necessity of indifference – on the analogy of the hair which, it is said, however great the tension, will not break under it, if it be evenly distributed, or of the man who, though exceedingly hungry and thirsty, and both equally, yet being equidistant from food and drink, is therefore bound to stay where he is – even so, it still remains to explain why fire stays at the extremities. It is strange, too, to ask about things staying still but not about their motion, – why, I mean, one thing, if nothing stops it, moves up, and another thing to the centre. Again, their statements are not true. It happens, indeed, to be the case that a thing to which movement this way and that is equally inappropriate is obliged to remain at the centre. But so far as their argument goes, instead of remaining there, it will move, only not as a mass but in fragments. For the argument applies equally to fire. Fire, if set at the centre, should stay there, like earth, since it will be indifferently related to every point on the extremity. Nevertheless it will move, as in fact it always does move when nothing stops it, away from the centre to the extremity. It will not, however, move in a mass to a single point on the circumference – the only possible result on the lines of the indifference theory – but rather each corresponding portion of fire to the corresponding part of the extremity, each fourth part, for instance, to a fourth part of the circumference. For since no body is a point, it will have parts. The expansion, when the body increased the place occupied, would be on the same principle as the contraction, in which the place was diminished. Thus, for all the indifference theory shows to the contrary, the earth also would have moved in this manner away from the centre, unless the centre had been its natural place.

We have now outlined the views held as to the shape, position, and rest or movement of the earth.

14 Let us first decide the question whether the earth moves or is at rest. For, as we said, there are some who make it one of the stars, and others who, setting it at the centre, suppose it to be rolled and in motion about the pole as axis. That both views are untenable will be clear if we take as our starting-point the fact that the earth's motion, whether the earth be at the centre or away from it, must needs be a constrained motion. It cannot be the movement of the earth itself. If it were, any portion of it would have this movement; but in fact every part moves in a straight line to the centre. Being, then, constrained and unnatural, the movement could not be eternal. But the order of the universe is eternal. Again, everything that moves with the circular movement, except the first sphere, is observed to be passed, and to move with more than one motion. The earth, then, also, whether it moves about the centre or is stationary at it, must necessarily move with two motions. But if this were so, there would have to be passings and turnings of the fixed stars. Yet no such thing is observed. The same stars always rise and set in the same parts of the earth.

Further, the natural movement of the earth, part and whole alike, is to the centre of the whole – whence the fact that it is now actually situated at the centre – but it might be questioned, since both centres are the same, which centre it is that portions of earth and other heavy things move to. Is this their goal because it is the centre of the earth or because it is the centre of the whole? The goal, surely, must be the centre of the whole. For fire and other light things move to the extremity of the area which contains the centre. It happens, however, that the centre of the earth and of the whole is the same. Thus they do move to the centre of the earth, but accidentally, in virtue of the fact that the earth's centre lies at the centre of the whole. That the centre of the earth is the goal of their movement is indicated by the fact that heavy bodies moving towards the earth do not move parallel but so as to make equal angles, and thus to a single centre, that of the earth. It is clear, then, that the earth must be at the centre and immovable, not only for the reasons already given, but also because heavy bodies forcibly thrown quite straight upward return to the point from which they started, even if they are thrown to an unlimited distance. From these considerations then it is clear that the earth does not move and does not lie elsewhere than at the centre.

From what we have said the explanation of the earth's immobility is also apparent. If it is the nature of earth, as observation shows, to move from any point to the centre, as of fire contrariwise to move from the centre to the extremity, it is impossible that any portion of earth should move away from the centre except by constraint. For a single thing has a single movement, and a simple thing a simple: contrary movements cannot belong to the same thing, and movement away from the centre is the contrary of movement to it. If then no portion of earth can move away from the centre, obviously still less can the earth as a whole so move. For it is the nature of the whole to move to the point to which the part naturally moves. Since, then, it would require a force greater than itself to move it, it must needs stay at the centre. This view is further supported by the contributions of mathematicians to astronomy, since the phenomena – the changes of the shapes by which the order of the stars is determined – are fully accounted for on the hypothesis that the earth lies at the centre. Of the position of the earth and of the manner of its rest or movement, our discussion may here end.

Its shape must necessarily be spherical. For every portion of earth has weight until it reaches the centre, and the jostling of parts greater and smaller would bring about not a waved surface, but rather compression and convergence of part and part until the centre is reached. The process should be conceived by supposing the earth to come into being in the way that some of the natural philosophers describe. Only they attribute the downward movement to constraint, and it is better to keep to the truth and say that the reason of this motion is that a thing which possesses weight is naturally endowed with a centripetal movement. When the mixture, then, was merely potential, the things that were separated off moved similarly from every side towards the centre. Whether the parts which came together at the centre were distributed at the extremities evenly, or in some other way, makes no difference. If, on the one hand, there were a similar movement from each quarter of the extremity to the single centre, it is obvious that the resulting mass would be similar on every side. For if an equal amount is added on every side the extremity of the mass will be everywhere equidistant from its centre, i.e. the figure will be spherical. But neither will it in any way affect the argument if there is not a similar accession of concurrent fragments from every side. For the greater quantity, finding a lesser in front of it, must necessarily drive it on, both having an impulse whose goal is the centre, and the greater weight driving the lesser forward till this goal is reached. In this we have also the solution of a possible difficulty. The earth, it might be argued, is at the centre and spherical in shape: if, then, a weight many times that of the earth were added to one hemisphere, the centre of the earth and of the whole will no longer be coincident. So that either the earth will not stay at the centre, or if it does, it might even now be at rest without being at the centre but at a place where it is its nature to move. Such is the difficulty. A short consideration will give us an easy answer, if we first give precision to our postulate that any body endowed with weight, of whatever size, moves towards the centre. Clearly it will not stop when its edge touches the centre. The greater quantity must prevail until its own centre occupies the centre. For that is the goal of its impulse. Now it makes no difference whether we apply this to a clod or arbitrary fragment of earth or to the earth as a whole. The fact indicated does not depend upon degrees of size but applies universally to everything that has the centripetal impulse. Therefore earth in motion whether in a mass or in fragments, necessarily continues to move until it occupies the centre equally every way, the less being forced to equalize itself by the greater owing to the forward drive of the impulse.

If the earth was generated, then, it must have been formed in this way, and so clearly its generation was spherical; and if it is ungenerated and has remained so always, its character must be that which the initial generation, if it had occurred, would have given it. But the spherical shape, necessitated by this argument, follows also from the fact that the motions of heavy bodies always make equal angles, and are not parallel. This would be the natural form of movement towards what is naturally spherical. Either then the earth is spherical or it is at least naturally spherical. And it is right to call anything that which nature intends it to be, and which belongs to it, rather than that which it is by constraint and contrary to nature. The evidence of the senses further corroborates this. How else would eclipses of the moon show segments shaped as we see them? As it is, the shapes which the moon itself each month shows are of every kind – straight, gibbous, and concave – but in eclipses the outline is always curved; and, since it is the interposition of the earth that makes the eclipse, the form of this line will be caused by the form of the earth's surface, which is therefore spherical. Again, our observations of the stars make it evident, not only that the earth is circular, but also that it is a circle of no great size. For quite a small change of position on our part to south or north causes a manifest alteration of the horizon. There is much change, I mean, in the stars which are overhead, and the stars seen are different, as one moves northward or southward. Indeed there are some stars seen in Egypt and in the neighbourhood of Cyprus which are not seen in the northerly regions; and stars, which in the north are never beyond the range of observation, in those regions rise and set. All of which goes to show not only that the earth is circular in shape, but also that it is a sphere of no great size; for otherwise the effect of so slight a change of place would not be so quickly apparent. Hence one should not be too sure of the incredibility of the view of those who conceive that there is continuity between the parts about the pillars of Hercules and the parts about India, and that in this way the ocean is one. As further evidence in favour of this they quote the case of elephants, a species occurring in each of these extreme regions, suggesting that the common characteristic of these extremes is explained by their continuity. Also, those mathematicians who try to calculate the size of the earth's circumference arrive at the figure 400,000 stades. This indicates not only that the earth's mass is spherical in shape, but also that as compared with the stars it is not of great size.

1.9

The Divisions of Nature and the Divisions of Knowledge

Aristotle

Aristotle was a keen observer of nature and had a great interest in unusual plants and animals. Knowing this, his pupil Alexander sent back to his old tutor unusual specimens of flora and fauna encountered on his military conquests. In this selection from the first book of *On the Parts of Animals*, Aristotle once again asks, and partly answers, fundamental questions: In order to learn about animals, should we examine them in classes or one by one as they come? What are the differences between the study of living things, such as a seed, and the study of abstract objects? How do the different sorts of causes give structure to our inquiry? How should one go about trying to grasp the real causes of things?

[...]

... Should one take each substantial being singly and define it independently, e.g. taking up one by one the nature of mankind, lion, ox, and any other animal as well; or should one first establish, according to something common, the attributes common to all? For many of the same attributes are present in many different kinds of animals, e.g. sleep, respiration, growth, deterioration, death, and in addition any remaining affections and dispositions such as these. (I add this because at the moment it is permissible to speak unclearly and indefinitely about these things.) It is apparent that, especially when speaking one by one, we shall repeatedly say the same things about many kinds; for instance, each of the attributes just mentioned belongs to horses, dogs, and human beings. So if one speaks of their attributes one by one, it will be necessary to speak repeatedly about the same things – whenever, that is, the same things are present in different forms of animal, yet themselves have no difference.

Yet there are probably other attributes which turn out to have the same predicate, but to differ by a difference in form, e.g. the locomotion of animals; it is apparent that locomotion is not one in form, because flying, swimming, walking,

From *Aristotle on the Parts of Animals*, trans. James G. Lennox (Oxford: Clarendon Press, 2001), pp. 1–8.

and crawling differ. Accordingly, the following question about how one is to carry out an examination should not be overlooked – I mean the question of whether one should study things in common according to kind first, and then later their distinctive characteristics, or whether one should study them one by one straight away. At present this matter has not been determined, nor has the question that will now be stated, namely, whether just as the mathematicians explain the phenomena in the case of astronomy, so the natural philosopher too, having first studied the phenomena regarding the animals and the parts of each, should then state the reason why and the causes, or whether he should proceed in some other way.

And in addition to these questions, since we see more than one cause of natural generation, e.g. both the cause for the sake of which and the cause from which comes the origin of motion, we need also to determine, about these causes, which sort is naturally first and which second. Now it is apparent that first is the one we call for the sake of which; for this is an account, and the account is an origin alike in things composed according to art and in things composed by nature. For once the doctor has defined health, and the builder has defined house, either by thought or perception, they provide the accounts and the causes of each of the things they produce, and the reason why it must be produced in this way. Yet that for the sake of which and the good are present more in the works of nature than in those of art.

What is of necessity is not present in all natural things in the same way; yet nearly everyone attempts to refer their accounts back to it without having distinguished in how many ways the necessary is said. That which is necessary without qualification is present in the eternal things, while that which is conditionally necessary is also present in all generated things, as it is in artefacts such as a house or any other such thing. It is necessary that a certain sort of matter be present if there is to be a house or any other end, and this must come to be and be changed first, then that, and so on continuously up to the end and that for the sake of which each comes to be and is. It is the same way too with things that come to be by nature.

However, the mode of demonstration and of necessity is different in natural science and the theoretical sciences. [...] For the origin is, in the latter cases, what is, but in the former, what will be. So: 'Since health or mankind is such, it is necessary for *this* to be or come to be', instead of 'Since *this* is or has come about, *that* from necessity is or will be'. Nor is it possible to connect the necessity in such a demonstration into eternity, as if to say, 'Since *this* is, therefore *that* is'. [...]

We should also not forget to ask whether it is appropriate to state, as those who studied nature before us did, how each thing has naturally come to be, rather than how it is. For the one differs not a little from the other. It seems we should begin, even with generation, precisely as we said before: first one should get hold of the phenomena concerning each kind, then state their causes. For even with house-building, it is rather that these things happen because the form of the house is such as it is, than that the house is such as it is because it comes to be in this way. For generation is for the sake of substantial being, rather than substantial being for the sake of generation. That is precisely why Empedocles misspoke when he said that many things are present in animals because of how things happened during generation – for example, that the backbone is such as it is because it happened to get broken through being twisted. He failed to understand, first, that seed already constituted with this sort of potential must be present, and second, that its producer was prior – not only in account but also in time. For one human being generates another; consequently, it is on account of *that* one being such as it is that *this* one's generation turns out a certain way. It is likewise both with things that seem to come to be spontaneously and with artefacts; for in some cases the same things produced by art also come to be spontaneously, e.g. health. Now in some of these cases there pre-exists a productive capacity like them, e.g. the art of sculpture; for a statue does not come to be spontaneously. The art is the account of the product without the matter. And it is likewise with the products of chance; for as the art has it, so they come to be.

Hence it would be best to say that, since this is what it is to be a human being, on account of this it has these things; for it cannot be without these parts. If one cannot say this, one should say the next best thing, i.e. either that in general it cannot be otherwise, or that at least it is good thus.

And these things follow. And since it is such, its generation necessarily happens in this way and is such as it is. (This is why this part comes to be first, then that one.) And in like manner one should speak in precisely this way about all of the things constituted by nature.

Now the ancients who first began philosophizing about nature were examining the material origin and that sort of cause: what matter is and what sort of thing it is, and how the whole comes to be from it and what moves it (e.g. whether strife, friendship, reason, or spontaneity). They also examined what sort of nature the underlying matter has of necessity, e.g. whether the nature of fire is hot, of earth cold, and whether the nature of fire is light, of earth heavy. In fact, even the cosmos they generate in this way. And they speak in a like manner too of the generation of animals and plants, saying, for example, that as water flowed into the body a stomach and every part that receives nourishment and residue came to be; and as the breath passed through, the nostrils were burst open.

Air and water are matter for bodies; that is, it is from such things that all the ancients constitute the nature of bodies. But if human beings, animals, and their parts exist by nature, one should speak about flesh, bone, blood, and all the uniform parts. Likewise too, about the non-uniform parts such as face, hand, and foot, one should say in virtue of what each of them is such as it is, and in respect of what sort of potential. For it is not enough to say from what things they are constituted, e.g. from fire or earth. It is just as if we were speaking about a bed or any other such thing; we would attempt to define its form rather than its matter, e.g. the bronze or the wood. And if we could not do this, we would at least attempt to define the matter of the composite; for a bed is a 'this-in-that' or 'this-such', so that we would have to mention its configuration as well, and what its visible character is. For the nature in respect of shape is more important than the material nature.

Now if it is by virtue of its configuration and colour that each of the animals and their parts is what it is, Democritus might be speaking correctly; for he appears to assume this. Note that he says it is clear to everyone what sort of thing a human being is in respect of shape, since it is known by way of its figure and its colour. And yet though the configuration of a corpse has the same shape, it is nevertheless *not* a human being. And further, it is impossible for something in any condition whatsoever, such as bronze or wooden, to be a hand, except homonymously (like a doctor in a picture). For such a hand will not be able to do its work, just as stone flutes will not be able to do theirs and the doctor in the picture his. Likewise none of the parts of a corpse is any longer such – I mean, for example, any longer an eye or a hand.

What Democritus has said, then, is too unqualified, and is said in the same way as a carpenter might speak about a wooden hand. Indeed this is also the way the natural philosophers speak of the generations and causes of configuration. Ask them by what potencies things were crafted. Well, no doubt the carpenter will say an axe or an auger, while the natural philosopher will say air and earth – albeit the carpenter's response is better; for it will be insufficient for him to say merely that when the tool fell this became a depression and that flat. Rather, he will state the cause, the reason why he made such a blow and for the sake of what, in order that it might then come to be this or that sort of shape.

It is clear, then, that these natural philosophers speak incorrectly. Clearly, one should state that the animal is of such a kind, noting about each of its parts what it is and what sort of thing it is, just as one speaks of the form of the bed. Suppose what one is thus speaking about is soul, or a part of soul, or is not without soul (at least when the soul has departed there is no longer an animal, nor do any of the parts remain the same, except in configuration, like those in myths that are turned to stone) – if these things are so, then it will be up to the natural philosopher to speak and know about the soul; and if not all of it, about that very part in virtue of which the animal is such as it is. He will state both what the soul or that very part of it is, and speak about the attributes it has in virtue of the sort of substantial being it is, especially since the nature of something is spoken of and is in two ways: as matter and as substantial being. And nature as substantial being is both nature as mover and nature as end. And it is the soul – either all of it or some part of it – that is such in the animal's case. So in this way too it will be requisite for the person studying nature to speak about soul more than the matter, inasmuch as it is more that the matter is nature because of soul than the reverse. And indeed, the wood is a bed or a stool because it is potentially these things.

In view of what was said just now, one might puzzle over whether it is up to natural science to speak about *all* soul, or some part, since if it speaks about all, no philosophy is left besides natural science. This is because reason is of the objects of reason, so that natural science would be knowledge about everything. For it is up to the same science to study reason and its objects, if they truly are correlative and the same study in every case attends to correlatives, as in fact is the case with perception and perceptible objects.

However, it is not the case that all soul is an origin of change, nor all its parts; rather, of growth the origin is the part which is present even in plants, of alteration the perceptive part, and of locomotion some other part, and not the rational; for locomotion is present in other animals too, but thought in none. So it is clear that one should not speak of all soul; for not all of the soul is a nature, but some part of it, one part or even more.

Further, none of the abstract objects can be objects of natural study, since nature does everything for the sake of something. For it is apparent that, just as in artefacts there is the art, so in things themselves there is an other sort of origin and cause, which we have as we do the hot and the cold – from the entire universe. This is why it is more likely that the heaven has been brought into being by such a cause – if it has come to be – and is due to such a cause, than that the mortal animals have been. Certainly the ordered and definite are far more apparent in the heavens than around us, while the fluctuating and random are more apparent in the mortal sphere. Yet some people say that each of the animals is and came to be by nature, while the heaven, in which there is not the slightest appearance of chance and disorder, was constituted in that way by chance and the spontaneous.

We say 'this is for the sake of that' whenever there appears to be some end towards which the change proceeds if nothing impedes it. So it is apparent that there is something of this sort, which is precisely what we call a nature. Surely it is not any chance thing that comes to be from each seed, nor a chance seed which comes from a chance body; rather, *this* one comes from *that* one. Therefore the seed is an origin and is productive of what comes from it. For these things are by nature; at least they grow from seed. But prior even to this is what the seed is the seed of; for while the seed is becoming, the end is being. And prior again to both of these is what the seed is from. For the seed is a seed in two ways, *from* which and *of* which; that is, it is a seed both of what it came from, e.g. from a horse, and it is a seed of what will be from it, e.g. of a mule, though not in the same way, but of each in the way mentioned. Further, the seed is in potentiality; and we know how potentiality is related to complete actuality.

Therefore there are these two causes, the cause for the sake of which and the cause from necessity; for many things come to be because it is a necessity. One might perhaps be puzzled about what sort of necessity those who say 'from necessity' mean; for it cannot be either of the two sorts defined in our philosophical discussions. But it is especially in things that partake of generation that the third sort is present; for we say nourishment is something necessary according to neither of those two sorts of necessity, but because it is not possible to be without it. And this is, as it were, conditionally necessary; for just as, since the axe must split, it is a necessity that it be hard, and if hard, then made of bronze or iron, so too since the body is an instrument (for each of the parts is for the sake of something, and likewise also the whole), it is therefore a necessity that it be of such a character and constituted from such things, if that is to be.

Clearly, then, there are two sorts of cause, and first and foremost one should succeed in stating both, but failing that, at least attempt to do so; and clearly all who do not state this say virtually nothing about nature. For nature is an origin more than matter. Even Empedocles occasionally stumbles upon this, led by the truth itself, and is forced to say that the substantial being and the nature is the account, e.g. when he says what bone is. He does not say that it is some one of the elements, or two or three, or all of them, but rather that it is an account of their mixture. Accordingly, it is clear that flesh too, and each of the other such parts, is what it is in the same way.

One reason our predecessors did not arrive at this way is that there was no 'what it is to be' and 'defining substantial being'. Democritus touched on this first, not however as necessary for the study of nature, but because he was carried away by the subject itself; while in Socrates' time interest in this grew, but research into the natural world ceased, and philosophers turned instead to practical virtue and politics.

One should explain in the following way, e.g. breathing exists for the sake of *this*, while *that* comes to be from necessity because of *these*. But 'necessity' sometimes signifies that if that – i.e. that for the sake of which – is to be, it is necessary for these things to obtain, while at other times it signifies that things are thus in respect of their character and nature. For it is necessary for the hot to go out and enter again upon meeting resistance, and for the air to flow in. This is directly necessary; and it is as the internal heat retreats during the cooling of the external air that inhalation and exhalation occur. This then is the way of investigation, and it is in relation to these things and things such as these that one should grasp the causes.

1.10

On Methods of Inference

Philodemus

Philodemus (c.110–c.40 BC) was an Epicurean philosopher and poet whose literary work influenced his famous pupil, Virgil. When the town of Herculaneum was flattened by the eruption of Vesuvius, it buried a villa containing a large number of papyri, flattening and carbonizing them but also embalming them. With infrared imaging and other techniques, we have now been able to recover much of the content of these scrolls. The fragment from Philodemus given here shows a surprisingly sophisticated appreciation of, and a surprisingly modern-sounding approach to, the problem of induction.

And further the Stoics err in so far as they have not taken the trouble to understand the right method of analogical inference. Whenever we say,

> Since things in our experience are of such a nature,
> Unperceived objects are also of this nature *in so far as* things in our experience are of this nature,

we judge that there is a necessary connection between an unperceived object and the objects of our experience. For example,

> Since men in our experience *as men* are mortal,
> If there are men anywhere,
> They are mortal.

There are four things that the words "as such," "according as," and "in so far as," signify; . . .

[Here Philodemus distinguishes four senses of these phrases.]

But those who attack the inference from analogy do not indicate the distinctions just mentioned, namely, howe we are to take the "according as," as in the statement, for example,

> Man as man is mortal.

From Phillip Howard DeLacy and Estell Allen DeLacy, "Philodemus: On Methods of Inference. A Study in Ancient Empiricism," *The American Journal of Philology* 68: 3 (1947): 321–2 (extracts).

Hence they say that if the "according as" is omitted, the argument will be inconclusive; if it is admitted, the method of contraposition is used. But we Epicureans take this to be necessarily connected with that from the fact that this has been observed to be a property of all cases that we have come upon, and because we have observed many varied living creatures of the same genus who have differences in all other respects from each other, but who all share in certain common qualities (e.g., mortality). According to this method we say that man according as and in so far as he is man is mortal, on the ground that we have examined systematically many diverse men, and have found no variation in respect to this characteristic and no evidence to the contrary . . .

1.11

The Explanatory Power of Atomism

Lucretius

Lucretius (c.94–c.49 BC) was a Roman philosopher and poet whose great work *On the Nature of Things* provides the fullest and most detailed exposition of atomism in antiquity. Lucretius' analysis of physical phenomena from the atomistic perspective is powerful and persuasive, but like other atomists he extended the reduction beyond physics, claiming that not just the body but the soul could be accounted for wholly in terms of atoms and the void. As a consequence, atomists of the scientific revolution like Galileo, Boyle, and Hooke had to persuade their contemporaries that atomism and Christianity were compatible. This selection is from the first book of *On the Nature of Things*.

For what is to follow, my Memmius, lay aside your cares and lend undistracted ears and an attentive mind to true reason. Do not scornfully reject, before you have understood them, the gifts I have marshaled for you with zealous devotion. I will set out to discourse to you on the ultimate realities of heaven and the gods. I will reveal those *atoms* from which nature creates all things and increases and feeds them and into which, when they perish, nature again resolves them. To these in my discourse I commonly give such names as the 'raw material', or 'generative bodies' or 'seeds' of things. Or I may call them 'primary particles', because they come first and everything else is composed of them.

[...]

[O]ur starting-point will be this principle: *Nothing can ever be created by divine power out of nothing.* The reason why all mortals are so gripped by fear is that they see all sorts of things happening on the earth and in the sky with no discernible cause, and these they attribute to the will of a god. Accordingly, when we have seen that nothing can be created out of nothing, we shall

From *On the Nature of the Universe*, trans. R. E. Latham, revised with an introduction by John Goodwin (Penguin, 1951, 1994), from book 1, pp. 11, 13–27.

then have a clearer picture of the path ahead, the problem of how things are created and occasioned without the aid of the gods.

First then, if things were made out of nothing, any species could spring from any source and nothing would require seed. Men could arise from the sea and scaly fish from the earth, and birds could be hatched out of the sky. Cattle and other domestic animals and every kind of wild beast, multiplying indiscriminately, would occupy cultivated and wastelands alike. The same fruits would not grow constantly on the same trees, but they would keep changing: any tree might bear any fruit. If each species were not composed of its own generative bodies, why should each be born always of the same kind of mother? Actually, since each is formed out of specific seeds, it is born and emerges into the sunlit world only from a place where there exists the right material, the right kind of atoms. This is why everything cannot be born of everything, but a specific power of generation inheres in specific objects.

Again, why do we see roses appear in spring, grain in summer's heat, grapes under the spell of autumn? Surely, because it is only after specific seeds have drifted together at their own proper time that every created thing stands revealed, when the season is favorable and the life-giving earth can safely deliver delicate growths into the sunlit world. If they were made out of nothing, they would spring up suddenly after varying lapses of time and at abnormal seasons, since there would of course be no primary bodies which could be prevented by the harshness of the season from entering into generative unions. Similarly, in order that things might grow, there would be no need of any lapse of time for the accumulation of seed. Tiny tots would turn suddenly into grown men, and trees would shoot up spontaneously out of the earth. But it is obvious that none of these things happens, since everything grows gradually, as is natural, from a specific seed and retains its specific character. It is a fair inference that each is increased and nourished by its own raw material.

Here is a further point. Without seasonable showers the earth cannot send up gladdening growths. Lacking food, animals cannot reproduce their kind or sustain life. This points to the conclusion that many elements are common to many things, as letters are to words, rather than to the theory that anything can come into existence without atoms.

[. . .]

The second great principle is this: *nature resolves everything into its component atoms and never reduces anything to nothing.* If anything were perishable in all its parts, anything might perish all of a sudden and vanish from sight. There would be no need of any force to separate its parts and loosen their links. In actual fact, since everything is composed of indestructible seeds, nature obviously does not allow anything to perish till it has encountered a force that shatters it with a blow or creeks into chinks and unknits it.

If the things that are banished from the scene by age are annihilated through the exhaustion of their material, from what source does Venus bring back the several races of animals into the light of life? And, when they are brought back, where does the inventive earth find for each the special food required for its sustenance and growth? From what fount is the sea replenished by its native springs and the streams that flow into it from afar? Whence does the ether draw nutriment for the stars? For everything consisting of a mortal body must have been exhausted by the long day of time, the illimitable past. If throughout this bygone eternity there have persisted bodies from which the universe has been perpetually renewed, they must certainly be possessed of immortality. Therefore things cannot be reduced to nothing.

Again, all objects would regularly be destroyed by the same force and the same cause, were it not that they are sustained by imperishable matter more or less tightly fastened together. Why, a mere touch would be enough to bring about destruction supposing there were no imperishable bodies whose union could be dissolved only by the appropriate force. Actually, because the fastenings of the atoms are of various kinds while their matter is imperishable, compound objects remain intact until one of them encounters a force that proves strong enough to break up its particular constitution. Therefore nothing returns to nothing, but everything is resolved into its constituent bodies. . . .

Well, Memmius, I have taught you that things cannot be created out of nothing nor, once born, be summoned back to nothing. Perhaps, however, you are becoming mistrustful of my words, because these atoms of mine are not visible to the eye. Consider, therefore, this further evidence of *bodies whose existence you must acknowledge though they cannot be seen*. First, wind, when its force is roused, whips up waves, founders tall ships and scatters clouds. Sometimes scouring plains with hurricane force it strews them with huge trees and batters mountain peaks with blasts that hew down forests. Such is wind in its fury, when it whoops aloud with a mad menace in its shouting. Without question, therefore, there must be invisible particles of wind that sweep sea, that sweep land, that sweep the clouds in the sky, swooping upon them and whirling them along in a headlong hurricane. In the way they flow and the havoc they spread they are no different from a torrential flood of water when it rushes down in a sudden spate from the mountain heights, swollen by heavy rains, and heaps together wreckage from the forest and entire trees. Soft though it is by nature, the sudden shock of oncoming water is more than even stout bridges can withstand, so furious is the force with which the turbid, storm-flushed torrent surges against their piers. With a mighty roar it lays them low, rolling huge rocks under its waves and brushing aside every obstacle from its course. Such, therefore, must be the movement of blasts of wind also. When they have come surging along some course like a rushing river, they push obstacles before them and buffet them with repeated blows; and sometimes, eddying round and round, they snatch them up and carry them along in a swiftly circling vortex. Here then is proof upon proof that winds have invisible bodies, since in their actions and behavior they are found to rival great rivers, whose bodies are plain to see.

Then again, we smell the various scents of things though we never see them approaching our nostrils. Similarly, we do not look upon scorching heat nor can we grasp cold in our eyes and we do not see sounds. Yet all these must be composed of physical bodies, since they are able to impinge upon our senses. For nothing can touch or be touched except bodies.

Again, clothes hung out on a surf-beaten shore grow moist. Spread in the sun they grow dry. But we do not see how the moisture has soaked into them, nor again how it has been dispelled by the heat. It follows that the moisture is split up into minute parts which the eye cannot possibly see.

Again, in the course of many annual revolutions of the sun a ring is worn thin next to the finger with continual rubbing. Dripping water hollows a stone. A curved ploughshare, iron though it is, dwindles imperceptibly in the furrow. We see the cobblestones of the highway worn by the feet of many wayfarers. The bronze statues by the city gates show their right hands worn thin by the touch of travelers who have greeted them in passing. We see that all these are being diminished, since they are worn away. But to perceive what particles drop off at any particular time is a power grudged to us by our ungenerous sense of sight.

To sum up, whatever is added to things gradually by nature and the passage of days, causing a cumulative increase, eludes the most attentive scrutiny of our eyes. Conversely, you cannot see what objects lose by the wastage of age – sheer sea cliffs, for instance, exposed to prolonged erosion by the mordant brine – or at what time the loss occurs. It follows that nature works through the agency of invisible bodies.

On the other hand, things are not hemmed in by the pressure of solid bodies in a tight mass. This is because *there is vacuity in things*. A grasp of this fact will be helpful to you in many respects and will save you from much bewildered doubting and questioning about the universe and from mistrust of my teaching. Well then, by vacuity I mean intangible and empty space. If it did not exist, things could not move at all. For the distinctive action of matter, which is counteraction and obstruction, would be in force always and everywhere. Nothing could move forward, because nothing would give it a starting-point by receding. As it is, we see with our eyes at sea and on land and high up in the sky that all sorts of things in all sorts of ways are on the move. If there were no empty space, these things would be denied the power of restless movement – or rather, they could not possibly have come into existence, embedded as they would have been in motionless matter.

Besides, there are clear indications that things that pass for solid are in fact porous. Even in rocky caves a trickle of water seeps through, and every surface weeps with brimming drops. Food percolates to every part of an animal's body. Trees grow and pour forth their fruit in season, because their food is distributed throughout their length from the tips of the roots through the trunk and along every branch. Noises pass through walls and fly into closed buildings. Freezing cold penetrates to the bones. If there were no vacancies through which the various bodies could make their way, none of these phenomena would be possible.

Again, why do we find some things outweigh others of equal volume? If there is as much matter in a ball of wool as in one of lead, it is natural that it should weigh as heavily, since it is the function of matter to press everything downwards, while it is the function of space on the other hand to remain weightless. Accordingly, when one thing is not less bulky than another but obviously lighter, it plainly declares that there is more vacuum in it, while the heavier object proclaims that there is more matter in it and much less empty space. We have therefore reached the goal of our diligent enquiry: there is in things an admixture of what we call vacuity.

[. . .]

To pick up the thread of my discourse, all nature as it is in itself consists of two things – bodies and the vacant space in which the bodies are situated and through which they move in different directions. The existence of bodies is vouched for by the agreement of the senses. If a belief resting directly on this foundation is not valid, there will be no standard to which we can refer any doubt on obscure questions for rational confirmation. If there were no place and space, which we call vacuity, these bodies could not be situated anywhere or move in any direction whatever. This I have just demonstrated. It remains to show that nothing exists that is distinct both from body and from vacuity and could be ranked with the others as a third substance. For whatever is must also be something. If it offers resistance to touch, however light and slight, it will increase the mass of body by such amount, great or small, as it may amount to, and will rank with it. If, on the other hand, it is intangible, so that it offers no resistance whatever to anything passing through it, then it will be that empty space which we call vacuity. Besides, whatever it may be in itself, either it will act in some way, or react to other things acting upon it, or else it will be such that things can be and happen in it. But without body nothing can act or react; and nothing can afford a place except emptiness and vacancy. Therefore, besides matter and vacuity, we cannot include in the number of things any third substance that can either affect our senses at any time or be grasped by the reasoning of our minds.

You will find that anything that can be named is either a property or an accident of these two. A property is something that cannot be detached or separated from a thing without destroying it, as weight is a property of rocks, heat of fire, fluidity of water, tangibility of all bodies, intangibility of vacuum. On the other hand, servitude, poverty and riches, freedom, war, peace and all other things whose advent or departure leaves the essence of a thing intact, all these it is our practice to call by their appropriate name, accidents.

Similarly, time by itself does not exist; but from things themselves there results a sense of what has already taken place, what is now going on and what is to ensue. It must not be claimed that anyone can sense time by itself apart from the movement of things or their restful immobility.

[. . .]

Material objects are of two kinds, atoms and compounds of atoms. The atoms themselves cannot be swamped by any force, for they are preserved indefinitely by their absolute solidity. Admittedly, it is hard to believe that anything can exist that is absolutely solid. The lightning stroke from the sky penetrates closed buildings, as do shouts and other noises. Iron glows white-hot in the fire, and rocks crack in savage scorching heat. Hard gold is softened and melted by heat; and the ice of bronze is liquefied by flame. Both heat and piercing cold seep through silver, since we feel both alike when a cooling shower of water is poured into a goblet that we hold ceremonially in our hands. All these facts point to the conclusion that nothing is really solid. But sound reasoning and nature itself drive us to the opposite conclusion. Pay attention, therefore, while I demonstrate in a few lines that there exist certain bodies that are absolutely solid and indestructible, namely those atoms which according to our teaching are the seeds of prime units of things from which the whole universe is built up.

In the first place, we have found that nature is twofold, consisting of two totally different things, matter and the space in which things happen. Hence each of these must exist by itself without admixture of the other. For, where there is empty space (what we call vacuity), there matter is not; where matter exists, there cannot be a vacuum. Therefore the prime units of matter are solid and free from vacuity.

Again, since composite things contain some vacuum, the surrounding matter must be solid. For you cannot reasonably maintain that anything can hide vacuity and hold it within its body unless you allow that the container itself is solid. And what contains the vacuum in things can only be an accumulation of matter. Hence matter, which possesses absolute solidity, can be everlasting when other things are decomposed.

Again, if there were no empty space, everything would be one solid mass; if there were no material objects with the property of filling the space they occupy, all existing space would be utterly void. It is clear, then, that there is an alternation of matter and vacuity, mutually distinct, since the whole is neither completely full nor completely empty. There are therefore solid bodies, causing the distinction between empty space and full. And these, as I have just shown, can be neither decomposed by blows from without nor invaded and unknit from within nor destroyed by any other form of assault. For it seems that a thing without vacuum can be neither knocked to bits nor snapped nor chopped in two by cutting; nor can it let in moisture or seeping cold or piercing fire, the universal agents of destruction. The more vacuum a thing contains within it, the more readily it yields to these assailants. Hence, if the units of matter are solid and without vacuity, as I have shown, they must be everlasting.

Yet again, if the matter in things had not been everlasting, everything by now would have gone back to nothing, and the things we see would be the product of rebirth out of nothing. But, since I have already shown that nothing can be created out of nothing nor any existing thing be summoned back to nothing, the atoms must be made of imperishable stuff into which everything can be resolved in the end, so that there may be a stock of matter for building the world anew. The atoms, therefore, are absolutely solid and unalloyed. In no other way could they have survived throughout infinite time to keep the world renewed.

Furthermore, if nature had set no limit to the breaking of things, the particles of matter in the course of ages would have been ground so small that nothing could be generated from them so as to attain from them in the fullness of time to the summit of its growth. For we see that anything can be more speedily disintegrated than put together again. Hence, what the long day of time, the bygone eternity, has already shaken and loosened to fragments could never in the residue of time be reconstructed. As it is, there is evidently a limit set to breaking, since we see that everything is renewed and each according to its kind has a fixed period in which to grow to its prime.

Here is a further argument. Granted that the particles of matter are absolutely solid, we can still explain the composition and behavior of soft things – air, water, earth, fire – by their intermixture with empty space. On the other hand, supposing the atoms to be soft, we cannot account for the origin of hard flint and iron. For there would be no foundation for nature to build on. Therefore there must be bodies strong in their unalloyed solidity by whose closer clustering things can be knit together and display unyielding toughness.

If we suppose that there is no limit set to the breaking of matter, we must still admit that material objects consist of particles which throughout eternity have resisted the forces of destruction. To say that these are breakable does not square with the fact that they have survived throughout eternity under a perpetual bombardment of innumerable blows.

Again, there is laid down for each thing a specific limit to its growth and its tenure of life, and the laws of nature ordain what each can do and what it cannot. No species is ever changed, but each remains so much itself that every kind of bird displays on its body its own specific markings. This is a further proof that their bodies are made of changeless matter. For, if the atoms could yield in any way to change, there would be no certainty as to what could arise and what could not, at what point the power of everything was limited by an immovable frontier post; nor could successive generations so regularly repeat the nature, behavior, habits and movements of their parents.

To proceed with our argument, there is an ultimate point in visible objects that represents the smallest thing that can be seen. So also there must be an ultimate point in objects that lie below the limit of perception by our senses. This point is without parts and is the smallest thing that can exist. It never has been and never will be able to exist by itself, but only as one primary part of something else. It is with a mass of such parts, solidly jammed together in formation, that matter is filled up. Since they cannot exist by themselves, they must needs stick together in a mass from which they cannot by any means be prized loose. The atoms, therefore, are absolutely solid and unalloyed, consisting of a mass of least parts tightly packed together. They are not compounds formed by the coalescence of their parts, but bodies of absolute and everlasting solidity. To these nature allows no loss or diminution, but guards them as seeds for things. If there are no such least parts, even the smallest bodies consist of an infinite number of parts, since they can always be halved and their halves halved again without limit. On this showing, what difference will there be between the whole universe and the very least of things? None at all. For, however endlessly infinite the universe may be, yet the smallest things will equally consist of an infinite number of parts. Since true reason cries out against this and denies that the mind can believe it, you must needs give in and admit that there are least parts which themselves are partless. Granted that these parts exist, you must needs admit that the atoms they compose are also solid and everlasting. But, if all things were compelled by all-creating nature to be broken up into these least parts, nature would lack the power to rebuild anything out of them. For partless objects cannot have the essential properties of generative matter – those varieties of attachment, weight, impetus, impact and movement on which everything depends.

[. . .]

The truth, as I maintain, is this: there are certain bodies whose impacts, movements, order, position and shapes produce fires. When their order is changed, they change their nature. In themselves they do not resemble fire or anything else that can bombard our senses with particles or impinge on our organs of touch.

To say, as Heraclitus does, that everything is fire, and nothing can be numbered among things as a reality except fire, seems utterly crazy. On the basis of the senses he attacks and unsettles the senses – the foundation of all belief and the only source of his knowledge of that which he calls fire. He believes that the senses clearly perceive fire, but not the other things that are in fact no less clear. This strikes me as not only pointless but mad. For what is to be our standard of reference? What can be a surer guide to the distinction of true from false than our own senses?

1.12

The Earth: Its Size, Shape, and Immobility

Claudius Ptolemy

Ptolemy (c.90–c.168) was a Roman astronomer whose major work, usually referred to by the title of its Arabic translation as the *Almagest*, provides a computationally workable model of the solar system in which the earth is motionless near (but not quite at) the center of the sun's orbit while the sun, moon, and other planets travel around the earth. The *Almagest* was the definitive treatise on astronomy through the Middle Ages until Copernicus. In this selection from the first book of the *Almagest*, Ptolemy argues that the earth is spherical, that in relation to the distance of the fixed stars it is so small that it should be considered to be a mere geometric point, and that it does not move.

The Heavens Move Like a Sphere

It is plausible to suppose that the ancients got their first notions on these topics from the following kind of observations. They saw that the sun, moon, and other stars moved from east to west along circular paths which were always parallel to each other, that they started by rising up from below the earth itself as it were, gradually achieving their ascent, and then kept circling in the same way and getting lower, until, seeming to fall to earth, they vanished completely. Then, after remaining invisible for some time, they rose and set once more. And they saw that the intervals between these motions, and also the locations of the rising and setting, were on the whole determined and regular.

The main phenomenon that led them to the idea of a sphere was the revolution of the ever-visible stars. They observed that this revolution was circular as well as continuous about a single common center. Naturally they considered that point to be the pole of the heavenly sphere. For they saw that the closer were stars to that point,

From *The Book of the Cosmos*, ed. Dennis Richard Danielson (Cambridge, MA: Perseus, 2000), extracts from pp. 69–74. Adapted from Claudius Ptolemy, *Almagest*, trans. G. J. Toomer (New York: Springer-Verlag, 1984).

the smaller were their circles. And the farther were stars from it, the greater were their circles – right out to the limit where stars became invisible. But here too they saw that some heavenly bodies near the ever-visible stars remained visible for only a short time, while some farther away remained invisible for a long time, again depending on how far away they were from the pole. So they arrived at the idea of the heavenly sphere merely from this kind of inference. But from then on, in subsequent investigations, they found that everything else fit with this notion, and that absolutely all appearances contradicted any alternative notion that was proposed.

For suppose that the stars' motion takes place in a straight line towards infinity, as some have thought. How then could one explain their appearing to set out from the same starting-point every day? How could the stars return if their motion were towards infinity? Or, if they did return, would not the straight-line hypothesis be obviously wrong? For according to it, the stars would gradually have to diminish in size until they disappeared, whereas in fact they appear greater at the very moment of their disappearance, at which point they are obstructed and cut off, as it were, by the earth's surface.

It is also absurd to imagine the stars ignited as they rise out of the earth and extinguished again as they fall to earth. Just suppose that the strict order in their size and number, their intervals, positions, and periods could be restored by such a random and chance process, and that one whole region of earth has igniting properties, and another has extinguishing properties – or rather that the same region ignites stars for one set of observers and extinguishes them for another set, and that the same stars are already ignited or extinguished for some observers while they are not yet for others! Even on this ridiculous supposition, what could we say about the ever-visible stars, which neither rise nor set? The stars that are ignited and extinguished ought to rise and set for observers everywhere, while those that are not ignited and extinguished should always be visible to observers everywhere. How would we explain the fact that this is not so? We can hardly say that stars that are ignited and extinguished for some observers never undergo this process for other observers. Yet it is utterly obvious that the very same stars that rise and set in certain regions of the earth neither rise nor set in other regions.

Finally, to assume any motion at all other than spherical motion would entail that the distances of stars measured from the earth upwards must vary, regardless of where or how we assume the earth itself is situated. Hence the apparent sizes of the stars and the distances between them would necessarily vary for the same observers during the course of each revolution, for their distances from the objects of observation would be now greater, now lesser. Yet we see that no such variation occurs. And the apparent increase in their sizes at the horizon is caused not by a decrease in their distances but by the exhalations of moisture surrounding the earth. These intervene between the place from which we observe and the heavenly bodies. In the same way, objects placed in water appear bigger than they really are, and the lower they sink, the bigger they appear.

[. . .]

The Earth Too, Taken as a Whole, Is Sensibly Spherical

That the earth, too, taken as a whole, is sensibly spherical can best be grasped from the following considerations. To repeat, we see that the sun, moon, and other stars do not rise and set simultaneously for everyone on earth, but do so earlier for those towards the east and later for those towards the west. And eclipses, especially lunar eclipses, take place simultaneously for all observers yet are not recorded by all observers as occurring at the same *hour* (that is, at an equal distance from noon). Rather, the hour recorded by observers in the east is always later than that recorded by those in the west. And we find that the differences in the recorded hour are proportional to the distances between the places of observation. Hence, one can reasonably conclude that the earth's surface is spherical, because its evenly curving surface (for so it is when considered as a whole) cuts off the heavenly bodies for each set of observers in a manner that is gradual and regular.

This would not happen if the earth's shape were other than spherical, as one can see from the following arguments. If the shape were concave, the stars would be seen rising first by those more towards the west; if it were a plane, they would rise and set simultaneously for everyone on earth; if it were triangular or square or any other polygonal shape, similarly they would rise and set simultaneously for all those living on the same planar surface. Yet clearly nothing like this takes place. Nor could the earth be cylindrical, with the curved surface in the east–west direction, and the flat sides towards the poles of the universe, as some might suppose more plausible. For to those living on the curved surface none of the stars would be ever-visible. Either all stars would rise and set for all observers, or the same stars, for an equal celestial distance from each of the poles, would always be invisible for all observers. In fact, however, the further we travel toward the north, the more of the southern stars disappear and the more of the northern stars become visible. Clearly, then, here too the curvature of the earth cuts off the heavenly bodies in a regular fashion in a north–south direction and demonstrates the sphericity of the earth in all directions.

Moreover, if we sail towards mountains or elevated places from whatever direction, north, south, east or west, we observe them to increase gradually in size as if rising up from the sea itself in which they had previously been submerged. This is due to the curvature of the surface of the water.

The Earth Has the Ratio of a Point to the Heavens

The earth has, to the senses, the ratio of a point to the distance of the sphere of the so-called fixed stars. This is strongly indicated by the fact that the sizes and distances of the stars at any given time appear equal and the same from any and every place on earth. Observations of the same celestial objects from different latitudes are found to have not the least discrepancy from each other. Moreover, gnomons set up in any part of the earth whatever, and likewise the centers of armillary spheres, operate like the real center of the earth. . . .

Another clear demonstration of the above proposition is that a plane drawn through the observer's line of sight at any point on earth – we call this plane one's "horizon" – always bisects the whole heavenly sphere. This would not happen if the earth were of perceptible size in relation to the distance of the heavenly bodies. In that case only the plane drawn through the center of the earth could exactly bisect the sphere, and a plane through any point on the surface of the earth would always make the section of the heavens below the plane greater than the section above it.

Neither Does the Earth Have Any Motion from Place to Place

One can show by arguments like the one above that the earth can have no motion in the directions mentioned, nor indeed can it ever move at all from its position at the center. For if it did move, the same phenomena would result as those that would follow from its having any position other than the central one. To me it seems pointless, therefore, to ask why objects move towards the center of the earth, once it has been so clearly established from actual phenomena that the earth occupies the middle place in the universe, and that all heavy objects are carried towards that place. The following fact alone amply supports this claim. Absolutely everywhere on the face of the earth – which has been shown to be spherical and in the middle of the universe – the direction and path of the motion (I mean proper, natural motion) of all heavy bodies is everywhere consistently at right angles to the plane that is tangent to the point of impact on the earth's surface. Clearly, therefore, if these falling objects were not stopped by the earth's surface, they would certainly reach the center of the earth itself, since any line drawn through the center of a sphere is always perpendicular to the tangent plane at the line's point of intersection with the sphere's surface.

Those who think it paradoxical that the earth, having such great weight, is not supported by anything and yet does not move, seem to me to be making the mistake of judging on the basis of their own experience instead of taking into account the peculiar nature of the universe. They would not, I think, consider this fact strange if they realized that the magnitude of the earth, when compared with the whole surrounding mass of the universe, has the ratio of a point to it. Given this way of thinking, it will seem quite consistent that (relatively speaking) the smallest of things should be overpowered and pressed in equally from all directions to a position of equilibrium by the greatest of things (which possess a uniform nature). For there is no up and down in the universe with respect to itself, any more than "up" and "down" make sense within a sphere. Rather, in the universe, the proper and natural motion of compound bodies is as follows: light and rarefied bodies drift outwards towards the circumference, but seem to move in the direction which is "up" for each observer, since the overhead direction for all of us, which we also call "up," points towards the surrounding surface. Heavy and dense bodies, on the contrary, are carried towards the middle and the center, but seem to fall downwards, again because the line of movement towards our feet, which we call "down," also points towards the center of the earth. These heavy bodies, as one would expect, settle about the center because of their mutual pressure and resistance, which is equal and uniform from all directions. For the same reason it is plausible that the earth, since its total mass is so great compared with the bodies which fall towards it, can remain motionless under the impact of these very small weights (for they strike it from all sides), and receive, as it were, the objects that fall upon it. . . .

Certain people, however, propose what they consider to be a more convincing model. They do not disagree with what I have said above, since they have no argument to bring against it. But they think no evidence prevents them from supposing, for example, that the heavens remain motionless and that the earth revolves from west to east about the same axis, making approximately one revolution each day. Or they suppose that both heaven and earth move by some amount, each about the same axis and in such a way as to preserve the overtaking of one by the other. However, they do not realize that, although there is perhaps nothing in the celestial phenomena to count against that simpler hypothesis, nevertheless what would occur here on earth and in the air would render such a notion quite ridiculous.

For the sake of argument, let us suppose that, contrary to nature, the most rare and light matter should either be motionless or else move in exactly the same way as matter with the opposite nature. . . . Suppose, too, that the densest and heaviest objects have a proper motion of the quick and uniform kind which they suppose (although, again, as everyone knows, earthly objects are sometimes not readily moved even by an external force). Even granted this supposition, they would have to admit that the revolving motion of the earth must be the most violent of all the motions they postulate, given that the earth makes one revolution in such a short time. Accordingly, all objects not actually standing on the earth would appear to have the same motion, opposite to that of the earth: neither clouds nor other flying or thrown objects would ever be seen moving towards the east, since the earth's motion towards the east would always outrun and overtake them, so that all other objects would seem to move backwards towards the west. Even if they claim that the air is carried around in the same direction and with the same speed as the earth, still the compound objects in the air would always seem to be left behind by the motion of both earth and air together. Or, if those objects too were carried around, fused as it were to the air, then they would never appear to have any motion either forwards or backwards. They would always appear still, neither wandering about nor changing position, whether they were things in flight or objects thrown. Yet we quite plainly see that they do undergo all these kinds of motion in such a way that they are not even slowed down or speeded up at all by any motion of the earth.

1.13

The Weaknesses of the Hypotheses

Proclus

Proclus (c.410–485), one of the last of the Greek Neoplatonist philosophers, wrote commentaries on several of Plato's dialogues and on the first book of Euclid's *Elements*. His work was influential on subsequent philosophical and scientific thought, not least among Arab philosophers. In this selection from his *Hypotyposis astronomicarum positionum*, Proclus expresses skepticism about the physical reality of the epicycles that astronomers following Ptolemy used to account for the apparent motions of the planets. His complaint that the Ptolemaic account lacks unity was revived over a millennium later by Copernicus, who called the Ptolemaic system monstrous; and his charge that astronomers have derived the causes of natural movements from something that does not exist in nature comes up again in the work of Kepler.

My dear friend: The great Plato thinks that the real philosopher ought to study the sort of astronomy that deals with entities more abstract than the visible heaven, without reference to either sense perception or ever-changing matter. In that world of abstract entities he will come to know slowness itself and speed itself in their true numerical relationships. Now, I think, you wish to bring us down from that contemplation of abstract truth to consideration of the orbits on the visible heaven, to the observations of professional astronomers and to the hypotheses which they have devised from these observations, hypotheses which people like Aristarchus, Hipparchus, Ptolemy and others like them are always writing about. I suppose you want to become acquainted with their theories because you wish to examine carefully all the theories, as far as that is possible, with which the ancients, in their speculations about the universe, have abundantly supplied us.

Last year, when I was staying with you in central Lydia, I promised you that when I had time,

I would work with you on these matters in my accustomed way. Now that I have arrived in Athens and heaven has freed me from those many unending troubles, I keep my promise to you and will . . . explain to you the real truth which those who are so eager to contemplate the heavenly bodies have come to believe by means of long and, indeed, endless chains of reasoning. In doing so I must, of course, pretend to myself to forget, for the moment at any rate, Plato's exhortations and the theoretical explanations which he taught us to maintain. Even so, I shall not be able to refrain from applying, as is my habit, a critical mind to their doctrines, though I shall do so sparingly, since I am convinced that the exposition of their doctrines will suggest to you quite clearly what the weaknesses of their hypotheses are, hypotheses of which they are so proud when developing their theories.

Before I end, I wish to add this: in their endeavor to demonstrate that the movements of the heavenly bodies are uniform, the astronomers have unwittingly shown the nature of these movements to be lacking in uniformity and to be the subject of outside influences. What shall we say of the eccentrics and the epicycles of which they speak so much? Are they only conceptual notions or do they have a substantial existence in the spheres with which they are connected? If they exist only as concepts, then the astronomers have passed, without noticing it, from bodies really existing in nature to mathematical notions and, again without noticing it, have derived the causes of natural movements from something that does not exist in nature. I will add further that there is absurdity also in the way in which they attribute particular kinds of movement to heavenly bodies. That we conceive of these movements, that is not proof that the stars which we conceive of moving in these circles really move anomalously.

On the other hand, if the astronomers say that the circles have a real, substantial existence, then they destroy the coherence of the spheres themselves on which the circles are situated. They attribute a separate movement to the circles and another to the spheres, and again, the movement they attribute to the circles is not the same for all of them; indeed, sometimes these movements take place in opposite directions. They vary the distances between them in a confused way; sometimes the circles come together in one plane, at other times they stand apart, and cut each other. There will, therefore, be all sorts of divisions, foldings and separations.

I want to make this further observation: the astronomers exhibit a very casual attitude in their exposition of these hypothetical devices. Why is it that, on any given hypothesis, the eccentric or, for that matter, the epicycle moves (or is stationary) in such and such a way while the star moves either in direct or retrograde motion? And what are the explanations (I mean the real explanations) of those planes and their separations? This they never explain in a way that would satisfy our yearning for complete understanding. They really go backwards: they do not derive their conclusions deductively from their hypotheses, as one does in the other sciences; instead, they attempt to formulate the hypotheses starting from the conclusions, which they ought to derive from the hypotheses. It is clear that they do not even solve such problems as could well be solved.

One must, however, admit that these are the simplest hypotheses and the most fitting for divine bodies, and that they have been constructed with a view to discovering the characteristic movements of the planets (which, in real truth, move in exactly the same way as they *seem* to move) and to formulating the quantitative measures applicable to them.

1.14
Projectile Motion

John Philoponus

John Philoponus (c.490–c.570), sometimes called John of Alexandria or John the Grammarian, was a Christian philosopher, scientist, and theologian who lived and worked in Alexandria. Though he was trained as a Neoplatonist by one of Proclus's students, he broke in fundamental ways with the Neoplatonic tradition. His work influenced Arab philosophers such as Avempace (Ibn Bajja), the medieval philosopher-scientists Buridan and Oresme, and Galileo. In the following selection from his commentary on Aristotle's *Physics*, Philoponus mercilessly critiques Aristotle's suggested account of projectile motion and introduces the idea of *impetus*, an "incorporeal motive force," as a more plausible explanation.

Such, then, is Aristotle's account in which he seeks to show that forced motion and motion contrary to nature could not take place if there were a void. But to me this argument does not seem to carry conviction. For in the first place really nothing has been adduced, sufficiently cogent to satisfy our minds, to the effect that motion contrary to nature or forced motion is caused in one of the ways enumerated by Aristotle. . . .

For in the case of *antiperistasis* [the process whereby P_1 pushes P_2 into P_3's place, P_2 pushes P_3 into P_4's place, . . . , P_{n-1} pushes P_n into P_1's place] there are two possibilities; (1) the air that has been pushed forward by the projected arrow or stone moves back to the rear and takes the place of the arrow or stone, and being thus behind it pushes it on, the process continuing until the impetus of the missile is exhausted, or, (2) it is not the air pushed ahead but the air from the sides that takes the place of the missile. . . .

Let us suppose that *antiperistasis* takes place according to the first method indicated above, namely, that the air pushed forward by the arrow gets to the rear of the arrow and thus pushes it from behind. On that assumption, one would be hard put to it to say what it is (since

From Morris R. Cohen and I. E. Drabkin (eds.), *A Source Book in Greek Science* (Cambridge, Mass.: Harvard University Press, 1966), pp. 221–3.

there seems to be no counter force) that causes the air, once it has been pushed forward, to move back, that is along the sides of the arrow, and, after it reaches the rear of the arrow, to turn around once more and push the arrow forward. For, on this theory, the air in question must perform three distinct motions: it must be pushed forward by the arrow, then move back, and finally turn and proceed forward once more. Yet air is easily moved, and once set in motion travels a considerable distance. How, then, can the air, pushed by the arrow, fail to move in the direction of the impressed impulse, but instead, turning about, as by some command, retrace its course? Furthermore, how can this air, in so turning about, avoid being scattered into space, but instead impinge precisely on the notched end of the arrow and again push the arrow on and adhere to it? Such a view is quite incredible and borders rather on the fantastic.

Again, the air in front that has been pushed forward by the arrow is, clearly, subjected to some motion, and the arrow, too, moves continuously. How, then, can this air, pushed by the arrow, take the place of the arrow, that is, come into the place which the arrow has left? For before this air moves back, the air from the sides of the arrow and from behind it will come together and, because of the suction caused by the vacuum, will instantaneously fill up the place left by the arrow, particularly so the air moving along with the arrow from behind it. Now one might say that the air pushed forward by the arrow moves back and pushes, in its turn, the air that has taken the place of the arrow, and thus getting behind the arrow pushes it into the place vacated by the very air pushed forward (by the arrow) in the first instance. But in that case the motion of the arrow would have to be discontinuous. For before the air from the sides, which has taken the arrow's place, is itself pushed, the arrow is not moved. For this air does not move it. But if, indeed, it does, what need is there for the air in front to turn about and move back? And in any case, how or by what force could the air that had been pushed forward receive an impetus for motion in the opposite direction? . . .

So much, then, for the argument which holds that forced motion is produced when air takes the place of the missile (*antiperistasis*). Now there is a second argument which holds that the air which is pushed in the first instance [i.e. when the arrow is first discharged] receives an impetus to motion, and moves with a more rapid motion than the natural [downward] motion of the missile, thus pushing the missile on while remaining always in contact with it until the motive force originally impressed on this portion of air is dissipated. This explanation, though apparently more plausible, is really no different from the first explanation by *antiperistasis*, and the following refutation will apply also to the explanation by *antiperistasis*.

In the first place we must address the following questions to those who hold the views indicated: "When one projects a stone by force, is it by pushing the air behind the stone that one compels the latter to move in a direction contrary to its natural direction? Or does the thrower impart a motive force to the stone, too?" Now if he does not impart any such force to the stone, but moves the stone merely by pushing the air, and if the bowstring moves the arrow in the same way, of what advantage is it for the stone to be in contact with the hand, or for the bowstring to be in contact with the notched end of the arrow?

For it would be possible, without such contact, to place the arrow at the top of a stick, as it were on a thin line, and to place the stone in a similar way, and then, with countless machines, to set a large quantity of air in motion behind these bodies. Now it is evident that the greater the amount of air moved and the greater the force with which it is moved the more should this air push the arrow or stone, and the further should it hurl them. But the fact is that even if you place the arrow or stone upon a line or point quite devoid of thickness and set in motion all the air behind the projectile with all possible force, the projectile will not be moved the distance of a single cubit.

If, then, the air, though moved with a greater force [than that used by one who hurls a projectile], could not impart motion to the projectile, it is evident, in the case of the hurling of missiles or the shooting of arrows, it is not the air set in motion by the hand or bowstring that produces the motion of the missile or arrow. For why would such a result be any more likely when the projector is in contact with the projectile than when he is not? And, again, if the arrow is in direct contact with the bowstring and the stone with the hand, and there is nothing between, what air behind the projectile could be moved? If it is the air from the sides that is moved, what has that to do with the projectile? For that air falls outside the [trajectory of the] projectile.

From these considerations and from many others we may see how impossible it is for forced motion to be caused in the way indicated. Rather is it necessary to assume that some incorporeal motive force is imparted by the projector to the projectile, and that the air set in motion contributes either nothing at all or else very little to this motion of the projectile. If, then, forced motion is produced as I have suggested, it is quite evident that if one imparts motion "contrary to nature" or forced motion to an arrow or a stone the same degree of motion will be produced much more readily in a void than in a plenum. And there will be no need of any agency external to the projector. . . .

1.15

Free Fall

John Philoponus

In the following selection, which is also taken from his commentary on Aristotle's *Physics*, Philoponus critiques Aristotle's account of free fall. Unlike Aristotle, Philoponus takes seriously the physical possibility of there being a void and asks how an object might move in a void. Where Aristotle's remarks suggest that (within limits) he is relating force, velocity, and resistance in a way we might be tempted to express as $V \propto F/R$, Philoponus' analysis suggests something like $V \propto F - R$. Though neither expression is correct by modern standards, Philoponus' critique of Aristotle's position represents a definite conceptual advance since it does not lead to absurd consequences when resistance goes to zero.

Weight, then, is the efficient cause of downward motion, as Aristotle himself asserts. This being so, given a distance to be traversed, I mean through a void where there is nothing to impede motion, and given that the efficient cause of the motion differs [i.e., that there are differences in weight], the resultant motions will inevitably be at different speeds, even through a void. . . . Clearly, then, it is the natural weights of bodies, one having a greater and another a lesser downward tendency, that causes differences in motion. For that which has a greater downward tendency divides a medium better. Now air is more effectively divided by a heavier body. To what other cause shall we ascribe this fact than that that which has greater weight has, by its own nature, a greater downward tendency, even if the motion is not through a plenum? . . .

And so, if a body cuts through a medium better by reason of its greater downward tendency, then, even if there is nothing to be cut, the body will none the less retain its greater downward tendency. . . . And if bodies possess a greater or a lesser downward tendency in and of themselves, clearly they will possess this difference in themselves even if they move through a void. The same space will consequently be traversed by the heavier body in shorter time and by the lighter body in longer time, even though the space be void. The result will be due not to greater or lesser interference with the motion [i.e. the resistance of the medium, since in a void there is none] but to the greater or lesser downward tendency, in proportion to the natural weight of the bodies in question. . . .

Sufficient proof has been adduced to show that if motion took place through a void, it would not follow that all bodies would move therein with equal speed. We have also shown that Aristotle's attempt to prove that they would so move does not carry conviction. Now if our reasoning up to this point has been sound it follows that our earlier proposition is also true, namely, that it is possible for motion to take place through a void in finite time. . . .

Thus, if a certain time is required for each weight, in and of itself, to accomplish a given motion, it will never be possible for one and the same body to traverse a given distance, on one occasion through a plenum and on another through a void, in the same time.

For if a body moves the distance of a stade through air, and the body is not at the beginning and at the end of the stade at one and the same instant, a definite time will be required, dependent on the particular nature of the body in question, for it to travel from the beginning of the course to the end (for, as I have indicated, the body is not at both extremities at the same instant), and this would be true even if the space traversed were a void. But a certain *additional time* is required because of the interference of the medium. For the pressure of the medium and the necessity of cutting through it make the motion through it more difficult.

Consequently, the thinner we conceive the air to be through which a motion takes place, the less will be the *additional time* consumed in dividing the air. And if we continue indefinitely to make this medium thinner, the additional time will also be reduced indefinitely, since time is indefinitely divisible. But even if the medium be thinned out indefinitely in this way, the total time consumed will never be reduced to the time which the body consumes in moving the distance of a stade through the void. I shall make my point clearer by examples.

If a stone moves the distance of a stade through a void, there will necessarily be a time, let us say an hour, which the body will consume in moving the given distance. But if we suppose this distance of a stade filled with water, no longer will the motion be accomplished in one hour, but a certain additional time will be necessary because of the resistance of the medium. Suppose that for the division of the water another hour is required, so that the same weight covers the distance through a void in one hour and through water in two. Now if you thin out the water, changing it into air, and if air is half as dense as water, the time which the body has consumed in dividing the water will be proportionately reduced. In the case of water the additional time was an hour. Therefore the body will move the same distance through air in an hour and a half [i.e., the hour it would take to go through a void, plus half an hour (half as much as the hour that would be added to pass through water) because air offers only half the resistance of water]. If, again, you make the air half as dense [as you already did], the motion will be accomplished in an hour and a quarter. And if you continue indefinitely to rarefy the medium, you will decrease indefinitely the time required for the division of the medium, for example, the additional hour required in the case of water. But you will never completely eliminate this additional time, for time is indefinitely divisible.

If, then, by rarefying the medium you will never eliminate this additional time, and if in the case of motion through a plenum there is always some portion of the second hour to be added, in proportion to the density of the medium, clearly the stade will never be traversed by a body through a void in the same time as through a plenum. . . .

But it is completely false and contrary to the evidence of experience to argue as follows: "If a stade is traversed through a plenum in two hours, and through a void in one hour, then if I take a medium half as dense as the first, the same distance will be traversed through this rarer medium in half the time, that is, in one hour: hence the same distance will be traversed through a plenum in the same time as through a void." *For Aristotle wrongly assumes that the ratio of the times required for motion through various media is equal to the ratio of the densities of the media. . . .*

Now this argument of Aristotle's seems convincing and the fallacy is not easy to detect because it is impossible to find the ratio which air bears to water, in its composition, that is, to find how much denser water is than air, or one specimen of air than another. But from a consideration of the moving bodies themselves we are able to refute Aristotle's contention. [Philoponus spends the rest of this paragraph drawing out a consequence of Aristotle's view before attacking it in the next paragraph.] For if, in the case of one and the same body moving through two different media, the ratio of the times required for the motions were equal to the ratio of the densities of the respective media, then, since differences of velocity are determined not only by the media but also by the moving bodies themselves, the following proposition would be a fair conclusion: "in the case of bodies differing in weight and moving through one and the same medium, the ratio of the times required for the motions is equal to the inverse ratio of the weights." For example, if the weight were doubled, the time would be halved. That is, if a weight of two pounds moved the distance of a stade through the air in one-half hour, a weight of one pound would move the same distance in one hour. Conversely, the ratio of the weights of the bodies would have to be equal to the inverse ratio of the times required for the motions.

But this is completely erroneous, and our view may be corroborated by actual observation more effectively than by any sort of verbal argument. *For if you let fall from the same height two weights of which one is many times as heavy as the other, you will see that the ratio of the times required for the motion does not depend on the ratio of the weights, but that the difference in time is a very small one.* And so, if the difference in the weights is not considerable, that is, if one is, let us say, double the other, there will be no difference, or else an imperceptible difference, in time, though the difference in weight is by no means negligible, with one body weighing twice as much as the other.

Now if, in the case of different weights in motion through the same medium, the ratio of the times required for the motions is not equal to the inverse ratio of the weights, and, conversely, the ratio of the weights is not equal to the inverse ratio of the times, the following proposition would surely be reasonable: "If identical bodies move through different media, like air and water, the ratio of the times required for the motions through the air and water, respectively, is not equal to the ratio of the densities of air and water, and conversely."

Now if the ratio of the times is not determined by the ratio of the densities of the media, it follows that a medium half as dense will not be traversed in half the time, but longer than half. Furthermore, as I have indicated above, in proportion as the medium is rarefied, the shorter is the *additional* time required for the division of the medium. But this additional time is never completely eliminated; it is merely decreased in proportion to the degree of rarefaction of the medium, as has been indicated. . . . And so, if the *total* time required is not reduced in proportion to the degree of rarefaction of the medium, and if the time added for the division of the medium is diminished in proportion to the rarefaction of the medium, but never entirely eliminated, it follows that a body will never traverse the same distance through a plenum in the same time as through a void.

1.16

Against the Reality of Epicycles and Eccentrics

Moses Maimonides

Moses Maimonides (1135–1204) was arguably the greatest Jewish thinker of the Middle Ages and made fundamental contributions to rabbinical doctrine as well as to philosophy. In this selection from Book II, chapter 24 of his *Guide of the Perplexed*, he addresses the question of the physical reality of the equants and epicycles used in Ptolemaic astronomy. Maimonides rejects these devices as incompatible with Aristotle's physics, which he considers to be firmly established. But he is perplexed by the fact that an astronomy constructed using these fictitious devices is empirically accurate.

You know of astronomical matters what you have read under my guidance and understood from the contents of the "Almagest." But there was not enough time to begin another speculative study with you. What you know already is that as far as the action of ordering the motions and making the course of the stars conform to what is seen is concerned, everything depends on two principles: either that of the epicycles or that of the eccentric spheres or on both of them. Now I shall draw your attention to the fact that both those principles are entirely outside the bounds of reasoning and opposed to all that has been made clear in natural science. In the first place, if one affirms as true the existence of an epicycle revolving round a certain sphere, positing at the same time that that revolution is not around the center of the sphere carrying the epicycles – and this has been supposed with regard to the moon and to the five planets – it follows necessarily that there is rolling, that is, that the epicycle rolls and changes its place completely. Now this is the impossibility that was to be avoided, namely, the assumption that there should be something in the heavens that changes its place. For this reason Abu Bakr Ibn al-Sa'igh states in his extant discourse on astronomy that the existence of epicycles is impossible. He points out the necessary inference already mentioned. In addition to this impossibility necessarily following from the assumption

The Guide of the Perplexed, Vol. 1, trans. Shlomo Pines (Chicago: University of Chicago Press, 1963). Reprinted with permission from The University of Chicago Press.

of the existence of epicycles, he sets forth there other impossibilities that also follow from that assumption. I shall explain them to you now.

The revolution of the epicycles is not around the center of the world. Now it is a fundamental principle of this world that there are three motions: a motion from the midmost point of the world, a motion toward that point, and a motion around that point. But if an epicycle existed, its motion would be neither from that point nor toward it nor around it.

Furthermore, it is one of the preliminary assumptions of Aristotle in natural science that there must necessarily be some immobile thing around which circular motion takes place. Hence it is necessary that the earth should be immobile. Now if epicycles exist, theirs would be a circular motion that would not revolve round an immobile thing. I have heard that Abu Bakr has stated that he had invented an astronomical system in which no epicycles figured, but only eccentric circles. However, I have not heard this from his pupils. And even if this were truly accomplished by him, he would not gain much thereby. For eccentricity also necessitates going outside the limits posed by the principles established by Aristotle, those principles to which nothing can be added. It was by me that attention was drawn to this point. In the case of eccentricity, we likewise find that the circular motion of the spheres does not take place around the midmost point of the world, but around an imaginary point that is other than the center of the world. Accordingly, that motion is likewise not a motion taking place around an immobile thing. If, however, someone having no knowledge of astronomy thinks that eccentricity with respect to these imaginary points may be considered – when these points are situated inside the sphere of the moon, as they appear to be at the outset – as equivalent to motion round the midmost point of the world, we would agree to concede this to him if that motion took place round a point in the zone of fire or of air, though in that case that motion would not be around an immobile thing. We will, however, make it clear to him that the measures of eccentricity have been demonstrated in the "Almagest" according to what is assumed there. And the latter-day scientists have given a correct demonstration, regarding which there is no doubt, of how great the measure of these eccentricities is compared with half the diameter of the earth, just as they have set forth all the other distances and dimensions. It has consequently become clear that the eccentric point around which the sun revolves must of necessity be outside the concavity of the sphere of the moon and beneath the convexity of the sphere of Mercury. Similarly the point around which Mars revolves, I mean to say the center of its eccentric sphere, is outside the concavity of the sphere of Mercury and beneath the convexity of the sphere of Venus. Again the center of the eccentric sphere of Jupiter is at the same distance – I mean between the sphere of Mercury and Venus. As for Saturn, the center of its eccentric sphere is between the spheres of Mars and Jupiter. See now how all these things are remote from natural speculation! All this will became clear to you if you consider the distances and dimensions, known to you, of every sphere and star, as well as the evaluation of all of them by means of half the diameter of the earth so that everything is calculated according to one and the same proportion and the eccentricity of every sphere is not evaluated in relation to the sphere itself.

Even more incongruous and dubious is the fact that in all cases in which one of two spheres is inside the other and adheres to it on every side, while the centers of the two are different, the smaller sphere can move inside the bigger one without the latter being in motion, whereas the bigger sphere cannot move upon any axis whatever without the smaller one being in motion. For whenever the bigger sphere moves, it necessarily, by means of its movement, sets the smaller one in motion, except in the case in which its motion is on an axis passing through the two centers. From this demonstrative premise and from the demonstrated fact that vacuum does not exist and from the assumptions regarding eccentricity, it follows necessarily that when the higher sphere is in motion it must move the sphere beneath it with the same motion and around its own center. Now we do not find that this is so. We find rather that neither of the two spheres, the containing and the contained, is set in motion by the movement of the other nor does it move around the other's center or poles, but that each of them has its own particular motion. Hence necessity obliges the belief that between every two spheres there are bodies other than those of the spheres. Now if this be so, how many obscure points remain? Where will you suppose the centers of those bodies existing between every two spheres to be? And those bodies should likewise have their own particular motion. Thabit has explained this in a treatise of his and has demonstrated what we have said, namely, that there must be the body of a sphere between every two spheres. All this I did not explain to you when you read under my guidance, for fear of confusing you with regard to that which it was my purpose to make you understand.

As for the inclination and deviation that are spoken of regarding the latitude of Venus and Mercury, I have explained to you by word of mouth and I have shown you that it is impossible to conceive their existence in those bodies. For the rest Ptolemy has said explicitly, as you have seen, that one was unable to do this, stating literally: No one should think that these principles and those similar to them may only be put into effect with difficulty, if his reason for doing this be that he regards that which we have set forth as he would regard things obtained by artifice and the subtlety of art and which may only be realized with difficulty. For human matters should not be compared to those that are divine. This is, as you know, the text of his statement. I have indicated to you the passages from which the true reality of everything I have mentioned to you becomes manifest, except for what I have told you regarding the examination of where the points lie that are the centers of the eccentric circles. For I have never come across anybody who has paid attention to this. However this shall become clear to you through the knowledge of the measure of the diameter of every sphere and what the distance is between the two centers as compared with half the diameter of the earth, according to what has been demonstrated by al-Qabisi in the "Epistle concerning the Distances." If you examine those distances, the truth of the point to which I have drawn your attention will become clear to you.

Consider now how great these difficulties are. If what Aristotle has stated with regard to natural science is true, there are no epicycles or eccentric circles and everything revolves round the center of the earth. But in that case how can the various motions of the stars come about? Is it in any way possible that motion should be on the one hand circular, uniform, and perfect, and that on the other hand the things that are observable should be observed in consequence of it, unless this be accounted for by making use of one of the two principles or of both of them? This consideration is all the stronger because of the fact that if one accepts everything stated by Ptolemy concerning the epicycle of the moon and its deviation toward a point outside the center of the world and also outside the center of the eccentric circle, it will be found that what is calculated on the hypothesis of the two principles is not at fault by even a minute. The truth of this is attested by the correctness of the calculations – always made on the basis of these principles – concerning the eclipses and the exact determination of their times as well as of the moment when it begins to be dark and of the length of time of the darkness. Furthermore, how can one conceive the retrogradation of a star, together with its other motions, without assuming the existence of an epicycle? On the other hand, how can one imagine a rolling motion in the heavens or a motion around a center that is not immobile? This is the true perplexity.

However, I have already explained to you by word of mouth that all this does not affect the astronomer. For his purpose is not to tell us in which way the spheres truly are, but to posit an astronomical system in which it would be possible for the motions to be circular and uniform and to correspond to what is apprehended through sight, regardless of whether or not things are thus in fact. You know already that in speaking of natural science, Abu Bakr Ibn al-Sa'igh expresses a doubt whether Aristotle knew about the eccentricity of the sun and passed over it in silence – treating of what necessarily follows from the sun's inclination, inasmuch as the effect of eccentricity is not distinguishable from that of inclination – or whether he was not aware of eccentricity. Now the truth is that he was not aware of it and had never heard about it, for in his time mathematics had not been brought to perfection. If, however, he had heard about it, he would have violently rejected it; and if it were to his mind established as true, he would have become most perplexed about all his assumptions on the subject. I shall repeat here what I have said before. All that Aristotle states about that which is beneath the sphere of the moon is in accordance with reasoning; these are things that have a known cause, that follow one upon the other, and concerning which it is clear and manifest at what points wisdom and natural providence are effective. However, regarding all that is in the heavens, man grasps nothing but a small measure of what is mathematical; and you know what is in it. I shall accordingly say in the manner of poetical preciousness: *The heavens are the heavens of the Lord, but the earth hath He given to the sons of man.* I mean thereby that the deity alone fully knows the true reality, the nature, the substance, the form, the motions, and the causes of the heavens. But He has enabled man to have knowledge of what is beneath the heavens, for that is his world and his dwelling-place in which he has been placed and of which he himself is a part. This is the truth. For it is impossible for us to accede to the points starting from which conclusions may be drawn about the heavens; for the latter are too far away from us and too high in place and in rank. And even the general conclusion that may be drawn from them, namely, that they prove the existence of their Mover, is a matter the knowledge of which cannot be reached by human intellects. And to fatigue the minds with notions that cannot be grasped by them and for the grasp of which they have no instrument, is a defect in one's inborn disposition or some sort of temptation. Let us then stop at a point that is within our capacity, and let us give over the things that cannot be grasped by reasoning to him who was reached by the mighty divine overflow so that it could be fittingly said of him: *With him do I speak mouth to mouth.* That is the end of what I have to say about this question. It is possible that someone else may find a demonstration by means of which the true reality of what is obscure for me will become clear to him. The extreme predilection that I have for investigating the truth is evidenced by the fact that I have explicitly stated and reported my perplexity regarding these matters as well as by the fact that I have not heard nor do I know a demonstration as to anything concerning them.

1.17

Impetus and Its Applications

Jean Buridan

Jean Buridan (1300–c.1358) was one of the great philosophers of the high Middle Ages. In this selection from his *Questions on the Eight Books of Aristotle's Physics*, Buridan criticizes Aristotle's theory of projectile motion by appealing our experience of common objects such as a top, a lance, and a ship – a method of persuasion that Galileo subsequently adopts. Picking up on the suggestion made by John Philoponus, Buridan develops a theory of impressed force, which he calls *impetus*, to account for the continued motion of an object when it is no longer in contact with the body that set it in motion. Such a theory, he claims, fits the appearances. Yet Buridan is somewhat tentative in his conclusion because he has arrived at it only by showing the inadequacy of the alternatives others have advanced.

Book VIII, Question 12

1 It is sought whether a projectile after leaving the hand of the projector is moved by the air, or by what it is moved.

It is argued that it is not moved by the air, because the air seems rather to resist, since it is necessary that it be divided. Furthermore, if you say that the projector in the beginning moved the projectile and the ambient air along with it, and then that air, having been moved, moves the projectile further to such and such a distance, the doubt will return as to by what the air is moved after the projector ceases to move. For there is just as much difficulty regarding this (the air) as there is regarding the stone which is thrown.

Aristotle takes the opposite position in the eighth [book] of this work (the *Physics*) thus: "Projectiles are moved further after the projectors are no longer in contact with them, either by antiperistasis, as some say, or by the fact that the air having been pushed, pushes with a movement swifter than the movement of impulsion by which it (the body) is carried towards its own

From Marshall Clagett, *The Science of Mechanics in the Middle Ages* (Madison, WI: University of Wisconsin Press, 1959), pp. 532–8.

[natural] place." He determines the same thing in the seventh and eighth [books] of this work (the *Physics*) and in the third [book] of the *De caelo*.

2 This question I judge to be very difficult because Aristotle, as it seems to me, has not solved it well. For he touches on two opinions. The first one, which he calls "antiperistasis," holds that the projectile swiftly leaves the place in which it was, and nature, not permitting a vacuum, rapidly sends air in behind to fill up the vacuum. The air moved swiftly in this way and impinging upon the projectile impels it along further. This is repeated continually up to a certain distance. . . . But such a solution notwithstanding, it seems to me that this method of proceeding was without value because of many experiences (*experientie*).

The first experience concerns the top (*trocus*) and the smith's mill (i.e. wheel – *mola fabri*) which are moved for a long time and yet do not leave their places. Hence, it is not necessary for the air to follow along to fill up the place of departure of a top of this kind and a smith's mill. So it cannot be said [that the top and the smith's mill are moved by the air] in this manner.

The second experience is this: A lance having a conical posterior as sharp as its anterior would be moved after projection just as swiftly as it would be without a sharp conical posterior. But surely the air following could not push a sharp end in this way, because the air would be easily divided by the sharpness.

The third experience is this: a ship drawn swiftly in the river even against the flow of the river, after the drawing has ceased, cannot be stopped quickly, but continues to move for a long time. And yet a sailor on deck does not feel any air from behind pushing him. He feels only the air from the front resisting [him]. Again, suppose that the said ship were loaded with grain or wood and a man were situated to the rear of the cargo. Then if the air were of such an impetus that it could push the ship along so strongly, the man would be pressed very violently between that cargo and the air following it. Experience shows this to be false. Or, at least, if the ship were loaded with grain or straw, the air following and pushing would fold over (*plico*) the stalks which were in the rear. This is all false.

3 Another opinion, which Aristotle seems to approve, is that the projector moves the air adjacent to the projectile [simultaneously] with the projectile and that air moved swiftly has the power of moving the projectile. He does not mean by this that the same air is moved from the place of projection to the place where the projectile stops, but rather that the air joined to the projector is moved by the projector and that air having been moved moves another part of the air next to it, and that [part] moves another (i.e., the next) up to a certain distance. Hence the first air moves the projectile into the second air, and the second [air moves it] into the third air, and so on. Aristotle says, therefore, that there is not one mover but many in turn. Hence he also concludes that the movement is not continuous but consists of succeeding or contiguous entities.

But this opinion and method certainly seems to me equally as impossible as the opinion and method of the preceding view. For this method cannot solve the problem of how the top or smith's mill is turned after the hand [which sets them into motion] has been removed. Because, if you cut off the air on all sides near the smith's mill by a cloth (*linteamine*), the mill does not on this account stop but continues to move for a long time. Therefore it is not moved by the air.

Also a ship drawn swiftly is moved a long time after the haulers have stopped pulling it. The surrounding air does not move it, because if it were covered by a cloth and the cloth with the ambient air were withdrawn, the ship would not stop its motion on this account. And even if the ship were loaded with grain or straw and were moved by the ambient air, then that air ought to blow exterior stalks toward the front. But the contrary is evident, for the stalks are blown rather to the rear because of the resisting ambient air.

Again, the air, regardless of how fast it moves, is easily divisible. Hence it is not evident as to how it would sustain a stone of weight of one thousand pounds projected in a sling or in a machine.

Furthermore, you could, by pushing your hand, move the adjacent air, if there is nothing in your hand, just as fast or faster than if you were holding in your hand a stone which you wish to project. If, therefore, that air by reason of the velocity of its motion is of a great enough impetus to move the stone swiftly, it seems that if I were to impel air toward you equally as fast, the air ought to push you impetuously and with sensible strength. [Yet] we would not perceive this.

Also, it follows that you would throw a feather farther than a stone and something less heavy farther than something heavier, assuming equal magnitudes and shapes. Experience shows this to be false. The consequence is manifest, for the air having been moved ought to sustain or carry or move a feather more easily than something heavier. . . .

4 Thus we can and ought to say that in the stone or other projectile there is impressed something which is the motive force (*virtus motiva*) of that projectile. And this is evidently better than falling back on the statement that the air continues to move that projectile. For the air appears rather to resist. Therefore, it seems to me that it ought to be said that the motor in moving a moving body impresses (*imprimit*) in it a certain impetus (*impetus*) or a certain motive force (*vis motiva*) of the moving body, [which impetus acts] in the direction toward which the mover was moving the moving body, either up or down, or laterally, or circularly. *And by the amount the motor moves that moving body more swiftly, by the same amount it will impress in it a stronger impetus.* It is by that impetus that the stone is moved after the projector ceases to move. But that impetus is continually decreased (*remittitur*) by the resisting air and by the gravity of the stone, which inclines it in a direction contrary to that in which the impetus was naturally predisposed to move it. Thus the movement of the stone continually becomes slower, and finally that impetus is so diminished or corrupted that the gravity of the stone wins out over it and moves the stone down to its natural place.

This method, it appears to me, ought to be supported because the other methods do not appear to be true and also because all the appearances (*apparentia*) are in harmony with this method.

5 For if anyone seeks why I project a stone farther than a feather, and iron or lead fitted to my hand farther than just as much wood, I answer that the cause of this is that the reception of all forms and natural dispositions is in matter and by reason of matter. *Hence by the amount more there is of matter, by that amount can the body receive more of that impetus and more intensely* (intensius). *Now in a dense and heavy body, other things being equal, there is more of prime matter than in a rare and light one. Hence a dense and heavy body receives more of that impetus and more intensely, just as iron can receive more calidity than wood or water of the same quantity.* Moreover, a feather receives such an impetus so weakly (*remisse*) that such an impetus is immediately destroyed by the resisting air. *And so also if light wood and heavy iron of the same volume and of the same shape are moved equally fast by a projector, the iron will be moved farther because there is impressed in it a more intense impetus, which is not so quickly corrupted as the lesser impetus would be corrupted. This also is the reason why it is more difficult to bring to rest a large smith's mill which is moving swiftly than a small one, evidently because in the large one, other things being equal, there is more impetus.* And for this reason you could throw a stone of one-half or one pound weight farther than you could a thousandth part of it. For the impetus in that thousandth part is so small that it is overcome immediately by the resisting air.

6 From this theory also appears the cause of why the natural motion of a heavy body downward is continually accelerated (*continue velocitatur*). For from the beginning only the gravity was moving it. Therefore, it moved more slowly, but in moving it impressed in the heavy body an impetus. This impetus now [acting] together with its gravity moves it. Therefore, the motion becomes faster; and by the amount it is faster, so the impetus becomes more intense. Therefore, the movement evidently becomes continually faster.

[The impetus then also explains why] one who wishes to jump a long distance drops back a way in order to run faster, so that by running he might acquire an impetus which would carry him a longer distance in the jump. Whence the person so running and jumping does not feel the air moving him, but [rather] feels the air in front strongly resisting him.

Also, since the Bible does not state that appropriate intelligences move the celestial bodies, it could be said that it does not appear necessary to posit intelligences of this kind, because it would be answered that God, when He created the world, moved each of the celestial orbs as He pleased, and in moving them He impressed in them impetuses which moved them without his having to move them any more except by the method of general influence whereby he concurs as a co-agent in all things which take place; "for thus on the seventh day He rested from all work which He had executed by committing to others the actions and the passions in turn." And these impetuses which He impressed in the celestial bodies were not decreased nor corrupted afterwards, because there was no inclination of the celestial bodies for other movements. Nor was there resistance which would be corruptive or repressive of that impetus. But this I do not say assertively, but [rather tentatively] so that I might seek from the theological masters what they might teach me in these matters as to how these things take place. . . .

7 The first [conclusion] is that that impetus is not the very local motion in which the projectile is moved, because that impetus moves the projectile and the mover produces motion. Therefore, the impetus produces that motion, and the same thing cannot produce itself. Therefore, etc.

Also since every motion arises from a motor being present and existing simultaneously with that which is moved, if the impetus were the motion, it would be necessary to assign some other motor from which that motion would arise. And the principal difficulty would return. Hence there would be no gain in positing such an impetus. But others cavil when they say that the prior part of the motion which produces the projection produces another part of the motion which is related successively and that produces another part and so on up to the cessation of the whole movement. But this is not probable, because the "producing something" ought to exist when the something is made, but the prior part of the motion does not exist when the posterior part exists, as was elsewhere stated. Hence, neither does the prior exist when the posterior is made. This consequence is obvious from this reasoning. For it was said elsewhere that motion is nothing else than "the very being produced" (*ipsum fieri*) and the "very being corrupted" (*ipsum corumpi*). Hence motion does not result when it *has been* produced (*factus est*) but when it *is being* produced (*fit*).

8 The second conclusion is that that impetus is not a purely successive thing (*res*), because motion is just such a thing and the definition of motion [as a successive thing] is fitting to it, as was stated elsewhere. And now it has just been affirmed that that impetus is not the local motion.

Also, since a purely successive thing is continually corrupted and produced, it continually demands a producer. But there cannot be assigned a producer of that impetus which would continue to be simultaneous with it.

9 The third conclusion is that that impetus is a thing of permanent nature (*res nature permanentis*), distinct from the local motion in which the projectile is moved. This is evident from the two aforesaid conclusions and from the preceding [statements]. And it is probable (*verisimile*) that that impetus is a quality naturally present and predisposed for moving a body in which it is impressed, just as it is said that a quality impressed in iron by a magnet moves the iron to the magnet. And it also is probable that just as that quality (the impetus) is impressed in the moving body along with the motion by the motor; so with the motion it is remitted, corrupted, or impeded by resistance or a contrary inclination.

10 And in the same way that a luminant generating light generates light reflexively because of an obstacle, so that impetus because of an obstacle acts reflexively. It is true, however, that other causes aptly concur with that impetus for greater or longer reflection. For example, the ball which we bounce with the palm in falling to earth is reflected higher than a stone, although the stone falls more swiftly and more impetuously (*impetuosius*) to the earth. This is because many things are curvable or intracompressible by violence which are innately disposed to return swiftly and by themselves to their correct position or to the disposition natural to them. In thus returning, they can impetuously push or draw something conjunct to them, as is evident in the case of the bow (*arcus*). Hence in this way the ball thrown to the hard ground is compressed into itself by the impetus of its motion; and immediately after striking, it returns swiftly to its sphericity by elevating itself upwards. From this elevation it acquires to itself an impetus which moves it upward a long distance.

Also, it is this way with a cither cord which, put under strong tension and percussion, remains a long time in a certain vibration (*tremulatio*) from which its sound continues a notable time. And this takes place as follows: As a result of striking [the chord] swiftly, it is bent violently in one direction, and so it returns swiftly toward its normal straight position. But on account of the impetus, it crosses beyond the normal straight position in the contrary direction and then again returns. It does this many times. For a similar reason a bell (*campana*), after the ringer ceases to draw [the chord], is moved a long time, first in one direction, now in another. And it cannot be easily and quickly brought to rest.

This, then, is the exposition of the question. I would be delighted if someone would discover a more probable way of answering it. And this is the end.

1.18

The Possibility of a Rotating Earth

Nicole Oresme

Nicole Oresme (c.1325–1382) was a student of Jean Buridan and extended the work of his great teacher. Like Buridan, Oresme made important contributions to numerous fields, but his most influential work was in mathematics (where he took important steps toward analytic geometry) and what we would today call physics. In this selection from his *Book of Heaven and Earth*, Oresme anticipates Copernicus by undermining both physical and scriptural objections to the daily rotation of the earth. He stops short of endorsing diurnal rotation, but the door had been opened; two centuries later, Copernicus would walk through it.

[Aristotle's text]: There are others who hold that the earth is at the center of the world and that it revolves and moves in a circuit around the pole established for this purpose, as is written in Plato's book called *Timaeus.*

[Oresme's commentary]: This was the opinion of a philosopher named Heraclides Ponticus, who maintained that the earth moves circularly and that the heavens remain at rest. Here Aristotle does not refute these theories, possibly because they seemed to him of slight probability and were, moreover, sufficiently criticized in philosophical and astrological writings.

However, subject, of course, to correction, it seems to me that it is possible to embrace the argument and consider with favor the conclusions set forth in the above opinion that the earth rather than the heavens has a diurnal or daily rotation. At the outset, I wish to state that it is impossible to demonstrate from any experience at all that the contrary is true; second, that no argument is conclusive; and third, I shall demonstrate why this is so.

As to the first point, let us examine one experience: we can see with our eyes the rising and setting of the sun, the moon, and several stars,

From Nicole Oresme, *Le Livre du Ciel et du Monde*, ed. Albert D. Menut and Alexander J. Denomy, trans. Albert D. Menut (Madison, Wis.: University of Wisconsin Press, 1968), pp. 519–39.

while other stars turn around the arctic pole. Such a thing is due only to the motion of the heavens, [. . .] and, therefore, the heavens move with daily motion. Another experience is this one: if the earth is so moved, it makes its complete course in a natural day with the result that we and the trees and the houses are moved very fast toward the east; thus, it should seem to us that the air and wind are always coming very strong from the east and that it should make a noise such as it makes against the arrow shot from a crossbow or an even louder one, but the contrary is evident from experience. The third argument is Ptolemy's – namely, that, if someone were in a boat moving rapidly toward the east and shot an arrow straight upward, it would not fall in the boat but far behind it toward the west. Likewise, if the earth moves so very fast turning from west to east and if someone threw a stone straight upward, it would not fall back to the place from which it was thrown, but far to the west; and the contrary appears to be the case.

It seems to me that what I shall say below about these experiences could apply to all other theories which might be brought forward in this connection. Therefore, I state, in the first place, that the whole corporeal machine or the entire mass of all the bodies in the universe is divided into two parts: one is the heavens with the sphere of fire and the higher region of the air; all this part, according to Aristotle in Book I of *Meteors*, moves in a circle or revolves each day. The other part of the universe is all the rest – that is, the middle and lower regions of the air; the water, the earth, and the mixed bodies – and, according to Aristotle, all this part is immobile and has no daily motion.

Now, I take as a fact that local motion can be perceived only if we can see that one body assumes a different position relative to another body. For example, if a man is in a boat *a*, which is moving very smoothly either at rapid or slow speed, and if this man sees nothing except another boat *b*, which moves precisely like boat *a*, the one in which he is standing, I maintain that to this man it will appear that neither boat is moving. If *a* rests while *b* moves, he will be aware that *b* is moving; if *a* moves and *b* rests, it will seem to the man in *a* that *a* is resting and *b* is moving, just as before. Thus, if *a* rested an hour and *b* moved, and during the next hour it happened conversely that *a* moved and *b* rested, this man would not be able to sense this change or variation; it would seem to him that all this time *b* was moving. This fact is evident from experience, and the reason is that the two bodies *a* and *b* have a continual relationship to each other so that, when *a* moves, *b* rests and, conversely, when *b* moves, *a* rests. It is stated in Book Four of *The Perspective* by Witelo that we do not perceive motion unless we notice that one body is in the process of assuming a different position relative to another.

I say, therefore, that, if the higher of the two parts of the world mentioned above were moved today in daily motion – as it is – and the lower part remained motionless and if tomorrow the contrary were to happen so that the lower part moved in daily motion and the higher – that is, the heavens, etc. – remained at rest, we should not be able to sense or perceive this change, and everything would appear exactly the same both today and tomorrow with respect to this mutation. We should keep right on assuming that the part where we are was at rest while the other part was moving continually, exactly as it seems to a man in a moving boat that the trees on shore move. In the same way, if a man in the heavens, moved and carried along by their daily motion, could see the earth distinctly and its mountains, valleys, rivers, cities, and castles, it would appear to him that the earth was moving in daily motion, just as to us on earth it seems as though the heavens are moving. Likewise, if the earth moved with daily motion and the heavens were motionless, it would seem to us that the earth was immobile and that the heavens appeared to move; and this can be easily imagined by anyone with clear understanding. This obviously answers the first experience, for we could say that the sun and stars appear to rise and set as they do and that the heavens seem to revolve on account of the motion of the earth in which we live together with the elements.

To the second experience, the reply seems to be that, according to this opinion, not only the earth moves, but also with it the water and the air, as we stated above, although the water and air here below may be moved in addition by the winds or other forces. In a similar manner, if the air were closed in on a moving boat, it would seem to a person in that air that it was not moving.

Concerning the third experience which seems more complicated and which deals with the case of an arrow or stone thrown up into the air, etc., one might say that the arrow shot upward is moved toward the east very rapidly with the air through which it passes, along with all the lower portion of the world which we have already defined and which moves with daily motion; for this reason the arrow falls back to the place from which it was shot into the air. Such a thing could be possible in this way for, if a man were in a ship moving rapidly eastward without his being aware of the movement and if he drew his hand in a straight line down along the ship's mast, it would seem to him that his hand were moving with a rectilinear motion; so, according to this theory it seems to us that the same thing happens with the arrow which is shot straight down or straight up. Inside the boat moved rapidly eastward, there can be all kinds of movements – horizontal, criss-cross, upward, downward, in all directions – and they seem to be exactly the same as those when the ship is at rest. Thus, if a man in this boat walked toward the west less rapidly than the boat was moving toward the east, it would seem to the man that he was approaching the west when actually he was going east; and similarly as in the preceding case, all the motions here below would seem to be the same as though the earth rested.

Now, in order to explain the reply to the third experience in which this artificial illustration was used, I should like to present an example taken from nature, which, according to Aristotle, is true.

He supposes that there is a portion of pure fire called *a* in the higher region of the air; this fire, being very light, rises as high as possible to a place called *b* near the concave surface of the heavens [see Fig. 4]. I maintain that, just as with the arrow above, the motion of *a* in this case also must be compounded of rectilinear and, in part, of circular motion, because the region of the air and the sphere of fire through which *a* passed have, in Aristotle's opinion, circular motion. If they were not thus moved, *a* would go straight upward along the line *ab*; but because *b* is meanwhile drawn toward *c* by circular and daily motion, it appears that *a* describes the line *ac* as it ascends and that, therefore, the movement of *a* is compounded of rectilinear and of circular motion, and the movement of the arrow would be of this kind of mixed or compound motion that we spoke of in Chapter Three of Book I. I conclude, then, that it is impossible to demonstrate by any experience that the heavens have daily motion and that the earth does not have the same.

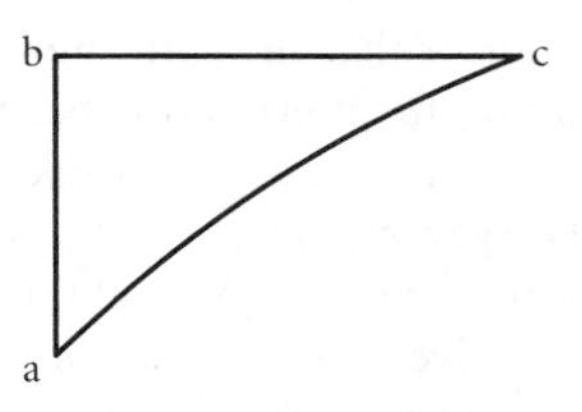

Figure 4 Mixed motion of fire according to Aristotle. A bit of fire, *a*, moves toward its natural place, *b*; but in the same span of time *b* is itself moved toward *c* as the heavens move in a vast circle. The combination of the two motions produces a curved path

[. . .]

. . . If neither experience nor reason indicates the contrary, it is much more reasonable, as stated above, that all the principal movements of the simple bodies in the world should go or proceed in one direction or manner. Now, according to the philosophers and astronomers, it cannot be that all bodies move from east to west; but, if the earth moves as we have indicated, then all proceed alike from west to east – that is, the earth by rotating once around the poles from west to east in one natural day and the heavenly bodies around the zodiacal poles: the moon in one month, the sun in one year, Mars in approximately two years, and so on with the other bodies. It is unnecessary to posit in the heavens other primary poles or two kinds of motion, one from the east to the west and the other on different poles in the opposite direction, but such an assumption is definitely necessary if the heavens move with diurnal motion. . . .

. . . [I]f we assume that the earth moves as stated above, . . . appearances can be saved in this way, as is evident from the reply to the seventh argument, presented against this opinion. . . . Thus, it is apparent that one cannot demonstrate by any experience whatever that the heavens move with diurnal motion; whatever the fact may be, assuming that the heavens move and the earth does not or that the earth moves and the heavens do not, to an eye in the heavens which could see the earth clearly, it would appear to move; if the eye were on the earth, the heavens would appear to move. Nor would the vision of this eye be deceived, for it can sense or see nothing but the process of the movement itself. But if the motion is relative to some particular body or object, this judgment is made by the senses from within that particular body, as Witelo explains in *The Perspective*; and the senses are often deceived in such cases, as was related above in the example of the man on the moving ship. Afterward, it was demonstrated how it cannot be proved conclusively by argument that the heavens move. In the third place, we offered arguments opposing their diurnal motion. However, everyone maintains, and I think myself, that the heavens do move and not the earth: For God hath established the world which shall not be moved, in spite of contrary reasons because they are clearly not conclusive persuasions. However, after considering all that has been said, one could then believe that the earth moves and not the heavens, for the opposite is not clearly evident. Nevertheless, at first sight, this seems as much against natural reason as, or more against natural reason than, all or many of the articles of our faith. What I have said by way of diversion or intellectual exercise can in this manner serve as a valuable means of refuting and checking those who would like to impugn our faith by argument.

Figure acknowledgment

Figure 4: Oresme, "The Compatibility of the Earth's Diurnal Rotation," from Edward Grant, *A Source Book in Medieval Science* (Cambridge: Harvard University Press, 1974), p. 67.

Unit 2

The Scientific Revolution

Historical periods are to some extent arbitrarily defined, but one could do worse than to place the beginning of the scientific revolution at 1543 and its end, or at least its climax, at 1687. The boundaries are fixed by the publication of two books, Copernicus's *On the Revolutions of the Heavenly Spheres* and Isaac Newton's masterpiece on the mathematical principles of natural philosophy, now universally known by a shortened part of its Latin title, *Principia Mathematica*. The story of the road from the one great work to the other is one of the central stories of the scientific revolution, for in it we see the interconnected development of modern astronomy and dynamics and the importance of some key questions about scientific method and inference.

1 The Copernican Revolution

The great Polish astronomer Nicholas Copernicus, whose model of the solar system tripped off the scientific revolution, was an unlikely revolutionary. The idea that the earth rotates on its axis and orbits the sun was not a new one, as Copernicus himself was at pains to point out; it had been advocated in the third century BC by Aristarchus of Samos. And although the new model contradicted Aristotle's view that the earth is stationary and could not be reconciled with Aristotle's dynamics, Copernicus was trying, not to overthrow Aristotle, but to return to Aristotle's principle of uniform circular motion and to produce an astronomy that would live up to Aristotle's lofty conception of scientific knowledge.

The central principle of Aristotle's astronomy is that the heavenly bodies move with uniform circular motion. This is not a crude blunder; Aristotle was aware of the complexities of retrograde motion and discussed the system of Eudoxus, which was designed to reproduce that motion, in detail. But Aristotle considered circular motion to be geometrically perfect, ideally suited to model phenomena that endlessly repeat their intricate patterns; and the simplest sort of circular motion is motion that is uniform as seen from the center, the one point in a circle that is uniquely determined by the geometry of the circle itself. In Copernicus's view, Ptolemy had abused this principle of celestial motion by judging uniformity not from the center but from the equant, a point not at the center of a circle and therefore not uniquely determined by the circle itself. In his own system, Copernicus rigorously avoids any use of the equant.

Another undesirable feature of the Ptolemaic system was the arbitrariness with which it treated the orbital distances of the planets. Is the deferent, the primary circle, for the orbit of Mercury, for example, closer to the earth than the deferent for Venus, or vice versa? Ptolemy gives an order, but he has no compelling argument for it. How large is the deferent for Mars? Whatever answer Ptolemy gives, the visible location of Mars would be essentially the same if the construction were scaled up by a constant and all of the angular velocities remained the same. The ordering and distances of the planets are essentially arbitrary in the Ptolemaic system. Nothing could be further from the Aristotelian ideal of knowledge of the reasoned fact.

Copernicus was also distressed by the physical impossibility of reconciling all of Ptolemy's constructions even with each other. In constructing a model for the speed of the moon's orbit, for example, Ptolemy uses so large an epicycle that the apparent size of the moon, seen from earth, would vary much more than it actually does. When he comes to deal with the distance of the moon from the earth, he ignores the earlier construction and uses a different, incompatible one. If one were to draw these geometric diagrams on sheets of clear plastic and then lay one over the other, there would be no way, even by magnifying or shrinking one of the sheets, to make them coincide.

If one's goal is merely to construct an ephemeris, a set of tables indicating the positions of the celestial bodies at any given time, such constructions may suffice. The components of a calculating device need not represent the underlying structures of reality in order to yield accurate predictions. But Copernicus was at heart a realist; he wanted not only foresight but scientific understanding of the actual structure of the solar system. A system that violated Aristotle's principle of celestial motion, that incorporated arbitrary constants at every point, and that employed mutually irreconcilable constructions could not yield that sort of understanding. So Ptolemy's system, notwithstanding its predictive accuracy, had to be scrapped.

Reading 2.3 is taken from the first book of Copernicus's *On the Revolutions*, and it lays the groundwork for the detailed mathematical work that occupies most of that work. Many of the tools Copernicus used in constructing his new system were the same as those used by Ptolemy. Though at first he had hopes that he could reduce the overall number of epicycles in the system, this proved to be too optimistic. In order to achieve predictive accuracy on a par with Ptolemy's, Copernicus had to fine-tune his system with one epicycle after another; by the time he was finished, mere counting of epicycles left little to choose between the two models.

The fact that in the Copernican system the earth is a planet just like the others has many attractive consequences. The most critical is that it provides a geometric explanation for retrograde motion. In a broadly heliocentric model, the earth is on the inside track with respect to the superior planets Mars, Jupiter, and Saturn. As a result, it will on a regular basis "lap" these other planets – with its shorter orbital circumference, it will pass them out in the race around the sun. When this happens, the planet being lapped will seem to move backwards against the stars. But this is an optical illusion: none of the planets ever actually reverses its course in three-dimensional space. The motion of the earth, rather than the use of epicycles, accounts for the phenomenon of retrograde motion. Maimonides' perplexity is resolved.

But this is only the beginning of the Copernican system's virtues. The geometry that accounts for retrograde motion also entails that a planet will retrogress when it is "in opposition" – when it is on the opposite side of the earth from the sun, which is to say, when it is directly overhead at midnight. Observation bears this out. What is in the Ptolemaic system just a coincidence brought about by the clever setting of independent parameters turns out to be inescapable, a geometric necessity, in the Copernican system. Again, in the Copernican model a planet must be in opposition at just the time of the earth's nearest approach to it, which will be the time when it shines most brightly in the sky. The Ptolemaic system contains enough carefully chosen independent parameters to predict this fact, but just because the parameters are independent any one of them could be changed without affecting other aspects of the system. In the Copernican system all of these phenomena are interrelated by geometric necessity and cannot be dissociated.

Perhaps the crowning achievement of the Copernican system is the use of retrograde motion to determine the relative orbital distances of the planets from the sun. A little thought shows that, in Copernicus's model, the further away a superior planet is, the smaller its retrogression will be. Using this bit of geometry and using the earth's own orbit as a yardstick, Copernicus is able to move beyond the arbitrary selection of orbital radii and give, for the first time, a compelling argument for expressing all of the planetary orbits as definite fractions or multiples of the earth's orbit. The explanatory power of the assumption that the earth orbits the sun was astonishing, causing Georg Rheticus, one of Copernicus's earliest advocates, to exclaim that "all these phenomena appear to be linked most nobly together, as by a golden chain; and each of the planets, by its position and order and very inequality of its motion, bears witness that the Earth moves . . ." Reading 2.1 is an excerpt from Rheticus's *Narratio Prima*.

How could so elegant a system fail to win universal assent? Sociological explanations are inadequate. Some authors have attributed the relatively slow acceptance of the Copernican view to the retarding influence of biblical literalism, and there was some of this; Luther famously dismissed Copernicus with the offhand remark that in Scripture we are told that Joshua made the sun, not the earth, to stand still (Joshua 10:12–13).[1] But such an attitude was by no means universal even among the devout. As Copernicus, himself a canon in the Catholic Church, was composing *On the Revolutions*, he received a very friendly and solicitous note from Cardinal Schönberg asking for a copy of his work on a new system of astronomy in which the earth moves.

The more fundamental problems were empirical. First, if Venus orbits the sun, it ought to show phases rather like those of the moon; but we never see such phases. Second, if we take the Copernican system literally, the visible disc of Venus should be about forty times as large when it is closest to us as it is when it is on the far side of the sun. Yet Venus appears to change hardly at all in brightness. Something similar goes for Mars.

A third problem concerns the absence of stellar parallax. The visible angle between two distant points grows as one approaches those points. If Copernicus is right, then the orbit of the earth is enormous: if we take the visible angle between two distant stars visible just before dawn in January and then look at them again at midnight months later, we should be much closer to them and should therefore expect to see a greater angle between them. Yet we do not. Again, if the sphere of the fixed stars is not much further away than the orbit of Saturn then the celestial North Pole should change with the seasons. But it does not. Copernicus countered that the stars were so far away as to render the diameter of the earth's orbit insignificant. He was right, but the answer must have seemed lame to his critics. For individual stars look too large to be that far away – too large, that is, if we assume that our sun is fairly typical of stars in general. If the distances required by Copernicus were put together with the apparent size of the stars, many of the stars would have diameters larger than our entire solar system.

Besides these observational problems, the Copernican view faced a challenge on the front of elegance. Ptolemy's system, for all its flaws, has one very attractive feature: there is only one center for celestial motion. Copernicus's system has two, since the earth and the other planets orbit the sun but the moon orbits the earth. Partisans of the opposing views might dispute endlessly on the relative merits of this consideration of simplicity, on the one hand, and the unification effected by heliocentrism on the other. But it is hard to see how either side could force the other to acknowledge its point.

And beyond all this, Copernicus ran afoul of Aristotle's dynamics. By setting the earth in motion, he destroyed the fundamental division Aristotle had made between the celestial and terrestrial realms. Ptolemy's argument against a moving earth worked forward from Aristotle's principle of forced motion as a premise; if Copernicus now wished to deny the conclusion, the onus was on him to explain how Aristotle could have gone so badly wrong and to propose a believable alternative. Copernicus labored to the end of his life to complete *On the Revolutions of the Heavenly Spheres*; one of the first printed copies was put into his hands while he lay on his deathbed in 1543. The creation of a new dynamics was a task beyond his failing strength.

For all of these reasons, the decision to accept or reject the Copernican view was difficult and required weighing up the significance of multiple factors that are not easily compared. The sorts of theoretical virtues exhibited in the Copernican system are compelling enough if considered alone. But how should one balance them against the fact that the earth does not seem, to the senses, to be moving? The Aristotelian assumption about motion that lies behind this natural objection is one that many people, even today, find intuitively compelling. And the appeal to experience shows just how complicated the epistemic situation was during the scientific revolution. Modern writers sometimes pillory the followers of Aristotle for failing to look at the actual behavior of objects around them, as if the principal force behind the scientific revolution were the invention of crude empiricism. But in many ways it was the Aristotelians who were appealing to the evidence of the senses and their opponents who were stressing the necessity of treating sensation critically and giving full weight to reason.

Not surprisingly, the initial reception of Copernicus's work was mixed. Georg Joachim Rheticus, a young mathematics professor who had just completed his degree, set off on a visit to Poland in 1539 to determine whether Copernicus was a genius or a crackpot. Four months later he was a complete convert, and he published his long and almost worshipful defense of Copernicus, the *Narratio Prima*, in the following year. But not all even of Copernicus's friends were equally enthusiastic. Andreas Osiander, who was entrusted with the task of seeing the manuscript of *On the Revolutions* into print, slipped a "Letter to the Reader," reprinted here as reading 2.2, into the front of the manuscript warning readers not to take the system literally but rather to use it as a mere calculating device. The letter, which was unsigned, appeared to come from the author. Copernicus may never have seen it, but when Rheticus saw it in print he immediately suspected Osiander of the deed and said that, if he could prove it, he would give Osiander such a thrashing that he would never again meddle in the affairs of scientific men. In his own copy of *On the Revolutions*, Rheticus crossed out the offending passage.

Astronomers in the next two generations puzzled over what to make of Copernicus's difficult and imposing book. Those with the skill to work through the mathematics openly admired his account of retrograde motion even when they did not accept the motion of the earth. The reaction of Thomas Blundeville, an English astronomer, is typical: "Copernicus . . . affirmeth that the earth turneth about and that the sun standeth still in the midst of the heavens, by help of which false supposition he hath made truer demonstrations of the motions and revolutions of the celestial spheres, than ever were made before."[2]

2 Tycho Brahe: The Gradual Breakdown of the Ptolemaic View

Between the publication of Copernicus's proposal and its enthusiastic defense by Galileo there is a gap of nearly three quarters of a century. In that time, astronomy made no great leaps but did take several steps forward. The person centrally responsible for a number of those steps was the Danish astronomer Tycho Brahe.

One of his advances was to accept a part of the Copernican hypothesis. Tycho found the Copernican explanation of retrograde motion so compelling that he created his own model of the solar system in which the planets orbit the sun. But in his system the earth is not a planet: the sun itself orbits the earth, which sits motionless just as in the Ptolemaic system. Because he has not set the earth in motion, Tycho has no problems squaring his system with Aristotelian dynamics and no problem with the absence of a measurable stellar parallax. He inherits only a subset of Copernicus's problems: the apparent lack of phases of Venus, the surprising lack of variation in planetary brightness, and the existence of two centers of motion as opposed to Ptolemy's one. For those who felt torn between the elegance of the Copernican explanation of retrograde motion and the established dynamics of Aristotle, Tycho's system offered an attractive halfway house.

Tycho was deeply attached to his geoheliocentric system, but it was not his greatest contribution to astronomy. The prediction of the eclipse of August 21, 1560 impressed him deeply, and he began to study the available charts of the planets and the stars. To his dismay, they all disagreed with each other. Using his contacts at the Danish court, he eventually won financial support for the construction of two fabulous observatories on the island of Hven between Denmark and Sweden. At these observatories Tycho developed techniques that vastly improved both the precision and the accuracy of astronomical measurement. In an era without telescopes, his data were unsurpassed and provided a much needed corrective to the bad data that circulated undetected in the sixteenth century. These data also set a high standard for all future astronomical theories, since astronomers had less excuse for a mismatch between their theories and the observed locations of the stars and planets.

Tycho's third contribution happened almost by accident. In November of 1572 he observed the appearance of a new star in the familiar constellation Cassiopeia. The phenomenon became a source of dispute, for according to Aristotle's cosmology the heavens are necessarily unchanging. Was the object truly in the heavens, beyond the sphere of the moon, or could it be (as some claimed) an atmospheric phenomenon instead? In *The New Star* (1573), Tycho brought a combination of accurate observation and simple geometry to the problem and showed that the new star had no measurable parallax and must therefore be well beyond the sphere of the moon (reading 2.4). The heavens were not unchanging after all.

Another opportunity came to Tycho in 1577, when a comet appeared. Once again using a combination of careful observation and basic geometry, Tycho showed that the comet must be a celestial phenomenon beyond the sphere of the moon. This posed an explicit challenge to the popular Renaissance idea that the heavenly orbs were solid crystalline spheres; for the comet, passing obliquely through these spheres, would shatter them.

These achievements would be enough to earn Tycho an honored place in the history of astronomy, yet for sheer impact on astronomy they pale in comparison with something that happened at the end of his life. In December of 1599, Tycho invited an impoverished young German mathematician to meet him in Prague and work on the orbit of Mars using his new and jealously guarded data. Two years later, Tycho died and his data passed into the hands of that assistant, Johannes Kepler, with results that transformed astronomy beyond anything Tycho could have dreamed.

3 Johannes Kepler: The Physics of the Skies

The German astronomer and mathematician Johannes Kepler is one of the loneliest figures in the history of science. His father, a soldier, was rarely home; when he was, the house was filled with marital strife. Johannes himself was a sickly child who was born two months prematurely and suffered a bout of smallpox in childhood that left his hands crippled and his eyesight weak. During his time in Catholic schools, to the consternation of his masters, he developed Protestant opinions. It is no great surprise that later in life he found himself almost entirely alone in his struggles with a problem that his teachers and peers, even Galileo himself, did not fully understand and that he produced a solution that they could not appreciate.

Kepler's early education contained almost no hint of astronomy, but his deep personal piety, his diligence in his studies, and his undeniable brilliance obviously made an impression on his teachers. When he applied in 1591 for the renewal of a fellowship, his professors at Tübingen strongly supported his request because, in their words, "Kepler has such a superior and magnificent mind that something special may be expected of him."[3]

Kepler's growing reputation as a mathematician brought him to the attention of Tycho Brahe. Tycho was critical of some of Kepler's earlier work, in part because Kepler had borrowed bad data from Copernicus, data that Tycho had long since corrected. Still, Tycho needed a good mathematician, and at about the same time Graz became inhospitable to Protestants. Though his personality clashed with Tycho's, Kepler took a position in Prague as Tycho's assistant and worked assiduously to construct a model of the orbit of Mars that would fit Tycho's accurate observations.

When Tycho died suddenly in 1601, Kepler took charge of his data and continued to work on the problem of the orbit of Mars. His work was constrained in more ways than astronomers at the time could understand. For Kepler was convinced that an adequate astronomy must give not only a description of the motions of the planets but also their causes, and he was also convinced an empty point in space has no causal properties. There must be some force in the sun that drives the planets around in their orbits, and somehow that force must also account for the actual paths the planets take. From his correspondence at the time we can see just how wide was the gap between Kepler's view and that of his contemporaries. To an Aristotelian, the very notion of a "physics of the skies" was contradictory: physics deals with terrestrial phenomena, whereas the business of astronomy is in another realm altogether. Even Copernicans like Mästlin did not appreciate the fact that the new astronomy required the breaking down of the old dichotomy between the heavens and the earth. Reading 2.5 is a collection of passages from Kepler's correspondence and minor works that illustrates just how much in advance of his time some of his ideas were.

The struggle to fit the orbit of Mars continued for years, as Kepler threw more and more Aristotelian ballast overboard in a desperate effort to find a workable model. The mathematical difficulties were staggering, and he appealed for help to leading mathematicians and to his old professor Mästlin. No one answered. When he finally solved the problem, he published both his results and the history of his struggles in the *New Astronomy*. Reading 2.6 is taken from the introduction to that work, and it gives a remarkable window on Kepler's vision of the physical world and his own desperate attempt to unlock the secret of Mars. At one point, as he relates in the *New Astronomy*, he achieved fit so close that the maximum error was a mere eight minutes of arc, sufficient to supersede anything Copernicus had ever dreamt of achieving. But given the accuracy of Brahe's observations, it was not sufficient. With remarkable intellectual integrity, as reading 2.7 shows, Kepler discarded his near miss and started over again. Only then, and only by an uncanny series of accidents and cross-cancelling errors, did he discover that Mars moves in an elliptical orbit with the sun at one focus. The discovery, generalized to all of the planetary orbits, is known today as Kepler's first law of planetary motion.

Lost in wonder at the marvelous simplicity of the orbital path, Kepler searched for further harmonies. His second law states that a radius from the sun to a planet sweeps out equal areas in equal times; his third, that the ratio of the squares of the periodic times of any two planets is equal to the ratio of the cubes of their mean distances from the sun. In each pattern Kepler found further confirmation of his conviction that God is the consummate geometer. In his calendar for 1604, when he was at the critical point in his "war on Mars," he expresses his conviction with eloquent simplicity:

> I may say with truth that whenever I consider in my thoughts the beautiful order, how one thing issues out of and is derived from another, then it is as though I had read a divine text, written into the world itself, not with letters but rather with essential objects, saying: "Man, stretch thy reason hither, so that thou mayest comprehend these things."[4]

Kepler lived two decades beyond the publication of the *New Astronomy*. He made fundamental contributions to optics and eagerly embraced the telescopic observations of Galileo. Yet his contributions to astronomy were almost entirely ignored by his professional colleagues. In 1616 Mästlin wrote to him expressing bewilderment over the idea that the irregularities in the orbit of the moon should be traced to physical causes. Galileo and Kepler exchanged some polite letters, but there is no evidence that Galileo ever actually read the *New Astronomy*; and if he had he would almost certainly have rejected the ellipse, a mathematically unwieldy object in the days of the infancy of algebra, as an insufficiently simple curve.

What is worse, Kepler stuffed his works with a mountain of irrelevant patterns, wild conjectures, and mystical speculations, prompting one historian to remark dryly that if Kepler had burnt three quarters of what he printed, we should in all probability have formed a higher opinion of him today.[5] What kept him from being a mere crackpot was his scrupulous willingness to submit every fanciful idea to the test of comparison with the facts and his almost unparalleled honesty in relinquishing cherished theories when they failed that test. But Kepler's overflowing imagination was too unfocused for him to lead the scientific revolution forward in his own day, and his enormous contributions could only be sifted from the dross and seen for their true worth by a relentlessly powerful synthesizing intellect. He did not live to see that intellect's birth.

4 Galileo: New Dynamics for the New Astronomy

The history of science has its share of giants, but even in their company the Italian astronomer and mathematician Galileo Galilei stands out. Even a portion of what he achieved in the course of his brilliant and tempestuous career would have guaranteed him a place among the greatest scientists of all time, and what he endured for his outspoken advocacy of the Copernican system is now quite literally the stuff of legend.

It is questionable whether Galileo ever dropped two objects, one heavier than the other, from the Tower of Pisa in an attempt to disprove the Aristotelian doctrine of free fall; in any event, that sort of experiment had been tried in his own time by the Flemish engineer Simon Stevin and a millennium earlier by John Philoponus, so the problems with the Aristotelian position were well known. We do know that quite early in his career Galileo became deeply skeptical of Aristotelian physics. Yet when a nova appeared in 1604, Galileo, then a professor of mathematics at Padua, was hesitant to render a public judgment on its significance for the Copernican view. He was well aware of the empirical difficulties the Copernican view faced and could not at that point see a way to surmount them.

A happy accident broke this impasse. In the fall of 1608, some mischievous boys playing in a Dutch lens maker's shop discovered that looking through two lenses at the proper distance greatly magnified the appearance of a tower across the street. Within months word of the discovery had spread throughout Europe, and by the summer hawkers at the fairs outside of Paris were selling crude telescopes. A friend sent word of the new invention to Galileo, who immediately worked out the mathematical principles of the telescope, improved on the design, and began constructing instruments himself. Late in the fall of 1609, he turned one of his telescopes to the skies.

What he observed provided spectacular vindication of Copernicus's views. The telescope disclosed four points of light moving back and forth around Jupiter. At first Galileo took these for stars. But observations on subsequent nights showed that they moved with Jupiter through the field of the fixed stars, and Galileo quickly realized that these were moons orbiting Jupiter just as our moon orbits the earth. The discovery undermined the "two centers of motion" objection to Copernicanism, for now even the Ptolemaic system would need to accommodate multiple centers of motion.

Turning the telescope to Venus, Galileo discovered to his delight that the planet showed a full cycle of phases just as the moon does; the problem is simply that the unaided human eye cannot resolve the disc of Venus. This showed that the Ptolemaic view, according to which Venus circulates on an epicycle between the earth and the sun, could not be right. It also resolved the mystery of the minimal changes in Venus's brightness. When it is almost behind the sun at its furthest visible point from earth, Venus displays a nearly full disc; when it is closest to earth it does indeed appear much larger, but it is mostly dark, showing only a thin illuminated crescent. What had been a strike against the Copernican view was transformed into a powerful argument in its favor.

The telescope also cleared up the problem of the apparent sizes of the stars. Galileo's telescope magnified by about 20 diameters, and thus it greatly increased the apparent distance between stars. But it scarcely changed their visible sizes at all: even in the telescope, the stars appeared to be mere points of light. When we view stars with the naked eye, they appear larger than geometry would indicate they should because the resolving power of the eye is limited by the diameter of the pupil: their images are blurred out and they look larger than their true sizes. With this discovery, another objection to the Copernican view collapsed.

Galileo published some of these discoveries immediately and went on tour, presenting noblemen with telescopes of his own construction and inviting the intelligentsia to look through the telescope and see for themselves. The reception was mixed. Some critics objected that, while the telescope clearly worked marvels on earth, in the heavens it deceived; for certain stars appear double in the telescope while anyone can see with the naked eye that they are plainly not double. Here again the eye's limited resolving power was to blame, but there was no way to disprove decisively the objections of someone determined to hold onto the Ptolemaic view. Most professional astronomers, however, did not doubt Galileo's observations. The astronomers at the Collegio Romano, asked by Cardinal Roberto Bellarmine to render their opinion on Galileo's observational claims, wrote back that Venus did indeed show phases and that most likely Galileo was right about Jupiter's moons. In 1611, Galileo received a triumphal welcome in Rome and enjoyed friendly conversation about his discoveries with astronomers. Most of them were not persuaded of the Copernican view, but there were few who were willing to deny the evidence of their own eyes.

His favorable reception in Rome seems to have emboldened Galileo. Twice in the next few years he returned there of his own accord and was well received, enjoying conversation with fellow astronomers, several cardinals, and even the Pope. But a shadow darkened his third visit in 1616. Galileo already had a reputation as a vigorous polemicist, and he had made some enemies among the Jesuits, particularly Fr. Orazio Grassi, whom Galileo had attacked mercilessly in *The Assayer* (reading 2.8). One of these enemies obtained a copy of a letter Galileo had written in 1614 to a friend in which he had ventured explanations for passages of Scripture that seem to indicate that the earth is motionless and the sun moves. This would have been innocuous enough a century earlier; no one had given Copernicus any trouble about his views. But in its reaction to the Protestant Reformation of the mid 1500s the Catholic Church had tightened its position on the interpretation of Scripture, reserving that right to the Church itself. Since the Church had interpreted Scripture in line with the best existing science of the time, the established interpretations of these passages favored the Ptolemaic view. At the same time, the Inquisition took a first close look at the Copernican view and determined, on the basis of those interpretations, that it was contrary to Scripture.

Galileo was instructed of this ruling and agreed – under the circumstances he could scarcely do otherwise – not to hold the Copernican view. But on the strength of Galileo's letter his enemies circulated a rumor that he was a heretic, and the rumors grew in the telling so that by the time they reached Galileo, who was still in Rome, they represented him as having been tried for heresy, convicted, forced to recant, and given penance. In frustration, Galileo asked Cardinal Roberto Bellarmine to refute the rumors. Bellarmine wrote a brief note in his own handwriting stating clearly that Galileo had not been forced to abjure any of his opinions or given any penances but that he had been warned that the Copernican doctrine was contrary to Holy Scripture and was therefore not to be "defended or held." Galileo took the note with him when he returned to Florence.

Galileo had been personally convinced of the truth of the Copernican view for nearly 20 years, but he knew that he would have to overcome the dynamical difficulties in order for the view to gain widespread acceptance. Mindful of the injunction not to defend the Copernican view, he began drafting the *Dialogue Concerning the Two Chief World Systems* in which three friends, a Copernican (Salviati), an Aristotelian (Simplicio), and an uncommitted but intelligent inquirer (Sagredo), discuss the relative merits of the two systems. Salviati comes off best in every dialectical exchange with Simplicio, but in good Renaissance fashion Galileo frames the entire work as an attempt simply to lay out the arguments on both sides, and he throws in periodic qualifications, perhaps not very convincing, to the effect that the discussion must necessarily be inconclusive since the Church has the final authority in this matter – something he had to say because of the clash between Copernicanism and the standard interpretation of certain Scriptures. Readings 2.9 through 2.11, from the second and third days of the *Dialogue*, give a good impression of Galileo's style and of his manner of arguing, which involves both undermining the Aristotelian objections to Copernicus and presenting positive evidence, including the evidence made available by the telescope, for the Copernican view.

Galileo submitted his manuscript, written in lively vernacular Italian rather than in formal scholarly Latin, to the Church's censors and willingly made all of the changes they required. Notably, they asked him to include a favorite argument of Pope Urban VIII, one that Urban had communicated to Galileo back in 1616 when he was merely a cardinal: since God could make a world in which all of our evidence is just as it now seems to be but in which the earth stands still, we can never be certain that He did not choose to do so. This argument from empirical equivalence is correct as far as it goes. But since Galileo was already moving away from the Aristotelian notion of science as demonstratively certain knowledge and forming a new conception of scientific reasoning based on considerations of simplicity and probable inference, a stricture against certainty carried very little weight with him. Unwisely, he put the argument at the end of the *Dialogue* in the mouth of Simplicio, whose name in Italian conveys the same connotation of "simpleton" that it suggests in English.

The *Dialogue* displays Galileo's brilliant gifts both as a scientist and as a controversialist. He is very careful to lay out a full range of empirical arguments against the Copernican view, even having Salviati help Simplicio to make the case against Copernicus stronger. Then in a *tour de force* he refutes the objections, demolishes Aristotle's dynamics and creates a new and empirically sounder one to take its place, lays out a positive case for the Copernican view using the new telescopic discoveries, and presents a theory of the tides that requires the motion of the earth. His treatment of motion is particularly important, for he articulates a theory of inertia that is very nearly correct and uses it to explain why we do not feel the motion of the earth even though Copernicus is correct. The interchanges between Salviati and Simplicio are laced with irony at Simplicio's expense, and Salviati's protestations against the stubbornness and ignorance of certain "Peripatetics" (another name for Aristotelians) span the full rhetorical range from gentle admonition to blistering invective. Notwithstanding Galileo's carefully worded introduction and the occasional disclaimers placed in the mouth of Salviati, the impression of the whole work is overwhelmingly pro-Copernican.

In February of 1632, the *Dialogue* was published in Florence with the *Imprimatur* of the Church, certifying that it was free of doctrinal or moral error. For several months it sold quite briskly. But Galileo's enemies, including those he had embarrassed in polemical exchanges in the previous decade and a half, seized on its publication as an opportunity to make the accusation of heresy stick. Some of them charged that by discussing and defending the Copernican view he had violated the injunction Bellarmine had laid on him; others, perhaps more shrewdly, drew the Pope's attention to the fact that his own favorite argument was put in the mouth of the *Dialogue*'s most foolish character and insinuated that the entire book was intended as a mocking caricature of the Pope and his views. Both charges struck home, but the former provided the grounds for the official summons, issued in the fall of 1632, demanding that the 68-year-old Galileo come to Rome.

In April of 1633, the officers of the Inquisition verbally interrogated Galileo regarding his publication of the *Dialogue* and particularly regarding his 1616 conversation with Bellarmine, who had died in 1621 and therefore could not give direct evidence. To Galileo's surprise, they produced a document from the files of the Inquisition purporting to be a record of what Galileo had been told by Bellarmine: this document stated that he had been told not to hold, defend, *or discuss, even hypothetically*, the Copernican view – this last being something of which he was surely guilty. To have legal force such a document would require Galileo's signature, but this document was unsigned. Galileo in turn astonished the Inquisitors by producing the letter Bellarmine had written at his request in 1616, which did contain the injunction neither to defend nor to hold the Copernican view but said nothing about discussing it hypothetically. Galileo's signed letter from Bellarmine was unquestionably genuine, and it not only trumped the unsigned document but cast serious doubt on the latter's authenticity. The officials assigned to the case withdrew in consternation.

Eventually they worked out something like a plea bargain. Galileo was to acknowledge that he had, unintentionally, defended the Copernican view by making the arguments in its favor seem more powerful than those against it; he would then ask the court's clemency for this marginal and inadvertent infraction of Bellarmine's injunction. In return, the Inquisition would drop the other charges and let him off lightly. Given the fact that Galileo undoubtedly had defended the Copernican view, this compromise was perhaps the best resolution of the whole affair one could have hoped for. On April 30, 1633 Galileo made his humble apology in private court and the Inquisition prepared to close the case.

But the decision was not left in their hands. Pope Urban was under pressure from the Spanish cardinals who accused him of being soft on heresy, and perhaps he was also nettled by the suspicion maliciously planted in his mind that Galileo had been mocking him by putting his words in the mouth of Simplicio. Whatever his reasons, the Pope remanded the decision to the Inquisition, requiring them to reexamine Galileo under the verbal threat of torture. Galileo held firmly to the position he had taken in April, but by the Pope's specific directive he was forced to abjure his errors in public, his book was banned, and he was placed under arrest. The 10 members of the Board of Inquisition cannot have been very happy with this turn of affairs; three of them, including the Pope's own nephew, did not sign the condemnation. Their secret sympathy for Galileo is further suggested by the fact that they mitigated the conditions of his sentence, allowing him to return to Florence and live out his days there under house arrest in his own home.

Galileo spent his final years receiving visitors, including the English poet John Milton, and finishing his last book, the *Discourse concerning Two New Sciences*, which does not mention the motion of the earth but draws together over 40 years' worth of his work on a new dynamics. His method of reasoning comes through clearly in this work, both in practice and in explicit formulation. A comment from a discussion of the rise and fall of a pendulum bob is typical. Galileo presents without proof the idea that, in the absence of impeding obstacles, an object will rise to the same height from which it has fallen; then he writes "Let us take this for the present as a postulate, of which the absolute truth will be later established for us by our seeing that other conclusions, built on this hypothesis, do indeed correspond with and exactly conform to experience."[6]

This methodology, so foreign to the Aristotelian conception of scientific practice, is an early version of the hypothetico-deductive model of scientific reasoning. It is perhaps the closest philosophical point of contact between Galileo and Kepler.

5 Isaac Newton

Isaac Newton was born, by the Julian calendar, in 1642, the year of Galileo's death. As a student he was brilliant but indifferent; when he chose, he could excel in any subject, but his mind often wandered. Though he does not seem to have sought trouble, it did occasionally find him. A school bully once made the mistake of misevaluating the silent and abstracted boy; Isaac responded to the provocation of a blow by thrashing his assailant and then smashing his face into the stone wall of the church.[7] In a way, this episode reveals the pattern of his life. Newton seems to have wanted nothing so much as to be left alone; but once he was roused – by an intellectual problem, by a perceived slight, by a practical challenge – he displayed a relentless, almost fanatical energy. He had countless admirers, several notable enemies, and very few friends.

Newton's earliest scientific interest was the theory of colors. Using prisms he separated white light into a spectrum and then, in a delicate and difficult bit of work, found a way to recombine the spectrum into white light again. In 1672 he sent some of his discoveries and his interpretation of them to certain members of the Royal Society, the earliest formal scientific society in England, where they were received with mild interest but also with some resistance. Young Newton proposed a particle theory of light; several prominent scientists, including Robert Hooke, preferred a wave theory. Hooke wrote a rather patronizing review of Newton's work, sifting out what he thought true and claiming priority for it himself. Meanwhile the French Jesuit Gaston Pardies wrote a letter to Newton in which he claimed that Newton's theory went beyond observation, though it was a "very ingenious hypothesis."[8] Both criticisms stung. Newton wrote up a rather defensive account of his inductive methodology to Henry Oldenburg, the secretary of the Royal Society, urging that careful experimentation be given priority over the mere framing of hypotheses (see reading 2.14). Regarding Hooke he had scarcely a good word to say, and the enmity between the two men smoldered for the rest of the decade.

Meanwhile the scientific revolution proceeded apace. The great Irish experimentalist Robert Boyle, whose book *The Skeptical Chymist* (1661) had given a vigorous defense of a version of atomism that he preferred to call the corpuscular hypothesis, was enormously productive in the 1670s, writing on fields as diverse as mineralogy, magnetism, electricity, cold air, the vacuum, and the salinity of the oceans. Boyle's approach, revealed in reading 2.12 from his *Philosophical Papers*, was empirical and explicitly corpuscularian. The unifying theme of the new science was the mechanical view of nature; what was missing was the set of laws governing the vast machine of the cosmos.

Scientists on the continent were not idle either. The Dutch mathematician and scientist Christiaan Huygens published his work on the pendulum in

1672 and wrote his *Treatise on Light* in 1678, though he did not publish it for another 12 years. The *Treatise on Light* is valuable not only for its elegant development of the wave theory of light but also for Huygens's explicit discussion of the nature of scientific method in the Preface, which is reproduced here as reading 2.13. Whereas Newton bristled at having his idea described as an "ingenious hypothesis," Huygens has no reservations about the employment of hypotheses as guides to inquiry. His discussion raises important issues regarding probability and confirmation that were in turn taken up by Herschel (reading 4.3) and have become key issues in contemporary philosophy of science.

In England, Hooke, Edmund Halley, and the architect Christopher Wren were all intuitively convinced that gravitation was a force that varied inversely with the square of the distance between two objects, but none of them was a good enough mathematician to link this intuition with Kepler's laws. In 1684, Halley, then a young man who had not yet discovered the periodicity of the comet that now bears his name, made a trip to Cambridge to consult with Newton. "What," he asked, "would be the curve described by the planets on the supposition that gravity diminished as the square of the distance?" Newton immediately answered, "An *ellipse*." Halley asked how he knew: "Why, I have calculated it," Newton replied. Halley asked in awe to see the calculation, but in the clutter of his papers Newton could not find it. In his solitude he had found and then absentmindedly mislaid the key to the scientific revolution.[9]

Over the course of the next two years Halley, with infinite tact, drew the reclusive genius out of himself and prevailed upon him to write up his monumental work on gravitation. The *Principia Mathematica* swelled to fill three forbidding books, and Halley, in an act as foresighted as it was generous, funded its publication from his own pocket. The structure of the work is deductive, much like that of Euclid's *Elements*. The beginning of the first Book, reproduced here in reading 2.15, summarizes, in definitions and axioms, virtually the whole of mechanics as it was then known, clarifying it and making it more precise in the process. Mass, momentum, inertia and force are each clearly and accurately defined, and in some cases the clarification amounts almost to the introduction of new concepts, as when Newton distinguishes accelerative, motive, and centripetal force. Euclidean space is the stage on which the drama of the cosmos is played out, structured and unchanging as objects come and go. Three laws govern all motions. Acceleration and deceleration are the evidence that a force is at work; for in the absence of a force, moving objects would continue on straight trajectories at the same velocity forever. At the beginning of the third Book of the *Principia* (reading 2.16), Newton lays down his rules of reasoning in the sciences without apology.

But the crowning achievement of the *Principia* is Newton's law of universal gravitation. Here is the key that Galileo lacked, the force that holds the universe together. In an interesting Scholium, or explanatory discussion, at the end of the third Book, Newton declines to "feign hypotheses" regarding the causes of gravitation. It is enough that the force exists and its effects can be treated by mathematical principles (reading 2.17). From the publication of the *Principia* onward, the search for causes was in considerable measure subordinated – particularly in England – to the task of giving an accurate description of natural phenomena. In a book called *The System of the World*, Newton illustrates how this task should be carried out, demonstrating that astronomical data on the positions of the planets are in excellent agreement with the predictions of his law of gravitation (reading 2.18).

What Newton achieved in the *Principia* is sometimes called the "Newtonian synthesis," since he drew together the scattered insights and half-glimpsed truths from the incompatible doctrines of his great predecessors and wove them into a single and nearly seamless fabric of scientific law and method. From Copernicus, he took the rotation and orbit of the earth; from Kepler, the three laws of planetary motion; from Galileo, the principle of inertia and the law of free fall, both corrected and generalized, and the composition of motions and forces generalized and made rigorous in the parallelogram of forces familiar to all modern students of physics. There was work yet to be done; Euler and Lagrange in particular reformulated Newton's laws in the language of the calculus and extended and refined his system. Still, the principal glory belonged to Newton. More than a century later, the French mathematician and physicist Lagrange remarked that Newton was the most fortunate of men, for there is only one universe, and Newton had the honor of discovering its laws.

With the publication of the *Principia*, Isaac Newton became an international icon. The Royal Society, from which he had once threatened to resign because of Hooke's treatment of his ideas, became Newton's domain; Hooke withdrew from active participation, and when Hooke died in 1703 Newton was elected president almost immediately. An international debate raged over the priority of Newton and Leibniz for the discovery of the calculus, but now Newton had no lack of partisans willing to take up his cause. In the event, he outlived Leibniz and died in his eighty-fifth year, the undisputed monarch of British intellectual life. A memorial statue in the ante-chapel of Trinity College, Cambridge bears an inscription that says simply, "He surpassed all men of genius."[10]

Yet Newton's theory raised many questions, both physical and philosophical. Did gravitation act at a distance? If so, was this not a violation of the very mechanical philosophy he supposedly advocated? What could one make of Newton's conception of absolute space and time? Could these be more than philosophical fictions? Newton's laws seemed so conclusively demonstrated that it seemed he had gained a priori insight into the structure of reality itself – but is such insight even possible? How could the observation of instances *justify* any general laws? What is meant by 'causation', and what role does our knowledge of cause and effect play in our knowledge of the world around us? All of these questions raised by Newton's monumental achievement drove philosophical inquiry for over a century.

Notes

1 Jaroslav Pelikan et al. (eds.), 1986, *Luther's Works*, Vol. 54. Philadelphia: Fortress Press, pp. 358–9.
2 Thomas Blundeville, 1594, *M. Blundeville His Exercises*. London: Iohn Windlet; quoted in Toulmin and Goodfield, 1961, *The Fabric of the Heavens*. Chicago: University of Chicago Press, p. 176.
3 Max Caspar, 1959, *Kepler: 1571–1630*. New York: Collier Books, p. 46.
4 Quoted in Caspar, *Kepler*, p. 157.
5 Arthur Berry, 1961, *A Short History of Astronomy from Earliest Times through the Nineteenth Century*. New York: Dover, p. 197.
6 Galileo, 1974, *Two New Sciences* (ed. and trans.), Stillman Drake. Madison: University of Wisconsin Press, p. 164.
7 The story of this encounter comes from a manuscript written by Newton's friend, John Conduitt, Newton's assistant at the mint and husband of Newton's niece. See Richard S. Westfall, 1980, *Never at Rest: A Biography of Isaac Newton*. London: Cambridge University Press, pp. 59–60.
8 Pardies had intended no offense, and when he realized that Newton had taken his remark amiss he went out of his way to be conciliatory. See Westfall, *Never at Rest*, pp. 241–2.
9 This account is given, in substantially the same terms, by both John Conduitt and Abraham de Moivre. For de Moivre's account, see Westfall, p. 403. For Conduitt's version, see David Brewster, 1855, *Memoirs of the Life, Writings, and Discoveries of Sir Isaac Newton*, Vol. I. Edinburgh: Thomas Constable and Co., pp. 296–7.
10 *Qui genus humanum ingenis superavit*. See H. D. Anthony, 1961, *Sir Isaac Newton*. New York: Collier Books, p. 171.

Suggestions for Further Reading

Items marked with an asterisk assume little or no background in the subject; these would be a good place to begin further reading.

Achinstein, Peter (ed.), 2004, *Science Rules: A Historical Introduction to Scientific Methods*. Baltimore: Johns Hopkins University Press, Parts I, II, and V.
Barbour, Julian, 2001, *The Discovery of Dynamics*. Oxford: Oxford University Press, chapters 5–13.
Ben-Chaim, Michael, 2004, *Experimental Philosophy and the Birth of Empirical Science: Boyle, Locke and Newton*. Burlington, VT: Ashgate Publishing Company.
Berlinski, David, 2000, *Newton's Gift*. New York: Simon and Schuster.
Butts, Robert E. and Davis, John W. (eds.), 1970, *The Methodological Heritage of Newton*. Toronto: University of Toronto Press.
Clatterbaugh, Kenneth, *The Causation Debate in Modern Philosophy, 1637–1739*. New York: Routledge, 1999.
*Cohen, I. Bernard, 1985, *The Birth of a New Physics*, rev. edn. New York: W. W. Norton.
Crowe, Michael, 2001, *Theories of the World from Antiquity to the Copernican Revolution*, 2nd edn. New York: Dover, chapters 6–9.
Densmore, Dana, 2003, *Newton's Principia: The Central Argument*, 3rd edn. Santa Fe: Green Lion Press.

Donahue, William H. (trans. and ed.), 2004, *Selections from Kepler's* Astronomia Nova. Santa Fe: Green Lion Press.

Drake, Stillman, 1957, *Discoveries and Opinions of Galileo*. New York: Anchor Books.

Finocchiaro, Maurice, 1989, *The Galileo Affair: A Documentary History*. Berkeley: University of California Press.

Finocchiaro, Maurice, 1997, *Galileo on the World Systems*. Berkeley: University of California Press.

Grant, Edward, 1996, *Planets, Stars, and Orbs: The Medieval Cosmos, 1200–1687*. Cambridge: Cambridge University Press.

Hall, A. Rupert, 1981, *From Galileo to Newton*. New York: Dover.

Koyré, Alexander, 1957, *From the Closed World to the Infinite Universe*. Baltimore: Johns Hopkins University Press.

Koyré, Alexander, 1968, *Metaphysics and Measurement: Essays in Scientific Revolution*. Cambridge, MA: Harvard University Press.

Kozhamthadam, Job, 1994, *The Discovery of Kepler's Laws*. Notre Dame: Notre Dame University Press.

Kuhn, Thomas, 1957, *The Copernican Revolution*. Cambridge, MA: Harvard University Press.

Lattis, James, 1994, *Between Copernicus and Galilieo: Christoph Clavius and the Collapse of Ptolemaic Cosmology*. Chicago: University of Chicago Press.

Madden, Edward H. (ed.), 1960, *Theories of Scientific Method: The Renaissance through the Nineteenth Century*. Seattle: University of Washington Press, chapters 3–7.

Sobel, Dava, 2000, *Galileo's Daughter*. New York: Penguin Books.

Strong, E. W., 1936, *Procedures and Metaphysics*. Berkeley: University of California Press.

Toulmin, Stephen and Goodfield, June, 1961, *The Fabric of the Heavens*. Chicago: University of Chicago Press, chapters 6–10.

Westfall, Richard S., 1978, *The Construction of Modern Science*. Cambridge: Cambridge University Press.

2.1

The Nature and Grounds of the Copernican System

Georg Joachim Rheticus

Georg Joachim Rheticus (1514–1574), whose given name was Georg Joachim von Lauchen, was an Austrian-born mathematician, cartographer, and astronomer. In May of 1539 he traveled to Poland in order to determine whether Copernicus, who was said to be developing a system of astronomy in which the earth both rotated and revolved around the sun, was more than a mere crackpot. Four months later he was a complete convert to the Copernican view. In this excerpt from his *Narratio Prima*, or First Letter (1540), Rheticus explains to his former teacher Schoner the nature and grounds of the Copernican system. The considerations Rheticus finds persuasive would also persuade Galileo at the dawn of the next century.

The Second Part of the Hypotheses: The Motions of the Five Planets

When I reflect on this truly admirable structure of new hypotheses wrought by my teacher, I frequently recall, most learned Schoner, the Platonic dialogue which indicates the qualities required in an astronomer and then adds "No nature except an extraordinary one could ever easily formulate a theory."

When I was with you last year and watched your work and that of other learned men in the improvement of the motions of Regiomontanus and his teacher Peurbach, I first began to understand what sort of task and how great a difficulty it was to recall this queen of mathematics, astronomy, to her palace, as she deserved, and to restore the boundaries of her kingdom. But from the time that I became, by God's will, a spectator and witness of the labors which my teacher performs with energetic mind and has in large measure already accomplished, I realized that I had not dreamed of even the shadow of so great a burden of work. And it is so great a labor that it is not any hero who can endure it and finally complete it. . . .

[. . .]

From Edward Rosen, trans. and ed., *Three Copernican Treatises*, 2nd edn. (New York: Dover, 1959), pp. 162, 164–8 (selections).

With regard to the apparent motions of the sun and moon, it is perhaps possible to deny what is said about the motion of the earth, although I do not see how the explanation of precession is to be transferred to the sphere of the stars. But if anyone desires to look either to the principal end of astronomy and the order and harmony of the system of the spheres or to ease and elegance and a complete explanation of the causes of the phenomena, by the assumption of no other hypotheses will he demonstrate the apparent motions of the remaining planets more neatly and correctly. For all these phenomena appear to be linked most nobly together, as by a golden chain; and each of the planets, by its position and order and every inequality of its motion, bears witness that the earth moves and that we who dwell upon the globe of the earth, instead of accepting its changes of position, believe that the planets wander in all sorts of motions of their own. . . .

The ancients attributed to the epicycles of the three superior planets the entire inequality of motion which they discovered that these planets had with respect to the sun. Then they saw that the remaining apparent inequality in these planets did not occur simply on the theory of an eccentric. The results obtained by calculating the motions of these planets in imitation of the hypotheses for Venus agreed with experience and the observations. Hence they decided to assume for the second apparent inequality a device like that which their analyses established for Venus. As in the case of Venus, the center of the epicycle of each planet was to move at a uniform distance from the center of the eccentric, but at a uniform rate with respect to the center of the equant; and this point was to be the center of uniform motion also for the planet, as it moved on the epicycle with its own motion, starting from the mean apogee. So long as the ancients strove to retain the earth in the center of the universe, they were compelled by the observations to affirm that, just as Venus revolved with its own special motion on the epicycle, but by reason of the eccentric advanced with the mean motion of the sun, so conversely the superior planets in the epicycle were related to the sun, but moved with special motions on the eccentric. But in the theory of Mercury, the ancients thought that they had to accept, in addition to the devices which they deemed adequate to save the appearances of Venus, a different position for the equant, and revolution on a small circle for the center from which the epicycle was equidistant. All these arrangements were shrewdly devised, like most of the work of antiquity, and would agree satisfactorily with the motions and appearances if we granted that the celestial circles admit an inequality about their centers – a relation which nature abhors – and if we regarded the especially notable first inequality of apparent motion as essential to the five planets, although it is clearly accidental.

[. . .]

. . . [A]s Aristotle points out at length in another connection, men by nature desire to know. Hence it is quite vexing that the causes of phenomena are nowhere else so hidden and wrapped, as it were, in Cimmerian darkness, a feeling which Ptolemy shares with us. . . . I sincerely cherish Ptolemy and his followers equally with my teacher, since I have ever in mind and memory that sacred precept of Aristotle: "We must esteem both parties but follow the more accurate." And yet somehow I feel more inclined to the hypotheses of my teacher. This is so perhaps partly because I am persuaded that now at last I have a more accurate understanding of that delightful maxim which on account of its weightiness and truth is attributed to Plato: "God ever geometrizes"; but partly because in my teacher's revival of astronomy I see, as the saying is, with both eyes and as though a fog had lifted and the sky were now clear, the force of that wise statement of Socrates in the *Phaedrus*: "If I think any other man is able to see things that can naturally be collected into one and divided into many, him I follow after and 'walk in his footsteps as if he were a god.'"

2.2
The Unsigned Letter

Andreas Osiander

Andreas Osiander (1498–1552), born Andreas Hosemann, was a Lutheran theologian and friend of Copernicus who was entrusted with seeing the manuscript of the latter's masterwork, *On the Revolutions of the Heavenly Spheres*, through the press. Osiander did so, but he inserted an unsigned letter at the beginning instructing the reader to take the system as a mere calculating device rather than as a representation of reality. The true authorship of the unsigned letter was disclosed to the astronomer Michael Mästlin and made public by his pupil Kepler, who described Osiander's letter as "the view of a jackass written for the consumption of other jackasses."

To the Reader Concerning the Hypotheses of This Work

Since the newness of the hypotheses of this work – which sets the earth in motion and puts an immovable sun at the centre of the universe – has already received a great deal of publicity, I have no doubt that certain of the savants have taken grave offense and think it wrong to raise any disturbance among liberal disciplines which have had the right set-up for a long time now. If, however, they are willing to weigh the matter scrupulously, they will find that the author of this work has done nothing which merits blame. For it is the job of the astronomer to use painstaking and skilled observation in gathering together the history of the celestial movements, and then – since he cannot by any line of reasoning reach the true causes of these movements – to think up or construct whatever causes or hypotheses he pleases such that, by the assumption of these causes, those same movements can be calculated from the principles of geometry for the past and for the future too. This artist is markedly outstanding in both of these respects: for it is not necessary that these hypotheses should be true, or even probably; but it is enough if they provide a calculus which fits the observations – unless by some chance there is anyone so ignorant of geometry and optics as to hold the epicycle of Venus as probable and to believe this to be a cause why Venus alternately precedes and follows the sun at an angular distance of up to 40° or more. For who does not see that it necessarily follows from this assumption that the diameter of the planet in its perigee should appear more than four times greater, and the body of the planet more than sixteen times greater, than in its apogee? Nevertheless the experience of all the ages is opposed to that. There are also other things in this discipline which are just as absurd, but it is not necessary to examine them right now. For it is sufficiently clear that this art is absolutely and profoundly ignorant of the causes of the apparent irregular movements. And if it constructs and thinks up causes – and it has certainly thought up a good many – nevertheless it does not think them up in order to persuade anyone of their truth but only in order that they may provide a correct basis for calculation. But since for one and the same movement varying hypotheses are proposed from time to time, as eccentricity or epicycle for the movement of the sun, the astronomer much prefers to take the one which is easiest to grasp. Maybe the philosopher demands probability instead; but neither of them will grasp anything certain or hand it on, unless it has been divinely revealed to him. Therefore let us permit these new hypotheses to make a public appearance among old ones which are themselves no more probable, especially since they are wonderful and easy and bring with them a vast storehouse of learned observations. And as far as hypotheses go, let no one expect anything in the way of certainty from astronomy, since astronomy can offer us nothing certain, lest, if anyone take as true that which has been constructed for another use, he go away from this discipline a bigger fool than when he came to it. Farewell.

2.3

The Motion of the Earth

Nicholas Copernicus

Nicholas Copernicus (1473–1543) was the great Polish astronomer whose work *On the Revolutions of the Heavenly Spheres* (1543), published when he was on his deathbed, challenged the Ptolemaic system and catalyzed the revolution in astronomy that was carried on by Galileo and Kepler and completed by Isaac Newton. In this selection from the preface and first book of that work, Copernicus dedicates the fruits of his labors to the Pope and then argues, in chapters that parallel those in Ptolemy's *Almagest*, that the earth moves and that its motion gives us the key to "the marvelous symmetry of the universe."

To His Holiness Pope Paul III

Holy Father, I can guess already that some people, as soon as they find out about this book I have written on the revolutions of the universal spheres, in which I ascribe a kind of motion to the earthly globe, will clamor to have me and my opinions shouted down. Nor am I so pleased with my own work that I disregard others' judgments concerning it. I know that a philosopher's thoughts are beyond the reach of common opinion, because his aim is to search out the truth in all things – so far as human reason, by God's permission, can do that. But I do think that completely false opinions are to be avoided.

My thoughts, then, were these: Those who know that the judgment of many centuries supports the view that the earth stands firm in the midst of the heavens – their center, as it were – will think it an absurd bit of theater if I on the contrary declare that the earth moves. So for a long time I wavered. Should I publish my argument showing that the earth moves? Or would I be better to follow the example of the Pythagoreans, and some others, who handed down the secrets of their philosophy only to relatives and friends

From Dennis Danielson (ed.), *The Book of the Cosmos: Imagining the Universe from Heraclitus to Hawking* (Cambridge, MA: Perseus, 2000), selections from pp. 104–17.

– orally, not in writing – as the letter of Lysis to Hipparchus indicates. They did so, it seems to me, not (as some think) out of mere unwillingness to share their teachings, but out of a desire to protect beauties and profundities discovered by great men from the contempt of those who refuse to give any effort to literary accomplishment unless it turns a profit – or who, even if by the advice and example of others they do apply themselves freely to the study of philosophy, are, as a result of their native stupidity, among philosophers like drones among bees. Accordingly, when I contemplated the contempt I would face on account of the novelty and absurdity of my opinion, I almost gave up completely the work I had started.

And yet, although for a long time I hesitated and even resisted, my friends drew me along. Foremost among them was Nicholas Schönberg, Cardinal of Capua, famous in all fields of learning. Next to him was my dear friend Tiedemann Giese, Bishop of Kulm, a great student of the sacred writings and all good literature. For repeatedly he encouraged me, commanded me, sometimes sharply, to publish this book and let it see the light of day after lying buried and hidden, not for nine years but going on four times nine. More than a few other eminent and learned men advised me to do the same. They urged me to set aside my anxieties, abandon my reluctance, and share my work for the common good of astronomical learning. According to them, the more absurd my doctrine of the earth's motion appeared to most people, the greater would be their amazement and gratitude once my book was published and the clouds of absurdity had been dispersed by radiant proofs. These persuasions and this hope, therefore, finally convinced me to allow my friends, as they had long requested, to prepare an edition of this work.

However, Your Holiness, perhaps your amazement that I would publish these findings – after having so exerted myself in the work of thinking them through, even to the point of deciding to commit to writing my conclusions concerning the earth's motion – perhaps your amazement is not so great as your desire to hear why it would occur to me, contrary to the received opinions of astronomers and almost contrary to common sense, to dare to imagine any motion of the earth.

Accordingly, Your Holiness, I would have you know that what moved me to conceive a different model for explaining the motions of the universal spheres was merely my realization that the astronomers are not consistent among themselves regarding this subject. In the first place, they are so uncertain concerning the motions of the sun and the moon that they can neither observe nor predict even the constant length of a tropical year. Secondly, in calculating the motions of these as well as the other five planets, they do not use the same principles and assumptions, nor the same explanations for their apparent revolutions and motions. For while some use only concentric circles, others employ eccentrics and epicycles, from which however the desired results do not quite follow. Those relying on concentrics, though they may use these for modelling diverse motions, nevertheless have not been successful in using them to obtain firm results in perfect accordance with the phenomena. Yet those who have invented eccentric circles, while they seem for the most part to have solved apparent motion in a manner that is arithmetically consistent, at the same time also seem to have introduced several ideas that contradict the first principles of uniform motion. Nor have they been able to discover or deduce by means of their eccentrics the main point, which is to describe the form of the universe and the sure symmetry of its parts. Instead they have been like someone attempting a portrait by assembling hands, feet, head, and other parts from different sources. These several bits may be well painted, but they do not fit together to make up a single body. Bearing no genuine relationship to each other, such components, joined together, would compose a monster, not a man.

Thus in their process of demonstration – "method," as they call it – those employing eccentrics have either omitted something essential or else admitted something extraneous and irrelevant. This would not have happened if they had observed sound principles. For unless the hypothesis they adopted were fallacious, all the predictions following from them would be verifiable beyond dispute. (Even if what I am saying here is obscure, it will become clearer in its proper place.)

Thus I pondered for a long time this lack of resolution within the astronomical tradition as far as the derivation of the motions of the universal spheres is concerned. It began to irritate me that the philosophers, who otherwise scrutinized so precisely the minutiae of this world, could not agree on a more reliable theory concerning the motions of the system of the universe, which the best and most orderly Artist of all framed for our sake. So I set myself the task of rereading all the philosophers whose books I could lay my hands on, to find out whether anyone had ever held another opinion concerning the motion of the universal spheres than those asserted by the teachers of astronomy in the schools. Indeed, I found, first in Cicero, that Nicetus supposed the earth to move. And later I discovered in Plutarch that some others held the same opinion. . . .

Following their example, therefore, I too began to contemplate the possibility that the earth moves. To be sure, it seemed an absurd idea. Yet I knew that others before me had been accorded the liberty to imagine whatever circles they chose in order to explain the astronomical phenomena. Thus I presumed that I likewise would surely be permitted to test, given some motion of the earth, whether a more solid explanation of the revolutions of the heavenly spheres were possible than had so far been provided.

Accordingly, I posited the motion which later in this volume I assign to the earth. And by deep and extensive investigation I finally found that if the motion of the other planets is viewed in relation to the circular motion of the earth, and if this calculation is made for the revolution of each planet, then not only do the phenomena follow consistently, but also the orders and magnitudes of all the orbs and spheres and heaven itself are so interconnected that not one of its parts could be removed without throwing the other parts and the whole universe into confusion.

In the arrangement of this work, therefore, I have observed the following order. In the first book I set out all the positions of the spheres along with the motions I ascribe to the earth, so that this book comprises as it were the overall structure of the universe. And in the remaining books I relate the motions of the rest of the planets and all the spheres to the movement of the earth in order to show to what extent their appearances, if we do relate them to the earth's motion, can be saved.

I have no doubt that astute and learned astronomers will agree with me if, in keeping with the chief requirement of this discipline, they will study and examine – not superficially but in depth – the evidence for these matters which I set forth in this work. However, so that learned and unlearned alike may see I am a person who flees the judgment of no one at all, I have chosen to dedicate these my late-night studies to you, Your Holiness, rather than to anyone else. For even here in this remote corner of the earth which I inhabit, you are held to be the highest authority by virtue of your exalted office and your love for all literature, even astronomy. Thus by your authority and discernment you may easily repress the malice of slanderers, even if (as the proverb says) there is no remedy against the teeth of a backbiter.

Perhaps some idle talkers, thinking they can judge astronomy though completely ignorant of it, and distorting some passage of Scripture twisted to their purposes, will dare to criticize and censure my teaching. I shall not waste time on them; I have only contempt for their audacity. As is well known, Lactantius, otherwise a distinguished writer but no astronomer, speaks quite immaturely about the shape of the earth when he mocks those who assert that the earth is spherical. No scholar need be surprised, therefore, if such persons ridicule me likewise. Astronomy is written for astronomers – and they, if I am not mistaken, will see the value that these efforts of mine have for the ecclesiastical community over which Your Holiness now holds dominion. For not long ago, under Leo X, the Lateran Council raised the issue of emending the church calendar. No decision was then arrived at merely because the Council concluded that the lengths of the year and the month and the motions of the sun and the moon were not yet measured accurately enough. Since then, at the urging of that most eminent man Dr Paulus, Bishop of Sempronia, who was in charge of the proceedings, I have concentrated on studying these matters with greater accuracy. What I have accomplished in this regard, however, I hand over to be judged by Your Holiness in particular and by all other learned astronomers. And lest Your Holiness should think I promise more regarding the usefulness of this volume than I can fulfill, I now proceed to the work itself.

Chapter 1 The Universe Is Spherical

The first thing for us to realize is that the universe is spherical. This is so either because, of all forms, the sphere is the most perfect, requiring no joins, and being an integrated whole; or because it is the most capacious of all forms, and so best fitted to enclose and preserve all things – or also because the most perfected parts of the universe such as the sun, the moon, and the stars display this shape; or because all things strive to be bounded thus, as we observe in drops of water and other liquids when they seek to be bounded within themselves. There can be no doubt, then, about the rightness of ascribing this shape to the heavenly bodies.

Chapter 2 The Earth Too Is Spherical

The earth also has the shape of a globe, because all of its parts tend towards its center. We do not immediately perceive it as a perfect sphere because the mountains are so high and the valleys so deep, and yet these hardly affect the overall sphericity of the earth. This is clear from the fact that if one travels northward, the pole of the diurnal rotation gradually rises, while the opposite pole sinks accordingly, and more stars in the northern sky seem never to set, while some in the south seem never to rise. Thus Italy does not see Canopus, which is visible from Egypt; and Italy does see the last star of the River, which up here in our frozen territory is unknown. Conversely, as one travels southward, such stars rise higher, while those which appear high to us sink lower. Also, the angle of elevation of the poles is everywhere constantly proportionate to the distance one thus travels across the earth, something that happens with no other shape than a sphere. Hence we see that the earth too is delimited by poles and thus shaped like a globe. Moreover, evening eclipses of sun or moon are not seen by inhabitants of the east, nor morning eclipses by inhabitants of the west; but those that occur in between are observed later by the former and earlier by the latter.

Seafarers know that the waters too conform to this shape, for land that is not visible from the ship is observed from the top of the mast. And if a bright light is placed at the top of the mast, then to those remaining on the shore it appears gradually to sink as the ship moves farther off from land. Finally, the light as it were sets and disappears. Also, like earth, water, in keeping with its nature as a fluid, always obviously seeks a lower level and so does not push farther inland than the curvature of the shore permits. Hence, one accepts that whatever land emerges from the ocean is higher than it is.

[. . .]

Chapter 4 The Motion of the Heavenly Bodies Is Uniform, Perpetual, and Circular, or Made Up of Circular Motions

Let us now recall that the motion of heavenly bodies is circular. For that movement which a sphere possesses is movement in a circle, by which action it expresses its own form as the simplest shape, in which no beginning or end is to be discerned, nor can these be distinguished from each other, while the sphere keeps on moving within its own bounds. Yet a multitude of motions applies to the various spheres. The most obvious of all is the daily rotation, which the Greeks call *nuchthemeron*, which measures the passage of a day and a night. By this motion, it is assumed the whole universe – except for the earth – glides round from east to west. This is recognized as the common measure of all motion, since we reckon even time itself mainly by counting days.

Next we see other, as it were contrary revolutions, moving from west to east, namely those of the sun, the moon, and the five planets. Thus the sun metes out our year and the moon our month, these being the other most common measures of time. In this way too each of the five planets completes its circuit. Yet there are differences among their various motions. First, these do not turn about the same axis as the primary motion but take a slantwise course through the zodiac. Secondly, they are not observed moving uniformly in their orbits. For we see that the sun and moon in their courses sometimes move slowly and sometimes more quickly. As for the other five planets, as we observe, sometimes they even come to a stop and retrace their steps. And while the sun always keeps strictly to its own pathway, these others wander in various ways, sometimes towards the south, sometimes towards the north – which is why they are called planets [from Greek *planetes*, "wanderer"]. Moreover, sometimes they are nearer the earth and said to be "in perigee"; at other times they are farther off and said to be "in apogee."

We must admit, nonetheless, that their motions are circular, or made up of several circles, because these nonuniformities conform to a consistent law and to the fact that the planets return to where they began, which could not be the case unless the motions were circular, for only a circle can replicate what has already taken place. For example, by a motion made up of circles the sun causes for us a repetition of unequal days and nights and of the four seasons. In this cycle we discern several motions, since no simple heavenly body can move irregularly in a single sphere. For such irregularity would have to result either from an inconstancy in the force of movement, whether arising internally or externally, or from some irregularity in the revolving body. But either alternative is abhorrent to reason. We must not ascribe any such indignity to things framed and governed optimally.

We must conclude, then, that their uniform motions appear to us as irregular either because they take place around different axes, or else because the earth is not at the center of their circles of revolution. For us on earth as we observe the movements of these planets, this is what happens: because of their nonuniform distances they appear larger when they are near us than when they are farther away (optics proves this principle). And similarly, across equal portions of their circumferences their motions over a given time will appear unequal because viewed from different distances.

Above all, then, I think we must examine carefully the relationship of the earth to the heavens. Otherwise, in our desire to investigate things of the highest order we may remain ignorant of what is nearest to us, likewise mistakenly attributing things that are earthly to things that are heavenly.

Chapter 5 Is Circular Motion Appropriate to the Earth? And What Is Earth's Location?

[. . .]

Every apparent change of place is caused by the movement either of the observer or of the thing observed, or indeed by some unequal alteration in the position of both. (When observer and observed move uniformly relative to each other, no motion is perceived.) Yet it is from earth that we behold the circuit of the heavens; it is here that its spectacle is represented to us. Therefore if any motion is predicated of the earth, the same motion will appear in all that is beyond the earth, but in a contrary direction, as if everything were moving about it. The prime example of this is the daily rotation, whereby apparently the whole universe except for the earth itself is driven round. However, if you grant that the heavens have no part in this rotation, but that the earth itself turns from west to east, then considering the matter seriously you will find this is actually the case as far as the rising and setting of sun, moon, and stars are concerned. And since everything is contained within the heavens, which serve as the location and setting of all things, it is not immediately apparent why motion should be attributed to the container rather than the contents, to the location rather than the thing located. Indeed, this was the opinion of Heraclides and Ecphantus the Pythagoreans and, according to Cicero, of Nicetus of Syracuse, who held that earth rotates in the middle of the universe. For their judgment was that the stars set when the earth comes in the way and rise when it ceases to be in the way.

Given this rotation, a further, equally important question follows concerning the earth's location. Admittedly, virtually everyone has been taught, and believes, that the earth is the center of the universe. However, anyone who denies that the earth occupies the center or midpoint may still assert that its distance from the center is negligible by comparison with that of the sphere of the fixed stars, yet noticeable and noteworthy relative to the spheres of the sun and other planets. He may consider that this is why their motions appear nonuniform, and that they are regular relative to some center other than that of the earth. In this way, perhaps, he can offer a not-so-inept explanation for the appearance of irregular motion. For the fact that we observe the planets sometimes nearer the earth and sometimes farther away is logical proof that the center of the earth is not the center of their orbits. . . .

Chapter 6 The Immensity of the Heavens Compared to the Size of the Earth

[. . .]

Things enclosed within a smaller orbit revolve more quickly than those turning in a larger circle. Thus Saturn, the highest of the planets, revolves in thirty years, while the moon, undoubtedly the nearest the earth, has a circuit of one month. Finally, one will presume, earth rotates within the space of a day and a night. Hence the question of the daily rotation arises once more.

And yet the question of earth's location remains uncertain, all the more so because of what was said above. For nothing has been proven except the indescribable size of the heavens compared to that of the earth. But the degree of that immensity remains unclear. (Consider that at the opposite extreme there are minuscule indivisible bodies called "atoms." Because they are imperceptible, when they are taken two or several at a time they do not make up any visible body. Yet they can be sufficiently multiplied to the point where there are enough of them to form a visible mass. The same thing applies to the position of the earth. Even though it is not in the center of the universe, its distance from the center is nevertheless inconsiderable when compared to the distance of the sphere of the fixed stars.)

[. . .]

Chapter 8 Explanation of These Reasons for Believing that the Earth Rests in the Middle of the Universe as if It Were Its Centre and of Their Insufficiency

[. . .]

Ptolemy therefore has no reason to fear that earth and all things terrestrial will fly apart on account of a rotation brought about by means of nature's own operation, which is very different from anything artificial or devised by human ingenuity.

Yet why is he not just as worried about the universe as a whole, whose swiftness of motion must be that much greater in proportion as the heavens are greater than the earth? Or are the heavens so immense precisely because the ineffable force of their motion impels them away from the center? Would they otherwise collapse if they did stand still? If this reasoning were sound, then surely the magnitude of the heavens must expand to infinity. For the higher they are impelled by the force of their motion, the faster their motion will be on account of the continuously expanding circumference which has to make its revolution every twenty-four hours. In turn, as the motion increased, so would the immensity of the heavens – speed thus increasing size, and size increasing speed, *ad infinitum*. Yet according to that axiom of physics, nothing that is infinite can be traversed nor moved by any means, and so the heavens are necessarily at rest.

But beyond the heavens, it is said, there is no body, no place, no vacuity, absolutely nothing, and so there is nowhere for the heavens to go. It is truly miraculous, then, if something can be contained within nothing. However, if the heavens are infinite, and finite only in their hollow interior, then perhaps there will be greater reason to believe that outside the heavens there is nothing, for in this case every single thing, no matter how much space it takes up, will be inside them. But the heavens will remain motionless. For the strongest piece of evidence produced in support of the earth's finitude is its motion. Whether the universe is finite or infinite, however – let us leave that question for the natural philosophers to dispute while we hold firmly to the belief that the earth is delimited by its poles and enclosed by a spherical surface.

Why, then, do we still hesitate to accept the earth's movement in keeping with the nature of its form instead of attributing motion to the whole universe, whose bounds are unknown and unknowable. As regards the daily rotation, why not grant that in the heavens is the appearance but in the earth is the reality? It is like the case spoken of by Vergil's Aeneas: "We sail forth from the harbor, and lands and cities draw backwards" [*Aeneid*, III.72]. For when a ship glides along smoothly, its passengers see its motion reflected by everything outside of the ship and, by contrast, suppose themselves and everything else on board to be motionless. No wonder, then, that the movement of the earth makes us think the whole universe is turning round.

[...]

Chapter 9 Can the Earth Be Said to Move in More Ways Than One? Where Is the Center of the Universe?

Since therefore nothing precludes the earth's movement, I propose we now consider whether it may be thought to move in more than one way: can it be regarded as one of the planets?

For earth is not the center of all the revolutions. This claim is demonstrated by the apparently nonuniform motion of the planets and by their variable distances from the earth, which cannot be conceived as implying circles concentric to the earth. Therefore, there being numerous centers, it is worth asking whether the center of the universe, or some other, is the center of earthly gravity. In my view gravity is nothing but a certain natural desire which by divine providence the Creator of all has infused into the parts, whereby they draw themselves into a unity and an integrity in the form of a globe. The same desire may be credibly predicated also of the sun, the moon, and the other luminous planets; by its efficacy they persist in the rounded shape in which we behold them, although they pursue their own various orbits.

Therefore, if the earth too moves in other ways – about a center, for example – then this must similarly be reflected in many external things. Among them, it would seem, is the annual revolution. For if, granting immobility to the sun, we exchange earthly movement for solar movement, then the risings and settings of the constellations and the fixed stars which accompany morning and evening will appear just as they do. Furthermore, the stations as well as both the backward and forward motions of the planets will be seen not as their own motions but as earthly motion transmuted into apparent planetary motions. Finally, it will be accepted that the sun occupies the center of the universe.

We learn all these things by discerning the order whereby the planets follow one another and by the harmony of the entire universe – if only we examine these matters (as they say) with both eyes open.

Chapter 10 The Order of the Heavenly Spheres

[...]

We should not be ashamed to admit that this whole domain encircled by the moon, with the center of the earth, traverses this great orbit amidst the other planets in an annual revolution around the sun, and that near the sun is the center of the universe; and moreover that, since the sun stands still, whatever motion the sun appears to have is instead actually attributable to the motion of the earth. Furthermore, although the distance between the earth and the sun is quite noticeable relative to the size of the other planetary orbits, it is imperceptible as compared with the sphere of the fixed stars – so great indeed is the size of the universe. I think it is a lot easier to accept this than to drive our minds to distraction multiplying spheres almost ad infinitum, as has been the compulsion of those who would detain earth in the center of the universe. Instead, it is better to follow the wisdom of nature, which just as it strongly avoids producing anything superfluous or useless, so it often prefers to endow a single thing with multiple effects.

This whole matter is difficult, almost paradoxical, and certainly contrary to many people's way of thinking. In what follows, however, God helping me, I shall make these things clearer than sunlight, at least to those not ignorant of the art of astronomy. And so, with the first principle firmly established (for nobody can propose one more fitting than that the magnitude of a planet's orbit is proportionate to its period of revolution), the order of the spheres is as follows, beginning with the highest:

First and highest of all is the sphere of the fixed stars, containing itself and all things, and therefore immovable, the very location of the universe, that to which the motion and position of all the other heavenly bodies is referred. . . . This is followed by the first of the planets, Saturn, which completes his circuit in thirty years. Then comes Jupiter, moving in a revolution with a twelve-year period. Next, the circuit of Mars is two years. Fourth comes the annual revolution in which, as mentioned earlier, the earth is carried along, with the moon as it were in an epicycle. Venus, in fifth place, circles round in nine months. And then in sixth place Mercury completes his course in the space of eighty days.

And behold, in the midst of all resides the sun. For who, in this most beautiful temple, would set this lamp in another or a better place, whence to illuminate all things at once? For aptly indeed do some call him the lantern – and others the mind or the ruler – of the universe. Hermes Trismegistus calls him the visible god, and Sophocles' Electra "the beholder" of all things. Truly indeed does the sun, as if seated upon a royal throne, govern his family of planets as they circle about him. Nor is the earth thus deprived of the moon's services; rather, as Aristotle asserts in his book on animals, the moon shares closest kinship with the earth. Meanwhile, the earth is impregnated by the sun, by whom is begotten her annual offspring.

Thus we discover in this orderly arrangement the marvelous symmetry of the universe and a firm harmonious connection between the motion and the size of the spheres such as can be discerned by no other means. For this model permits anyone who is diligent to comprehend why the progressions and regressions of Jupiter appear greater than those of Saturn and smaller than those of Mars, and again greater for Venus than for Mercury. And the reversals appear more frequently in Saturn than in Jupiter, and even more rarely in Mars and in Venus than in Mercury. . . . All these phenomena appear for the same reason: that the earth moves.

However, that none of these phenomena appears in the fixed stars proves that these are immensely distant, for which reason even the motion of the annual revolution, or the appearance thereof, vanishes from sight. For each visible thing has a certain limit of distance beyond which it becomes invisible, as demonstrated in optics. The sparkling of the stars shows what an enormous distance remains between their sphere and that of the highest planet, Saturn. This is principally what distinguishes them from the planets, for there had to be an enormous difference between that which moves and that which does not. So great, certainly, is the divine handiwork of Him who is himself the greatest and the best.

2.4

The New Star

Tycho Brahe

Tycho Brahe (1546–1601) was a Danish astronomer who set new standards of precision for the measurement of stellar and planetary positions. In this excerpt from *The New Star* (1573), Tycho describes the appearance of the nova of 1572 in Cassiopeia. Since the official Aristotelian position was that there is never any change in the heavens, Tycho had to establish that the nova was indeed a heavenly phenomenon and not something beneath the sphere of the moon. This he does with the aid of a bit of trigonometry. The Aristotelian foundations were not yet wholly undermined, but Tycho's argument helped to shift the ground beneath them.

On a New Star, Not Previously Seen within the Memory of Any Age since the Beginning of the World

Its First Appearance in 1572. – Last year [1572], in the month of November, on the eleventh day of that month, in the evening, after sunset, when, according to my habit, I was contemplating the stars in a clear sky, I noticed that a new and unusual star, surpassing the other stars in brilliancy, was shining almost directly above my head; and since I had, almost from boyhood, known all the stars of the heavens perfectly (there is no great difficulty in attaining that knowledge), it was quite evident to me that there had never before been any star in that place in the sky, even the smallest, to say nothing of a star so conspicuously bright as this. I was so astonished at this sight that I was not ashamed to doubt the trustworthiness of my own eyes. But when I observed that others, too, on having the place pointed out to them, could see that there was really a star there, I had no further doubts. A miracle indeed, either the greatest of all that have occurred in the whole range of nature since the beginning of the world, or one certainly that is to be classed with those attested by the Holy Oracles, the staying of the Sun in its course in answer to the prayers of

Tycho Brahe from *De Nova Stella*, "Opera Omnia, Tomus I," ed. J. L. E. Dreyer, 1913, translated by John H. Walden, 1928. From Harlow Shapley and Helen E. Howarth (eds.), *A Source Book on Astronomy* (Cambridge, MA: Harvard University Press, 1969), selections from pp. 233–9.

Joshua, and the darkening of the Sun's face at the time of the Crucifixion. For all philosophers agree, and facts clearly prove it to be the case, that in the ethereal region of the celestial world no change, in the way either of generation or of corruption, takes place; but that the heavens and the celestial bodies in the heavens are without increase or diminution, and that they undergo no alteration, either in number or in size or in light or in any other respect; that they always remain the same, like unto themselves in all respects, no years wearing them away. Furthermore, the observations of all the founders of the science, made some thousands of years ago, testify that all the stars have always retained the same number, position, order, motion, and size as they are found, by careful observation on the part of those who take delight in heavenly phenomena, to preserve even in our own day. Nor do we read that it was ever before noted by any one of the founders that a new star had appeared in the celestial world, except only by Hipparchus, if we are to believe Pliny. For Hipparchus, according to Pliny, (Book II of his Natural History) noticed a star different from all others previously seen, one born in his own age . . .

Its Position with Reference to the Diameter of the World and Its Distance from the Earth, the Center of the Universe. – It is a difficult matter, and one that requires a subtle mind, to try to determine the distances of the stars from us, because they are so incredibly far removed from the earth; nor can it be done in any way more conveniently and with greater certainty than by the measure of the parallax [diurnal], if a star have one. For if a star that is near the horizon is seen in a different place than when it is at its highest point and near the vertex, it is necessarily found in some orbit with respect to which the Earth has a sensible size. How far distant the said orbit is, the size of the parallax compared with the semi-diameter of the Earth will make clear. If, however, a [circumpolar] star, that is as near to the horizon [at lower culmination] as to the vertex [at upper culmination], is seen at the same point of the Primum Mobile, there is no doubt that it is situated either in the eighth sphere or not far below it, in an orbit with respect to which the whole Earth is as a point.

In order, therefore, that I might find out in this way whether this star was in the region of the Element or among the celestial orbits, and what its distance was from the Earth itself, I tried to determine whether it had a parallax, and, if so, how great a one; and this I did in the following way: I observed the distance between this star and Schedir of Cassiopeia (for the latter and the new star were both nearly on the meridian), when the star was at its nearest point to the vertex, being only 6 degrees removed from the zenith itself (and for that reason, though it were near the Earth, would produce no parallax in that place, the visual position of the star and the real position then uniting in one point, since the line from the center of the Earth and that from the surface nearly coincide.) I made the same observation when the star was farthest from the zenith and at its nearest point to the horizon, and in each case I found that the distance from the abovementioned fixed star was exactly the same, without the variation of a minute: namely 7 degrees and 55 minutes. Then I went through the same process, making numerous observations with other stars. Whence I conclude that this new star has no diversity of aspect, even when it is near the horizon. For otherwise in its least altitude it would have been farther away from the abovementioned star in the breast of Cassiopeia than when in its greatest altitude. Therefore, we shall find it necessary to place this star, not in the region of the Element, below the Moon, but far above, in an orbit with respect to which the Earth has no sensible size.

[Brahe here gives a geometric argument proving that, if the new star had been an object closer to the earth than the sphere of the moon, its apparent position against the stars would vary noticeably, depending on whether it were seen nearly overhead or near the horizon.]

[. . .]

Therefore, this new star is neither in the region of the Element, below the Moon, nor among the orbits of the seven wandering stars, but it is in the eighth sphere, among the other fixed stars, which was what we had to prove. Hence it follows that it is not some peculiar kind of comet or some other kind of fiery meteor become visible. For none of these are generated in the heavens themselves, but they are below the Moon, in the upper region of the air, as all philosophers testify; unless one would believe with Albategnius that comets are produced, not in the air, but in the heavens. For he believes that he has observed a comet above the Moon, in the sphere of Venus. That this can be the case, is not yet clear to me. But, please God, sometime, if a comet shows itself in our age, I will investigate the truth of the matter. Even should we assume that it can happen (which I, in company with other philosophers, can hardly admit), still it does not follow that this star is a kind of comet; first, by reason of its very form, which is the same as the form of the real stars and different from the form of all the comets hitherto seen, and then because, in such a length of time, it advances neither latitudinally nor longitudinally by any motion of its own, as comets have been observed to do. For, although these sometimes seem to remain in one place several days, still, when the observation is made carefully by exact instruments, they are seen not to keep the same position for so very long or so very exactly. I conclude, therefore, that this star is not some kind of comet or a fiery meteor, whether these be generated beneath the Moon or above the Moon, but that it is a star shining in the firmament itself – one that has never previously been seen before our time, in any age since the beginning of the world.

2.5

A Man Ahead of His Time

Johannes Kepler

Johannes Kepler (1571–1630) was a German mathematician and astronomer. He studied under the Copernican astronomer Michael Mästlin and was working as Tycho Brahe's assistant at the time of the latter's death in 1601. In sharp contrast to the other astronomers of his day, Kepler was convinced that the heavens and the earth form one unified realm governed by one set of physical laws. In this selection from his correspondence and other works, we can see the great distance between Kepler's understanding of the task of astronomy and that of his peers. Because they did not see the need to bring astronomy and physics together, Kepler had to create a "physics of the sky" on his own.

Kepler on the Purpose of Astronomy

May God make it come to pass that my delightful speculation [the *Mysterium Cosmographicum*] have everywhere among reasonable men fully the effect which I strove to obtain in the publication; namely, that the belief in the creation of the world be fortified through this external support, that thought of the creator be recognized in its nature, and that His inexhaustible wisdom shine forth daily more brightly. Then man will at last measure the power of his mind on the true scale, and will realize that God, who founded everything in the world according to the norm of quantity, also has endowed man with a mind which can comprehend these norms. For as the eye for color, the ear for musical sounds, so is the mind of man created for the perception not of any arbitrary entities, but rather of quantities; the mind comprehends a thing the more correctly the closer the thing approaches toward pure quantity as its origin.

– Kepler to his teacher Mästlin,
Apr. 19, 1597

From Gerald Holton, "Johannes Kepler's Universe: Its Physics and Metaphysics," *American Journal of Physics*, Vol. 24, No. 5 (1956), selections from pp. 340–51.

Those laws [which govern the material world] lie within the power of understanding of the human mind; God wanted us to perceive them when He created us in His image in order that we may take part in His own thoughts. . . . Our knowledge [of numbers and quantities] is of the same kind as God's, at least insofar as we can understand something of it in this mortal life.

– Kepler to Herwart von Hohenburg, Apr. 9/10, 1599

The Clockwork Image

I am much occupied with the investigation of the physical causes. My aim in this is to show that the celestial machine is to be likened not to a divine organism but rather to a clockwork . . . , insofar as nearly all the manifold movements are carried out by means of a single, quite simple magnetic force, as in the case of a clockwork all motions [are caused] by a simple weight. Moreover I show how this physical conception is to be presented through calculation and geometry.

– Kepler to Herwart, Feb. 10, 1605

Against the Potency of Place

Das Mittele is nur ein Düpfflin. ["The center is just a little point."]

How can the Earth, or its nature, notice, recognize and seek after the center of the world which is only a little point – and then go toward it? The Earth is not a hawk, and the center of the world not a little bird; it [the center] is also not a magnet which could attract the Earth, for it has no substance and therefore cannot exert a force.

– from the *Objections* to his translation of Aristotle's *On the Heavens*

Mästlin Misses the Point

Concerning the motion of the moon you write you have traced all the inequalities to physical causes; I do not quite understand this. I think rather that here one should leave physical causes out of account, and should explain astronomical matters only according to astronomical method with the aid of astronomical, not physical, causes and hypotheses. That is, the calculation demands astronomical bases in the field of geometry and arithmetic. . . .

– Mästlin to Kepler, Oct. 1, 1616

I call my hypotheses physical for two reasons. . . . My aim is to assume only those things of which I do not doubt they are real and consequently physical, where one must refer to the nature of the heavens, not the elements. When I dismiss the perfect eccentric and the epicycle, I do so because they are purely geometric assumptions, for which a corresponding body in the heavens does not exist. The second reason for my calling my hypotheses physical is this. . . . I prove that the irregularity of the motion [of the planets] corresponds to the nature of the planetary sphere; i.e., is physical.

– note in Kepler's handwriting in the margin of this letter from Mästlin

Kepler on Hypotheses in Astronomy

The difference consists only in this, that you use circles, I use bodily forces. . . .

When you say it is not to be doubted that all motions occur on a perfect circle, then this is fase for the composite, i.e., the real motions. According to Copernicus, as explained, they occur on an orbit distended at the sides, whereas according to Ptolemy and Brahe on spirals. But if you speak of components of motion, then you speak of something existing in thought; i.e., something that is not there in reality. For nothing courses on the heavens except the planetary bodies themselves – no orbs, no epicycles. . . .

– Kepler to Fabricius, Aug. 1, 1607

2.6

On Arguments about a Moving Earth

Johannes Kepler

Kepler made fundamental advances in the theory of the solar system, showing that the orbits of the planets were not circles or complex constructs of circles and epicycles but rather ellipses with the sun at one focus. This extension of the Copernican view was driven by the accurate data Tycho left behind and by Kepler's own restless search for heavenly harmonies. In this excerpt from the introduction to the *New Astronomy* (1609), Kepler lays out the alternatives he is considering and deals directly with various challenges – physical, conceptual, and theological – to the idea of a moving earth.

The Introduction to This Work Is Aimed at Those Who Study the Physical Sciences

. . . [T]here are many points that should be brought together here at the beginning which are presented bit by bit throughout the work, and are therefore not so easy to attend to in passing. Furthermore, I shall reveal, especially for the sake of those professors of the physical sciences who are irate with me, as well as with Copernicus and even with the remotest antiquity, on account of our having shaken the foundations of the sciences with the motion of the earth – I shall, I say, reveal faithfully the intent of the principal chapters which deal with this subject, and shall propose for inspection all the principles of the proofs upon which my conclusions, so repugnant to them, are based.

For when they see that this is done faithfully, they will then have the free choice either of reading through and understanding the proofs themselves with much exertion, or of trusting me, a professional mathematician, concerning the sound and geometrical method presented. Meanwhile, they, for their part, will turn to the principles of the proofs thus gathered for their inspection, and will examine them thoroughly, knowing that unless they are refuted the proof erected upon them will not topple. I shall also do the same where, as is customary in the physical sciences, I mingle the probable with the necessary

From William H. Donahue, *Selections from Kepler's* Astronomia Nova (Santa Fe, NM: Green Lion Press, 2004), pp. 3–28, selections.

and draw a plausible conclusion from the mixture. For since I have mingled celestial physics with astronomy in this work, no one should be surprised at a certain amount of conjecture. This is the nature of physics, of medicine, and of all the sciences which make use of other axioms besides the most certain evidence of the eyes.

On the Schools of Thought in Astronomy

The reader should be aware that there are two schools of thought among astronomers, one distinguished by its chief, Ptolemy, and by the assent of the large majority of the ancients, and the other attributed to more recent proponents, although it is the most ancient. The former treats the individual planets separately and assigns causes to the motions of each in its own orb, while the latter relates the planets to one another, and deduces from a single common cause those characteristics that are found to be common to their motions. The latter school is again subdivided. Copernicus, with Aristarchus of remotest antiquity, ascribes to the translational motion of our home the earth the cause of the planets' appearing stationary and retrograde. Tycho Brahe, on the other hand, ascribes this cause to the sun, in whose vicinity he says the eccentric circles of all five planets are connected as if by a kind of knot (not physical, of course, but only quantitative). Further, he says that this knot, as it were, revolves about the motionless earth, along with the solar body.

For each of these three opinions concerning the world there are several other peculiarities which themselves also serve to distinguish these schools, but these peculiarities can each be easily altered and amended in such a way that, so far as astronomy, or the celestial appearances, are concerned, the three opinions are for practical purposes equivalent to a hair's breadth, and produce the same results.

The Twofold Aim of the Work

My aim in the present work is chiefly to reform astronomical theory (especially of the motion of Mars) in all three forms of hypotheses, so that what we compute from the tables may correspond to the celestial phenomena. Hitherto, it has not been possible to do this with sufficient certainty. In fact, in August of 1608, Mars was a little less than four degrees beyond the position given by calculation from the Prutenic tables. In August and September of 1593 this error was a little less than five degrees, while in my new calculation the error is entirely suppressed.

On the Physical Causes of the Motions

Meanwhile, although I place this goal first and pursue it cheerfully, I also make an excursion into Aristotle's *Metaphysics*, or rather, I inquire into celestial physics and the natural causes of the motions. The eventual result of this consideration is the formulation of very clear arguments showing that only Copernicus's opinion concerning the world (with a few small changes) is true, that the other two are false, and so on.

Indeed, all things are so interconnected, involved, and intertwined with one another that after trying many different approaches to the reform of astronomical calculations, some well trodden by the ancients and others constructed in emulation of them and by their example, none other could succeed than the one founded upon the motions' physical causes themselves, which I establish in this work.

The First Step: The Planes of the Eccentrics Intersect in the Sun

Now my first step in investigating the physical causes of the motions was to demonstrate that [the planes of] all the eccentrics intersect in no other place than the very center of the solar body (not some nearby point), contrary to what Copernicus and Brahe thought. If this correction of mine is carried over into the Ptolemaic theory, Ptolemy will have to investigate not the motion of the center of the epicycle, about which the epicycle proceeds uniformly, but the motion of some point whose distance from that center, in proportion to the diameter, is the same as the distance of the center of the solar orb from the earth for Ptolemy, which point is also on the same line, or one parallel to it. . . .

In the second part of the work I take up the main subject, and describe the positions of Mars at apparent opposition to the sun, not worse, but indeed much better, with my method than they expressed the positions of Mars at mean opposition to the sun with the old method.

Meanwhile, throughout the entire second part (as far as concerns geometrical demonstrations from the observations) I leave in suspense the question of whose procedure is better, theirs or mine, seeing that we both match a great many observations (this is, indeed, a basic requirement for our theorizing). However, my method is in agreement with physical causes, and their old one is in disagreement, as I have partly shown in the first part, especially Chapter 6.

But finally in the fourth part of the work, in Chapter 52, I consider certain other observations, no less trustworthy than the previous ones were, which their old method could not match, but which mine matches most beautifully. I thereby demonstrate most soundly that Mars's eccentric is so situated that the center of the solar body lies upon its line of apsides, and not any nearby point, and hence, that all the [planes of the] eccentrics intersect in the sun itself. . . .

[There follows a section entitled "The second step: there is an equant in the theory of the sun," which is omitted here.]

The Earth Is Moved and the Sun Stands Still. Physico-astronomical Arguments

With these things thus demonstrated by a reliable method, the previous step towards the physical causes is now confirmed, and a new step is taken towards them, most clearly in the theories of Copernicus and Brahe, and more obscurely but at least plausibly in the Ptolemaic theory. For whether it is the earth or the sun that is moved, it has certainly been demonstrated that the body that is moved is moved in a nonuniform manner, that is, slowly when it is farther from the body at rest, and more swiftly when it has approached this body.

Thus the physical difference is now immediately apparent – by way of conjecture, it is true, but yielding nothing in certainty to conjectures of doctors on physiology or to any other natural science.

First, Ptolemy is certainly condemned. For who would believe that there are as many theories of the sun (so closely resembling one another that they are in fact equal) as there are planets, when he sees that for Brahe a single solar theory suffices for the same task, and it is the most widely accepted axiom in the natural sciences that Nature makes use of the fewest possible means? That Copernicus is better able than Brahe to deal with celestial physics is proven in many ways.

1. First, although Brahe did indeed take up those five solar theories from the theories of the planets, bringing them down to the centers of the eccentrics, hiding them there, and conflating them into one, he nevertheless left in the world the effects produced by those theories. For Brahe no less than for Ptolemy, besides that motion which is proper to it, each planet is still actually moved with the sun's motion, the two being mixed into one, the result being a spiral. That it results from this that there are no solid orbs, Brahe has demonstrated most firmly. Copernicus, on the other hand, entirely removed this extrinsic motion from the five planets, assigning its cause to a deception arising from the circumstances of observation. Thus the motions are still multiplied to no purpose by Brahe, as they were before by Ptolemy.
2. Second, if there are no orbs, the conditions under which the intelligences and moving souls must operate are made very difficult, since they have to attend to so many things to introduce to the planet two intermingled motions. They would at least have to attend at one and the same time to the principles, centers, and periods of the two motions. But if the earth is moved, I show that most of this can be done with physical rather than animate faculties, namely, magnetic ones. But these are more general points. There follow others arising specifically from demonstrations, upon which we now begin.
3. For if the earth is moved, it has been demonstrated that the increases and decreases of its velocity are governed by its approaching towards and receding from the sun. And in fact the same happens with the rest of the planets: they are urged on or held back according to the approach toward or recession

from the sun. So far, the demonstration is geometrical.

And now, from this very reliable demonstration, the conclusion is drawn, using a physical conjecture, that the source of the five planets' motion is in the sun itself. It is therefore very likely that the source of the earth's motion is in the same place as the source of the other five planets' motion, namely, in the sun as well. It is therefore likely that the earth is moved, since a likely cause of its motion is apparent.

4. That, on the other hand, the sun remains in place in the center of the world, is most probably shown by (among other things) its being the source of motion for at least five planets. For whether you follow Copernicus or Brahe, the source of motion for five of the planets is in the sun, and in Copernicus, for a sixth as well, namely, the earth. And it is more likely that the source of all motion should remain in place rather than move.
5. But if we follow Brahe's theory and say that the sun moves, this first conclusion still remains valid, that the sun moves slowly when it is more distant from the earth and swiftly when it approaches, and this not only in appearance, but in fact. For this is the effect of the circle of the equant, which, by an inescapable demonstration, I have introduced into the theory of the sun.

   Upon this most valid conclusion, making use of the physical conjecture introduced above, might be based the following theorem of natural philosophy: the sun, and with it the whole huge burden (to speak coarsely) of the five eccentrics, is moved by the earth; or, the source of the motion of the sun and the five eccentrics attached to the sun is in the earth.

   Now let us consider the bodies of the sun and the earth, and decide which is better suited to being the source of motion for the other body. Does the sun, which moves the rest of the planets, move the earth, or does the earth move the sun, which moves the rest, and which is so many times greater? Unless we are to be forced to admit the absurd conclusion that the sun is moved by the earth, we must allow the sun to be fixed and the earth to move.
6. What shall I say of the motion's periodic time of 365 days, intermediate in quantity between the periodic time of Mars of 687 days and that of Venus of 225 days? Does not the nature of things cry out with a great voice that the circuit in which these 365 days are used up also occupies a place intermediate between those of Mars and Venus about the sun, and thus itself also encircles the sun, and hence, that this circuit is a circuit of the earth about the sun, and not of the sun about the earth? . . .

Objections to the Earth's Motion

I trust the reader's indulgence if I take this opportunity to present a few brief replies to a number of objections which, capturing people's minds, use the following arguments to shed darkness. For these replies are by no means irrelevant to matters that concern the physical causes of the planets' motion, which I discuss chiefly in parts three and four of the present work.

I On the motion of heavy bodies

Many are prevented by the motion of heavy bodies from believing that the earth is moved by an animate motion, or better, by a magnetic one. They should ponder the following propositions.

The theory of gravity is in error

A mathematical point, whether or not it is the center of the world, can neither effect the motion of heavy bodies nor act as a object towards which they tend. Let the physicists prove that this force is in a point which neither is a body nor is grasped otherwise than through mere relation.

It is impossible that, in moving its body, the form of a stone seek out a mathematical point (in this instance, the center of the world), without respect to the body in which this point is located. Let the physicists prove that natural things have a sympathy for that which is nothing. . . .

[. . .]

True theory of gravity

The true theory of gravity rests upon the following axioms.

Every corporeal substance, to the extent that it is corporeal, has been so made as to be suited to rest in every place in which it is put by itself, outside the sphere of influence of a kindred body.

Gravity is a mutual corporeal disposition among kindred bodies to unite or join together, thus, the earth attracts a stone much more than the stone seeks the earth. (The magnetic faculty is another example of this sort.)

Heavy bodies (most of all if we establish the earth in the center of the world) are not drawn towards the center of the world because it is the center of the world, but because it is the center of a kindred spherical body, namely, the earth. Consequently, wherever the earth be established, or whithersoever it be carried by its animate faculty, heavy bodies are drawn towards it.

If the earth were not round, heavy bodies would not everywhere be drawn in straight lines towards the middle point of the earth, but would be drawn towards different points from different sides.

If two stones were set near one another in some place in the world outside the sphere of influence of a third kindred body, these stones, like two magnetic bodies, would come together in an intermediate place, each approaching the other by an interval proportional to the bulk of the other.

If the moon and the earth were not each held back in its own circuit by an animate force or something else equivalent to it, the earth would ascend towards the moon by one fifty-fourth part of the interval, and the moon would descend towards the earth about fifty-three parts of the interval, and there they would be joined together; provided, that is, that the substance of each is of the same density.

If the earth should cease to attract its waters to itself, all the sea water would be lifted up, and would flow onto the body of the moon.

Reason for the ebb and flow of the sea

The sphere of influence of the attractive power in the moon is extended all the way to the earth, and in the torrid zone calls the waters forth, particularly when it comes to be overhead in one or another of its passages. This is imperceptible in enclosed seas, but noticeable where the beds of the ocean are widest and there is much free space for the waters' reciprocation. . . .

[. . .]

Although these things are appropriate to a different topic, I wanted to present them all in one context in order to make more credible the ocean tide and through it the moon's attractive power.

For it follows that if the moon's power of attraction extends to the earth, the earth's power of attraction will be much more likely to extend to the moon and far beyond, and accordingly, that nothing that consists to any extent whatever of terrestrial material, carried up on high, ever escapes the grasp of this mighty power of attraction.

True theory of levity

Nothing that consists of corporeal material is absolutely light. It is only comparatively lighter, because it is less dense, either by its own nature or through an influx of heat. By "less dense" I do not just mean that which is porous and divided into many cavities, but in general that which, while occupying a place of the same magnitude as that occupied by some heavier body, contains a lesser quantity of corporeal material.

The motion of light things also follows from their definition. For it should not be thought that they flee all the way to the surface of the world when they are carried upwards, or that they are not attracted by the earth. Rather, they are less attracted than heavy bodies, and are thus displaced by heavy bodies, whereupon they come to rest and are kept in their place by the earth.

To the objection that objects projected vertically fall back to their places

But even if the earth's power of attraction is extended very far upwards, as was said, nevertheless, if a stone were at a distance that was perceptible in relation to the earth's diameter, it is true that, the earth being moved, such a stone would not simply follow, but its forces of resistance would mingle with the earth's forces of attraction, and it would thus detach itself somewhat from the earth's grasp. In just the same way, violent motion detaches projectiles somewhat from the earth's grasp, so that they either run on ahead if they are shot eastwards, or are left behind if shot westwards, thus leaving the place from which they are shot, under the compulsion of force. Nor can the earth's revolving effect impede this violent motion all at once, as long as the violent motion is at its full strength.

But no projectile is separated from the surface of the earth by even a hundred thousandth part of the earth's diameter, and not even the clouds themselves, or smoke, which partake of earthy matter to the very least extent, achieve an altitude of a thousandth part of the semidiameter. Therefore, none of the clouds, smokes, or objects shot vertically upwards can make any resistance, nor, I say, can the natural inclination to rest do anything to impede this grasp of the earth's, at least where this resistance is negligible in proportion to that grasp. Consequently, anything shot vertically upwards falls back to its place, the motion of the earth notwithstanding. For the earth cannot be pulled out from under it, since the earth carries with it anything sailing through the air, linked to it by the magnetic force no less firmly than if those bodies were actually in contact with it.

When these propositions have been grasped by the understanding and pondered carefully, not only do the absurdity and falsely conceived physical impossibility of the earth's motion vanish, but it also becomes clear how to reply to the physical objections, however they are framed. . . .

[Kepler here mentions two other sorts of objections, numbered II and III (those arising from the swiftness of the earth's motion and those arising from the immensity of the heavens), referring the reader in both cases to his discussion in *On the New Star*.]

IV To objections concerning the dissent of holy scripture, and its authority

There are, however, many more people who are moved by piety to withhold assent from Copernicus, fearing that falsehood might be charged against the Holy Spirit speaking in the scriptures if we say that the earth is moved and the sun stands still.

But let them consider that since we acquire most of our information, both in quality and quantity, through the sense of sight, it is impossible for us to abstract our speech from this ocular sense. Thus, many times each day we speak in accordance with the sense of sight, although we are quite certain that the truth of the matter is otherwise. . . .

A generation passes away (says Ecclesiastes [1:4]), and a generation comes, but the earth stands forever. Does it seem here as if Solomon wanted to argue with the astronomers? No; rather, he wanted to warn people of their own mutability, while the earth, home of the human race, remains always the same, the motion of the sun perpetually returns to the same place, the wind blows in a circle and returns to its starting point, rivers flow from their sources into the sea, and from the sea return to the sources, and finally, as these people perish, others are born. Life's tale is ever the same; there is nothing new under the sun.

You do not hear any physical dogma here. The message is a moral one, concerning something self-evident and seen by all eyes but seldom pondered. Solomon therefore urges us to ponder. Who is unaware that the earth is always the same? Who does not see the sun return daily to its place of rising, rivers perennially flowing towards the sea, the winds returning in regular alternation, and men succeeding one another? But who really considers that the same drama of life is always being played, only with different characters, and that not a single thing in human affairs is new? So Solomon, by mentioning what is evident to all, warns of that which almost everyone wrongly neglects. . . .

Advice to astronomers

I, too, implore my reader, when he departs from the temple and enters astronomical studies, not to forget the divine goodness conferred upon men, to the consideration of which the psalmodist chiefly invites. I hope that, with me, he will praise and celebrate the Creator's wisdom and greatness, which I unfold for him in the more perspicacious explanation of the world's form, the investigation of causes, and the detection of errors of vision. Let him not only extol the Creator's divine beneficence in His concern for the well-being of all living things, expressed in the firmness and stability of the earth, but also acknowledge His wisdom expressed in its motion, at once so well hidden and so admirable.

Advice for idiots

But whoever is too stupid to understand astronomical science, or too weak to believe Copernicus without affecting his faith, I would advise him that, having dismissed astronomical studies and having damned whatever philosophical opinions he pleases, he mind his own business and betake himself home to scratch in his own dirt patch, abandoning this wandering about the world. He should raise his eyes (his only means of vision) to this visible heaven and with his whole heart burst forth in giving thanks and praising God the Creator. He can be sure that he worships God no less than the astronomer, to whom God has granted the more penetrating vision of the mind's eye, and an ability and desire to celebrate his God above those things he has discovered.

Commendation of the Brahean hypothesis

At this point, a modest (though not too modest) commendation to the learned should be made on behalf of Brahe's opinion of the form of the world, since in a way it follows a middle path. On the one hand, it frees the astronomers as much as possible from the useless apparatus of so many epicycles and, with Copernicus, it includes the causes of motion, unknown to Ptolemy, giving some place to physical theory in accepting the sun as the center of the planetary system. And on the other hand, it serves the mob of literalists and eliminates the motion of the earth, so hard to believe, although many difficulties are thereby insinuated into the theories of the planets in astronomical discussions and demonstrations, and the physics of the heavens is no less disturbed.

V To objections concerning the authority of the pious

So much for the authority of holy scripture. As for the opinions of the pious on these matters of nature, I have just one thing to say: while in theology it is authority that carries the most weight, in philosophy it is reason. Therefore, Lactantius is pious, who denied that the earth is round, Augustine is pious, who, though admitting the roundness, denied the antipodes, and the Inquisition nowadays is pious, which, though allowing the earth's smallness, denies its motion. To me, however, the truth is more pious still, and (with all due respect for the Doctors of the Church) I prove philosophically not only that the earth is round, not only that it is inhabited all the way around at the antipodes, not only that it is contemptibly small, but also that it is carried along among the stars.

But enough about the truth of the Copernican hypothesis. Let us return to the plan I proposed at the beginning of this introduction.

I had begun to say that in this work I treat all of astronomy by means of physical causes rather than fictitious hypotheses, and that I had taken two steps in my effort to reach this central goal: first, that I had discovered that the planetary eccentrics all intersect in the body of the sun, and second, that I had understood that in the theory of the earth there is an equant circle, and that its eccentricity is to be bisected.

The Third Step Towards the Physical Explanation. The Eccentricity of Mars's Equant Is to Be Precisely Bisected

Now we come to the third step, namely, that it has been demonstrated with certainty, by a comparison of the conclusions of Parts 2 and 4, that the eccentricity of Mars's equant is also to be precisely bisected, a fact long held in doubt by Brahe and Copernicus.

Therefore, by induction extending to all the planets . . . since there are (of course) no solid orbs, as Brahe demonstrated from the paths of comets, the body of the sun is the source of the power that drives all the planets around. Moreover, I have specified the manner [in which this occurs] as follows: that the sun, although it stays in one place, rotates as if on a lathe, and out of itself sends into the space of the world an immaterial *species* of its body, analogous to the immaterial species of its light. This *species* itself, as a consequence of the rotation of the solar body, also rotates like a very rapid whirlpool throughout the whole breadth of the world, and carries the bodies of the planets along with itself in a gyre, its grasp stronger or weaker according to the greater density or rarity it acquires through the law governing its diffusion. . . .

[. . .]

Fourth Step to the Physical Explanation. The Planet Describes an Oval Path

But my exhausting task was not complete: I had a fourth step yet to make towards the physical hypotheses. By most laborious proofs and by computations on a very large number of observations, I discovered that the course of a planet in the heavens is not a circle, but an oval path, perfectly elliptical.

Geometry gave assent to this, and taught that such a path will result if we assign to the planet's own movers the task of making the planet's body reciprocate along a straight line extended towards the sun. Not only this, but also the correct eccentric equations, agreeing with the observations, resulted from such a reciprocation.

Finally, the pediment was added to the structure, and proven geometrically: that it is in the order of things for such a reciprocation to be the result of a magnetic corporeal faculty. Consequently, these movers belonging to the planets individually are shown with great probability to be nothing but properties of the planetary bodies themselves, like the magnet's property of seeking the pole and catching up iron. As a result, every detail of the celestial motions is caused and regulated by faculties of a purely corporeal nature, that is, magnetic, with the sole exception of the whirling of the solar body as it remains fixed in its space. For this, a vital faculty seems required.

2.7

Eight Minutes of Arc

Johannes Kepler

After a great deal of computational labor, Kepler arrived at a construction of the orbit of Mars that fit Tycho's observational data to within eight minutes of arc (about 1/8 of one degree), an achievement that surpassed anything Ptolemy or Copernicus even attempted. Yet Kepler knew that Tycho's data – data which were unpublished and in his care – were so accurate that this minute mismatch indicated the falsehood of his construction. In a breathtaking display of intellectual honesty, he abandoned his theory and started over. In this excerpt from chapter 19 of the *New Astronomy*, we see the candid admission of failure that paved the way for his discovery of the elliptical orbit.

Chapter 19

Who would have thought it possible? This hypothesis, so closely in agreement with the acronychal observations, is nonetheless false, . . .

And from this difference of eight minutes, so small that it is, the reason is clear why Ptolemy, when he made use of bisection, was satisfied with a fixed equalizing point. For if the eccentricity of the equant, whose magnitude the very large equations in the middle longitudes fix indubitably, be bisected, you see that the very greatest error from the observations reaches 8′, and this in Mars, which has the greatest eccentricity; it is therefore less for the rest.

Now Ptolemy professed not to go below 10′, or the sixth part of a degree, in his observation. The uncertainty or (as they say) the 'latitude' of the observations exceeds the error in this Ptolemaic computation.

Since the divine benevolence has vouchsafed us Tycho Brahe, a most diligent observer, from whose observations the 8′ error in this Ptolemaic computation is shown, it is fitting that we with thankful mind both acknowledge and honour this benefit of God. For it is in this that we shall

From William H. Donahue and Owen Gingerich (trans. and ed.), *Johannes Kepler's New Astronomy* (Cambridge University Press, 1993), ch. 19, selections.

carry on, to find at length the true form of the celestial motions, supported as we are by these arguments showing our suppositions to be fallacious. In what follows, I shall myself, to the best of my ability, lead the way for others on this road. For if I had thought I could ignore eight minutes of longitude, in bisecting the eccentricity I would already have made enough of a correction in the hypothesis found in ch. 16. Now, because they could not have been ignored, these eight minutes alone will have led the way to the reformation of all of astronomy, and have constituted the material for a great part of the present work.

2.8

Tradition and Experience

Galileo Galilei

Galileo Galilei (1564–1642) was an Italian astronomer, mathematician, and physicist who did more than anyone else of his generation to complete the Copernican revolution. In the course of his career he engaged in many controversies and made powerful enemies. One of those enemies was the Jesuit Grassi, who published an attack on some of Galileo's work under the pen name "Sarsi" – ostensibly the name of a student, but really a rearrangement of some of the letters of Grassi's own name. Galileo shot back with *The Assayer* (1623), a blistering critique in which he pillories "Sarsi" and articulates a tough-minded empiricism as an alternative to the mere citation of venerable authority.

In Sarsi I seem to discern the firm belief that in philosophizing one must support oneself upon the opinion of some celebrated author, as if our minds ought to remain completely sterile and barren unless wedded to the reasoning of some other person. Possibly he thinks that philosophy is a book of fiction by some writer, like the *Iliad* or *Orlando Furioso*, productions in which the least important thing is whether what is written there is true. Well, Sarsi, that is not how matters stand. Philosophy is written in this grand book, the universe, which stands continually open to our gaze. But the book cannot be understood unless one first learns to comprehend the language and read the letters in which it is composed. It is written in the language of mathematics, and its characters are triangles, circles, and other geometric figures without which it is humanly impossible to understand a single word of it; without these, one wanders about in a dark labyrinth.

[. . .]

To put aside hints and speak plainly, and dealing with science as a method of demonstration and reasoning capable of human pursuit, I hold that

From "The Assayer," in *Discoveries and Opinions of Galileo*, trans. with an introduction and notes by Stillman Drake (New York: Anchor Books, Doubleday, 1957), selections from pp. 237–41, 258, 271–2.

the more this partakes of perfection the smaller the number of propositions it will promise to teach, and fewer yet will it conclusively prove. Consequently the more perfect it is the less attractive it will be, and the fewer its followers. On the other hand magnificent titles and many grandiose promises attract the natural curiosity of men and hold them forever involved in fallacies and chimeras, without ever offering them one single sample of that sharpness of true proof by which the taste may be awakened to know how insipid is the ordinary fare of philosophy. Such things will keep an infinite number of men occupied, and that man will indeed be fortunate who, led by some unusual inner light, can turn from dark and confused labyrinths in which he might have gone perpetually winding with the crowd and becoming ever more entangled.

Hence I consider it not very sound to judge a man's philosophical opinions by the number of his followers. Yet though I believe the number of disciples of the best philosophy may be quite small, I do not conclude conversely that those opinions and doctrines are necessarily perfect which have few followers, for I know well enough that some men hold opinions so erroneous as to be rejected by everyone else . . .

If I accept Sarsi's charge of negligence because various motions that might have been attributed to the comet did not occur to me, I fail to see how he can free his teacher from the same criticism for not considering the possibility of motion in a straight line. . . . There is no doubt whatever that by introducing irregular lines one may save not only the appearance in question but any other. Yet I warn Sarsi that far from being of any assistance to his teacher's case, this would only prejudice it more seriously; not only because he did not mention this, and on the contrary accepted the most regular line there is (the circular), but because it would have been very flippant to propose such a thing. Sarsi himself may understand this if he will consider what is meant by an irregular line. Lines are called regular when, having a fixed and definite description, they are susceptible of definition and of having their properties demonstrated. Thus the spiral is regular, and its definition originates in two uniform motions, one straight and the other circular. So is the ellipse, which originates from the cutting of a cone or a cylinder. Irregular lines are those which have no determinacy whatever, but are indefinite and casual and hence undefinable; no property of such lines can be demonstrated, and in a word nothing can be known about them. Hence to say, "Such events take place thanks to an irregular path" is the same as to say, "I do not know why they occur." The introduction of such lines is in no way superior to the "sympathy," "antipathy," "occult properties," "influences," and other terms employed by some philosophers as a cloak for the correct reply, which would be: "I do not know." That reply is as much more tolerable than the others as candid honesty is more beautiful than deceitful duplicity.

[. . .]

I could illustrate with many more examples Nature's bounty in producing her effects, as she employs means we could never think of without our senses and our experiences to teach them to us – and sometimes even these are insufficient to remedy our lack of understanding. So I should not be condemned for being unable to determine precisely the way in which comets are produced, especially in view of the fact that I have never boasted that I could do this, knowing that they may originate in some manner that is far beyond our power of imagination. The difficulty of comprehending how the cicada forms its song even when we have it singing to us right in our hands ought to be more than enough to excuse us for not knowing how comets are formed at such immense distances. Let us therefore go no further than our original intention, which was to set forth the questions that appeared to upset the old theories, and to propose a few new ideas.

[. . .]

I cannot but be astonished that Sarsi should persist in trying to prove by means of witnesses something that I may see for myself at any time by means of experiment. Witnesses are examined in doubtful matters which are past and transient, not in those which are actual and present. A judge must seek by means of witnesses to determine whether Peter injured John last night, but not whether John was injured, since the judge can see that for himself. But even in conclusions which can be known only by reasoning, I say that the testimony of many has little more value than that of few, since the number of people who reason well in complicated matters is much smaller than that of those who reason badly. If reasoning were like hauling I should agree that several reasoners would be worth more than one, just as several horses can haul more sacks of grain than one can. But reasoning is like racing and not like hauling, and a single Arabian steed can outrun a hundred plowhorses. So when Sarsi brings in this multitude of authors it appears to me that instead of strengthening his conclusion he merely ennobles our case by showing that we have outreasoned many men of great reputation.

If Sarsi wants me to believe with Suidas that the Babylonians cooked their eggs by whirling them in slings, I shall do so; but I must say that the cause of this effect was very different from what he suggests. To discover the true cause I reason as follows: "If we do not achieve an effect which others formerly achieved, then it must be that in our operations we lack something that produced their success. And if there is just one single thing we lack, then that alone can be the true cause. Now we do not lack eggs, nor slings, nor sturdy fellows to whirl them; yet our eggs do not cook, but merely cool down faster if they happen to be hot. And since nothing is lacking to us except being Babylonians, then being Babylonians is the cause of the hardening of eggs, and not friction of the air." And this is what I wished to discover. Is it possible that Sarsi has never observed the coolness produced on his face by the continual change of air when he is riding post? If he has, then how can he prefer to believe things related by other men as having happened two thousand years ago in Babylon rather than present events which he himself experiences? . . .

Sarsi says he does not wish to be numbered among those who affront the sages by disbelieving and contradicting them. I say I do not wish to be counted as an ignoramus and an ingrate toward Nature and toward God; for it they have given me my senses and my reason, why should I defer such great gifts to the errors of some man? Why should I believe blindly and stupidly what I wish to believe, and subject the freedom of my intellect to someone else who is just as liable to error as I am? . . .

2.9

A Moving Earth Is More Probable Than the Alternative

Galileo Galilei

Galileo's most influential work was undoubtedly his *Dialogue Concerning the Two Chief World Systems* (1632), in which he systematically demolishes a wide range of Aristotelian ("Peripatetic") arguments against the Copernican view. In this selection from the second day of the dialogue, Galileo's spokesman Salviati offers his companions, Sagredo (the open-minded inquirer) and Simplicio (the rather closed-minded Aristotelian) seven reasons why it is more probable that the earth moves and the heavens do not than that the rest of the universe moves while the earth is motionless.

Salv. Then let the beginning of our reflections be the consideration that whatever motion comes to be attributed to the earth must necessarily remain imperceptible to us and as if nonexistent, so long as we look only at terrestrial objects; for as inhabitants of the earth, we consequently participate in the same motion. But on the other hand it is indeed just as necessary that it display itself very generally in all other visible bodies and objects which, being separated from the earth, do not take part in this movement. So the true method of investigating whether any motion can be attributed to the earth, and if so what it may be, is to observe and consider whether bodies separated from the earth exhibit some appearance of motion which belongs equally to all. For a motion which is perceived only, for example, in the moon, and which does not affect Venus or Jupiter or the other stars, cannot in any way be the earth's or anything but the moon's.

Now there is one motion which is most general and supreme over all, and it is that by which the sun, moon, and all other planets and fixed stars – in a word, the whole universe, the earth alone excepted – appear to be moved as a unit from east to west in the space of twenty-four hours. This, in so far as first appearances are concerned, may just as logically belong to the earth alone as to the rest of the universe, since the same appearances would prevail as much in the one situation as in

From Galileo Galilei, *Dialogue Concerning the Two Chief World Systems: Ptolemaic and Copernican.* Trans. with notes by Stillman Drake (New York: The Modern Library, 1953), pp. 132–41.

the other. Thus it is that Aristotle and Ptolemy, who thoroughly understood this consideration, in their attempt to prove the earth immovable do not argue against any other motion than this diurnal one, though Aristotle does drop a hint against another motion ascribed to it by an ancient writer, of which we shall speak in the proper place.

Sagr. I am quite convinced of the force of your argument, but it raises a question for me from which I do not know how to free myself, and it is this: Copernicus attributed to the earth another motion than the diurnal. By the rule just affirmed, this ought to remain imperceptible to all observations on the earth, but be visible in the rest of the universe. It seems to me that one may deduce as a necessary consequence either that he was grossly mistaken in assigning to the earth a motion corresponding to no appearance in the heavens generally, or that if the correspondent motion does exist, then Ptolemy was equally at fault in not explaining it away, as he explained away the other.

Salv. This is very reasonably questioned, and when we come to treat of the other movement you will see how greatly Copernicus surpassed Ptolemy in acuteness and penetration of mind by seeing what the latter did not – I mean the wonderful correspondence with which such a movement is reflected in all the other heavenly bodies. But let us postpone this for the present and return to the first consideration, with respect to which I shall set forth, commencing with the most general things, those reasons which seem to favor the earth's motion, so that we may then hear their refutation from Simplicio.

First, let us consider only the immense bulk of the starry sphere in contrast with the smallness of the terrestrial globe, which is contained in the former so many millions of times. Now if we think of the velocity of motion required to make a complete rotation in a single day and night, I cannot persuade myself that anyone could be found who would think it the more reasonable and credible thing that it was the celestial sphere which did the turning, and the terrestrial globe which remained fixed.

Sagr. If, throughout the whole variety of effects that could exist in nature as dependent upon these motions, all the same consequences followed indifferently to a hairsbreadth from both positions, still my first general impression of them would be this: I should think that anyone who considered it more reasonable for the whole universe to move in order to let the earth remain fixed would be more irrational than one who should climb to the top of your cupola just to get a view of the city and its environs, and then demand that the whole countryside should revolve around him so that he would not have to take the trouble to turn his head. Doubtless there are many and great advantages to be drawn from the new theory and not from the previous one (which to my mind is comparable with or even surpasses the above in absurdity), making the former more credible than the latter. But perhaps Aristotle, Ptolemy, and Simplicio ought to marshal their advantages against us and set them forth, too, if such there are; otherwise it will be clear to me that there are none and cannot be any.

Salv. Despite much thinking about it, I have not been able to find any difference, so it seems to me I have found that there can be no difference; hence I think it vain to seek one further. For consider: Motion, in so far as it is and acts as motion, to that extent exists relatively to things that lack it; and among things which all share equally in any motion, it does not act, and is as if it did not exist. Thus the goods with which a ship is laden leaving Venice, pass by Corfu, by Crete, by Cyprus and go to Aleppo. Venice, Corfu, Crete, etc. stand still and do not move with the ship; but as to the sacks, boxes, and bundles with which the boat is laden and with respect to the ship itself, the motion from Venice to Syria is as nothing, and in no way alters their relation among themselves. This is so because it is common to all of them and all share equally in it. If, from the cargo in the ship, a sack were shifted from a chest one single inch, this alone would be more of a movement for it than the two-thousand-mile journey made by all of them together.

Simp. This is good, sound doctrine, and entirely Peripatetic.

Salv. I should have thought it somewhat older. And I question whether Aristotle entirely understood it when selecting it from some good school of thought, and whether he has not, by altering it in his writings, made it a source of confusion among those who wish to maintain everything he said. When he wrote that everything which is moved is moved upon something immovable,

I think he only made equivocal the saying that whatever moves, moves with respect to something motionless. This proposition suffers no difficulties at all, whereas the other has many.

Sagr. Please do not break the thread, but continue with the argument already begun.

Salv. It is obvious, then, that motion which is common to many moving things is idle and inconsequential to the relation of these movables among themselves, nothing being changed among them, and that it is operative only in the relation that they have with other bodies lacking that motion, among which their location is changed. Now, having divided the universe into two parts, one of which is necessarily movable and the other motionless, it is the same thing to make the earth alone move, and to move all the rest of the universe, so far as concerns any result which may depend upon such movement. For the action of such a movement is only in the relation between the celestial bodies and the earth, which relation alone is changed. Now if precisely the same effect follows whether the earth is made to move and the rest of the universe stay still, or the earth alone remains fixed while the whole universe shares one motion, who is going to believe that nature (which by general agreement does not act by means of many things when it can do so by means of few) has chosen to make an immense number of extremely large bodies move with inconceivable velocities, to achieve what could have been done by a moderate movement of one single body around its own center?

Simp. I do not quite understand how this very great motion is as nothing for the sun, the moon, the other planets, and the innumerable host of the fixed stars. Why do you say it is nothing for the sun to pass from one meridian to the other, rise above this horizon and sink beneath that, causing now the day and now the night; and for the moon, the other planets, and the fixed stars to vary similarly?

Salv. Every one of these variations which you recite to me is nothing except in relation to the earth. To see that this is true, remove the earth; nothing remains in the universe of rising and setting of the sun and moon, nor of horizons and meridians, nor day and night, and in a word from this movement there will never originate any changes in the moon of sun or any stars you please, fixed or moving. All these changes are in relation to the earth, all of them meaning nothing except that the sun shows itself now over China, then to Persia, afterward to Egypt, to Greece, to France, to Spain, to America, etc. And the same holds for the moon and the rest of the heavenly bodies, this effect taking place in exactly the same way if, without embroiling the biggest part of the universe, the terrestrial globe is made to revolve upon itself.

And let us redouble the difficulty with another very great one, which is this. If this great motion is attributed to the heavens, it has to be made in the opposite direction from the specific motion of all the planetary orbs, of which each one incontrovertibly has its own motion from west to east, this being very gentle and moderate, and must then be made to rush the other way; that is, from east to west, with this very rapid diurnal motion. Whereas by making the earth itself move, the contrariety of motions is removed, and the single motion from west to east accommodates all the observations and satisfies them all completely.

Simp. As to the contrariety of motions, that would matter little, since Aristotle demonstrates that circular motions are not contrary to one another, and their opposition cannot be called true contrariety.

Salv. Does Aristotle demonstrate that, or does he just say it because it suits certain designs of his? If, as he himself declares, contraries are those things which mutually destroy each other, I cannot see how two movable bodies meeting each other along a circular line conflict any less than if they had met along a straight line.

Sagr. Please stop a moment. Tell me, Simplicio, when two knights meet tilting in an open field, or two whole squadrons, or two fleets at sea go to attack and smash and sink each other, would you call their encounters contrary to one another?

Simp. I should say they were contrary.

Sagr. Then why are two circular motions not contrary? Being made upon the surface of the land or sea, which as you know is spherical, these motions become circular. Do you know what circular motions are not contrary to each other, Simplicio? They are those of two circles which touch from the outside; one being turned, the other naturally moves the opposite way. But if one circle should be inside the other, it is impossible that their motions should be made in opposite directions without their resisting each other.

Salv. "Contrary" or "not contrary," these are quibbles about words, but I know that with facts it is a much simpler and more natural thing to keep everything with a single motion than to introduce two, whether one wants to call them contrary or opposite. But I do not assume the introduction of two to be impossible, nor do I pretend to draw a necessary proof from this; merely a greater probability. The improbability is shown for a third time in the relative disruption of the order which we surely see existing among those heavenly bodies whose circulation is not doubtful, but most certain. This order is such that the greater orbits complete their revolutions in longer times, and the lesser in shorter; thus Saturn, describing a greater circle than the other planets, completes it in thirty years; Jupiter revolves in its smaller one in twelve years, Mars in two; the moon covers its much smaller circle in a single month. And we see no less sensibly that of the satellites of Jupiter (*stelle Medicee*), the closest one to that planet makes its revolution in a very short time, that is in about forty-two hours; the next, in three and a half days; the third in seven days and the most distant in sixteen. And this very harmonious trend will not be a bit altered if the earth is made to move on itself in twenty-four hours. But if the earth is desired to remain motionless, it is necessary, after passing from the brief period of the moon to the other consecutively larger ones, and ultimately to that of Mars in two years, and the greater one of Jupiter in twelve, and from this to the still larger one of Saturn, whose period is thirty years – it is necessary, I say, to pass on beyond to another incomparably larger sphere, and make this one finish an entire revolution in twenty-four hours. Now this is the minimum disorder that can be introduced, for if one wished to pass from Saturn's sphere to the stellar, and make the latter so much greater than Saturn's that it would proportionally be suited to a very slow motion of many thousands of years, a much greater leap would be required to pass beyond that to a still larger one and then make that revolve in twenty-four hours. But by giving mobility to the earth, order becomes very well observed among the periods; from the very slow sphere of Saturn one passes on to the entirely immovable fixed stars, and manages to escape a fourth difficulty necessitated by supposing the stellar sphere to be movable. This difficulty is the immense disparity between the motions of the stars, some of which would be moving very rapidly in vast circles, and others very slowly in little tiny circles, according as they are located farther from or closer to the poles. This is indeed a nuisance, for just as we see that all those bodies whose motion is undoubted move in large circles, so it would not seem to have been good judgment to arrange bodies in such a way that they must move circularly at immense distances from the center, and then make them move in little tiny circles.

Not only will the size of the circles and consequently the velocities of motion of these stars be very diverse from the orbits and motions of some others, but (and this shall be the fifth difficulty) the same stars will keep changing their circles and their velocities, since those which two thousand years ago were on the celestial equator, and which consequently described great circles with their motion, are found in our time to be many degrees distant, and must be made slower in motion and reduced to moving in smaller circles. Indeed, it is not impossible that a time will come when some of the stars which in the past have always been moving will be reduced, by reaching the pole, to holding fast, and then after that time will start moving once more; whereas all those stars which certainly do move describe, as I said, very large circles in their orbits and are unchangeably preserved in them.

For anyone who reasons soundly, the unlikelihood is increased – and this is the sixth difficulty – by the incomprehensibility of what is called the "solidity" of that very vast sphere in whose depths are firmly fixed so many stars which, without changing place in the least among themselves, come to be carried around so harmoniously with such a disparity of motions. If, however, the heavens are fluid (as may much more reasonably be believed) so that each star roves around in it by itself, what law will regulate their motion so that as seen from the earth they shall appear as if made into a single sphere? For this to happen, it seems to me that it is as much more effective and convenient to make them immovable than to have them roam around, as it is easier to count the myriad tiles set in a courtyard than to number the troop of children running around on them.

Finally, for the seventh objection, if we attribute the diurnal rotation to the highest heaven, then this has to be made of such strength and power as to carry with it the innumerable host of fixed stars, all of them vast bodies and much larger than the earth, as well as to carry along the planetary orbs despite the fact that the two move naturally in opposite ways. Besides this, one must grant that the element of fire and the greater part of the air are likewise hurried along, and that only the little body of the earth remains defiant and resistant to such power. This seems to me to be most difficult; I do not understand why the earth, a suspended body balanced on its center and indifferent to motion or to rest, placed in and surrounded by an enclosing fluid, should not give in to such force and be carried around too. We encounter no such objections if we give the motion to the earth, a small and trifling body in comparison with the universe, and hence unable to do it any violence.

Sagr. I am aware of some ideas whirling around in my own imagination which have been confusedly roused in me by these arguments. If I wish to keep my attention on the things about to be said, I shall have to try to get them in better order and to place the proper construction upon them, if possible. Perhaps it will help me to express myself more easily if I proceed by interrogation. Therefore I ask Simplicio, first, whether he believes that the same simple movable body can naturally partake of diverse movements, or whether only a single motion suits it, this being its own natural one?

Simp. For a simple movable body there can be but a single motion, and no more, which suits it naturally; any others it can possess only incidentally and by participation. Thus when a man walks along the deck of a ship, his own motion is that of walking, while the motion which takes him to port is his by participation; for he could never arrive there by walking if the ship did not take him there by means of its motion.

Sagr. Second, tell me about this motion which is communicated to a movable body by participation, when it itself is moved by some other motion different from that in which it participates. Must this shared motion in turn reside in some subject, or can it indeed exist in nature without other support?

Simp. Aristotle answers all these questions for you. He tells you that just as there is only one motion for one movable body, so there is but one movable body for that motion. Consequently no motion can either exist or even be imagined except as inhering in its subject.

Sagr. Now in the third place I should like you to tell me whether you believe that the moon and the other planets and celestial bodies have their own motions, and what these are.

Simp. They have, and they are those motions in accordance with which they run through the zodiac – the moon in a month, the sun in a year, Mars in two, the stellar sphere in so many thousands. These are their own natural motions.

Sagr. Now as to that motion with which the fixed stars, and with them all the planets, are seen rising and setting and returning to the east every twenty-four hours. How does that belong to them?

Simp. They have that by participation.

Sagr. Then it does not reside in them; and neither residing in them, nor being able to exist without some subject to reside in, it must be made the proper and natural motion of some other sphere.

Simp. As to this, astronomers and philosophers have discovered another very high sphere, devoid of stars, to which the diurnal rotation naturally belongs. To this they have given the name *primum mobile*; this speeds along with it all the inferior spheres, contributing to and sharing with them its motion.

Sagr. But when all things can proceed in most perfect harmony without introducing other huge and unknown spheres; without other movements or imparted speedings; with every sphere having only its simple motion, unmixed with contrary movements, and with everything taking place in the same direction, as must be the case if all depend upon a single principle, why reject the means of doing this, and give assent to such outlandish things and such labored conditions?

2.10

The Ship and the Tower

Galileo Galilei

In the following selection, also from the second day of Galileo's *Dialogue*, Salviati attacks the argument that we can tell from watching the fall of a stone that the earth does not move. He first tricks Simplicio into accepting the analogy between a stone dropped from the mast of a moving ship and a stone dropped from a tower on the moving earth; then he springs the trap, agreeing that they are analogous but stunning Simplicio with the claim that the stone falls at the base of the mast even on a moving ship. The upshot is that the observation in question, that falling stones strike the ground at the base of the tower from which they are dropped, proves nothing either way. Because of the relativity of observed motion, the argument from the fall of a stone to an unmoving earth is undermined.

Salv. [. . .] We may go on therefore to the fourth, with which it will be proper to deal at length, this being founded upon that experience from which most of the remaining arguments derive their force. Aristotle says, then, that a most certain proof of the earth's being motionless is that things projected perpendicularly upward are seen to return by the same line to the same place from which they were thrown, even though the movement is extremely high. This, he argues, could not happen if the earth moved, since in the time during which the projectile is moving upward and then downward it is separated from the earth, and the place from which the projectile began its motion would go a long way toward the east, thanks to the revolving of the earth, and the falling projectile would strike the earth that distance away from the place in question. Thus we can accommodate here the argument of the cannon ball as well as the other argument, used by Aristotle and Ptolemy, of seeing heavy bodies falling from great heights along a straight line perpendicular to the surface of the earth. Now, in order to begin to untie these knots, I ask Simplicio by

From Galileo Galilei, *Dialogue Concerning the Two Chief World Systems: Ptolemaic and Copernican.* Trans. with notes by Stillman Drake (New York: The Modern Library, 1953), pp. 161–73.

what means he would prove that freely falling bodies go along straight and perpendicular lines directed toward the center, should anyone refuse to grant this to Aristotle and Ptolemy.

Simp. By means of the senses, which assure us that the tower is straight and perpendicular, and which show us that a falling stone goes along grazing it, without deviating a hairsbreadth to one side or the other, and strikes at the foot of the tower exactly under the place from which it was dropped.

Salv. But if it happened that the earth rotated, and consequently carried along the tower, and if the falling stone were seen to graze the side of the tower just the same, what would its motion then have to be?

Simp. In that case one would have to say "its motions," for there would be one with which it went from top to bottom, and another one needed for following the path of the tower.

Salv. The motion would then be a compound of two motions; the one with which it measures the tower, and the other with which it follows it. From this compounding it would follow that the rock would no longer describe that simple straight perpendicular line, but a slanting one, and perhaps not straight.

Simp. I don't know about its not being straight, but I understand well enough that it would have to be slanting, and different from the straight perpendicular line it would describe with the earth motionless.

Salv. Hence just from seeing the falling stone graze the tower, you could not say for sure that it described a straight and perpendicular line, unless you first assumed the earth to stand still.

Simp. Exactly so; for if the earth were moving, the motion of the stone would be slanting and not perpendicular.

Salv. Then here, clear and evident, is the paralogism of Aristotle and of Ptolemy, discovered by you yourself. They take as known that which is intended to be proved.

Simp. In what way? It looks to me like a syllogism in proper form, and not a *petitio principii.*

Salv. In this way: Does he not, in his proof, take the conclusion as unknown?

Simp. Unknown, for otherwise it would be superfluous to prove it.

Salv. And the middle term; does he not require that to be known?

Simp. Of course; otherwise it would be an attempt to prove *ignotum per aeque ignotum.*

Salv. Our conclusion, which is unknown and is to be proved; is this not the motionlessness of the earth?

Simp. That is what it is.

Salv. Is not the middle term, which must be known, the straight and perpendicular fall of the stone?

Simp. That is the middle term.

Salv. But wasn't it concluded a little while ago that we could not have any knowledge of this fall being straight and perpendicular unless it was first known that the earth stood still? Therefore in your syllogism, the certainty of the middle term is drawn from the uncertainty of the conclusion. Thus you see how, and how badly, it is a paralogism.

Sagr. On behalf of Simplicio I should like, if possible, to defend Aristotle, or at least to be better persuaded as to the force of your deduction. You say that seeing the stone graze the tower is not enough to assure us that the motion of the rock is perpendicular (and this is the middle term of the syllogism) unless one assumes the earth to stand still (which is the conclusion to be proved). For if the tower moved along with the earth and the rock grazed it, the motion of the rock would be slanting, and not perpendicular. But I reply that if the tower were moving, it would be impossible for the rock to fall grazing it; therefore, from the scraping fall is inferred the stability of the earth.

Simp. So it is. For to expect the rock to go grazing the tower if that were carried along by the earth would be requiring the rock to have two natural motions; that is, a straight one toward the center, and a circular one about the center, which is impossible.

Salv. So Aristotle's defense consists in its being impossible, or at least in his having considered it impossible, that the rock might move with a motion mixed of straight and circular. For if he had not held it to be impossible that the stone might move both toward and around the center at the same time, he would have understood how it could happen that the falling rock might go grazing the tower whether that was moving or was standing still, and consequently he would have been able to perceive that this grazing could imply nothing as to the motion or rest of the earth.

Nevertheless this does not excuse Aristotle, not only because if he did have this idea he ought to have said so, it being such an important point in the argument, but also, and more so, because it cannot be said either that such an effect is impossible or that Aristotle considered it impossible. The former cannot be said because, as I shall shortly prove to you, this is not only possible but necessary; and the latter cannot be said either, because Aristotle himself admits that fire moves naturally upward in a straight line and also turns in the diurnal motion which is imparted by the sky to all the element of fire and to the greater part of the air. Therefore if he saw no impossibility in the mixing of straight-upward with circular motion, as communicated to fire and to the air up as far as the moon's orbit, no more should he deem this impossible with regard to the rock's straight-downward motion and the circular motion natural to the entire globe of the earth, of which the rock is a part.

Simp. It does not look that way to me at all. If the element of fire goes around together with the air, this is a very easy and even a necessary thing for a particle of fire, which, rising high from the earth, receives that very motion in passing through the moving air, being so tenuous and light a body and so easily moved. But it is quite incredible that a very heavy rock or a cannon ball which is dropping without restraint should let itself be budged by the air or by anything else. Besides which, there is the very appropriate experiment of the stone dropped from the top of the mast of a ship, which falls to the foot of the mast when the ship is standing still, but falls as far from that same point when the ship is sailing as the ship is perceived to have advanced during the time of the fall, this being several yards when the ship's course is rapid.

Salv. There is a considerable difference between the matter of the ship and that of the earth under the assumption that the diurnal motion belongs to the terrestrial globe. For it is quite obvious that just as the motion of the ship is not its natural one, so the motion of all the things in it is accidental; hence it is no wonder that this stone which was held at the top of the mast falls down when it is set free, without any compulsion to follow the motion of the ship. But the diurnal rotation is being taken as the terrestrial globe's own and natural motion, and hence that of all its parts, as a thing indelibly impressed upon them by nature. Therefore the rock at the top of the tower has as its primary tendency a revolution about the center of the whole in twenty-four hours, and it eternally exercises this natural propensity no matter where it is placed. To be convinced of this, you have only to alter an outmoded impression made upon your mind, saying: "Having thought until now that it is a property of the earth's globe to remain motionless with respect to its center, I have never had any difficulty in or resistance to understanding that each of its particles also rests naturally in the same quiescence. Just so, it ought to be that if the natural tendency of the earth were to go around its center in twenty-four hours, each of its particles would also have an inherent and natural inclination not to stand still but to follow that same course."

And thus without encountering any obstacle you would be able to conclude that since the motion conferred by the force of the oars upon a boat, and through the boat upon all things contained in it, is not natural but foreign to them, then it might well be that this rock, once separated from the boat, is restored to its natural state and resumes its exercise of the simple tendency natural to it.

I might add that at least that part of the air which is lower than the highest mountains must be swept along and carried around by the roughness of the earth's surface, or must naturally follow the diurnal motion because of being a mixture of various terrestrial vapors and exhalations. But the air around a boat propelled by oars is not moved by them. So arguing from the boat to the tower has no inferential force. The rock coming from the top of the mast enters a medium which does not have the motion of the boat; but that which leaves the top of the tower finds itself in a medium which has the same motion as the entire terrestrial globe, so that far from being impeded by the air, it rather follows the general course of the earth with assistance from the air.

Simp. I am not convinced that the air could impress its own motion upon a huge stone or a large ball of iron or lead weighing, say, two hundred pounds, as it might upon feathers, snow, and other very light bodies. In fact, I can see that a weight of that sort does not move a single inch from its place even when exposed to the wildest wind you please; now judge whether the air alone would carry it along.

Salv. There is an enormous difference between this experience of yours and our example. You make the wind arrive upon this rock placed at rest, and we are exposing to the already moving air a rock which is also moving with the same speed, so that the air need not confer upon it some new motion, but merely maintain – or rather, not impede – what it already has. You want to drive the rock with a motion foreign and unnatural to it; we, to conserve its natural motion in it. If you want to present a more suitable experiment, you ought to say what would be observed (if not with one's actual eyes, at least with those of the mind) if an eagle, carried by the force of the wind, were to drop a rock from its talons. Since this rock was already flying equally with the wind, and thereafter entered into a medium moving with the same velocity, I am pretty sure that it would not be seen to fall perpendicularly, but, following the course of the wind and adding to this that of its own weight, would move in a slanting path.

Simp. It would be necessary to be able to make such an experiment and then to decide according to the result. Meanwhile, the result on shipboard confirms my opinion up to this point.

Salv. You may well say "up to this point," since perhaps in a very short time it will look different. And to keep you no longer on tenterhooks, as the saying goes, tell me, Simplicio: Do you feel convinced that the experiment on the ship squares so well with our purpose that one may reasonably believe that whatever is seen to occur there must also take place on the terrestrial globe?

Simp. So far, yes; and though you have brought up some trivial disparities, they do not seem to me of such moment as to suffice to shake my conviction.

Salv. Rather, I hope that you will stick to it, and firmly insist that the result on the earth must correspond to that on the ship, so that when the latter is perceived to be prejudicial to your case you will not be tempted to change your mind.

You say, then, that since when the ship stands still the rock falls to the foot of the mast, and when the ship is in motion it falls apart from there, then conversely, from the falling of the rock at the foot it is inferred that the ship stands still, and from its falling away it may be deduced that the ship is moving. And since what happens on the ship must likewise happen on the land, from the falling of the rock at the foot of the tower one necessarily infers the immobility of the terrestrial globe. Is that your argument?

Simp. That is exactly it, briefly stated, which makes it easy to understand.

Salv. Now tell me: If the stone dropped from the top of the mast when the ship was sailing rapidly fell in exactly the same place on the ship to which it fell when the ship was standing still, what use could you make of this falling with regard to determining whether the vessel stood still or moved?

Simp. Absolutely none; just as by the beating of the pulse, for instance, you cannot know whether a person is asleep or awake, since the pulse beats in the same manner in sleeping as in waking.

Salv. Very good. Now, have you ever made this experiment of the ship?

Simp. I have never made it, but I certainly believe that the authorities who adduced it had carefully observed it. Besides, the cause of the difference is so exactly known that there is no room for doubt.

Salv. You yourself are sufficient evidence that those authorities may have offered it without having performed it, for you take it as certain without having done it, and commit yourself to the good faith of their dictum. Similarly it not only may be, but must be that they did the same thing too – I mean, put faith in their predecessors, right on back without ever arriving at anyone who had performed it. For anyone who does will find that the experiment shows exactly the opposite of what is written; that is, it will show that the stone always falls in the same place on the ship, whether the ship is standing still or moving with any speed you please. Therefore, the same cause holding good on the earth as on the ship, nothing can be inferred about the earth's motion or rest from the stone falling always perpendicularly to the foot of the tower.

Simp. If you had referred me to any other agency that experiment, I think that our dispute would not soon come to an end; for this appears to me to be a thing so remote from human reason that there is no place in it for credulity or probability.

Salv. For me there is, just the same.

Simp. So you have not made a hundred tests, or even one? And yet you so freely declare it to be certain? I shall retain my incredulity, and my own confidence that the experiment has been made by the most important authors who make use of it, and that it shows what they say it does.

Salv. Without experiment, I am sure that the effect will happen as I tell you, because it must happen that way; and I might add that you yourself also know that it cannot happen otherwise, no matter how you may pretend not to know it – or give that impression. But I am so handy at picking people's brains that I shall make you confess this in spite of yourself.

Sagredo is very quiet; it seemed to me that I saw him move as though he were about to say something.

Sagr. I was about to say something or other, but the interest aroused in me by hearing you threaten Simplicio with this sort of violence in order to reveal the knowledge he is trying to hide has deprived me of any other desire; I beg you to make good your boast.

Salv. If only Simplicio is willing to reply to my interrogation, I cannot fail.

Simp. I shall reply as best I can, certain that I shall be put to little trouble; for of the things I hold to be false, I believe I can know nothing, seeing that knowledge is of the true and not of the false.

Salv. I do not want you to declare or reply anything that you do not know for certain. Now tell me: Suppose you have a plane surface as smooth as a mirror and made of some hard material like steel. This is not parallel to the horizon, but somewhat inclined, and upon it you have placed a ball which is perfectly spherical and of some hard and heavy material like bronze. What do you believe this will do when released? Do you not think, as I do, that it will remain still?

Simp. If that surface is tilted?

Salv. Yes, that is what was assumed.

Simp. I do not believe that it would stay still at all; rather, I am sure that it would spontaneously roll down.

Salv. Pay careful attention to what you are saying, Simplicio, for I am certain that it would stay wherever you placed it.

Simp. Well, Salviati, so long as you make use of assumptions of this sort I shall cease to be surprised that you deduce such false conclusions.

Salv. Then you are quite sure that it would spontaneously move downward?

Simp. What doubt is there about this?

Salv. And you take this for granted not because I have taught it to you – indeed, I have tried to persuade you to the contrary – but all by yourself, by means of your own common sense.

Simp. Oh, now I see your trick; you spoke as you did in order to get me out on a limb, as the common people say, and not because you really believed what you said.

Salv. That was it. Now how long would the ball continue to roll, and how fast? Remember that I said a perfectly round ball and a highly polished surface, in order to remove all external and accidental impediments. Similarly I want you to take away any impediment of the air caused by its resistance to separation, and all other accidental obstacles, if there are any.

Simp. I completely understood you, and to your question I reply that the ball would continue to move indefinitely, as far as the slope of the surface extended, and with a continually accelerated motion. For such is the nature of heavy bodies, which *vires acquirunt eundo*; and the greater the slope, the greater would be the velocity.

Salv. But if one wanted the ball to move upward on this same surface, do you think it would go?

Simp. Not spontaneously, no; but drawn or thrown forcibly, it would.

Salv. And if it were thrust along with some impetus impressed forcibly upon it, what would its motion be, and how great?

Simp. The motion would constantly slow down and be retarded, being contrary to nature, and would be of longer or shorter duration according to the greater or lesser impulse and the lesser or greater slope upward.

Salv. Very well; up to this point you have explained to me the events of motion upon two different planes. On the downward inclined plane, the heavy moving body spontaneously descends and continually accelerates, and to keep it at rest requires the use of force. On the upward slope, force is needed to thrust it along or even to hold it still, and motion which is impressed upon it continually diminishes until it is entirely annihilated. You say also that a difference in the two instances arises from the greater or lesser upward or downward slope of the plane, so that from a greater slope downward there follows a greater speed, while on the contrary upon the upward slope a given movable body thrown with a given force moves farther according as the slope is less.

Now tell me what would happen to the same movable body placed upon a surface with no slope upward or downward.

Simp. Here I must think a moment about my reply. There being no downward slope, there can be no natural tendency toward motion; and there being no upward slope, there can be no resistance to being moved, so there would be an indifference between the propensity and the resistance to motion. Therefore it seems to me that it ought naturally to remain stable. But I forgot; it was not so very long ago that Sagredo gave me to understand that that is what would happen.

Salv. I believe it would do so if one set the ball down firmly. But what would happen if it were given an impetus in any direction?

Simp. It must follow that it would move in that direction.

Salv. But with what sort of movement? One continually accelerated, as on the downward plane, or increasingly retarded as on the upward one?

Simp. I cannot see any cause for acceleration or deceleration, there being no slope upward or downward.

Salv. Exactly so. But if there is no cause for the ball's retardation, there ought to be still less for its coming to rest; so how far would you have the ball continue to move?

Simp. As far as the extension of the surface continued without rising or falling.

Salv. Then if such a space were unbounded, the motion on it would likewise be boundless? That is, perpetual?

Simp. It seems so to me, if the movable body were of durable material.

Salv. That is of course assumed, since we said that all external and accidental impediments were to be removed, and any fragility on the part of the moving body would in this case be one of the accidental impediments.

Now tell me, what do you consider to be the cause of the ball moving spontaneously on the downward inclined plane, but only by force on the one tilted upward?

Simp. That the tendency of heavy bodies is to move toward the center of the earth, and to move upward from its circumference only with force; now the downward surface is that which gets closer to the center, while the upward one gets farther away.

Salv. Then in order for a surface to be neither downward nor upward, all its parts must be equally distant from the center. Are there any such surfaces in the world?

Simp. Plenty of them; such would be the surface of our terrestrial globe if it were smooth, and not rough and mountainous as it is. But there is that of the water, when it is placid and tranquil.

Salv. Then a ship, when it moves over a calm sea, is one of these movables which courses over a surface that is tilted neither up nor down, and if all external and accidental obstacles were removed, it would thus be disposed to move incessantly and uniformly from an impulse once received?

Simp. It seems that it ought to be.

Salv. Now as to that stone which is on top of the mast; does it not move, carried by the ship, both of them going along the circumference of a circle about its center? And consequently is there not in it an ineradicable motion, all external impediments being removed? And is not this motion as fast as that of the ship?

Simp. All this is true, but what next?

Salv. Go on and draw the final consequence by yourself, if by yourself you have known all the premises.

Simp. By the final conclusion you mean that the stone, moving with an indelibly impressed motion, is not going to leave the ship, but will follow it, and finally will fall at the same place where it fell when the ship remained motionless. And I, too, say that this would follow if there were no external impediments to disturb the motion of the stone after it was set free. But there are two such impediments; one is the inability of the movable body to split the air with its own impetus alone, once it has lost the force from the oars which it shared as part of the ship while it was on the mast; the other is the new motion of falling downward, which must impede its other, forward, motion.

Salv. As for the impediment of the air, I do not deny that to you, and if the falling body were of very light material, like a feather or a tuft of wool, the retardation would be quite considerable. But in a heavy stone it is insignificant, and if, as you yourself just said a little while ago, the force of the wildest wind is not enough to move a large stone from its place, just imagine how much the quiet air could accomplish upon meeting a rock which moved no faster than the ship! All the same, as I said, I concede to you the small effect which may depend upon such an impediment, just as I know you will concede to me that if the air were moving at the same speed as the ship and the rock, this impediment would be absolutely nil.

As for the other, the supervening motion downward, in the first place it is obvious that these two motions (I mean the circular around the center and the straight motion toward the center) are not contraries, nor are they destructive of one another, nor incompatible. As to the moving body, it has no resistance whatever to such a motion, for you yourself have already granted the resistance to be against motion which increases the distance from the center, and the tendency to be toward motion which approaches the center. From this it follows necessarily that the moving body has neither a resistance nor a propensity to motion which does not approach toward or depart from the center, and in consequence no cause for diminution in the property impressed upon it. Hence the cause of motion is not a single one which must be weakened by the new action, but there exist two distinct causes. Of these, heaviness attends only to the drawing of the movable body toward the center, and impressed force only to its being led around the center, so no occasion remains for any impediment.

2.11

The Copernican View Vindicated

Galileo Galilei

In this selection from the third day of Galileo's *Dialogue*, Salviati lays out three observational difficulties for the Copernican view and then answers them. The passage where he praises Copernicus, who did not have a telescope at his disposal, for following reason against the evidence of sensible experience is particularly interesting. It shows that Galileo is not a naive empiricist and that he understands that even a true theory may for a time be beset with anomalies. The selection concludes with a geometric discussion of how the Copernican view accounts elegantly for the appearance of retrograde motion – the same feature Rheticus singled out in the *Narratio Prima* when he described the "golden chain."

Salv. A while ago I sketched for you an outline of the Copernican system, against the truth of which the planet Mars launches a ferocious attack. For if it were true that the distances of Mars from the earth varied as much from minimum to maximum as twice the distance from the earth to the sun, then when it is closest to us its disc would have to look sixty times as large as when it is most distant. Yet no such difference is to be seen. Rather, when it is in opposition to the sun and close to us, it shows itself as only four or five times as large as when, at conjunction, it becomes hidden behind the rays of the sun.

Another and greater difficulty is made for us by Venus, which, if it circulates around the sun as Copernicus says, would be now beyond it and now on this side of it, receding from and approaching toward us by as much as the diameter of the circle it describes. Then when it is beneath the sun and very close to us, its disc ought to appear to us a little less than forty times as large as when it is beyond the sun and near conjunction. Yet the difference is almost imperceptible.

Add to these another difficulty; for if the body of Venus is intrinsically dark, and like the moon it shines only by illumination from the sun,

From Galileo Galilei, *Dialogue Concerning the Two Chief World Systems: Ptolemaic and Copernican.* Trans. with notes by Stillman Drake (New York: The Modern Library, 1953), pp. 388–400.

which seems reasonable, then it ought to appear horned when it is beneath the sun, as the moon does when it is likewise near the sun – a phenomenon which does not make itself evident in Venus. For that reason, Copernicus declared that Venus was either luminous in itself or that its substance was such that it could drink in the solar light and transmit this through its entire thickness in order that it might look resplendent to us. In this manner Copernicus pardoned Venus its unchanging shape, but he said nothing about its small variation in size; much less of the requirements of Mars. I believe this was because he was unable to rescue to his own satisfaction an appearance so contradictory to his view; yet being persuaded by so many other reasons, he maintained that view and held it to be true.

Besides these things, to have all the planets move around together with the earth, the sun being the center of their rotations, then the moon alone disturbing this order and having its own motion around the earth (going around the sun in a year together with the earth and the whole elemental sphere) seems in some way to upset the whole order and to render it improbable and false.

These are the difficulties which make me wonder at Aristarchus and Copernicus. They could not have helped noticing them, without having been able to resolve them; nevertheless they were confident of that which reason told them must be so in the light of many other remarkable observations. Thus they confidently affirmed that the structure of the universe could have no other form than that which they had described. Then there are other very serious but beautiful problems which are not easy for ordinary minds to resolve, but which were seen through and explained by Copernicus; these we shall put off until we have answered the objections of people who show themselves hostile to this position.

Coming now to the explanations and replies to the three grave objections mentioned, I say that the first two are not only not contrary to the Copernican system, but that they absolutely favor it, and greatly. For both Mars and Venus do show themselves variable in the assigned proportions, and Venus does appear horned when beneath the sun, and changes her shape in exactly the same way as the moon.

Sagr. But if this was concealed from Copernicus, how is it revealed to you?

Salv. These things can be comprehended only through the sense of sight, which nature has not granted so perfect to men that they can succeed in discerning such distinctions. Rather, the very instrument of seeing introduces a hindrance of its own. But in our time it has pleased God to concede to human ingenuity an invention so wonderful as to have the power of increasing vision four, six, ten, twenty, thirty, and forty times, and an infinite number of objects which were invisible, either because of distance or extreme minuteness, have become visible by means of the telescope.

Sagr. But Venus and Mars are not objects which are invisible because of any distance or small size. We perceive these by simple natural vision. Why, then, do we not discern the differences in their sizes and shapes?

Salv. In this the impediment of our eyes plays a large part, as I have just hinted to you. On account of that, bright distant objects are not represented to us as simple and plain, but are festooned with adventitious and alien rays which are so long and dense that the bare bodies are shown as expanded ten, twenty, a hundred, or a thousand times as much as would appear to us if the little radiant crown which is not theirs were removed.

Sagr. Now I recall having read something of the sort, but I don't remember whether it was in the *Solar Letters* or in *Il Saggiatore* by our friend. It would be a good thing, in order to refresh my memory as well as to inform Simplicio, who perhaps has not read those works, to explain to us in more detail how the matter stands. For I should think that a knowledge of this would be most essential to an understanding of what is now under discussion.

Simp. Everything that Salviati is presently setting forth is truly new to me. Frankly, I had no interest in reading those books, nor up till now have I put any faith in the newly introduced optical device. Instead, following in the footsteps of other Peripatetic philosophers of my group, I have considered as fallacies and deceptions of the lenses those things which other people have admired as stupendous achievements. If I have been in error, I shall be glad to be lifted out of it; and, charmed by the other new things I have heard from you, I shall listen most attentively to the rest.

Salv. The confidence which men of that stamp have in their own acumen is as unreasonable as the small regard they have for the judgments of others. It is a remarkable thing that they should think themselves better able to judge such an instrument without ever having tested it, than those who have made thousands and thousands of experiments with it and make them every day. But let us forget about such headstrong people, who cannot even be censured without doing them more honor than they deserve.

Getting back to our purpose, I say that shining objects, either because their light is refracted in the moisture that covers the pupil, or because it is reflected from the edges of the eyelids and these reflected rays are diffused over the pupil, or for some other reason, appear to our eyes as if surrounded by new rays. Hence these bodies look much larger than they would if they were seen by us deprived of such irradiations. This enlargement is made in greater and greater proportion as such luminous objects become smaller and smaller, in exactly such a manner as if we were to suppose a growth of shining hair, say four inches long, to be added around a circle four inches in diameter, which would increase its apparent size nine times; but . . .

Simp. I think you meant to say "three times," since four inches added on each side of a circle four inches in diameter would amount to tripling its magnitude and not to enlarging it nine times.

Salv. A little geometry, Simplicio; it is true that the diameter increases three times, but the surface (which is what we are talking about) grows nine times. For the surfaces of circles, Simplicio, are to each other as the squares of their diameters, and a circle four inches in diameter has to another of twelve inches the same ratio which the square of four has to the square of twelve; that is, 16 to 144. Therefore it will be nine times as large, not three. This is for your information, Simplicio.

Now, to continue, if we add this coiffure of four inches to a circle of only two inches in diameter, the diameter of the crown will be ten inches and the ratio of the circle to the bare body will be as 100 to 4 (for such are the squares of 10 and of 2), so the enlargement would be twenty-five times. And finally, the four inches of hair added to a tiny circle of one inch in diameter would enlarge this eighty-one times. Thus the increase is continually made larger and larger proportionately, according as the real objects which are increased become smaller and smaller.

Sagr. The question which gave Simplicio trouble did not really bother me, but there are some other things about which I desire a clearer explanation. In particular I should like to learn the basis upon which you affirm such a growth to be always equal in all visible objects.

Salv. I have already partly explained by saying that only luminous objects increase; not dark ones. Now I shall add the rest. Of shining objects, those which are brightest in light make the greatest and strongest reflections upon our pupils, thereby showing themselves as much more enlarged than those less bright. And so as not to go on too long about this detail, let us resort to what is shown us by our greatest teacher; this evening, when the sky is well darkened, let us look at Jupiter; we shall see it very radiant and large. Then let us cause our vision to pass through a tube, or even through a tiny opening which we may leave between the palm of our hand and our fingers, clenching the fist and bringing it to the eye; or through a hole made by a fine needle in a card. We shall see the disc of Jupiter deprived of rays and so very small that we shall indeed judge it to be even less than one-sixtieth of what had previously appeared to us to be a great torch when seen with the naked eye. Afterwards, we may look at the Dog Star, a very beautiful star and larger than any other fixed star. To the naked eye it looks to be not much smaller than Jupiter, but upon taking away its headdress in the manner described above, its disc will be seen to be so small that one would judge it to be no more than one-twentieth the size of Jupiter. Indeed, a person lacking perfect vision will be able to find it only with great difficulty, from which it may reasonably be inferred that this star is one which has a great deal more luminosity than Jupiter, and makes larger irradiations.

Next, the irradiations of the sun and of the moon are as nothing because of the size of these bodies, which by themselves take up so much room in our eye as to leave no place for adventitious rays, so that their discs are seen as shorn and bounded.

We may assure ourselves of the same fact by another experiment which I have made many times – assure ourselves, I mean, that the resplendent bodies of more vivid illumination give out many more rays than those which have only a pale light. I have often seen Jupiter and Venus together, twenty-five or thirty degrees from the sun, the sky being very dark. Venus would appear eight or even ten times as large as Jupiter when looked at with the naked eye. But seen afterward through a telescope, Jupiter's disc would be seen to be actually four or more times as large as Venus. Yet the liveliness of Venus's brilliance was incomparably greater than the pale light of Jupiter, which comes about only because Jupiter is very distant from the sun and from us, while Venus is close to us and to the sun.

These things having been explained, it will not be difficult to understand how it might be that Mars, when in opposition to the sun and therefore seven or more times as close to the earth as when it is near conjunction, looks to us scarcely four or five times as large in the former state as in the latter. Nothing but irradiation is the cause of this. For if we deprive it of the adventitious rays we shall find it enlarged in exactly the proper ratio. And to remove its head of hair from it, the telescope is the unique and supreme means. Enlarging its disc nine hundred or a thousand times, it causes this to be seen bare and bounded like that of the moon, and in the two positions varying in exactly the proper proportion.

Next in Venus, which at its evening conjunction when it is beneath the sun ought to look almost forty times as large as in its morning conjunction, and is seen as not even doubled, it happens in addition to the effects of irradiation that it is sickle-shaped, and its horns, besides being very thin, receive the sun's light obliquely and therefore very weakly. So that because it is small and feeble, it makes its irradiations less ample and lively than when it shows itself to us with its entire hemisphere lighted. But the telescope plainly shows us its horns to be as bounded and distinct as those of the moon, and they are seen to belong to a very large circle, in a ratio almost forty times as great as the same disc when it is beyond the sun, toward the end of its morning appearances.

Sagr. O Nicholas Copernicus, what a pleasure it would have been for you to see this part of your system confirmed by so clear an experiment!

Salv. Yes, but how much less would his sublime intellect be celebrated among the learned! For as I said before, we may see that with reason as his guide he resolutely continued to affirm what sensible experience seemed to contradict. I cannot get over my amazement that he was constantly willing to persist in saying that Venus might go around the sun and be more than six times as far from us at one time as at another, and still look always equal, when it should have appeared forty times larger.

Sagr. I believe then that in Jupiter, Saturn, and Mercury one ought also to see differences of size corresponding exactly to their varying distances.

Salv. In the two outer planets I have observed this with precision in almost every one of the past twenty-two years. In Mercury no observations of importance can be made, since it does not allow itself to be seen except at its maximum angles with the sun, in which the inequalities of its distances from the earth are imperceptible. Hence such differences are unobservable, and so are its changes of shape, which must certainly take place as in Venus. But when we do see it, it would necessarily show itself to us in the shape of a semicircle, just as Venus does at its maximum angles, though its disc is so small and its brilliance so lively that the power of the telescope is not sufficient to strip off its hair so that it may appear completely shorn.

It remains for us to remove what would seem to be a great objection to the motion of the earth. This is that though all the planets turn about the sun, the earth alone is not solitary like the others, but goes together in the company of the moon and the whole elemental sphere around the sun in one year, while at the same time the moon moves around the earth every month. Here one must once more exclaim over and exalt the admirable perspicacity of Copernicus, and simultaneously regret his misfortune at not being alive in our day. For now Jupiter removes this apparent anomaly of the earth and moon moving conjointly. We see Jupiter, like another earth, going around the sun in twelve years accompanied not by one but by four moons, together with everything that may be contained within the orbits of its four satellites.

Sagr. And what is the reason for your calling the four Jovian planets "moons"?

Salv. That is what they would appear to be to anyone who saw them from Jupiter. For they are dark in themselves, and receive their light from the sun; this is obvious from their being eclipsed when they enter into the cone of Jupiter's shadow. And since only that hemisphere of theirs is illuminated which faces the sun, they always look entirely illuminated to us who are outside their orbits and closer to the sun; but to anyone on Jupiter they would look completely lighted only when they were at the highest points of their circles. In the lowest part – that is, when between Jupiter and the sun – they would appear horned from Jupiter. In a word, they would make for Jovians the same changes of shape which the moon makes for us Terrestrials.

Now you see how admirably these three notes harmonize with the Copernican system, when at first they seemed so discordant with it. From this, Simplicio will be much better able to see with what great probability one may conclude that not the earth, but the sun, is the center of rotation of the planets. And since this amounts to placing the earth among the world bodies which indubitably move about the sun (above Mercury and Venus but beneath Saturn, Jupiter, and Mars), why will it not likewise be probable, or perhaps even necessary, to admit that it also goes around?

Simp. These events are so large and so conspicuous that it is impossible for Ptolemy and his followers not to have had knowledge of them. And having had, they must also have found a way to give reasons sufficient to account for such sensible appearances; congruous and probable reasons, since they have been accepted for so long by so many people.

Salv. You argue well, but you must know that the principal activity of pure astronomers is to give reasons just for the appearances of celestial bodies, and to fit to these and to the motions of the stars such a structure and arrangement of circles that the resulting calculated motions correspond with those same appearances. They are not much worried about admitting anomalies which might in fact be troublesome in other respects. Copernicus himself writes, in his first studies, of having rectified astronomical science upon the old Ptolemaic assumptions, and corrected the motions of the planets in such a way that the computations corresponded much better with the appearances, and vice versa. But this was still taking them separately, planet by planet. He goes on to say that when he wanted to put together the whole fabric from all individual constructions, there resulted a monstrous chimera composed of mutually disproportionate members, incompatible as a whole. Thus however well the astronomer might be satisfied merely as a calculator, there was no satisfaction and peace for the astronomer as a scientist. And since he very well understood that although the celestial appearances might be saved by means of assumptions essentially false in nature, it would be very much better if he could derive them from true suppositions, he set himself to inquiring diligently whether any one among the famous men of antiquity had attributed to the universe a different structure from that of Ptolemy's which is commonly accepted. Finding that some of the Pythagoreans had in particular attributed the diurnal rotation to the earth, and others the annual revolution as well, he began to examine under these two new suppositions the appearances and peculiarities of the planetary motions, all of which he had readily at hand. And seeing that the whole then corresponded to its parts with wonderful simplicity, he embraced this new arrangement, and in it he found peace of mind.

Simp. But what anomalies are there in the Ptolemaic arrangement which are not matched by greater ones in the Copernican?

Salv. The illnesses are in Ptolemy, and the cures for them in Copernicus. First of all, do not all philosophical schools hold it to be a great impropriety for a body having a natural circular movement to move irregularly with respect to its own center and regularly around another point? Yet Ptolemy's structure is composed of such uneven movements, while in the Copernican system each movement is equable around its own center. With Ptolemy it is necessary to assign to the celestial bodies contrary movements, and make everything move from east to west and at the same time from west to east, whereas with Copernicus all celestial revolutions are in one direction, from west to east. And what are we to say of the apparent movement of a planet, so uneven that it not only goes fast at one time and slow at another, but sometimes stops entirely and even goes backward a long way after doing so? To save these appearances, Ptolemy introduces vast epicycles, adapting them one by one to each planet, with certain rules about incongruous motions – all of which can be done away with by one very simple motion of the earth. Do you not think it extremely absurd, Simplicio, that in Ptolemy's construction where all planets are assigned their own orbits, one above another, it should be necessary to say that Mars, placed above the sun's sphere, often falls so far that it breaks through the sun's orb, descends below this and gets closer to the earth than the body of the sun is, and then a little later soars immeasurably above it? Yet these and other anomalies are cured by a single and simple annual movement of the earth.

Sagr. I should like to arrive at a better understanding of how these stoppings, retrograde motions, and advances, which have always seemed to me highly improbable, come about in the Copernican system.

Salv. Sagredo, you will see them come about in such a way that the theory of this alone ought to be enough to gain assent for the rest of the doctrine from anyone who is neither stubborn nor unteachable. I tell you, then, that no change occurs in the movement of Saturn in thirty years, in that of Jupiter in twelve, that of Mars in two, Venus in nine months, or in that of Mercury in about eighty days. The annual movement of the earth alone, between Mars and Venus, causes all the apparent irregularities of the five stars named. For an easy and full understanding of this, I wish to draw you a picture of it. Now suppose the sun to be located in the center O, around which we shall designate the orbit described by the earth with its annual movement, BGM. The circle described by Jupiter (for example) in 12 years will be *BGM* here, and in the stellar sphere we shall take the circle of the zodiac to be *PUA*. In addition, in the earth's annual orbit we shall take a few equal arcs, BC, CD, DE, EF, FG, GH, HI, IK, KL, and LM, and in the circle of Jupiter we shall indicate these other arcs passed over in the same times in which the earth is passing through these. These are *BC*, *CD*, *DE*, *EF*, *FG*, *GH*, *HI*, *IK*, *KL*, and *LM*, which will be proportionately smaller than those noted on the earth's orbit, as the motion of Jupiter through the zodiac is slower than the annual celestial motion.

Now suppose that when the earth is at B, Jupiter is at *B*; then it will appear to us as being in the zodiac at *P*, along the straight line B*BP*. Next let the earth move from B to C and Jupiter from *B* to *C* in the same time; to us, Jupiter will appear to have arrived at *Q* in the zodiac, having advanced in the order of the signs from *P* to *Q*. The earth then passing to D and Jupiter to *D*, it will be seen in the zodiac at *R*; and from E, Jupiter being at *E*, it will appear in the zodiac at *S*, still advancing. But now when the earth begins to get directly between Jupiter and the sun (having arrived at F and Jupiter at *F*), to us Jupiter will appear to be ready to commence returning backward through the zodiac, for during the time in which the earth will have passed through the arc EF, Jupiter will have been slowed down between the points *S* and *T*, and will look to us almost stationary. Later the earth coming to G, Jupiter at *G* (in opposition to the sun) will be seen in the zodiac at *U*, turned far back through the whole arc *TU* in the zodiac; but in reality, following always its uniform course, it has advanced not only in its own circle but in the zodiac too, with respect to the center of the zodiac and to the sun which is located there.

The earth and Jupiter then continuing their movements, when the earth is at H and Jupiter is at *H*, it will be seen as having returned far back through the zodiac by the whole arc *UX*; but the earth having arrived at I and Jupiter at *I*, it will apparently have moved in the zodiac by only the small space *XY*, and will there appear stationary. Then when the earth shall have progressed to K and Jupiter to *K*, Jupiter will have advanced through the arc *YN*, in the zodiac; and, continuing

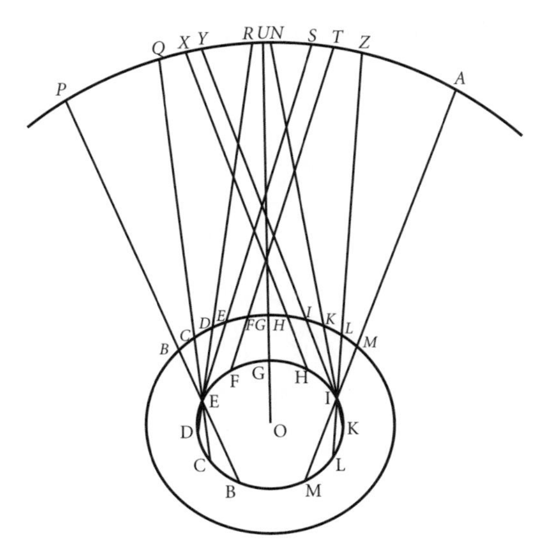

Figure 5 Diagram of the sun and planets

its course, from L the earth will see Jupiter at *L* in the point Z. Finally, Jupiter at *M* will be seen from the earth at M to have passed to *A*, still advancing. And its whole apparent retrograde motion in the zodiac will be as much as the arc *TX*, made by Jupiter while it is passing in its own circle through the arc *FH*, the earth going through FH in its orbit.

Now what is said here of Jupiter is to be understood of Saturn and Mars also. In Saturn these retrogressions are somewhat more frequent than in Jupiter, because its motion is slower than Jupiter's, so that the earth overtakes it in a shorter time. In Mars they are rarer, its motion being faster than that of Jupiter, so that the earth spends more time in catching up with it.

Next, as to Venus and Mercury, whose circles are included within that of the earth, stoppings and retrograde motions appear in them also, due not to any motion that really exists in them, but to the annual motion of the earth. This is acutely demonstrated by Copernicus, enlisting the aid of Apollonius of Perga, in chapter 35 of Book V in his *Revolutions.*

You see, gentlemen, with what ease and simplicity the annual motion – if made by the earth – lends itself to supplying reasons for the apparent anomalies which are observed in the movements of the five planets, Saturn, Jupiter, Mars, Venus, and Mercury. It removes them all and reduces these movements to equable and regular motions; and it was Nicholas Copernicus who first clarified for us the reasons for this marvelous effect.

Figure acknowledgment

Figure 5: from *Dialogue Concerning the Two Chief World Systems, Epigraph and Day 3* from Galileo Galilei, *Dialogue Concerning the Two Chief World Systems: Ptolemaic and Copernican*, trans. with notes by Stillman Drake (New York: The Modern Library), p. 153, copyright renewed 1981 by Stillman Drake, p. 399.

2.12
The "Corpuscular" Philosophy

Robert Boyle

Robert Boyle (1627–1691), an Irish scientist and philosopher, was a child prodigy (by the age of eight he had learned French, Latin and Greek and was sent to Eton College) whose deep contributions to physics and chemistry gave a distinctive shape to the scientific revolution. In this excerpt from his philosophical papers, Boyle gives a clear statement of the principles of the "corpuscular philosophy" – a name he preferred to "atomism" since the latter had connotations he rejected – as applied to the material constitution of bodies. He argues that not everything that we are inclined to call a "quality" of a body is a real, distinct physical entity; many qualities of bodies are due to the arrangement of their parts and to the effects upon them of surrounding bodies.

The Origin of Forms and Qualities According to the Corpuscular Philosophy

[. . .]

That before I descend to particulars I may, Pyrophilus, furnish you with some general apprehension of the doctrine (or rather the *hypothesis*) which is to be collated with, and to be either confirmed or disproved by, the historical truths that will be delivered concerning particular qualities (and forms), I will assume the person of a Corpuscularian, and here at the entrance give you (in a general way) a brief account of the hypothesis itself, as it concerns the origin of qualities (and forms); and for distinction's sake, I shall comprise it in the eight following particulars, which, that the whole scheme may be the better comprehended and, as it were, surveyed under one prospect, I shall do little more than barely propose them that either seem evident enough by their own light, or may without prejudice have divers of their proofs reserved for proper places in the following part of this treatise.

From M. A. Stewart (ed.), *Selected Philosophical Papers of Robert Boyle* (Indianapolis, IN: Hackett, 1991), selections from pp. 18–20, 23–30 (all footnotes cut).

I. I agree with the generality of philosophers, so far as to allow that there is one catholic or universal matter common to all bodies, by which I mean a substance extended, divisible, and impenetrable.

II. But because, this matter being in its own nature but one, the diversity we see in bodies must necessarily arise from somewhat else than the matter they consist of, and since we see not how there could be any change in matter if all its (actual or designable) parts were perpetually at rest among themselves, it will follow that, to discriminate the catholic matter into variety of natural bodies, it must have motion in some or all its designable parts; and that motion must have various tendencies, that which is in this part of the matter tending one way, and that which is in that part tending another: as we plainly see in the universe or general mass of matter there is really a great quantity of motion, and that variously determined, and that yet divers portions of matter are at rest.

That there is local motion in many parts of matter is manifest to sense, but how matter came by this motion was of old, and is still, hotly disputed of. For the ancient Corpuscularian philosophers (whose doctrine in most other points, though not in all, we are the most inclinable to), not acknowledging an Author of the universe, were thereby reduced to make motion congenite to matter, and consequently coeval with it: but since local motion, or an endeavour at it, is not included in the nature of matter, which is as much matter when it rests as when it moves, and since we see that the same portion of matter may from motion be reduced to rest, and, after it hath continued at rest as long as other bodies do not put it out of that state, may by external agents be set a-moving again, I, who am not wont to think a man the worse naturalist for not being an atheist, shall not scruple to say with an eminent philosopher of old – whom I find to have proposed among the Greeks that opinion (for the main) that the excellent Descartes hath revived amongst us – that the origin of motion in matter is from God; and not only so, but that thinking it very unfit to be believed that matter, barely put into motion and then left to itself, should casually constitute this beautiful and orderly world, I think also further that the wise Author of things did, by establishing the laws of motion among bodies, and by guiding the first motions of the small parts of matter, bring them to convene after the manner requisite to compose the world, and especially did contrive those curious and elaborate engines, the bodies of living creatures, endowing most of them with a power of propagating their species. But though these things are my persuasions, yet, because they are not necessary to be supposed here – where I do not pretend to deliver any complete discourse of the principles of natural philosophy, but only to touch upon such notions as are requisite to explicate the origin of qualities and forms – I shall pass on to what remains, as soon as I have taken notice that *local motion seems to be indeed the principal amongst second causes, and the grand agent of all that happens in nature.* For though bulk, figure, rest, situation, and texture, do concur to the phenomena of nature, yet in comparison of motion they seem to be in many cases effects, and in many others little better than *conditions*, or *requisites*, or causes *sine quibus non*, which modify the operation that one part of matter by virtue of its motion hath upon another: as in a watch the number, the figure, and coaptation, of the wheels and other parts, is requisite to the showing the hour and doing the other things that may be performed by the watch, but till these parts be actually put into motion all their other affections remain inefficacious; and so in a key, though if it were too big or too little, or if its shape were incongruous to that of the cavity of the lock, it would be unfit to be used as a key though it were put into motion, yet, let its bigness and figure be never so fit, unless actual motion intervene it will never lock or unlock anything, as without the like actual motion neither a knife nor razor will actually cut, how much soever their shape and other qualities may fit them to do so. And so brimstone, what disposition of parts soever it have to be turned into flame, would never be kindled unless some actual fire or other parcel of vehemently and variously agitated matter should put the sulphureous corpuscles into a very brisk motion.

III. These two grand and most catholic principles of bodies, matter and motion, being thus established, it will follow both that matter must be actually divided into parts, that being the genuine effect of variously determined motion; and that each of the primitive fragments, or other distinct and entire masses of matter, must have two attributes – its own magnitude, or rather *size*, and its own *figure* or *shape*. And since experience shows us (especially that which is afforded us by chemical operations, in many of which matter is divided into parts too small to be singly sensible) that this division of matter is frequently made into insensible corpuscles or particles, we may conclude that the minutest fragments, as well as the biggest masses, of the universal matter are likewise endowed each with its peculiar bulk and shape. For, being a finite body, its dimensions must be terminated and measurable; and though it may change its figure, yet for the same reason it must necessarily have *some figure* or other. So that now we have found out and must admit three essential properties of each entire or undivided, though insensible, part of matter: namely, *magnitude* (by which I mean not quantity in general, but a determined quantity, which we in English oftentimes call the *size* of a body), *shape*, and either *motion* or *rest* (for betwixt them two there is no mean), the two first of which may be called *inseparable accidents* of each distinct part of matter – *inseparable* because, being extended and yet finite, it is physically impossible that it should be devoid of some bulk or other and some determinate shape or other, and yet *accidents* because that, whether or no the shape can by physical agents be altered, or the body subdivided, yet mentally both the one and the other may be done, the whole essence of matter remaining undestroyed.

[...]

We may consider, then, that when Tubal Cain, or whoever else were the smith that invented *locks* and *keys*, had made his first lock (for we may reasonably suppose him to have made that before the *key*, though the comparison may be made use of without that supposition), that was only a piece of iron contrived into such a shape; and when afterwards he made a key to that lock, that also in itself considered was nothing but a piece of iron of such a determinate figure. But in regard that these two pieces of iron might now be applied to one another after a certain manner, and that there was a congruity betwixt the wards of the lock and those of the key, the lock and the key did each of them now obtain a new capacity; and it became a main part of the notion and description of a *lock* that it was capable of being made to lock or unlock by that other piece of iron we call a *key*, and it was looked upon as a peculiar faculty and power in the key that it was fitted to open and shut the lock: and yet by these new attributes there was not added any real or physical entity either to the lock or to the key, each of them remaining indeed nothing but the same piece of iron, just so shaped as it was before. And when our smith made other keys of differing bignesses or with differing wards, though the first lock was not to be opened by any of those keys, yet that indisposition, however it might be considered as a peculiar power of resisting this or that key, and might serve to discriminate it sufficiently from the locks those keys belonged to, was nothing new in the lock, or distinct from the figure it had before those keys were made. To carry this comparison a little further, let me add that, though one that would have defined the first lock and the first key would have given them distinct definitions with reference to each other, yet (as I was saying) these definitions, being given but upon the score of certain respects which the defined bodies had one to another, would not infer that these two iron instruments did physically differ otherwise than in the figure, size, or contrivement, of the iron whereof each of them consisted. And proportionably hereunto, I do not see why we may not conceive that, as to those qualities (for instance) which we call *sensible*, though, by virtue of a certain congruity or incongruity in point of figure or texture (or other mechanical attributes) to our sensories, the portions of matter they modify are enabled to produce various effects upon whose account we make bodies to be endowed with qualities, yet they are not in the bodies that are endowed with them any real or distinct entities, or differing from the matter itself furnished with such a determinate bigness, shape, or other mechanical modifications. Thus, though the modern goldsmiths and refiners reckon amongst the most distinguishing qualities of gold, by which men may be certain of its being true and not sophisticated, that it is easily dissoluble in *aqua regis*, and that *aqua fortis* will not work upon it, yet these attributes are not in the gold anything distinct from its peculiar texture; nor is the gold we have now of any other nature than it was in Pliny's time, when *aqua fortis* and *aqua regis* had not been found out (at least in these parts of the world), and were utterly unknown to the Roman goldsmiths.

And this example I have the rather pitched upon, because it affords me an opportunity to represent that, unless we admit the doctrine I have been proposing, we must admit that a body may have an almost infinite number of new real entities accruing to it without the intervention of any physical change in the body itself: as, for example, gold was the same natural body immediately before *aqua regis* and *aqua fortis* were first made, as it was immediately after, and yet now it is reckoned amongst its principal properties that it is dissoluble by the former of those two menstruums, and that it is not, like other metals, dissoluble or corrodible by the latter. And if one should invent another menstruum (as possibly I may think myself master of such a one), that will but in part dissolve pure gold, and change some part of it into another metalline body, there will then arise another new property whereby to distinguish that from other metals; and yet the nature of gold is not a whit other now than it was before this last menstruum was first made. There are some bodies not cathartic nor sudorific, with some of which gold being joined acquires a purgative virtue, and with others a power to procure sweat. And, in a word, nature herself doth sometimes otherwise, and sometimes by chance, produce so many things that have new relations unto others; and art, especially assisted by chemistry, may, by variously dissipating natural bodies, or compounding either them or their constituent parts with one another, make such an innumerable company of new productions, that will each of them have new operations either immediately upon our sensories or upon other bodies whose changes we are able to perceive, that no man can know but that the most familiar bodies may have multitudes of qualities that he dreams not of: and a considering man will hardly imagine that so numerous a crowd of real physical entities can accrue to a body, whilst in the judgement of all our senses it remains unchanged and the same that it was before.

[...]

And this puts me in mind to add that the multiplicity of qualities that are sometimes to be met with in the same natural bodies needs not make men reject the opinion we have been proposing, by persuading them that so many differing attributes as may be sometimes found in one and the same natural body cannot proceed from the bare texture and other mechanical affections of its matter. For we must consider each body not barely as it is in itself an entire and distinct portion of matter, but as it is a part of the universe, and consequently placed among a great number and variety of other bodies, upon which it may act and by which it may be acted on in many ways (or upon many accounts), each of which men are wont to fancy as a distinct power or quality in the body by which those actions, or in which those passions, are produced. For if we thus consider things, we shall not much wonder that a portion of matter, that is indeed endowed but with a very few mechanical affections – as such a determinate texture and motion – but is placed among a multitude of other bodies that differ in those attributes from it and one another, should be capable of having a great number and variety of relations to those other bodies, and consequently should be thought to have many distinct inherent qualities, by such as look upon those several relations or respects it may have to bodies without it as real and distinct entities implanted in the body itself. When a curious watch is going, though the spring be that which puts all the parts into motion, yet we do not fancy (as an Indian or Chinese would perchance do) in this spring one faculty to move the index uniformly round the dial-plate, another to strike the hour, and perhaps a third to give an alarm, or show the age of the moon or the tides: all the action of the spring (which is but a flexible piece of steel forcibly coiled together) being but an endeavour to dilate or unbind itself, and the rest being performed by the various respects it hath to the several bodies (that compose the watch) among which it is placed, and which they have one to another. We all know that the sun hath a power to harden clay, and soften wax, and melt butter, and thaw ice, and turn water into vapours, and make air expand itself in weather-glasses, and contribute to blanch linen, and make the white skin of the face swarthy and mowed grass yellow, and ripen fruit, hatch the eggs of silk-worms, caterpillars and the like insects, and perform I know not how many other things, divers of which seem contrary effects; and yet these are not distinct powers or faculties in the sun, but only the productions of its heat (which itself is but the brisk and confused local motion of the minute parts of a body), diversified by the differing textures of the body that it chances to work upon, and the condition of the other bodies that are concerned in the operation. And therefore, whether the sun in some cases have any influence at all distinct from its light and heat, we see that all those phenomena we have thought fit to name are producible by the heat of the common culinary fire duly applied and regulated. . . .

I have said thus much, Pyrophilus, to remove the mistake that *everything men are wont to call a quality* must needs be a real and physical entity, because of the importance of the subject. And yet I have omitted some things that might have been pertinently added, partly because I may hereafter have opportunity to take them in, and partly because I would not any farther lengthen this *excursion*, which yet I must not conclude till I have added this short advertisement: that I have chosen to declare what I mean by *qualities* rather by examples than definitions, partly because, being immediately or reductively the objects of sense, men generally understand pretty well what one another mean when they are spoken of (as to say that the taste of such a thing is saline or sour, or that such a sound is melodious, shrill, or jarring (especially if, when we speak of sensible qualities, we add some enumeration of particular subjects wherein they do the most eminently reside), will make a man as soon understood as if he should go about to give logical definitions of those qualities); and partly because the notions of things are not yet so well stated and agreed on but that it is many times difficult to assign their true genuses. And Aristotle himself doth not only define *accidents* without setting down their genus, but when he comes to define *qualities*, he tells us that *quality is that by which a thing is said to be qualis* – where I would have you take notice both that in his definition he omits the genus, and that it is no such easy thing to give a very good definition of qualities, since he that is reputed the great master of logic, where he pretends to give us one, doth but upon the matter define the thing by the same thing; for it is supposed to be as little known what *qualis* is as what *qualitas* is, and methinks he does just as if I should define whiteness to be that for which a thing is called white, or virtue that for which a man is said to be virtuous. Besides that I much doubt whether his definition be not untrue as well as obscure. For to the question '*qualis res est?*' answer may be returned out of *some*, if not *all*, of the other *predicaments of accidents*: which some of the modern logicians being aware of, they have endeavoured to salve the matter with certain cautions and limitations, which however they may argue the devisers to be ingenious, do, for aught I can discern, leave us still to seek for a right and intelligible definition of quality in general; though to give such a one be probably a much easier task than to define many qualities that may be named in particular, as saltness, sourness, green, blue, and many others, which when we hear named, every man knows what is meant by them, though no man (that I know of) hath been able to give accurate definitions of them.

2.13

Successful Hypotheses and High Probability

Christiaan Huygens

Christiaan Huygens (1629–1695) was a Dutch mathematician, astronomer, and physicist who was an older contemporary of both Newton and Leibniz. In the Preface to his *Treatise on Light* (1690), Huygens gives a revealing picture of his conception of scientific reasoning, which does not produce the same sort of certainty as one finds in geometry and yet may rise to a "degree of probability" that is "scarcely less than complete proof." The method he outlines is broadly hypothetico-deductive, but he singles out predictive success for special notice.

Preface

I wrote this Treatise during my sojourn in France twelve years ago, and I communicated it in the year 1678 to the Royal Academy of Science, to the membership of which the King had done me the honour of calling me. Several of that body who are still alive will remember having been present when I read it, and above the rest those amongst them who applied themselves particularly to the study of Mathematics; of whom I cannot cite more than the celebrated gentlemen Cassini, Roemer, and De la Hire. And although I have since corrected and changed some parts, the copies which I had made of it at that time may serve for proof that I have yet added nothing to it save some conjectures touching the formation of Iceland Crystal, and a novel observation on the refraction of Rock Crystal. I have desired to relate these particulars to make known how long I have meditated the things which now I publish, and not for the purpose of detracting from the merit of those who, without having seen anything that I have written, may be found to have treated of like matters: as has in fact occurred in two eminent Geometricians, Messieurs Newton and Leibnitz, with respect to the problem of the figure of glasses for collecting rays when one of the surfaces is given.

One may ask why I have so long delayed to bring this work to the light. The reason is that I wrote it rather carelessly in the Language in which it appears, with the intention of translating it into Latin, so doing in order to obtain greater

From Christiaan Huygens, *Treatise on Light*, trans. Sylvanus P. Thompson (University of Chicago Press, 1912).

attention to the thing. After which I proposed to myself to give it out along with another Treatise on Dioptrics, in which I explain the effects of Telescopes and those things which belong more to that Science. But the pleasure of novelty being past, I have put off from time to time the execution of this design, and I know not when I shall ever come to an end of it, being often turned aside either by business or by some new study. Considering which I have finally judged that it was better worth while to publish this writing, such as it is, than to let it run the risk, by waiting longer, of remaining lost.

There will be seen in it demonstrations of those kinds which do not produce as great a certitude as those of Geometry, and which even differ much therefrom, since whereas the Geometers prove their Propositions by fixed and incontestable Principles, here the Principles are verified by the conclusions drawn from them; the nature of these things not allowing of this being done otherwise. It is always possible to attain thereby to a degree of probability which very often is scarcely less than complete proof. To wit, when things which have been demonstrated by the Principles that have been assumed correspond perfectly to the phenomena which experiment has brought under observation; especially when there are a great number of them, and further, principally, when one can imagine and foresee new phenomena which ought to follow from the hypotheses which one employs, and when one finds that therein the fact corresponds to our prevision. But if all these proofs of probability are met within that which I propose to discuss, as it seems to me there are, this ought to be a very strong confirmation of the success of my inquiry; and it must be ill if the facts are not pretty much as I represent them. I would believe then that those who love to know the Causes of things and who are able to admire the marvels of Light, will find some satisfaction in these various speculations regarding it, and in the new explanation of its famous property which is the main foundation of the construction of our eyes and of those great inventions which extend so vastly the use of them. I hope also that there will be some who by following these beginnings will penetrate much further into this question than I have been able to do, since the subject must be far from being exhausted. This appears from the passages which I have indicated where I leave certain difficulties without having resolved them, and still more from matters which I have not touched at all, such as Luminous Bodies of several sorts, and all that concerns Colours; in which no one until now can boast of having succeeded. Finally, there remains much more to be investigated touching the nature of Light which I do not pretend to have disclosed, and I shall owe much in return to him who shall be able to supplement that which is here lacking to me in knowledge.

The Hague
The 8 January 1690.

2.14

Inductive Methodology

Isaac Newton

Sir Isaac Newton (1642–1727), an English mathematician and physicist, is regarded by many as the greatest scientist of all time. In these selections from the *Optiks* (1706) and from his correspondence we can see that Newton thought explicitly about the proper method in scientific reasoning. The July 6, 1762 note to Henry Oldenburg is particularly noteworthy, since Newton expressly rejects a Baconian account ("'tis thus because not otherwise") of his optical discoveries. His rejection of "conjectures about the truth of things from the mere possibility of hypotheses" in the June 10 note set the stage for a century-long methodological dispute between Newtonian "inductivists" and those scientists, principally European, who were more attracted to the method of hypothesis described in Huygens.

From Query 31

As in Mathematicks, so in Natural Philosophy, the Investigation of difficult Things by the Method of Analysis, ought ever to precede the Method of Composition. This Analysis consists in making Experiments and Observations, and in drawing general Conclusions from them by Induction, and admitting of no Objections against the Conclusions, but such as are taken from Experiments, or other certain Truths. For Hypotheses are not to be regarded in experimental Philosophy. And although the arguing from Experiments and Observations by Induction be no Demonstration of general Conclusions; yet it is the best way of arguing which the Nature of Things admits of, and may be looked upon as so much the stronger, by how much the Induction is more general. And if no Exception occur from Phæanomena, the Conclusion may be pronounced generally.

From Newton, *Optiks*, 4th edn., "From Query 31" (New York: Dover Publications, 1952), pp. 404–5. © 1952. Reprinted with permission from Dover Publications. And from *The Correspondence of Isaac Newton*, ed. H. W. Turnbull, J. F. Scott, A. Rupert Hall, and Laura Tilling, 7 vols. (Cambridge: Cambridge University Press, 1959–77), vol. 5, pp. 396–7, Vol. 1, pp. 96–7, Vol. 1, p. 169, Vol. 1, pp. 209–10. © 1959–77. Reprinted with permission from The Royal Society.

But if at any time afterwards and Exception shall occur from Experiments, it may then begin to be pronounced with such Exceptions as occur. By this way of Analysis we may proceed from Compounds to Ingredients, and from Motions to the Forces producing them; and in general, from Effects to their Causes, and from particular Causes to more general ones, till the Argument end in the most general. This is the Method of Analysis: And the Synthesis consists in assuming the Causes discover'd and establish'd as Principles, and by them explaining the Phæanomena proceeding from them, and proving the Explanations.

[. . .]

From Newton to Cotes, March 28, 1713

Sir

I had yours of Feb 18th, and the Difficulty you mention which lies in these words "Et cum Attractio omnis mutua sit" is removed by considering that as in Geometry the word Hypothesis is not taken in so large a sense as to include the Axiomes and Postulates, so in experimental Philosophy it is not to be taken in so large a sense as to include the first Principles or Axiomes which I call the laws of motion. These Principles are deduced from Phænomena, and made general by Induction: which is the highest evidence that a Proposition can have in this philosophy. And the word Hypothesis is here used by me to signify only such a Proposition as is not a Phænomenon nor deduced from any Phænomena but assumed or supposed without an experimental proof. Now the mutual and mutally equal attraction of bodies is a branch of the third Law of motion and how this branch is deduced from Phænomena you may see in the end of the Corollaries of the Laws of Motion, pag. 22. If a body attracts another body contiguous to it and is not mutually attracted by the other: the attracted body will drive the other before it and both will go away together with an accelerated motion in infinitum, as it were by a self moving principle, contrary to the first law of motion, whereas there is no such phænomenon in all nature.

From Newton to Oldenburg, February 6, 1672

A naturalist would scearce expect to see the science of those become mathematicall, and yet I dare affirm that there is as much certainty in it as in any other part of Opticks. For what I shall tell concerning them is not an Hypothesis but most rigid consequence, not conjectured by barely inferring 'tis thus because not otherwise or because it satisfies all phænomena (the Philosophers universall Topick,) but evinced by the mediation of experiments concluding directly and without any suspicion of doubt. To continue the historicall narration of these experiments would make a discourse too tedious and confused, and therefore I shall rather lay down the *Doctrine* first, and then, for its examination, give you an instance or two of the *Experiments*, as a specimen of the rest.

From Newton to Oldenburg, June 10, 1672

For the best and safest method of philosophizing seems to be, first, to inquire diligently into the properties of things and to establish those properties by experiments, and to proceed later to hypotheses for the explanation of things themselves. For hypotheses ought to be applied only in the explanation of the properties of things, and not made use of in determining them; except in so far as they may furnish experiments. And if anyone offers conjectures about the truth of things from the mere possibility of hypotheses, I do not see by what stipulation anything certain can be determined in any science; since one or another set of hypotheses may always be devised which will appear to supply new difficulties. Hence I judged that one should abstain from contemplating hypotheses, as from improper argumentation.

From Newton to Oldenburg, July 6, 1672

In the meane while give me leave to insinuate that I cannot think it effectuall for determining truth to examin the severall ways by which Phænomena may be explained, unlesse where there can be a perfect enumeration of all those ways. You know the proper Method for inquiring after the properties of things is to deduce them from Experiments. And I told you that the Theory which I propounded was evinced to me, *not by inferring tis thus because not otherwise*, that is not by deducing it onely from a confutation of contrary suppositions, but *by deriving it from Experiments concluding positively and directly.* The way therefore to examin it is by considering whether the experiments which I propound do prove those parts of the Theory to which they are applyed, or by prosecuting other experiments which the Theory may suggest for its examination. And this I would have done in a due Method; the Laws of Refraction being throughly inquired into and determined before the nature of colours be taken into consideration. It may not be amiss to proceed according to the series of these Queries: The decision of which I could wish to be stated, and the events declared by those that may have the curiosity to examin them.

1. Whether rays that are alike incident on the same Medium have unequall refractions, and how great are the inequalities of their refractions at any incidence?
2. What is the law according to which each ray is more or lesse refracted, whether it be that the same ray is ever refracted according to the same ratio of the sines of incidence and refraction; and divers rays, according to divers ratios; Or that the refraction of each ray is greater or lesse without any certain rule? That is, whether each ray have a certain degree of refrangibility according to which its refraction is performed, or is refracted without that regularity?
3. Whether rays which are indued with particular degrees of refrangibility, when they are by any meanes separated, have particular colours constantly belonging to them: viz, the least refrangible, scarlet; the most refrangible, deep violet; the middle, Sea-green; and others, other colours? And on the contrary?
4. Whether the colour of any sort of rays apart may be changed by refraction?
5. Whether colours by coalescing do really change one another to produce a new colour, or produce it by mixing onely?
6. Whether a due mixture of rays, indued with all variety of colours, produces light perfectly like that of the Sun, and which hath all the same properties and exhibits the same Phænomena?
8. Whether there be any other colours produced by refractions then such, as ought to result from the colours belonging to the diversly refrangible rays by their being separated or mixed by that refraction?

To determin by experiments these and such like Queries which involve the propounded Theory seemes the most proper and direct way to a conclusion. And therefore I could wish all objections were suspended, taken from Hypotheses or any other Heads then these two; Of showing the insufficiency of experiments to determin these Queries or prove any other parts of my Theory, by assigning the flaws and defects in my Conclusions drawn from them; Or of producing other Experiments which directly contradict me, if any such may seem to occur. For if the Experiments, which I urge be defective it cannot be difficult to show the defects, but if valid, then by proving the Theory they must render all other Objections invalid.

2.15

Space, Time, and the Elements of Physics

Isaac Newton

Newton's masterwork is the *Philosophiae Naturalis Principia Mathematica* (1687), or *Mathematical Principles of Natural Philosophy* – a title he may have chosen for the contrast it makes with Descartes's *Principles of Philosophy*. In this selection from the first book of the *Principia*, Newton lays down the definitions of the fundamental concepts he will employ and then gives a Scholium, or explanatory discussion, in which he articulates his theory of absolute space and time. His discussion in the final paragraph of the Scholium regarding the detection of inertial forces poses a key issue for all subsequent discussion of the nature of space.

Definition 1

Quantity of matter is a measure of matter that arises from its density and volume jointly.

If the density of air is doubled in a space that is also doubled, there is four times as much air, and there is six times as much if the space is tripled. The case is the same for snow and powders condensed by compression or liquefaction, and also for all bodies that are condensed in various ways by any causes whatsoever. For the present, I am not taking into account any medium, if there should be any, freely pervading the interstices between the parts of bodies. Furthermore, I mean this quantity whenever I use the term "body" or "mass" in the following pages. It can always be known from a body's weight, for – by making very accurate experiments with pendulums – I have found it to be proportional to the weight, as will be shown below.

Definition 2

Quantity of motion is a measure of motion that arises from the velocity and the quantity of matter jointly.

From Isaac Newton, *The* Principia: *Mathematical Principles of Natural Philosophy*, ed. I. Bernard Cohen and Anne Whitman (Berkeley: University of California Press, 1999), pp. 403–15.

The motion of a whole is the sum of the motions of the individual parts, and thus if a body is twice as large as another and has equal velocity there is twice as much motion, and if it has twice the velocity there is four times as much motion.

Definition 3

Inherent force of matter is the power of resisting by which every body, so far as it is able, perseveres in its state either of resting or of moving uniformly straight forward.

This force is always proportional to the body and does not differ in any way from the inertia of the mass except in the manner in which it is conceived. Because of the inertia of matter, every body is only with difficulty put out of its state either of resting or of moving. Consequently, inherent force may also be called by the very significant name of force of inertia. Moreover, a body exerts this force only during a change of its state, caused by another force impressed upon it, and this exercise of force is, depending on the viewpoint, both resistance and impetus: resistance insofar as the body, in order to maintain its state, strives against the impressed force, and impetus insofar as the same body, yielding only with difficulty to the force of a resisting obstacle, endeavors to change the state of that obstacle. Resistance is commonly attributed to resting bodies and impetus to moving bodies; but motion and rest, in the popular sense of the terms, are distinguished from each other only by point of view, and bodies commonly regarded as being at rest are not always truly at rest.

Definition 4

Impressed force is the action exerted on a body to change its state either of resting or of moving uniformly straight forward.

This force consists solely in the action and does not remain in a body after the action has ceased. For a body perseveres in any new state solely by the force of inertia. Moreover, there are various sources of impressed force, such as percussion, pressure, or centripetal force.

Definition 5

Centripetal force is the force by which bodies are drawn from all sides, are impelled, or in any way tend, toward some point as to a center.

One force of this kind is gravity, by which bodies tend toward the center of the earth; another is magnetic force, by which iron seeks a lodestone; and yet another is that force, whatever it may be, by which the planets are continually drawn back from rectilinear motions and compelled to revolve in curved lines.

A stone whirled in a sling endeavors to leave the hand that is whirling it, and by its endeavor it stretches the sling, doing so the more strongly the more swiftly it revolves; and as soon as it is released, it flies away. The force opposed to that endeavor, that is, the force by which the sling continually draws the stone back toward the hand and keeps it in an orbit, I call centripetal, since it is directed toward the hand as toward the center of an orbit. And the same applies to all bodies that are made to move in orbits. They all endeavor to recede from the centers of their orbits, and unless some force opposed to that endeavor is present, restraining them and keeping them in orbits and hence called by me centripetal, they will go off in straight lines with uniform motion. If a projectile were deprived of the force of gravity, it would not be deflected toward the earth but would go off in a straight line into the heavens and do so with uniform motion, provided that the resistance of the air were removed. The projectile, by its gravity, is drawn back from a rectilinear course and continually deflected toward the earth, and this is so to a greater or lesser degree in proportion to its gravity and its velocity of motion. The less its gravity in proportion to its quantity of matter, or the greater the velocity with which it is projected, the less it will deviate from a rectilinear course and the farther it will go. If a lead ball were projected with a given velocity along a horizontal line from the top of some mountain by the force of gunpowder and went in a curved line for a distance of two miles before falling to the earth, then the same ball projected with twice the velocity would go about twice as far and with ten times the velocity about ten times as far, provided that the resistance of the air were removed. And by increasing the velocity, the distance to which it would be projected could be increased at will and the curvature of the line that it would describe could be decreased, in such a way that it would finally fall at a distance of 10 or 30 or 90 degrees or even go around the whole earth or, lastly, go off into the heavens and continue indefinitely in this motion. And in the same way that a projectile could, by the force of gravity, be deflected into an orbit and go around the whole earth, so too the moon, whether by the force of gravity – if it has gravity – or by any other force by which it may be urged toward the earth, can always be drawn back toward the earth from a rectilinear course and deflected into its orbit; and without such a force the moon cannot be kept in its orbit. If this force were too small, it would not deflect the moon sufficiently from a rectilinear course; if it were too great, it would deflect the moon excessively and draw it down from its orbit toward the earth. In fact, it must be of just the right magnitude, and mathematicians have the task of finding the force by which a body can be kept exactly in any given orbit with a given velocity and, alternatively, to find the curvilinear path into which a body leaving any given place with a given velocity is deflected by a given force.

The quantity of centripetal force is of three kinds: absolute, accelerative, and motive.

Definition 6

The absolute quantity of centripetal force is the measure of this force that is greater or less in proportion to the efficacy of the cause propagating it from a center through the surrounding regions.
An example is magnetic force, which is greater in one lodestone and less in another, in proportion to the bulk or potency of the lodestone.

Definition 7

The accelerative quantity of centripetal force is the measure of this force that is proportional to the velocity which it generates in a given time.
One example is the potency of a lodestone, which, for a given lodestone, is greater at a smaller distance and less at a greater distance. Another example is the force that produces gravity, which is greater in valleys and less on the peaks of high mountains and still less (as will be made clear below) at greater distances from the body of the earth, but which is everywhere the same at equal distances, because it equally accelerates all falling bodies (heavy or light, great or small), provided that the resistance of the air is removed.

Definition 8

The motive quantity of centripetal force is the measure of this force that is proportional to the motion which it generates in a given time.
An example is weight, which is greater in a larger body and less in a smaller body; and in one and the same body is greater near the earth and less out in the heavens. This quantity is the centripetency, or propensity toward a center, of the whole body, and (so to speak) its weight, and it may always be known from the force opposite and equal to it, which can prevent the body from falling.

These quantities of forces, for the sake of brevity, may be called motive, accelerative, and absolute forces, and, for the sake of differentiation, may be referred to bodies seeking a center, to the places of the bodies, and to the center of the forces: that is, motive force may be referred to a body as an endeavor of the whole directed toward a center and compounded of the endeavors of all the parts; accelerative force, to the place of the body as a certain efficacy diffused from the center through each of the surrounding places in order to move the bodies that are in those places; and absolute force, to the center as having some cause without which the motive forces are not propagated through the surrounding regions, whether this cause is some central body (such as a lodestone in the center of a magnetic force or the earth in the center of a force that produces gravity) or whether it is some other cause which is not apparent. This concept is purely mathematical, for I am not now considering the physical causes and sites of forces.

Therefore, accelerative force is to motive force as velocity to motion. For quantity of motion arises from velocity and quantity of matter jointly, and motive force from accelerative force and quantity of matter jointly. For the sum of the actions of the accelerative force on the individual particles of a body is the motive force of the whole body. As a consequence, near the surface of the earth, where the accelerative gravity, or the force that produces gravity, is the same in all bodies universally, the motive gravity, or weight, is as the body, but in an ascent to regions where the accelerative gravity becomes less, the weight will decrease proportionately and will always be as the body and the accelerative gravity jointly. Thus, in regions where the accelerative gravity is half as great, a body one-half or one-third as great will have a weight four or six times less.

Further, it is in this same sense that I call attractions and impulses accelerative and motive. Moreover, I use interchangeably and indiscriminately words signifying attraction, impulse, or any sort of propensity toward a center, considering these forces not from a physical but only from a mathematical point of view. Therefore, let the reader beware of thinking that by words of this kind I am anywhere defining a species or mode of action or a physical cause or reason, or that I am attributing forces in a true and physical sense to centers (which are mathematical points) if I happen to say that centers attract or that centers have forces.

Scholium

Thus far it has seemed best to explain the senses in which less familiar words are to be taken in this treatise. Although time, space, place, and motion are very familiar to everyone, it must be noted that these quantities are popularly conceived solely with reference to the objects of sense perception. And this is the source of certain preconceptions; to eliminate them it is useful to distinguish these quantities into absolute and relative, true and apparent, mathematical and common.

1. Absolute, true, and mathematical time, in and of itself and of its own nature, without reference to anything external, flows uniformly and by another name is called duration. Relative, apparent, and common time is any sensible and external measure (precise or imprecise) of duration by means of motion; such a measure – for example, an hour, a day, a month, a year – is commonly used instead of true time.
2. Absolute space, of its own nature without reference to anything external, always remains homogeneous and immovable. Relative space is any movable measure or dimension of this absolute space; such a measure or dimension is determined by our senses from the situation of the space with respect to bodies and is popularly used for immovable space, as in the case of space under the earth or in the air or in the heavens, where the dimension is determined from the situation of the space with respect to the earth. Absolute and relative space are the same in species and in magnitude, but they do not always remain the same numerically. For example, if the earth moves, the space of our air, which in a relative sense and with respect to the earth always remains the same, will now be one part of the absolute space into which the air passes, now another part of it, and thus will be changing continually in an absolute sense.
3. Place is the part of space that a body occupies, and it is, depending on the space, either absolute or relative. I say the part of space, not the position of the body or its outer surface. For the places of equal solids are always equal, while their surfaces are for the most part unequal because of the dissimilarity of shapes; and positions, properly speaking, do not have quantity and are not so much places as attributes of places. The motion of a whole is the same as the sum of the motions of the parts; that is, the change in position of a whole from its place is the same as the sum of the changes in position of its parts from their places, and thus the place of a whole is the same as the sum of the places of the parts and therefore is internal and in the whole body.
4. Absolute motion is the change of position of a body from one absolute place to another; relative motion is change of position from one relative place to another. Thus, in a ship under sail, the relative place of a body is that region of the ship in which the body happens to be or that part of the whole interior of the ship which the body fills and which accordingly moves along with the ship, and relative rest is the continuance of the body in that same region of the ship or same part of its interior. But true rest is the continuance of a body in the same part of that unmoving space in which the ship itself, along with its interior and

all its contents, is moving. Therefore, if the earth is truly at rest, a body that is relatively at rest on a ship will move truly and absolutely with the velocity with which the ship is moving on the earth. But if the earth is also moving, the true and absolute motion of the body will arise partly from the true motion of the earth in unmoving space and partly from the relative motion of the ship on the earth. Further, if the body is also moving relatively on the ship, its true motion will arise partly from the true motion of the earth in unmoving space and partly from the relative motions both of the ship on the earth and of the body on the ship, and from these relative motions the relative motion of the body on the earth will arise. For example, if that part of the earth where the ship happens to be is truly moving eastward with a velocity of 10,010 units, and the ship is being borne westward by sails and wind with a velocity of 10 units, and a sailor is walking on the ship toward the east with a velocity of 1 unit, then the sailor will be moving truly and absolutely in unmoving space toward the east with a velocity of 10,001 units and relatively on the earth toward the west with a velocity of 9 units.

In astronomy, absolute time is distinguished from relative time by the equation of common time. For natural days, which are commonly considered equal for the purpose of measuring time, are actually unequal. Astronomers correct this inequality in order to measure celestial motions on the basis of a truer time. It is possible that there is no uniform motion by which time may have an exact measure. All motions can be accelerated and retarded, but the flow of absolute time cannot be changed. The duration or perseverance of the existence of things is the same, whether their motions are rapid or slow or null; accordingly, duration is rightly distinguished from its sensible measures and is gathered from them by means of an astronomical equation. Moreover, the need for using this equation in determining when phenomena occur is proved by experience with a pendulum clock and also by eclipses of the satellites of Jupiter.

Just as the order of the parts of time is unchangeable, so, too, is the order of the parts of space. Let the parts of space move from their places, and they will move (so to speak) from themselves. For times and spaces are, as it were, the places of themselves and of all things. All things are placed in time with reference to order of succession and in space with reference to order of position. It is of the essence of spaces to be places, and for primary places to move is absurd. They are therefore absolute places, and it is only changes of position from these places that are absolute motions.

But since these parts of space cannot be seen and cannot be distinguished from one another by our senses, we use sensible measures in their stead. For we define all places on the basis of the positions and distances of things from some body that we regard as immovable, and then we reckon all motions with respect to these places, insofar as we conceive of bodies as being changed in position with respect to them. Thus, instead of absolute places and motions we use relative ones, which is not inappropriate in ordinary human affairs, although in philosophy abstraction from the senses is required. For it is possible that there is no body truly at rest to which places and motions may be referred.

Moreover, absolute and relative rest and motion are distinguished from each other by their properties, causes, and effects. It is a property of rest that bodies truly at rest are at rest in relation to one another. And therefore, since it is possible that some body in the regions of the fixed stars or far beyond is absolutely at rest, and yet it cannot be known from the position of bodies in relation to one another in our regions whether or not any of these maintains a given position with relation to that distant body, true rest cannot be defined on the basis of the position of bodies in relation to one another.

It is a property of motion that parts which keep given positions in relation to wholes participate in the motions of such wholes. For all the parts of bodies revolving in orbit endeavor to recede from the axis of motion, and the impetus of bodies moving forward arises from the joint impetus of the individual parts. Therefore, when bodies containing others move, whatever is relatively at rest within them also moves. And thus true and absolute motion cannot be determined by means of change of position from the vicinity of bodies that are regarded as being at rest. For the exterior bodies ought to be regarded not only as being at rest but also as being truly at rest. Otherwise all contained bodies, besides being subject to change of position from the vicinity of the containing bodies, will participate in the true motions of the containing bodies and, if there is no such change of position, will not be truly at rest but only be regarded as being at rest. For containing bodies are to those inside them as the outer part of the whole to the inner part or as the shell to the kernel. And when the shell moves, the kernel also, without being changed in position from the vicinity of the shell, moves as a part of the whole.

A property akin to the preceding one is that when a place moves, whatever is placed in it moves along with it, and therefore a body moving away from a place that moves participates also in the motion of its place. Therefore, all motions away from places that move are only parts of whole and absolute motions, and every whole motion is compounded of the motion of a body away from its initial place, and the motion of this place away from its place, and so on, until an unmoving place is reached, as in the above-mentioned example of the sailor. Thus, whole and absolute motions can be determined only by means of unmoving places, and therefore in what has preceded I have referred such motions to unmoving places and relative motions to movable places. Moreover, the only places that are unmoving are those that all keep given positions in relation to one another from infinity to infinity and therefore always remain immovable and constitute the space that I call immovable.

The causes which distinguish true motions from relative motions are the forces impressed upon bodies to generate motion. True motion is neither generated nor changed except by forces impressed upon the moving body itself, but relative motion can be generated and changed without the impression of forces upon this body. For the impression of forces solely on other bodies with which a given body has a relation is enough, when the other bodies yield, to produce a change in that relation which constitutes the relative rest or motion of this body. Again, true motion is always changed by forces impressed upon a moving body, but relative motion is not necessarily changed by such forces. For if the same forces are impressed upon a moving body and also upon other bodies with which it has a relation, in such a way that the relative position is maintained, the relation that constitutes the relative motion will also be maintained. Therefore, every relative motion can be changed while the true motion is preserved, and can be preserved while the true one is changed, and thus true motion certainly does not consist in relations of this sort.

The effects distinguishing absolute motion from relative motion are the forces of receding from the axis of circular motion. For in purely relative circular motion these forces are null, while in true and absolute circular motion they are larger or smaller in proportion to the quantity of motion. If a bucket is hanging from a very long cord and is continually turned around until the cord becomes twisted tight, and if the bucket is thereupon filled with water and is at rest along with the water and then, by some sudden force, is made to turn around in the opposite direction and, as the cord unwinds, perseveres for a while in this motion; then the surface of the water will at first be level, just as it was before the vessel began to move. But after the vessel, by the force gradually impressed upon the water, has caused the water also to begin revolving perceptibly, the water will gradually recede from the middle and rise up the sides of the vessel, assuming a concave shape (as experience has shown me), and, with an ever faster motion, will rise further and further until, when it completes its revolutions in the same times as the vessel, it is relatively at rest in the vessel. The rise of the water reveals its endeavor to recede from the axis of motion, and from such an endeavor one can find out and measure the true and absolute circular motion of the water, which here is the direct opposite of its relative motion. In the beginning, when the relative motion of the water in the vessel was greatest, that motion was not giving rise to any endeavor to recede from the axis; the water did not seek the circumference by rising up the sides of the vessel but remained level, and therefore its true circular motion had not yet begun. But afterward, when the relative motion of the water decreased, its rise up the sides of the vessel revealed its endeavor to recede from the axis, and this endeavor showed the true circular motion of the water to be continually increasing and finally becoming greatest when the water was relatively at rest in the vessel. Therefore, that endeavor does not depend on the change of position of the water with respect to surrounding bodies, and thus true circular motion cannot be determined by means of such changes of position. The truly circular motion of each revolving body is unique, corresponding to a unique endeavor as its proper and sufficient effect, while relative motions are innumerable in accordance with their varied relations to external bodies and, like relations, are completely lacking in true effects except insofar as they participate in that true and unique motion. Thus, even in the system of those who hold that our heavens revolve below the heavens of the fixed stars and carry the planets around with them, the individual parts of the heavens, and the planets that are relatively at rest in the heavens to which they belong, are truly in motion. For they change their positions relative to one another (which is not the case with things that are truly at rest), and as they are carried around together with the heavens, they participate in the motions of the heavens and, being parts of revolving wholes, endeavor to recede from the axes of those wholes.

Relative quantities, therefore, are not the actual quantities whose names they bear but are those sensible measures of them (whether true or erroneous) that are commonly used instead of the quantities being measured. But if the meanings of words are to be defined by usage, then it is these sensible measures which should properly be understood by the terms "time," "space," "place," and "motion," and the manner of expression will be out of the ordinary and purely mathematical if the quantities being measured are understood here. Accordingly those who there interpret these words as referring to the quantities being measured do violence to the Scriptures. And they no less corrupt mathematics and philosophy who confuse true quantities with their relations and common measures.

It is certainly very difficult to find out the true motions of individual bodies and actually to differentiate them from apparent motions, because the parts of that immovable space in which the bodies truly move make no impression on the senses. Nevertheless, the case is not utterly hopeless. For it is possible to draw evidence partly from apparent motions, which are the differences between the true motions, and partly from the forces that are the causes and effects of the true motions. For example, if two balls, at a given distance from each other with a cord connecting them, were revolving about a common center of gravity, the endeavor of the balls to recede from the axis of motion could be known from the tension of the cord, and thus the quantity of circular motion could be computed. Then, if any equal forces were simultaneously impressed upon the alternate faces of the balls to increase or decrease their circular motion, the increase or decrease of the motion could be known from the increased or decreased tension of the cord, and thus, finally, it could be discovered which faces of the balls the forces would have to be impressed upon for a maximum increase in the motion, that is, which were the posterior faces, or the ones that are in the rear in a circular motion. Further, once the faces that follow and the opposite faces that precede were known, the direction of the motion would be known. In this way both the quantity and the direction of this circular motion could be found in any immense vacuum, where nothing external and sensible existed with which the balls could be compared. Now if some distant bodies were set in that space and maintained given positions with respect to one another, as the fixed stars do in the regions of the heavens, it could not, of course, be known from the relative change of position of the balls among the bodies whether the motion was to be attributed to the bodies or to the balls. But if the cord was examined and its tension was discovered to be the very one which the motion of the balls required, it would be valid to conclude that the motion belonged to the balls and that the bodies were at rest, and then, finally, from the change of position of the balls among the bodies, to determine the direction of this motion. But in what follows, a fuller explanation will be given of how to determine true motions from their causes, effects, and apparent differences, and, conversely, of how to determine from motions, whether true or apparent, their causes and effects. For this was the purpose for which I composed the following treatise.

2.16

Four Rules of Reasoning

Isaac Newton

At the beginning of book 3 of the *Principia*, Newton briefly lays down four methodological rules. Together with his correspondence, these rules illuminate Newton's conception of his own enterprise. His preference for experiment and solid data over unfounded hypotheses comes across very clearly here, as does his appreciation for Ockham's Razor – the principle that explanatory causes should not be multiplied without necessity. These are more than mere afterthoughts: Newton appeals to them explicitly, just as he does to his axioms and theorems, in the derivation of the law of gravitation.

Rule 1

No more causes of natural things should be admitted than are both true and sufficient to explain their phenomena.

As the philosophers say: Nature does nothing in vain, and more causes are in vain when fewer suffice. For nature is simple and does not indulge in the luxury of superfluous causes.

Rule 2

Therefore, the causes assigned to natural effects of the same kind must be, so far as possible, the same.

Examples are the cause of respiration in man and beast, or of the falling of stones in Europe and America, or of the light of a kitchen fire and the sun, or of the reflection of light on our earth and the planets.

Rule 3

Those qualities of bodies that cannot be intended and remitted [i.e., qualities that cannot be increased and diminished] and that belong to all bodies on which experiments can be made should be taken as qualities of all bodies universally.

For the qualities of bodies can be known only through experiments; and therefore qualities that

From Newton, *The* Principia: *Mathematical Principles of Natural Philosophy*, trans. I. Bernard Cohen and Anne Whitman (Berkeley: University of California Press, 1999), selections from pp. 794–6.

square with experiments universally are to be regarded as universal qualities; and qualities that cannot be diminished cannot be taken away from bodies. Certainly idle fancies ought not to be fabricated recklessly against the evidence of experiments, nor should we depart from the analogy of nature, since nature is always simple and ever consonant with itself. The extension of bodies is known to us only through our senses, and yet there are bodies beyond the range of these senses; but because extension is found in all sensible bodies, it is ascribed to all bodies universally. We know by experience that some bodies are hard. Moreover, because the hardness of the whole arises from the hardness of its parts, we justly infer from this not only the hardness of the undivided particles of bodies that are accessible to our senses, but also of all other bodies. That all bodies are impenetrable we gather not by reason but by our senses. We find those bodies that we handle to be impenetrable, and hence we conclude that impenetrability is a property of all bodies universally. That all bodies are movable and persevere in motion or in rest by means of certain forces (which we call forces of inertia) we infer from finding these properties in the bodies that we have seen. The extension, hardness, impenetrability, mobility, and force of inertia of the whole arise from the extension, hardness, impenetrability, mobility, and force of inertia of each of the parts; and thus we conclude that every one of the least parts of all bodies is extended, hard, impenetrable, movable, and endowed with a force of inertia. And this is the foundation of all natural philosophy. Further, from phenomena we know that the divided, contiguous parts of bodies can be separated from one another, and from mathematics it is certain that the undivided parts can be distinguished into smaller parts by our reason. But it is uncertain whether those parts which have been distinguished in this way and not yet divided can actually be divided and separated from one another by the forces of nature. But if it were established by even a single experiment that in the breaking of a hard and solid body, any undivided particle underwent division, we should conclude by the force of this third rule not only that divided parts are separable but also that undivided parts can be divided indefinitely.

Finally, if it is universally established by experiments and astronomical observations that all bodies on or near the earth gravitate [*lit.* are heavy] toward the earth, and do so in proportion to the quantity of matter in each body, and that the moon gravitates [is heavy] toward the earth in proportion to the quantity of its matter, and that our sea in turn gravitates [is heavy] toward the moon, and that all planets gravitate [are heavy] toward one another, and that there is a similar gravity [heaviness] of comets toward the sun, it will have to be concluded by this third rule that all bodies gravitate toward one another. Indeed, the argument from phenomena will be even stronger for universal gravity than for the impenetrability of bodies, for which, of course, we have not a single experiment, and not even an observation, in the case of the heavenly bodies. Yet I am by no means affirming that gravity is essential to bodies. By inherent force I mean only the force of inertia. This is immutable. Gravity is diminished as bodies recede from the earth.

Rule 4

In experimental philosophy, propositions gathered from phenomena by induction should be considered either exactly or very nearly true notwithstanding any contrary hypotheses, until yet other phenomena make such propositions either more exact or liable to exceptions.

This rule should be followed so that arguments based on induction may not be nullified by hypotheses.

2.17
General Scholium

Isaac Newton

In the General Scholium at the end of the third book of the *Principia*, Newton draws together the threads of his work in ways that are sometimes surprising. He opens with a dismissal of the Cartesian theory that the planets are moved in vortices, but he moves from this directly, even abruptly, to a theological discussion (largely omitted here) of the majesty, dominion, and nature of God. Of more direct scientific interest is Newton's refusal to assign a cause to gravity. In keeping with his methodological principles, he declares that he does not "feign hypotheses" about gravity. Yet he does not say that it has no cause, but merely that he has not yet been able to discover it.

The hypothesis of vortices is beset with many difficulties. If, by a radius drawn to the sun, each and every planet is to describe areas proportional to the time, the periodic times of the parts of the vortex must be as the squares of the distances from the sun. If the periodic times of the planets are to be as the $^3/_2$ powers of the distances from the sun, the periodic times of the parts of the vortex must be as the $^3/_2$ powers of the distances. If the smaller vortices revolving about Saturn, Jupiter, and the other planets are to be preserved and are to float without agitation in the vortex of the sun, the periodic times of the parts of the solar vortex must be the same. The axial revolutions [i.e., rotations] of the sun and planets, which would have to agree with the motions of their vortices, differ from all these proportions. The motions of comets are extremely regular, observe the same laws as the motions of planets, and cannot be explained by vortices. Comets go with very eccentric motions into all parts of the heavens, which cannot happen unless vortices are eliminated.

The only resistance which projectiles encounter in our air is from the air. With the air removed, as it is in Boyle's vacuum, resistance ceases, since

From Isaac Newton, *The* Principia: *Mathematical Principles of Natural Philosophy*, trans. I. Bernard Cohen and Anne Whitman (Berkeley: University of California Press, 1999), selections from pp. 939–40, 943.

a tenuous feather and solid gold fall with equal velocity in such a vacuum. And the case is the same for the celestial spaces, which are above the atmosphere of the earth. All bodies must move very freely in these spaces, and therefore planets and comets must revolve continually in orbits given in kind and in position, according to the laws set forth above. They will indeed persevere in their orbits by the laws of gravity, but they certainly could not originally have acquired the regular position of the orbits by these laws.

The six primary planets revolve about the sun in circles concentric with the sun, with the same direction of motion, and very nearly in the same plane. Ten moons revolve about the earth, Jupiter, and Saturn in concentric circles, with the same direction of motion, very nearly in the planes of the orbits of the planets. And all these regular motions do not have their origin in mechanical causes, since comets go freely in very eccentric orbits and into all parts of the heavens. And with this kind of motion the comets pass very swiftly and very easily through the orbits of the planets; and in their aphelia, where they are slower and spend a longer time, they are at the greatest possible distance from one another, so as to attract one another as little as possible.

This most elegant system of the sun, planets, and comets could not have arisen without the design and dominion of an intelligent and powerful being. And if the fixed stars are the centers of similar systems, they will all be constructed according to a similar design and subject to the dominion of *One*, especially since the light of the fixed stars is of the same nature as the light of the sun, and all the systems send light into all the others. And so that the systems of the fixed stars will not fall upon one another as a result of their gravity, he has placed them at immense distances from one another.

[. . .]

Thus far I have explained the phenomena of the heavens and of our sea by the force of gravity, but I have not yet assigned a cause to gravity. Indeed, this force arises from some cause that penetrates as far as the centers of the sun and planets without any diminution of its power to act, and that acts not in proportion to the quantity of the *surfaces* of the particles on which it acts (as mechanical causes are wont to do) but in proportion to the quantity of *solid* matter, and whose action is extended everywhere to immense distances, always decreasing as the squares of the distances. Gravity toward the sun is compounded of the gravities toward the individual particles of the sun, and at increasing distances from the sun decreases exactly as the squares of the distances as far out as the orbit of Saturn, as is manifest from the fact that the aphelia of the planets are at rest, and even as far as the farthest aphelia of the comets, provided that those aphelia are at rest. I have not as yet been able to deduce from phenomena the reason for these properties of gravity, and I do not feign hypotheses. For whatever is not deduced from the phenomena must be called a hypothesis; and hypotheses, whether metaphysical or physical, or based on occult qualities, or mechanical, have no place in experimental philosophy. In this experimental philosophy, propositions are deduced from the phenomena and are made general by induction. The impenetrability, mobility, and impetus of bodies, and the laws of motion and the law of gravity have been found by this method. And it is enough that gravity really exists and acts according to the laws that we have set forth and is sufficient to explain all the motions of the heavenly bodies and of our sea.

2.18

The System of the World

Isaac Newton

The third book of Newton's *Principia* is entitled "The System of the World," but it is densely mathematical – deliberately so, as Newton said he wrote it in a mathematical way so that his rival Hooke would not be able to read it. But Newton left in the library at Cambridge a manuscript of some of his lectures, also entitled *The System of the World*, which covers the same material in a much more accessible fashion. In this selection from those lectures, Newton lays out the theory of universal gravitation as an inverse square force between masses at specified distances and then shows that it accounts for Kepler's third law – that the cubes of the distances of the planets are proportional to the squares of their periodic times.

[. . .]

2 The Principle of Circular Motion in Free Spaces

After this time, we do not know in what manner the ancients explained the question, how the planets came to be retained within certain bounds in these free spaces, and to be drawn off from the rectilinear courses, which, left to themselves, they should have pursued, into regular revolutions in curvilinear orbits. Probably it was to give some sort of satisfaction to this difficulty that solid orbs had been introduced.

Modern [i.e., more recent] philosophers want either vortices to exist, as Kepler and Descartes, or some other principle whether of impulse or attraction, as Borelli, Hooke, and others of our countrymen. From the first law of motion it is most certain that some force is required. Our purpose is to bring out its quantity and properties and to investigate mathematically its effects in moving bodies. Further, in order not to determine the type hypothetically, we have called by the general name 'centripetal' that [force] which

From *Newton*, Norton Critical Edition, ed. I. Bernard Cohen and Richard Westfall (New York: W. W. Norton & Co., 1995), selections from pp. 259–66.

tends toward the sun, 'circumterrestrial' [that force] which [tends] toward the earth, 'circumjovial' [that force] which tends toward Jupiter, and so in the others.

3 The Action of Centripetal Forces

That by means of centripetal forces the planets may be retained in certain orbits, we may easily understand, if we consider the motions of projectiles; for a stone that is projected is by the pressure of its own weight forced out of the rectilinear path, which by the initial projection alone it should have pursued, and made to describe a curved line in the air; and through that crooked way is at last brought down to the ground; and the greater the velocity is with which it is projected, the farther it goes before it falls to the earth. We may therefore suppose the velocity to be so increased, that it would describe an arc of 1, 2, 5, 10, 100, 1000 miles before it arrived at the earth, till at last, exceeding the limits of the earth, it should pass into space without touching it.

Let AFB represent the surface of the earth, C its centre, VD, VE, VF the curved lines which a body would describe, if projected in an horizontal direction from the top of an high mountain successively with more and more velocity; and, because the celestial motions are scarcely retarded by the little or no resistance of the spaces in which they are performed, to keep up the parity of cases, let us suppose either that there is no air about the earth, or at least that it is endowed with little or no power of resisting; and for the same reason that the body projected with a less velocity describes the lesser arc VD, and with a greater velocity the greater arc VE, and, augmenting the velocity, it goes farther and farther to F and G, if the velocity was still more and more augmented, it would reach at last quite beyond the circumference of the earth, and return to the mountain from which it was projected.

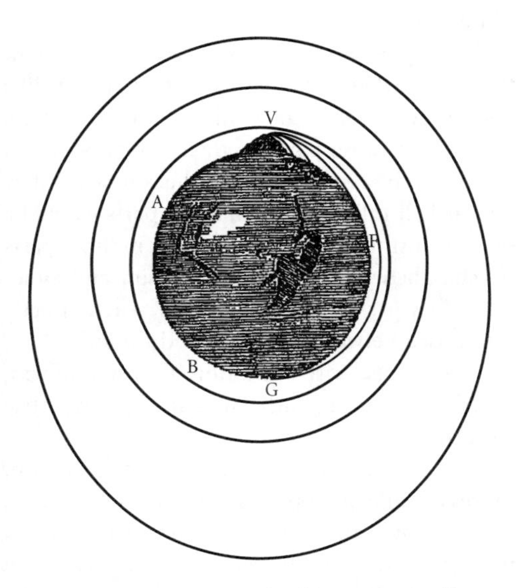

Figure 6 Gravitation and orbits. A projectile launched horizontally from a mountain will go farther as it is given greater force. If it is launched with sufficient force, it will not come down but rather go into orbit around the earth

And since the areas which by this motion it describes by a radius drawn to the centre of the earth are proportional to the times in which they are described, its velocity, when it returns to the mountain, will be no less than it was at first; and, retaining the same velocity, it will describe the same curve over and over, by the same law.

But if we now imagine bodies to be projected in the directions of lines parallel to the horizon from greater heights, as of 5, 10, 100, 1000, or more miles, or rather as many semidiameters of the earth, those bodies, according to their different velocity, and the different force of gravity in different heights, will describe arcs either concentric with the earth, or variously eccentric, and go on revolving through the heavens in those orbits just as the planets do in their orbits.

4 The Certainty of the Proof

As when a stone is projected obliquely, that is, any way but in the perpendicular direction, the continual deflection thereof towards the earth from the right line in which it was projected is a proof of its gravitation to the earth, no less certain than its direct descent when suffered to fall freely from rest; so the deviation of bodies moving in free spaces from rectilinear paths and continual deflection therefrom towards any place, is a sure indication of the existence of some force which from all quarters impels those bodies towards that place.

And as, from the supposed existence of gravity, it necessarily follows that all bodies about the earth must press downwards, and therefore must either descend directly to the earth, if they are let fall from rest, or at least continually deviate from right lines towards the earth, if they are projected obliquely; so, from the supposed existence of a force directed to any centre, it will follow, by the like necessity, that all bodies upon which this force acts must either descend directly to that centre, or at least deviate continually towards it from right lines, if otherwise they should have moved obliquely in these right lines.

And how from the motions given we may infer the forces, or from the forces given we may determine the motions, is shown in the first two Books of our *Principles of Philosophy*.

If the earth is supposed to stand still, and the fixed stars to be revolved in free spaces in the space of 24 hours, it is certain the forces by which the fixed stars are held in their orbits are not directed to the earth, but to the centres of those orbits, that is, of the several parallel circles, which the fixed stars, declining to one side and the other from the equator, describe daily; also that by radii drawn to the centres of the orbits the fixed stars describe areas exactly proportional to the times of description. Then, because the periodic times are equal, it follows that the centripetal forces are as the radii of the several orbits, and that they will continually revolve in the same orbits. And the like consequences may be drawn from the supposed diurnal motion of the planets.

That forces should be directed to no body on which they physically depend, but to innumerable imaginary points in the axis of the earth, is an hypothesis too incongruous. It is more incongruous still that those forces should increase exactly in proportion of the distances from this axis; for this is an indication of an increase to immensity, or rather to infinity; whereas the forces of natural things commonly decrease in receding from the fountain from which they flow. But, what is yet more absurd, neither are the areas described by the same star proportional to the times, nor are its revolutions performed in the same orbit; for as the star recedes from the neighboring pole, both areas and orbits increase; and from the increase of the area it is demonstrated that the forces are not directed to the axis of the earth. And this difficulty arises from the twofold motion that is observed in the fixed stars, one diurnal round the axis of the earth, the other exceedingly slow round the axis of the ecliptic. And the explication thereof requires a composition of forces so involved and so variable, that it is hardly to be reconciled with any physical theory.

5 Centripetal Forces Are Directed to the Individual Centres of the Planets

That there are centripetal forces actually directed to the bodies of the sun, of the earth, and other planets, I thus infer.

The moon revolves about our earth, and by radii drawn to its centre describes areas nearly proportional to the times in which they are described, as is evident from its velocity compared with its apparent diameter; for its motion is slower when its diameter is less (and therefore its distance greater), and its motion is swifter when its diameter is greater.

The revolutions of the satellites of Jupiter about that planet are more regular; for they describe circles concentric with Jupiter by uniform motions, as exactly as our senses can perceive.

And so the satellites of Saturn are revolved about this planet with motions nearly circular and uniform, scarcely disturbed by any eccentricity hitherto observed.

That Venus and Mercury are revolved about the sun, is demonstrable from their moon-like appearances; when they shine with a full face, they are in those parts of their orbits which in respect of the earth lie beyond the sun; when they appear half full, they are in those parts which lie over against the sun; when horned, in those parts which lie between the earth and the sun; and sometimes they pass over the sun's disk, when directly interposed between the earth and the sun.

And Venus, with a motion almost uniform, describes an orbit nearly circular and concentric with the sun.

But Mercury, with a more eccentric motion, makes remarkable approaches to the sun, and goes off again by turns; but it is always swifter as it is near to the sun, and therefore by a radius drawn to the sun still describes areas proportional to the times.

Lastly, that the earth describes about the sun, or the sun about the earth, by a radius from the one to the other, areas exactly proportional to the times, is demonstrable from the apparent diameter of the sun compared with its apparent motion.

These are astronomical experiments; from which it follows, by Prop. 1, 2, 3, in the first Book of our *Principles*, and their Corollaries, that there are centripetal forces actually directed (either accurately or without considerable error) to the centres of the earth, of Jupiter, of Saturn, and of the sun. In Mercury, Venus, Mars, and the lesser planets, where experiments are wanting, the arguments from analogy must be allowed in their place.

6 Centripetal Forces Decrease Inversely as the Square of the Distances from the Centres of the Planets

[...]

7 The Superior Planets Revolve about the Sun, and the Radii Drawn to the Sun Describe Areas Proportional to the Times

[...]

8 The Force Which Controls the Superior Planets Is Not Directed to the Earth, but to the Sun

[...]

9 The Circumsolar Force Decreases in All Planetary Spaces Inversely as the Square of the Distances from the Sun

The distances of the planets from the sun come out the same, whether, with Tycho, we place the earth in the centre of the system, or the sun with Copernicus: and we have already proved that these distances are true in Jupiter.

Kepler and Boulliau have, with great care, determined the distances of the planets from the sun; and hence it is that their tables agree best with the heavens. And in all the planets, in Jupiter and Mars, in Saturn and the earth, as well as in Venus and Mercury, the cubes of their distances are as the squares of their periodic times; and therefore the centripetal circumsolar force throughout all the planetary regions decreases as the inverse square of the distances from the sun. . . .

10 The Circumterrestrial Force Decreases Inversely as the Square of the Distances from the Earth. This Is Shown on the Hypothesis That the Earth Is at Rest

[...]

11 The Same Proved on the Hypothesis That the Earth Moves

[...]

12 The Decrease of the Forces Inversely as the Square of the Distances from the Earth and Planets Is Proved also from the Eccentricities of the Planets and the Very Slow Motion of the Apsides

[...]

18 Another Analogy between Forces and Bodies Attracted Is Shown in the Heavens

Of kin to the analogy we have been describing is another observed between the forces and the bodies attracted. Because the action of the centripetal force upon the planets decreases inversely as the square of the distance, and the periodic time increases as the $^3/_2$th power of the distance, it is evident that the actions of the centripetal force, and therefore the periodic times, would be equal in equal planets at equal distances from the sun; and in equal distances of unequal planets the total actions of the centripetal force would be as the bodies of the planets; for if the actions were not proportional to the bodies to be moved, they could not equally retract these bodies from the tangents of their orbs in equal times: nor could the motions of the satellites of Jupiter be so regular, if it was not that the circumsolar force was equally exerted upon Jupiter and all its satellites in proportion of their several weights. And the same thing is to be said of Saturn in respect of its satellites, and of our earth in respect of the moon. . . . And, therefore, at equal distances, the actions of the centripetal force are equal upon all the planets in proportion to their bodies, or to the quantities of matter in their several bodies; and for the same reason must be the same upon all the particles of the same size of which the planet is composed; for if the action was greater upon some sort of particles than upon others, in proportion to their quantity of matter, it would be also greater or less upon the whole planets, not in proportion to the quantity only, but likewise to the sort of matter more copiously found in one and more sparingly in another.

19 It Is Found Also in Terrestrial Bodies

In such bodies as are found on our earth of very different sorts, I examined this analogy with great care.

If the action of the circumterrestrial force is proportional to the bodies to be moved, it will (by the second Law of Motion) move them with equal velocity in equal times, and will make all bodies, let fall, to descend through equal spaces in equal times, and all bodies, hung by equal threads, to vibrate in equal times. If the action of the force was greater, the times would be less; if that was less, these would be greater.

But it has been long ago observed by others, that (allowance being made for the small resistance of the air) all bodies descend through equal spaces in equal times; and, by the help of pendulums, that equality of times may be observed with great exactness.

I tried the thing in gold, silver, lead, glass, sand, common salt, wood, water, and wheat. I provided two equal wooden boxes. I filled the one with wood, and suspended an equal weight of gold (as exactly as I could) in the centre of oscillation of the other. The boxes, hung by equal threads of 11 feet, made a couple of pendulums perfectly equal in weight and figure, and equally exposed to the resistance of the air: and, placing the one by the other, I observed them to play together forwards and backwards for a long while, with equal vibrations. And therefore the quantity of matter in the gold was to the quantity of matter in the wood as the action of the motive force upon all the gold to the action of the same upon all the wood; that is, as the weight of the one to the weight of the other.

And by these experiments, in bodies of the same weight, one could have discovered a difference of matter less than the thousandth part of the whole.

20 The Agreement of Those Analogies

Since the action of the centripetal force upon the bodies attracted is, at equal distances, proportional to the quantities of matter in those bodies, reason requires that it should be also proportional to the quantity of matter in the body attracting.

For all action is mutual, and makes the bodies approach one to the other, and therefore must be the same in both bodies. It is true that we may consider one body as attracting, another as attracted; but this distinction is more mathematical than natural. The attraction resides really in each body towards the other, and is therefore of the same kind in both.

Figure acknowledgment

Figure 6: from Newton, selections from "System of the World," from *Newton: Texts, Backgrounds, Commentaries*, selected and edited by I. B. Cohen and R. S. Westfall (New York: W. W. Norton & Co., 1995), selections from pp. 259–66, p. 260.

Unit 3
The Modern Philosophers

As the scientific revolution unfolded, philosophy underwent a parallel upheaval. The reaction against Aristotle, once begun, swept across the intellectual landscape of the seventeenth century like a wildfire. The philosophical positions that grew up in its wake were quite diverse and not always consistent with each other, but in varying degrees they all represented a departure from Aristotle.

Several of the key figures were both philosophers and scientists in their own right, and it is to some extent a matter of arbitrary classification to place them in one group or the other. René Descartes and Gottfried Wilhelm von Leibniz, in particular, could just as easily have been classified among the scientists. But the two disciplines were not yet clearly separated, and the word 'scientist' would not be coined until the early nineteenth century. What matters most are not the labels we put on them but the contents of their works.

1 Francis Bacon: Naive Empiricism and Scientific Method

The name of Francis Bacon is now firmly established in the history of ideas as the philosophical spokesman for the scientific revolution. Certainly it is a distinction he actively sought; in his *Novum Organum* and *New Atlantis* he set out very self-consciously to reform scientific thought. But although Bacon was many things in his checkered career – lawyer, counselor to Kings, Lord Chancellor of England, and disgraced ex-Chancellor – he was not a major figure in either science or philosophy during his own lifetime. Indeed, he had a very imperfect grasp of the revolution happening around him, and the reaction of scientists to Bacon's earnest works on scientific method was decidedly mixed.

What chiefly motivated Bacon was exasperation at the apparent stagnation of science. His diagnosis of the causes of this stagnation is in part a diagnosis of the frailty of the human mind, which is incapable of its greatest achievements so long as it remains captive to four "Idols." Idols of the Tribe arise from the nature of the human understanding, which distorts reality as a warped mirror distorts an image. Idols of the Cave are those cherished falsehoods that distort the judgment of an individual who clings to them without evidence. Idols of the Marketplace arise from the use of language and our tendency to reason regarding words rather than things. Finally, Idols of the Theater are distortions of thinking induced by traditional philosophical systems. The human mind, according to Bacon, must be cleansed of all four sorts of idols in order to function as it should.

No doubt the largest target in the fourth group of idols was the Aristotelian philosophy, which Bacon attacks savagely. Yet his suggested replacement is in many ways quite similar to the view he is trying so desperately to overcome. Part of the explanation for this irony is the fact that Aristotle's own empiricism had been to some extent submerged in the hands of his most enthusiastic followers, so that a philosophical system that had originally grown up almost too closely linked to everyday experience had by Bacon's time become a matter of deduction from first principles laid down by Aristotle or abstracted from his writings.

To overcome the stagnation of the sciences, Bacon proposes an almost mechanical method for generating discoveries. In broad outline, Bacon's procedure for discovering the true cause of some phenomenon is to list the possible causes and then, through careful experimentation, to eliminate them one by one until at last only the true cause remains. This is, of course, easier than it sounds; there is some evidence that Bacon came to realize that a comprehensive list is not nearly as easy to create as he had hoped. Still, the idea of a mechanical procedure for the production of scientific discoveries has a certain attraction. It minimizes the significance of individual genius; if discoveries can be achieved through plodding hard work, what need is there for bold and imaginative thinkers? Bacon himself seems to have had little need for them and little understanding of their worth. With spectacular bad judgment he rejected some of the central discoveries of the great minds of the scientific revolution – Nicholas Copernicus about the solar system, William Gilbert about magnetism, and William Harvey about the circulation of the blood. For their part, those scientists who took notice of the methodological portions of Bacon's works do not seem to have thought much of them. Harvey famously quipped that Bacon "wrote philosophy like a Lord Chancellor,"[1] and when Isaac Newton outlined his own methodology to Henry Oldenburg he dismissed the Baconian approach with the remark that he had not arrived at his theories "by barely inferring 'tis thus because not otherwise" (reading 2.14).

In his second proposal, Bacon was more fortunate. He saw the future of scientific work as a collaborative effort in which individuals need not have mastery of the full range of ideas and methods in order for their contributions to mesh. To a significant extent, this vision was borne out within the next century, which saw the founding of scientific societies and the publication and dissemination of their works. The purpose of scientific work, moreover, was for the betterment of the human condition; on these terms scientific labor was honorable, and scientists could see their activities as part of a collective humanitarian endeavor even when their individual results might not seem to be tending toward such ends. It was Bacon who first saw clearly the structure and utility of such work, which Thomas Kuhn would later christen "normal science"[2] (see Reading 7.2).

Bacon's reputation has fluctuated wildly in the nearly four centuries since his death. Despite his misevaluation of the work of several of the great minds of the scientific revolution, he had a gift for articulating important truths in memorable form, and his thought-provoking aphorisms are widely quoted even today. "All true and fruitful Natural Philosophy," he wrote, "has a double scale or ladder, ascendant and descendant, ascending from experiments to axioms and descending from axioms to the invention of new experiments."[3] The admirable balance between top-down and bottom-up reasoning expressed in this quotation was upset in the next century, as the followers of Newton in England and of Leibniz and Christiaan Huygens on the continent fell into a protracted dispute regarding the relative importance of inferring data from hypotheses and ascending to the laws of nature from experimental evidence. Reading 3.1 presents a collection of some of Bacon's most memorable aphorisms from the *Novum Organum*.

2 René Descartes: Radical Rationalism

If Bacon lacked credentials as a scientist and a philosopher, Descartes could boast of both. His "La Géométrie" changed the face of mathematics by bringing geometry and algebra together with the coordinate axes now universally known as "Cartesian," and his *Principles of Philosophy* exerted a great influence on the next several generations of scientists even though in the end many of his specific doctrines proved false. In philosophy, Descartes's reputation is secure for all time; the *Meditations on First Philosophy*

(1641) would by itself suffice to earn him the title "father of modern philosophy."

Like Bacon, Descartes was deeply mistrustful of Aristotelian science. But where Bacon advocated an extreme form of empiricism, Descartes moved in the opposite direction, arguing that knowledge both general and certain could not be founded on the evidence of the senses alone but required, rather, that we apprehend with reason the most general features of physical body. This emphasis on reason shows up in reading 3.2, taken from Descartes's *Discourse on the Method of Rightly Conducting One's Reason and Seeking the Truth in the Sciences*, where he argues that what we most need in approaching scientific problems is the proper method. The contrast with Bacon's proposal is striking: whereas Bacon advocates acquiring a great mass of observations and then ruling out alternative hypotheses through observation and experiment, Descartes proposes that what we really need to do is to break down complex problems into their simplest parts and then approach these parts in the proper order. Bacon's method is designed to help us find things that are hidden; Descartes's is suited for solving mathematical problems.

This is not an accident, for on Descartes's view much of physical science is very nearly reducible to an exercise in armchair reasoning. This comes out clearly in reading 3.3 from the *Principles of Philosophy*, where he rejects the form of atomism that was gaining popularity among other scientists on largely a priori grounds. This section of that work must have annoyed Isaac Newton greatly; it is a paradigm of reasoning from preconceived hypotheses rather than ascending carefully from observation and experiment, and most of Descartes's substantive proposals are false. Nevertheless, he gets one thing exactly right: Descartes realizes that an object moving without any forces acting upon it will travel at a constant rate in a straight line. In this respect he sees more clearly than Kepler, who thought that a moving object would have a natural tendency to slow down, and than Galileo, who believed that unimpeded objects would move forever in a circle. When Newton brought together the best ideas of his predecessors in the great synthesis of the *Principia Mathematica*, he had harsh words for many of Descartes's ideas. But the Cartesian principle of inertia is one of the foundation stones of Newtonian dynamics.

3 John Locke: Moderate Empiricism

If anyone deserves to be called the philosopher of the scientific revolution, it is the physician and philosopher John Locke. Unlike Bacon, Locke knew some of the greatest minds of the revolution at first hand, and his *Essay Concerning Human Understanding* developed out of meetings of some of the members of the Royal Society in which they explicitly discussed the epistemological implications of the new science – and the scientific implications of the new epistemology. Though Locke made no noteworthy contributions to science and describes himself modestly as merely an "underlabourer" clearing a little of the ground for the great architects of the revolution, his *Essay* gives us one of the best views of the philosophy that lay behind the work of the best scientific minds in England toward the end of the seventeenth century.

Locke is often called an empiricist, and the label is no doubt appropriate. But it does not tell us much. Aristotle was certainly an empiricist in some sense of the term, yet Locke's views are very different from Aristotle's. It is more revealing to characterize Locke's position as an example of *concept empiricism* and *judgment empiricism*. Locke is a concept empiricist because he maintains that all of our concepts either are given in experience or arise from reflection on concepts given in experience. His judgment empiricism is reflected in the position that all of our knowledge of the world depends, ultimately, on information we obtain through sensory experience. Since knowledge of the essential natures of objects cannot be obtained directly in sensation, it would at best have to be inferred from experience.

Locke is, however, rather pessimistic about the possibility of our discovering the essential natures of things, and in reading 3.4 he gives three reasons for this. First, in many cases we simply lack the ability to obtain the relevant ideas, as a dim-sighted mole could not conceive of the telescopic vision of the eagle; in fact, Locke suggests, our direct experience is so narrowly bounded by our own limitations that it bears no significant proportion to the range of what might be known. Second, even those ideas we might in principle obtain through experience are limited by the reach of our experience. Remote stars and the minute realm of the microscopically small are both inaccessible to our direct gaze. Third, the actual mechanical causes of our ideas in some cases bear no resemblance to the sensations they produce in us. For example, a matrix of molecules at the surface of an object reflects certain rays of light but not others. We perceive the surface as red, but the color, as we experience it, is a secondary property; the individual molecules are not themselves red. The connection, Locke says, is one that we are simply incapable of seeing. It must be attributed to the decision of God, who knows better than we do and has given us such limited knowledge of nature as is necessary for life but has not given us deep scientific insight into the intrinsic natures of things.

Some of Locke's pessimistic arguments were destined to be overcome by the refinement of instruments being made by his colleagues in the Royal Society and their correspondents abroad. In 1672 Newton invented a reflecting telescope that circumvented the problem of chromatic aberration; around the same time Robert Hooke and Anton van Leeuwenhoek were probing the world of the minute with a succession of ever-improving microscopes. Later discoveries would permit scientists to probe the world inaccessible to our senses. But it remains true that our experience, though vastly wider than it was in Locke's time, is still very limited. Some means of inference is required to move us beyond what we experience to conclusions about the unobservable. The difficulty of determining how such inferences might go and how to justify them remains one of the central problems of philosophy.

4 Gottfried Wilhelm von Leibniz: Sophisticated Rationalism

The sharp contrast between English and Continental science is most clearly visible in the clash between Newton and the German polymath Leibniz. Both men were brilliant mathematicians. Newton invented calculus somewhat earlier than Leibniz but did not publish his discovery at first; Leibniz independently hit upon the same ideas and published them. There ensued a bitter and protracted priority dispute that diverted the intellectual energies of some of the best minds of the time and did credit to neither side. The dispute over the calculus set the tone for their subsequent, and more interesting, disputes concerning physical science.

Leibniz's approach to nature is profoundly mathematical, but in his metaphysics he is prepared to contemplate things only suggested by the mathematics. Reading 3.5, an excerpt from one of his unpublished papers, shows this clearly. After extolling the wisdom of God in the structure of creation, Leibniz gives a geometric analysis of reflection from a curved surface and shows that the path taken by reflected light, though not forced upon us by any known principle of geometry, is nevertheless determined by certain constraints that can be geometrically described. In a surprising move that would never have occurred to a seventeenth-century Englishman, Leibniz suggests that the explanation for this version of the principle of least action may lie in the Aristotelian doctrine of final causes – in an appeal to the purposes for which things happen. While the English scientific establishment was busy exorcising the last vestiges of Aristotelianism from science, Leibniz was actively seeking insights from that tradition. These two philosophical poles continue to exert an attraction on scientific minds as different as Ernst Mach and Charles Sanders Peirce down through the twentieth century.

One of the most interesting and revealing examples of Leibniz's reasoning and his view of science came about in a curious fashion. In November of 1715, Leibniz wrote a letter to Caroline of Ansbach, the Princess of Wales, to whom he had acted as a philosophical mentor, expressing some criticisms of Newton's views. The chief burden of Leibniz's criticisms was that religion was decaying in England since Locke and Newton were leading their followers into materialism, denying the immateriality of the soul and suggesting that God had need of an organ with which to perceive physical bodies. Caroline shared the letter with her friend Samuel Clarke, who was a close friend of Newton. There followed an exchange of letters, mediated and preserved by Caroline, between Leibniz and Clarke; but it is plain that some of the content of Clarke's letters came from Newton himself.

The first few exchanges feature some sparring over the supposedly atheistical tendencies of atomism, but in the selections reproduced here as reading 3.6 they move into some deep questions regarding the principle of sufficient reason and the nature of space. Newton had maintained in the *Principia* that space is a thing that has a nature and structure independent of the material things passing through it; Leibniz protests that space is merely relative, the ordered co-existence of material things, and that the Newtonian view violates the principle of sufficient reason. The exchange continued, the letters becoming longer and more nuanced, with neither party willing to give way to the other; it was terminated by Leibniz's death in 1716. From a contemporary perspective it is quite difficult to adjudicate the controversy; modern physics appears to endorse some components of each of the opposing views. We may be fairly certain that neither Leibniz nor Newton would have been satisfied.

5 David Hume: Radical Empiricism

If Leibniz and Clarke were concerned to defend their systems against the charge that they encouraged atheism, that issue was the furthest thing from the mind of David Hume. The brilliant Scottish philosopher had abandoned revealed religion in his youth and never looked backward. He was, however, deeply interested in the empiricist account of human understanding offered by Locke, and in his *Enquiry Concerning Human Understanding* he modified Locke's views – he would have called it a consistent outworking of Locke's own principles – until he arrived at an extreme form of empiricism with radically skeptical consequences.

On Locke's view, it is difficult for us to arrive at knowledge of matters of fact that lie beyond our direct experience. In Hume's hands, this difficulty becomes an impossibility. For in order to know matters of fact beyond our experience, we must reason from our experience. The reasoning cannot be deductive, for our experience does not *entail* matters of fact beyond it; it is always logically possible that our experience be as we find it but that any given matter of fact be true or be false. This does not sound too serious. Logical possibility is fairly cheap, and granting that something is logically possible does not commit one to saying that it is plausible. But here Hume unleashes his famous problem of induction. What is the nature of the reasoning that takes us beyond our experience? How can we arrive even at reasonable beliefs about things of which we are not directly aware? Starting with some of the same examples Locke used regarding the limitations of our knowledge, Hume draws sweeping skeptical conclusions. Not only do we not know the reasons that the microstructure of bread is such that it nourishes us, we do not even know, on the basis of our past experience, that bread which has nourished us at one time will do so at another. For to make such an inference, we would have to know that bread causes health in the body. But to know that, we have to know that what has caused health in the past will do so in the future – and that the future will resemble the past is the very thing we are trying to show.

Hume's problem of induction, presented here in reading 3.7 from the *Enquiry Concerning Human Understanding* (1748), is perhaps the most celebrated problem in the whole of philosophy. The conclusion seems absurd; of course experience gives us justified beliefs about matters that go beyond experience. Yet the reasoning Hume uses to reach that conclusion is disturbingly plausible, at least to those who have given up on Aristotle's idea that the mind can abstract the essential natures of things in some manner other than inference from bare sensory data. It is a measure of Hume's success in stating the problem that even at the dawn of the twenty-first century there is nothing remotely like a consensus among philosophers regarding the proper solution to the problem of induction.

If the problem of induction is an epistemological puzzle, Hume's treatment of causality in reading 3.8, also drawn from the *Enquiry*, poses a problem in concept analysis. What is a causal law – what do we mean when we say that events of type A cause events of type B? Hume answers that it is an unbroken correlation of instances in which A causes B; causal laws are simply shorthand statements for collections of particular causal facts. But what does it mean to say that, in a particular instance, A causes B? Here again Hume's view is reductive: to say that A causes B is simply to say that B invariably follows A and the two events are spatially contiguous – they happen next to one another. We may naively think that there is something deeper behind the scenes, perhaps invisible to us but nevertheless real, that binds things together causally. Hume's response is that we never see such "hidden connections" and have no reason to infer them; there is nothing of value in our ordinary notion of causation that is not captured by his doubly reductive analysis, nothing we needed to say that we cannot say in terms of his account. The challenge for anyone who would disagree with Hume's radical proposal is to articulate just what is wrong with it and to supply a defensible alternative. Here again, there is no consensus on whether Hume was right or, if he was not, on what goes wrong with his analysis.

Hume himself was happy to accept the results of Newtonian science. But he did so, as he would say, merely out of custom and habit; what he denied was that there is any sound chain of reasoning that leads from the meager data with which we are acquainted at first hand to conclusions about matters beyond our experience. Many subsequent critics have found his blend of sophisticated skepticism incompatible with his complacent acquiescence in both scientific and everyday conclusions that could not, by Hume's own admission, be justified on his own terms. But his skeptical challenges have resisted easy solution, and strands of his thought run like rhizomes under the philosophical landscape, cropping up again and again.

6 Immanuel Kant's "Copernican Revolution"

Hume's writings struck a nerve almost immediately. The German professor Immanuel Kant described Hume's skeptical challenges as rousing him from his "dogmatic slumber" and forcing him to rethink the foundations of metaphysics and epistemology. The result was a massive and difficult work, the *Critique of Pure Reason*, in which Kant attempted to meet and overcome Hume's challenges by a profound reorientation of some of our most fundamental ideas regarding reality and experience – a reorientation that Kant himself referred to as a "Copernican Revolution" in philosophy.

To understand Kant's dilemma, it is important to see how he viewed the new science. Newton, Kant believed, had not merely discovered the laws of nature: he had achieved *certainty* regarding the laws of nature. Yet on Hume's empiricist principles, this was utterly impossible; the laws of nature go far beyond the experience of any individual, and Hume's empiricism provides no resources for bridging the gap in a rational way. Under the pressure of his need to reconcile these two beliefs, Kant opted for the conclusion that, contrary to Hume, reason must be capable of grasping some truths about nature a priori. Yet Kant accepted Hume's argument that we can never have a priori knowledge of any matter of fact independent of our minds. Again, he found a way to reconcile all of these beliefs: nature is not wholly independent of our minds but is partly constituted by the structures of our understanding.

The details of Kant's *Critique* are intricate, but a few examples give some sense of how he employed this idea. Newton's conception of absolute space is Euclidean – lines and planes in space have just the properties that Euclid's geometry ascribes to them. Kant suggests that this is because we are creatures who could not experience space in any other way. We could not, according to Kant, experience objects otherwise than in Euclidean space, because our minds are bound to represent the world that way; in fact, creatures of any kind who are capable of sensory experience would be subject to the same condition. We may therefore have certainty regarding the structure of space. Similarly, other concepts such as cause and effect are categories of the understanding; we could not experience the world except as a causally connected system because the notion of causation is built into the very structure of our understanding. This is why we may legitimately be certain that every event has a cause.

Kant's *Critique* is punishing reading even by philosophers' standards. But in a shorter work, the *Metaphysical Foundations of Natural Science* (1786), he gives a more lucid argument for his position as it applies to science. The heart of his argument appears here as reading 3.9. Echoing Leibniz, Kant says that natural science is not merely a haphazard piling up of knowledge; it requires principles, not merely mathematical or geometric principles, but *metaphysical* principles that cannot be justified by inference from experience since they are part of the conditions that make experience itself possible. This form of argument, that X must exist because it is a necessary condition for the possibility of experience, is sometimes called a *transcendental argument*, and it is a major weapon in Kant's philosophical arsenal.

Like Hume, Kant cast a long shadow over all subsequent philosophers, and his work frequently influenced scientists and historians of science as well. Relatively few thinkers accepted Kant's views without modification. But his influence shows itself in William Whewell's resistance to John Stuart Mill's empiricism, in Peirce's pragmatism, and even in Albert Einstein's philosophical writings and correspondence. It has become part of the cultural heritage of both science and philosophy.

Notes

1 The comment is related in John Aubrey, *Minutes of Lives*, subsequently published under the title *Brief Lives, Edited from The Original Manuscripts And with A Life of John Aubrey*, by Oliver Lawson Dick (Ann Arbor, MI: University of Michigan Press, 1957).

2 See selection 7.3 and §6 of the commentary to Unit 7.

3 *The Advancement of Learning* (1605) II.7.1.

Suggestions for Further Reading

Items marked with an asterisk assume little or no background in the subject; these would be a good place to begin further reading.

*Buchdahl, Gerd, 1969, *Metaphysics and the Philosophy of Science. The Classical Origins: Descartes to Kant.* Cambridge, MA: MIT Press.

Cottingham, John, 1988, *The Rationalists.* Oxford: Oxford University Press.

Garber, Daniel, 1992, *Descartes' Metaphysical Physics.* Chicago: University of Chicago Press.

Gilbert, Neal W., 1960, *Renaissance Concepts of Method.* New York: Columbia University Press.

Hall, A. Rupert, 1980, *Philosophers at War: The Quarrel between Newton and Leibniz.* Cambridge: Cambridge University Press.

Lowe, E. J., 1995, *Locke on Human Understanding.* New York: Routledge.

*Popkin, Richard, 2003, *The History of Skepticism from Erasmus to Descartes.* Oxford: Oxford University Press.

Tlumak, Jeffrey, 2006, *Classical Modern Philosophy: A Contemporary Introduction.* New York: Routledge.

Vailati, Ezio, 1997, *Leibniz and Clarke: A Study of Their Correspondence.* Oxford: Oxford University Press.

Williams, Bernard, 2006, *Descartes: The Project of Pure Inquiry.* New York: Routledge.

Woolhouse, R. S., 1988, *The Empiricists.* Oxford: Oxford University Press.

3.1

The Inductive Method

Francis Bacon

Francis Bacon (1561–1626) was a statesman, essayist, and philosopher whose vision of the systematic interrogation of nature and the collective nature of the scientific enterprise crystallized the vision of the scientific revolution of the seventeenth century into crisp prose. In this selection from his *New Organon*, Bacon lamints the lack of sure progress in the understanding of nature and outlines a new "inductive" method – the elimination of faulty rival explanations until only the true are left – that will render individual talent unimportant.

Book I

Aphorism I

Man is Nature's agent and interpreter; he does and understands only as much as he has observed of the order of nature in fact or by inference; he does not know and cannot do more.

II

Neither the bare hand nor the unaided intellect has much power; the work is done by tools and assistance, and the intellect needs them as much as the hand. As the hand's tools either prompt or guide its motions, so the mind's tools either prompt or warn the intellect.

III

Human knowledge and human power come to the same thing, because ignorance of cause frustrates effect. For Nature is conquered only by obedience; and that which in thought is a cause, is like a rule in practice.

[. . .]

From *Francis Bacon: The New Organon*, ed. Lisa Jardine and Michael Silverthorne (West Nyack, NY: Cambridge University Press, 2000), pp. 33, 34, 36–7, 39, 40–2, 45–6, 49–50, 83–4.

VIII

Even the results which have been discovered already are due more to chance and experience than to sciences; for the sciences we now have are no more than elegant arrangements of things previously discovered, not methods of discovery or pointers to new results.

IX

The cause and root of nearly all the deficiencies of the sciences is just this: that while we mistakenly admire and praise the powers of the human mind, we do not seek its true supports.

[. . .]

XVIII

The things that have hitherto been discovered in the sciences all fit nicely into common notions; in order to penetrate to the more inward and remote parts of nature, both notions and axioms must be abstracted from things in a more certain, better-grounded way; and a more certain and altogether better intellectual procedure must come into use.

XIX

There are, and can be, only two ways to investigate and discover truth. The one leaps from sense and particulars to the most general axioms, and from these principles and their settled truth, determines and discovers intermediate axioms; this is the current way. The other elicits axioms from sense and particulars, rising in a gradual and unbroken ascent to arrive at last at the most general axioms; this is the true way, but it has not been tried.

XX

Left to itself the intellect goes the same way as it does when it follows the order of dialectic (i.e. the first of the two ways above). The mind loves to leap to generalities, so that it can rest; it only takes it a little while to get tired of experience. These faults have simply been magnified by dialectic, for ostentatious disputes.

XXI

In a sober, grave and patient character the intellect left to itself (especially if unimpeded by received doctrines) makes some attempt on that other way, which is the right way, but with little success; since without guidance and assistance it is a thing inadequate and altogether incompetent to overcome the obscurity of things.

XXII

Both ways start from sense and particulars, and come to rest in the most general; but they are vastly different. For one merely brushes experience and particulars in passing, the other deals fully and properly with them; one forms certain abstract and useless generalities from the beginning, the other rises step by step to what is truly better known by nature.

[. . .]

XXXI

It is futile to expect a great advancement in the sciences from overlaying and implanting new things on the old; a new beginning has to be made from the lowest foundations, unless one is content to go round in circles for ever, with meagre, almost negligible, progress.

[. . .]

XXXVI

There remains one simple way of getting our teaching across, namely to introduce men to actual particulars and their sequences and orders, and for men in their turn to pledge to abstain for a while from notions, and begin to get used to actual things.

XXXVII

In its initial positions our way agrees to some extent with the method of the supporters of lack of conviction; but in the end our ways are far apart and strongly opposed. They assert simply that nothing can be known; but we say that not much can be known in nature by the way which is now in use. They thereupon proceed to destroy the authority of sense and intellect; but we devise and provide assistance to them.

XXXVIII

The illusions and false notions which have got a hold on men's intellects in the past and are now profoundly rooted in them, not only block their minds so that it is difficult for truth to gain access, but even when access has been granted and allowed, they will once again, in the very renewal of the sciences, offer resistance and do mischief unless men are forewarned and arm themselves against them as much as possible.

XXXIX

There are four kinds of illusions which block men's minds. For instruction's sake, we have given them the following names: the first kind are called idols of the tribe; the second idols of the cave; the third idols of the marketplace; the fourth idols of the theatre.

XL

Formation of notions and axioms by means of true induction is certainly an appropriate way to banish idols and get rid of them; but it is also very useful to identify the idols. Instruction about idols has the same relation to the interpretation of nature as teaching the sophistic refutations has to ordinary logic.

XLI

The idols of the tribe are founded in human nature itself and in the very tribe or race of mankind. The assertion that the human senses are the measure of things is false; to the contrary, all perceptions, both of sense and mind, are relative to man, not to the universe. The human understanding is like an uneven mirror receiving rays from things and merging its own nature with the nature of things, which thus distorts and corrupts it.

XLII

The idols of the cave are the illusions of the individual man. For (apart from the aberrations of human nature in general) each man has a kind of individual cave or cavern which fragments and distorts the light of nature. This may happen either because of the unique and particular nature of each man; or because of his upbringing and the company he keeps; or because of his reading of books and the authority of those whom he respects and admires; or because of the different impressions things make on different minds, preoccupied and prejudiced perhaps, or calm and detached, and so on. The evident consequence is that the human spirit (in its different dispositions in different men) is a variable thing, quite irregular, almost haphazard. Heraclitus well said that men seek knowledge in lesser, private worlds, not in the great or common world.

XLIII

There are also illusions which seem to arise by agreement and from men's association with each other, which we call idols of the marketplace; we take the name from human exchange and community. Men associate through talk; and words are chosen to suit the understanding of the common people. And thus a poor and unskilful code of words incredibly obstructs the understanding. The definitions and explanations with which learned men have been accustomed to protect and in some way liberate themselves, do not restore the situation at all. Plainly words do violence to the understanding, and confuse everything; and betray men into countless empty disputes and fictions.

[. . .]

L

But much the greatest obstacle and distortion of human understanding comes from the dullness, limitations and deceptions of the senses; so that things that strike the senses have greater influence than even powerful things which do not directly strike the senses. And therefore thought virtually stops at sight; so that there is little or no notice taken of things that cannot be seen. And so all operation of spirits enclosed in tangible bodies remains hidden and escapes men's notice. And all the more subtle structural change in the parts of dense objects (which is commonly called alteration, although in truth it is movement of particles) is similarly hidden. Yet unless the two things mentioned are investigated and brought to the light, nothing important can be done in nature as far as results are concerned. Again, the very nature of the common air and of all the bodies which surpass air in rarity (of which there are many) is virtually unknown. For by itself sense is weak and prone to error, nor do instruments for amplifying and sharpening the senses do very much. And yet every interpretation of nature which has a chance to be true is achieved by instances, and suitable and relevant experiments, in which sense only gives a judgement on the experiment, while the experiment gives a judgement on nature and the thing itself.

LI

The human understanding is carried away to abstractions by its own nature, and pretends that things which are in flux are unchanging. But it is better to dissect nature than to abstract; as the school of Democritus did, which penetrated more deeply into nature than the others. We should study matter, and its structure (schematismus), and structural change (meta-schematismus), and pure act, and the law of act or motion; for forms are figments of the human mind, unless one chooses to give the name of forms to these laws of act.

[. . .]

LXI

Idols of the theatre are not innate or stealthily slipped into the understanding; they are openly introduced and accepted on the basis of fairytale theories and mistaken rules of proof. It is not at all consistent with our argument to attempt or undertake to refute them.

There is no possibility of argument, since we do not agree either about the principles or about the proofs. It is a happy consequence that the ancients may keep their reputation. I take nothing from them, since the question is simply about the way. As the saying goes, a lame man on the right road beats the runner who misses his way. It is absolutely clear that if you run the wrong way, the better and faster you are, the more you go astray.

Our method of discovery in the sciences is designed not to leave much to the sharpness and strength of the individual talent; it more or less equalises talents and intellects. In drawing a straight line or a perfect circle, a good deal depends on the steadiness and practice of the hand, but little or nothing if a ruler or a compass is used. Our method is exactly the same. But though there is no point in specific refutations, something must be said about the sects and kinds of such theories; and then of the external signs that the situation is bad; and lastly of the reasons for so much failure, and such persistent and general agreement in error; so that there may be easier access to true things, and the human understanding may be more willing to cleanse itself and dismiss its idols.

[. . .]

CV

In forming an axiom we need to work out a different form of induction from the one now in use; not only to demonstrate and prove so-called principles, but also lesser and intermediate axioms, in fact all axioms. For the induction which proceeds by simple enumeration is a childish thing, its conclusions are precarious, and it is exposed to the danger of the contrary instance; it normally bases its judgement on fewer instances than is appropriate, and merely on available instances. But the induction which will be useful for the discovery and proof of sciences and arts should separate out a nature, by appropriate rejections and exclusions; and then, after as many negatives as are required, conclude on the affirmatives. This has not yet been done, nor even certainly tried except only by Plato, who certainly makes use of this form of induction to some extent in settling on definitions and ideas. But any number of things need to be included in a true, legitimate account of this kind of induction or demonstration, which have never occurred to anyone to think about, so that more effort needs to be put into this than has ever been spent on the syllogism. It is this kind of induction whose help we must have not only to discover axioms but also to define concepts. And we may certainly have the greatest hopes for this kind of induction.

3.2

Rules for the Discovery of Scientific Truth

René Descartes

René Descartes (1596–1650), a French mathematician and scientist, is best known as the "father of modern philosophy." His work in the theory of knowledge was motivated by his strong desire to find certainty in the sciences, something he thought he could achieve by beginning with certainties and following an appropriate procedure. In this selection from his early work, *Discourse on the Method of Rightly Conducting One's Reason and Seeking the Truth in the Sciences* (composed about 1626–28 but first published posthumously in 1684), Descartes lays out four rules that he believes will, if rigidly obeyed, enable him to discover scientific truth.

Discourse on the Method of Rightly Conducting One's Reason and Seeking the Truth in the Sciences

Part One

Good sense is the best distributed thing in the world: for everyone thinks himself so well endowed with it that even those who are the hardest to please in everything else do not usually desire more of it than they possess. In this it is unlikely that everyone is mistaken. It indicates rather that the power of judging well and of distinguishing the true from the false – which is what we properly call 'good sense' or 'reason' – is naturally equal in all men, and consequently that the diversity of our opinions does not arise because some of us are more reasonable than others but solely because we direct our thoughts along different paths and do not attend to the same things. For it is not enough to have a good mind; the main thing is to apply it well. The greatest souls are capable of the greatest vices as well as the greatest virtues; and those who proceed but very slowly can make much greater progress, if they always follow the right path, than those who hurry and stray from it.

From *The Philosophical Writings of Descartes*, Vol. 1. Trans. by J. Cottingham, R. Stoothoff and D. Murdoch (Cambridge: Cambridge University Press, 1985), pp. 111, 118–20.

[. . .]

Part Two

[. . .]

. . . I cannot by any means approve of those meddlesome and restless characters who, called neither by birth nor by fortune to the management of public affairs, are yet forever thinking up some new reform. And if I thought this book contained the slightest ground for suspecting me of such folly, I would be very reluctant to permit its publication. My plan has never gone beyond trying to reform my own thoughts and construct them upon a foundation which is all my own. If I am sufficiently pleased with my work to present you with this sample of it, this does not mean that I would advise anyone to imitate it. Those on whom God has bestowed more of his favours will perhaps have higher aims; but I fear that even my aim may be too bold for many people. The simple resolution to abandon all the opinions one has hitherto accepted is not an example that everyone ought to follow. The world is largely composed of two types of minds for whom it is quite unsuitable. First, there are those who, believing themselves cleverer than they are, cannot avoid precipitate judgements and never have the patience to direct all their thoughts in an orderly manner; consequently, if they once took the liberty of doubting the principles they accepted and of straying from the common path, they could never stick to the track that must be taken as a short-cut, and they would remain lost all their lives. Secondly, there are those who have enough reason or modesty to recognize that they are less capable of distinguishing the true from the false than certain others by whom they can be taught; such people should be content to follow the opinions of these others rather than seek better opinions themselves.

For myself, I would undoubtedly have been counted among the latter if I had had only one teacher or if I had never known the differences that have always existed among the opinions of the most learned. But in my college days I discovered that nothing can be imagined which is too strange or incredible to have been said by some philosopher; and since then I have recognized through my travels that those with views quite contrary to ours are not on that account barbarians or savages, but that many of them make use of reason as much or more than we do. I thought, too, how the same man, with the same mind, if brought up from infancy among the French or Germans, develops otherwise than he would if he had always lived among the Chinese or cannibals; and how, even in our fashions of dress, the very thing that pleased us ten years ago, and will perhaps please us again ten years hence, now strikes us as extravagant and ridiculous. Thus it is custom and example that persuade us, rather than any certain knowledge. And yet a majority vote is worthless as a proof of truths that are at all difficult to discover; for a single man is much more likely to hit upon them than a group of people. I was, then, unable to choose anyone whose opinions struck me as preferable to those of all others, and I found myself as it were forced to become my own guide.

But, like a man who walks alone in the dark, I resolved to proceed so slowly, and to use such circumspection in all things, that even if I made but little progress I should at least be sure not to fall. Nor would I begin rejecting completely any of the opinions which may have slipped into my mind without having been introduced there by reason, until I had first spent enough time in planning the work I was undertaking and in seeking the true method of attaining the knowledge of everything within my mental capabilities.

[. . .] Now a multiplicity of laws often provides an excuse for vices, so that a state is much better governed when it has but few laws which are strictly observed; in the same way, I thought, in place of the large number of rules that make up logic, I would find the following four to be sufficient, provided that I made a strong and unswerving resolution never to fail to observe them.

The first was never to accept anything as true if I did not have evident knowledge of its truth: that is, carefully to avoid precipitate conclusions and preconceptions, and to include nothing more in my judgements than what presented itself to my mind so clearly and so distinctly that I had no occasion to doubt it.

The second, to divide each of the difficulties I examined into as many parts as possible and as may be required in order to resolve them better.

The third, to direct my thoughts in an orderly manner, by beginning with the simplest and most easily known objects in order to ascend little by little, step by step, to knowledge of the most complex, and by supposing some order even among objects that have no natural order of precedence.

And the last, throughout to make enumerations so complete, and reviews so comprehensive, that I could be sure of leaving nothing out.

Those long chains composed of very simple and easy reasonings, which geometers customarily use to arrive at their most difficult demonstrations, had given me occasion to suppose that all the things which can fall under human knowledge are interconnected in the same way. And I thought that, provided we refrain from accepting anything as true which is not, and always keep to the order required for deducing one thing from another, there can be nothing too remote to be reached in the end or too well hidden to be discovered. I had no great difficulty in deciding which things to begin with, for I knew already that it must be with the simplest and most easily known. Reflecting, too, that of all those who have hitherto sought after truth in the sciences, mathematicians alone have been able to find any demonstrations – that is to say, certain and evident reasonings – I had no doubt that I should begin with the very things that they studied.

3.3

Rationalism and Scientific Method

René Descartes

In his more mature work, *The Principles of Philosophy* (1644), Descartes grapples directly with fundamental physical questions. Yet even here, his thinking is characterized by a downplaying of the senses and an emphasis on what reason alone can discover. In this selection from the *Principles*, Descartes applies this a priori approach to such topics as the vacuum (which he denied outright) and atomism (which he rejected in one form) and the nature and causes of motion and rest. Curiously, although he tries to uphold a circulatory theory of projectile motion, he hits upon the right conception of inertia. This was the only component of Descartes's scientific thought that found its way into Newton's magnificent synthesis; but it was critical.

The Principles of Material Things

1 The arguments that lead to the certain knowledge of the existence of material things

Everyone is quite convinced of the existence of material things. But earlier on we cast doubt on this belief and counted it as one of the preconceived opinions of our childhood. So it is necessary for us to investigate next the arguments by which the existence of material things may be known with certainty. Now, all our sensations undoubtedly come to us from something that is distinct from our mind. For it is not in our power to make ourselves have one sensation rather than another; this is obviously dependent on the thing that is acting on our senses. Admittedly one can raise the question of whether this thing is God or something different from God. But we have sensory awareness of, or rather as a result of sensory stimulation we have a clear and distinct perception of, some kind of matter, which is extended in length, breadth and depth,

From *The Philosophical Writings of Descartes*, Vol. 1. Trans. by J. Cottingham, R. Stoothoff and D. Murdoch (Cambridge: Cambridge University Press, 1985), pp. 223–5, 230–8, 240–5, 247.

and has various differently shaped and variously moving parts which give rise to our various sensations of colours, smells, pain and so on. And if God were himself immediately producing in our mind the idea of such extended matter, or even if he were causing the idea to be produced by something which lacked extension, shape and motion, there would be no way of avoiding the conclusion that he should be regarded as a deceiver. For we have a clear understanding of this matter as something that is quite different from God and from ourselves or our mind; and we appear to see clearly that the idea of it comes to us from things located outside ourselves, which it wholly resembles. And we have already noted that it is quite inconsistent with the nature of God that he should be a deceiver. The unavoidable conclusion, then, is that there exists something extended in length, breadth and depth and possessing all the properties which we clearly perceive to belong to an extended thing. And it is this extended thing that we call 'body' or 'matter'.

[...]

3 Sensory perception does not show us what really exists in things, but merely shows us what is beneficial or harmful to man's composite nature

It will be enough, for the present, to note that sensory perceptions are related exclusively to this combination of the human body and mind. They normally tell us of the benefit or harm that external bodies may do to this combination, and do not, except occasionally and accidentally, show us what external bodies are like in themselves. If we bear this in mind we will easily lay aside the preconceived opinions acquired from the senses, and in this connection make use of the intellect alone, carefully attending to the ideas implanted in it by nature.

4 The nature of body consists not in weight, hardness, colour, or the like, but simply in extension

If we do this, we shall perceive that the nature of matter, or body considered in general, consists not in its being something which is hard or heavy or coloured, or which affects the senses in any way, but simply in its being something which is extended in length, breadth and depth. For as regards hardness, our sensation tells us no more than that the parts of a hard body resist the motion of our hands when they come into contact with them. If, whenever our hands moved in a given direction, all the bodies in that area were to move away at the same speed as that of our approaching hands, we should never have any sensation of hardness. And since it is quite unintelligible to suppose that, if bodies did move away in this fashion, they would thereby lose their bodily nature, it follows that this nature cannot consist in hardness. By the same reasoning it can be shown that weight, colour, and all other such qualities that are perceived by the senses as being in corporeal matter, can be removed from it, while the matter itself remains intact; it thus follows that its nature does not depend on any of these qualities.

5 This truth about the nature of body is obscured by preconceived opinions concerning rarefaction and empty space

But there are still two possible reasons for doubting that the true nature of body consists solely in extension. The first is the widespread belief that many bodies can be rarefied and condensed in such a way that when rarefied they possess more extension than when condensed. Indeed, the subtlety of some people goes so far that they distinguish the substance of a body from its quantity, and even its quantity from its extension. The second reason is that if we understand there to be nothing in a given place but extension in length, breadth and depth, we generally say not that there is a body there, but simply that there is a space, or even an empty space; and almost everyone is convinced that this amounts to nothing at all.

6 How rarefaction occurs

But with regard to rarefaction and condensation, anyone who attends to his own thoughts, and is willing to admit only what he clearly perceives, will not suppose that anything happens in these processes beyond a change of shape. Rarefied bodies, that is to say, are those which have many gaps between their parts – gaps which are occupied by other bodies; and they become denser simply in virtue of the parts coming together and reducing or completely closing the gaps. In this last eventuality a body becomes so dense that it would be a contradiction to suppose that it could be made any denser. Now in this condition, the extension of a body is no less than when it occupies more space in virtue of the mutual separation of its parts; for whatever extension is comprised in the pores or gaps left between the parts must be attributed not to the body itself but to the various other bodies which fill the gaps. In just the same way, when we see a sponge filled with water or some other liquid, we do not suppose that in terms of its own individual parts it has a greater extension than when it is squeezed dry; we simply suppose that its pores are open wider, so that it spreads over a greater space.

[. . .]

17 The ordinary use of the term 'empty' does not imply the total absence of bodies

In its ordinary use the term 'empty' usually refers not to a place or space in which there is absolutely nothing at all, but simply to a place in which there is none of the things that we think ought to be there. Thus a pitcher made to hold water is called 'empty' when it is simply full of air; a fishpond is called 'empty', despite all the water in it, if it contains no fish; and a merchant ship is called 'empty' if it is loaded only with sand ballast. And similarly a space is called 'empty' if it contains nothing perceivable by the senses, despite the fact that it is full of created, self-subsistent matter; for normally the only things we give any thought to are those which are detected by our senses. But if we subsequently fail to keep in mind what ought to be understood by the terms 'empty' and 'nothing', we may suppose that a space we call empty contains not just nothing perceivable by the senses but nothing whatsoever; that would be just as mistaken as thinking that the air in a jug is not a subsistent thing on the grounds that a jug is usually said to be empty when it contains nothing but air.

18 How to correct our preconceived opinion regarding an absolute vacuum

Almost all of us fell into this error in our early childhood. Seeing no necessary connection between a vessel and the body contained in it, we reckoned there was nothing to stop God, at least, removing the body which filled the vessel, and preventing any other body from taking its place. But to correct this error we should consider that, although there is no connection between a vessel and this or that particular body contained in it, there is a very strong and wholly necessary connection between the concave shape of the vessel and the extension, taken in its general sense, which must be contained in the concave shape. Indeed, it is no less contradictory for us to conceive of a mountain without a valley than it is for us to think of the concavity apart from the extension contained within it, or the extension apart from the substance which is extended; for, as I have often said, nothingness cannot possess any extension. Hence, if someone asks what would happen if God were to take away every single body contained in a vessel, without allowing any other body to take the place of what had been removed, the answer must be that the sides of the vessel would, in that case, have to be in contact. For when there is nothing between two bodies they must necessarily touch each other. And it is a manifest contradiction for them to be apart, or to have a distance between them, when the distance in question is nothing; for every distance is a mode of extension, and therefore cannot exist without an extended substance.

[. . .]

20 The foregoing results also demonstrate the impossibility of atoms

We also know that it is impossible that there should exist atoms, that is, pieces of matter that are by their very nature indivisible as some philosophers have imagined. For if there were any atoms, then no matter how small we imagined them to be, they would necessarily have to be extended; and hence we could in our thought divide each of them into two or more smaller parts, and hence recognize their divisibility. For anything we can divide in our thought must, for that very reason, be known to be divisible; so if we were to judge it to be indivisible, our judgement would conflict with our knowledge. Even if we imagine that God has chosen to bring it about that some particle of matter is incapable of being divided into smaller particles, it will still not be correct, strictly speaking, to call this particle indivisible.

For, by making it indivisible by any of his creatures, God certainly could not thereby take away his own power of dividing it, since it is quite impossible for him to diminish his own power, as has been noted above. Hence, strictly speaking, the particle will remain divisible, since it is divisible by its very nature.

21 Similarly, the extension of the world is indefinite

What is more we recognize that this world, that is, the whole universe of corporeal substance, has no limits to its extension. For no matter where we imagine the boundaries to be, there are always some indefinitely extended spaces beyond them, which we not only imagine but also perceive to be imaginable in a true fashion, that is, real. And it follows that these spaces contain corporeal substance which is indefinitely extended. For, as has already been shown very fully, the idea of the extension which we conceive to be in a given space is exactly the same as the idea of corporeal substance.

22 Similarly, the earth and the heavens are composed of one and the same matter; and there cannot be a plurality of worlds

It can also easily be gathered from this that celestial matter is no different from terrestrial matter. And even if there were an infinite number of worlds, the matter of which they were composed would have to be identical; hence, there cannot in fact be a plurality of worlds, but only one. For we very clearly understand that the matter whose nature consists simply in its being an extended substance already occupies absolutely all the imaginable space in which the alleged additional worlds would have to be located; and we cannot find within us an idea of any other sort of matter.

23 All the variety in matter, all the diversity of its forms, depends on motion

The matter existing in the entire universe is thus one and the same, and it is always recognized as matter simply in virtue of its being extended. All the properties which we clearly perceive in it are reducible to its divisibility and consequent mobility in respect of its parts, and its resulting capacity to be affected in all the ways which we perceive as being derivable from the movement of the parts. If the division into parts occurs simply in our thought, there is no resulting change; any variation in matter or diversity in its many forms depends on motion. This seems to have been widely recognized by the philosophers, since they have stated that nature is the principle of motion and rest. And what they meant by 'nature' in this context is what causes all corporeal things to take on the characteristics of which we are aware in experience.

24 What is meant by 'motion' in the ordinary sense of the term

Motion, in the ordinary sense of the term, is simply *the action by which a body travels from one place to another*. By 'motion', I mean local motion; for my thought encompasses no other kind, and hence I do not think that any other kind should be imagined to exist in nature. Now I pointed out above that the same thing can be said to be changing and not changing its place at the same time; and similarly the same thing can be said to be moving and not moving. For example, a man sitting on board a ship which is leaving port considers himself to be moving relative to the shore which he regards as fixed; but he does not think of himself as moving relative to the ship, since his position is unchanged relative to its parts. Indeed, since we commonly think all motion involves action, while rest consists in the cessation of action, the man sitting on deck is more properly said to be at rest than in motion, since he does not have any sensory awareness of action in himself.

25 What is meant by 'motion' in the strict sense of the term

If, on the other hand, we consider what should be understood by *motion*, not in common usage but in accordance with the truth of the matter, and if our aim is to assign a determinate nature to it, we may say that *motion is the transfer of one piece of matter, or one body, from the vicinity of the other bodies which are in immediate contact with it, and which are regarded as being at rest, to the vicinity of other bodies*. By 'one body' or 'one piece of matter' I mean whatever is transferred at a given time, even though this may in fact consist of many parts which have different motions relative to each other. And I say 'the transfer' as opposed to the force or action which brings about the transfer, to show that motion is always in the moving body as opposed to the body which brings about the movement. The two are not normally distinguished with sufficient care; and I want to make it clear that the motion of something that moves is, like the lack of motion in a thing which is at rest, a mere mode of that thing and not itself a subsistent thing, just as shape is a mere mode of the thing which has shape.

26 No more action is required for motion than for rest

It should be noted that in this connection we are in the grip of a strong preconceived opinion, namely the belief that more action is needed for motion than for rest. We have been convinced of this since early childhood owing to the fact that our bodies move by our will, of which we have inner awareness, but remain at rest simply in virtue of sticking to the earth by gravity, the force of which we do not perceive through the senses. And because gravity and many other causes of which we are unaware produce resistance when we try to move our limbs, and make us tired, we think that a greater action or force is needed to initiate a motion than to stop it; for we take *action* to be the effort we expend in moving our limbs and moving other bodies by the use of our limbs. We will easily get rid of this preconceived opinion if we consider that it takes an effort on our part not only to move external bodies, but also, quite often, to stop them, when gravity and other causes are insufficient to arrest their movement. For example, the action needed to move a boat which is at rest in still water is no greater than that needed to stop it suddenly when it is moving – or rather it is not much greater, for one must subtract the weight of the water displaced by the ship and the viscosity of the water, both of which could gradually bring it to a halt.

27 Motion and rest are merely various modes of a body in motion

We are dealing here not with the action which is understood to exist in the body which produces or arrests the motion, but simply with the transfer of a body, and with the absence of a transfer, i.e. rest. So it is clear that this transfer cannot exist outside the body which is in motion, and that when there is a transfer of motion, the body is in a different state from when there is no transfer, i.e. when it is at rest. Thus motion and rest are nothing else but two different modes of a body.

28 Motion in the strict sense is to be referred solely to the bodies which are contiguous with the body in motion

In my definition I specified that the transfer occurs from the vicinity of contiguous bodies to the vicinity of other bodies; I did not say that there was a transfer from one place to another. This is because, as explained above, the term 'place' has various meanings, depending on how we think of it; but when we understand motion as a transfer occurring from the vicinity of contiguous bodies, then, given that only one set of bodies can be contiguous with the same moving body at any one time, we cannot assign several simultaneous motions to this body, but only one.

29 And it is to be referred only to those contiguous bodies which are regarded as being at rest

I further specified that the transfer occurs from the vicinity not of *any* contiguous bodies but from the vicinity of those which 'are regarded as being at rest'. For transfer is in itself a reciprocal process: we cannot understand that a body AB is transferred from the vicinity of a body CD without simultaneously understanding that CD is transferred from the vicinity of AB. Exactly the same force and action is needed on both sides. So if we wished to characterize motion strictly in terms of its own nature, without reference to anything else, then in the case of two contiguous bodies being transferred in opposite directions, and thus separated, we should say that there was just as much motion in the one body as in the other. But this would clash too much with our ordinary way of speaking. For we are used to standing on the earth and regarding it as at rest; so although we may see some of its parts, which are contiguous with other smaller bodies, being transferred out of their vicinity, we do not for that reason think of the earth itself as in motion.

[. . .]

31 *How there may be countless different motions in the same body*

Each body has only one proper motion, since it is understood to be moving away from only one set of bodies, which are contiguous with it and at rest. But it can also share in countless other motions, namely in cases where it is a part of other bodies which have other motions. For example, if someone walking on board ship has a watch in his pocket, the wheels of the watch have only one proper motion, but they also share in another motion because they are in contact with the man who is taking his walk, and they and he form a single piece of matter. They also share in an additional motion through being in contact with the ship tossing on the waves; they share in a further motion through contact with the sea itself; and lastly, they share in yet another motion through contact with the whole earth, if indeed the whole earth is in motion. Now all the motions will really exist in the wheels of the watch, but it is not easy to have an understanding of so many motions all at once, nor can we have knowledge of all of them. So it is enough to confine our attention to that single motion which is the proper motion of each body.

[. . .]

33 *How in every case of motion there is a complete circle of bodies moving together*

I noted above that every place is full of bodies, and that the same portion of matter always takes up the same amount of space, so that it is impossible for it to fill a greater or lesser space, or for any other body to occupy its place while it remains there. It follows from this that each body can move only in a complete circle of matter, or ring of bodies which all move together at the same time: a body entering a given place expels another, and the expelled body moves on and expels another, and so on, until the body at the end of the sequence enters the place left by the first body at the precise moment when the first body is leaving it. We can easily understand this in the case of a perfect circle, since we see that no vacuum and no rarefaction or condensation is needed to enable part A of the circle to move towards B, provided that B simultaneously moves towards C, C towards D and D towards A. But the same thing is intelligible even in the case of an imperfect circle however irregular it may be, provided we notice how all the variations in the spaces can be compensated for by variations in speed. . . .

[. . .]

36 *God is the primary cause of motion; and he always preserves the same quantity of motion in the universe*

After this consideration of the nature of motion, we must look at its cause. This is in fact twofold: first, there is the universal and primary cause – the general cause of all the motions in the world; and second there is the particular cause which produces in an individual piece of matter some motion which it previously lacked. Now as far as the general cause is concerned, it seems clear to me that this is no other than God himself. In the beginning in his omnipotence he created matter, along with its motion and rest; and now, merely by his regular concurrence, he preserves the same amount of motion and rest in the material universe as he put there in the beginning. Admittedly motion is simply a mode of the matter which is moved. But nevertheless it has a certain determinate quantity; and this, we easily understand, may be constant in the universe as a whole while varying in any given part. Thus if one part of matter moves twice as fast as another which is twice as large, we must consider that there is the same quantity of motion in each part; and if one part slows down, we must suppose that some other part of equal size speeds up by the same amount. For we understand that God's perfection involves not only his being immutable in himself, but also his operating in a manner that is always utterly constant and immutable. Now there are some changes whose occurrence is guaranteed either by our own plain experience or by divine revelation, and either our perception or our faith shows us that these take place without any change in the creator; but apart from these we should not suppose that any other changes occur in God's works, in case this suggests some inconstancy in God. Thus, God imparted various motions to the parts of matter when he first created them, and he now preserves all this matter in the same way, and by the same process by which he originally created it; and it follows from what we have said that this fact alone makes it most reasonable to think that God likewise always preserves the same quantity of motion in matter.

37 The first law of nature: each and every thing, in so far as it can, always continues in the same state; and thus what is once in motion always continues to move

From God's immutability we can also know certain rules or laws of nature, which are the secondary and particular causes of the various motions we see in particular bodies. The first of these laws is that each thing, in so far as it is simple and undivided, always remains in the same state, as far as it can, and never changes except as a result of external causes. Thus, if a particular piece of matter is square, we can be sure without more ado that it will remain square for ever, unless something coming from outside changes its shape. If it is at rest, we hold that it will never begin to move unless it is pushed into motion by some cause. And if it moves, there is equally no reason for thinking it will ever lose this motion of its own accord and without being checked by something else. Hence we must conclude that what is in motion always, so far as it can, continues to move. But we live on the Earth, whose composition is such that all motions occurring near it are soon halted, often by causes undetectable by our senses. Hence from our earliest years we have often judged that such motions, which are in fact stopped by causes unknown to us, come to an end of their own accord. And we tend to believe that what we have apparently experienced in many cases holds good in all cases – namely that it is in the very nature of motion to come to an end, or to tend towards a state of rest. This, of course, is a false preconceived opinion which is utterly at variance with the laws of nature; for rest is the opposite of motion, and nothing can by its own nature tend towards its opposite, or towards its own destruction.

38 The motion of projectiles

Indeed, our everyday experience of projectiles completely confirms this first rule of ours. For there is no other reason why a projectile should persist in motion for some time after it leaves the hand that throws it, except that what is once in motion continues to move until it is slowed down by bodies that are in the way. And it is clear that projectiles are normally slowed down, little by little, by the air or other fluid bodies in which they are moving, and that this is why their motion cannot persist for long. The fact that air offers resistance to other moving bodies may be confirmed either by our own experience, through the sense of touch if we beat the air with a fan, or by the flight of birds. And in the case of any other fluid, the resistance offered to the motion of a projectile is even more obvious than in the case of air.

39 The second law of nature: all motion is in itself rectilinear; and hence any body moving in a circle always tends to move away from the centre of the circle which it describes

The second law is that every piece of matter, considered in itself, always tends to continue moving, not in any oblique path but only in a straight line. This is true despite the fact that many particles are often forcibly deflected by the impact of other bodies; and, as I have said above, in any motion the result of all the matter moving simultaneously is a kind of circle. The reason for this second rule is the same as the reason for the first rule, namely the immutability and simplicity of the operation by which God preserves motion in matter. For he always preserves the motion in the precise form in which it is occurring at the very moment when he preserves it, without taking any account of the motion which was occurring a little while earlier. It is true that no motion takes place in a single instant of time; but clearly whatever is in motion is determined, at the individual instants which can be specified as long as the motion lasts, to continue moving in a given direction along a straight line, and never in a curve . . .

40 The third law: if a body collides with another body that is stronger than itself, it loses none of its motion; but if it collides with a weaker body, it loses a quantity of motion equal to that which it imparts to the other body

The third law of nature is this: when a moving body collides with another, if its power of continuing in a straight line is less than the resistance of the other body, it is deflected so that, while the quantity of motion is retained, the direction is altered; but if its power of continuing is greater than the resistance of the other body, it carries that body along with it, and loses a quantity of motion equal to that which it imparts to the other body. Thus we find that when hard projectiles strike some other hard body, they do not stop, but rebound in the opposite direction; when, by contrast, they encounter a soft body, they are immediately halted because they readily transfer all their motion to it. . . .

41 The proof of the first part of this rule

The first part of this law is proved by the fact that there is a difference between motion considered in itself the motion of a thing and its determination in a certain direction; for the determination of the direction can be altered, while the motion remains constant. As I have said above, everything that is not composite but simple, as motion is, always persists in being as it is in itself and not in relation to other things, so long as it is not destroyed by an external cause by meeting another object. Now if one body collides with a second, hard body in its path which it is quite incapable of pushing, there is an obvious reason why its motion should not remain fixed in the same direction, namely the resistance of the body which deflects its path; but there is no reason why its motion should be stopped or diminished, since it is not removed by the other body or by any other cause, and since one motion is not the opposite of another motion. Hence it follows that the motion in question ought not to diminish at all.

42 The proof of the second part of this rule

The second part of the law is proved from the immutability of the workings of God, by means of which the world is continually preserved through an action identical with its original act of creation. For the whole of space is filled with bodies, and the motion of every single body is rectilinear in tendency; hence it is clear that when he created the world in the beginning God did not only impart various motions to different parts of the world, but also produced all the reciprocal impulses and transfers of motion between the parts. Thus, since God preserves the world by the selfsame action and in accordance with the selfsame laws as when he created it, the motion which he preserves is not something permanently fixed in given pieces of matter, but something which is mutually transferred when collisions occur. The very fact that creation is in a continual state of change is thus evidence of the immutability of God.

43 The nature of the power which all bodies have to act on, or resist, other bodies

In this connection we must be careful to note what it is that constitutes the power of any given body to act on, or resist the action of, another body. This power consists simply in the fact that everything tends, so far as it can, to persist in the same state, as laid down in our first law. Thus what is joined to another thing has some power of resisting separation from it; and what is separated has some power of remaining separate. Again, what is at rest has some power of remaining at rest and consequently of resisting anything that may alter the state of rest; and what is in motion has some power of persisting in its motion, i.e. of continuing to move with the same speed and in the same direction. An estimate of this last power must depend firstly on the size of the body in question and the size of the surface which separates it from other bodies, and secondly on the speed of the motion, and on the various ways in which different bodies collide, and the degree of opposition involved.

44 The opposite of motion is not some other motion but a state of rest; and the opposite of the determination of a motion in a given direction is its determination in the opposite direction

It should be noted that one motion is in no way contrary to another of equal speed. Strictly speaking there are only two sorts of opposition to be found here. One is the opposition between motion and rest, together with the opposition between swiftness and slowness of motion (in so far, that is, as such slowness shares something of the nature of rest). And the second sort is the opposition between the determination of motion in a given direction and an encounter somewhere in that direction with another body which is at rest or moving in another direction. The degree of this opposition varies in accordance with the direction in which a body is moving when it collides with another.

[. . .]

52 *The seventh rule*

. . . These matters do not need proof since they are self-evident the demonstrations are so certain that even if our experience seemed to show us the opposite, we should still be obliged to have more faith in our reason than in our senses.

53 *The application of these rules is difficult because each body is simultaneously in contact with many others*

In fact it often happens that experience may appear to conflict with the rules I have just explained, but the reason for this is evident. Since no bodies in the universe can be so isolated from all others, and no bodies in our vicinity are normally perfectly hard, the calculation for determining how much the motion of a given body is altered by collision with another body is much more difficult than those given above. . . .

[. . .]

64 *The only principles which I accept, or require, in physics are those of geometry and pure mathematics; these principles explain all natural phenomena, and enable us to provide quite certain demonstrations regarding them*

I will not here add anything about shapes or about the countless different kinds of motions that can be derived from the infinite variety of different shapes. These matters will be quite clear in themselves when the time comes for me to deal with them. I am assuming that my readers know the basic elements of geometry already, or have sufficient mental aptitude to understand mathematical demonstrations. For I freely acknowledge that I recognize no matter in corporeal things apart from that which the geometers call quantity, and take as the object of their demonstrations, i.e. that to which every kind of division, shape and motion is applicable. Moreoever, my consideration of such matter involves absolutely nothing apart from these divisions, shapes and motions; and even with regard to these, I will admit as true only what has been deduced from indubitable common notions so evidently that it is fit to be considered as a mathematical demonstration. And since all natural phenomena can be explained in this way, as will become clear in what follows, I do not think that any other principles are either admissible or desirable in physics.

3.4

Human Knowledge: Its Scope and Limits

John Locke

John Locke (1632–1704), an English physician and philosopher, was an early member of the Royal Society for the Improvement of Natural Knowledge, a group that self-consciously advanced the scientific revolution from its chartering in 1662 onward. In his *Essay Concerning Human Understanding* (1689), Locke clarified the brand of empiricism characteristic of the Royal Society and sought to understand the scope and limits of our investigation of nature. In this selection from book IV of the *Essay*, he reflects on the causes of ignorance. The modesty of Locke's conclusion should be compared with Newton's refusal to speculate on the cause of gravity in the General Scholium to the *Principia* (Reading 2.17).

[. . .]

22 Our knowledge being so narrow, as I have shown, it will perhaps give us some light into the present state of our minds if we look a little into the dark side, and take a view of *our ignorance*; which, being infinitely larger than our knowledge, may serve much to the quieting of disputes, and improvement of useful knowledge; if, discovering how far we have clear and distinct ideas, we confine our thoughts within the contemplation of those things that are within the reach of our understandings, and launch not out into that abyss of darkness, (where we have not eyes to see, nor faculties to perceive anything), out of a presumption that nothing is beyond our comprehension. But to be satisfied of the folly of such a conceit, we need not go far. He that knows anything, knows this, in the first place, that he need not seek long for instances of his ignorance. The meanest and most obvious things that come in our way have dark sides, that the quickest sight cannot penetrate into. The clearest and most enlarged understandings of thinking men find themselves puzzled and at a loss in every particle of matter. We shall the less wonder to find it so, when we consider the *causes of our ignorance*; which, from what has been said, I suppose will be found to be these three:

From John Locke, *An Essay Concerning Human Understanding*, Vol. 2 (New York: Dover Publications, 1959). Selections from pp. 212–18, 220–3.

First, Want of ideas.
Secondly, Want of a discoverable connection between the ideas we have.
Thirdly, Want of tracing and examining our ideas.

23 First, There are some things, and those not a few, that we are ignorant of, for want of ideas.

First, all the simple ideas we have are confined (as I have shown) to those we receive from corporeal objects by sensation, and from the operations of our own minds as the objects of reflection. But how much these few and narrow inlets are disproportionate to the vast whole extent of all beings, will not be hard to persuade those who are not so foolish as to think their span the measure of all things. What other simple ideas it is possible the creatures in other parts of the universe may have, by the assistance of senses and faculties more or perfecter than we have, or different from ours, it is not for us to determine. But to say or think there are no such, because we conceive nothing of them, is no better an argument than if a blind man should be positive in it, that there was no such thing as sight and colours, because he had no manner of idea of any such thing, nor could by any means frame to himself any notions about seeing. The ignorance and darkness that is in us no more hinders nor confines the knowledge that is in others, than the blindness of a mole is an argument against the quicksightedness of an eagle. He that will consider the infinite power, wisdom, and goodness of the Creator of all things will find reason to think it was not all laid out upon so inconsiderable, mean, and impotent a creature as he will find man to be; who in all probability is one of the lowest of all intellectual beings. What faculties, therefore, other species of creatures have to penetrate into the nature and inmost constitutions of things; what ideas they may receive of them far different from ours, we know not. This we know and certainly find, that we want several other views of them besides those we have, to make discoveries of them more perfect. And we may be convinced that the ideas we can attain to by our faculties are very disproportionate to things themselves, when a positive, clear, distinct one of substance itself, which is the foundation of all the rest, is concealed from us. But want of ideas of this kind, being a part as well as cause of our ignorance, cannot be described. Only this I think I may confidently say of it, That the intellectual and sensible world are in this perfectly alike: that that part which we see of either of them holds no proportion with what we see not; and whatsoever we can reach with our eyes or our thoughts of either of them is but a point, almost nothing in comparison of the rest.

24 Secondly, Another great cause of ignorance is the want of ideas we are capable of. As the want of ideas which our faculties are not able to give us shuts us wholly from those views of things which it is reasonable to think other beings, perfecter than we, have, of which we know nothing; so the want of ideas I now speak of keeps us in ignorance of things we conceive capable of being known to us. Bulk, figure, and motion we have ideas of. But though we are not without ideas of these primary qualities of bodies in general, yet not knowing what is the particular bulk, figure, and motion, of the greatest part of the bodies of the universe, we are ignorant of the several powers, efficacies, and ways of operation, whereby the effects which we daily see are produced. These are hid from us, in some things by being too remote, and in others by being too minute. When we consider the vast distance of the known and visible parts of the world, and the reasons we have to think that what lies within our ken is but a small part of the universe, we shall then discover a huge abyss of ignorance. What are the particular fabrics of the great masses of matter which make up the whole stupendous frame of corporeal beings; how far they are extended; what is their motion, and how continued or communicated; and what influence they have one upon another, are contemplations that at first glimpse our thoughts lose themselves in. If we narrow our contemplations, and confine our thoughts to this little canton – I mean this system of our sun, and the grosser masses of matter that visibly move about it, What several sorts of vegetables, animals, and intellectual corporeal beings, infinitely different from those of our little spot of earth, may there probably be in the other planets, to the knowledge of which, even of their outward figures and parts, we can no way attain whilst we are confined to this earth; there being no natural means, either by sensation or reflection, to convey their certain ideas into our minds? They are out of the reach of those inlets

of all our knowledge: and what sorts of furniture and inhabitants those mansions contain in them we cannot so much as guess, much less have clear and distinct ideas of them.

25 If a great, nay, far the greatest part of the several ranks of bodies in the universe escape our notice by their remoteness, there are others that are no less concealed from us by their minuteness. These *insensible corpuscles*, being the active parts of matter, and the great instruments of nature, on which depend not only all their secondary qualities, but also most of their natural operations, our want of precise distinct ideas of their primary qualities keeps us in an incurable ignorance of what we desire to know about them. I doubt not but if we could discover the figure, size, texture, and motion of the minute constituent parts of any two bodies, we should know without trial several of their operations one upon another; as we do now the properties of a square or a triangle. Did we know the mechanical affections of the particles of rhubarb, hemlock, opium, and a man, as a watchmaker does those of a watch, whereby it performs its operations; and of a file, which by rubbing on them will alter the figure of any of the wheels; we should be able to tell beforehand that rhubarb will purge, hemlock kill, and opium make a man sleep: as well as a watchmaker can, that a little piece of paper laid on the balance will keep the watch from going till it be removed; or that, some small part of it being rubbed by a file, the machine would quite lose its motion, and the watch go no more. The dissolving of silver in *aqua fortis*, and gold in *aqua regia*, and not *vice versa*, would be then perhaps no more difficult to know than it is to a smith to understand why the turning of one key will open a lock, and not the turning of another. But whilst we are destitute of senses acute enough to discover the minute particles of bodies, and to give us ideas of their mechanical affections, we must be content to be ignorant of their properties and ways of operation; nor can we be assured about them any further than some few trials we make are able to reach. But whether they will succeed again another time, we cannot be certain. This hinders our certain knowledge of universal truths concerning natural bodies: and our reason carries us herein very little beyond particular matter of fact.

26 And therefore I am apt to doubt that, how far soever human industry may advance useful and experimental philosophy in physical things, *scientifical* will still be out of our reach: because we want perfect and adequate ideas of those very bodies which are nearest to us, and most under our command. Those which we have ranked into classes under names, and we think ourselves best acquainted with, we have but very imperfect and incomplete ideas of. Distinct ideas of the several sorts of bodies that fall under the examination of our senses perhaps we may have: but adequate ideas, I suspect, we have not of any one amongst them. And though the former of these will serve us for common use and discourse, yet whilst we want the latter, we are not capable of scientifical knowledge; nor shall ever be able to discover general, instructive, unquestionable truths concerning them. Certainty and demonstration are things we must not, in these matters, pretend to. By the colour, figure, taste, and smell, and other sensible qualities, we have as clear and distinct ideas of sage and hemlock, as we have of a circle and a triangle: but having no ideas of the particular primary qualities of the minute parts of either of these plants, nor of other bodies which we would apply them to, we cannot tell what effects they will produce; nor when we see those effects can we so much as guess, much less know, their manner of production. Thus, having no ideas of the particular mechanical affections of the minute parts of bodies that are within our view and reach, we are ignorant of their constitutions, powers, and operations: and of bodies more remote we are yet more ignorant, not knowing so much as their very outward shapes, or the sensible and grosser parts of their constitutions.

[. . .]

28 Secondly, What a small part of the substantial beings that are in the universe the want of ideas leaves open to our knowledge, we have seen. In the next place, another cause of ignorance, of no less moment, is a want of a discoverable connection between those ideas we have. For wherever we want that, we are utterly incapable of universal and certain knowledge; and are, in the former case, left only to observation and experiment: which, how narrow and confined it is, how far from general knowledge we need not be told. I shall give some few instances of this cause

of our ignorance, and so leave it. It is evident that the bulk, figure, and motion of several bodies about us produce in us several sensations, as of colours, sounds, tastes, smells, pleasure, and pain, etc. These mechanical affections of bodies having no affinity at all with those ideas they produce in us, (there being no conceivable connection between any impulse of any sort of body and any perception of a colour or smell which we find in our minds,) we can have no distinct knowledge of such operations beyond our experience; and can reason no otherwise about them, than as effects produced by the appointment of an infinitely Wise Agent, which perfectly surpass our comprehensions. As the ideas of sensible secondary qualities which we have in our minds, can by us be no way deduced from bodily causes, nor any correspondence or connection be found between them and those primary qualities which (experience shows us) produce them in us; so, on the other side, the operation of our minds upon our bodies is as inconceivable. How any thought should produce a motion in body is as remote from the nature of our ideas, as how any body should produce any thought in the mind. That it is so, if experience did not convince us, the consideration of the things themselves would never be able in the least to discover to us. These, and the like, though they have a constant and regular connection in the ordinary course of things; yet that connection being not discoverable in the ideas themselves, which appearing to have no necessary dependence one on another, we can attribute their connection to nothing else but the arbitrary determination of that All-wise Agent who has made them to be, and to operate as they do, in a way wholly above our weak understandings to conceive.

29 In some of our ideas there are certain relations, habitudes, and connections, so visibly included in the nature of the ideas themselves, that we cannot conceive them separable from them by any power whatsoever. And in these only we are capable of certain and universal knowledge. Thus the idea of a right-lined triangle necessarily carries with it an equality of its angles to two right ones. Nor can we conceive this relation, this connection of these two ideas, to be possibly mutable, or to depend on any arbitrary power, which of choice made it thus, or could make it otherwise. But the coherence and continuity of the parts of matter; the production of sensation in us of colours and sounds, etc., by impulse and motion; nay, the original rules and communication of motion being such, wherein we can discover no natural connection with any ideas we have, we cannot but ascribe them to the arbitrary will and good pleasure of the Wise Architect. . . . The things that, as far as our observation reaches, we constantly find to proceed regularly, we may conclude do act by a law set them; but yet by a law that we know not: whereby, though causes work steadily, and effects constantly flow from them, yet their connections and dependencies being not discoverable in our ideas, we can have but an experimental knowledge of them. From all which it is easy to perceive what a darkness we are involved in, how little it is of Being, and the things that are, that we are capable to know. And therefore we shall do no injury to our knowledge, when we modestly think with ourselves, that we are so far from being able to comprehend the whole nature of the universe and all the things contained in it, that we are not capable of a philosophical knowledge of the bodies that are about us, and make a part of us: concerning their secondary qualities, powers, and operations, we can have no universal certainty. Several effects come every day within the notice of our senses, of which we have so far sensitive knowledge: but the causes, manner, and certainty of their production, for the two foregoing reasons, we must be content to be very ignorant of. In these we can go no further than particular experience informs us matter of fact, and by analogy to guess what effects the like bodies are, upon other trials, like to produce. But as to a perfect science of natural bodies, (not to mention spiritual beings,) we are, I think, so far from being capable of any such thing, that I conclude it lost labour to seek after it.

3.5

The Principle of Least Action

Gottfried Wilhelm Leibniz

Gottfried Wilhelm von Leibniz (1646–1716) was a German polymath who shares with Newton the honor of having discovered the calculus. No short summary can do justice to the scope of Leibniz's many contributions to mathematics, science, and philosophy. In this excerpt from a manuscript unpublished until after his death, Leibniz celebrates the structure of the beauty of the laws of nature as a proof of God's wisdom as well as His power. Leibniz's particular illustration is a version of the principle of least action. He argues that it is neither a mere contingent mechanical fact nor a consequence of geometry alone but is best seen from the standpoint of the doctrine of final causes.

I have shown on several occasions that the final analysis of the laws of nature leads us to the most sublime principles of order and perfection, which indicate that the universe is the effect of a universal intelligent power. As the ancients already held, this truth is the chief fruit of our investigations; without mentioning Pythagoras and Plato, whose primary aim was such an analysis, even Aristotle sought to demonstrate a prime mover in his works, particularly in his *Metaphysics*. It is true that these ancient thinkers were not informed about the laws of nature as are we, since they lacked many of the methods which we have and of which we ought to take advantage. The knowledge of nature gives birth to the arts, it gives us many means of conserving life, and it even provides us with conveniences; but the satisfaction of spirit which comes from wisdom and virtue, in addition to being the greatest ornament of life, raises us to what is eternal, whereas this life, in contrast, is most brief. As a result, whatever serves to establish maxims which locate happiness in virtue and show that everything follows the principle of perfection is infinitely more useful to man, and even to the state, than all that serves the arts. Discoveries useful to life, moreover, are very often merely the corollaries of more important insights; it is true

From Gottfried Wilhelm Leibniz, *Philosophical Papers and Letters*, Vol. II, ed. and trans. by L. E. Loemker (Chicago: University of Chicago Press, 1956), pp. 777–83, 785–8.

here too that those who seek the kingdom of God find the rest on their way.

The inquiry into final causes in physics is precisely the application of the method which I think ought to be used, and those who have sought to banish it from their philosophy have not adequately considered its usefulness. For I do not wish to do them the injury of thinking that they have evil designs in doing this. Others followed them, however, who have abused their position, and who, not content with excluding final causes from physics but restoring them elsewhere, have tried to destroy them entirely and to show that the Creator of the universe is most powerful, indeed, but without any intelligence. There have been still others who have not admitted any universal cause, like the ancients who recognized nothing in the universe but a concourse of corpuscles. This seems plausible to those minds in whom the imaginative faculty predominates, because they believe that they need to use only mathematical principles, without having any need either for metaphysical principles, which they treat as illusory, or for principles of the good, which they reduce to human morals; as if perfection and the good were only a particular result of our thinking and not to be found in universal nature.

I recognize that it is rather easy to fall into this error, especially when one's thinking stops at what imagination alone can supply, namely, at magnitudes and figures and their modifications. But when one pushes forward his inquiry after reasons, it is found that the laws of motion cannot be explained through purely geometric principles or by imagination alone. This is also why some very able philosophers of our day have held that the laws of motion are purely arbitrary. They are right in this if they take *arbitrary* to mean coming from choice and not from geometric necessity, but it is wrong to extend this concept to mean that laws are entirely indifferent, since it can be shown that they originate in the wisdom of their Author or in the principle of greatest perfection, which has led to their choice.

This consideration gives us the true middle term that is needed for satisfying truth as well as piety. We know that while there have been, on the one hand, able philosophers who recognized nothing except what is material in the universe, there are, on the other hand, learned and zealous theologians who, shocked at the corpuscular philosophy and not content with checking its misuse, have felt obliged to maintain that there are phenomena in nature which cannot be explained by mechanical principles; as, for example, light, weight, and elastic force. But since they do not reason with exactness in this matter, and it is easy for the corpuscular philosophers to reply to them, they injure religion in trying to render it a service, for they merely confirm those in their error who recognize only material principles. The true middle term for satisfying both truth and piety is this: all natural phenomena could be explained mechanically if we understood them well enough, but the principles of mechanics themselves cannot be explained geometrically, since they depend on more sublime principles which show the wisdom of the Author in the order and perfection of his work.

The most beautiful thing about this view seems to me to be that the principle of perfection is not limited to the general but descends also to the particulars of things and of phenomena and that in this respect it closely resembles the method of *optimal forms*, that is to say, of forms *which provide a maximum or minimum*, as the case may be – a method which I have introduced into geometry in addition to the ancient method of *maximal and minimal quantities*. For in these forms or figures the *optimum* is found not only in the whole but also in each part, and it would not even suffice in the whole without this. For example, if in the case of the curve of shortest descent between two given points, we choose any two points on this curve at will, the part of the line intercepted between them is also necessarily the line of shortest descent with regard to them. It is in this way that the smallest parts of the universe are ruled in accordance with the order of greatest perfection; otherwise the whole would not be so ruled.

It is for this reason that I usually say that there are, so to speak, two kingdoms even in corporeal nature, which interpenetrate without confusing or interfering with each other – the realm of power, according to which everything can be explained *mechanically* by efficient causes when we have sufficiently penetrated into its interior, and the realm of wisdom, according to which everything can be explained architectonically, so to speak, or by final causes when we understand its ways sufficiently. In this sense one can say with

Lucretius not only that animals see because they have eyes but also that eyes have been given them in order to see, though I know that some people, in order the better to pass as free thinkers, admit only the former. Those who enter into the details of natural machines, however, must have need of a strong bias to resist the attractions of their beauty. Even Galen, after learning something about the function of the parts of animals, was so stirred with admiration that he held that to explain them was essentially to sing hymns to the honor of divinity. I have often wished that an able physician would undertake to prepare a special work whose title – or whose aim, at least – would be *The Hymn of Galen*.

What is more, our thinking sometimes furnishes us with considerations revealing the value of final causes, not merely in increasing our admiration for the supreme Author, but also in making discoveries among his works. Some day I shall show this in a special case in which I shall propose as a general principle of optics that a ray of light moves from one point to another by the path which is found to be easiest in relation to the plane surfaces which must serve as the rule for other surfaces. For it must be kept in mind that if we claimed to use this principle as an efficient cause, and as if the easiest path would prevail among all the possible competing rays, it would be necessary to consider the whole surface as it is, without considering the plane tangent to it, and then the principle would not always work out successfully, as I shall show presently. But far from concealing that there is a certain final cause involved in this principle – an objection which was once made against Mr Fermat, who had used it in his *Dioptrics* – I have found it more beautiful and more important than that of mechanism for a more sublime application. And an able author who has published a work on optics in England has expressed his indebtedness to my view. Order demands that curved lines and surfaces be treated as composed of straight lines and planes, and a ray is determined by the plane on which it falls, which is considered as forming the curved surface at that point. But the same order demands that the effect of the greatest ease be obtained in relation to the planes, at least those which serve as elements to other surfaces, since it cannot be obtained with regard to these surfaces also. This is all the more true since it thus satisfies, with respect to these curves, another principle which now supersedes the preceding one, and which holds that in the absence of a minimum it is necessary to hold to the *most determined*, which can be the *simplest* even when it is a *maximum*.

Now we find that the ancients, and among others Ptolemy, already used this hypothesis of the easiest path of a ray which falls on a plane, to account for the equality of the angles of incidence and reflection, the principle at the basis of catoptrics. It is by this same hypothesis that Mr Fermat provided a reason for the law of refraction according to the sines, or to formulate it in other terms, as Snell did, according to the secants. But, what is more, I have no doubt whatever that this law was first discovered by this method. It is known that Willebrord Snell, one of the greatest geometricians of his time and well versed in the methods of the ancients, invented it, having even written a work which was not published because of its author's death. But, since he had taught it to his disciples, all appearances point to the conclusion that Descartes, who had come to Holland a little later and who was most interested in this problem, learned it there. For the way in which Descartes has tried to explain the law of refraction by efficient causes or by the composition of directions in imitation of the reflection of bullets is extremely forced and not intelligible enough. To say no more about it here, it shows clearly that it is an afterthought adjusted somehow to the conclusion and was not discovered by the method he gives. So we may well believe that we should not have had this beautiful discovery so soon without the method of final causes.

I recall that capable writers have frequently objected that this principle does not seem to work in reflection itself when applied to curved surfaces and that in concave mirrors the path of reflection happens sometimes to be the longest. But in addition to what I have already said, that, according to architectonic principles, curved surfaces must be ruled by the planes tangent to them, I shall now explain how it remains always universally true that the ray is directed in the most determined or unique path, even in relation to curves. It is also worth noting that, in the method of analysis by maxima and minima, the same operation suffices for the problems of the greatest and the smallest, without distinguishing between them except in applying the method to different cases, since we seek the most determined magnitude in both cases, which is sometimes the greatest and sometimes the smallest in its order, the analysis being based solely on the disappearance of a difference or on the unique result of reuniting twins, and not at all on a comparison of the greatest and smallest with all other magnitudes. For given a curve *AB*, concave or convex, and an axis *ST* to which the ordinates of the curve are referred; then it is seen that to each ordinate, like *Q* or *R*, there corresponds another one equal to it, its twin, *q* or *r* (Fig. 7). But there is one particular ordinate *EC* which is unique, or the only determinate one of its magnitude, and has no twin, since the two twins *EC* and *ec* coincide in it and make but one. And this *EC* is the greatest ordinate of the concave curve and the smallest of the convex curve. So instead of two infinitely near ordinates in all other cases, having a difference of *dm* if the ordinate is called m, whose ratio to *Ee*, a correspondingly small part of the axis, would give the angle of the curve or of its tangent to the axis *ST*, the infinitely close ordinates or twins become coincident in this case at *C* and have no difference; *dm* becomes 0, and the tangent at *C* is parallel to the axis. Thus the basis of the analysis is this uniqueness caused by the union of the twins, without any concern as to whether the ordinate is the greatest or smallest. . . . [Leibniz here gives proofs, using calculus, of two related theorems, that the angles of incidence and reflection will be equal for any reflective surface, regardless of its shape, and that the sine of the angle of refraction of a ray of light passing through a refractive medium like water or glass will be proportional to the sine of the angle of incidence.]

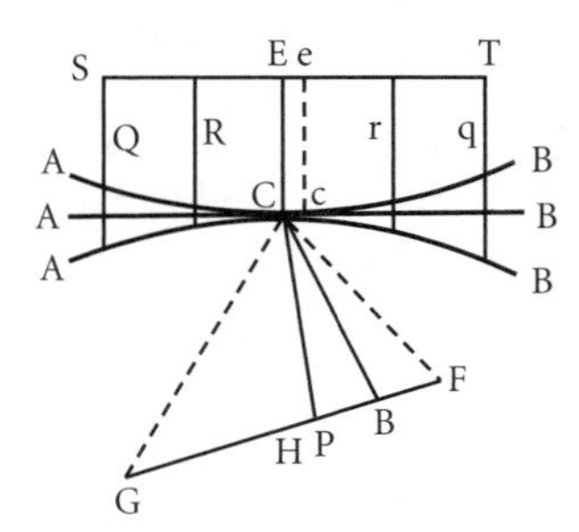

Figure 7 Leibniz's geometric illustration of the uniqueness of a local maximum or minimum

This makes us see, finally, that the rule of the unique path, or the path most determined in length of time, applies generally to the direct and the broken ray, whether reflected or refracted, whether by plane or by curved surfaces, whether convex or concave, without distinguishing in the process whether the time is the longest or the shortest, though it is in fact the shortest when that which should provide the rule is taken into consideration, that is, the tangent plane; nature being governed, as it is, by sovereign wisdom, shows the general design throughout of controlling curves by straight lines or planes tangent to them, as if the curves were composed of these, although this is not strictly true.

[. . .]

This principle of nature, that it acts in the most determined ways which we may use, is purely architectonic in fact, yet it never fails to be observed. Assume the case that nature were obliged in general to construct a triangle and that for this purpose only the perimeter or the sum of the sides were given, and nothing else; then nature would construct an equilateral triangle. This example shows the difference between architectonic and geometric determinations. Geometric determinations introduce an absolute necessity, the contrary of which implies a contradiction, but architectonic determinations introduce only a necessity of choice whose contrary means imperfection – a little like the saying in jurisprudence: *Quae contra bonos mores sunt, ea nec facere nos posse credendum est.* So there is even in the algebraic calculus what I call the law of justice, which greatly aids us in finding good solutions. If nature were brutish, so to speak, that is, purely material or geometrical, the above case would be impossible, and, unless something more determinative were given than merely the perimeter, nature would not produce a triangle. But since nature is governed architectonically, the half-determinations of geometry are sufficient for it to achieve its work; otherwise it would most often have been stopped. And this is particularly true with regard to the laws of nature. Perhaps someone will deny that what I have said above applies to the laws of motion and will maintain that an entirely geometric demonstration can be given of them. I reserve the proof of the contrary for another discourse, where I shall show that they cannot be derived from their sources without assuming architectonic grounds. One of the most important of these, which I believe I am the first to have introduced into physics, is the law of continuity, which I discussed many years ago in the *Nouvelles de la republique des lettres*, where I showed with examples how it serves as the touchstone of theories. It serves not merely to test, however, but also as a very fruitful principle of discovery, as I plan to show some day. But I have also found other very beautiful and extended laws of nature, quite different, however, from those usually employed, yet always depending on architectonic principles. Nothing seems to me to be more effectual in proving and admiring the sovereign wisdom of the Author of things as shown in the very principles of things themselves.

Figure acknowledgment

Figure 7: from Leibniz, selections from *Philosophical Papers and Letters*, vol. II, ed. and trans. L. E. Loemker (Chicago: University of Chicago Press, 1956), p. 783.

3.6

Space, Time, and Symmetry

Gottfried Wilhelm Leibniz

One of the most famous controversies of the early eighteenth century was an exchange of letters – really a set of increasingly long papers – between Leibniz and Samuel Clarke. The dispute served as a proxy for a debate between Leibniz and Isaac Newton, Clarke's friend and the undoubted source for some of the questions and criticisms Clarke passed on. Both national and professional pride were at stake, and neither side made any concessions to the other. In this excerpt from the correspondence, we see the sharp difference between the Newtonian and Leibnizian conceptions of space and time and the role of symmetry considerations in Leibniz's scientific thought.

[...]

IV Clarke's Second Reply

When I said that the *Mathematical Principles of Philosophy* are opposite to those of the materialists, the meaning was that, whereas materialists suppose the frame of nature to be such as could have arisen from mere mechanical principles of matter and motion, of necessity and fate, the *Mathematical Principles of Philosophy* show on the contrary that the state of things (the constitution of the sun and planets) is such as could not arise from anything but an intelligent and free cause. As to the propriety of the name, so far as metaphysical consequences follow demonstratively from mathematical principles, so far the mathematical principles may (if it be thought fit) be called metaphysical principles.

'Tis very true that nothing is without a sufficient reason why it is and why it is thus rather than otherwise. And, therefore, where there is no cause, there can be no effect. But this sufficient reason is ofttimes no other than the mere will of God. For instance, why this particular system of matter should be created in one particular place, when (all place being absolutely indifferent to all matter) it would have been exactly the same thing vice versa, supposing the two systems (or

From Gottfried Wilhelm Leibniz, *Philosophical Papers and Letters*, Vol. II, ed. and trans. by L. E. Loemker (Chicago: University of Chicago Press, 1956), pp. 1104–9, 1145–8.

the particles) of matter to be alike; there can be no other reason but the mere will of God. Which if it could in no case act without a predetermining cause, any more than a balance can move without a preponderating weight, this would tend to take away all power of choosing and to introduce fatality.

[. . .]

3 The word "sensory" does not properly signify the organ but the place of sensation. The eye, the ear, etc., are organs but not sensoria. Besides, Sir Isaac Newton does not say that space is the sensory but that it is, by way of similitude only, as it were, the sensory, etc.

4 It was never supposed that the presence of the soul was sufficient but only that it is necessary in order to perception. Without being present to the images of the things perceived, it could not possibly perceive them. But being present is not sufficient, without it be also a living substance. Any inanimate substance, tho' present, perceives nothing, and a living substance can only there perceive where it is present either to the things themselves (as the omnipresent God is to the whole universe) or the images of things (as the soul of man is in its proper sensory). Nothing can any more act or be acted upon where it is not present than it can be where it is not. The soul's being indivisible does not prove it to be present only in a mere point. Space, finite or infinite, is absolutely indivisible, even so much as in thought (to imagine its parts moved from each other is to imagine them moved out of themselves), and yet space is not a mere point.

[. . .]

6 and 7 'Tis very true that the excellency of God's workmanship does not consist in its showing the power only but in its showing the wisdom also of its author. But then this wisdom of God appears not in making nature (as an artificer makes a clock) capable of going on without him (for that's impossible, there being no powers of nature independent upon God, as the powers of weights and springs are independent upon men). But the wisdom of God consists in framing originally the perfect and complete idea of a work which began and continues according to that original perfect idea by the continual uninterrupted exercise of his power and government.

8 The word "correction" or "amendment" is to be understood, not with regard to God, but to us only. The present frame of the solar system, for instance, according to the present laws of motion, will in time fall into confusion and perhaps, after that, will be amended or put into a new form. But this amendment is only relative with regard to our conceptions. In reality and with regard to God, the present frame, and the consequent disorder, and the following renovation are all equally parts of the design framed in God's original perfect idea. 'Tis in the frame of the world, as in the frame of man's body; the wisdom of God does not consist in making the present frame of either of them eternal but to last so long as he thought fit.

[. . .]

Mr Leibniz's Third Paper

[. . .]

2 The author grants me this important principle, that nothing happens without a sufficient reason why it should be so rather than otherwise. But he grants it only in words and in reality denies it. Which shows that he does not fully perceive the strength of it. And therefore he makes use of an instance which exactly falls in with one of my demonstrations against real absolute space, which is an idol of some modern Englishmen. I call it an idol, not in a theological sense, but in a philosophical one, as Chancellor Bacon says that there are *idola tribus, idola specus.*

3 These gentlemen maintain, therefore, that space is a real absolute being. But this involves them in great difficulties, for such a being must needs be eternal and infinite. Hence some have believed it to be God himself, or one of his attributes, his immensity. But since space consists of parts, it is not a thing which can belong to God.

4 As for my own opinion, I have said more than once that I hold space to be something merely relative, as time is; that I hold it to be an order of coexistences as time is an order of successions. For space denotes, in terms of possibility, an order

of things which exist at the same time, considered as existing together, without inquiring into their particular manner of existing. And when many things are seen together, one perceives that order of things among themselves.

5 I have many demonstrations to confute the fancy of those who take space to be a substance or at least an absolute being. But I shall only use, at the present, one demonstration, which the author here gives me occasion to insist upon. I say, then, that if space was an absolute being, there would something happen for which it would be impossible there should be a sufficient reason. Which is against my axiom. And I can prove it thus. Space is something absolutely uniform, and, without the things placed in it, one point of space does not absolutely differ in any respect whatsoever from another point of space. Now from hence it follows (supposing space to be something in itself, besides the order of bodies among themselves) that 'tis impossible there should be a reason why God, preserving the same situations of bodies among themselves, should have placed them in space after one certain particular manner and not otherwise; why everything was not placed the quite contrary way, for instance, by changing east into west. But if space is nothing else but that order or relation, and is nothing at all without bodies but the possibility of placing them, then those two states, the one such as it now is, the other supposed to be the quite contrary way, would not at all differ from one another. Their difference therefore is only to be found in our chimerical supposition of the reality of space in itself. But in truth the one would exactly be the same thing as the other, they being absolutely indiscernible, and consequently there is no room to inquire after a reason of the preference of the one to the other.

6 The case is the same with respect to time. Supposing anyone should ask why God did not create everything a year sooner, and the same person should infer from thence that God has done something concerning which 'tis not possible there should be a reason why he did it so and not otherwise; the answer is that his inference would be right if time was anything distinct from things existing in time. For it would be impossible there should be any reason why things should be applied to such particular instants rather than to others, their succession continuing the same. But then the same argument proves that instants, considered without the things, are nothing at all and that they consist only in the successive order of things, which order remaining the same, one of the two states, viz., that of a supposed anticipation, would not at all differ, nor could be discerned from the other which now is.

[...]

47 I will here show how men come to form to themselves the notion of space. They consider that many things exist at once, and they observe in them a certain order of coexistence, according to which the relation of one thing to another is more or less simple. This order is their situation or distance. When it happens that one of those coexistent things changes its relation to a multitude of others which do not change their relations among themselves, and that another thing, newly come, acquires the same relation to the others as the former had, we then say it is come into the *place* of the former; and this change we call a *motion* in that body wherein is the immediate cause of the change. And though many, or even all, the coexistent things should change according to certain known rates of direction and swiftness, yet one may always determine the relation of situation which every coexistent acquires with respect to every other coexistent, and even that relation which any other coexistent would have to this, or which this would have to any other, if it had not changed or if it had changed any otherwise. And supposing or feigning that among those coexistents there is a sufficient number of them which have undergone no change, then we may say that those which have such a relation to those fixed existents as others had to them before have now the same *place* which those others had. And that which comprehends all those places is called *space*. Which shows that in order to have an idea of place, and consequently of space, it is sufficient to consider these relations and the rules of their changes, without needing to fancy any absolute reality out of the things whose situation we consider; and to give a kind of definition, *place* is that which we say is the same to *A* and to *B*, when the relation of the coexistence of *B*, with *C*, *E*, *F*, *G*, etc., agrees perfectly with the relation of the coexistence which *A* had with the same *C*, *E*, *F*, *G*, etc., supposing there has

been no cause of change in *C*, *E*, *F*, *G*, etc. It might be said also, without entering into any further particularity, that place is that which is the same in different moments to different existent things when their relations of coexistence with certain other existents which are supposed to continue fixed from one of those moments to the other agree entirely together. And *fixed existents* are those in which there has been no cause of any change of the order of their coexistence with others, or (which is the same thing) in which there has been no motion. Lastly, *space* is that which results from places taken together. And here it may not be amiss to consider the difference between place and the relation of situation which is in the body that fills up the place. For the place of *A* and *B* is the same, whereas the relation of *A* to fixed bodies is not precisely and individually the same as the relation which *B* (that comes into its place) will have to the same fixed bodies; but these relations agree only. For two different subjects, as *A* and *B*, cannot have precisely the same individual affection, it being impossible that the same individual accident should be in two subjects or pass from one subject to another. But the mind, not contented with an agreement, looks for an identity, for something that should be truly the same, and conceives it as being extrinsic to the subject; and this is what we here call *place* and *space*. But this can only be an ideal thing, containing a certain order, wherein the mind conceives the application of relations. In like manner as the mind can fancy to itself an order made up of genealogical lines whose bigness would consist only in the number of generations wherein every person would have his place; and if to this one should add the fiction of a metempsychosis and bring in the same human souls again, the persons in those lines might change place; he who was a father or a grandfather might become a son or a grandson, etc. And yet those genealogical places, lines, and spaces, though they should express real truths, would only be ideal things. I shall allege another example to show how the mind uses, upon occasion of accidents which are in subjects, to fancy to itself something answerable to those accidents out of the subjects. The ratio or proportion between two lines *L* and *M* may be conceived three several ways: as a ratio of the greater *L* to the lesser *M*, as a ratio of the lesser *M* to the greater *L*, and, lastly, as something abstracted from both, that is, the ratio between *L* and *M* without considering which is the antecedent or which the consequent, which the subject and which the object. And thus it is that proportions are considered in music. In the first way of considering them, *L* the greater, in the second, *M* the lesser, is the subject of that accident which philosophers call "relation." But which of them will be the subject in the third way of considering them? It cannot be said that both of them, *L* and *M* together, are the subject of such an accident; for, if so, we should have an accident in two subjects, with one leg in one and the other in the other, which is contrary to the notion of accidents. Therefore we must say that this relation, in this third way of considering it, is indeed out of the subjects; but, being neither a substance nor an accident, it must be a mere ideal thing, the consideration of which is nevertheless useful. To conclude, I have here done much like Euclid, who, not being able to make his readers well understand what *ratio* is absolutely in the sense of geometricians, defines what are the *same ratios*. Thus, in like manner, in order to explain what *place* is, I have been content to define what is the *same place*. Lastly, I observe that the traces of movable bodies, which they leave sometimes upon the immovable ones on which they are moved, have given men occasion to form in their imagination such an idea, as if some trace did still remain, even when there is nothing unmoved. But this is a mere ideal thing and imports only that *if there was any unmoved thing there, the trace might be marked out upon it*. And 'tis this analogy which makes men fancy places, traces, and spaces, though these things consist only in the truth of relations and not at all in any absolute reality.

3.7

The Problem of Induction

David Hume

David Hume (1711–1776) was a brilliant historian and philosopher whose crisp formulation of skeptical problems set the agenda for subsequent philosophical discussion. In this selection from *An Enquiry Concerning Human Understanding* (1748), Hume divides all of human knowledge into two parts, *relations of ideas* and *matters of fact*, and argues that inductive extrapolations of our experience cannot be justified by appeal to either. To reason inductively is to reason in a customary or habitual way, but this has, Hume claims, no rational or normative force. This skepticism has profound implications for the rationality of science. Hume's challenge still occupies philosophers of science today.

Section IV Sceptical Doubts Concerning the Operations of the Understanding

Part I

All the objects of human reason or inquiry may naturally be divided into two kinds, to wit, *Relations of Ideas*, and *Matters of Fact*. Of the first kind are the sciences of Geometry, Algebra, and Arithmetic, and, in short, every affirmation which is either intuitively or demonstratively certain. *That the square of the hypothenuse is equal to the square of the two sides*, is a proposition which expresses a relation between these figures. *That three times five is equal to the half of thirty*, expresses a relation between these numbers. Propositions of this kind are discoverable by the mere operation of thought, without dependence on what is anywhere existent in the universe. Though there never were a circle or triangle in nature, the truths demonstrated by Euclid would for ever retain their certainty and evidence.

Matters of fact, which are the second objects of human reason, are not ascertained in the same manner; nor is our evidence of their truth, however great, of a like nature with the foregoing. The contrary of every matter of fact is still possible, because it can never imply a contradiction, and is conceived by the mind with the

From *The Philosophical Works of David Hume*, Vol. 4 (London: Little, Brown and Co., 1854), pp. 30–46.

same facility and distinctness, as if ever so conformable to reality. *That the sun will not rise to-morrow,* is no less intelligible a proposition, and implies no more contradiction, than the affirmation, *that it will rise.* We should in vain, therefore, attempt to demonstrate its falsehood. Were it demonstratively false, it would imply a contradiction, and could never be distinctly conceived by the mind.

It may therefore be a subject worthy of curiosity, to inquire what is the nature of that evidence, which assures us of any real existence and matter of fact, beyond the present testimony of our senses, or the records of our memory. This part of philosophy, it is observable, has been little cultivated, either by the ancients or moderns; and therefore our doubts and errors, in the prosecution of so important an inquiry, may be the more excusable, while we march through such difficult paths without any guide or direction. They may even prove useful, by exciting curiosity, and destroying that implicit faith and security which is the bane of all reasoning and free inquiry. The discovery of defects in the common philosophy, if any such there be, will not, I presume, be a discouragement, but rather an incitement, as is usual, to attempt something more full and satisfactory than has yet been proposed to the public.

All reasonings concerning matter of fact seem to be founded on the relation of *Cause and Effect.* By means of that relation alone we can go beyond the evidence of our memory and senses. If you were to ask a man why he believes any matter of fact which is absent, for instance, that his friend is in the country or in France, he would give you a reason, and this reason would be some other fact: as a letter received from him, or the knowledge of his former resolutions and promises. A man, finding a watch or any other machine in a desert island, would conclude that there had once been men in that island. All our reasonings concerning fact are of the same nature. And here it is constantly supposed, that there is a connection between the present fact and that which is inferred from it. Were there nothing to bind them together, the inference would be entirely precarious. The hearing of an articulate voice and rational discourse in the dark assures us of the presence of some person: why? because these are the effects of the human make and fabric, and closely connected with it. If we anatomize all the other reasonings of this nature, we shall find, that they are founded on the relation of cause and effect, and that this relation is either near or remote, direct or collateral. Heat and light are collateral effects of fire, and the one effect may justly be inferred from the other.

If we would satisfy ourselves, therefore, concerning the nature of that evidence which assures us of matters of fact, we must inquire how we arrive at the knowledge of cause and effect.

I shall venture to affirm, as a general proposition which admits of no exception, that the knowledge of this relation is not, in any instance, attained by reasonings a priori; but arises entirely from experience, when we find, that any particular objects are constantly conjoined with each other. Let an object be presented to a man of ever so strong natural reason and abilities; if that object be entirely new to him, he will not be able, by the most accurate examination of its sensible qualities, to discover any of its causes or effects. Adam, though his rational faculties be supposed, at the very first, entirely perfect, could not have inferred from the fluidity and transparency of water, that it would suffocate him; or from the light and warmth of fire that it would consume him. No object ever discovers, by the qualities which appear to the senses, either the causes which produced it, or the effects which will arise from it; nor can our reason, unassisted by experience, ever draw any inference concerning real existence and matter of fact.

This proposition, *that causes and effects are discoverable, not by reason but by experience,* will readily be admitted with regard to such objects as we remember to have once been altogether unknown to us; since we must be conscious of the utter inability which we then lay under of foretelling what would arise from them. Present two smooth pieces of marble to a man who has no tincture of natural philosophy; he will never discover that they will adhere together in such a manner as to require great force to separate them in a direct line, while they make so small a resistance to a lateral pressure. Such events as bear little analogy to the common course of nature, are also readily confessed to be known only by experience; nor does any man imagine that the explosion of gunpowder, or the attraction of a loadstone, could ever be discovered by arguments a priori. In like manner, when an effect is supposed to depend upon an intricate machinery or secret structure of parts, we make no difficulty in attributing all our knowledge of it to experience. Who will assert that he can give the ultimate reason, why milk or bread is proper nourishment for a man, not for a lion or a tiger?

But the same truth may not appear at first sight to have the same evidence with regard to events, which have become familiar to us from our first appearance in the world, which bear a close analogy to the whole course of nature, and which are supposed to depend on the simple qualities of objects, without any secret structure of parts. We are apt to imagine, that we could discover these effects by the mere operation of our reason without experience. We fancy, that were we brought on a sudden into this world, we could at first have inferred that one billiard-ball would communicate motion to another upon impulse; and that we needed not to have waited for the event, in order to pronounce with certainty concerning it. Such is the influence of custom, that where it is strongest, it not only covers our natural ignorance, but even conceals itself, and seems not to take place, merely because it is found in the highest degree.

But to convince us, that all the laws of nature, and all the operations of bodies, without exception, are known only by experience, the following reflections may perhaps suffice. Were any object presented to us, and were we required to pronounce concerning the effect which will result from it, without consulting past observation; after what manner, I beseech you, must the mind proceed in this operation? It must invent or imagine some event which it ascribes to the object as its effect; and it is plain that this invention must be entirely arbitrary. The mind can never possibly find the effect in the supposed cause, by the most accurate scrutiny and examination. For the effect is totally different from the cause, and consequently can never be discovered in it. Motion in the second billiard-ball is a quite distinct event from motion in the first; nor is there anything in the one to suggest the smallest hint of the other. A stone or piece of metal raised into the air, and left without any support, immediately falls: but to consider the matter a priori, is there anything we discover in this situation which can beget the idea of a downward, rather than an upward, or any other motion, in the stone or metal?

And as the first imagination or invention of a particular effect, in all natural operations, is arbitrary, where we consult not experience; so must we also esteem the supposed tie or connection between the cause and effect, which binds them together, and renders it impossible, that any other effect could result from the operation of that cause. When I see, for instance, a billiard-ball moving in a straight line towards another; even suppose motion in the second ball should by accident be suggested to me, as the result of their contact or impulse; may I not conceive that a hundred different events might as well follow from that cause? May not both these balls remain at absolute rest? May not the first ball return in a straight line, or leap off from the second in any line or direction? All these suppositions are consistent and conceivable. Why then should we give the preference to one, which is no more consistent or conceivable than the rest? All our reasonings a priori will never be able to show us any foundation for this preference.

In a word, then, every effect is a distinct event from its cause. It could not, therefore, be discovered in the cause; and the first invention or conception of it, a priori, must be entirely arbitrary. And even after it is suggested, the conjunction of it with the cause must appear equally arbitrary; since there are always many other effects, which, to reason, must seem fully as consistent and natural. In vain, therefore, should we pretend to determine any single event, or infer any cause or effect, without the assistance of observation and experience.

Hence we may discover the reason, why no philosopher, who is rational and modest, has ever pretended to assign the ultimate cause of any natural operation, or to show distinctly the action of that power, which produces any single effect in the universe. It is confessed, that the utmost effort of human reason is, to reduce the principles productive of natural phenomena to a greater simplicity, and to resolve the many particular effects into a few general causes, by means of reasonings from analogy, experience, and observation. But as to the causes of these general causes, we should in vain attempt their discovery; nor shall we ever be able to satisfy ourselves by any particular explication of them. These ultimate springs and principles are totally shut up from human curiosity and inquiry. Elasticity, gravity, cohesion of parts, communication of motion by impulse; these are probably the ultimate causes and principles which we shall ever discover in nature; and we may esteem ourselves sufficiently happy, if, by accurate enquiry and reasoning, we can trace up the particular phenomena to, or near to, these general principles. The most perfect philosophy of the natural kind only staves off our ignorance a little longer; as perhaps the most perfect philosophy of the moral or metaphysical kind serves only to discover larger portions of it. Thus the observation of human blindness and weakness is the result of all philosophy, and meets us, at every turn, in spite of our endeavors to elude or avoid it.

Nor is geometry, when taken into the assistance of natural philosophy, ever able to remedy this defect, or lead us into the knowledge of ultimate causes, by all that accuracy of reasoning for which it is so justly celebrated. Every part of mixed mathematics proceeds upon the supposition, that certain laws are established by nature in her operations; and abstract reasonings are employed, either to assist experience in the discovery of these laws, or to determine their influence in particular instances, where it depends upon any precise degree of distance and quantity. Thus, it is a law of motion, discovered by experience, that the moment or force of any body in motion, is in the compound ratio or proportion of its solid contents and its velocity; and consequently, that a small force may remove the greatest obstacle, or raise the greatest weight, if by any contrivance or machinery, we can increase the velocity of that force, so as to make it an overmatch for its antagonist. Geometry assists us in the application of this law, by giving us the just dimensions of all the parts and figures which can enter into any species of machine; but still the discovery of the law itself is owing merely to experience; and all the abstract reasonings in the world could never lead us one step towards the knowledge of it. When we reason a priori, and consider merely any object or cause, as it appears to the mind, independent of all observation, it never could suggest to us the notion of any distinct object, such as its effect; much less show us the inseparable and inviolable connection between them. A man must be very sagacious who could discover by reasoning, that crystal is the effect of heat, and ice of cold, without being previously acquainted with the operation of these qualities.

Part II

But we have not yet attained any tolerable satisfaction with regard to the question first proposed. Each solution still gives rise to a new question as difficult as the foregoing, and leads us on to further inquiries. When it is asked, *What is the nature of all our reasonings concerning matter of fact?* the proper answer seems to be, That they are founded on the relation of cause and effect. When again it is asked, *What is the foundation of all our reasonings and conclusions concerning that relation?* it may be replied in one word, EXPERIENCE. But if we still carry on our sifting humor, and ask, *What is the foundation of all conclusions from experience?* this implies a new question, which may be of more difficult solution and explication. Philosophers that give themselves airs of superior wisdom and sufficiency, have a hard task when they encounter persons of inquisitive dispositions, who push them from every corner to which they retreat, and who are sure at last to bring them to some dangerous dilemma. The best expedient to prevent this confusion, is to be modest in our pretensions, and even to discover the difficulty ourselves before it is objected to us. By this means we may make a kind of merit of our very ignorance.

I shall content myself in this section with an easy task, and shall pretend only to give a negative answer to the question here proposed. I say then, that even after we have experience of the operations of cause and effect, our conclusions from that experience are *not* founded on reasoning, or any process of the understanding. This answer we must endeavor both to explain and to defend.

It must certainly be allowed, that nature has kept us at a great distance from all her secrets, and has afforded us only the knowledge of a few superficial qualities of objects; while she conceals from us those powers and principles on which the influence of those objects entirely depends. Our senses inform us of the color, weight, and consistence of bread; but neither sense nor reason can ever inform us of those qualities which fit it for the nourishment and support of a human body.

Sight or feeling conveys an idea of the actual motion of bodies, but as to that wonderful force or power which would carry on a moving body for ever in a continued change of place, and which bodies never lose but by communicating it to others; of this we cannot form the most distant conception. But notwithstanding this ignorance of natural powers and principles, we always presume when we see like sensible qualities, that they have like secret powers, and expect that effects similar to those which we have experienced will follow from them. If a body of like color and consistence with that bread which we have formerly eat, be presented to us, we make no scruple of repeating the experiment, and foresee, with certainty, like nourishment and support. Now, this is a process of the mind or thought, of which I would willingly know the foundation. It is allowed on all hands that there is no known connection between the sensible qualities and the secret powers; and consequently, that the mind is not led to form such a conclusion concerning their constant and regular conjunction, by anything which it knows of their nature. As to past *Experience*, it can be allowed to give *direct* and *certain* information of those precise objects only, and that precise period of time which fell under its cognizance: but why this experience should be extended to future times, and to other objects, which, for aught we know, may be only in appearance similar, this is the main question on which I would insist. The bread which I formerly eat nourished me; that is, a body of such sensible qualities was, at that time, endued with such secret powers: but does it follow, that other bread must also nourish me at another time, and that like sensible qualities must always be attended with the like secret powers? The consequence seems nowise necessary. At least, it must be acknowledged, that there is here a consequence drawn by the mind, that there is a certain step taken, a process of thought, and an inference which wants to be explained. These two propositions are far from being the same, *I have found that such an object has always been attended with such an effect*, and *I foresee, that other objects, which are, in appearance, similar, will be attended with similar effects.* I shall allow, if you please, that the one proposition may justly be inferred from the other: I know, in fact, that it always is inferred. But if you insist that the inference is made by a chain of reasoning, I desire you to produce that reasoning. The connection between these propositions is not intuitive. There is required a medium, which may enable the mind to draw such an inference, if indeed it be drawn by reasoning and argument. What that medium is, I must confess passes my comprehension; and it is incumbent on those to produce it who assert that it really exists, and is the original of all our conclusions concerning matter of fact.

This negative argument must certainly, in process of time, become altogether convincing, if many penetrating and able philosophers shall turn their enquiries this way; and no one be ever able to discover any connecting proposition or intermediate step which supports the understanding in this conclusion. But as the question is yet new, every reader may not trust so far to his own penetration as to conclude, because an argument escapes his enquiry, that therefore it does not really exist. For this reason it may be requisite to venture upon a more difficult task; and enumerating all the branches of human knowledge, endeavor to show, that none of them can afford such an argument.

All reasonings may be divided into two kinds, namely, demonstrative reasoning, or that concerning relations of ideas; and moral reasoning, or that concerning matter of fact and existence. That there are no demonstrative arguments in the case, seems evident, since it implies no contradiction, that the course of nature may change, and that an object, seemingly like those which we have experienced, may be attended with different or contrary effects. May I not clearly and distinctly conceive, that a body, falling from the clouds, and which in all other respects resembles snow, has yet the taste of salt or feeling of fire? Is there any more intelligible proposition than to affirm, that all the trees will flourish in December and January, and will decay in May and June? Now, whatever is intelligible, and can be distinctly conceived, implies no contradiction, and can never be proved false by any demonstrative argument or abstract reasoning a priori.

If we be, therefore, engaged by arguments to put trust in past experience, and make it the standard of our future judgment, these arguments must be probable only, or such as regard matter of fact and real existence, according to the division above mentioned. But that there is no argument of this kind, must appear, if our explication of that species of reasoning be admitted as solid and satisfactory. We have said that all arguments concerning existence are founded on the relation of cause and effect; that our knowledge of that relation is derived entirely from experience; and that all our experimental conclusions proceed upon the supposition, that the future will be conformable to the past. To endeavor, therefore, the proof of this last supposition by probable arguments, or arguments regarding existence, must be evidently going in a circle, and taking that for granted which is the very point in question.

In reality, all arguments from experience are founded on the similarity which we discover among natural objects, and by which we are induced to expect effects similar to those which we have found to follow from such objects. And though none but a fool or madman will ever pretend to dispute the authority of experience, or to reject that great guide of human life, it may surely be allowed a philosopher to have so much curiosity at least as to examine the principle of human nature which gives this mighty authority to experience, and makes us draw advantage from that similarity which nature has placed among different objects. From causes which appear similar, we expect similar effects. This is the sum of all our experimental conclusions. Now it seems evident, that if this conclusion were formed by reason, it would be as perfect at first, and upon one instance, as after ever so long a course of experience: but the case is far otherwise. Nothing so like as eggs; yet no one, on account of this appearing similarity expects the same taste and relish in all of them. It is only after a long course of uniform experiments in any kind, that we attain a firm reliance and security with regard to a particular event. Now, where is that process of reasoning, which from one instance, draws a conclusion so different from that which it infers from a hundred instances that are nowise different from that single one? This question I propose, as much for the sake of information, as with an intention of raising difficulties. I cannot find, I cannot imagine, any such reasoning. But I keep my mind still open to instruction, if any one will vouchsafe to bestow it on me.

Should it be said, that, from a number of uniform experiments, we *infer* a connection between the sensible qualities and the secret powers, this, I must confess, seems the same difficulty, couched in different terms. The question still recurs, on what process of argument is this *inference* founded? Where is the medium, the interposing ideas, which join propositions so very wide of each other? It is confessed, that the color, consistence, and other sensible qualities of bread, appear not of themselves to have any connection with the secret powers of nourishment and support: for otherwise we could infer these secret powers from the first appearance of these sensible qualities, without the aid of experience, contrary to the sentiment of all philosophers, and contrary to plain matter of fact. Here then is our natural state of ignorance with regard to the powers and influence of all objects. How is this remedied by experience? It only shows us a number of uniform effects resulting from certain objects, and teaches us that those particular objects, at that particular time, were endowed with such powers and forces. When a new object, endowed with similar sensible qualities, is produced, we expect similar powers and forces, and look for a like effect. From a body of like color and consistence with bread, we expect like nourishment and support. But this surely is a step or progress of the mind which wants to be explained. When a man says, *I have found, in all past instances, such sensible qualities conjoined with such secret powers*; and when he says, *similar sensible qualities will always be conjoined with similar secret powers*; he is not guilty of a tautology, nor are these propositions in any respect the same. You say that the one proposition is an inference from the other: but you must confess that the inference is not intuitive, neither is it demonstrative. Of what nature is it, then? To say it is experimental, is begging the question. For all inferences from experience suppose, as their foundation, that the future will resemble the past, and that similar powers will be conjoined with similar sensible qualities. If there be any suspicion that the course of nature may change, and that the past may be no rule for the future, all experience becomes useless, and can give rise to no inference or conclusion. It is impossible, therefore, that any arguments from experience can prove this resemblance of the past to the future: since all these arguments are founded on the supposition of that resemblance. Let the course of things be allowed hitherto ever so regular, that alone, without some new argument or inference, proves not that for the future it will continue so. In vain do you pretend to have learned the nature of bodies from your past experience. Their secret nature, and consequently all their effects and influence, may change, without any change in their sensible qualities. This happens sometimes, and with regard to some objects: why may it not happen always, and with regard to all objects? What logic, what process or argument, secures you against this supposition? My practice, you say, refutes my doubts. But you mistake the purport of my question. As an agent, I am quite satisfied in the point; but as a philosopher, who has some share of curiosity, I will not say scepticism, I want to learn the foundation of this inference. No reading, no enquiry, has yet been able to remove my difficulty, or give me satisfaction in a matter of such importance. Can I do better than propose the difficulty to the public, even though, perhaps, I have small hopes of obtaining a solution? We shall at least, by this means, be sensible of our ignorance, if we do not augment our knowledge.

I must confess that a man is guilty of unpardonable arrogance, who concludes, because an argument has escaped his own investigation, that therefore it does not really exist. I must also confess that, though all the learned, for several ages, should have employed themselves in fruitless search upon any subject, it may still, perhaps, be rash to conclude positively, that the subject must therefore pass all human comprehension. Even though we examine all the sources of our knowledge, and conclude them unfit for such a subject, there may still remain a suspicion, that the enumeration is not complete, or the examination not accurate. But with regard to the present subject, there are some considerations which seem to remove all this accusation of arrogance or suspicion of mistake.

It is certain, that the most ignorant and stupid peasants, nay infants, nay even brute beasts, improve by experience, and learn the qualities of natural objects, by observing the effects which result from them. When a child has felt the sensation of pain from touching the flame of a candle, he will be careful not to put his hand near any candle, but will expect a similar effect from a cause which is similar in its sensible qualities and appearance. If you assert, therefore, that the understanding of the child is led into this conclusion by any process of argument or ratiocination, I may justly require you to produce that argument; nor have you any pretence to refuse so equitable a demand. You cannot say that the argument is abstruse, and may possibly escape your inquiry; since you confess that it is obvious to the capacity of a mere infant. If you hesitate, therefore, a moment, or if, after reflection, you produce any intricate or profound argument, you, in a manner, give up the question, and confess, that it is not reasoning which engages us to suppose the past resembling the future, and to expect similar effects from causes which are to appearance similar. This is the proposition which I intended to enforce in the present section. If I be right, I pretend not to have made any mighty discovery. And if I be wrong, I must acknowledge myself to be indeed a very backward scholar, since I cannot now discover an argument which, it seems, was perfectly familiar to me long before I was out of my cradle.

3.8

The Nature of Cause and Effect

David Hume

Hume's views on causation, illustrated in this second selection from his *Enquiry*, provide an excellent illustration of thoroughgoing empiricism. Neither sensation nor introspection gives us any indication of the hidden powers of objects; we can never discover by thought the necessary connection between certain properties of bodies and others, between causes and effects. If we have any knowledge of these things, therefore, it must arise from experience. But in experience we never see the connection either; we see only one event (B) happening on the occasion of another (A). Hume's ruthless solution to the problem is reductive: our concept of causation is nothing but the concept of a constant conjunction of A with B, one after the other, in spatial contiguity. The illusion that we can detect a necessary connection arises from our expectation that B will follow A. In fact, Hume argues, we have no *conception* of causation above constant conjunction.

Section VII Of the Idea of Necessary Connection

Part I

[. . .]

There are no ideas, which occur in metaphysics, more obscure and uncertain, than those of *power, force, energy*, or *necessary connection*, of which it is every moment necessary for us to treat in all our disquisitions. We shall, therefore, endeavor, in this section, to fix, if possible, the precise meaning of these terms, and thereby remove some part of that obscurity, which is so much complained of in this species of philosophy.

It seems a proposition, which will not admit of much dispute, that all our ideas are nothing but copies of our impressions, or, in other words, that it is impossible for us to *think* of anything,

From *The Philosophical Works of David Hume*, Vol. 4 (London: Little, Brown and Co., 1854), pp. 71–7, 79–80, 84–90.

which we have not antecedently *felt*, either by our external or internal senses. I have endeavored to explain and prove this proposition, and have expressed my hopes, that, by a proper application of it, men may reach a greater clearness and precision in philosophical reasonings, than what they have hitherto been able to attain. . . .

To be fully acquainted, therefore, with the idea of power or necessary connection, let us examine its impression; and, in order to find the impression with greater certainty, let us search for it in all the sources from which it may possibly be derived.

When we look about us towards external objects, and consider the operation of causes, we are never able, in a single instance, to discover any power or necessary connection; any quality, which binds the effect to the cause, and renders the one an infallible consequence of the other. We only find that the one does actually in fact follow the other. The impulse of one billiard-ball is attended with motion in the second. This is the whole that appears to the *outward* senses. The mind feels no sentiment or *inward* impression from this succession of objects: consequently there is not, in any single, particular instance of cause and effect, any thing which can suggest the idea of power or necessary connection.

From the first appearance of an object, we never can conjecture what effect will result from it. But were the power or energy of any cause discoverable by the mind, we could foresee the effect, even without experience; and might, at first, pronounce with certainty concerning it, by mere dint of thought and reasoning.

In reality, there is no part of matter that does ever, by its sensible qualities, discover any power or energy, or give us ground to imagine that it could produce any thing, or be followed by any other object which we could denominate its effect. Solidity, extension, motion; these qualities are all complete in themselves, and never point out any other event which may result from them. The scenes of the universe are continually shifting, and one object follows another in an uninterrupted succession; but the power of force, which actuates the whole machine, is entirely concealed from us, and never discovers itself in any of the sensible qualities of body. We know that, in fact, heat is a constant attendant of flame; but what is the connection between them we have no room so much as to conjecture or imagine. It is impossible, therefore, that the idea of power can be derived from the contemplation of bodies, in single instances of their operation; because no bodies ever discover any power, which can be the original of this idea.

Since, therefore, external objects, as they appear to the senses, give us no idea of power or necessary connection, by their operation in particular instances, let us see, whether this idea be derived from reflection on the operations of our own minds, and be copied from any internal impression. It may be said, that we are every moment conscious of internal power while we feel, that, by the simple command of our will, we can move the organs of our body, or direct the faculties of our mind. An act of volition produces motion in our limbs, or raises a new idea in our imagination. This influence of the will we know by consciousness. Hence we acquire the idea of power or energy; and are certain, that we ourselves and all other intelligent beings are possessed of power. This idea, then, is an idea of reflection, since it arises from reflecting on the operations of our own mind, and on the command which is exercised by will, both over the organs of the body and faculties of the soul.

We shall proceed to examine this pretension: and first, with regard to the influence of volition over the organs of the body. This influence, we may observe, is a fact which, like all other natural events, can be known only by experience, and can never be foreseen from any apparent energy or power in the cause, which connects it with the effect, and renders the one an infallible consequence of the other. The motion of our body follows upon the command of our will. Of this we are every moment conscious. But the means by which this is effected, the energy by which the will performs so extraordinary an operation; of this we are so far from being immediately conscious, that it must for ever escape our most diligent enquiry.

For, *first*: Is there any principle in all nature more mysterious than the union of soul with body; by which a supposed spiritual substance acquires such an influence over a material one, that the most refined thought is able to actuate the grossest matter? Were we empowered, by a secret wish, to remove mountains, or control the planets in their orbit, this extensive authority would not be more extraordinary, nor more beyond our comprehension. But if, by consciousness, we perceived any power or energy in the will, we must know this power; we must know its connection with the effect; we must know the secret union of soul and body, and the nature of both these substances, by which the one is able to operate, in so many instances, upon the other.

Secondly, We are not able to move all the organs of the body with a like authority, though we cannot assign any reason besides experience, for so remarkable a difference between one and the other. Why has the will an influence over the tongue and fingers, not over the heart or liver? This question would never embarrass us, were we conscious of a power in the former case, not in the latter. We should then perceive, independent of experience, why the authority of will, over the organs of the body, is circumscribed within such particular limits. Being in that case fully acquainted with the power or force by which it operates, we should also know why its influence reaches precisely to such boundaries, and no further.

A man, suddenly struck with palsy in the leg or arm, or who had newly lost those members, frequently endeavors, at first, to move them, and employ them in their usual offices. Here he is as much conscious of power to command such limbs as a man in perfect health is conscious of power to actuate any member which remains in its natural state and condition. But consciousness never deceives. Consequently, neither in the one case nor in the other are we ever conscious of any power. We learn the influence of our will from experience alone. And experience only teaches us how one event constantly follows another, without instructing us in the secret connection which binds them together, and renders them inseparable.

Thirdly, We learn from anatomy, that the immediate object of power in voluntary motion is not the member itself which is moved, but certain muscles, and nerves, and animal spirits, and, perhaps, something still more minute and more unknown, through which the motion is successively propagated, ere it reach the member itself whose motion is the immediate object of volition. Can there be a more certain proof that the power by which this whole operation is performed, so far from being directly and fully known by an inward sentiment or consciousness, is to the last degree mysterious and unintelligible? Here the mind wills a certain event: immediately another event, unknown to ourselves, and totally different from the one intended, is produced: this event produces another, equally unknown: till at last, through a long succession, the desired event is produced. But if the original power were felt, it must be known: were it known, its effect also must be known, since all power is relative to its effect. And, *vice versa*, if the effect be not known, the power cannot be known nor felt. How indeed can we be conscious of a power to move our limbs, when we have no such power, but only that to move certain animal spirits, which, though they produce at last the motion of our limbs, yet operate in such a manner as is wholly beyond our comprehension?

We may therefore conclude from the whole, I hope, without any temerity, though with assurance, that our idea of power is not copied from any sentiment or consciousness of power within ourselves, when we give rise to animal motion, or apply our limbs to their proper use and office. That their motion follows the command of the will, is a matter of common experience, like other natural events: but the power or energy by which this is effected, like that in the other natural events, is unknown and inconceivable.

[. . .]

The generality of mankind never find any difficulty in accounting for the more common and familiar operations of nature; such as the descent of heavy bodies, the growth of plants, the generation of animals, or the nourishment of bodies by food: but suppose that, in all these cases, they perceive the very force or energy of the cause by which it is connected with its effect, and is for ever infallible in its operation. They acquire, by long habit, such a turn of mind, that upon the appearance of the cause, they immediately expect, with assurance, its usual attendant, and hardly conceive it possible that any other event could result from it. It is only on the discovery of extraordinary phaenomena, such as earthquakes, pestilence, and prodigies of any kind, that they find themselves at a loss to assign a proper cause, and to explain the manner in which the effect is produced by it. It is usual for men, in such difficulties, to have recourse to some invisible intelligent principle, as the immediate cause of that event, which surprises them, and which they think cannot be accounted for from the common powers of nature. But philosophers, who carry their scrutiny a little further, immediately perceive, that, even in the most familiar events, the energy of the cause is as unintelligible as in the most unusual, and that we only learn by experience the frequent conjunction of objects, without being ever able to comprehend anything like connection between them.

[...]

Part II

But to hasten to a conclusion of this argument, which is already drawn out to too great a length; we have sought in vain for an idea of power or necessary connection, in all the sources from which we could suppose it to be derived. It appears, that in single instances of the operation of bodies, we never can, by our utmost scrutiny, discover any thing but one event following another; without being able to comprehend any force or power by which the cause operates, or any connection between it and its supposed effect. The same difficulty occurs in contemplating the operations of mind on body; where we observe the motion of the latter to follow upon the volition of the former; but are not able to observe or conceive the tie which binds together the motion and volition, or the energy by which the mind produces this effect. The authority of the will over its own faculties and ideas, is not a whit more comprehensible: so that, upon the whole, there appears not, throughout all nature, any one instance of connection, which is conceivable by us. All events seem entirely loose and separate. One event follows another, but we never can observe any tie between them. They seem *conjoined*, but never *connected*. But as we can have no idea of any thing which never appeared to our outward sense or inward sentiment, the necessary conclusion *seems* to be, that we have no idea of connection or power at all, and that these words are absolutely without any meaning, when employed either in philosophical reasonings or common life.

But there still remains one method of avoiding this conclusion, and one source which we have not yet examined. When any natural object or event is presented, it is impossible for us, by any sagacity or penetration, to discover, or even conjecture, without experience, what event will result from it, or to carry our foresight beyond that object, which is immediately present to the memory and senses. Even after one instance or experiment, where we have observed a particular event to follow upon another, we are not entitled to form a general rule, or foretell what will happen in like cases; it being justly esteemed an unpardonable temerity to judge of the whole course of nature from one single experiment, however accurate or certain. But when one particular species of event has always, in all instances, been conjoined with another, we make no longer any scruple of foretelling one upon the appearance of the other, and of employing that reasoning, which can alone assure us of any matter of fact or existence. We then call the one object *Cause*, the other *Effect*. We suppose that there is some connection between them; some power in the one, by which it infallibly produces the other, and operates with the greatest certainty and strongest necessity.

It appears, then, that this idea of a necessary connection among events arises from a number of similar instances which occur, of the constant conjunction of these events; nor can that idea ever be suggested by any one of these instances, surveyed in all possible lights and positions. But there is nothing in a number of instances, different from every single instance, which is supposed to be exactly similar; except only, that after a repetition of similar instances, the mind is carried by habit, upon the appearance of one event, to expect its usual attendant, and to believe that it will exist. This connection, therefore, which we *feel* in the mind, this customary transition of the imagination from one object to its usual attendant, is the sentiment or impression from which we form the idea of power or necessary connection. Nothing further is in the case. Contemplate the subject on all sides, you will never find any other origin of that idea. This is the sole difference between one instance, from which we can never receive the idea of connection, and a number of similar instances, by which it is suggested. The first time a man saw the communication of motion by impulse, as by the shock of two billiard-balls, he could not pronounce that the one event was *connected*, but only that it was *conjoined* with the other. After he has observed several instances of this nature, he then pronounces them to be *connected*. What alteration has happened to give rise to this new idea of *connection*? Nothing but that he now *feels* these events to be *connected* in his imagination, and can readily foretell the existence of one from the appearance of the other. When we say, therefore, that one object is connected with another, we mean only that they have acquired a connection in our thought, and give rise to this inference, by which they become proofs of each other's existence; a conclusion which is somewhat extraordinary, but which seems founded on sufficient evidence. Nor will its evidence be weakened by any general diffidence of the understanding, or sceptical suspicion concerning every conclusion which is new and extraordinary. No conclusions can be more agreeable to scepticism than such as make discoveries concerning the weakness and narrow limits of human reason and capacity.

And what stronger instance can be produced of the surprising ignorance and weakness of the understanding than the present? For surely, if there be any relation among objects, which it imports to us to know perfectly, it is that of cause and effect. On this are founded all our reasonings concerning matter of fact or existence. By means of it alone, we attain any assurance concerning objects, which are removed from the present testimony of our memory and senses. The only immediate utility of all sciences is to teach us how to control and regulate future events by their causes. Our thoughts and enquiries are, therefore, every moment, employed about this relation: yet so imperfect are the ideas which we form concerning it, that it is impossible to give any just definition of cause, except what is drawn from something extraneous and foreign to it. Similar objects are always conjoined with similar. Of this we have experience. Suitably to this experience, therefore, we may define a cause to be *an object, followed by another, and where all the objects, similar to the first, are followed by objects similar to the second*. Or, in other words, *where, if the first object had not been, the second never had existed*. The appearance of a cause always conveys the mind, by a customary transition, to the idea of the effect. Of this also we have experience. We may, therefore, suitably to this experience, form another definition of cause, and call it, *an object followed by another, and whose appearance always conveys the thought to that other*. But though both these definitions be drawn from circumstances foreign to the cause, we cannot remedy this inconvenience, or attain any more perfect definition, which may point out that circumstance in the cause, which gives it a connection with its effect. We have no idea of this connection; nor even any distant notion what it is we desire to know, when we endeavor at a conception of it. We say, for instance, that the vibration of this string is the cause of this particular sound. But what do we mean by that affirmation? We either mean, *that this vibration is followed by this sound, and that all similar vibrations have been followed by similar sounds*: or, *that this vibration is followed by this sound, and that, upon the appearance of one, the mind anticipates the senses, and forms immediately an idea of the other*. We may consider the relation of cause and effect in either of these two lights; but beyond these, we have no idea of it.

To recapitulate, therefore, the reasonings of this Section: every idea is copied from some preceding impression or sentiment; and where we cannot find any impression, we may be certain that there is no idea. In all single instances of the operation of bodies or minds, there is nothing that produces any impression, nor consequently can suggest any idea of power or necessary connection. But when many uniform instances appear, and the same object is always followed by the same event, we then begin to entertain the notion of cause and connection. We then *feel* a new sentiment or impression, to wit, a customary connection in the thought or imagination between one object and its usual attendant; and this sentiment is the original of that idea which we seek for. For as this idea arises from a number of similar instances, and not from any single instance, it must arise from that circumstance in which the number of instances differ from every individual instance. But this customary connection or transition of the imagination is the only circumstance in which they differ. In every other particular they are alike. The first instance which we saw of motion, communicated by the shock of two billiard-balls (to return to this obvious illustration), is exactly similar to any instance that may, at present, occur to us, except only that we could not at first *infer* one event from the other, which we are enabled to do at present, after so long a course of uniform experience. I know not whether the reader will readily apprehend this reasoning. I am afraid, that, should I multiply words about it, or throw it into a greater variety of lights, it would only become more obscure and intricate. In all abstract reasonings, there is one point of view, which, if we can happily hit, we shall go farther towards illustrating the subject than by all the eloquence and copious expression in the world. This point of view we should endeavor to reach, and reserve the flowers of rhetoric for subjects which are more adapted to them.

3.9

The Metaphysical Foundations of Natural Science

Immanuel Kant

Immanuel Kant (1724–1804) was a German philosopher who was roused by Hume's skeptical challenges into an attempt to account for the possibility of human knowledge in general and scientific knowledge in particular. In this selection from the Preface to his *Metaphysical Foundations of Natural Science* (1786), Kant gives a detailed taxonomy of the study of nature in order to locate natural science in its proper place. He insists that natural science requires metaphysical principles that cannot be grounded in experience but are nevertheless required for the intelligibility of the very concept of nature and that the doctrine of any determinate nature must have mathematical principles.

Preface

If the word nature is taken simply in its *formal* meaning, where it means the first inner principle of all that belongs to the existence of a thing, then there can be as many different natural sciences as there are specifically different things, each of which must contain its own peculiar inner principle of the determinations belonging to its existence. But nature is also taken otherwise in its *material* meaning, not as a constitution, but as the sum total of all things, insofar as they can be *objects of our senses*, and thus also of experience. Nature, in this meaning, is therefore understood as the whole of all appearances, that is, the sensible world, excluding all nonsensible objects. Now nature, taken in this meaning of the word, has two principal parts, in accordance with the principal division of our senses, where the one contains the objects of the *outer* senses, the other the object of *inner* sense. In this meaning, therefore, a twofold doctrine of nature is possible, the *doctrine of body* and the *doctrine of the soul*, where the first considers *extended* nature, the second *thinking* nature.

Every doctrine that is supposed to be a system, that is, a whole of cognition ordered according

From Kant, *Metaphysical Foundations of Natural Science*, trans. and ed. Michael Friedman (Cambridge: Cambridge University Press, 2004), pp. 3–14.

to principles, is called a science. And, since such principles may be either principles of *empirical* or of *rational* connection of cognitions into a whole, then natural science, be it the doctrine of body or the doctrine of the soul, would have to be divided into *historical* or *rational* natural science, were it not that the word *nature* (since this signifies a derivation of the manifold belonging to the existence of things from their inner *principle*) makes necessary a cognition through reason of the interconnection of natural things, insofar as this cognition is to deserve the name of a science. Therefore, the doctrine of nature can be better divided into *historical doctrine of nature*, which contains nothing but systematically ordered facts about natural things (and would in turn consist *of natural description*, as a system of classification for natural things in accordance with their similarities, and *natural history*, as a systematic presentation of natural things at various times and places), and natural science. Natural science would now be either *properly* or *improperly* so-called natural science, where the first treats its object wholly according to a priori principles, the second according to laws of experience.

What can be called *proper* science is only that whose certainty is apodictic; cognition that can contain mere empirical certainty is only *knowledge* improperly so-called. Any whole of cognition that is systematic can, for this reason, already be called *science*, and, if the connection of cognition in this system is an interconnection of grounds and consequences, even *rational* science. If, however, the grounds or principles themselves are still in the end merely empirical, as in chemistry, for example, and the laws from which the given facts are explained through reason are mere laws of experience, then they carry with them no consciousness of their *necessity* (they are not apodictally certain), and thus the whole of cognition does not deserve the name of a science in the strict sense; chemistry should therefore be called a systematic art rather than a science.

A rational doctrine of nature thus deserves the name of a natural science, only in case the fundamental natural laws therein are cognized a priori, and are not mere laws of experience. One calls a cognition of nature of the first kind *pure*, but that of the second kind is called *applied* rational cognition. Since the word nature already carries with it the concept of laws, and the latter carries with it the concept of the *necessity* of all determinations of a thing belonging to its existence, one easily sees why natural science must derive the legitimacy of this title only from its pure part – namely, that which contains the a priori principles of all other natural explanations – and why only in virtue of this pure part is natural science to be proper science. Likewise, [one sees] that, in accordance with demands of reason, every doctrine of nature must finally lead to natural science and conclude there, because this necessity of laws is inseparably attached to the concept of nature, and therefore makes claim to be thoroughly comprehended. Hence, the most complete explanation of given appearances from chemical principles still always leaves behind a certain dissatisfaction, because one can adduce no a priori grounds for such principles, which, as contingent laws, have been learned merely from experience.

All *proper* natural science therefore requires a *pure* part, on which the apodictic certainty that reason seeks therein can be based. And because this pure part is wholly different, in regard to its principles, from those that are merely empirical, it is also of the greatest utility to expound this part as far as possible in its entirety, separated and wholly unmixed with the other part; indeed, in accordance with the nature of the case it is an unavoidable duty with respect to method. This is necessary in order that one may precisely determine what reason can accomplish for itself, and where its power begins to require the assistance of principles of experience. Pure rational cognition from mere *concepts* is called pure philosophy or metaphysics; by contrast, that which grounds its cognition only on the *construction* of concepts, by means of the presentation of the object in an a priori intuition, is called mathematics.

Properly so-called natural science presupposes, in the first place, metaphysics of nature. For laws, that is, principles of the necessity of that which belongs to the *existence* of a thing, are concerned with a concept that cannot be constructed, since existence cannot be presented a priori in any intuition. Thus proper natural science presupposes metaphysics of nature. Now this latter must always contain solely principles that are not empirical (for precisely this reason it bears the name of a metaphysics), but it can still either: *first*, treat the laws that make possible the concept of a nature in general, even without relation to any determinate object of experience, and thus undetermined with respect to the nature of this or that thing in the sensible world, in which case it is the *transcendental* part of the metaphysics of nature; or *second*, concern itself with a particular nature of this or that kind of thing, for which an empirical concept is given, but still in such a manner that, outside of what lies in this concept, no other empirical principle is used for its cognition (for example, it takes the empirical concept of matter or of a thinking being as its basis, and it seeks that sphere of cognition of which reason is capable a priori concerning these objects), and here such a science must still always be called a metaphysics of nature, namely, of corporeal or of thinking nature. However, [in this second case] it is then not a general, but a *special* metaphysical natural science (physics or psychology), in which the above transcendental principles are applied to the two species of objects of our senses.

I assert, however, that in any special doctrine of nature there can be only as much *proper* science as there is *mathematics* therein. For, according to the preceding, proper science, and above all proper natural science, requires a pure part lying at the basis of the empirical part, and resting on a priori cognition of natural things. Now to cognize something a priori means to cognize it from its mere possibility. But the possibility of determinate natural things cannot be cognized from their mere concepts; for from these the possibility of the thought (that it does not contradict itself) can certainly be cognized, but not the possibility of the object, as a natural thing that can be given outside the thought (as existing). Hence, in order to cognize the possibility of determinate natural things, and thus to cognize them a priori, it is still required that the *intuition* corresponding to the concept be given a priori, that is, that the concept be constructed. Now rational cognition through construction of concepts is mathematical. Hence, although a pure philosophy of nature in general, that is, that which investigates only what constitutes the concept of a nature in general, may indeed be possible even without mathematics, a pure doctrine of nature concerning *determinate* natural things (doctrine of body or doctrine of soul) is only possible by means of mathematics. And, since in any doctrine of nature there is only as much proper science as there is a priori knowledge therein, a doctrine of nature will contain only as much proper science as there is mathematics capable of application there.

So long, therefore, as there is still for chemical actions of matters on one another no concept to be discovered that can be constructed, that is, no law of the approach or withdrawal of the parts of matter can be specified according to which, perhaps in proportion to their density or the like, their motions and all the consequences thereof can be made intuitive and presented a priori in space (a demand that will only with great difficulty ever be fulfilled), then chemistry can be nothing more than a systematic art or experimental doctrine, but never a proper science, because its principles are merely empirical, and allow of no a priori presentation in intuition. Consequently, they do not in the least make the principles of chemical appearances conceivable with respect to their possibility, for they are not receptive to the application of mathematics.

Yet the empirical doctrine of the soul must remain even further from the rank of a properly so-called natural science than chemistry. In the first place, because mathematics is not applicable to the phenomena of inner sense and their laws, the only option one would have would be to take the *law of continuity* in the flux of inner changes into account – which, however, would be an extension of cognition standing to that which mathematics provides for the doctrine of body approximately as the doctrine of the properties of the straight line stands to the whole of geometry. For the pure inner intuition in which the appearances of the soul are supposed to be constructed is *time*, which has only one dimension. [In the second place,] however, the empirical doctrine of the soul can also never approach chemistry even as a systematic art of analysis or experimental doctrine, for in it the manifold of inner observation can be separated only by mere division in thought, and cannot then be held separate and recombined at will (but still less does another thinking subject suffer himself to be experimented upon to suit our purpose), and even observation by itself already changes and displaces the state of the observed object. Therefore, the empirical doctrine of the soul can never become anything more than an historical doctrine of nature, and, as such, a natural doctrine of inner sense which is as systematic as possible, that is, a natural description of the soul, but never a science of the soul, nor even, indeed, an experimental psychological doctrine. This is also the reason for our having used, in accordance with common custom, the general title of natural science for this work, which actually contains the principles of the doctrine of body, for only to it does this title belong in the proper sense, and so no ambiguity is thereby produced.

But in order to make possible the application of mathematics to the doctrine of body, which only through this can become natural science, principles for the *construction* of the concepts that belong to the possibility of matter in general must be introduced first. Therefore, a complete analysis of the concept of a matter in general will have to be taken as the basis, and this is a task for pure philosophy – which, for this purpose, makes use of no particular experiences, but only that which it finds in the isolated (although intrinsically empirical) concept itself, in relation to the pure intuitions in space and time, and in accordance with laws that already essentially attach to the concept of nature in general, and is therefore a genuine *metaphysics of corporeal nature.*

Hence all natural philosophers who have wished to proceed mathematically in their occupation have always, and must have always, made use of metaphysical principles (albeit unconsciously), even if they themselves solemnly guarded against all claims of metaphysics upon their science. Undoubtedly they have understood by the latter the folly of contriving possibilities at will and playing with concepts, which can perhaps not be presented in intuition at all, and have no other certification of their objective reality than that they merely do not contradict themselves. All true metaphysics is drawn from the essence of the faculty of thinking itself, and is in no way fictitiously invented on account of not being borrowed from experience. Rather, it contains the pure actions of thought, and thus a priori concepts and principles, which first bring the manifold of *empirical representations* into the law-governed connection through which it can become *empirical* cognition, that is, experience. Thus these mathematical physicists could in no way avoid metaphysical principles, and, among them, also not those that make the concept of their proper object, namely, matter, a priori suitable for application to outer experience, such as the concept of motion, the filling of space, inertia, and so on. But they rightly held that to let merely empirical principles govern these concepts would in no way be appropriate to the apodictic certainty they wished their laws of nature to possess, so they preferred to postulate such [principles], without investigating them with regard to their a priori sources.

Yet it is of the greatest importance to separate heterogeneous principles from one another, for the advantage of the sciences, and to place each in a special system so that it constitutes a science of its own kind, in order to guard against the uncertainty arising from mixing things together, where one finds it difficult to distinguish to which of the two the limitations, and even mistakes, that might occur in their use may be assigned. For this purpose I have considered it necessary [to isolate] the former from the pure part of natural science (*physica generalis*), where metaphysical and mathematical constructions customarily run together, and to present them, together with principles of the construction of these concepts (and thus principles of the possibility of a mathematical doctrine of nature itself), in a system. Aside from the already mentioned advantage that it provides, this isolation has also a special charm arising from the unity of cognition, when one takes care that the boundaries of the sciences do not run together, but rather each takes in its own separated field.

The following can serve as still another ground for commending this procedure. In everything that is called metaphysics one can hope for the *absolute completeness* of the sciences, of such a kind one may expect in no other type of cognition. Therefore, just as in the metaphysics of nature in general, here also the completeness of the metaphysics of corporeal nature can confidently be expected. The reason is that in metaphysics the object is only considered in accordance with the general laws of thought, whereas in other sciences it must be represented in accordance with data of intuition (pure as well as empirical), where the former, because here the object has to be compared always with *all* the necessary laws of thought, must yield a determinate number of cognitions that may be completely exhausted, but the latter, because they offer an infinite manifold of intuitions (pure or empirical), and thus an infinite manifold of objects of thought, never attain absolute completeness, but can always be extended to infinity, as in pure mathematics and empirical doctrine of nature. I also take myself to have completely exhausted this metaphysical doctrine of body, so far as it may extend, but not to have thereby accomplished any great [piece of] work.

But the schema for completeness of a metaphysical system, whether it be of nature in general, or of corporeal nature in particular, is the table of categories, For there are no more pure concepts of the understanding which can be concerned with the nature of things. All determinations of the general concept of a matter in general must be able to be brought under the four classes of [pure concepts of the understanding], those of *quantity*, of *quality*, of *relation*, and finally *of modality* – and so, too, [must] all that may be either thought a priori in this concept, or presented in mathematical construction, or given as a determinate object of experience. There is no more to be done, or to be discovered, or to be added here, except, if need be, to improve it where it may lack in clarity or exactitude.

The concept of matter had therefore to be carried through all four of the indicated functions of the concepts of the understanding (in four chapters), where in each a new determination of this concept was added. The basic determination of something that is to be an object of the outer senses had to be motion, because only thereby can these senses be affected. The understanding traces back all other predicates of matter belonging to its nature to this, and so natural science, therefore, is either a pure or applied *doctrine of motion*. The *metaphysical* foundations of natural science are therefore to be brought under *four* chapters. The *first* considers *motion* as a pure *quantum* in accordance with its composition, without any quality of the movable, and may be called **phoronomy**. The *second* takes into consideration motion as belonging to the *quality* of matter, under the name of an original moving force, and is therefore called **dynamics**. The *third* considers matter with this quality as in *relation* to another through its own inherent motion, and therefore appears under the name of **mechanics**. The *fourth* chapter, however, determines matter's motion or rest merely in relation to the mode of representation or *modality*, and thus as appearance of the outer senses, and is called **phenomenology**.

Yet aside from the inner necessity to isolate the metaphysical foundations of the doctrine of body, not only from physics, which needs empirical principles, but even from the rational premises of physics that concern the use of mathematics therein, there is still an external, certainly only accidental, but nonetheless important reason for detaching its detailed treatment from the general system of metaphysics, and presenting it systematically as a special whole. For if it is permissible to draw the boundaries of a science, not simply according to the constitution of the object and its specific mode of cognition, but also according to the end that one has in mind for this science itself in uses elsewhere; and if one finds that metaphysics has busied so many heads until now, and will continue to do so, not in order thereby to extend natural knowledge (which takes place much more easily and surely through observation, experiment, and the application of mathematics to outer appearances), but rather so as to attain cognition of that which lies wholly beyond all boundaries of experience, of God, Freedom, and Immortality; then one gains in the advancement of this goal if one frees it from an offshoot that certainly springs from its root, but nonetheless only hinders its regular growth, and one plants this offshoot specially, yet without failing to appreciate the origin of [this offshoot] from it, and without omitting the mature plant from the system of general metaphysics. This does not impair the completeness of general metaphysics, and in fact facilitates the uniform progress of this science towards its end, if, in all instances where one requires the general doctrine of body, one may call only upon the isolated system, without swelling this greater system with the latter. It is also indeed very remarkable (but cannot be expounded in detail here) that general metaphysics, in all instances where it requires examples (intuitions) in order to provide meaning for its pure concepts of the understanding, must always take them from the general doctrine of body, and thus from the form and principles of outer intuition; and, if these are not exhibited completely, it gropes uncertainly and unsteadily among mere meaningless concepts. This is the source of the well-known disputes, or at least obscurity, in the questions concerning the possibility of a conflict of realities, of intensive magnitude, and so on, in which the understanding is taught only by examples from corporeal nature what the conditions are under which such concepts can alone have objective reality, that is, meaning and truth. And so a separated metaphysics of corporeal nature does excellent and indispensable service for *general* metaphysics, in that the former furnishes examples (instances *in concreto*) in which to realize the concepts and propositions of the latter (properly speaking, transcendental philosophy), that is, to give a mere form of thought sense and meaning.

In this treatise, although I have not followed the mathematical method with thoroughgoing rigor (which would have required more time than I had to spend thereon), I have nonetheless imitated that method – not in order to obtain a better reception for the treatise, through an ostentatious display of exactitude, but rather because I believe that such a system would certainly be capable of this rigor, and also that such perfection could certainly be reached in time by a more adept hand, if, stimulated by this sketch, mathematical natural scientists should find it not unimportant to treat the metaphysical part, which they cannot leave out in any case, as a special fundamental part in their general physics, and to bring it into union with the mathematical doctrine of motion.

Newton, in the preface to his *Mathematical First Principles of Natural Science*, says (after he had remarked that geometry requires only two of the mechanical operations that it postulates, namely, to describe a straight line and a circle): *Geometry is proud of the fact that with so little derived from without it is able to produce so much.* By contrast, one can say of metaphysics: *it is dismayed that with so much offered to it by pure mathematics it can still accomplish so little.* Nevertheless, this small amount is still something that even mathematics unavoidably requires in its application to natural science; and thus, since it must here necessarily borrow from metaphysics, need also not be ashamed to let itself be seen in community with the latter.

Unit 4

Methodology and Revolution

Part of the fascination of the history of science is the way that philosophy and the various sciences cross-pollinate each other. A century after Newton's grand synthesis, physics and astronomy were going strong and philosophers were posing increasingly sophisticated questions regarding the scope and methods of the sciences – by which they meant, of course, those sciences that had come of age in the scientific revolution. Between the 1780s and the end of the nineteenth century, other disciplines such as chemistry, geology, and biology began to catch up. At every step along the way, philosophical discussions of proper scientific methodology shaped the development of the emerging sciences.

1 Methodology in France

The breakthrough in chemistry came from the work of Antoine Lavoisier, whose *Elements of Chemistry* appeared in 1789. Part of what enabled Lavoisier to reform the chemistry of his time was his use of exquisitely crafted experimental equipment with which he carried out delicate measurements. But in the preface to his book, from which reading 4.1 is taken, Lavoisier draws attention not to his equipment but to the use he made of it. In particular, he lays stress on a few simple maxims that should govern scientific inquiry: we should proceed from known facts to what is unknown, never searching for truth except "by the natural road of experiment and observation," never forming a conclusion that is not "an immediate consequence necessarily flowing from observation and experiment." These sentiments are thoroughly Newtonian, recalling Newton's famous statements on methodology in the correspondence with Oldenburg (reading 2.14) and his Rules at the opening of the third book of the *Principia* (reading 2.16). Lavoisier endeavored to apply them to chemistry.

He also introduced a reform in the nomenclature of chemistry. Names for compounds, for example, should be combinations of the names of the things being combined. The insight is so simple that it is difficult for us now to imagine a chemistry without it. Yet at the time it was a revolutionary idea – one that Lavoisier owed, in turn, to the empiricist philosopher Étienne Condillac, whose work he quotes repeatedly in his preface. Lavoisier found Condillac's views on language particularly inspiring, for they held out the promise of reducing the difficulty of analytical thinking. "The art of reasoning," Condillac wrote, "is nothing more than a language well arranged." By his reform of the language of chemistry, Lavoisier made basic chemical relationships and transformations sound obvious and self-explanatory.

Lavoisier lost his life in the French Revolution, but his countryman, the mathematical physicist Pierre-Simon, Marquis de Laplace, shared his concern with the art of reasoning. What Laplace offers in reading 4.2, however, is an even grander methodological vision, the vision of "good sense reduced to a calculus." The calculus in question was the calculus of probability; Laplace, who had made good use of probability in his scientific work, saw no reason that its merits should not be felt in everyday reasoning as well. In proposing the calculus of probability as a model for our reasoning, Laplace shows just how thoroughly the scientific world had divested itself of the Aristotelian ideal of knowledge. To give up certainty in exchange for probability was to take a risk that could be justified only on the ground that certainty, in scientific matters and in everyday life, was never attainable in the first place.

2 Methodology in England

On the other side of the English Channel, discussions of methodology were bound up with the problem of inductive inference. The term 'induction', like many philosophical terms, has a wide and a narrow meaning – or rather, several narrow meanings. In the broadest sense, inductive reasoning is just non-deductive reasoning: inductive arguments are simply those in which the premises, though offering some support for the conclusion, fall short of entailing it. In a narrower sense, 'induction' can be used to refer to the act of reasoning from an observed uniformity to its holding in unobserved cases, sometimes called 'enumerative induction', or to the sort of process outlined by Bacon in which one narrows down the possible causes for a phenomenon, sometimes called 'eliminative induction'. In England, the early nineteenth century saw lively discussions not only of instances of inductive reasoning but also of the meaning of 'induction'.

Sir John Herschel's *Preliminary Discourse on the Study of Natural Philosophy* (1830) gives an articulate statement of a moderate empiricist point of view. In reading 4.3, Herschel starts out with a summary of the points John Locke had made a century and a half earlier in the *Essay Concerning Human Understanding* (reading 3.4). Our senses, even extended by the microscope and telescope, are insufficiently sensitive to give us insight into the hidden processes of nature; the true causes of things are revealed only indirectly through their visible effects. But in some cases there are two or more hypotheses that could be used to explain the observable data; and as long as the hypotheses are actually suggested by the evidence and not merely frivolously proposed, they may prove very fruitful.

Herschel uses the example of light, which had been explained both in terms of a particle model (favored by Newtonians) and in terms of waves (the model favored by European scientists such as Christiaan Huygens and Leonard Euler). Throughout the eighteenth century, English scientists had as a group stood steadfastly by the particle model and scorned the "mere hypotheses" of their continental colleagues. But in the early nineteenth century, the English prodigy Thomas Young and the French mathematical physicist Augustin Fresnel successfully overturned the consensus, providing through their studies of interference and polarization "such a weight of analogy and probability" in favor of the wave theory that we must admit that it is either true or very nearly so.

In cases where two theories explain a large number of facts equally well, we must look for a crucial experiment that will enable us to discriminate between them. The best way to find a theory that will fit the facts is to reflect on established general laws. But when only one theory explains all of the known facts, Herschel argues, it is very probable that it will explain others – and how we arrived at it is of no importance. If we were therefore to reject theories on the grounds of their origins, we would lose all of the discoveries to which they would probably lead us. There is no absolute requirement to build one's theories up from the observational data step by inductive step.

One of the major philosophers to write on induction was the English polymath John Stuart Mill. Eloquent, erudite, and opinionated, Mill dominated philosophical discourse in England in the mid-nineteenth century. His first major work, *A System of Logic: Ratiocinative and Inductive*, appeared in 1843 and went through a number of editions in his own lifetime. The treatment of names, propositions, and the Aristotelian syllogism in the first two books of the *System of Logic* is not remarkable, but in the third book Mill turns his attention to induction, and here his views really are original. In reading 4.4 he defines induction narrowly as the process by which we infer that what holds true in particular cases will hold in all cases resembling those particular cases in certain respects. To determine which respects are appropriate, we must determine what the true causes of phenomena are and separate them from factors only accidentally connected with the effects of interest. Mill proposed five canons of inference by which to eliminate the accidental factors. It is perhaps the closest that any great philosopher came to working out Francis Bacon's account of scientific reasoning.

David Hume had, of course, argued that there was no rational bridge that could take us across the gap between our limited experience and the truths of nature that transcend that experience (see reading 3.7). But Mill counters that the gap can be bridged by the uniformity of nature, and we may be sure enough that nature is uniform in those respects that have been found invariable in our experience. Past inductions, in short, have worked out well; on this ground we may expect them to continue working out well in the future. If it occurred to Mill that there is something odd about using an induction to justify induction, the thought does not seem to have disturbed him overmuch.

Mill's great antagonist in the English methodology debates was another polymath, William Whewell, Master of Trinity College, Cambridge. Whewell was an eminent authority on scientific reasoning in his own right, and his *History of the Inductive Sciences* (1837) had been a major influence on Mill. But the two powerful personalities clashed, and successive editions of Mill's *Logic* are full of references to their controversy. At almost every point their thinking diverged. Where Mill embraces the tradition of British empiricism from Locke to Hume, Whewell finds empiricism inadequate. Its theory of concepts is wrong, he argues in reading 4.5, for we have a concept of *cause* that is not (*pace* Hume) derived from experience. And the empiricist theory of judgment is also inadequate, since it cannot supply a justification for certain scientific truths we know perfectly well, such as Newton's third law that for every reaction there is an equal and opposite reaction. Whewell's thought shows clearly the influence of Immanuel Kant; the chief problem for him is not the empiricist's problem of trying to find enough information somehow to justify our scientific beliefs, but rather the problem of finding a philosophy adequate to account for what we undoubtedly know. Where Mill seems almost Baconian in his methods for sifting through accidental circumstances to determine true causes, Whewell sets out to reform the canons of scientific inference, deliberately titling one of his books *Novum Organum Renovatum* – complete with aphorisms – to make the point that Bacon's *Novum Organum* required revision (reading 4.6).

A test case for both Mill and Whewell was Johannes Kepler's discovery of the elliptical orbit of Mars. Whewell argued that Kepler had made a genuine inference from Tycho Brahe's data to the ellipse. Mill retorted that the ellipse was in the observations already and that Kepler merely "colligated" the data rather than inferring anything from them. There was an induction, Mill granted, in the inference that the planet's orbital path was the result of a centrally directed force. Neither philosopher could ever see any merit in the position of the other. Whewell died in 1866, Mill in 1873. Their controversy was taken up after their departure and adjudicated by the brilliant American logician Charles Sanders Peirce.

3 Geology and the Uniformity of Nature

While the philosophers were wrestling with the uniformity of nature as a theoretical postulate underwriting induction, geologists were grappling with the question of uniformity on a practical level. The world as we find it around us today is on the whole not subject to vast catastrophes. The forces of nature are, of course, very powerful; earthquakes damage cities, hurricanes devastate homes, floods cut channels through rock, and volcanic eruptions change the landscape. But within recorded history, all of these destructive phenomena occur with approximately the same frequency from century to century; and on the whole their effects are local rather than global. Can we extrapolate this observed uniformity backward in time, or was the earth in earlier ages subjected to vast catastrophes? And what sort of picture of the earth's past will emerge from these opposing assumptions?

One of the most articulate spokesmen for the catastrophists was the French naturalist Georges Cuvier (1769–1832), who founded the field of vertebrate paleontology. In reading 4.7, Cuvier explains the evidence for a catastrophe, or a series of catastrophes, in the early history of the earth. Observation of geological strata made it obvious that at one point lands now well above sea level had been under water, for marine fossils are preserved in layer after layer. Near mountain ranges, we find strata that run nearly vertically, great geological slabs tilted nearly upright, some of them with still further horizontal layers on top of the vertical ones. Clearly the sedimentary layers were laid down horizontally and then twisted upright in some vast upheaval. And at least some catastrophes must have happened very rapidly, causing mass extinctions. For all of these reasons, Cuvier concluded that the earlier history of the earth was marked by multiple catastrophes that shaped and changed it much more rapidly than the forces we see at work today.

To infer the history of the earth by induction, however, one would take the present as the key to the past; and in the present we see no such vast catastrophes. James Hutton and Charles Lyell developed the uniformitarian position on just such inductive grounds. Instead of multiple rapid catastrophes, they proposed a combination of gradualism and a vastly expanded scale of geological time for gradual processes to work; the earth's crust, gradually rising and falling over time, could buckle and break to create the sorts of formations that had attracted Cuvier's attention. In reading 4.8, Lyell argues that such an extrapolation does not have the consequence some critics leveled against it – that it would compel us to say that the earth had never had a beginning, a position advocated by Democritus that had atheistic overtones. Lyell's response is that the inference has been misunderstood. The claim is not that the earth had no beginning but rather that in tracing back the earth's history we have not yet found any decisive evidence of a beginning; and it would be arbitrary to say that the earth must have existed only as far back as we have so far been able to trace its history.

The uniformitarian, gradualist hypothesis could accommodate many of the same data that Cuvier had noted, provided that sufficient time was available. If the earth had been created only a few thousand years ago, as a literal reading of the opening chapters of Genesis would suggest, the required time was unavailable; if Lyell was right, on the other hand, then at the very least the most natural reading of Genesis would have to be reconsidered. Thus geology unexpectedly became bound up in the religious controversies of the nineteenth century.

4 The Darwinian Synthesis

The first volume of Lyell's *Principles of Geology* came off the presses in 1830 just in time to be put into the hands of the young naturalist Charles Darwin as he set off on HMS *Beagle* for a two-year survey trip to South America. In the event the trip took almost five years, and from his observation of living creatures in isolated regions and of geological formations, which he saw through Lyell's eyes, Darwin was himself gradually persuaded of the uniformitarian point of view. Extending this idea, he conceived of the alterations of living things by the gradual accumulation of small variations.

In the nature of the case, an argument of this sort cannot be made briefly. *On the Origin of Species by Natural Selection* (1859) is, in Darwin's own phrase, "one long argument,"[1] a cumulative case for evolution by natural selection that cannot be adequately appreciated in snippets. But in the Introduction to *The Variation of Animals and Plants under Domestication* (1868), from which reading 4.9 is taken, Darwin gives a summary of his argument and a justification for his adopting the hypothesis of natural selection. His discussion shows that he is very familiar with the methodological debates of the early to mid nineteenth century. He was a personal friend of Mill, and in his *Autobiography* he states explicitly that Herschel's work was one of the two major influences on his scientific development.[2] That influence shows through clearly in the very terms in which Darwin makes his defense: "In scientific investigations it is permitted to invent any hypothesis, and if it explains various large and independent classes of facts it rises to the rank of a well-grounded theory."

The "large and independent classes of facts" Darwin had in mind were the succession of creatures in the geological column, the geographical distribution of creatures both past and present, and the homology of their structures. Each of these sets of facts, Darwin argued, was explained well on the hypothesis of the gradual descent of organisms with modification by natural selection.

But on the hypothesis of the divine creation of separate species, these facts could not be explained at all; for merely saying that it pleased God to create things in this manner connects no facts with laws. The explanatory unification brought about by the hypothesis of natural selection was itself all the justification Darwin needed.

But if it was all Darwin needed, it was not all that his critics demanded. The hypothesis of descent by modification through natural selection met with some dissent on the grounds that it contradicted Scripture, but there were also empirical challenges for it to overcome. Darwin had a grand idea but could say nothing of the mechanism by which the modifications were acquired and passed on. Still worse, the timescales required by uniformitarian geology ran into problems on several fronts. William Thompson, Lord Kelvin calculated that it would take at most 400 million years – and perhaps as little as 20 million years – for the earth to cool from a molten ball of rock to its present temperature. Hermann von Helmholtz and Simon Newcomb independently calculated that the sun, condensing from nebular dust and glowing from the heat of gravitational attraction, would take about 100 million years to reach its present brightness and diameter, a result consistent with Kelvin's calculation. The Irish scientist John Joly calculated the rate at which the oceans would accumulate salt from erosion; it came out at 90 million years, confirming the other calculations.

Each of these arguments was in its time a serious empirical challenge to the gradual evolution of species. The comparison with the anomalies confronting the Copernican hypothesis is striking, as are the parallels in the gradual acceptance of the two views. For a time, the uniformitarian geologists and the Darwinians simply stuck to the fact that their arguments were strong; there must, they said, be something wrong with the calculations from other branches of science. It took the work of several generations to vindicate their stubbornness; the empirical objections were overcome only by new discoveries that amounted, in aggregate, to a second scientific revolution. The problem of the mechanism of inheritance awaited the work of Gregor Mendel and the subsequent development of molecular genetics. Kelvin's computation did not take into account the effect of radioactive decay on the earth's temperature, which had not been discovered at that time; when Kelvin learned in 1905 about the work of Pierre and Marie Curie on radioactive heating, he withdrew his estimate. The calculations regarding the sun's radius and brightness were also based on a mistaken assumption since nuclear fusion was unknown until 1938. Joly's calculations depended on several mistaken assumptions including, ironically, the idea that salination occurs at a uniform rate.

Darwin's triumph was in part the triumph of uniformitarianism, which in turn cast Mill's assumption of the uniformity of nature in a favorable light. But at a deeper level it was the triumph of explanatory inference. In the twentieth century, questions about the nature and force of explanatory reasoning would continue to attract the close attention of both philosophers and scientists.

5 Holism, Induction, and Explanatory Reasoning

A critical component of Herschel's scientific methodology, and one that goes back at least to Francis Bacon's *Novum Organum*, is the search for crucial experiments that can discriminate between two rival views. But in *The Aim and Structure of Physical Theory*, the French physicist and historian of science Pierre Duhem offers a penetrating critique of this idea. Duhem's central argument, presented in reading 4.13, is that experiments never test a hypothesis in isolation; an experiment tests a system of beliefs, a whole group of hypotheses. If the system as a whole does not line up satisfactorily with experience, something is wrong, no doubt; but there is no way to tell a priori where the correction needs to be made. It is a further consequence of this holism that what appear to be crucial experiments – say, between the wave theory and the particle theory of light – are actually experiments that can show the inadequacy of clusters of hypotheses. We cannot assume, as Herschel assumes in reading 4.3, that the cluster including the wave theory of light is true simply because the cluster containing the particle theory is faulty.

Duhem's critique strikes another blow against the Baconian conception of scientific method. But it also raises another, more disturbing problem.

If experience, even that carefully controlled sort of experience we call experiment, cannot decisively refute a hypothesis in isolation, then it seems we can never decisively get rid of extravagant hypotheses; we can without logical contradiction maintain that the earth is flat, to take an extreme example, if we are willing to make sufficiently radical adjustments in the rest of our beliefs. This extension of Duhem's point was articulated by the American logician Willard van Orman Quine and is sometimes known as the Duhem-Quine thesis.

It might be wondered whether much is lost in conceding Duhem's point. After all, from the fact that we cannot decisively disprove a hypothesis it does not follow that we cannot be pretty confident that it is false. If we have lowered our aim from the certainty required in Aristotelian science to well-grounded rational confidence, Duhem's pronouncement need not concern us. But if we are going to appeal to good but non-decisive reasons, we are back to the problem of induction, broadly conceived. Hume's skeptical challenge arises once again.

The philosopher who did the deepest work on non-deductive inference around the turn of the twentieth century was the American logician Charles Sanders Peirce. In reading 4.10, Peirce gives a taxonomy of the forms of non-deductive inference and argues that they are all, at bottom, inferences from sampling. In the sorts of inductions Hume considered, A is invariably associated with B in our experience; the question then is how we are to justify the inference that A and B are associated in cases lying outside our experience. Peirce generalizes the form of the inference to include cases where A and B are often, but not invariably associated, as when most but not all of the wheat in a sample is of a particular quality. Acknowledging that we have from the sample no knowledge of the essence of wheat and that our inference regarding the unexamined wheat is provisional, Peirce argues that we are nevertheless justified in our provisional inference and that without the need for any "postulates" of the sort Mill envisaged.

Peirce was a strong advocate of the method of hypothesis in scientific procedure, and he gives a fascinating defense of it in reading 4.11. Hypotheses are not mere "colligations of facts," to use Mill's term; they go beyond our direct experience and suggest other facts not yet observed. At first, a hypothesis might be entertained interrogatively, as a mere conjecture: "I wonder whether . . . ?" Peirce gives a delightful example of this mode of reasoning in action. There is, he says, a certain element of guesswork to the inference, since we cannot readily quantify the sorts of characteristics in terms of which we make our judgments. But it is not all *mere* guesswork; even when we are uncertain how much weight should be ascribed to a particular piece of evidence, we can tell which direction it points – can tell that it ought to incline us to believe the hypothesis in question.

6 Hypothesis and Method in Physics

One of the characteristics of physics in the early twentieth century was an increase in the attention given by physicists to the philosophical underpinnings of their work. In his essay "Hypotheses in Physics," from which reading 4.12 is taken, the French mathematician and physicist Henri Poincaré sets out to answer a deceptively simple question: if experiment is the sole source of truth, what role is left for mathematical physics? The answer, simply put, is that experiment alone does not teach us what we wish to know. Error infects our measurements, and we know this and correct for it. But fitting a curve to data points is not a dot-to-dot procedure; the curve will pass between and near the data points but not generally through them.

But what could possibly justify our treating the actual experimental data so casually? According to Poincaré it is that we know, or think that we know, that the true laws of nature are simple. In fact, he argues, we have no choice but to fall back upon simplicity, even though the history of science shows that there is usually complexity beneath the simplicity and yet deeper simplicity beneath the complexity in a seemingly unending pattern. But this does not mean that we should despair. On the contrary, in discovering that our hypotheses are nearly but not quite correct we often obtain an unexpected insight into nature that could be had in no other way. Poincaré's discussion of this point echoes Herschel's remarks in reading 4.3 regarding the value of hypotheses.

Thirty years later, in a lecture here reproduced as Reading 4.14, Albert Einstein revisited the same question that Poincaré had raised. Reason and experience, Einstein says, are the two poles of scientific thinking. Reason provides the structure of the system of theoretical physics, but without experience reason is empty. Experience provides the data of physics. But experience, though it may guide us in our theorizing, cannot be the source of the mathematical concepts required for an adequate physics. Here Einstein diagnoses a subtle error in the thought of Isaac Newton, who seems to have thought that the basic concepts and laws of physics could be derived from experience. On the contrary, Einstein argues, theories are free creations of the human mind.

But if our theories are free creations of the mind and not forced on us by the data of experience, what assurance can we have that our theories will correspond to reality? Here Einstein expresses his conviction that Nature – the capitalization is his – really does operate by simple laws that are in principle discoverable. More than this, he argues that our success in the investigation of nature so far justifies this conviction. There are still unsolved problems, Einstein admits, particularly regarding quantum theory. But throughout the remainder of his life he pressed forward under the conviction he expresses here that at the deepest level the laws of nature govern events directly, not probabilistically, and that those laws are at bottom beautifully simple.

Notes

1 Charles Darwin, *On the Origin of Species: A Facsimile of the First Edition*, 1964, edited by Ernst Mayr, Cambridge: Harvard University Press, p. 459.

2 Francis Darwin (ed.), 1896, *The Life and Letters of Charles Darwin: Including an Autobiographical Chapter*, Vol. 1, New York: D. Appleton and Co., p. 47.

Suggestions for Further Reading

Items marked with an asterisk assume little or no background in the subject; these would be a good place to begin further reading.

*Albritton, Jr., Claude C., 2002, *The Abyss of Time*. New York: Dover.

Bowler, Peter J., 1984, *Evolution: The History of an Idea*. Berkeley: University of California Press.

Cantor, Geoffrey, 1984, *Optics After Newton*. Manchester: Manchester University Press.

Desmond, Adrian and Moore, James, 1991, *Darwin*. New York: Warner Books.

*Eiseley, Loren, 1961, *Darwin's Century: Evolution and the Men who Discovered It*. New York: Doubleday.

Giere, Ronald N. and Westfall, Richard S. (eds.), 1973, *Foundations of Scientific Method: The Nineteenth Century*. Bloomington: Indiana University Press.

*Harman, P. M., 1982, *Energy, Force, and Matter: The Conceptual Development of Nineteenth-Century Physics*. Cambridge: Cambridge University Press.

Hull, David L., 1973, *Darwin and His Critics*. Chicago: University of Chicago Press.

Kohn, David (ed.), 1985, *The Darwinian Heritage*. Princeton: Princeton University Press.

Madden, Edward H. (ed.), 1960, *Theories of Scientific Method: The Renaissance through the Nineteenth Century*. Seattle: University of Washington Press, chapters 8–12.

Olson, Richard, 1975, *Scottish Philosophy and British Physics 1750–1880: A Study in the Foundations of the Victorian Scientific Style*. Princeton: Princeton University Press.

Singer, Charles, 1989, *A History of Biology to About the Year 1900*. Ames, IA: Iowa State University Press, Part III.

*Toulmin, Stephen and Goodfield, June, 1982, *The Discovery of Time*. Chicago: University of Chicago Press.

4.1

The Nature of Scientific Explanation

Antoine Lavoisier

Antoine Lavoisier (1743–1794) was a French nobleman whose accurate experiments and methodological insights earned him the title of "father of modern chemistry." He articulated an early version of the principle of the conservation of mass, discovered several of the elements, and overthrew the phlogiston theory of combustion. In this selection from the introduction to his *Elements of Chemistry* (1789), Lavoisier explains why he felt the need to reform the discipline and teaching of chemistry and articulates what he considers fundamental principles of scientific explanation: proceed from the known to the unknown, and do not make up for the lack of facts with speculative theories.

Preface of the Author: Elements of Chemistry

When I began the following Work, my only object was to extend and explain more fully the Memoir which I read at the public meeting of the Academy of Science in the month of April 1787, on the necessity of reforming and completing the Nomenclature of Chemistry. While engaged in this employment, I perceived, better than I had ever done before, the justice of the following maxims of the Abbé de Condillac, in his *System of Logic*, and some other of his works.

"We think only through the medium of words. – Languages are true analytical methods. – Algebra, which is adapted to its purpose in every species of expression, in the most simple, most exact, and best manner possible, is at the same time a language and an analytical method. – The art of reasoning is nothing more than a language well arranged."

Thus, while I thought myself employed only in forming a Nomenclature, and while I proposed to myself nothing more than to improve the chemical language, my work transformed itself by degrees, without my being able to

From Antoine Lavoisier, *Elements of Chemistry*, trans. Robert Kerr (New York: Dover, 1965), pp. xiii–xxxvii.

prevent it, into a treatise upon the Elements of Chemistry.

The impossibility of separating the nomenclature of a science from the science itself, is owing to this, that every branch of physical science must consist of three things; the series of facts which are the objects of the science, the ideas which represent these facts, and the words by which these ideas are expressed. Like three impressions of the same seal, the word ought to produce the idea, and the idea to be a picture of the fact. And, as ideas are preserved and communicated by means of words, it necessarily follows that we cannot improve the language of any science without at the same time improving the science itself; neither can we, on the other hand, improve a science, without improving the language or nomenclature which belongs to it. However certain the facts of any science may be, and, however just the ideas we may have formed of these facts, we can only communicate false impressions to others, while we want words by which these may be properly expressed.

[. . .]

It is a maxim universally admitted in geometry, and indeed in every branch of knowledge, that, in the progress of investigation, we should proceed from known facts to what is unknown. In early infancy, our ideas spring from our wants; the sensation of want excites the idea of the object by which it is to be gratified. In this manner, from a series of sensations, observations, and analyses, a successive train of ideas arises, so linked together, that an attentive observer may trace back to a certain point the order and connection of the whole sum of human knowledge.

When we begin the study of any science, we are in a situation, respecting that science, similar to that of children; and the course by which we have to advance is precisely the same which Nature follows in the formation of their ideas. In a child, the idea is merely an effect produced by a sensation; and, in the same manner, in commencing the study of a physical science, we ought to form no idea but what is a necessary consequence, and immediate effect, of an experiment or observation. Besides, he that enters upon the career of science, is in a less advantageous situation than a child who is acquiring his first ideas. To the child, Nature gives various means of rectifying any mistakes he may commit respecting the salutary or hurtful qualities of the objects which surround him. On every occasion his judgments are corrected by experience; want and pain are the necessary consequences arising from false judgment; gratification and pleasure are produced by judging aright. Under such masters, we cannot fail to become well informed; and we soon learn to reason justly, when want and pain are the necessary consequences of a contrary conduct.

In the study and practice of the sciences it is quite different; the false judgments we form neither affect our existence nor our welfare; and we are not forced by any physical necessity to correct them. Imagination, on the contrary, which is ever wandering beyond the bounds of truth, joined to self-love and that self-confidence we are so apt to indulge, prompt us to draw conclusions which are not immediately derived from facts; so that we become in some measure interested in deceiving ourselves. Hence it is by no means to be wondered, that, in the science of physics in general, men have often made suppositions, instead of forming conclusions. These suppositions, handed down from one age to another, acquire additional weight from the authorities by which they are supported, till at last they are received, even by men of genius, as fundamental truths.

The only method of preventing such errors from taking place, and of correcting them when formed, is to restrain and simplify our reasoning as much as possible. This depends entirely upon ourselves, and the neglect of it is the only source of our mistakes. We must trust to nothing but facts: These are presented to us by Nature, and cannot deceive. We ought, in every instance, to submit our reasoning to the test of experiment, and never to search for truth but by the natural road of experiment and observation. Thus mathematicians obtain the solution of a problem by the mere arrangement of data, and by reducing their reasoning to such simple steps, to conclusions so very obvious, as never to lose sight of the evidence which guides them.

Thoroughly convinced of these truths, I have imposed upon myself, as a law, never to advance but from what is known to what is unknown; never to form any conclusion which is not an immediate consequence necessarily flowing from observation and experiment; and always to arrange the fact, and the conclusions which are drawn from them, in such an order as shall render it most easy for beginners in the study of chemistry thoroughly to understand them. Hence I have been obliged to depart from the usual order of courses of lectures and of treatises upon chemistry, which always assume the first principles of the science, as known, when the pupil or the reader should never be supposed to know them till they have been explained in subsequent lessons. In almost every instance, these begin by treating of the elements of matter, and by explaining the table of affinities, without considering, that, in so doing, they must bring the principal phenomena of chemistry into view at the very outset: They make use of terms which have not been defined, and suppose the science to be understood by the very persons they are only beginning to teach. It ought likewise to be considered, that very little of chemistry can be learned in a first course, which is hardly sufficient to make the language of the science familiar to the ears, or the apparatus familiar to the eyes. It is almost impossible to become a chemist in less than three or four years of constant application.

These inconveniencies are occasioned not so much by the nature of the subject, as by the method of teaching it; and, to avoid them, I was chiefly induced to adopt a new arrangement of chemistry, which appeared to me more consonant to the order of Nature. I acknowledge, however, that in thus endeavouring to avoid difficulties of one kind, I have found myself involved in others of a different species, some of which I have not been able to remove; but I am persuaded, that such as remain do not arise from the nature of the order I have adopted, but are rather consequences of the imperfection under which chemistry still labours. This science still has many chasms, which interrupt the series of facts, and often render it extremely difficult to reconcile them with each other: It has not, like the elements of geometry, the advantage of being a complete science, the parts of which are all closely connected together: Its actual progress, however, is so rapid, and the facts, under the modern doctrine, have assumed so happy an arrangement, that we have ground to hope, even in our own times, to see it approach near to the highest state of perfection of which it is susceptible.

The rigorous law from which I have never deviated, of forming no conclusions which are not fully warranted by experiment, and of never supplying the absence of facts, has prevented me from comprehending in this work the branch of chemistry which treats of affinities, although it is perhaps the best calculated of any part of chemistry for being reduced into a completely systematic body. . . . This science of affinities, or elective attractions, holds the same place with regard to the other branches of chemistry, as the higher or transcendental geometry does with respect to the simpler and elementary part; and I thought it improper to involve those simple and plain elements, which I flatter myself the greatest part of my readers will easily understand, in the obscurities and difficulties which still attend that other very useful and necessary branch of chemical science.

[. . .]

It will, no doubt, be a matter of surprise, that in a treatise upon the elements of chemistry, there should be no chapter on the constituent and elementary parts of matter; but I shall take occasion, in this place, to remark, that the fondness for reducing all the bodies in nature to three or four elements, proceeds from a prejudice which has descended to us from the Greek Philosophers. The notion of four elements, which, by the variety of their proportions, compose all the known substances in nature, is a mere hypothesis, assumed long before the first principles of experimental philosophy or of chemistry had any existence. In those days, without possessing facts, they framed systems; while we, who have collected facts, seem determined to reject them, when they do not agree with our prejudices. The authority of these fathers of human philosophy still carry great weight, and there is reason to fear that it will even bear hard upon generations yet to come.

It is very remarkable, that, notwithstanding of the number of philosophical chemists who have supported the doctrine of the four elements, there is not one who has not been led by the evidence of facts to admit a greater number of elements into their theory. The first chemists that wrote after the revival of letters, considered sulphur and salt as elementary substances entering into the composition of a great number of substances; hence, instead of four, they admitted the existence of six elements. Beccher assumes the existence of three kinds of earth, from the combination of which, in different proportions, he supposed all the varieties of metallic substances to be produced. Stahl gave a new modification to this system; and succeeding chemists have taken the liberty to make or to imagine changes and additions of a similar nature. All these chemists were carried along by the influence of the genius of the age in which they lived, which contented itself with assertions without proofs; or, at least, often admitted as proofs the slightest degrees of probability, unsupported by that strictly rigorous analysis required by modern philosophy.

All that can be said upon the number and nature of elements is, in my opinion, confined to discussions entirely of a metaphysical nature. The subject only furnishes us with indefinite problems, which may be solved in a thousand different ways, not one of which, in all probability, is consistent with nature. I shall therefore only add upon this subject, that if, by the term *elements*, we mean to express those simple and indivisible atoms of which matter is composed, it is extremely probable we know nothing at all about them; but, if we apply the term *elements*, or *principles of bodies*, to express our idea of the last point which analysis is capable or reaching, we must admit, as elements, all the substances into which we are capable, by any means, to reduce bodies by decomposition. Not that we are entitled to affirm, that these substances we consider as simple may not be compounded of two, or even of a greater number of principles; but, since these principles cannot be separated, or rather since we have not hitherto discovered the means of separating them, they act with regard to us as simple substances, and we ought never to suppose them compounded until experiment and observation has proved them to be so.

. . . I have endeavoured, as much as possible, to denominate simple bodies by simple terms, and I was naturally led to name these first. It will be recollected, that we were obliged to retain that name of any substance by which it had been long known in the world, and that in two cases only we took the liberty of making alterations; first, in the case of those which were but newly discovered, and had not yet obtained names, or at least which had been known but for a short time, and the names of which had not yet received the sanction of the public; and, secondly, when the names which had been adopted, whether by the ancients or the moderns, appeared to us to express evidently false ideas, when they confounded the substances, to which they were applied, with others possessed of different, or perhaps opposite qualities. We made no scruple, in this case, of substituting other names in their room, and the greatest number of these were borrowed from the Greek language. We endeavoured to frame them in such a manner as to express the most general and the most characteristic quality of the substances; and this was attended with the additional advantage both of assisting the memory of beginners, who find it difficult to remember a new word which has no meaning, and of accustoming them early to admit no word without connecting with it some determinate idea.

To those bodies which are formed by the union of several simple substances we gave new names, compounded in such a manner as the nature of the substances directed; but, as the number of double combinations is already very considerable, the only method by which we could avoid confusion, was to divide them into classes. In the natural order of ideas, the name of the class or genus is that which expresses a quality common to a great number of individuals: The name of the species, on the contrary, expresses a quality peculiar to certain individuals only.

These distinctions are not, as some may imagine, merely metaphysical, but are established by Nature. "A child," says the Abbé de Condillac, "is taught to give the name *tree* to the first one which is pointed out to him. The next one he sees presents the same idea, and he gives it the same name. This he does likewise to a third and a fourth, till at last the word *tree*, which he first applied to an individual, comes to be employed by him as the name of a class or a genus, an abstract idea, which comprehends all trees in general. But, when he learns that all trees serve not the same purpose, that they do not all produce the same kind of fruit, he will soon learn to distinguish them by specific and particular names." This is the logic of all the sciences, and is naturally applied of chemistry.

The acids, for example, are compounded of two substances, of the order of those which we consider as simple; the one constitutes acidity, and is common to all acids, and, from this substance, the name of the class or the genus ought to be taken; the other is peculiar to each acid, and distinguishes it from the rest, and from this substance is to be taken the name of the species. But, in the greatest number of acids, the two constituent elements, the acidifying principle, and that which it acidifies, may exist in different proportions, constituting all the possible points of equilibrium or of saturation. This is the case in the *sulphuric* and the *sulphurous acids*; and these two states of the same acid we have marked by varying the termination of the specific name.

[. . .]

I shall conclude this preface by transcribing, literally, some observations of the Abbé de Condillac, which I think describe, with a good deal of truth, the state of chemistry at a period not far distant from our own. These observations were made on a different subject; but they will not, on this account, have less force, if the application of them be thought just.

> Instead of applying observation to the things we wished to know, we have chosen rather to imagine them. Advancing from one ill founded supposition to another, we have at last bewildered ourselves amidst a multitude of errors. These errors becoming prejudices, are, of course, adopted as principles, and we thus bewilder ourselves more and more. The method, too, by which we conduct our reasonings is as absurd; we abuse words which we do not understand, and call this the art of reasoning. When matters have been brought this length, when errors have been thus accumulated, there is but one remedy by which order can be restored to the faculty of thinking; this is, to forget all that we have learned, to trace back our ideas to their source, to follow the train in which they rise, and, as my Lord Bacon says, to frame the human understanding anew.
>
> This remedy becomes the more difficult in proportion as we think ourselves more learned. Might it not be thought that works which treated of the sciences with the utmost perspicuity, with great precision and order, must be understood by every body? The fact is, those who have never studied any thing will understand them better than those who have studied a great deal, and especially those who have written a great deal.

At the end of the fifth chapter, the Abbé de Condillac adds: "But, after all, the sciences have made progress, because philosophers have applied themselves with more attention to observe, and have communicated to their language that precision and accuracy which they have employed in their observations: In correcting their language they reason better."

4.2

Determinism, Ignorance, and Probability

Pierre-Simon Laplace

Pierre-Simon, Marquis de Laplace (1749–1827) was a French astronomer and mathematician who reformulated the celestial mechanics of Newton in the language of the calculus. In this excerpt from *A Philosophical Essay on Probabilities* (1819), Laplace brings together a powerful set of ideas: the principle of sufficient reason, physical determinism, and a conception of probability as relative in part to our ignorance, in part to our knowledge. Though the notion of "equipossible events" subsequently came in for heavy criticism, we should remember that this was Laplace's attempt to make the theory of probability, which he had himself developed in great mathematical detail, more widely accessible.

On Probability

All events, even those that on account of their rarity (or insignificance) seem not to obey the great laws of nature, are as necessary a consequence of these laws as the revolutions of the sun. Ignorant of the bonds that link them to the entire system of the universe, we have made them depend on final causes, or on chance, according as they occur and succeed each other in a regular fashion, or without apparent order. But these fancied causes have been successively moved back as the boundaries of our knowledge have expanded, and they vanish entirely in the face of a sound philosophy, which sees in them only the expression of our ignorance of the true causes.

The connexion between present and preceding events is based on the evident principle that a thing cannot come into existence without there being a cause to produce it. This axiom, known as the *principle of sufficient reason*, extends even to actions between which one is indifferent. The freest will is unable to give rise to them without a specific reason; for if, all circumstances of two situations being exactly the same, it (i.e. the will) were acting in the one but not in the other, its choice would be an effect without cause; it would then, says Leibniz, be the blind chance of

From Pierre-Simon Laplace, *Philosophical Essay on Probabilities*, trans. Andrew I. Dale (New York: Springer-Verlag, 1995), selections from pp. 2–6.

the Epicureans. The contrary opinion is an illusion of the mind that, losing sight of the fleeting reasons for the choice of the will in matters between which we are indifferent, persuades itself that it (i.e. the choice) is determined of its own accord and without motives.

We ought then to consider the present state of the universe as the effect of its previous state and as the cause of that which is to follow. An intelligence that, at a given instant, could comprehend all the forces by which nature is animated and the respective situation of the beings that make it up, if moreover it were vast enough to submit these data to analysis, would encompass in the same formula the movements of the greatest bodies of the universe and those of the lightest atoms. For such an intelligence nothing would be uncertain, and the future, like the past, would be open to its eyes. The human mind affords, in the perfection that it has been able to give to astronomy, a feeble likeness of this intelligence. Its discoveries in mechanics and in geometry, joined to the discovery of universal gravitation, have enabled it to comprehend in the same analytical expressions the past and future states of the system of the world. In applying the same method to some other objects of its knowledge, it has succeeded in relating observed phenomena to general laws, and in anticipating those that given circumstances ought to bring to light. All these efforts in the search for truth tend to lead the mind continually towards the intelligence we have just mentioned, although it will always remain infinitely distant from this intelligence. This tendency, peculiar to the human race, is what makes it superior to the animals; and their progress in this respect distinguishes nations and ages, and constitutes their real glory.

Let us recall that formerly, and indeed not too long ago, torrential rain or severe drought, a comet with a very long tail in train, eclipses, the aurora borealis and generally all extraordinary phenomena were regarded as so many signs of divine anger. Heaven was implored to avert their disastrous influence, though no one used to pray that the planets and the sun might be halted in their courses: observation would soon have shown the futility of such prayers. But as these phenomena, coming and going at long intervals, seemed contrary to the order of nature, it was supposed that heaven, incensed by the crimes of the earth, had created them to give warning of its vengeance. Thus the long tail of the comet of 1456 spread terror in a Europe already disheartened by the rapid success of the Turks, who had just overthrown the Byzantine Empire. This heavenly body, after four (further) revolutions, has excited amongst us a completely different interest. The knowledge of the laws of the system of the world, acquired in this period, has dissipated the fears engendered by ignorance of the true relationship of man to the universe; and Halley, having recognized that this comet was identical to those of the years 1531, 1607 and 1682, predicted its next return for the end of 1758 or the beginning of 1759. The learned world impatiently awaited this return, which would confirm one of the greatest discoveries ever made in science, and which would realize the prediction of Seneca when he said, in speaking of the orbits of those heavenly bodies coming from a great distance, "The day will come when, after continual study for several centuries, the things now hidden will appear with clearness, and posterity will be astonished that truths so clear could have escaped us". Clairaut then attempted to analyse the perturbations that the comet had undergone owing to the motion of the two largest planets, Jupiter and Saturn. After immense calculation he fixed its next passage to the perihelion about the beginning of April 1759, and this was soon verified by observation. Without any doubt, the regularity that astronomy shows us in the movement of the comets takes place in all phenomena. The trajectory of a simple molecule of air or vapour is regulated in a manner as certain as that of the planetary orbits; the only difference between them is that which is contributed by our ignorance.

Probability is relative in part to this ignorance and in part to our knowledge. Suppose we know that, of three or more events, one alone must occur, but that nothing leads us to believe that one of them will happen rather than the others. In this state of indecision, it is impossible for us to say anything with certainty about their occurrence. However, it is probable that one of these events chosen at will (or at random), will not occur, because there are several equally possible cases that exclude its occurrence, while only a single one favours it.

The theory of chances consists in reducing all events of the same kind to a certain number of equally possible cases, that is to say, to cases whose existence we are equally uncertain of, and in determining the number of cases favourable to the event whose probability is sought. The ratio of this number to that of all possible cases is the measure of this probability, which is thus only a fraction whose numerator is the number of favourable cases, and whose denominator is the number of all possible cases.

The preceding notion of probability implies that the probability remains the same when the number of favourable cases and the number of all possible cases are increased in the same ratio. To convince oneself of this, consider two urns *A* and *B*, the first containing four white and two black balls, and the second only two white and one black ball. Suppose that the two black balls of the first urn are connected by a thread that breaks at the moment one of the balls is taken hold of before being drawn, and suppose further that the four white balls form two similar systems. All the chances that will lead to one of the balls of the black system being taken hold of will result in the drawing of a black ball. If one supposes now that the threads which join the balls do *not* break, it is clear that the number of possible chances will not change, and no more will that of the chances favourable to the drawing of black balls; the only difference is that two balls will together be drawn from the urn. The probability of drawing a black ball from the urn will then be the same as before. But then one clearly has the case of urn *B*, with the single difference that the three balls of this last urn are replaced by three systems of two balls inseparably linked.

When all the cases are favourable to an event, its probability changes into certainty, which is expressed by the number 1. In this respect certainty and probability are comparable, although there may be an essential difference between the two states of the mind, when a truth is rigorously demonstrated to it, or when a small source of error is still seen.

In matters that are only probable, the difference in the data that each man has on them is one of the principal causes of the diversity of opinions found to hold on such matters. Let us suppose, for example, that we have three urns *A*, *B* and *C*, one of which contains only black balls, while the two other contain only white balls. A ball is to be drawn from urn *C*: what is the probability that this ball will be black? If we do not know which of the three urns contains only black balls, so that we have no reason to believe that it is *C* rather than *B* or *A*, these three hypotheses will seem equally possible; and as a black ball can be drawn only under the first hypothesis (i.e. only if it is *C* that contains balls of this colour), the probability of such a drawing will be equal to a third. If one knows that urn *A* contains only white balls, then the uncertainty extends only to the urns *B* and *C*, and the probability that the ball drawn from urn *C* will be black is a half. Finally, this probability is changed into certainty if one is assured that the urns *A* and *B* contain only white balls.

It is thus that the same matter recounted before a large crowd of people, finds various degrees of belief according to the extent of the listeners' knowledge. If the man who reports it is deeply convinced of it (i.e. of its truth), and if by his calling and character he inspires great confidence, his account, however extraordinary it may be, will have the same degree of likelihood (or plausibility) for ignorant listeners as an ordinary matter reported by the same man, and they will believe it implicitly. However, if any one of the listeners knows that the same matter is denied by other equally respectable men, he will doubt the truth of the report; and the matter will be judged false by well-informed listeners who deem it inconsistent, either with well-authenticated matters or with the immutable laws of nature.

It is to the influence of the viewpoint of those whom popular opinion judges best informed, and to whom it has been accustomed to give its trust in the most important matters of life, that the propagation of those errors is due that, in times of ignorance, have covered the face of the earth. Magic and Astrology offer us two prime examples. These fallacies, inculcated in infancy, sanctioned without scrutiny, and based only on universal credulity, have maintained their position for a very long time. At last the march of science has effaced them from the minds of enlightened men, whose opinion in turn has caused them to vanish among the common people, through the same power of imitation and custom that had spread them so widely. This power, the most potent resource of the moral (or mental) world, establishes and preserves in an entire nation, ideas completely contrary to those maintained elsewhere with the same conviction. What forbearance ought we not then to have for opinions different from ours, seeing that this difference often depends only on the multifarious points of view to which circumstances have led us! Let us enlighten those whom we do not judge sufficiently well informed; but before doing so let us severely examine our own opinions, and weigh up their respective probabilities impartially.

Difference of opinion depends moreover on the manner in which each of us determines the influence of data known to him. Probability theory depends on such delicate considerations that it is not surprising that, with the same data, two people will arrive at different results, especially in very complicated questions. Let us now consider the general principles of this theory.

4.3

Hypotheses, Data, and Crucial Experiments

John Herschel

Sir John Herschel (1792–1871), son of the famous astronomer Sir William Herschel, was an English mathematician and astronomer. In this selection from his *Preliminary Discourse on the Study of Natural Philosophy* (London, 1830), Herschel plots a middle course between the brittle inductivism that had dominated British science in the eighteenth century and the more speculative method of hypothesis advocated by many continental scientists. He recommends a flexible approach in which data may guide the formation of hypotheses but, regardless of how they are formed, in the end the test of their worth is extensive in comparison with a great mass of facts. His observations on crucial experiments provide a foil for the later arguments of Pierre Duhem and W. V. O. Quine.

The immediate object we propose to ourselves in physical theories is the analysis of phenomena, and the knowledge of the hidden processes of nature in their production, so far as they can be traced by us. An important part of this knowledge consists in a discovery of the actual structure or mechanism of the universe and its parts, through which, and by which, those processes are executed; and of the agents which are concerned in their performance. Now, the mechanism of nature is for the most part either on too large or too small a scale to be immediately cognizable by our senses; and her agents in like manner elude direct observation, and become known to us only by their effects. It is in vain therefore that we desire to become witnesses to the processes carried on with such means, and to be admitted into the secret recesses and laboratories where they are effected. Microscopes have been constructed which magnify more than a thousand times in *linear* dimension, so that the smallest visible grain of sand may be enlarged to the appearance of one a thousand million times more bulky; yet the only impression we receive by viewing it through such a magnifier is, that it reminds us of some vast fragment of a rock, while the intimate structure

From John Herschel, *Preliminary Discourse on the Study of Natural Philosophy* (London: Longman, Kees, Orme, Brown, and Green, 1830), pp. 191–3, 195–9, 201–2, 206–9.

on which depend its colour, its hardness, and its chemical properties, remains still concealed: we do not seem to have made even an approach to a closer analysis of it by any such scrutiny. . . .

[T]he agents employed by nature to act on material structures are invisible, and only to be traced by the effects they produce. Heat dilates matter with an irresistible force; but what heat *is*, remains yet a problem. A current of electricity passing along a wire moves a magnetized needle at a distance; but except from this effect we perceive no difference between the condition of the wire when it conveys and when it does not convey the stream: and we apply the terms current, or stream, to the electricity only because in some of its relations it reminds us of something we have observed in a stream of air or water. In like manner we see that the moon circulates about the earth; and because we believe it to be a solid mass, and have never seen one solid substance revolve round another within our reach to handle and examine unless retained by a force or united by a tie, we conclude that there *is* a force, and a mode of connection, between the moon and the earth; though, what that mode can be, we have no conception, nor can we imagine *how* such a force can be exerted at a distance, and with empty space, or at most an invisible fluid, between. [. . .]

Now, nothing is more common in physics than to find two, or even many, *theories* maintained as to the origin of a natural phenomenon. For instance, in the case of heat itself, one considers it as a really existing material fluid, of such exceeding subtlety as to penetrate all bodies, and even to be capable of combining with them chemically; while another regards it as nothing but a rapid vibratory or rotatory motion in the ultimate particles of the bodies heated; and produces a singularly ingenious train of mechanical reasoning to show, that there is nothing contradictory to sound dynamical principles in such a doctrine. Thus, again, with light: one considers it as consisting in actual particles darted forth from luminous bodies, and acted upon in their progress by forces of extreme intensity residing in the substances on which they strike; another, in the vibratory motion of the particles of luminous bodies, communicated to a peculiar subtle and highly elastic etherial medium, filling all space, and conveyed through it into our eyes, as sounds are to our ears, by the undulations of air.

Now, are we to be deterred from framing hypotheses and constructing theories, because we meet with such dilemmas, and find ourselves frequently beyond our depth? Undoubtedly not. *Est quodam prodire tenus si non datur ultra.* ["One can go at least this far, even if one cannot go further" – a saying from Horace that Dalton used as the epigraph for his *Meteorological Observations and Essays* (1793).] Hypotheses, with respect to theories, are what presumed proximate causes are with respect to particular inductions: they afford us motives for searching into analogies; grounds of citation to bring before us all the cases which seem to bear upon them, for examination. A well imagined hypothesis, if it have been suggested by a fair inductive consideration of general laws, can hardly fail at least of enabling us to generalize a step farther, and group together several such laws under a more universal expression. But this is taking a very limited view of the value and importance of hypotheses: it may happen (and it has happened in the case of the undulatory doctrine of light) that such a weight of analogy and probability may become accumulated on the side of an hypothesis, that we are compelled to admit one of two things; either that it is an actual statement of what really passes in nature, or that the reality, whatever it be, must run so close a parallel with it, as to admit of some mode of expression common to both, at least in so far as the phenomena actually known are concerned. Now, this is a very great step, not only for its own sake, as leading us to a high point in philosophical speculation, but for its applications; because whatever conclusions we deduce from an hypothesis so supported must have at least a strong presumption in their favour: and we may be thus led to the trial of many curious experiments, and to the imagining of many useful and important contrivances, which we should never otherwise have thought of, and which, at all events, *if* verified in practice, are real additions to our stock of knowledge and to the arts of life.

In framing a theory which shall render a rational account of any natural phenomenon, we have *first* to consider the agents on which it depends, or the causes to which we regard it as ultimately referable. These agents are not to be arbitrarily assumed; they must be such as we have good inductive grounds to believe do exist in nature, and do perform a part in phenomena analogous to those we would render an account of; or such, whose presence in the actual case can be demonstrated by unequivocal signs. They must be *verae causae* ["real causes"], in short, which we can not only show to exist and to act, but the laws of whose action we can derive independently, by direct induction, from experiments purposely instituted; or at least make such suppositions respecting them as shall not be contrary to our experience, and which will remain to be verified by the coincidence of the conclusions we shall deduce from them, with facts. . . . [S]ince it is a fact that the moon does circulate about the earth, it must be drawn towards the earth by a force; for if there were no force acting upon it, it would go on in a straight line without turning aside to circulate in an orbit, and would, therefore, soon go away and be lost in space. This force, then, which we call the *force* of gravity, is a real cause.

We have next to consider the laws which regulate the action of these our primary agents; and these we can only arrive at in three ways: 1st, By inductive reasoning; that is, by examining all the cases in which we know them to be exercised, inferring, as well as circumstances will permit, its amount or intensity in each particular case, and then piecing together, as it were, these *disjecta membra* ["scattered fragments," e.g., bits and pieces] generalizing from them, and so arriving at the laws desired; 2dly, By forming at once a bold hypothesis, particularizing the law, and trying the truth of it by following out its consequences and comparing them with facts; or, 3dly, By a process partaking of both these, and combining the advantages of both without their defects, viz. by assuming indeed the laws we would discover, but so generally expressed, that they shall include an unlimited variety of particular laws; – following out the consequences of this assumption, by the application of such general principles as the case admits; – comparing them in succession with all the particular cases within our knowledge; and, lastly, *on this comparison*, so modifying and restricting the general enunciation of our laws as to *make the results agree*. . . .

To return to our example: particular inductions drawn from the motions of the several planets about the sun, and of the satellites round their primaries, &c. having led us to the general conception of an attractive force exerted by every particle of matter in the universe on every other according to the law to which we attach the name of gravitation; when we would verify this induction, we must set out with assuming this law, considering the whole system as subjected to its influence and implicitly obeying it, and nothing interfering with its action; we then, for the first time, perceive a train of modifying circumstances which had not occurred to us when reasoning upwards from particulars to obtain the fundamental law; we perceive that *all the planets* must attract *each other*, must therefore draw each other out of the orbits which they would have if acted on only by the sun; and as this was never contemplated in the inductive process, its validity becomes a question, which can only be determined by ascertaining precisely how great a deviation this new class of mutual actions will produce. To do this is no easy task, or rather, it is the most difficult task which the genius of man has ever yet accomplished: still, it *has* been accomplished by the mere application of the general laws of dynamics; and the result (undoubtedly a most beautiful and satisfactory one) is, that all those observed deviations in the motions of our system which stood out as exceptions . . . , or were noticed as residual phenomena and reserved for further enquiry . . . , in that imperfect view of the subject which we got in the subordinate process by which we rose to our general conclusion, prove to be the immediate consequences of the above-mentioned mutual actions. As such, they are neither exceptions nor residual facts, but fulfilments of general rules, and essential features in the statement of the case, *without* which our induction would be invalid, and the law of gravitation positively untrue. . . .

When two theories run parallel to each other, and each explains a great many facts in common with the other, any experiment which affords a crucial instance to decide between them, or by which one or other must fall, is of great importance. In thus verifying theories, since they are grounded on general laws, we may appeal, not merely to particular cases, but to whole classes of facts; and we therefore have a great range among the individuals of these for the selection of some particular effect which ought to take place oppositely in the event of one of the two suppositions at issue being right and the other wrong. A curious example is given by M. Fresnel, as decisive, in his mind, of the question between the two great opinions on the nature of light, which, since the time of Newton and Huyghens, have divided philosophers. . . . When two very clean glasses are laid one on the other, if they be not perfectly flat, but one or both in an almost imperceptible degree convex or prominent, beautiful and vivid colours will be seen between them; and if these be viewed through a red glass, their appearance will be that of alternate dark and bright stripes. These stripes are formed *between* the two surfaces in apparent contact, as any one may satisfy himself by using, instead of a flat *plate* of glass for the upper one, a triangular-shaped piece, called a prism, like a three-cornered stick, and looking through the inclined side of it next the eye, by which arrangement the reflection of light from the upper surface is prevented from intermixing with that from the surfaces in contact. Now, the coloured stripes thus produced are explicable on both theories, and are appealed to by both as strong confirmatory facts; but there is a difference in one circumstance according as one or the other theory is employed to explain them. In the case of the Huyghenian doctrine, the intervals between the bright stripes ought to appear *absolutely black*; in the other, *half bright*, when so viewed through a prism. This curious case of difference was tried as soon as the opposing consequences of the two theories were noted by M. Fresnel, and the result is stated by him to be decisive in favour of that theory which makes light to consist in the vibrations of an elastic medium.

Theories are best arrived at by the consideration of general laws; but most securely verified by comparing them with particular facts, because this serves as a verification of the whole train of induction, from the lowest term to the highest. But then, the comparison must be made with facts purposely selected so as to include every variety of case, not omitting extreme ones, and in sufficient number to afford every reasonable probability of detecting error. A single numerical coincidence in a final conclusion, however striking the coincidence or important the subject, is not sufficient. Newton's theory of sound, for example, leads to a numerical expression for the actual velocity of sound, differing but little from that afforded by the correct theory afterwards explained by La Grange, and (when certain conditions not contemplated by him are allowed for) agreeing with fact; yet this coincidence is no verification of Newton's view of the general subject of sound, which is defective in an essential point, as the great geometer last named has very satisfactorily shown. This example is sufficient to inspire caution in resting the verification of theories upon any thing but a very extensive comparison with a great mass of observed facts.

But, on the other hand, when a theory will bear the test of such extensive comparison, it matters little how it has been originally framed. However strange and, at first sight, inadmissible its postulates may appear, or however singular it may seem that such postulates should have been fixed upon, – if they only lead us, by legitimate reasonings, to conclusions in exact accordance with numerous observations purposely made under such a variety of circumstances as fairly to embrace the whole range of the phenomena which the theory is intended to account for, we cannot refuse to admit them; or if we still hesitate to regard them as demonstrated truths, we cannot, at least, object to receive them as temporary substitutes for such truths, until the latter shall become known. If they suffice to explain all the phenomena known, it becomes highly improbable that they will not explain more; and if all their conclusions we have tried have proved correct, it is probable that others yet untried will be found so too; so that in rejecting them altogether, we should reject all the discoveries to which they may lead.

4.4

An Empiricist Account of Scientific Discovery

John Stuart Mill

John Stuart Mill (1806–1873) was an English philosopher whose wide-ranging work left a mark on fields as diverse as economics, moral theory, and philosophy of science. In this excerpt from his *System of Logic* (1843), Mill attempts to give an account of scientific reasoning that accords with his version of empiricism. His description of Johannes Kepler's discovery of the laws of planetary motion reduces them to a mere descriptive summary of the facts, an analysis sharply opposed to the more rationalist proposal of William Whewell. Mill goes on to contrast Kepler's achievement with "induction, properly so called," which in his view is grounded on the principle of the uniformity of nature.

Book III: Of Induction

Chapter I. Preliminary observations on induction in general

[. . .]

§1 [. . .] We have found that all Inference, consequently all Proof, and all discovery of truths not self-evident, consists of inductions, and the inter-pretation of inductions; that all our knowledge, not intuitive, comes to us exclusively from that source. What Induction is, therefore, and what conditions render it legitimate, cannot but be deemed the main question of the science of logic – the question which includes all others. It is, however, one which professed writers on logic have almost entirely passed over. [. . .]

§2. For the purposes of the present inquiry, Induction may be defined, the operation of discovering and proving general propositions. It is true that (as already shown) the process of indirectly ascertaining individual facts is as truly inductive as that by which we establish general truths. But it is not a different kind of induction; it is a form of the very same process: since, on the one hand, generals are but collections of particulars, definite in kind but indefinite in number; and on the other

From John Stuart Mill, *A System of Logic, Ratiocinative and Inductive*, Vol. 2 (London: John W. Parker, 1843), selections from Book III, pp. 185–7, 191–4, 197–8, 200–3.

hand, whenever the evidence which we derive from observation of known cases justifies us in drawing an inference respecting even one unknown case, we should on the same evidence be justified in drawing a similar inference with respect to a whole class of cases. The inference either does not hold at all, or it holds in all cases of a certain description; in all cases which, in certain definable respects, resemble those we have observed.

If these remarks are just; if the principles and rules of inference are the same whether we infer general propositions or individual facts; it follows that a complete logic of the sciences would be also a complete logic of practical business and common life. Since there is no case of legitimate inference from experience, in which the conclusion may not legitimately be a general proposition, an analysis of the process by which general truths are arrived at is virtually an analysis of all induction whatever. Whether we are inquiring into a scientific principle or into an individual fact, and whether we proceed by experiment or by ratiocination, every step in the train of inferences is essentially inductive, and the legitimacy of the induction depends in both cases on the same conditions.

. . . Success is here dependent on natural or acquired sagacity, aided by knowledge of the particular subject and of subjects allied with it. Invention, though it can be cultivated, cannot be reduced to rule; there is no science which will enable a man to bethink himself of that which will suit his purpose.

But when he *has* thought of something, science can tell him whether that which he has thought of will suit his purpose or not. The inquirer or arguer must be guided by his own knowledge and sagacity in the choice of the inductions out of which he will construct his argument. But the validity of the argument when constructed depends on principles and must be tried by tests which are the same for all descriptions of inquiries, whether the result be to give A an estate, or to enrich science with a new general truth. In the one case and in the other, the senses, or testimony, must decide on the individual facts; the rules of the syllogism will determine whether, those facts being supposed correct, the case really falls within the formulæ of the different inductions under which it has been successively brought; and finally, the legitimacy of the inductions themselves must be decided by other rules, and these it is now our purpose to investigate. . . .

If the identity of the logical processes which prove particular facts and those which establish general scientific truths required any additional confirmation, it would be sufficient to consider that in many branches of science single facts have to be proved, as well as principles; facts as completely individual as any that are debated in a court of justice, but which are proved in the same manner as the other truths of the science, and without disturbing in any degree the homogeneity of its method. A remarkable example of this is afforded by astronomy. The individual facts on which that science grounds its most important deductions, such facts as the magnitudes of the bodies of the solar system, their distances from one another, the figure of the earth, and its rotation, are scarcely any of them accessible to our means of direct observation: they are proved indirectly by the aid of inductions founded on other facts which we can more easily reach. For example, the distance of the moon from the earth was determined by a very circuitous process. The share which direct observation had in the work consisted in ascertaining, at one and the same instant, the zenith distances of the moon, as seen from two points very remote from one another on the earth's surface. The ascertainment of these angular distances ascertained their supplements; and since the angle at the earth's centre subtended by the distance between the two places of observation was deducible by spherical trigonometry from the latitude and longitude of those places, the angle at the moon subtended by the same line became the fourth angle of a quadrilateral of which the other three angles were known. The four angles being thus ascertained, and two sides of the quadrilateral being radii of the earth; the two remaining sides and the diagonal, or in other words, the moon's distance from the two places of observation, and from the centre of the earth, could be ascertained, at least in terms of the earth's radius, from elementary theorems of geometry. At each step in this demonstration a new induction is taken in, represented in the aggregate of its results by a general proposition.

[. . .]

§3. There remains a third improper use of the term Induction, which it is of real importance to clear up, because the theory of Induction has been, in no ordinary degree, confused by it, and because the confusion is exemplified in the most recent and elaborate treatise on the inductive philosophy which exists in our language. The error in question is that of confounding a mere description, by general terms, of a set of observed phenomena, with an induction from them.

Suppose that a phenomenon consists of parts, and that these parts are only capable of being observed separately, and as it were piecemeal. When the observations have been made, there is a convenience (amounting for many purposes to a necessity) in obtaining a representation of the phenomenon as a whole, by combining, or, as we may say, piecing these detached fragments together.

. . . [I]t is the facts themselves; it is a summary of those facts; the description of a complex fact, to which those simpler ones are as the parts of a whole.

Now there is, I conceive, no difference in kind between this simple operation, and that by which Kepler ascertained the nature of the planetary orbits; and Kepler's operation, all at least that was characteristic in it, was not more an inductive act than that of our supposed navigator.

The object of Kepler was to determine the real path described by each of the planets, or let us say by the planet Mars (since it was of that body that he first established the two of his three laws which did not require a comparison of planets). To do this there was no other mode than that of direct observation; and all which observation could do was to ascertain a great number of the successive places of the planet, or rather, of its apparent places. That the planet occupied successively all these positions, or at all events, positions which produced the same impressions on the eye, and that it passed from one of these to another insensibly, and without any apparent breach of continuity; thus much the senses, with the aid of the proper instruments, could ascertain. What Kepler did more than this, was to find what sort of a curve these different points would make, supposing them to be all joined together. He expressed the whole series of the observed places of Mars by what Dr Whewell calls the general conception of an ellipse. This operation was far from being as easy as that of the navigator who expressed the series of his observations on successive points of the coast by the general conception of an island. But it is the very same sort of operation; and if the one is not an induction but a description, this must also be true of the other.

The only real induction concerned in the case consisted in inferring that because the observed places of Mars were correctly represented by points in an imaginary ellipse, therefore Mars would continue to revolve in that same ellipse; and in concluding (before the gap had been filled up by further observations) that the positions of the planet during the time which intervened between two observations, must have coincided with the intermediate points of the curve. For these were facts which had not been directly observed. They were inferences from the observations; facts inferred, as distinguished from facts seen. But these inferences were so far from being a part of Kepler's philosophical operation, that they had been drawn long before he was born. Astronomers had long known that the planets periodically returned to the same places. When this had been ascertained, there was no induction left for Kepler to make, nor did he make any further induction. He merely applied his new conception to the facts inferred, as he did to the facts observed. Knowing already that the planets continued to move in the same paths; when he found that an ellipse correctly represented the past path he knew that it would represent the future path. In finding a compendious expression for the one set of facts, he found one for the other: but he found the expression only, not the inference; nor did he (which is the true test of a general truth) add anything to the power of prediction already possessed.

[. . .]

It is true that for these simply descriptive operations, as well as for the erroneous inductive one, a conception of the mind was required. The conception of an ellipse must have presented itself to Kepler's mind before he could identify the planetary orbits with it. According to Dr Whewell, the conception was something added to the facts. He expresses himself as if Kepler had put something into the facts by his mode of conceiving them. But Kepler did no such thing. The ellipse was in the facts before Kepler recognised it;

just as the island was an island before it had been sailed round. Kepler did not *put* what he had conceived into the facts, but *saw* it in them. A conception implies, and corresponds to, something conceived: and though the conception itself is not in the facts, but in our mind, yet if it is to convey any knowledge relating to them it must be a conception *of* something which really is in the facts, some property which they actually possess, and which they could manifest to our senses if our senses were able to take cognisance of it. If, for instance, the planet left behind it in space a visible track, and if the observer were in a fixed position at such a distance from the plane of the orbit as would enable him to see the whole of it at once, he would see it to be an ellipse; and if gifted with appropriate instruments and powers of locomotion, he would prove it to be such by measuring its different dimensions. Nay, further: if the track were visible, and he were so placed that he could see all parts of it in succession, but not all of them at once, he might be able, by piecing together his successive observations, to discover both that it was an ellipse and that the planet moved in it. The case would then exactly resemble that of the navigator who discovers the land to be an island by sailing round it. If the path was visible, no one I think would dispute that to identify it with an ellipse is to describe it: and I cannot see why any difference should be made by its not being directly an object of sense, when every point in it is as exactly ascertained as if it were so.

[. . .]

In every way, therefore, it is evident that to explain induction as the colligation of facts by means of appropriate conceptions, that is, conceptions which will really express them, is to confound mere descriptions of the observed facts with inference from those facts, and ascribe to the latter what is a characteristic property of the former.

There is, however, between Colligation and Induction a real correlation, which it is important to conceive correctly. Colligation is not always induction; but induction is always colligation. The assertion that the planets move in ellipses was but a mode of representing observed facts; it was but a colligation; while the assertion that they are drawn or tend towards the sun was the statement of a new fact, inferred by induction. But the induction, once made, accomplishes the purposes of colligation likewise. It brings the same facts, which Kepler had connected by his conception of an ellipse, under the additional conception of bodies acted upon by a central force, and serves therefore as a new bond of connection for those facts; a new principle for their classification.

Further, the descriptions which are improperly confounded with induction are nevertheless a necessary preparation for induction; no less necessary than correct observation of the facts themselves. Without the previous colligation of detached observations by means of one general conception, we could never have obtained any basis for an induction, except in the case of phenomena of very limited compass. We should not be able to affirm any predicates at all of a subject incapable of being observed otherwise than piecemeal: much less could we extend those predicates by induction to other similar subjects. Induction, therefore, always presupposes, not only that the necessary observations are made with the necessary accuracy, but also that the results of these observations are, so far as practicable, connected together by general descriptions, enabling the mind to represent to itself as wholes whatever phenomena are capable of being so represented.

[. . .]

Chapter III. Of the ground of induction

§1. Induction, properly so called, as distinguished from those mental operations, sometimes though improperly designated by the name, which I have attempted in the preceding chapter to characterise, may, then, be summarily defined as Generalisation from Experience. It consists in inferring from some individual instances in which a phenomenon is observed to occur, that it occurs in all instances of a certain class; namely, in all which *resemble* the former, in what are regarded as the material circumstances.

In what way the material circumstances are to be distinguished from those which are immaterial, or why some of the circumstances are material and others not so, we are not yet ready to point out. We must first observe that there is a principle implied in the very statement of what

Induction is; an assumption with regard to the course of nature and the order of the universe; namely, that there are such things in nature as parallel cases; that what happens once will, under a sufficient degree of similarity of circumstances, happen again, and not only again, but as often as the same circumstances recur. This, I say, is an assumption involved in every case of induction. And if we consult the actual course of nature, we find that the assumption is warranted. The universe, so far as known to us, is so constituted, that whatever is true in any one case, is true in all cases of a certain description; the only difficulty is, to find what description.

This universal fact, which is our warrant for all inferences from experience, has been described by different philosophers in different forms of language; that the course of nature is uniform; that the universe is governed by general laws; and the like. One of the most usual of those modes of expression, but also one of the most inadequate, is that which has been brought into familiar use by the metaphysicians of the school of Reid and Stewart. The disposition of the human mind to generalise from experience, – a propensity considered by these philosophers as an instinct of our nature, – they usually describe under some such name as "our intuitive conviction that the future will resemble the past." Now it has been well pointed out by Mr Bailey, that (whether the tendency be or not an original and ultimate element of our nature) Time, in its modifications of past, present, and future, has no concern either with the belief itself, or with the grounds of it. We believe that fire will burn to-morrow, because it burned to-day and yesterday; but we believe, on precisely the same grounds, that it burned before we were born, and that it burns this very day in Cochin-China. It is not from the past to the future, as past and future, that we infer, but from the known to the unknown; from facts observed to facts unobserved; from what we have perceived, or been directly conscious of, to what has not come within our experience. In this last predicament is the whole region of the future; but also the vastly greater portion of the present and of the past.

Whatever be the most proper mode of expressing it, the proposition that the course of nature is uniform is the fundamental principle, or general axiom, of Induction. It would yet be a great error to offer this large generalisation as any explanation of the inductive process. On the contrary, I hold it to be itself an instance of induction, and induction by no means of the most obvious kind. Far from being the first induction we make, it is one of the last, or at all events one of those which are latest in attaining strict philosophical accuracy. As a general maxim, indeed, it has scarcely entered into the minds of any but philosophers; nor even by them, as we shall have many opportunities of remarking, have its extent and limits been always very justly conceived. The truth is, that this great generalisation is itself founded on prior generalisations. The obscurer laws of nature were discovered by means of it, but the more obvious ones must have been understood and assented to as general truths before it was ever heard of. We should never have thought of affirming that all phenomena take place according to general laws, if we had not first arrived, in the case of a great multitude of phenomena, at some knowledge of the laws themselves; which could be done no otherwise than by induction. In what sense, then, can a principle, which is so far from being our earliest induction, be regarded as our warrant for all the others? In the only sense in which (as we have already seen) the general propositions which we place at the head of our reasonings when we throw them into syllogisms ever really contribute to their validity. As Archbishop Whately remarks, every induction is a syllogism with the major premise suppressed; or (as I prefer expressing it) every induction may be thrown into the form of a syllogism by supplying a major premise. If this be actually done, the principle which we are now considering, that of the uniformity of the course of nature, will appear as the ultimate major premise of all inductions, and will, therefore, stand to all inductions in the relation in which, as has been shown at so much length, the major proposition of a syllogism always stands to the conclusion; not contributing at all to prove it, but being a necessary condition of its being proved; since no conclusion is proved for which there cannot be found a true major premise.

The statement that the uniformity of the course of nature is the ultimate major premise in all cases of induction may be thought to require some explanation. The immediate major premise in every inductive argument it certainly is not. Of that Archbishop Whately's must be held to be the correct account. The induction, "John, Peter, &c., are mortal, therefore all mankind are mortal," may, as he justly says, be thrown into a syllogism by prefixing as a major premise, (what is at any rate a necessary condition of the validity of the argument,) namely, that what is true of John, Peter, &c., is true of all mankind. But how came we by this major premise? It is not self-evident; nay, in all cases of unwarranted generalisation it is not true. How, then, is it arrived at? Necessarily either by induction or ratiocination; and if by induction, the process, like all other inductive arguments, may be thrown into the form of a syllogism. This previous syllogism it is, therefore, necessary to construct. There is, in the long run, only one possible construction. The real proof that what is true of John, Peter, &c., is true of all mankind, can only be, that a different supposition would be inconsistent with the uniformity which we know to exist in the course of nature. Whether there would be this inconsistency or not, may be a matter of long and delicate inquiry; but unless there would, we have no sufficient ground for the major of the inductive syllogism. It hence appears, that if we throw the whole course of any inductive argument into a series of syllogisms, we shall arrive by more or fewer steps at an ultimate syllogism, which will have for its major premise the principle or axiom of the uniformity of the course of nature.

It was not to be expected that in the case of this axiom, any more than of other axioms, there should be unanimity among thinkers with respect to the ground on which it is to be received as true. I have already stated that I regard it as itself a generalisation from experience. Others hold it to be a principle which, antecedently to any verification by experience, we are compelled by the constitution of our thinking faculty to assume as true.

4.5

Against Pure Empiricism

William Whewell

William Whewell (1794–1866) was a British polymath and longtime master of Trinity College, Cambridge. In this excerpt from his *Philosophy of the Inductive Sciences* (1840), Whewell argues, against the empiricism of philosophers like John Locke and David Hume, that our idea of cause is not something that we obtain from experience but is something that has its origin in the mind itself and that various maxims regarding causality have a necessity that transcends, and does not rest on, the evidence of the senses. In this respect, Whewell's thought owes much to Immanuel Kant.

Of the Idea of Cause

We see in the world around us a constant succession of causes and effects connected with each other. The laws of this connexion we learn in a great measure from experience, by observation of the occurrences which present themselves to our notice, succeeding one another. But in doing this, and in attending to this succession of appearances, of which we are aware by means of our senses, we supply from our own minds the Idea of Cause. This Idea . . . is not derived from experience, but has its origin in the mind itself; – is introduced into our experience by the active, and not by the passive part of our nature.

By Cause we mean some quality, power, or efficacy, by which a state of things produces a succeeding state. Thus the motion of bodies from rest is produced by a cause which we call *Force*: and in the particular case in which bodies fall to the earth, this force is termed *Gravity*. In these cases, the Conceptions of Force and Gravity receive their meaning from the Idea of Cause which they involve: for Force is conceived as the Cause of Motion. That this Idea of Cause is not derived from experience, we prove . . . by this consideration; that we can make assertions, involving this idea, which are rigorously necessary and universal; whereas knowledge derived from experience can only be true as far as experience goes, and can

From Whewell, *The Philosophy of the Inductive Sciences Founded Upon Their History*, Vol. 1 (London: John W. Parker, 1847), pp. 165–70.

never contain in itself any evidence whatever of its necessity. We assert that "Every event must have a cause:" and this proposition we know to be true, not only probably, and generally, and as far as we can see; but we cannot suppose it to be false in any single instance. We are as certain of it as of the truths of arithmetic or geometry. We cannot doubt that it must apply to all events past and future, in every part of the universe, just as truly as to those occurrences which we have ourselves observed. *What* causes produce what effects; – what is the cause of any particular event; – what will be the effect of any peculiar process; – these are points on which experience may enlighten us. Observation and experience may be requisite, to enable us to judge respecting such matters. But that every event has *some* cause, Experience cannot prove any more than she can disprove. She can add nothing to the evidence of the truth, however often she may exemplify it. This doctrine, then, cannot have been acquired by her teaching; and the Idea of Cause which the doctrine involves, and on which it depends, cannot have come into our minds from the region of observation.

That we do, in fact, apply the Idea of Cause in a more extensive manner than could be justified, if it were derived from experience only, is easily shown. For from the principle that everything must have a cause, we not only reason concerning the succession of the events which occur in the progress of the world, and which form the course of experience; but we infer that the world itself must have a cause; – that the chain of events connected by common causation, must have a First Cause of a nature different from the events themselves. This we are entitled to do, if our Idea of Cause be independent of, and superior to, experience: but if we have no Idea of Cause except such as we gather from experience, this reasoning is altogether baseless and unmeaning.

Again; by the use of our powers of observation, we are aware of a succession of appearances and events. But none of our senses or powers of external observation, can detect in these appearances the power or quality which we call Cause. Cause is that which connects one event with another; but no sense or perception discloses to us, or can disclose, any connexion among the events which we observe. We see that one occurrence follows another, but we can never see anything which shows that one occurrence *must* follow another. . . . One ball strikes another and causes it to move forward. But by what compulsion? Where is the necessity? If the mind can see any circumstance in this case which makes the result inevitable, let this circumstance be pointed out. But, in fact, there is no such discoverable necessity; for we can conceive this event not to take place at all. The struck ball may stand still, for aught we can see. "But the laws of motion will not allow it to do so." Doubtless they will not. But the laws of motion are learnt from experience, and therefore can prove no necessity. Why should not the laws of motion be other than they are? Are they necessarily true? That they are necessarily such as do actually regulate the impact of bodies, is at least no obvious truth; and therefore this necessity cannot be, in common minds, the ground of connecting the impact of one ball with the motion of another. And assuredly, if this fails, no other ground of such necessary connexion can be shown. In this case, then, the events are not seen to be necessarily connected. But if this case, where one ball moves another by impulse, be not an instance of events exhibiting a necessary connexion, we shall look in vain for any example of such a connexion. There is, then, no case in which events can be observed to be necessarily connected: our idea of causation, which implies that the event is necessarily connected with the cause, cannot be derived from observation.

But it may be said, we have not any such Idea of Cause, implying necessary connexion with effect, and a quality by which this connexion is produced. We see nothing but the succession of events; and by *cause* we mean nothing but a certain succession of events; – namely, a constant, unvarying succession. Cause and effect are only two events of which the second invariably follows the first. We delude ourselves when we imagine that our idea of causation involves anything more than this.

To this I reply by asking, what then is the meaning of the maxim above quoted, and allowed by all to be universally and necessarily true, that every event must have a cause? Let us put this maxim into the language of the explanation just noticed; and it becomes this: – "Every event must have a certain other event invariably preceding it." But why must it? Where is the necessity? Why must like events always be preceded by like, except so far as other events interfere? That there is such a necessity, no one can doubt. All will allow that if a stone ascend because it is thrown upwards in one case, a stone which ascends in another case has also been thrown upwards, or has undergone some equivalent operation. All will allow that in this sense, every kind of event must have some other specific kind of event preceding it. But this turn of men's thoughts shows that they see in events a connexion which is not mere succession. They see in cause and effect, not merely what does, often or always, precede and follow, but what *must* precede and follow. The events are not only conjoined, they are connected. The cause is more than the prelude, the effect is more than the sequel, of the fact. The cause is conceived not as a mere occasion; it is a power, an efficacy, which has a real operation.

Thus we have drawn from the maxim, that Every Effect must have a Cause, arguments to show that we have an Idea of Cause which is not borrowed from experience, and which involves more than mere succession. Similar arguments might be derived from any other maxims of universal and necessary validity, which we can obtain concerning Cause: as for example, the maxims that Causes are measured by their Effects, and that Reaction is equal and opposite to Action. These maxims we shall soon have to examine; but we may observe here, that the necessary truth which belongs to them, shows that they, and the Ideas which they involve, are not the mere fruits of observation; while their meaning, including, as it does, something quite different from the mere conception of succession of events, proves that such a conception is far from containing the whole import and signification of our Idea of Cause. . . .

4.6

The Causes Behind the Phenomena

William Whewell

In his *Novum Organum Renovatum* (1858), Whewell explicitly seeks to revise Francis Bacon's crudely empiricist philosophy of science, even writing his own aphorisms to replace those of his distinguished countryman. In this selection from his renovation of Bacon's *Novum Organum*, Whewell makes an eloquent plea for scientists to move beyond a mere articulation of natural laws, however indispensable those might be, and inquire into the causes of phenomena.

Of Laws of Phenomena and of Causes

Aphorism XVIII

> Inductive truths are of two kinds, Laws of Phenomena, and Theories of Causes. It is necessary to begin in every science with the Laws of Phenomena; but it is impossible that we should be satisfied to stop short of a Theory of Causes. In Physical Astronomy, Physical Optics, Geology, and other sciences, we have instances showing that we can make a great advance in inquiries after true Theories of Causes.

In the first attempts at acquiring an exact and connected knowledge of the appearances and operations which nature presents, men went no further than to learn *what* takes place, not *why* it occurs. They discovered an Order which the phenomena follow, Rules which they obey; but they did not come in sight of the Powers by which these rules are determined, the Causes of which this order is the effect. Thus, for example, they found that many of the celestial motions took place as if the sun and stars were carried round by the revolutions of certain celestial spheres; but what causes kept these spheres in constant motion, they were never able to explain. In like manner in modern times, Kepler discovered that the planets describe ellipses, before Newton explained why they select this particular curve, and

From Whewell, *The Philosophy of the Inductive Sciences Founded Upon Their History*, Vol. 2 (London: John W. Parker, 1847), selections from pp. 471, 95–8.

describe it in a particular manner. The laws of reflection, refraction, dispersion, and other properties of light have long been known; the causes of these laws are at present under discussion. And the same might be said of many other sciences. The discovery of *the Laws of Phenomena* is, in all cases, the first step in exact knowledge; these Laws may often for a long period constitute the whole of our science; and it is always a matter requiring great talents and great efforts, to advance to a knowledge of the *Causes* of the phenomena.

Hence the larger part of our knowledge of nature, at least of the certain portion of it, consists of the knowledge of the Laws of Phenomena. In Astronomy indeed, besides knowing the rules which guide the appearances, and resolving them into the real motions from which they arise, we can refer these motions to the forces which produce them. In Optics, we have become acquainted with a vast number of laws by which varied and beautiful phenomena are governed; and perhaps we may assume, since the evidence of the Undulatory Theory has been so fully developed, that we know also the Causes of the Phenomena. But in a large class of sciences, while we have learnt many Laws of Phenomena, the causes by which these are produced are still unknown or disputed. Are we to ascribe to the operation of a fluid or fluids, and if so, in what manner, the facts of heat, magnetism, electricity, galvanism? What are the forces by which the elements of chemical compounds are held together? What are the forces, of a higher order, as we cannot help believing, by which the course of vital action in organized bodies is kept up? In these and other cases, we have extensive departments of science; but we are as yet unable to trace the effects to their causes; and our science, so far as it is positive and certain, consists entirely of the laws of phenomena.

In those cases in which we have a division of the science which teaches us the doctrine of the causes, as well as one which states the rules which the effects follow, I have, in the *History*, distinguished the two portions of the science by certain terms. I have thus spoken of *Formal* Astronomy and *Physical* Astronomy. The latter phrase has long been commonly employed to describe that department of Astronomy which deals with those forces by which the heavenly bodies are guided in their motions; the former adjective appears well suited to describe a collection of rules depending on those ideas of space, time, position, number, which are, as we have already said, the *forms* of our apprehension of phenomena. The laws of phenomena may be considered as *formulæ*, expressing results in terms of those ideas. In like manner, I have spoken of Formal Optics and Physical Optics; the latter division including all speculations concerning the machinery by which the effects are produced. Formal Acoustics and Physical Acoustics may be distinguished in like manner, although these two portions of science have been a good deal mixed together by most of those who have treated of them. Formal Thermotics, the knowledge of the laws of the phenomena of heat, ought in like manner to lead to Physical Thermotics, or the Theory of Heat with reference to the cause by which its effects are produced; – a branch of science which as yet can hardly be said to exist.

What *kinds of cause* are we to admit in science? This is an important, and by no means an easy question. In order to answer it, we must consider in what manner our progress in the knowledge of causes has hitherto been made. By far the most conspicuous instance of success in such researches, is the discovery of the causes of the motions of the heavenly bodies. In this case, after the formal laws of the motions, – their conditions as to space and time, – had become known, men were enabled to go a step further; to reduce them to the familiar and general cause of motion – mechanical force; and to determine the laws which this force follows. That this was a step in addition to the knowledge previously possessed, and that it was a real and peculiar truth, will not be contested. And a step in any other subject which should be analogous to this in astronomy; – a discovery of causes and forces as certain and clear as the discovery of universal gravitation; – would undoubtedly be a vast advance upon a body of science consisting only of the laws of phenomena.

4.7

Catastrophist Geology

Georges Cuvier

Georges Cuvier (1769–1832) was a French naturalist and paleontologist who defended the position known as catastrophism, according to which geological phenomena have been shaped by a series of violent events in earth's history caused by forces that are not seen, or not on that scale, today. In this selection from his *Discourse on the Revolutionary Upheavals on the Surface of the Earth* (1826), Cuvier points to various geological phenomena that seem to require catastrophism for their explanation.

Introduction

In my work on *Fossil Bones*, I set myself the task of identifying the animals whose fossilized remains fill the surface strata of the earth. This project meant I had to travel along a path where we had so far taken only a few tentative steps. As a new sort of antiquarian, I had to learn to restore these memorials to past upheavals and, at the same time, to decipher their meaning. I had to collect and put together in their original order the fragments which made up these animals, to reconstruct the ancient creatures to which these fragments belonged, to recreate their proportions and characteristics, and finally to compare them to those alive today on the surface of the earth. This was an almost unknown art, which assumed a science hardly touched upon up until now, that of the laws which govern the formal coexistence of the various parts in organic beings. Thus, I had to prepare myself for these studies through a much longer research into animals which presently exist. Only an almost universal review of present creation could provide some proof for my results concerning created life long ago. But at the same time such a study had to provide me with a large collection of equally demonstrable rules and interconnections. In the course of this exploration into a small part of the theory of the earth, I would have to

From Cuvier, *Discourse on the Revolutionary Upheavals on the Surface of the Globe and on the Changes which they have Produced in the Animal Kingdom*, trans. Ian Johnston, pp. 1–9 http://www.mala.bc.ca/~johnstoi/cuvier/cuvier-e.htm.

be able to subject the entire animal kingdom in some way to new laws.

I was sustained in this double task by the constant interest which it promised to have and by service to the universal science of anatomy, the essential basis of all those sciences dealing with organic entities, and to the physical history of the earth, the foundation of mineralogy, geography, and, we can say, even of human history and everything really important for human beings to know about themselves.

If one finds it interesting to follow in the infancy of our species the almost eradicated traces of so many extinct nations, how could one not also find it interesting to search in the shadows of the earth's infancy for the traces of revolutionary upheavals which have preceded the existence of all nations? We admire the force with which the human spirit has measured the movements of planets which nature seemed to have concealed for ever from our view; human genius and science have stepped beyond the limits of space; some observations developed by reasoning have unveiled the mechanical workings of the world. Would there not also be some glory for human beings to know how to step beyond the limits of time and to recover, through some observations, the history of this earth and a succession of events which have preceded the birth of the human genus? No doubt the astronomers have proceeded more rapidly than the naturalists. The theory of the earth at the present time is rather like the one in which some philosophers believed that the sky was made of freestone and the moon was as big as the Peloponnese. But, following Anaxagoras, Copernicus and Kepler opened up the road to Newton. And why one day should natural history not also have its own Newton?

[. . .]

The Geological Record of Ancient Upheavals

The first appearance of the earth

When the traveler goes through fertile plains where tranquil waters nourish with their regular flow an abundant vegetation and where the ground, trodden by numerous people and decorated with flourishing villages, rich cities, and superb monuments, is never troubled except by ravages of war or by the oppression of men in power, he is not tempted to believe that nature has also had its internal wars and that the surface of the earth has been overthrown by revolutions and catastrophes. But his ideas change as soon as he seeks to dig through this soil, today so calm, or when he takes himself up into the hills which border the plain; his ideas expand, so to speak, with what he is looking at. They begin to embrace the extent and the grandeur of the ancient events as soon as he climbs up the higher mountains of which these are the foothills, or when he follows the stream beds which descend from these mountains and moves into their interior.

The first proofs of upheavals

The lowest and most level land areas show us, especially when we dig there to very great depths, nothing but horizontal layers of material more or less varied, which almost all contain innumerable products of the sea. Similar layers, with similar products, form the hills up to quite high elevations. Sometimes the shells are so numerous that they make up the entire mass of soil by themselves. They occur at elevations higher than the level of all seas, where no sea could be carried today by present causes. Not only are these shells encased in loose sand, but the hardest rocks often encrust them and are penetrated by them throughout. All the parts of the world, both hemispheres, all continents, and all islands of any size provide evidence of the same phenomenon. The time is past when ignorance could continue to maintain that these remains of organic bodies were simple games of nature, products conceived in the bosom of the earth by its creative forces, and the renewed efforts of certain metaphysicians will probably not be enough to make these old opinions acceptable. A scrupulous comparison of the shapes of these deposits, of their make up and often even their chemical composition shows not the slightest difference between these fossil shells and those which the sea nourishes. Their preservation is no less perfect. Very often one observes there neither shattering nor fractures, nothing which signifies a violent movement. The smallest of them keep their most delicate parts, their most subtle crests, their slenderest features.

Thus, not only have they lived in the sea, but they have been deposited by the sea, which has left them in the places where we find them. Moreover, this sea has remained in these locations, with a sufficient calm and duration to form deposits so regular, so thick, so extensive, and in places so solid, that they are full of the remains of marine animals. The sea basin therefore has provided evidence of at least one change, whether in extent or location. See what results already from the first inspections and the most superficial observation.

The traces of upheavals become more impressive when one moves a little higher, when one gets even closer to the foot of the great mountain ranges. There are still plenty of shell layers. We notice them, even thicker and more solid ones. The shells there are just as numerous and just as well preserved. But they are no longer the same species. Also, the strata which contain them are no longer generally horizontal. They lie obliquely, sometimes almost vertically. In contrast to the plains and the low hills, where it was necessary to dig deep to recognize the succession of layers, here we see them on the mountain flank, as we follow the valleys produced by their tearing apart. At the foot of the escarpments, immense masses of debris form rounded hillocks, whose height is increased by each thawing and each storm.

And those upright layers which form the crests of secondary mountains do not rest on the horizontal layers of hills which serve as their lower stages. By contrast, they sink under these hills, which rest on the slopes of these oblique strata. When we bore into the horizontal strata near mountains with oblique layers, we find these oblique layers deep down. Sometimes when the oblique layers are not very high, their summits are even crowned with horizontal strata. The oblique layers are therefore older than the horizontal layers. Since it is impossible, at least for most of them, not to have been formed horizontally, evidently they have been lifted up again and were in existence before the others which rest on top of them.

Thus, before forming these horizontal layers, the sea had formed other strata. These were for some reason or other broken, raised up, and overturned in thousands of ways. As several of these oblique layers which the sea formed in a previous age rise higher than the horizontal layers which succeeded them and which surrounded them, the causes which gave these layers their oblique orientation also made them protrude above the level of the sea and turned them into islands or at least reefs and uneven structures, whether they were raised again by an extreme condition or whether a contrasting subsidence made the waters sink. The second result is no less clear or less proven than the first for anyone who will take the trouble to study the monuments which provide evidence for these results.

[. . .]

Proofs that these revolutions have been sudden

But it is also really important to note that these eruptions and repeated retreats were not at all slow and did not all take place gradually. On the contrary, most of the catastrophes which brought them on have been sudden. That is especially easy to demonstrate for the last of these catastrophes, which by a double movement inundated and later left dry our present continents or at least a great part of the land which forms them today. That catastrophe also left in the northern countries the cadavers of great quadrupeds locked in the ice, preserved right up to our time with their skin, hair, and flesh. If they had not been frozen as soon as they were killed, decay would have caused them to decompose. On the other hand, this permanent freezing was not a factor previously in the places where these animals were trapped. For they would not have been able to live in such a temperature. Hence the same instant which killed the animals froze the country where they lived. This event was sudden, instantaneous, without any gradual development. What is so clearly demonstrated for this most recent catastrophe is hardly less so for the earlier ones. The rending, rearranging, and overturning of more ancient layers leave no doubt that sudden and violent causes placed them in the state in which we see them. The very force of the movements which the bodies of water experienced is still attested to by the mountain of remains and rounded pebbles interposed in many places between the solid layers. Thus, life on this earth has often been disturbed by dreadful events. Innumerable living creatures have been victims of these catastrophes. Some inhabitants of dry land have seen themselves swallowed up by floods; others living in the ocean depths when the bottom of the sea was lifted up suddenly were placed on dry land. Their very races were extinguished for ever, leaving behind nothing in the world but some hardly recognizable debris for the natural scientist.

Such are the conclusions to which we are necessarily led by the objects which we meet at every step and which we can verify at every instant in almost every country. These huge and terrible events are clearly printed everywhere for the eye which knows how to read the story in their monuments.

But what is even more astonishing and what is no less certain is that life has not always existed on the earth and that it is easy for the observer to recognize the point where life began to deposit her productions.

[. . .]

History of Geology and Geological Systems

Ancient systems of geologists

For a long time we have accepted only two events, two periods of changes on the earth: the Creation and the Flood. All the efforts of geologists have tended to explain the present state of the earth by imagining a certain original state, later modified by the Flood. Each of them has speculated also about the nature of the causes, the actions, and effects of these events.

Thus, according to one, the earth was first given a smooth and light crust which covered seas in the depths and which broke open to produce the Flood. The debris formed the mountains. According to another, the Flood was brought about by a momentary suspension of mineral cohesion. The mass of earth was entirely dissolved, and the mixture penetrated by shellfish. According to a third, God raised the mountains in order to make the waters which had produced the Flood flow out, and put the waters in places where there were the most rocks, because otherwise it would have been impossible to hold them in. A fourth created the earth with the atmosphere of a comet and had it overwhelmed by the tail of another comet. The heat which remained from its first origin excited all the living creatures to sin. Thus, they were all drowned, except the fish, who had apparently less excitable passions.

We see that, while entrenching themselves entirely within the limits set by the Book of Genesis, naturalists still gave themselves a large enough goal. They found themselves soon at an impasse. When they succeeded in seeing the six days of the Creation as so many indefinite periods, discounting the centuries, their systems took flight in proportion to the lapses of time which they were able to deal with.

Even the great Leibnitz amused himself, like Descartes, by making the earth an extinguished star, a glazed globe, on which vapours, trapped at the time of its cooling, formed the seas which later deposited calcified earth. Demaillet covered the entire globe with water for thousands of years. He had the waters gradually ebb. All the land animals at first lived in the sea. Even man started as a fish. And the author asserts that it is not rare to meet in the ocean fish which are only half human, but from them the species will become completely human one fine day.

Buffon's system is merely a development of Leibnitz's, only with the addition of a comet which by a violent shock caused the sun to emit the liquid mass of the earth at the same time as all the planets. From this theory one result is firm dating. For, by the present temperature of the earth, we can know how long it has been cooling. And since the other planets left the sun at the same time as the earth, we can calculate how many centuries the large ones still have to cool and at what point the small ones were already frozen.

More recent systems

In our time, freer spirits than ever before have also wished to busy themselves with this important subject. Certain writers have reproduced and enormously extended Demaillet's ideas. They claim that all was liquid at the beginning, that the liquid engendered at first very simple animals, like monads or other microscopic infusorian species, and that, with the passage of time and the development of different habits, the animal races became more complex and diversified to the point where we see them today. All these animal races have converted the water of the sea by degrees into calcified earth. The plants (on the origin and changes of which no one tells us anything) for their part turned this water into clay. But these two earths, by force of being stripped of the characters which life had imprinted on them, resolved themselves, in the last analysis, into silica. And lo and behold, for this reason the oldest mountains contain more silica than the others. All solid parts of the earth therefore owe their origin to living things, and without that life the earth would be still entirely liquid.

Some other writers have preferred Kepler's ideas. Like this great astronomer, they give the earth itself vital faculties. According to them, a fluid circles in the earth, and an assimilation takes place just as in animated bodies. Each of its parts is alive. Every elementary molecule has instinct and will; they attract and repel each other according to antipathies and sympathies; each sort of mineral can change immense masses into its own nature, as we convert our food into flesh and blood. The mountains are the respiratory organs of the earth, and the schists are the secretary organs. Through them sea water is decomposed to create the volcanic eruptions. The seams finally are the decaying teeth, the abscesses of the mineral kingdom, and the metals a product of decay and illness. That is why almost all of them feel unpleasant.

Even more recently, a philosophy which substitutes metaphors for rational argument, starting with the system of absolute identity or pantheism, ascribes the origin of all phenomena or, what in its eyes is the same thing, of all beings to polarization, like the two electricities, by calling all opposition and difference polarization. Whether we consider situation, nature, or function, this belief sees opposition in the following: God and the world, in the universe the sun and the planets, in each planet the solid and the liquid, and following this course, changing as necessary its tropes and its allegories, it reaches even the final details of organic species.

I must admit, however, that above we have chosen extreme examples and that not all geologists have carried the airing of their conceptions as far as those we have just cited. However, among those who have proceeded with more reserve and who have not looked for methods outside ordinary physics or chemistry, how much diversity and contradiction still rule!

4.8

Uniformitarian Geology

Charles Lyell

Charles Lyell (1797–1875) was a Scottish lawyer and geologist who popularized the position known as uniformitarianism, according to which geological phenomena have been shaped by the operation of the same sorts of processes as can be seen to be operating today. In this selection from his *Principles of Geology* (1830), Lyell argues that the sorts of geological phenomena pointed out by Georges Cuvier could be accommodated by a gradual, uniformitarian approach, provided that the timescale was sufficiently long.

Chapter I

Geology is the science which investigates the successive changes that have taken place in the organic and inorganic kingdoms of nature; it enquires into the causes of these changes, and the influence which they have exerted in modifying the surface and external structure of our planet.

By these researches into the state of the earth and its inhabitants at former periods, we acquire a more perfect knowledge of its *present* condition, and more comprehensive views concerning the laws *now* governing its animate and inanimate productions. When we study history, we obtain a more profound insight into human nature, by instituting a comparison between the present and former states of society. We trace the long series of events which have gradually led to the actual posture of affairs; and by connecting effects with their causes, we are enabled to classify and retain in the memory a multitude of complicated relations – the various peculiarities of national character – the different degrees of moral and intellectual refinement, and numerous other circumstances, which, without historical associations, would be uninteresting or imperfectly understood. As the present condition of nations is the result of many antecedent changes, some extremely remote and others recent, some gradual, others sudden and violent, so the state of the natural world is the result

From Charles Lyell, *The Principles of Geology* (London: John Murray, 1830), selections from pp. 1–4, 382–5.

of a long succession of events, and if we would enlarge our experience of the present economy of nature, we must investigate the effects of her operations in former epochs.

We often discover with surprise, on looking back into the chronicles of nations, how the fortune of some battle has influenced the fate of millions of our contemporaries, when it has long been forgotten by the mass of the population. With this remote event we may find inseparably connected the geographical boundaries of a great state, the language now spoken by the inhabitants, their peculiar manners, laws, and religious opinions. But far more astonishing and unexpected are the connexions brought to light, when we carry back our researches into the history of nature. The form of a coast, the configuration of the interior of a country, the existence and extent of lakes, valleys, and mountains, can often be traced to the former prevalence of earthquakes and volcanoes, in regions which have long been undisturbed. To these remote convulsions the present fertility of some districts, the sterile character of others, the elevation of land above the sea, the climate, and various peculiarities, may be distinctly referred. On the other hand, many distinguishing features of the surface may often be ascribed to the operation at a remote era of slow and tranquil causes – to the gradual deposition of sediment in a lake or in the ocean, or to the prolific growth in the same of corals and testacea. To select another example, we find in certain localities subterranean deposits of coal, consisting of vegetable matter, formerly drifted into seas and lakes. These seas and lakes have since been filled up, the lands whereon the forests grew have disappeared or changed their form, the rivers and currents which floated the vegetable masses can no longer be traced, and the plants belonged to species which for ages have passed away from the surface of our planet. Yet the commercial prosperity, and numerical strength of a nation, may now be mainly dependant on the local distribution of fuel determined by that ancient state of things.

Geology is intimately related to almost all the physical sciences, as is history to the moral. An historian should, if possible, be at once profoundly acquainted with ethics, politics, jurisprudence, the military art, theology; in a word, with all branches of knowledge, whereby any insight into human affairs, or into the moral and intellectual nature of man, can be obtained. It would be no less desirable that a geologist should be well versed in chemistry, natural philosophy, mineralogy, zoology, comparative anatomy, botany; in short, in every science relating to organic and inorganic nature. With these accomplishments the historian and geologist would rarely fail to draw correct and philosophical conclusions from the various monuments transmitted to them of former occurrences. They would know to what combination of causes analogous effects were referrible, and they would often be enabled to supply by inference, information concerning many events unrecorded in the defective archives of former ages. But the brief duration of human life, and our limited powers, are so far from permitting us to aspire to such extensive acquisitions, that excellence even in one department is within the reach of few, and those individuals most effectually promote the general progress, who concentrate their thoughts on a limited portion of the field of inquiry. As it is necessary that the historian and the cultivators of moral or political science should reciprocally aid each other, so the geologist and those who study natural history or physics stand in equal need of mutual assistance. A comparative anatomist may derive some accession of knowledge from the bare inspection of the remains of an extinct quadruped, but the relic throws much greater light upon his own science, when he is informed to what relative era it belonged, what plants and animals were its contemporaries, in what degree of latitude it once existed, and other historical details. A fossil shell may interest a conchologist, though he be ignorant of the locality from which it came; but it will be of more value when he learns with what other species it was associated, whether they were marine or fresh-water, whether the strata containing them were at a certain elevation above the sea, and what relative position they held in regard to other groups of strata, with many other particulars determinable by an experienced geologist alone. On the other hand, the skill of the comparative anatomist and conchologist are often indispensable to those engaged in geological research, although it will rarely happen that the geologist will himself combine these different qualifications in his own person.

Some remains of former organic beings, like the ancient temple, statue, or picture, may have both their intrinsic and their historical value, while there are others which can never be expected to attract attention for their own sake. A painter, sculptor, or architect, would often neglect many curious relics of antiquity, as devoid of beauty and uninstructive with relation to their own art, however illustrative of the progress of refinement in some ancient nation. It has therefore been found desirable that the antiquary should unite his labours to those of the historian, and similar co-operation has become necessary in geology. The field of inquiry in living nature being inexhaustible, the zoologist and botanist can rarely be induced to sacrifice time in exploring the imperfect remains of lost species of animals and plants, while those still existing afford constant matter of novelty. They must entertain a desire of promoting *geology* by such investigations, and some knowledge of its objects must guide and direct their studies. According to the different opportunities, tastes, and talents of individuals, they may employ themselves in collecting particular kinds of minerals, rocks, or organic remains, and these, when well examined and explained, afford data to the geologist, as do coins, medals, and inscriptions to the historian.

It was long ere the distinct nature and legitimate objects of geology were fully recognized, and it was at first confounded with many other branches of inquiry, just as the limits of history, poetry, and mythology were ill-defined in the infancy of civilization. Werner appears to have regarded geology as little other than a subordinate department of mineralogy, and Desmarest included it under the head of Physical Geography. But the identification of its objects with those of Cosmogony has been the most common and serious source of confusion. The first who endeavoured to draw a clear line of demarcation between these distinct departments, was Hutton, who declared that geology was in no ways concerned "with questions as to the origin of things." But his doctrine on this head was vehemently opposed at first, and although it has gradually gained ground, and will ultimately prevail, it is yet far from being established. We shall attempt in the sequel of this work to demonstrate that geology differs as widely from cosmogony, as speculations concerning the creation of man differ from history. . . .

[. . .]

Concluding Remarks

In our history of the progress of geology, in the first volume, we stated that the opinion originally promulgated by Hutton, 'that the strata called *primitive* were mere altered sedimentary rocks,' was vehemently opposed for a time, the main objection to the theory being its supposed tendency to promote a belief in the past eternity of our planet. Previously the absence of animal and vegetable remains in the so-called primitive strata, had been appealed to, as proving that there had been a period when the planet was uninhabited by living beings, and when, as was also inferred, it was uninhabitable, and, therefore, probably in a nascent state.

The opposite doctrine, that the oldest visible strata might be the monuments of an antecedent period, when the animate world was already in existence, was declared to be equivalent to the assumption, that there never was a beginning to the present order of things. The unfairness of this charge was clearly pointed out by Playfair, who observed, 'that it was one thing to declare that we had not yet discovered the traces of a beginning, and another to deny that the earth ever had a beginning.'

We regret, however, to find that the bearing of our arguments in the first volume has been misunderstood in a similar manner, for we have been charged with endeavouring to establish the proposition, that 'the existing causes of change have operated with absolute uniformity from all eternity.'

It is the more necessary to notice this misrepresentation of our views, as it has proceeded from a friendly critic whose theoretical opinions coincide in general with our own, but who has, in this instance, strangely misconceived the scope of our argument. With equal justice might an astronomer be accused of asserting, that the works of creation extend throughout *infinite* space, because he refuses to take for granted that the remotest stars now seen in the heavens are on the utmost verge of the material universe. Every improvement of the telescope has brought thousands of new worlds into view, and it would, therefore, be rash and unphilosophical to imagine that we already survey the whole extent of the vast scheme, or that it will ever be brought within the sphere of human observation.

But no argument can be drawn from such premises in favour of the infinity of the space that has been filled with worlds; and if the material universe has any limits, it then follows that it must occupy a minute and infinitessimal point in infinite space. So, if in tracing back the earth's history, we arrive at the monuments of events which may have happened millions of ages before our times, and if we still find no decided evidence of a commencement, yet the arguments from analogy in support of the probability of a beginning remain unshaken; and if the past duration of the earth be finite, then the aggregate of geological epochs, however numerous, must constitute a mere moment of the past, a mere infinitessimal portion of eternity.

It has been argued, that as the different states of the earth's surface, and the different species by which it has been inhabited, have had each their origin, and many of them their termination, so the entire series may have commenced at a certain period. It has also been urged, that as we admit the creation of man to have occurred at a comparatively modern epoch – as we concede the astonishing fact of the first introduction of a moral and intellectual being, so also we may conceive the first creation of the planet itself.

We are far from denying the weight of this reasoning from analogy; but although it may strengthen our conviction, that the present system of change has not gone on from eternity, it cannot warrant us in presuming that we shall be permitted to behold the signs of the earth's origin, or the evidences of the first introduction into it of organic beings.

In vain do we aspire to assign limits to the works of creation in *space*, whether we examine the starry heavens, or that world of minute animalcules which is revealed to us by the microscope. We are prepared, therefore, to find that in *time* also, the confines of the universe lie beyond the reach of mortal ken. But in whatever direction we pursue our researches, whether in time or space, we discover everywhere the clear proofs of a Creative Intelligence, and of His foresight, wisdom, and power.

As geologists, we learn that it is not only the present condition of the globe that has been suited to the accommodation of myriads of living creatures, but that many former states also have been equally adapted to the organization and habits of prior races of beings. The disposition of the seas, continents, and islands, and the climates have varied; so it appears that the species have been changed, and yet they have all been so modelled, on types analogous to those of existing plants and animals, as to indicate throughout a perfect harmony of design and unity of purpose. To assume that the evidence of the beginning or end of so vast a scheme lies within the reach of our philosophical inquiries, or even of our speculations, appears to us inconsistent with a just estimate of the relations which subsist between the finite powers of man and the attributes of an Infinite and Eternal Being.

4.9

The Explanatory Scope of the Evolutionary Hypothesis

Charles Darwin

Charles Darwin (1809–1882) was an English naturalist who developed and defended the theory of natural selection. In this selection from his book *Variation of Animals and Plants under Domestication*, Darwin summarizes the argument he had made in *Origin of Species*, stressing the explanatory scope of the theory of evolution and endorsing the position expressed by Sir John Herschel that a hypothesis, no matter what its origins, may rise to the rank of a well-grounded theory if it explains various large and independent classes of facts.

In scientific investigations it is permitted to invent any hypothesis, and if it explains various large and independent classes of facts it rises to the rank of a well-grounded theory. The undulations of the ether and even its existence are hypothetical, yet every one now admits the undulatory theory of light. The principle of natural selection may be looked at as a mere hypothesis, but rendered in some degree probable by what we positively know of the variability of organic beings in a state of nature, – by what we positively know of the struggle for existence, and the consequent almost inevitable preservation of favourable variations, – and from the analogical formation of domestic races. Now this hypothesis may be tested, – and this seems to me the only fair and legitimate manner of considering the whole question, – by trying whether it explains several large and independent classes of facts; such as the geological succession of organic beings, their distribution in past and present times, and their mutual affinities and homologies. If the principle of natural selection does explain these and other large bodies of facts, it ought to be received. On the ordinary view of each species having been independently created, we gain no scientific explanation of any one of these facts. We can only say that it has so pleased the Creator to command that the past and present inhabitants of the world should appear in a certain order and in certain

From Darwin, *The Variation of Animals and Plants under Domestication*, 2nd ed., Vol. 1 (New York: D. Appleton and Co., 1876), selections from pp. 9–14.

areas; that He has impressed on them the most extraordinary resemblances, and has classed them in groups subordinate to groups. But by such statements we gain no new knowledge; we do not connect together facts and laws; we explain nothing.

It was the consideration of such large groups of facts as these which first led me to take up the present subject. When I visited during the voyage of HMS *Beagle*, the Galapagos Archipelago, situated in the Pacific Ocean about 500 miles from South America, I found myself surrounded by peculiar species of birds, reptiles, and plants, existing nowhere else in the world. Yet they nearly all bore an American stamp. In the song of the mocking-thrush, in the harsh cry of the carrion-hawk, in the great candlestick-like opuntias, I clearly perceived the neighbourhood of America, though the islands were separated by so many miles of ocean from the mainland, and differed much in their geological constitution and climate. Still more surprising was the fact that most of the inhabitants of each separate island in this small archipelago were specifically different, though most closely related to each other. The archipelago, with its innumerable craters and bare streams of lava, appeared to be of recent origin; and thus I fancied myself brought near to the very act of creation. I often asked myself how these many peculiar animals and plants had been produced: the simplest answer seemed to be that the inhabitants of the several islands had descended from each other, undergoing modification in the course of their descent; and that all the inhabitants of the archipelago were descended from those of the nearest land, namely America, whence colonists would naturally have been derived. But it long remained to me an inexplicable problem how the necessary degree of modification could have been effected, and it would have thus remained for ever, had I not studied domestic productions, and thus acquired a just idea of the power of Selection. As soon as I had fully realized this idea, I saw, on reading Malthus on Population, that Natural Selection was the inevitable result of the rapid increase of all organic beings; for I was prepared to appreciate the struggle for existence by having long studied the habits of animals.

[. . .]

The innumerable past and present inhabitants of the world are connected together by the most singular and complex affinities, and can be classed in groups under groups, in the same manner as varieties can be classed under species and subvarieties under varieties, but with much higher grades of difference. These complex affinities and the rules for classification, receive a rational explanation on the theory of descent, combined with the principle of natural selection, which entails divergence of character and the extinction of intermediate forms. How inexplicable is the similar pattern of the hand of a man, the foot of a dog, the wing of a bat, the flipper of a seal, on the doctrine of independent acts of creation! how simply explained on the principle of the natural selection of successive slight variations in the diverging descendants from a single progenitor! So it is with certain parts or organs in the same individual animal or plant, for instance, the jaws and legs of a crab, or the petals, stamens, and pistils of a flower. During the many changes to which in the course of time organic beings have been subjected, certain organs or parts have occasionally become at first of little use and ultimately superfluous; and the retention of such parts in a rudimentary and useless condition is intelligible on the theory of descent. It can be shown that modifications of structure are generally inherited by the offspring at the same age at which each successive variation appeared in the parents; it can further be shown that variations do not commonly supervene at a very early period of embryonic growth, and on these two principles we can understand that most wonderful fact in the whole circuit of natural history, namely, the close similarity of the embryos within the same great class – for instance, those of mammals, birds, reptiles, and fish.

It is the consideration and explanation of such facts as these which has convinced me that the theory of descent with modification by means of natural selection is in the main true. These facts have as yet received no explanation on the theory of independent Creation; they cannot be grouped together under one point of view, but each has to be considered as an ultimate fact. As the first origin of life on this earth, as well as the continued life of each individual, is at present quite beyond the scope of science, I do not wish to lay much stress on the greater simplicity of the view of a few forms or of only one form having been originally created, instead of innumerable miraculous creations having been necessary at innumerable periods; though this more simple view accords well with Maupertuis's philosophical axiom of "least action."

In considering how far the theory of natural selection may be extended, – that is, in determining from how many progenitors the inhabitants of the world have descended, – we may conclude that at least all the members of the same class have descended from a single ancestor. A number of organic beings are included in the same class, because they present, independently of their habits of life, the same fundamental type of structure, and because they graduate into each other. Moreover, members of the same class can in most cases be shown to be closely alike at an early embryonic age. These facts can be explained on the belief of their descent from a common form; therefore it may be safely admitted that all the members of the same class are descended from one progenitor. But as the members of quite distinct classes have something in common in structure and much in common in constitution, analogy would lead us one step further, and to infer as probable that all living creatures are descended from a single prototype.

I hope that the reader will pause before coming to any final and hostile conclusion on the theory of natural selection. The reader may consult my 'Origin of Species' for a general sketch of the whole subject; but in that work he has to take many statements on trust. In considering the theory of natural selection, he will assuredly meet with weighty difficulties, but these difficulties relate chiefly to subjects – such as the degree of perfection of the geological record, the means of distribution, the possibility of transitions in organs, &c. – on which we are confessedly ignorant; nor do we know how ignorant we are. If we are much more ignorant than is generally supposed, most of these difficulties wholly disappear. Let the reader reflect on the difficulty of looking at whole classes of facts from a new point of view. Let him observe how slowly, but surely, the noble views of Lyell on the gradual changes now in progress on the earth's surface have been accepted as sufficient to account for all that we see in its past history. The present action of natural selection may seem more or less probable: but I believe in the truth of the theory, because it collects, under one point of view, and gives a rational explanation of, many apparently independent classes of facts.

4.10

Induction as a Self-Correcting Process

Charles Sanders Peirce

Charles Sanders Peirce (1839–1914) was a brilliant American philosopher, logician, and scientist whose voluminous writings broke new ground in logic, semiotics, and other fields. In this selection from the sixth volume of his *Collected Papers* (6.40–6.42), he identifies the inference from samples as the core of all forms of non-deductive reasoning and argues that when we proceed in an inductive manner our provisional inferences are self-correcting. He then considers various objections to his account and argues that, while we may have to rest content with more modest conclusions than we would wish, there are no decisive objections to his account.

Non-deductive or ampliative inference is of three kinds: induction, hypothesis, and analogy. If there be any other modes, they must be extremely unusual and highly complicated, and may be assumed with little doubt to be of the same nature as those enumerated. For induction, hypothesis, and analogy, as far as their ampliative character goes, that is, so far as they conclude something not implied in the premisses, depend upon one principle and involve the same procedure. All are essentially inferences from sampling. Suppose a ship arrives at Liverpool laden with wheat in bulk. Suppose that by some machinery the whole cargo be stirred up with great thoroughness. Suppose that twenty-seven thimblefuls be taken equally from the forward, midships, and aft parts, from the starboard, center, and larboard parts, and from the top, half depth, and lower parts of her hold, and that these being mixed and the grains counted, four-fifths of the latter are found to be of quality *A*. Then we infer, experientially and provisionally, that approximately four-fifths of all the grain in the cargo is of the same quality. I say we infer this *experientially* and *provisionally*. By saying that we infer it *experientially*, I mean that our conclusion makes no pretension to knowledge of wheat-in-itself, our ἀλήθεια, as the derivation of that word implies, has nothing to do with *latent*

From Peirce, *Collected Papers of Charles Sanders Peirce*, Vol. 6, ed. Charles Hartshorne and Paul Weiss (Cambridge, MA: The Belknap Press of Harvard University Press, 1960), 6.40–6.42.

wheat. We are dealing only with the matter of possible experience – experience in the full acceptation of the term as something not merely affecting the senses but also as the subject of thought. If there be any wheat hidden on the ship, so that it can neither turn up in the sample nor be heard of subsequently from purchasers – or if it be half-hidden, so that it may, indeed, turn up, but is less likely to do so than the rest – or if it can affect our senses and our pockets, but from some strange cause or causelessness cannot be reasoned about – all such wheat is to be excluded (or have only its proportional weight) in calculating that true proportion of quality *A*, to which our inference seeks to approximate. By saying that we draw the inference *provisionally*, I mean that we do not hold that we have reached any assigned degree of approximation as yet, but only hold that if our experience be indefinitely extended, and if every fact of whatever nature, as fast as it presents itself, be duly applied, according to the inductive method, in correcting the inferred ratio, then our approximation will become indefinitely close in the long run; that is to say, close to the experience *to come* (not merely close by the exhaustion of a finite collection) so that if experience in general is to fluctuate irregularly to and fro, in a manner to deprive the ratio sought of all definite value, we shall be able to find out approximately within what limits it fluctuates, and if, after having one definite value, it changes and assumes another, we shall be able to find that out, and in short, whatever may be the variations of this ratio in experience, experience indefinitely extended will enable us to detect them, so as to predict rightly, at last, what its ultimate value may be, if it have any ultimate value, or what the ultimate law of succession of values may be, if there be any such ultimate law, or that it ultimately fluctuates irregularly within certain limits, if it do so ultimately fluctuate. Now our inference, claiming to be no more than thus experiential and provisional, manifestly involves no postulate whatever.

41. For what is a postulate? It is the formulation of a material fact which we are not entitled to assume as a premiss, but the truth of which is requisite to the validity of an inference. Any fact, then, which might be supposed postulated, must either be such that it would ultimately present itself in experience, or not. If it will present itself, we need not postulate it now in our provisional inference, since we shall ultimately be entitled to use it as a premiss. But if it never would present itself in experience, our conclusion is valid but for the possibility of this fact being otherwise than assumed, that is, it is valid as far as possible experience goes, and that is all that we claim. Thus, every postulate is cut off, either by the provisionality or by the experientiality of our inference. For instance, it has been said that induction postulates that, if an indefinite succession of samples be drawn, examined, and thrown back each before the next is drawn, then in the long run every grain will be drawn as often as any other, that is to say, postulates that the ratio of the numbers of times in which any two are drawn will indefinitely approximate to unity. But no such postulate is made; for if, on the one hand, we are to have no other experience of the wheat than from such drawings, it is the ratio that presents itself in those drawings and not the ratio which belongs to the wheat in its latent existence that we are endeavoring to determine; while if, on the other hand, there is some other mode by which the wheat is to come under our knowledge, equivalent to another kind of sampling, so that after all our care in stirring up the wheat some experiential grains will present themselves in the first sampling operation more often than others in the long run, this very singular fact will be sure to get discovered by the inductive method, which must avail itself of every sort of experience; and our inference, which was only provisional, corrects itself at last. Again, it has been said, that induction postulates that under like circumstances like events will happen, and that this postulate is at bottom the same as the principle of universal causation. But this is a blunder, or *bévue*, due to thinking exclusively of inductions where the concluded ratio is either 1 or 0. If any such proposition were postulated, it would be that under like circumstances (the circumstances of drawing the different samples) different events occur in the same proportions in all the different sets – a proposition which is false and even absurd. But in truth no such thing is postulated, the experiential character of the inference reducing the condition of validity to this, that if a certain result does not occur, the opposite result will be manifested, a condition assured by the provisionality of the inference. But it may be asked

whether it is not conceivable that every instance of a certain class destined to be ever employed as a datum of induction should have one character, while every instance destined not to be so employed should have the opposite character. The answer is that, in that case, the instances excluded from being subjects of reasoning would not be experienced in the full sense of the word, but would be among these *latent* individuals of which our conclusion does not pretend to speak.

42. To this account of the rationale of induction I know of but one objection worth mention: it is that I thus fail to deduce the full degree of force which this mode of inference in fact possesses; that according to my view, no matter how thorough and elaborate the stirring and mixing process had been, the examination of a single handful of grain would not give me any assurance, sufficient to risk money upon, that the next handful would not greatly modify the concluded value of the ratio under inquiry, while, in fact, the assurance would be very high that this ratio was not greatly in error. If the true ratio of grains of quality *A* were 0.80 and the handful contained a thousand grains, nine such handfuls out of every ten would contain from 780 to 820 grains of quality *A*. The answer to this is that the calculation given is correct when we know that the units of this handful and the quality inquired into have the normal independence of one another, if for instance the stirring has been complete and the character sampled for has been settled upon in advance of the examination of the sample. But in so far as these conditions are not known to be complied with, the above figures cease to be applicable. Random sampling and predesignation of the character sampled for should always be striven after in inductive reasoning, but when they cannot be attained, so long as it is conducted honestly, the inference retains some value. When we cannot ascertain how the sampling has been done or the sample-character selected, induction still has the essential validity which my present account of it shows it to have.

4.11

The Nature of Abduction

Charles Sanders Peirce

One of Peirce's deepest contributions to the analysis of scientific reasoning is his development of the notion of "abduction," or reasoning by hypothesis. In the selection that follows, we see a brief outline of the structure of abductive reasoning (*Collected Papers* 5.189) and then, from a separate work (*Collected Papers* 6.522–6.528), an extended description of abduction as applied to a particular case.

§1 The Nature of Hypothesis

[. . .]

All our knowledge may be said to rest upon *observed facts*. It is true that there are psychological states which antecede our observing facts as such. Thus, it is a fact that I see an inkstand before me; but before I can say that I am obliged to have impressions of sense into which no idea of an inkstand, or of any separate object, or of an "I," or of seeing, enter at all; and it is true that my judging that I see an inkstand before me is the product of mental operations upon these impressions of sense. But it is only when the cognition has become worked up into a proposition, or judgment of a fact, that I can exercize any direct control over the process; and it is idle to discuss the "legitimacy" of that which cannot be controlled. Observations of fact have, therefore, to be accepted as they occur.

523 But observed facts relate exclusively to the particular circumstances that happened to exist when they were observed. They do not relate to any future occasions upon which we may be in doubt how we ought to act. They, therefore, do not, in themselves, contain any practical knowledge.

Such knowledge must involve additions to the facts observed. The making of those additions is an operation which we can control; and it is evidently a process during which error is liable to creep in.

524 Any proposition added to observed facts, tending to make them applicable in any way to other circumstances than those under which they were observed, may be called a hypothesis. A

From Peirce, *Collected Papers of Charles Sanders Peirce*, Vols. 5–6, ed. Charles Hartshorne and Paul Weiss (Cambridge, MA: The Belknap Press of Harvard University Press, 1960), 5.189 and 6.522–8.

hypothesis ought, at first, to be entertained interrogatively. Thereupon, it ought to be tested by experiment so far as practicable. There are two distinct processes, both of which may be performed rightly or wrongly. We may go wrong and be wasting time in so much as entertaining a hypothesis, even as a question. That is a subject for criticism in every case. There are some hypotheses which are of such a nature that they never can be tested at all. Whether such hypotheses ought to be entertained at all, and if so in what sense, is a serious question; but it hardly concerns our present inquiry. The hypotheses with which we shall have in this paper to deal are capable of being put to the test. How this is to be done is a question of extreme importance; but my intention is to consider it only in a very cursory manner, at present. There are, moreover, many hypotheses in regard to which knowledge already in our possession may, at once, quite justifiably either raise them to the rank of opinions, or even positive beliefs, or cause their immediate rejection. This also is a matter to be considered. But it is the first process, that of entertaining the question, which will here be of foremost importance.

525 Before we go further, let us get the points stated above quite clear. By a *hypothesis*, I mean, not merely a supposition about an observed object, as when I suppose that a man is a Catholic priest because that would explain his dress, expression of countenance, and bearing, but also any other supposed truth from which would result such facts as have been observed, as when van't Hoff, having remarked that the osmotic pressure of one per cent solutions of a number of chemical substances was inversely proportional to their atomic weights, thought that perhaps the same relation would be found to exist between the same properties of any other chemical substance. The first starting of a hypothesis and the entertaining of it, whether as a simple interrogation or with any degree of confidence, is an inferential step which I propose to call *abduction*. This will include a preference for any one hypothesis over others which would equally explain the facts, so long as this preference is not based upon any previous knowledge bearing upon the truth of the hypotheses, nor on any testing of any of the hypotheses, after having admitted them on probation. I call all such inference by the peculiar name, *abduction*, because its legitimacy depends upon altogether different principles from those of other kinds of inference.

§2 The Testing of Hypotheses

526 The operation of testing a hypothesis by experiment, which consists in remarking that, if it is true, observations made under certain conditions ought to have certain results, and then causing those conditions to be fulfilled, and noting the results, and, if they are favorable, extending a certain confidence to the hypothesis, I call *induction*. For example, suppose that I have been led to surmise that among our colored population there is a greater tendency toward female births than among our whites. I say, if that be so, the last census must show it. I examine the last census report and find that, sure enough, there was a somewhat greater proportion of female births among colored births than among white births in that census year. To accord a certain faith to my hypothesis on that account is legitimate. It is a strong induction. I have taken all the births of that year as a sample of all the births of years in general, so long as general conditions remain as they were then. It is a very large sample, quite unnecessarily so, were it not that the excess of the one ratio over the other is quite small. All induction whatever may be regarded as the inference that throughout a whole class a ratio will have about the same value that it has in a random sample of that class, provided the nature of the ratio for which the sample is to be examined is specified (or virtually specified) in advance of the examination. So long as the class sampled consists of units, and the ratio in question is a ratio between counts of occurrences, induction is a comparatively simple affair. But suppose we wish to test the hypothesis that a man is a Catholic priest, that is, has all the characters that are common to Catholic priests and peculiar to them. Now characters are not units, nor do they consist of units, nor can they be counted, in such a sense that one count is right and every other wrong. Characters have to be estimated according to their significance. The consequence is that there will be a certain element of guess-work in such an induction; so that I call it an *abductory induction*. I might say to myself, let me think of some other character that belongs to Catholic priests,

beside those that I have remarked in this man, a character which I can ascertain whether he possesses or not. All Catholic priests are more or less familiar with Latin pronounced in the Italian manner. If, then, this man is a Catholic priest, and I make some remark in Latin which a person not accustomed to the Italian pronunciation would not at once understand, and I pronounce it in that way, then if that man is a Catholic priest he will be so surprised that he cannot but betray his understanding of it. I make such a remark; and I notice that he does understand it. But how much weight am I to attach to that test? After all, it does not touch an essential characteristic of a priest or even of a Catholic. It must be acknowledged that it is but a weak confirmation, and all the more so, because it is quite uncertain how much weight should be attached to it. Nevertheless, it does and ought to incline me to believe that the man is a Catholic priest. It is an induction, because it is a test of the hypothesis by means of a prediction, which has been verified. But it is only an abductory induction, because it was a sampling of the characters of priests to see what proportion of them this man possessed, when characters cannot be counted, nor even weighed, except by guess-work. It also partakes of the nature of abduction in involving an original suggestion; while typical induction has no originality in it, but only tests a suggestion already made.

527 In induction, it is not the fact predicted that in any degree necessitates the truth of the hypothesis or even renders it probable. It is the fact that it has been predicted successfully and that it is a haphazard specimen of all the predictions which might be based on the hypothesis and which constitute its practical truth. But it frequently happens that there are facts which, merely as facts, apart from the manner in which they have presented themselves, necessitate the truth, or the falsity, or the probability in some definite degree, of the hypothesis. For example, suppose the hypothesis to be that a man believes in the infallibility of the Pope. Then, if we ascertain in any way that he believes in the immaculate conception, in the confessional, and in prayers for the dead, or on the other hand that he disbelieves all or some of these things, either fact will be almost decisive of the truth or falsity of the proposition. Such inference is *deduction*. So if we ascertain that the man in question is a violent partisan in politics and in many other subjects. If, then, we find that he has given money toward a Catholic institution, we may fairly reason that such a man would not do that unless he believed in the Pope's infallibility. Or again, we might learn that he is one of five brothers whose opinions are identical on almost all subjects. If, then, we find that the other four all believe in the Pope's infallibility or all disbelieve it, this will affect our confidence in the hypothesis. This consideration will be strengthened by our general experience that while different members of a large family usually differ about most subjects, yet it mostly happens that they are either all Catholics or all Protestants. Those are four different varieties of deductive considerations which may legitimately influence our belief in a hypothesis.

528 These distinctions are perfectly clear in principle, which is all that is necessary, although it might sometimes be a nice question to say to which class a given inference belongs. It is to be remarked that, in pure abduction, it can never be justifiable to accept the hypothesis otherwise than as an interrogation. But as long as that condition is observed, no positive falsity is to be feared; and therefore the whole question of what one out of a number of possible hypotheses ought to be entertained becomes purely a question of economy.

4.12

The Role of Hypotheses in Physical Theory

Henri Poincaré

Henri Poincaré (1854–1912) was a French mathematician, physicist, and philosopher of science. He made fundamental contributions to virtually every branch of mathematics, and he anticipated parts of Einstein's theory of relativity. In this excerpt from *Science and Hypothesis*, Poincaré addresses the tension between experiment and hypothesis in physics, arguing that although experimental data are indispensable, detached facts unordered by hypotheses are unable to give us science worth the name. The use of hypotheses carries risk. Yet they often prove fruitful even if, perhaps especially if, they turn out to be flawed.

The Rôle of Experiment and Generalisation

Experiment is the sole source of truth. It alone can teach us something new; it alone can give us certainty. These are two points that cannot be questioned. But then, if experiment is everything, what place is left for mathematical physics? What can experimental physics do with such an auxiliary – an auxiliary, moreover, which seems useless, and even may be dangerous?

However, mathematical physics exists. It has rendered undeniable service, and that is a fact which has to be explained. It is not sufficient merely to observe; we must use our observations, and for that purpose we must generalise. This is what has always been done, only as the recollection of past errors has made man more and more circumspect, he has observed more and more and generalised less and less. Every age has scoffed at its predecessor, accusing it of having generalised too boldly and too naively. Descartes used to commiserate the Ionians. Descartes in his turn makes us smile, and no doubt some day our children will laugh at us. Is there no way of getting at once to the gist of the matter, and thereby escaping the raillery which we foresee? Cannot we be content with experiment alone? No, that is impossible; that would be a complete misunderstanding of the true character of science. The man of science must work with method.

From Henri Poincaré, *Science and Hypothesis* (Dover Publications, Inc., 1952), pp. 140–53.

Science is built up of facts, as a house is built of stones; but an accumulation of facts is no more a science than a heap of stones is a house. Most important of all, the man of science must exhibit foresight. . . .

[. . .]

We all know that there are good and bad experiments. The latter accumulate in vain. Whether there are a hundred or a thousand, one single piece of work by a real master – by a Pasteur, for example – will be sufficient to sweep them into oblivion. Bacon would have thoroughly understood that, for he invented the phrase *experimentum crucis*; . . . What, then, is a good experiment? It is that which teaches us something more than an isolated fact. It is that which enables us to predict, and to generalise. Without generalisation, prediction is impossible. The circumstances under which one has operated will never again be reproduced simultaneously. The fact observed will never be repeated. All that can be affirmed is that under analogous circumstances an analogous fact will be produced. To predict it, we must therefore invoke the aid of analogy – that is to say, even at this stage, we must generalise. However timid we may be, there must be interpolation. Experiment only gives us a certain number of isolated points. They must be connected by a continuous line, and this is a true generalisation. But more is done. The curve thus traced will pass between and near the points observed; it will not pass through the points themselves. Thus we are not restricted to generalising our experiment, we correct it; and the physicist who would abstain from these corrections, and really content himself with experiment pure and simple, would be compelled to enunciate very extraordinary laws indeed. Detached facts cannot therefore satisfy us, and that is why our science must be ordered, or, better still, generalised.

It is often said that experiments should be made without preconceived ideas. That is impossible. Not only would it make every experiment fruitless, but even if we wished to do so, it could not be done. Every man has his own conception of the world, and this he cannot so easily lay aside. We must, for example, use language, and our language is necessarily steeped in preconceived ideas. Only they are unconscious preconceived ideas, which are a thousand times the most dangerous of all. Shall we say, that if we cause others to intervene of which we are fully conscious, that we shall only aggravate the evil? I do not think so. I am inclined to think that they will serve as ample counterpoises – I was almost going to say antidotes. They will generally disagree, they will enter into conflict one with another, and *ipso facto*, they will force us to look at things under different aspects. This is enough to free us. He is no longer a slave who can choose his master.

Thus, by generalisation, every fact observed enables us to predict a large number of others; only, we ought not to forget that the first alone is certain, and that all the others are merely probable. However solidly founded a prediction may appear to us, we are never *absolutely* sure that experiment will not prove it to be baseless if we set to work to verify it. But the probability of its accuracy is often so great that practically we may be content with it. It is far better to predict without certainty, than never to have predicted at all. We should never, therefore, disdain to verify when the opportunity presents itself. But every experiment is long and difficult, and the labourers are few, and the number of facts which we require to predict is enormous; and besides this mass, the number of direct verifications that we can make will never be more than a negligible quantity. Of this little that we can directly attain we must choose the best. Every experiment must enable us to make a maximum number of predictions having the highest possible degree of probability. The problem is, so to speak, to increase the output of the scientific machine. I may be permitted to compare science to a library which must go on increasing indefinitely; the librarian has limited funds for his purchases, and he must, therefore, strain every nerve not to waste them. Experimental physics has to make the purchases, and experimental physics alone can enrich the library. As for mathematical physics, her duty is to draw up the catalogue. If the catalogue is well done the library is none the richer for it; but the reader will be enabled to utilise its riches; and also by showing the librarian the gaps in his collection, it will help him to make a judicious use of his funds, which is all the more important, inasmuch as those funds are entirely inadequate. That is the rôle of mathematical physics. It must direct generalisation, so as to increase what I called just now the output of science. By what means it does this, and how it may do it without danger, is what we have now to examine.

The Unity of Nature

Let us first of all observe that every generalisation supposes in a certain measure a belief in the unity and simplicity of Nature. As far as the unity is concerned, there can be no difficulty. If the different parts of the universe were not as the organs of the same body, they would not re-act one upon the other; they would mutually ignore each other, and we in particular should only know one part. We need not, therefore, ask if Nature is one, but how she is one.

As for the second point, that is not so clear. It is not certain that Nature is simple. Can we without danger act as if she were?

There was a time when the simplicity of Mariotte's law was an argument in favour of its accuracy: when Fresnel himself, after having said in a conversation with Laplace that Nature cares naught for analytical difficulties, was compelled to explain his words so as not to give offence to current opinion. Nowadays, ideas have changed considerably; but those who do not believe that natural laws must be simple, are still often obliged to act as if they did believe it. They cannot entirely dispense with this necessity without making all generalisation, and therefore all science, impossible. It is clear that any fact can be generalised in an infinite number of ways, and it is a question of choice. The choice can only be guided by considerations of simplicity. Let us take the most ordinary case, that of interpolation. We draw a continuous line as regularly as possible between the points given by observation. Why do we avoid angular points and inflexions that are too sharp? Why do we not make our curve describe the most capricious zigzags? It is because we know beforehand, or think we know, that the law we have to express cannot be so complicated as all that. The mass of Jupiter may be deduced either from the movements of his satellites, or from the perturbations of the major planets, or from those of the minor planets. If we take the mean of the determinations obtained by these three methods, we find three numbers very close together, but not quite identical. This result might be interpreted by supposing that the gravitation constant is not the same in the three cases; the observations would be certainly much better represented. Why do we reject this interpretation? Not because it is absurd, but because it is uselessly complicated. We shall only accept it when we are forced to, and it is not imposed upon us yet. To sum up, in most cases every law is held to be simple until the contrary is proved.

This custom is imposed upon physicists by the reasons that I have indicated, but how can it be justified in the presence of discoveries which daily show us fresh details, richer and more complex? How can we even reconcile it with the unity of nature? For if all things are interdependent, the relations in which so many different objects intervene can no longer be simple.

If we study the history of science we see produced two phenomena which are, so to speak, each the inverse of the other. Sometimes it is simplicity which is hidden under what is apparently complex; sometimes, on the contrary, it is simplicity which is apparent, and which conceals extremely complex realities. What is there more complicated than the disturbed motions of the planets, and what more simple than Newton's law? There, as Fresnel said, Nature playing with analytical difficulties, only uses simple means, and creates by their combination I know not what tangled skein. Here it is the hidden simplicity which must be disentangled. Examples to the contrary abound. In the kinetic theory of gases, molecules of tremendous velocity are discussed, whose paths, deformed by incessant impacts, have the most capricious shapes, and plough their way through space in every direction. The result observable is Mariotte's simple law. Each individual fact was complicated. The law of great numbers has re-established simplicity in the mean. Here the simplicity is only apparent, and the coarseness of our senses alone prevents us from seeing the complexity.

[...]

And Newton's law itself? Its simplicity, so long undetected, is perhaps only apparent. Who knows if it be not due to some complicated mechanism, to the impact of some subtle matter animated by irregular movements, and if it has not become simple merely through the play of averages and large numbers? In any case, it is difficult not to suppose that the true law contains complementary terms which may become sensible at small distances. If in astronomy they are negligible, and if the law thus regains its simplicity, it is solely on account of the enormous distances of the celestial bodies. No doubt, if our means of investigation became more and more penetrating, we should discover the simple beneath the complex, and then the complex from the simple, and then again the simple beneath the complex, and so on, without ever being able to predict what the last term will be. We must stop somewhere, and for science to be possible we must stop where we have found simplicity. That is the only ground on which we can erect the edifice of our generalisations. But, this simplicity being only apparent, will the ground be solid enough? That is what we have now to discover.

For this purpose let us see what part is played in our generalisations by the belief in simplicity. We have verified a simple law in a considerable number of particular cases. We refuse to admit that this coincidence, so often repeated, is a result of mere chance, and we conclude that the law must be true in the general case.

Kepler remarks that the positions of a planet observed by Tycho are all on the same ellipse. Not for one moment does he think that, by a singular freak of chance, Tycho had never looked at the heavens except at the very moment when the path of the planet happened to cut that ellipse. What does it matter then if the simplicity be real or if it hide a complex truth? Whether it be due to the influence of great numbers which reduces individual differences to a level, or to the greatness or the smallness of certain quantities which allow of certain terms to be neglected – in no case is it due to chance. This simplicity, real or apparent, has always a cause. We shall therefore always be able to reason in the same fashion, and if a simple law has been observed in several particular cases, we may legitimately suppose that it still will be true in analogous cases. To refuse to admit this would be to attribute an inadmissible rôle to chance. However, there is a difference. If the simplicity were real and profound it would bear the test of the increasing precision of our methods of measurement. If, then, we believe Nature to be profoundly simple, we must conclude that it is an approximate and not a rigorous simplicity. This is what was formerly done, but it is what we have no longer the right to do. The simplicity of Kepler's laws, for instance, is only apparent; but that does not prevent them from being applied to almost all systems analogous to the solar system, though that prevents them from being rigorously exact.

Rôle of Hypothesis

Every generalisation is a hypothesis. Hypothesis therefore plays a necessary rôle, which no one has ever contested. Only, it should always be as soon as possible submitted to verification. It goes without saying that, if it cannot stand this test, it must be abandoned without any hesitation. This is, indeed, what is generally done; but sometimes with a certain impatience. Ah well! this impatience is not justified. The physicist who has just given up one of his hypotheses should, on the contrary, rejoice, for he found an unexpected opportunity of discovery. His hypothesis, I imagine, had not been lightly adopted. It took into account all the known factors which seem capable of intervention in the phenomenon. If it is not verified, it is because there is something unexpected and extraordinary about it, because we are on the point of finding something unknown and new. Has the hypothesis thus rejected been sterile? Far from it. It may be even said that it has rendered more service than a true hypothesis. Not only has it been the occasion of a decisive experiment, but if this experiment had been made by chance, without the hypothesis, no conclusion could have been drawn; nothing extraordinary would have been seen; and only one fact the more would have been catalogued, without deducing from it the remotest consequence.

Now, under what conditions is the use of hypothesis without danger? The proposal to submit all to experiment is not sufficient. Some hypotheses are dangerous, – first and foremost those which are tacit and unconscious. And since we make them without knowing them, we cannot get rid of them. Here again, there is a service that mathematical physics may render us. By the precision which is its characteristic, we are compelled to formulate all the hypotheses that we would unhesitatingly make without its aid. Let us also notice that it is important not to multiply hypotheses indefinitely. If we construct a theory based upon multiple hypotheses, and if experiment condemns it, which of the premisses must be changed? It is impossible to tell. Conversely, if the experiment succeeds, must we suppose that it has verified all these hypotheses at once? Can several unknowns be determined from a single equation?

We must also take care to distinguish between the different kinds of hypotheses. First of all, there are those which are quite natural and necessary. It is difficult not to suppose that the influence of very distant bodies is quite negligible, that small movements obey a linear law, and that effect is a continuous function of its cause. I will say as much for the conditions imposed by symmetry. All these hypotheses affirm, so to speak, the common basis of all the theories of mathematical physics. They are the last that should be abandoned. There is a second category of hypotheses which I shall qualify as indifferent. In most questions the analyst assumes, at the beginning of his calculations, either that matter is continuous, or the reverse, that it is formed of atoms. In either case, his results would have been the same. On the atomic supposition he has a little more difficulty in obtaining them – that is all. If, then, experiment confirms his conclusions, will he suppose that he has proved, for example, the real existence of atoms?

In optical theories two vectors are introduced, one of which we consider as a velocity and the other as a vortex. This again is an indifferent hypothesis, since we should have arrived at the same conclusions by assuming the former to be a vortex and the latter to be a velocity. The success of the experiment cannot prove, therefore, that the first vector is really a velocity. It only proves one thing – namely, that it is a vector; and that is the only hypothesis that has really been introduced into the premisses. To give it the concrete appearance that the fallibility of our minds demands, it was necessary to consider it either as a velocity or as a vortex. In the same way, it was necessary to represent it by an x or a y, but the result will not prove that we were right or wrong in regarding it as a velocity; nor will it prove we are right or wrong in calling it x and not y.

These indifferent hypotheses are never dangerous provided their characters are not misunderstood. They may be useful, either as artifices for calculation, or to assist our understanding by concrete images, to fix the ideas, as we say. They need not therefore be rejected. The hypotheses of the third category are real generalisations. They must be confirmed or invalidated by experiment. Whether verified or condemned, they will always be fruitful; but, for the reasons I have given, they will only be so if they are not too numerous.

4.13

Against Crucial Experiments

Pierre Duhem

Pierre Duhem (1861–1916), a French physicist and philosopher of science, was one of the first scholars to undertake a serious study of the history of science, paying particular and sympathetic attention to the science of the Middle Ages. In this selection from *The Aim and Structure of Physical Theory* (1906), Duhem argues through an extended consideration of examples that an experiment in physics cannot strictly refute an isolated hypothesis; all that it can do is to show us that something is wrong with a cluster of hypotheses. His argument, if successful, would undermine the emphasis that earlier scientists like Herschel placed on "crucial experiments." W. V. O. Quine later advanced a more general form of this holism, and Quine's stronger claim is sometimes known as the Duhem-Quine thesis.

[. . .]

[I]n the mind of the physicist there are constantly present two sorts of apparatus: one is the concrete apparatus in glass and metal, manipulated by him, the other is the schematic and abstract apparatus which theory substitutes for the concrete apparatus and on which the physicist does his reasoning. For these two ideas are indissolubly connected in his intelligence, and each necessarily calls on the other; the physicist can no sooner conceive the concrete apparatus without associating with it the idea of the schematic apparatus than a Frenchman can conceive an idea without associating it with the French word expressing it. This radical impossibility, preventing one from dissociating physical theories from the experimental procedures appropriate for testing these theories, complicates this test in a singular way, and obliges us to examine the logical meaning of it carefully.

Of course, the physicist is not the only one who appeals to theories at the very time he is experimenting or reporting the results of his experiments.

From Duhem, *The Aim and Structure of Physical Theory* (New York: Atheneum, 1962) (originally published by Princeton University Press), pp. 182–90, 208–12.

The chemist and the physiologist when they make use of physical instruments, e.g., the thermometer, the manometer, the calorimeter, the galvanometer, and the saccharimeter, implicitly admit the accuracy of the theories justifying the use of these pieces of apparatus as well as of the theories giving meaning to the abstract ideas of temperature, pressure, quantity of heat, intensity of current, and polarized light, by means of which the concrete indications of these instruments are translated. But the theories used, as well as the instruments employed, belong to the domain of physics; by accepting with these instruments the theories without which their readings would be devoid of meaning, the chemist and the physiologist show their confidence in the physicist, whom they suppose to be infallible. The physicist, on the other hand, is obliged to trust his own theoretical ideas or those of his fellow-physicists. From the standpoint of logic, the difference is of little importance; for the physiologist and chemist as well as for the physicist, the statement of the result of an experiment implies, in general, an act of faith in a whole group of theories.

2 An Experiment in Physics Can Never Condemn an Isolated Hypothesis but Only a Whole Theoretical Group

The physicist who carries out an experiment, or gives a report of one, implicitly recognizes the accuracy of a whole group of theories. Let us accept this principle and see what consequences we may deduce from it when we seek to estimate the role and logical import of a physical experiment.

In order to avoid any confusion we shall distinguish two sorts of experiments: experiments of *application*, which we shall first just mention, and experiments of *testing*, which will be our chief concern.

You are confronted with a problem in physics to be solved practically; in order to produce a certain effect you wish to make use of knowledge acquired by physicists; you wish to light an incandescent bulb; accepted theories indicate to you the means for solving the problem; but to make use of these means you have to secure certain information; you ought, I suppose, to determine the electromotive force of the battery of generators at your disposal; you measure this electromotive force: that is what I call an experiment of application. This experiment does not aim at discovering whether accepted theories are accurate or not; it merely intends to draw on these theories. In order to carry it out, you make use of instruments that these same theories legitimize; there is nothing to shock logic in this procedure.

But experiments of application are not the only ones the physicist has to perform; only with their aid can science aid practice, but it is not through them that science creates and develops itself; besides experiments of application, we have experiments of testing.

A physicist disputes a certain law; he calls into doubt a certain theoretical point. How will he justify these doubts? How will he demonstrate the inaccuracy of the law? From the proposition under indictment he will derive the prediction of an experimental fact; he will bring into existence the conditions under which this fact should be produced; if the predicted fact is not produced, the proposition which served as the basis of the prediction will be irremediably condemned.

F. E. Neumann assumed that in a ray of polarized light the vibration is parallel to the plane of polarization, and many physicists have doubted this proposition. How did O. Wiener undertake to transform this doubt into a certainty in order to condemn Neumann's proposition? He deduced from this proposition the following consequence: If we cause a light beam reflected at 45° from a plate of glass to interfere with the incident beam polarized perpendicularly to the plane of incidence, there ought to appear alternately dark and light interference bands parallel to the reflecting surface; he brought about the conditions under which these bands should have been produced and showed that the predicted phenomenon did not appear, from which he concluded that Neumann's proposition is false, viz., that in a polarized ray of light the vibration is not parallel to the plane of polarization.

Such a mode of demonstration seems as convincing and as irrefutable as the proof by reduction to absurdity customary among mathematicians; moreover, this demonstration is copied from the reduction to absurdity, experimental contradiction playing the same role in one as logical contradiction plays in the other.

Indeed, the demonstrative value of experimental method is far from being so rigorous or absolute: the conditions under which it functions are much more complicated than is supposed in what we have just said; the evaluation of results is much more delicate and subject to caution.

A physicist decides to demonstrate the inaccuracy of a proposition; in order to deduce from this proposition the prediction of a phenomenon and institute the experiment which is to show whether this phenomenon is or is not produced, in order to interpret the results of this experiment and establish that the predicted phenomenon is not produced, he does not confine himself to making use of the proposition in question; he makes use also of a whole group of theories accepted by him as beyond dispute. The prediction of the phenomenon, whose nonproduction is to cut off debate, does not derive from the proposition challenged if taken by itself, but from the proposition at issue joined to that whole group of theories; if the predicted phenomenon is not produced, not only is the proposition questioned at fault, but so is the whole theoretical scaffolding used by the physicist. The only thing the experiment teaches us is that among the propositions used to predict the phenomenon and to establish whether it would be produced, there is at least one error; but where this error lies is just what it does not tell us. The physicist may declare that this error is contained in exactly the proposition he wishes to refute, but is he sure it is not in another proposition? If he is, he accepts implicitly the accuracy of all the other propositions he has used, and the validity of his conclusion is as great as the validity of his confidence.

Let us take as an example the experiment imagined by Zenker and carried out by O. Wiener. In order to predict the formation of bands in certain circumstances and to show that these did not appear, Wiener did not make use merely of the famous proposition of F. E. Neumann, the proposition which he wished to refute; he did not merely admit that in a polarized ray vibrations are parallel to the plane of polarization; but he used, besides this, propositions, laws, and hypotheses constituting the optics commonly accepted: he admitted that light consists in simple periodic vibrations, that these vibrations are normal to the light ray, that at each point the mean kinetic energy of the vibratory motion is a measure of the intensity of light, that the more or less complete attack of the gelatine coating on a photographic plate indicates the various degrees of this intensity. By joining these propositions, and many others that would take too long to enumerate, to Neumann's proposition, Wiener was able to formulate a forecast and establish that the experiment belied it. If he attributed this solely to Neumann's proposition, if it alone bears the responsibility for the error this negative result has put in evidence, then Wiener was taking all the other propositions he invoked as beyond doubt. But this assurance is not imposed as a matter of logical necessity; nothing stops us from taking Neumann's proposition as accurate and shifting the weight of the experimental contradiction to some other proposition of the commonly accepted optics; as H. Poincaré has shown, we can very easily rescue Neumann's hypothesis from the grip of Wiener's experiment on the condition that we abandon in exchange the hypothesis which takes the mean kinetic energy as the measure of the light intensity; we may, without being contradicted by the experiment, let the vibration be parallel to the plane of polarization, provided that we measure the light intensity by the mean potential energy of the medium deforming the vibratory motion.

These principles are so important that it will be useful to apply them to another example; again we choose an experiment regarded as one of the most decisive ones in optics.

We know that Newton conceived the emission theory for optical phenomena. The emission theory supposes light to be formed of extremely thin projectiles, thrown out with very great speed by the sun and other sources of light; these projectiles penetrate all transparent bodies; on account of the various parts of the media through which they move, they undergo attractions and repulsions; when the distance separating the acting particles is very small these actions are very powerful, and they vanish when the masses between which they act are appreciably far from each other. These essential hypotheses joined to several others, which we pass over without mention, lead to the formulation of a complete theory of reflection and refraction of light; in particular, they imply the following proposition: The index of refraction of light passing from one medium into another is equal to the velocity of the light projectile within the medium it penetrates, divided by the velocity of the same projectile in the medium it leaves behind.

This is the proposition that Arago chose in order to show that the theory of emission is in contradiction with the facts. From this proposition a second follows: Light travels faster in water than in air. Now Arago had indicated an appropriate procedure for comparing the velocity of light in air with the velocity of light in water; the procedure, it is true, was inapplicable, but Foucault modified the experiment in such a way that it could be carried out; he found that the light was propagated less rapidly in water than in air. We may conclude from this, with Foucault, that the system of emission is incompatible with the facts.

I say the *system* of emission and not the *hypothesis* of emission; in fact, what the experiment declares stained with error is the whole group of propositions accepted by Newton, and after him by Laplace and Biot, that is, the whole theory from which we deduce the relation between the index of refraction and the velocity of light in various media. But in condemning this system as a whole by declaring it stained with error, the experiment does not tell us where the error lies. Is it in the fundamental hypothesis that light consists in projectiles thrown out with great speed by luminous bodies? Is it in some other assumption concerning the actions experienced by light corpuscles due to the media through which they move? We know nothing about that. It would be rash to believe, as Arago seems to have thought, that Foucault's experiment condemns once and for all the very hypothesis of emission, i.e., the assimilation of a ray of light to a swarm of projectiles. If physicists had attached some value to this task, they would undoubtedly have succeeded in founding on this assumption a system of optics that would agree with Foucault's experiment.

In sum, the physicist can never subject an isolated hypothesis to experimental test, but only a whole group of hypotheses; when the experiment is in disagreement with his predictions, what he learns is that at least one of the hypotheses constituting this group is unacceptable and ought to be modified; but the experiment does not designate which one should be changed.

We have gone a long way from the conception of the experimental method arbitrarily held by persons unfamiliar with its actual functioning. People generally think that each one of the hypotheses employed in physics can be taken in isolation, checked by experiment, and then, when many varied tests have established its validity, given a definitive place in the system of physics. In reality, this is not the case. Physics is not a machine which lets itself be taken apart; we cannot try each piece in isolation and, in order to adjust it, wait until its solidity has been carefully checked. Physical science is a system that must be taken as a whole; it is an organism in which one part cannot be made to function except when the parts that are most remote from it are called into play, some more so than others, but all to some degree. If something goes wrong, if some discomfort is felt in the functioning of the organism, the physicist will have to ferret out through its effect on the entire system which organ needs to be remedied or modified without the possibility of isolating this organ and examining it apart. The watchmaker to whom you give a watch that has stopped separates all the wheelworks and examines them one by one until he finds the part that is defective or broken. The doctor to whom a patient appears cannot dissect him in order to establish his diagnosis; he has to guess the seat and cause of the ailment solely by inspecting disorders affecting the whole body. Now, the physicist concerned with remedying a limping theory resembles the doctor and not the watchmaker.

3 A "Crucial Experiment" Is Impossible in Physics

Let us press this point further, for we are touching on one of the essential features of experimental method, as it is employed in physics.

Reduction to absurdity seems to be merely a means of refutation, but it may become a method of demonstration: in order to demonstrate the truth of a proposition it suffices to corner anyone who would admit the contradictory of the given proposition into admitting an absurd consequence. We know to what extent the Greek geometers drew heavily on this mode of demonstration.

Those who assimilate experimental contradiction to reduction to absurdity imagine that in physics we may use a line of argument similar to the one Euclid employed so frequently in geometry. Do you wish to obtain from a group of phenomena a theoretically certain and indisputable explanation? Enumerate all the hypotheses that can be made to account for this group of phenomena; then, by experimental contradiction eliminate all except one; the latter will no longer be a hypothesis, but will become a certainty.

Suppose, for instance, we are confronted with only two hypotheses. Seek experimental conditions such that one of the hypotheses forecasts the production of one phenomenon and the other the production of quite a different effect; bring these conditions into existence and observe what happens; depending on whether you observe the first or the second of the predicted phenomena, you will condemn the second or the first hypothesis; the hypothesis not condemned will be henceforth indisputable; debate will be cut off, and a new truth will be acquired by science. Such is the experimental test that the author of the *Novum Organum* called the "*fact of the cross*, borrowing this expression from the crosses which at an intersection indicate the various roads."

We are confronted with two hypotheses concerning the nature of light; for Newton, Laplace, or Biot light consisted of projectiles hurled with extreme speed, but for Huygens, Young, or Fresnel light consisted of vibrations whose waves are propagated within an ether. These are the only two possible hypotheses as far as one can see: either the motion is carried away by the body it excites and remains attached to it, or else it passes from one body to another. Let us pursue the first hypothesis; it declares that light travels more quickly in water than in air; but if we follow the second, it declares that light travels more quickly in air than in water. Let us set up Foucault's apparatus; we set into motion the turning mirror; we see two luminous spots formed before us, one colorless, the other greenish. If the greenish band is to the left of the colorless one, it means that light travels faster in water than in air, and that the hypothesis of vibrating waves is false. If, on the contrary, the greenish band is to the right of the colorless one, that means that light travels faster in air than in water, and that the hypothesis of emissions is condemned. We look through the magnifying glass used to examine the two luminous spots, and we notice that the greenish spot is to the right of the colorless one; the debate is over; light is not a body, but a vibratory wave motion propagated by the ether; the emission hypothesis has had its day; the wave hypothesis has been put beyond doubt, and the crucial experiment has made it a new article of the scientific credo.

What we have said in the foregoing paragraph shows how mistaken we should be to attribute to Foucault's experiment so simple a meaning and so decisive an importance; for it is not between two hypotheses, the emission and wave hypotheses, that Foucault's experiment judges trenchantly; it decides rather between two sets of theories each of which has to be taken as a whole, i.e., between two entire systems, Newton's optics and Huygens' optics.

But let us admit for a moment that in each of these systems everything is compelled to be necessary by strict logic, except a single hypothesis; consequently, let us admit that the facts, in condemning one of the two systems, condemn once and for all the single doubtful assumption it contains. Does it follow that we can find in the "crucial experiment" an irrefutable procedure for transforming one of the two hypotheses before us into a demonstrated truth? Between two contradictory theorems of geometry there is no room for a third judgment; if one is false, the other is necessarily true. Do two hypotheses in physics ever constitute such a strict dilemma? Shall we ever dare to assert that no other hypothesis is imaginable? Light may be a swarm of projectiles, or it may be a vibratory motion whose waves are propagated in a medium; is it forbidden to be anything else at all? Arago undoubtedly thought so when he formulated this incisive alternative: Does light move more quickly in water than in air? "Light is a body. If the contrary is the case, then light is a wave." But it would be difficult for us to take such a decisive stand; Maxwell, in fact, showed that we might just as well attribute light to a periodical electrical disturbance that is propagated within a dielectric medium.

Unlike the reduction to absurdity employed by geometers, experimental contradiction does not have the power to transform a physical hypothesis into an indisputable truth; in order to confer this power on it, it would be necessary to enumerate completely the various hypotheses which may cover a determinate group of phenomena; but the physicist is never sure he has exhausted all the imaginable assumptions. The truth of a physical theory is not decided by heads or tails.

[. . .]

8 Are Certain Postulates of Physical Theory Incapable of Being Refuted by Experiment?

We recognize a correct principle by the facility with which it straightens out the complicated difficulties into which the use of erroneous principles brought us.

If, therefore, the idea we have put forth is correct, namely, that comparison is established necessarily between the *whole* of theory and the *whole* of experimental facts, we ought in the light of this principle to see the disappearance of the obscurities in which we should be lost by thinking that we are subjecting each isolated theoretical hypothesis to the test of facts.

Foremost among the assertions in which we shall aim at eliminating the appearance of paradox, we shall place one that has recently been often formulated and discussed. Stated first by G. Milhaud in connection with the "*pure bodies*" of chemistry, it has been developed at length and forcefully by H. Poincaré with regard to principles of mechanics; Edouard Le Roy has also formulated it with great clarity.

That assertion is as follows: Certain fundamental hypotheses of physical theory cannot be contradicted by any experiment, because they constitute in reality *definitions*, and because certain expressions in the physicist's usage take their meaning only through them.

Let us take one of the examples cited by Le Roy:

When a heavy body falls freely, the acceleration of its fall is constant. Can such a law be contradicted by experiment? No, for it constitutes the very definition of what is meant by "falling freely." If while studying the fall of a heavy body we found that this body does not fall with uniform acceleration, we should conclude not that the stated law is false, but that the body does not fall freely, that some cause obstructs its motion, and that the deviations of the observed facts from the law as stated would serve to discover this cause and to analyze its effects.

Thus, M. Le Roy concludes, "laws are verifiable, taking things strictly . . . , because they constitute the very criterion by which we judge appearances as well as the methods that it would be necessary to utilize in order to submit them to an inquiry whose precision is capable of exceeding any assignable limit."

Let us study again in greater detail, in the light of the principles previously set down, what this comparison is between the law of falling bodies and experiment.

Our daily observations have made us acquainted with a whole category of motions which we have brought together under the name of motions of heavy bodies; among these motions is the falling of a heavy body when it is not hindered by any obstacle. The result of this is that the words "free fall of a heavy body" have a meaning for the man who appeals only to the knowledge of common sense and who has no notion of physical theories.

On the other hand, in order to classify the laws of motion in question the physicist has created a theory, the theory of weight, an important application of rational mechanics. In that theory, intended to furnish a symbolic representation of reality, there is also the question of "free fall of a heavy body," and as a consequence of the hypotheses supporting this whole scheme free fall must necessarily be a uniformly accelerated motion.

The words "free fall of a heavy body" now have two distinct meanings. For the man ignorant of physical theories, they have their *real* meaning, and they mean what common sense means in pronouncing them; for the physicist they have a *symbolic* meaning, and mean "uniformly accelerated motion." Theory would not have realized its aim if the second meaning were not the sign of the first, if a fall regarded as free by common sense were not also regarded as uniformly accelerated, or *nearly* uniformly accelerated, since common-sense observations are essentially devoid of precision, according to what we have already said.

This agreement, without which the theory would have been rejected without further examination, is finally arrived at: a fall declared by common sense to be nearly free is also a fall whose acceleration is nearly constant. But noticing this crudely approximate agreement does not satisfy us; we wish to push on and surpass the degree of precision which common sense can claim. With the aid of the theory that we have imagined, we put together apparatus enabling us to recognize with sensitive accuracy whether the fall of a body is or is not uniformly accelerated; this apparatus shows us that a certain fall regarded by common sense as a free fall has a slightly variable acceleration. The proposition which in our theory gives its symbolic meaning to the words "free fall" does not represent with sufficient accuracy the properties of the real and concrete fall that we have observed.

Two alternatives are then open to us.

In the first place, we can declare that we were right in regarding the fall studied as a free fall and in requiring that the theoretical definition of these words agree with our observations. In this case, since our theoretical definition does not satisfy this requirement, it must be rejected; we must construct another mechanics on new hypotheses, a mechanics in which the words "free fall" no longer signify "uniformly accelerated motion," but "fall whose acceleration varies according to a certain law."

In the second alternative, we may declare that we were wrong in establishing a connection between the concrete fall we have observed and the symbolic free fall defined by our theory, that the latter was too simplified a scheme of the former, that in order to represent suitably the fall as our experiments have reported it the theorist should give up imagining a weight falling freely and think in terms of a weight hindered by certain obstacles like the resistance of the air, that in picturing the action of these obstacles by means of appropriate hypotheses he will compose a more complicated scheme than a free weight but one more apt to reproduce the details of the experiment; in short, in accord with the language we have previously established, we may seek to eliminate by means of suitable "corrections" the "causes of error," such as air resistance, which influenced our experiment.

M. Le Roy asserts that we shall prefer the second to the first alternative, and he is surely right in this. The reasons dictating this choice are easy to perceive. By taking the first alternative we should be obliged to destroy from top to bottom a very vast theoretical system which represents in a most satisfactory manner a very extensive and complex set of experimental laws. The second alternative, on the other hand, does not make us lose anything of the terrain already conquered by physical theory; in addition, it has succeeded in so large a number of cases that we can bank with interest on a new success. But in this confidence accorded the law of fall of weights, we see nothing analogous to the certainty that a mathematical definition draws from its very essence, that is, to the kind of certainty we have when it would be foolish to doubt that the various points on a circumference are all equidistant from the center.

We have here nothing more than a particular application of the principle set down in Section 2 of this chapter. A disagreement between the concrete facts constituting an experiment and the symbolic representation which theory substitutes for this experiment proves that some part of this symbol is to be rejected. But which part? This the experiment does not tell us; it leaves to our sagacity the burden of guessing. Now among the theoretical elements entering into the composition of this symbol there is always a certain number which the physicists of a certain epoch agree in accepting without test and which they regard as beyond dispute. Hence, the physicist who wishes to modify this symbol will surely bring his modification to bear on elements other than those just mentioned.

But what impels the physicist to act thus is *not* logical necessity. It would be awkward and ill inspired for him to do otherwise, but it would not be doing something logically absurd; he would not for all that be walking in the footsteps of the mathematician mad enough to contradict his own definitions. More than this, perhaps some day by acting differently, by refusing to invoke causes of error and take recourse to corrections in order to reestablish agreement between the theoretical scheme and the fact, and by resolutely carrying out a reform among the propositions declared untouchable by common consent, he will accomplish the work of a genius who opens a new career for a theory.

Indeed, we must really guard ourselves against believing forever warranted those hypotheses which have become universally adopted conventions, and whose certainty seems to break through experimental contradiction by throwing the latter back on more doubtful assumptions. The history of physics shows us that very often the human mind has been led to overthrow such principles completely, though they have been regarded by common consent for centuries as inviolable axioms, and to rebuild its physical theories on new hypotheses.

Was there, for instance, a clearer or more certain principle for thousands of years than this one: In a homogeneous medium, light is propagated in a straight line? Not only did this hypothesis carry all former optics, catoptrics, and dioptrics, whose elegant geometric deductions represented at will an enormous number of facts, but it had become, so to speak, the physical definition of a straight line. It is to this hypothesis that any man wishing to make a straight line appeals, the carpenter who verifies the straightness of a piece of wood, the surveyor who lines up his sights, the geodetic surveyor who obtains a direction with the help of the pinholes of his alidade, the astronomer who defines the position of stars by the optical axis of his telescope. However, the day came when physicists tired of attributing to some cause of error the diffraction effects observed by Grimaldi, when they resolved to reject the law of the rectilinear propagation of light and to give optics entirely new foundations; and this bold resolution was the signal of remarkable progress for physical theory.

4.14

On the Method of Theoretical Physics

Albert Einstein

Albert Einstein (1879–1955), a German-born theoretical physicist who won the Nobel Prize for physics in 1921, is perhaps the best-known scientist of all time. Throughout his life Einstein reflected deeply on the interplay between philosophical issues and the practice of science. In the following selection, which was the Herbert Spencer Lecture delivered at Oxford in 1933, Einstein offers his own perspective on the question that concerned Poincaré: how shall we conceive the relation between empirical data and scientific theories? Einstein's lucid discussion emphasizes his conception of theories as free creations of the human mind, his commitment to realism, and his reliance on simplicity as evidence of truth.

If you wish to learn from the theoretical physicist anything about the methods which he uses, I would give you the following piece of advice: Don't listen to his words, examine his achievements. For to the discoverer in that field, the constructions of his imagination appear so necessary and so natural that he is apt to treat them not as the creations of his thoughts but as given realities.

This statement may seem to be designed to drive my audience away without more ado. For you will say to yourselves, 'The lecturer is himself a constructive physicist; on his own showing therefore he should leave the consideration of the structure of theoretical science to the epistemologist'.

So far as I personally am concerned, I can defend myself against an objection of this sort by assuring you that it was no suggestion of mine but the generous invitation of others which has placed me on this dais, which commemorates a man who spent his life in striving for the unification of knowledge.

But even apart from that, I have this justification for my pains, that it may possibly interest

From Einstein, "On the Method of Theoretical Physics," *Philosophy of Science* (Chicago: University of Chicago Press), Vol. 1, No. 2 (Apr. 1934), pp. 163–9.

you to know how a man thinks about his science after having devoted so much time and energy to the clarification and reform of its principles.

Of course his view of the past and present history of his subject is likely to be unduly influenced by what he expects from the future and what he is trying to realize to-day. But this is the common fate of all who have adopted a world of ideas as their dwelling-place.

He is in just the same plight as the historian, who also, even though unconsciously, disposes events of the past around ideals that he has formed about human society.

I want now to glance for a moment at the development of the theoretical method, and while doing so especially to observe the relation of pure theory to the totality of the data of experience. Here is the eternal antithesis of the two inseparable constituents of human knowledge, Experience and Reason, within the sphere of physics. We honour ancient Greece as the cradle of western science. She for the first time created the intellectual miracle of a logical system, the assertions of which followed one from another with such rigor that not one of the demonstrated propositions admitted of the slightest doubt – Euclid's geometry. This marvellous accomplishment of reason gave to the human spirit the confidence it needed for its future achievements. The man who was not enthralled in youth by this work was not born to be a scientific theorist. But yet the time was not ripe for a science that could comprehend reality, was not ripe until a second elementary truth had been realized, which only became the common property of philosophers after Kepler and Galileo. Pure logical thinking can give us no knowledge whatsoever of the world of experience; all knowledge about reality begins with experience and terminates in it.

Conclusions obtained by purely rational processes are, so far as Reality is concerned, entirely empty. It was because he recognized this, and especially because he impressed it upon the scientific world that Galileo became the father of modern physics and in fact of the whole of modern natural science.

But if experience is the beginning and end of all our knowledge about reality, what role is there left for reason in science? A complete system of theoretical physics consists of concepts and basic laws to interrelate those concepts and of consequences to be derived by logical deduction. It is these consequences to which our particular experiences are to correspond, and it is the logical derivation of them which in a purely theoretical work occupies by far the greater part of the book. This is really exactly analogous to Euclidean geometry, except that in the latter the basic laws are called 'axioms'; and, further, that in this field there is no question of the consequences having to correspond with any experiences. But if we conceive Euclidean geometry as the science of the possibilities of the relative placing of actual rigid bodies and accordingly interpret it as a physical science, and do not abstract from its original empirical content, the logical parallelism of geometry and theoretical physics is complete.

We have now assigned to reason and experience their place within the system of theoretical physics. Reason gives the structure to the system; the data of experience and their mutual relations are to correspond exactly to consequences in the theory. On the possibility alone of such a correspondence rests the value and the justification of the whole system, and especially of its fundamental concepts and basic laws. But for this, these latter would simply be free inventions of the human mind which admit of no a priori justification either through the nature of the human mind or in any other way at all.

The basic concepts and laws which are not logically further reducible constitute the indispensable and not rationally deducible part of the theory. It can scarcely be denied that the supreme goal of all theory is to make the irreducible basic elements as simple and as few as possible without having to surrender the adequate representation of a single datum of experience.

The conception here outlined of the purely fictitious character of the basic principles of theory was in the eighteenth and nineteenth centuries still far from being the prevailing one. But it continues to gain more and more ground because of the ever-widening logical gap between the basic concepts and laws on the one side and the consequences to be correlated with our experiences on the other – a gap which widens progressively with the developing unification of the logical structure, that is with the reduction in the number of the logically independent conceptual elements required for the basis of the whole system.

Newton, the first creator of a comprehensive and workable system of theoretical physics, still believed that the basic concepts and laws of his system could be derived from experience; his phrase 'hypotheses non fingo' can only be interpreted in this sense. In fact at that time it seemed that there was no problematical element in the concepts, Space and Time. The concepts of mass, acceleration, and force and the laws connecting them, appeared to be directly borrowed from experience. But if this basis is assumed, the expression for the force of gravity seems to be derivable from experience; and the same derivability was to be anticipated for the other forces.

One can see from the way he formulated his views that Newton felt by no means comfortable about the concept of absolute space, which embodied that of absolute rest; for he was alive to the fact that nothing in experience seemed to correspond to this latter concept. He also felt uneasy about the introduction of action at a distance. But the enormous practical success of his theory may well have prevented him and the physicists of the eighteenth and nineteenth centuries from recognizing the fictitious character of the principles of his system.

On the contrary the scientists of those times were for the most part convinced that the basic concepts and laws of physics were not in a logical sense free inventions of the human mind, but rather that they were derivable by abstraction, i.e. by a logical process, from experiments. It was the general Theory of Relativity which showed in a convincing manner the incorrectness of this view. For this theory revealed that it was possible for us, using basic principles very far removed from those of Newton, to do justice to the entire range of the data of experience in a manner even more complete and satisfactory than was possible with Newton's principles. But quite apart from the question of comparative merits, the fictitious character of the principles is made quite obvious by the fact that it is possible to exhibit two essentially different bases, each of which in its consequences leads to a large measure of agreement with experience. This indicates that any attempt logically to derive the basic concepts and laws of mechanics from the ultimate data of experience is doomed to failure.

If then it is the case that the axiomatic basis of theoretical physics cannot be an inference from experience, but must be free invention, have we any right to hope that we shall find the correct way? Still more – does this correct approach exist at all, save in our imagination? Have we any right to hope that experience will guide us aright, when there are theories (like classical mechanics) which agree with experience to a very great extent, even without comprehending the subject in its depths? To this I answer with complete assurance, that in my opinion there is *the* correct path and, moreover, that it is in our power to find it. Our experience up to date justifies us in feeling sure that in Nature is actualized the ideal of mathematical simplicity. It is my conviction that pure mathematical construction enables us to discover the concepts and the laws connecting them which give us the key to the understanding of the phenomena of Nature. Experience can of course guide us in our choice of serviceable mathematical concepts; it cannot possibly be the source from which they are derived; experience of course remains the sole criterion of the serviceability of a mathematical construction for physics, but the truly creative principle resides in mathematics. In a certain sense, therefore, I hold it to be true that pure thought is competent to comprehend the real, as the ancients dreamed.

To justify this confidence of mine, I must necessarily avail myself of mathematical concepts. The physical world is represented as a four-dimensional continuum. If in this I adopt a Riemannian metric, and look for the simplest laws which such a metric can satisfy, I arrive at the relativistic gravitation-theory of empty space. If I adopt in this space a vector-field, or in other words, the antisymmetrical tensor-field derived from it, and if I look for the simplest laws which such a field can satisfy, I arrive at the Maxwell equations for free space.

Having reached this point we have still to seek a theory for those parts of space in which the electrical density does not vanish. De Broglie surmised the existence of a wave-field, which could be used to explain certain quantum properties of matter. Dirac found in the 'spinor-field' quantities of a new kind, whose simplest equations make it possible to deduce a great many of the properties of the electron, including its quantum properties. I and my colleague discovered that these 'spinors' constitute a special case of a field of a new sort which is mathematically connected with the metrical continuum of four dimensions, and it seems that they are naturally fitted to describe important properties of the electrical elementary particles.

It is essential for our point of view that we can arrive at these constructions and the laws relating them one with another by adhering to the principle of searching for the mathematically simplest concepts and their connections. In the paucity of the mathematically existent simple field-types and of the relations between them, lies the justification for the theorist's hope that he may comprehend reality in its depths.

The most difficult point for such a field-theory at present is how to include the atomic structure of matter and energy. For the theory in its basic principles is not an atomic one in so far as it operates exclusively with continuous functions of space, in contrast to classical mechanics whose most important feature, the material point, squares with the atomistic structure of matter.

The modern quantum theory, as associated with the names of de Broglie, Schrödinger, and Dirac, which of course operates with continuous functions, has overcome this difficulty by means of a daring interpretation, first given in a clear form by Max Born: – the space functions which appear in the equations make no claim to be a mathematical model of atomic objects. These functions are only supposed to determine in a mathematical way the probabilities of encountering those objects in a particular place or in a particular state of motion, if we make a measurement. This conception is logically unexceptionable, and has led to important successes. But unfortunately it forces us to employ a continuum of which the number of dimensions is not that of previous physics, namely 4, but which has dimensions increasing without limit as the number of the particles constituting the system under examination increases. I cannot help confessing that I myself accord to this interpretation no more than a transitory significance. I still believe in the possibility of giving a model of reality, a theory, that is to say, which shall represent events themselves and not merely the probability of their occurrence. On the other hand, it seems to me certain that we have to give up the notion of an absolute localization of the particles in a theoretical model. This seems to me to be the correct theoretical interpretation of Heisenberg's indeterminacy relation. And yet a theory may perfectly well exist, which is in a genuine sense an atomistic one (and not merely on the basis of a particular interpretation), in which there is no localizing of the particles in a mathematical model. For example, in order to include the atomistic character of electricity, the field equations only need to involve that a three-dimensional volume of space on whose boundary the electrical density vanishes everywhere, contains a total electrical charge of an integral amount. Thus in a continuum theory, the atomistic character could be satisfactorily expressed by integral propositions without localizing the particles which constitute the atomistic system.

Only if this sort of representation of the atomistic structure be obtained could I regard the quantum problem within the framework of a continuum theory as solved.

Part II

Introduction to Part II

1 Logical Positivism

The first half of the twentieth century saw the rise of a philosophical movement called *logical positivism* ("positivism" hereafter). The positivist's conception of science – even now called the *received view*[1] – stands to contemporary philosophy of science much as the Copernican revolution did to subsequent scientific inquiry: much more than a revolution within the discipline, to a large extent it marks the beginning of the discipline itself as an independent field of inquiry. As a result the received view determined, at least for some time thereafter, the issues and problems with which the philosophy of science became concerned as well as the methodology used to investigate them.

Unlike the Copernican revolution, however, from about the 1950s onward just about every facet of the received view has been subjected to withering critique; it is now commonly understood to be more of historical interest than a live philosophical contender. Since positivism's downfall, the philosophy of science has splintered into a variety of competing camps with no single orientation dominating the field. The received view therefore remains the background against which contemporary alternatives have been formulated and developed, and by reference to which those views are often distinguished from one another. In at least this sense, it has dominated much of the philosophy of science in the latter half of the twentieth century as well.

In this introduction to Part II, we will discuss the origin and doctrines of positivism. In Unit 5 we will review the received view of scientific theories in detail. Subsequent criticism and development of alternatives will be the subject of the commentary for Units 6 through 9, which contain works representative of those criticisms and alternatives.

2 Origins of Positivism: The Vienna Circle

Positivism primarily originated in the *Vienna Circle*, a group of like-minded philosophers and scientists meeting in Vienna in the 1920s and 1930s.[2] The circle was led by Moritz Schlick and included Gustav Bergmann, Rudolf Carnap, Herbert Feigl, Philipp Frank, Kurt Gödel, Hans Hahn, Victor Kraft, Karl Menger, Marcel Natkin, Otto Neurath, Olga Hahn-Neurath, Theodor Radakovic, and Friedrich Waismann. The positivists shared certain central tenets that we will discuss in §§5–8 below. But it is worth noting that, as recent historical investigation is making increasingly clear, the Circle (and, in general, those philosophers subsequently identified as positivists) was not a homogenous group in complete theoretical accord as it has often been represented. For example, Schlick, whose views most closely approximate the traditional conception of positivist philosophy, represented the "right wing"; Otto

Neurath, representing the "left wing," differed sharply from Schlick on a number of issues.[3]

In the center was Rudolf Carnap. Carnap was, arguably, the most sophisticated positivist and, unarguably, the most influential. This was, in part, because he showed himself most open to theoretical elaboration and change in response to critique (which often originated with Carnap himself). His early work, *Der Logische Aufbau der Welt* (*The Logical Constitution of the World*),[4] has often been seen as an expression of classical positivist doctrine (e.g., by his most famous critic, W. V. O. Quine).[5] But the *Aufbau* was written before the Circle was formed (and published the year in which the Circle was founded); soon after its publication, Carnap rejected a number of its purportedly positivist doctrines, including definitional reductionism and phenomenalism (see §8 below). More to our present concern, the received view of science, including its development and alteration over the years, is to a great extent Carnap's creation.

3 Origins of Positivism: Other Sources

The members of the Vienna Circle were not, however, the only philosophers that advocated recognizably positivist views. The Berlin Circle,[6] formed by the philosopher Hans Reichenbach, was positivism's second epicenter. Reichenbach preferred "Logical Empiricism" to "Logical Positivism" by way of distinguishing his group from the Vienna Circle, but it is a testament to the doctrinal affinity between the two circles that these two terms are now often used interchangeably.[7] Its members included the philosopher Carl Hempel, the primary author of the received view of scientific explanation (reading 5.2). Hempel also formulated the famous "raven paradox" in confirmation theory (reading 6.2). Finally, notwithstanding his sympathies with positivism, he authored one of the most influential critiques of the verification criterion (reading 6.1).[8] The Berlin group also included the mathematician David Hilbert, the logician Kurt Grelling, and the scientist and probability theorist Richard von Mises. The Vienna and Berlin Circles interacted considerably and co-founded the journal *Erkenntnis* (edited by Carnap and Reichenbach).

The Polish Lvov-Warsaw School of logicians, whose members included Alfred Tarski, also espoused views close to those of the positivists. The English philosopher A. J. Ayer's immensely popular *Language, Truth, and Logic*[9] introduced positivism to the English-speaking philosophical community (and did much to shape its standard characterization). And Ernst Nagel's *The Structure of Science*[10] is arguably the work that launched analytic philosophy of science in the United States.

4 Positivism's Influences

The work of philosopher, physicist, and perceptual psychologist Ernst Mach (of "Mach speed" and "Mach bands") had a great impact on the Vienna Circle (which began its life as the *Ernst Mach Society*). Mach's characterization of science as a system for the economical ordering of sensory experience, rather than as a description of an underlying unobservable reality, is a common theme in the positivists' conception of science, as was his advocacy of a "scientific" philosophy freed of pernicious metaphysics. Mach's philosophy of science also dovetails nicely with the conventionalist theme in Henri Poincaré and Pierre Duhem (readings 4.12 and 4.13, respectively) with which the positivists were very sympathetic (see §6).

The philosopher Ludwig Wittgenstein was not a member of the Vienna Circle, but he met with some of its members (Schlick, Waismann, and Carnap). The Circle read his *Tractatus Logico-Philosophicus*,[11] finding much to their liking in both Wittgenstein's verificationism (see §5) and his view of logic as merely tautological and so empty of factual content (see §6).

The positivists were deeply impressed by the achievements in logic by Gottlieb Frege and Bertrand Russell (and in particular by the seminal work in logic and mathematics by Russell and A. N. Whitehead, *Principia Mathematica*).[12] They also endorsed Frege's and Russell's *logicist* approach in the foundations of mathematics, according to which mathematical truths can be reduced to logical truths. The trajectory of the positivist's philosophical development (particularly that of Carnap) was later heavily influenced by the stunning discoveries concerning the relation between logic and mathematics made by Kurt Gödel, whose incompleteness theorem seriously threatened the logicist program, as well as the formal semantics and "correspondence" theory of truth developed by Tarski of the Warsaw School.

Many of the positivists were also well-versed in recent developments in mathematics and science (some were themselves mathematicians and scientists). They were impressed by the development of non-Euclidean geometries, and particularly by their recent application in Albert Einstein's theory of relativity. (Einstein knew a number of the positivists personally, including Schlick and Reichenbach, and discussed many issues in the philosophy of science with them.) The positivists saw in his (empirical) theory a decisive refutation of Kant's claim that knowledge of the Euclidean geometry of space is synthetic a priori – that is, factual information known independently of experience.[13]

The positivists are also traditionally seen as heavily influenced by the empiricism of J. S. Mill and the earlier *British Empiricists* John Locke, George Berkeley, and David Hume. Indeed, positivism has often been treated as a kind of logico-linguistic re-formulation of British Empiricist philosophy. There is no question that the British Empiricists were well-respected among the positivists. The positivists shared the British Empiricists' denial of synthetic a priori knowledge. The positivists also endorsed certain of the British Empiricists' (especially Hume's) more specific doctrines, such as the account of causation as constant conjunction and of laws as empirical generalizations (reading 3.8). The characterization of positivism as little more than British Empiricism in linguistic clothing – a view propagated primarily by Ayer's *Language, Truth, and Logic* – has, however, been seriously criticized recently by historians of the movement (see §8).

5 Verificationism and the Rejection of Metaphysics

The positivists were empiricists: they recognized only sensory experience as a source of factual information, and so denied the synthetic a priori knowledge affirmed by the Rationalists (René Descartes, Gottfried Wilhelm Leibniz, and Baruch Spinoza) and Kant. Empirical science, they insisted, is the means by which the information from this source is best exploited. Any claim to knowledge of metaphysical facts – facts that purportedly concern realities inaccessible to empirical inquiry – is therefore rejected, and with it much philosophical system-building.

The positivists did not, however, reject metaphysics by denying it. That would be to call it false, which presupposes that it is meaningful. But metaphysics, they claimed, is not merely false; it is meaningless. They framed their empiricism and their rejection of metaphysics (and ultimately their entire philosophical outlook), not as a doctrine about the truth concerning the way things are, but instead as a doctrine about language, concerning what can be meaningfully said.

They thereby did much to usher in the *linguistic turn*. The linguistic turn was (and, arguably, still is to a great extent) a prominent philosophical approach in which philosophical issues are interpreted as concerning language and meaning, and in which philosophical method is conceptual or linguistic analysis, often accomplished by applying the analytical tools provided by developments in symbolic logic.[14]

Sometimes, the positivists said, metaphysics is meaningless because it violates the syntactic and semantic rules of language (a charge Carnap famously leveled at Martin Heidegger's claims concerning "nothingness").[15] Such violations occasionally arise through a confusion of levels or categories. "Caesar is a prime number," for example, is meaningless rather than false because it incoherently applies a property that is only properly applicable to numbers (viz., being prime) to a person.

At other times, the metaphysician might be misled through over-attention to grammatical form in natural language. "Santa Claus does not exist," for example, has been taken by some philosophers to resemble "The Eiffel Tower is not short," both of which seem to deny that something has a certain property. But that something – Santa or Tower – must somehow be available, it seems, in order to fail to possess that property. Santa must then in some sense be, or 'subsist', albeit while not existing. There are therefore subsistent but non-existent things, some of which could not possibly exist (e.g., a round square).

But this all takes the subject-predicate form too seriously; the original claim could have been more clearly expressed as "There is no such thing as Santa Claus," which avoids the suggestion of a shadowy subsisting but non-existent Santa.[16] In a logically perspicuous artificial language, such "pseudo-sentences" could not be formulated. The engineering of such languages was considered, especially by Carnap, to be a fundamental task of the philosopher.

The charge of meaninglessness was, however, most famously (or infamously) leveled against metaphysics by appeal to the verification criterion of cognitive significance. This criterion required of a cognitively significant expression that its truth or falsehood makes an empirically detectable difference.[17] The criterion underwent considerable revision over the years, but it remained central to the positivist's program throughout.

In light of this criterion, empirical science presents a dilemma. On the one hand, there is – by this time, at any rate – no question that scientific theorizing is not a mere cataloguing of empirical data. Nor does it only consist in inductive generalizations from observations of particular As that are Bs to the claim that all (or most) As are B. Scientists have found it absolutely indispensable, it seems, to posit various seemingly unobservable realities – atoms, electromagnetic waves, tectonic plates, and so on – behind the empirically discernible phenomena, by appeal to which realities the phenomena can be predicted and explained. In light of the high esteem in which the positivists held the sciences, they could hardly propose that such seemingly metaphysical posits be eliminated; doing so would have crippled the development and application of the theories that display such spectacular empirical success. On the other hand, the positivists' criterion of meaningfulness seems to cast suspicion on the positing of such theoretical entities so far removed from the reach of the senses. Therefore, a condition of acceptability for an adequate criterion of empirical significance – a criterion of the criterion, as it were – was that such apparent positing be respectable by its lights, so that the scientific baby is not thrown out with the metaphysical bathwater.

This dilemma played a crucial role in the development of the received view, which development amounts to a series of attempts to explain how the abstract theoretical terminology (the "theoretical vocabulary") in advanced scientific theories enjoyed empirical significance in virtue of its relation to the "observation vocabulary" in which experimental results are couched, relations supposedly not enjoyed by the metaphysical doctrines the positivists repudiated.

Indeed, an early liberalization of the criterion was an immediate result of the dilemma. Scientific laws are universal in scope, typically with a very large, and potentially infinite, domain of application. "Copper expands when heated," for example, concerns all copper at all times and places. But the samples of copper actually examined for the property of thermal expansion amount to far less than the quantity of that metal that the law concerns. So no observations that have, or ever will, be made can verify the law, if by "verify" we mean "demonstrates the truth of". So verification could not mean conclusive verification in this sense, but only confirmation. (The term "verification" is sometimes reserved for the conclusive sense; so understood, the verification criterion was liberalized to a confirmation criterion.) The empiricist criterion and its development within the received view will be discussed further in the Unit 5 commentary.

6 Analyticity, Conventionalism, Logic, and Mathematics

Logic and mathematics also present the empiricist with a dilemma: they are indispensable for advanced empirical scientific theorizing – mathematics is, as Galileo said, the language of science – and so cannot be dismissed as metaphysical nonsense. But mathematics and logic seem to display an independence from experience that is difficult to make sense of, especially for the empiricist who insists that the only information is empirical information. This independence is manifested in their subject matter: mathematics seems to concern otherworldly abstract mathematical objects, and both mathematics and logic seem to deliver necessary truths instead of the merely contingent truths presented to us in experience. This independence is also manifested in the means by which we apparently come to know the necessary truths of logic and mathematics, characterized as an inner grasping or a priori intuition of "clear and distinct" or "self-evident" truths discerned by the "light of reason."

The positivists' resolution of this dilemma was very much in line with their general linguistic orientation. They first adopted a logicist conception of mathematics, according to which mathematics is reducible to logic.[18] Second, they proposed a characterization of logical (and, therefore, mathematical) propositions as *analytic*, that is, as true solely in virtue of their meaning. Finally, they endorsed a conventionalist treatment of analyticity, according to which analytic truth originates in (implicit or explicit) linguistic conventions.[19]

The end result is that all truths of logic and mathematics turn out to be merely consequences of linguistic convention, and therefore devoid of factual content. They are necessary, but only because their analytic status ensures that they can only be coherently denied if their meaning is altered. The locus of their necessity is in the linguistic convention and its consequences rather than in a necessitating fact somehow out there in the world (or in an abstract world of its own). They are also a priori, but only because the truth of an analytic sentence is a product solely of its meaning, and so requires no further experience (other than that required for grasp of its meaning). Notwithstanding the indispensability of logic and mathematics to science, experience alone remains the sole source of (factual) information, which information is contingent and a posteriori.[20]

The positivists' conventionalism – inherited in part from Poincaré and Duhem (readings 4.12 and 4.13, respectively) – played a crucial role in the character and development of the positivists' views in general, and of their view of science in particular. If much of the abstract theorizing in science amounts to the setting up of conventions instead of being answerable to an independent state of affairs, it becomes possible to recognize more than one set of conventions that, while competing on grounds of simplicity and economy of expression and calculation, do not compete as possible representations of matters of fact. The positivists could, as a result, view with equanimity the development of both multiple geometries and even multiple systems of logic, and reject the seemingly metaphysical question as to which is really correct. Of course, once a system was settled on (e.g., a space–time metric), specific facts within the *linguistic framework* (e.g., the distance from one point to another) have *internal* determinate answers. But the *external* question concerning which such framework was right was denounced as meaningless.[21]

This pluralism of alternative frameworks played an increasingly prominent role in Carnap's work in particular (although the theme can be traced back to his *Aufbau*). It was expressed in his *principle of tolerance* of alternative frameworks:

> it is not our business to set up prohibitions, but to arrive at conventions . . . In logic there are no morals. Everyone is at liberty to build up his own logic, i.e., his own language, as he wishes. All that is required of him is that, if he wishes to discuss it, he must state his methods clearly, and give syntactical rules instead of philosophical arguments."[22]

A famous expression of this attitude later in his career is his essay, "Empiricism, Semantics, and Ontology" (reading 5.3).

7 Philosophy of Philosophy

Another consequence of the positivists' logico-linguistic formulation of empiricism was their view of philosophy itself. They denied that philosophy has any aspect of reality as its particular concern. They also denied that philosophy employed a unique non-empirical method of inquiry into matters of fact; empirical scientific inquiry alone delivers factual information. They correspondingly denied philosophy a legislative role as overseer of the scientific enterprise; the philosopher, they insisted, is in no position to challenge the methods or second-guess the results of scientific inquiry.

So far, their position is entirely in accord with the naturalism that has come to dominate much of the contemporary philosophical scene. Naturalists also reject the *first-philosophical* hierarchical relation between philosophy and science advocated by Descartes, according to which philosophy is prior to science and lays its foundations.

But unlike contemporary naturalists, the positivists did not view philosophy as continuous with, or an extension of, scientific inquiry. Instead, they assigned the philosopher a meta-theoretical role. Philosophy investigates the language of science, with attention to the syntactic, semantic, logical, and evidential structures and relations within scientific theories and between those theories and the observations that constitute their evidence. Along the way it resolves various philosophical disputes by showing that they are either really empirical issues to be handed over to the sciences, or transformable into questions concerning the logical structure of the language of science (and so answerable by logical analysis), or proposals of alternative linguistic structures or frameworks that empirical science might find of use, or dismissible as pseudo-questions for one or the other of the reasons described in §5 above.

Although philosophy's task so understood appears to be, for the most part, a straightforward description of the science that is its subject matter, the "explications" or "rational reconstructions" that the positivists produced often resembled actual science only through positivist-colored glasses. In fact, a common criticism, wielded with explosive effect by Thomas Kuhn (see reading 7.2 and Unit 7 commentary) is that the positivists rarely examined actual scientific history and practice at all, choosing instead to pursue their logical investigations in splendid isolation from historical and social context.

Whether this charge is appropriately applied to every positivist at all times is subject to dispute. Virtually all members of the Vienna Circle were, after all, either trained scientists or scientifically very well informed, and shaped their views explicitly in light of recent developments in relativity theory and quantum mechanics. But it cannot be denied that the work of their more formalist-oriented members – particularly Carnap – demonstrated little contact with historical record, thereby laying them open to the charge of having ignored crucial characteristics of scientific inquiry that are revealed when that record is examined.

8 Contested Themes

Positivism was advertised within the Anglo-American philosophical community primarily through Ayer's *Language, Truth, and Logic.* Ayer's characterization of the movement highlighted a number of characteristics of positivism that have come to be associated with the movement in addition to those described above, all of which became the subject of scathing critique. One such characteristic was *phenomenalistic reductionism*, the view that all meaningful discourse must be capable of being translated into language that describes sensory experiences (e.g., "There is now an experience of a red expanse in the visual field"). Phenomenalism is now as dead as a philosophical view can be, in part because no adequate scheme of translation into such a language was successfully developed. Carnap attempted it with great ingenuity and technical sophistication in the *Aufbau*; and even he conceded the failure of that project.

However, precisely because he did so, and did so well before the heyday of the Vienna Circle, it is at best misleading to suggest that either phenomenalism or definitional reductionism were an integral part of the positivists' philosophical project. Even in the *Aufbau*, Carnap recognized legitimate alternative systems with different linguistic bases, including the "physicalistic" language of ordinary discourse concerning mid-sized observable everyday objects (e.g., "The pointer on the dial is pointing at '3' "). And, indeed, Carnap came to favor the physicalistic over the phenomenalistic language soon after the *Aufbau*. The crucial aspect of Carnap's view is that he viewed the choice between the phenomenalistic and physicalistic bases as a pragmatic matter concerning the most convenient linguistic framework for science. He did not view the choice as necessitated by a framework-independent recognition of the phenomenalistic language as somehow closer to true experiential reality, and so as more epistemologically fundamental.[23]

Much the same has been said concerning the idea that the positivists were reductionists, at least in the sense of a definitional translation that exhausts the meaning of the translated expression. Not since the *Aufbau* did Carnap suggest that it was possible to translate, without remainder, all meaningful discourse into the basis (whatever language the basis might be expressed in).[24] But he never saw this as demonstrating the impossibility of representing the empirical significance of scientific theories, a project which occupied much of the rest of his career.

Much the same has also been said of the suggestion, implicit in Ayer's work, that the positivists were epistemic foundationalists. Carnap's *The Logical Syntax of Language*, arguably his most important philosophical work, is explicitly coherentist, as were the views of the other members of the 'left wing' of the circle. Neurath's famous parable of the mariner afloat on the seas while making repairs to his ship is in fact now taken to be a classic expression of coherentism, used to great rhetorical effect by positivism's most famous critic, W. V .O. Quine (see reading 6.3 and Unit 6 commentary).

9 Positivism as History

These interpretive issues are still being debated. Fortunately we need not attempt to resolve them for two reasons: first, our primary concern here is with the positivists' philosophy of science, which allows us to sidestep at least some of these debates; second, we are interested in the reactions of subsequent philosophers of science to what they thought to be the received view, and perception is reality at least for that purpose. We will discuss the received view and its developments in the Unit 5 commentary, and the criticisms that brought the downfall of the received view, as well as some subsequent developments and alternatives, in the commentary to Units 6–9.

Perhaps the received view of the received view, and of positivism in general, is oversimplified or misleading in various respects. But however these interpretive debates get resolved, the overall verdict is in: neither positivism in general nor the received view in particular constitutes a viable philosophical program.

Such a status has a silver lining: for when a philosophical standpoint is viewed as an increasingly distant historical epoch rather than a proximate threat, its treatment is inevitably kinder. Historians can reveal the complexities of the issues with which the movement's participants dealt, and the hitherto unrecognized sophistication of their responses to those issues. They can discern the differences and divisions among those participants that have been pasted over by reference to *the* movement and doctrine. And they can acknowledge the wide variety of motives and purposes generating both the participants' early views and their later articulation and refinement. Such a historically nuanced re-examination of positivism is currently flowering; contemporary philosophers might, in the many details thereby revealed, even now find inspiration.

Notes

1 According to Fredrick Suppe, the name "the received view" originated in Putnam (reading 6.5), although Suppe's own justly popular, comprehensive discussion of the received view and its development in Suppe ("The Search for Philosophic Understanding of Scientific Theories," in Suppe, F., 1977, *The Structure of Scientific Theories*, Chicago: University of Illinois Press, pp. 3–232) standardized its use.
2 See next section for positivism's other points of origin.
3 These differences concerned primarily how much the empiricist criterion of significance should be weakened or "liberalized". See §5 below.
4 Carnap 1928.
5 See reading 6.3.
6 Also known as the "Berlin Society" or "Berlin School."
7 Sometimes, however, "Logical Empiricism" is used as a more encompassing classification, with 'Logical Positivism' reserved for the more radical doctrines of the early Vienna Circle.
8 For discussion of Hempel on explanation, see §§3 and 4 of Unit 5 commentary; the raven paradox and his critique of the verification criterion are discussed in §§1 and 2 of Unit 6 commentary.
9 Ayer 1946.
10 Nagel, E., 1979, *The Structure of Science: Problems in the Logic of Scientific Explanation*, 2nd edn., Indianapolis: Hackett.
11 Wittgenstein, Ludwig, 1921, "Logisch-Philosophische Abhandlung," *Annalen der Nat. u. K. Philosophie* 14, pp. 185–262, trans. C. Ogden, *Tractatus Logico-Philosophicus*, London, 1922; rev. edn. 1933; repr. London: Routledge, Kegan Paul, 1983.
12 Whitehead, Alfred North, and Bertrand Russell, 1910, 1912, 1913, *Principia Mathematica*, 3 vols., Cambridge: Cambridge University Press; 2nd edn, 1925 (Vol. 1), 1927 (Vols. 2, 3). Abridged as *Principia Mathematica to *56*, Cambridge: Cambridge University Press, 1962.
13 *Synthetic* propositions are contrasted with *analytic* propositions. An analytic proposition is true solely in virtue of meaning. A synthetic proposition is not; it therefore requires factual knowledge of some sort to establish its truth or falsehood. A priori knowledge is knowledge that is available independently of experience. The most common examples of putative a priori knowledge are logic and mathematics (see §6 below). A priori knowledge is contrasted with a posteriori knowledge, which is only available through experience.

14 Another point of origin of the linguistic turn was the *ordinary language* philosophy of the later Wittgenstein, Gilbert Ryle, J. L. Austin, Peter Strawson, and Norman Malcolm. The positivists differed from the ordinary language philosophers primarily in that the latter were concerned with the correct understanding of natural language and the dispelling of conceptual confusions arising within it, whereas the positivists were at least as interested in the development of pristine artificial languages in which such confusions cannot be expressed.
15 See Rudolf Carnap, 1932, in Ayer 1959, pp. 60–81.
16 At least, it avoids such a suggestion as it arises from the existence predicate. The name "Santa Claus" itself, however, still has to be contended with, since one might think that meaningful names require referents. There are various options here, discussion of which is unfortunately well beyond the scope of this book.
17 This is intentionally vague in an attempt to accommodate the many forms the criterion took.
18 There are some exceptions, such as Ayer 1946 [1936].
19 These last two steps are not often distinguished, but they are distinct. One might say that analytic truths are true in virtue of meaning relations, but that meanings and their relations are independent Platonic existences rather than products of human acts of convention.
20 On the distinction between a priori and a posteriori propositions, see note 13.
21 See §9 of Unit 5 commentary.
22 Carnap 1937, §17.
23 This latter conception of the choice between phenomenalism and physicalism – one favoring phenomenalism – does, however, more closely describe the position of the "right wing" of the Circle, particularly that of Schlick.
24 See §7 of Unit 5 commentary.

Suggestions for Further Reading

Carnap's early attempt at phenomenalistic reductionism is Carnap 1928. His most influential work is arguably Carnap 1937[1934]; an accessible summary is Carnap 1935. The collection of essays by various positivists in Ayer 1959 includes the English translation of Carnap's most widely-cited polemic against metaphysics, "The Elimination of Metaphysics through the Logical Analysis of Language" (Carnap 1932). Ayer's collection also includes Moritz Schlick's classic "Positivism and Realism" (Schlick 1932) and a number of other important essays by the positivists. Ayer's *Language, Truth, and Logic* (Ayer 1946) is a very influential and accessible popularization of positivism. A number of Michael Friedman's essays collected in Friedman 1999 played a significant role in initiating the "revisionist" reading of positivism (which rejects the traditional phenomenalistic foundationalist interpretation encouraged, in part, by Ayer 1946). Another important revisionist work, with particular attention to the interpretation of Carnap 1928, is Richardson 1988.

Ayer, A. J., 1946[1936], *Language, Truth and Logic*, 2nd edn. London: Gollancz.
Ayer, A. J. (ed.), 1959, *Logical Positivism*. New York: Free Press.
Carnap, R., 1928, *Der logische Aufbau der Welt*, Berlin: Bernary. Trans. R.A. George, *The Logical Structure of the World*, 1967. Berkeley: University of California Press.
Carnap, R., 1932, "Überwindung der Metaphysik durch logische Analyse der Sprache," *Erkenntnis* 2: 219–41. Trans. A. Pap, "The Elimination of Metaphysics through Logical Analysis of Language", in Ayer 1959, pp. 60–81.
Carnap, R., 1935, *Philosophy and Logical Syntax*. London: Kegan Paul.
Carnap, R., 1937[1934], *Logische Syntax der Sprache*. Wien: Springer. Rev. edn. Trans. A. Smeaton, 1937, *The Logical Syntax of Language*. London: Kegan Paul.
Friedman, M., 1999, *Reconsidering Logical Positivism*. Cambridge: Cambridge University Press.
Richardson, A., 1988, *Carnap's Construction of the World*. Cambridge: Cambridge University Press.
Schlick, Moritz, 1932, "Positivismus und Realismus", *Erkenntnis* 3: 1–31. Trans. P. Heath, "Positivism and Realism" in Ayer 1959, pp. 82–107.

Unit 5

The Received View

1 Introduction

The received view is the conception of scientific theories developed by the logical positivists, which dominated the philosophy of science during the first half of the twentieth century. It was subsequently subjected to devastating critique, with the result that it is no longer viable. No one standpoint has risen as its successor; instead, a number of alternative approaches are now on the table, distinguishable by what in the received view they reject.

The history sketched in the last paragraph is commonly accepted, and there is certainly much truth in it. But it does obscure two crucial points. First, the received view underwent considerable elaboration and alteration during its tenure, so that criticisms applicable to early versions do not apply to later ones. Second, the title "the received view" is misleading in that it suggests one unified package of inseparable doctrines. But at least many of those doctrines are separable and were, in fact, separated. Certain advocates of the view endorsed only some of them, and not always the same ones. And a number of those doctrines survived long after the received view's downfall; some, indeed, are accepted by many present-day philosophers of science.

The received view is represented primarily by the first reading by Rudolf Carnap, excerpted from his *Philosophical Foundations of Physics*. The second reading by Carl Hempel, excerpted from his *Aspects of Scientific Explanation*, contains Hempel's classic presentation of the fundamentals of the received view of explanation including the *deductive-nomological* (D-N) and *inductive-statistical* (I-S) models. It also includes Hempel's discussion of the problem of ambiguity afflicting the I-S model and his *requirement of maximal specificity* that, he hoped, would solve that problem.

The third reading is Carnap's essay, "Empiricism, Semantics, and Ontology," which represents his attitude toward a variety of metaphysical disputes, including that between realists and anti-realists in the philosophy of science (see Unit 9). The views expressed in this essay are also the primary targets of one of the most famous critics of positivism, W. V. O. Quine.[1]

The fourth and fifth readings contain two of the most famous responses to David Hume's problem of induction.[2] There is no standard positivist response to Hume's problem; they more typically just dismissed it as a pseudo-problem. But these two responses are arguably those most compatible with the overall positivist standpoint. The first is by Hans Reichenbach, who was himself a positivist (specifically, the founder of the Berlin Society; see §3 of the Part II commentary). The author of the second response, Peter Strawson, was not a positivist. But he was one of the ordinary-language philosophers that shared the positivist's inclination to characterize many philosophical disputes as resulting from confusions concerning language and resolvable by conceptual analysis.[3]

His response to the problem of induction is indeed virtually identical to (although more detailed than) that of the English positivist, A. J. Ayer in Ayer's *Language, Truth, and Logic.*[4]

2 Empirical Laws and Confirmation

In keeping with their general logico-linguistic orientation (cf. §§4–8 of Part II commentary), the positivists understood a scientific theory to consist in a collection of sentences. These sentences exhibit the structure of an axiom system from which theorems can be derived, much as one might derive various theorems concerning, for example, the properties of a triangle from the axioms of Euclidean geometry. Unlike (pure) geometry, however, a scientific theory has empirical consequences; the scientist derives predictions from the axioms that can then be checked by observation.[5]

The view that a scientific theory is a linguistic entity, in particular a formalized axiom system, was a tenet of the received view that survived long after that view was abandoned. It is now called the *syntactic account* of scientific theories, and is contrasted with the *semantic account*, according to which a scientific theory is a set of models picked out by a linguistic formulation, rather than the linguistic formulation itself. The relative merits of these two approaches continue to be assessed.[6]

An axiom system requires a logical scaffolding in order to permit the derivation of theorems. The system employed by almost all versions of the received view[7] is first-order predicate logic with identity, now standard among philosophers and logicians. In addition to the logical system, a vocabulary is required for the description of observations. The positivists assumed (without much argument) that there is an isolable *observation-language* that refers exclusively to objects and properties that can be "directly perceived by the senses."[8] This vocabulary was essentially borrowed from everyday speech, and included such concepts as 'blue', 'hard', 'hot', and so on.[9] At least some of the theorems derived from the axioms of a scientific theory must be couched exclusively in the observation vocabulary in order to be capable of empirical check, without which the system is not an empirical scientific theory at all.

As Carnap points out in reading 5.1, scientists do not usually employ such a restrictive conception of observation, applying the term also to the results of simple measurements, such as temperature, weight, length, time-duration, and so on. There is, Carnap insisted, no question which vocabulary is the right observation-vocabulary; there is a continuum here, and any observable/unobservable line through it is bound to be "highly arbitrary."[10] Carnap was happy to extend the observation vocabulary to terms that referred, not just to directly observable properties, but also to the quantitative results of simple measurements.

The theory does not, however, imply observation-sentences concerning specific dateable and locatable observations or measurements. The kinetic theory of gases, for example, does not on its own imply that the temperature, pressure, and volume of a particular gas measured at a specific time and place will have such-and-such values. Instead, the theory implies empirical *laws*, such as the combined gas law $P_1V_1/T_1 = P_2V_2/T_2$ (where the variables stand for pressure, volume, and temperature, respectively). Although such laws are typically formulated as mathematical equations, most are taken by the positivists to be universal generalizations.[11] A universal generalization is an expression of the form "All P are Q" (such as "all ravens are black" or "all copper expands when heated"). In predicate logic this is written as "$(x)(Px \supset Qx)$," which reads "Every x is such that, if x is a P, then x is a Q." ("Px" is called the *antecedent* and "Qx" the *consequent.*) Such a law will apply to a specific observation so long as the antecedent is satisfied; its being true will then require that its consequent is satisfied as well. If the gas law is correct, for example, then any particular isolated quantity of gas will satisfy that equation.

Empirical laws are tested by deriving predictions from them. We begin by observing cases where the antecedent is realized. If the consequent is realized as well, then the law is *confirmed*; if not, it is *disconfirmed*. Carnap suggests that there is an important difference between confirmation and disconfirmation. We can never decisively establish that a law is correct because the law concerns an infinite number of cases. The gas law, for example, concerns every quantity of gas, past, present, and future. It even concerns possible but not actual gases: although I am not boiling the water in my cup, if I had done so, thereby producing water vapor, the law would apply to that vapor as well. But the number of observations that we have made or ever will make is finite (and of only actual cases). So it is always possible that another observation that we do not in fact make would, if we had made it, disconfirm the law (the next volume of gas would not satisfy the equation, the next raven would not be black, and so on). So empirical laws cannot be proven by observation to be correct, at least not if "proof" is meant in the mathematical sense that rules out the possibility of being incorrect. However, so long as the observation report itself is correct, it seems that an empirical law can be decisively refuted, since so much as one non-black raven will demonstrate that the law "all ravens are black" is incorrect.[12]

This is called the *hypothetico-deductive* (H-D) account of confirmation: a prediction is deduced from an hypothesis (or theory or law) and the prediction is then checked; if the prediction is correct the hypothesis is confirmed, and if not it is disconfirmed.[13]

3 Explanation

Empirical laws are also used to *explain* observable phenomena. The positivists considered prediction and explanation to be structurally identical. In both cases, one presents an argument in which one logically deduces an observation statement from premises that include at least one law and a set of initial conditions. If the deduced observation statement has not already been empirically verified, then the conclusion of the argument is a prediction. (A prediction will typically concern an event that has not yet taken place; but it might concern a past or present event that is not known, at the time of the prediction, to have occurred.) But if the deduced observation statement has been empirically verified, then the argument provides an *explanation* of that observed phenomenon. For example, from the law "all copper expands when heated" we might predict that if we heat a particular copper bar it will expand. But, if the bar has already been heated and expanded, then that expansion can be explained by citing the law and the condition of its having been heated. The phenomenon to be explained (that the bar expands, in this case) is called the *explanandum*; the law and the initial conditions that explain the phenomenon are called the *explanans*. This is called the *deductive-nomological* (D-N) model of explanation ('nomos' is Greek for 'law', and the inference is deductive). The structure of a D-N explanation is:

1) L_1-L_n (the laws)
2) C_1-C_k (the initial conditions)
3) E (observation statement)

The D-N model was elaborated and defended primarily by Carl Hempel,[14] and constitutes the received-view account of explanation. The claim that an explanation-after-the-observed-fact could equally well have served as a prediction-before-the-observed-fact is called the *prediction/explanation symmetry thesis*.

The laws we have considered so far have been universal, that is, they apply in all cases (to all ravens, all quantities of gas, and so on). The positivists recognized, however, that not all laws are universal; some report only statistical regularities. Such laws will therefore be formulated, not as "All P are Q," but "the probability of a P's being a Q is r." Statistical laws can still be used to generate predictions, so long as the value of r is fairly high. For example, if most weeds sprayed with a particular herbicide die, then I can reasonably expect that a particular weed that is sprayed will die. I cannot, however, be certain of it; given the premises, it is only *likely* that the conclusion will be true. The argument from which the prediction is derived is *inductive*, rather than deductive.

Hempel also thought that statistical laws can be used in explanations. Since such laws cannot guarantee the phenomenon's occurrence, the structure of such explanations will be inductive rather than deductive. He therefore called them *inductive-statistical* (I-S) explanations (because the argument structure is inductive and the law used is statistical). The structure of an I-S explanation is:

1 $p(Q|P) = r$.
2 i is a P.
Therefore it is probable to degree q that:
3 i is a Q.

Here, "p(Q|P)" reads "the probability of something's being a Q given that it is a P" (the probability that a weed will die given that it is sprayed, for example). "i is a P" reads "this particular case is an instance of P" ("this is a sprayed weed"), and similarly with "i is a Q". In most cases, the probabilities r and q will be the same.

Hempel insisted that q – the probability of the explanandum given the explanans – must be high enough that the explanandum is expectable. As a result, the prediction/explanation symmetry thesis holds for I-S as well as D-N explanations: an I-S explanation would provide rational grounds for a prediction of the explanandum.

4 The Requirement of Maximal Specificity

Notwithstanding their similarities, there is a problem that afflicts I-S explanation from which D-N explanations are immune. Consider one of Hempel's examples: suppose that you are surprised at the warm and sunny weather on November 27. You are offered the explanation that in fact the weather is very typically (95 percent of the time) warm and sunny in November. That explanation conforms to the I-S pattern as follows:

1 The probability that it is warm and sunny, given that today is in November, is 0.95.
2 Today is in November.
Therefore it is 0.95 probable that:
3 Today is warm and sunny.

But now suppose that it was cold and rainy yesterday, and that when the previous day was cold and rainy it is very rare (only 20 percent of the time) that the next day is warm and sunny. This can also be presented as an I-S "explanation":

1 The probability that it is not warm and sunny today, given that it was cold and rainy yesterday, is 0.8.
2 It was cold and rainy yesterday.
Therefore, it is 0.8 probable that:
3 Today is not warm and sunny.

This of course could not really be an explanation, since it is in fact warm and sunny (so that the conclusion of this last argument is false). But Hempel nevertheless considered the fact that we can construct two I-S explanations with contradictory explananda to be intolerable.[15] For it means that, regardless of what the weather is like, we have available an I-S explanation that will explain it. Moreover, assuming the prediction/explanation symmetry thesis (that every explanation is a potential prediction and vice versa) the result is incoherent: we could yesterday have predicted both that it would be warm and sunny today and that it would not be warm and sunny today. But to endorse that prediction would be to accept a contradiction. Hempel called this the *problem of ambiguity* in I-S explanation.

The problem is that the case we are interested in – today's weather, in the example – is, according to our background knowledge, a member of a great many classes: days that are in November, days that follow rainy and cold days, days that are in the last two months of the year, and so on. And the probability (again, according to our background knowledge) that a day that falls in one of these classes will be warm and sunny may well be different from the probability that a day that falls in another of these classes will be warm and sunny. So to which of these classes are we to refer in estimating the probability that it will be warm and sunny today?

Hempel's answer, in essence, is that we should pick the most specific reference class (according to our background knowledge). Return to the weather example. Should we pick days in November, or days that follow cold and rainy days? Answer: neither. We should pick days in November that *also* follow cold and rainy days. We then determine the probability that such days are warm and sunny. In general, if there is a more specific class whose probability of warm and sunny days is different from the more general class, then we must always pick that more specific class. Hempel called this requirement – that we use the most specific class that generates a change in probability – the *requirement of maximal specificity* (RMS).

However, if the probability of the property of interest (being warm and sunny) is no different for the more specific reference class than it is for the more general class according to our background knowledge, then the more general class is acceptable for use in an I-S explanation. For example, suppose that the probability that a day that falls in November and that also follows a cold and rainy day is warm and sunny is 0.19. Suppose also that the probability that a day that falls in November and that follows a cold and rainy day *and* that is a Tuesday, and which is warm and sunny, is also 0.19. Then the more general class (without specifying that the day is Tuesday) is acceptable for use as a premise in I-S explanations.[16]

This problem does not arise for D-N explanations because it is impossible to have two deductively valid arguments, both with true premises, which imply contradictory conclusions. One of the contradictory conclusions must be false (P and ~P cannot both be true).[17] So consider the argument (whichever it is) that has the false conclusion. Since it is valid, it is impossible for the premises to be true and the conclusion false. So at least one of the premises of that argument is false. So it cannot be true that all the premises of both arguments are true. So there cannot be two D-N explanations, both with true explanans, one of which explains P and the other of which explains ~P. (The same applies if the premises are drawn from our background knowledge, so long as that background knowledge is consistent.)

You might have noticed the frequent references to our background knowledge. Hempel maintained that the RMS applies, not to the *actual* various classes of which the particular case is a member, but to the classes of which our best scientific information at the time *says* the particular case is a member, and the probabilities that that information assigns to members of those classes having the relevant property. That is, we gather together all of the currently accepted scientific claims. Call the set of all those claims K. K will imply that the particular case in question – November 27, in our example – is a member of a variety of different classes (is in November, follows a cold and rainy day, etc.). It will also (typically) assign a probability that members of those classes have the relevant property (being warm and sunny). The RMS then applies to those classes, with those probabilities.

That the RMS applies to K in this way is very significant. For it means that what counts as an acceptable I-S explanation can change with changes in the body of accepted scientific claims. There is therefore no such thing as the right I-S explanation, except *relative* to such a body; change in that body means a change in what the right explanation is. This is radically different from Hempel's conception of D-N explanations, which are not relative in this way. I-S explanations are therefore subjective in a way that D-N explanations are not, on Hempel's view.

5 Two Kinds of Probability

Notice that two probability statements occur in the I-S explanation structure: first, within the statistical law that occurs in (at least) one of the premises; and second, the probability that the conclusion is true given that the premises are true. Although these are both formulated as statements of probability, the positivists (Carnap in particular) were careful to distinguish them. They understood the first to constitute a *statistical probability*: it indicates the proportion of Ps among the Qs (e.g., the proportion of sprayed weeds that die). Sometimes this is the actual frequency of the Ps among the Qs, as when the effectiveness of a particular drug is measured by the proportion of successful outcomes to all cases in which the drug is ingested. But, as we saw earlier, most scientific laws concern an infinite domain; the probability is then taken to concern the statistical frequency "in the long run," or "in the limit," that is, as the proportion to which the observed proportions tend as observations continue to be made. A statistical-frequency interpretation of the claim that the probability of heads on a coin toss is 0.5, for example, is not that any finite sequence will result in precisely one half heads (which is unlikely), but that, as the tosses increase, the proportion will more closely approximate that value.

The probability statement that concerns the support given to the conclusion by the premises was, however, taken to be a quasi-logical relation akin to that of deductive validity, indicating the degree of rational support that the premises provide for the conclusion. Carnap accordingly called it *logical probability*. Unlike statistical probability, logical probability is always a relation between sentences (premises and conclusion, or evidence and hypothesis). Just as it makes no sense to say that a conclusion of a deductive argument is valid (since validity is a relation between the premises and the conclusion rather than just a property of the conclusion on its own), it makes no sense to say that the logical probability, for example, of a sentence is 0.8, since 0.8 is a measure of the support that other sentences provide for that sentence rather than a property of that sentence on its own.

Statements of logical probability are statements in the *meta-language* since they report a relation between sentences, whereas statistical probabilities are *object-language* statements since they concern frequencies with which certain properties are realized. Statistical probabilities, as reported in statistical laws, are synthetic a posteriori claims about the world, but Carnap thought that statements of logical probability are analytic a priori claims (just as he, and the positivists generally, understood logical – and mathematical – statements to be analytic a priori claims).

Logical probabilities play three roles in the received view. First, they describe the relation between the explanans and explanandum in an I-S explanation. Second, they indicate the degree to which a prediction from a statistical law should be expected to be successful given that the law is true (that a sprayed weed will die, for example). Third, they measure the extent to which a collection of successful predictions confirm a theory. Notice that the last applies even when the law so confirmed is universal; many observations of black ravens confer a certain logical probability on the universal generalization that all ravens are black. Not surprisingly, Carnap spent much of his time later in life attempting to formulate a system of logical probability (or "inductive logic," or "logic of confirmation").

6 Explaining Laws and the Theoretical Language

Scientific theories do not only explain particular empirical facts; sometimes they provide explanations for the laws themselves. For example, I might derive the law of expansion of copper when heated from a law of thermal expansion for metals in general. In accordance with the account of explanation sketched above, the laws are themselves deduced from other laws (which might, in turn, be themselves deduced from further laws). When the explained law is universal, then the explanation is counted as an instance of the D-N model; when it is a statistical law, the explanation is what Hempel called a "D-S" explanation because it is a deductive derivation of the statistical law from other laws (which might themselves be statistical laws).

One law might explain another if the explaining law is more general (as in the derivation of the law of expansion of copper from a law of thermal expansion for metals). However, no matter how general the law is, it will always employ terms from the observation vocabulary. This is because the generalization will always lead to broader categories of observable properties (e.g., from copper to metal to solid body). But the positivists were well aware of the fact that scientific theories very often include laws that refer to putative unobservable, theoretical entities and processes (e.g., electrons, electromagnetic waves, tectonic plates). They also recognized that these theoretical laws were routinely taken to be confirmed in scientific practice and that they played an integral role in the prediction and explanation of observable phenomena.

The positivists therefore understood the language of a scientific theory to contain, in addition to the observation language, a theoretical language in which such laws were couched. The received view, therefore, divides the scientific language into three parts: the language of the underlying system of logic (which is also taken to include any mathematics employed in the theory's expression); the observation language, used to report specific observations as well as to formulate empirical generalizations; and the theoretical language, used to formulate theoretical laws.

The scientific theory was understood by the positivists to be composed of three levels. The top level consists in the theoretical laws which are used to derive, and thereby both predict and explain, empirical generalizations, which generalizations constitute the second level. Those generalizations are then used to derive, and thereby both predict and explain, specific empirical facts, which constitute the third level. And in the other direction, those specific empirical facts confirm the empirical generalizations, and the (confirmed) generalizations in turn confirm the theoretical laws from which they derive.[18]

The theoretical entities and laws, and the theoretical vocabulary used to refer to them, however, presented the positivists with a dilemma. On the one hand, given their verificationist criterion of cognitive significance, the apparent remoteness of such putatively unobservable entities from sensory experience might seem to require their banishment from science.[19] But on the other hand, the positivists well knew that, for whatever reason, appeal to such entities and laws has proven to be utterly indispensable in the development of the tremendously empirically successful theories that so impressed them. The development of the different versions of the received view was primarily an attempt to solve the "problem of theoretical terms" that this dilemma generates.

7 Theoretical Terms and Correspondence Rules

Theoretical laws contain only terms from the theoretical vocabulary, whereas empirical laws contain only those from the observation vocabulary. But it is not possible to deduce a sentence which employs one vocabulary from other sentences that exclusively employ another vocabulary. There must in addition be "mixed" sentences that contain terms from both. Such sentences, called *correspondence rules* by Carnap,[20] had two roles. First, they provided the link between the theoretical and empirical strata of the theory, allowing prediction and explanation in the downward direction and confirmation in the upward. Second, they were to provide the solution to the problem of theoretical terms.

An early, straightforward attempt at characterizing correspondence rules was to simply define the theoretical terms directly by equating them with expressions in the observation vocabulary. Such definitions proceeded by identifying the theoretical property with the operation by which that property was measured; these were called *operational definitions*. For example, an operational definition of "has a mass of 3 grams" might be "if placed on a spring scale, the scale's pointer will coincide with the number 3."[21]

But *operationism* (or *operationalism*), as this approach is called, faces a number of serious problems. First, according to it, no one property can be measured by more than one operation, since the operation itself defines the property. For example, mass-as-measured-by-a-balance-scale must be an entirely different property than mass-as-measured-by-a-spring-scale, which is different than mass-as-measured-by-water-displacement. For the same reason, it is not possible to discover a new method of measuring a property (the new method implies a *new property*, which is therefore distinct from the old). Such a proliferation of different concepts of mass is both unintuitive and contrary to scientific practice, in which these different operations are routinely taken to measure the same property. Second, while some theoretical properties (e.g., mass, temperature, and length) might seem sufficiently close to the empirical ground, as it were, to be characterizable in terms of the methods by which they are measured, the prospects for definitions for 'electron', 'tectonic plate', and other concepts more deeply embedded within the theoretical system seem dim indeed.

Finally, a certain logical problem afflicts operational definitions. Such definitions treat a theoretical property as a disposition (viz. to produce such-and-such a result under such-and-such measurement conditions). But consider an everyday disposition, such as the fragility of a pane of glass. An obvious definition of this disposition is "Something is fragile if and only if it breaks when it is struck." In predicate logic this is written (x)(Fx if and only if (if Sx then Bx)); that is, "for all x, x is fragile if and only if, if it is struck, then it breaks." The problem is that the right-hand side of the biconditional (viz. "if it is struck, then it breaks") was understood to be a *material conditional*, and a material conditional is true if its antecedent is false.[22] So it follows from this definition of fragility (and the definition of the material conditional) that so long as an object is never struck – even if it is made of lead – it will be fragile; obviously the wrong result.

One possible response is to employ subjunctive conditionals, like "were it to be struck, then it would break," rather than the material conditional "if it is struck, then it will break." The logic of such conditionals was, however, far too underdeveloped when the received view was formulated to rely on them. (In fact, controversy concerning their correct interpretation continues to this day.)

Instead, Carnap formulated an alternative, called *reduction sentences*. A reduction sentence for fragility is: "If x is struck, then it is fragile if and only if it breaks." When the antecedent is true this tells us that it will break if and only if it is fragile, which is intuitively correct. But if the antecedent is false then the conditional will count as true regardless of whether the consequent is true or false (again, a material conditional is true if the antecedent is false). So the fact that it did not break will not imply that it is fragile. But neither will it imply that it is not fragile; neither follows from the fact that the antecedent is false, since that leaves open whether the consequent is true or false.

The upshot is that a reduction sentence provides a definition for a dispositional term whenever the operation is performed or the test-condition realized (e.g., when the glass is struck). But it is silent concerning whether the term applies when the test-condition is *not* realized (e.g., when the glass is not struck). It is therefore what Carnap called a *partial definition*, since it provides necessary, but not sufficient, conditions for application of the concept. Other reduction sentences may be added to minimize this; we might add, for example, "If x is dropped onto a ceramic floor, then it is fragile if and only if it breaks." But so long as there are other conditions under which that disposition would be realized, no collection of such definitions will be complete.

In addition to avoiding the logical problem, reduction sentences avoid the problem of multiple methods of measurement facing operationism since there is no reason why we cannot think of the different reduction sentences as each specifying part (but never all) of the definition of the same property. They apply under different test-conditions, after all, and so do not compete with each other. But reduction sentences are still unlikely to be applicable to deeply theoretical terms like "electron", since it is hard to imagine what test-condition we could use in the formulation of either an operational definition or a reduction sentence.

Carnap's eventual solution was to suggest that most theoretical terms receive empirical content indirectly, through the network of theoretical laws in which they are embedded and, finally, by their relation to the empirical laws and facts derived from that network. To understand this approach, first consider the system of axioms-plus-derived-theorems that constitutes a scientific theory, not as a collection of meaningful sentences, but just as a formal, uninterpreted calculus or game of symbol-manipulations whose rules dictate only that certain sequences of symbols can be written down when other symbol-sequences are given. The question is then how to provide the expressions in the system with empirical significance so that the result is a scientific theory rather than just a formal calculus.

The obvious way to do this is to first assign a meaning to the basic or "primitive" theoretical terms (that is, those that are not defined by other terms) that occur in the axioms or basic laws of the theory. Then the meaning of defined expressions (and of the theorems that contain those defined expressions) will flow from those initial assignments, much as "triangle" in the system of Euclidean geometry might earn its meaning by definition from the already-meaningful terms "line segment" and "point" (and "three"). Carnap referred to this as the "top-down" approach, since meaning is assigned to the primitive terms at the top and then flows downward to the defined expressions. Both operationism and the method of reduction sentences are top-down approaches since, according to them, primitive theoretical terms are directly assigned an empirical meaning by the operational definition or reduction sentence.

Carnap, however, proposed another "bottom-up" method. We can set out the entire formal system, using the primitive theoretical terms (and the axioms containing those primitive terms) to define other expressions (and derive theorems containing those defined expressions). So far, no empirical meaning has been assigned to any expression. Then we could assign a meaning, not to the primitive terms at the top of the system, but instead to the defined expressions at the bottom, by correspondence rules (which will typically be reduction sentences). For example, measurements made by a thermometer of the temperature of a gas would be used to assign a meaning, not to a primitive term like "molecule", but rather to the expression "mean molecular kinetic energy" (which is obviously constructed using the primitive term "molecule").

In this bottom-up approach, no empirical significance is assigned directly to the primitive theoretical terms; "molecule", for example, is not directly provided with any empirical meaning as it is in the top-down approach. These terms acquire the only meaning they have in virtue of their role in generating the defined expressions, which are directly provided with empirical meaning. Empirical significance is therefore drawn up from the defined expressions at the bottom to the primitive theoretical terms at the top by a kind of semantic analogue to osmosis.

A complete interpretation of an expression would provide necessary and sufficient conditions for the application of that expression. But in the bottom-up approach the theoretical terms only inherit their empirical meaning in the indirect fashion described above rather than directly by the specification of necessary and sufficient conditions. Their interpretation is therefore only partial. This *partial interpretation* solution to the problem of theoretical terms is now standardly associated with the received view.[23]

A consequence of this bottom-up approach is that the theoretical terms do not enjoy empirical significance in isolation from one another, but only in virtue of their place in the system. So the only legitimate answer to questions concerning, for example, the meaning of "electron" or "molecule" is to present the entire theoretical system itself to which those terms belong. Appeal to pictures, models, or analogies (such as picturing molecules impacting the walls of the container as like little billiard balls hitting the walls of a pool table) is therefore at best unnecessary for a complete understanding of the meaning of theoretical terms, and, at worst, misleading. This is because such surplus content would be independent of the theoretical term's role in generating empirical consequences. But such surplus content would then contribute no empirical significance; it would therefore introduce transcendent metaphysics into the heart of theoretical science.

Another consequence is that the only way to distinguish a cognitively significant theoretical term from a cognitively insignificant metaphysical expression is to notice that removal of the theoretical term from the system in which it is embedded will result in an impoverishment of the empirical consequences that are derivable from that system, whereas removal of a metaphysical expression will have no such impact (the latter came to be called *isolated* sentences). The partial interpretation approach therefore introduces a kind of *semantic holism* into the received view, at least for the theoretical terms (although not for the observation terms).

8 Ancillary Doctrines

There are three other doctrines widely advocated among the positivists to consider. These are not obvious consequences of the central tenets of the received view as presented above. They are, however, strongly associated with it, and subsequent criticisms raised against them have inevitably been seen to attack the received view itself.

The first is the doctrine of the *unity of science*. This is the view that, notwithstanding the variety of scientific disciplines ranging from physics to biology to economics, there is only one scientific method employed (with varying success, perhaps) across all such disciplines. There is, moreover, ultimately one domain of objects and properties that are the concern of all of the sciences. That domain is most directly characterized within physics; the subject matter of the other sciences, if legitimate, must in the end be *reducible* to (translatable into) the language of physics. Call the first claim concerning scientific method the doctrine of *synchronic methodological reductionism* and the second concerning object-domains *synchronic ontological reductionism* ("synchronic" means "at the same time").

The positivists also typically advocated *diachronic reductionism* of both the ontological and methodological varieties ("diachronic" means "over time"). According to diachronic ontological reductionism, when successor theories replaced reasonably successful earlier ones, this did not amount to wholesale replacement of one theory, then considered false, by an entirely different theory. Instead, the positivists saw it in one of two ways. First, the old theory might be *extended* into new domains to which it was not originally applied, as when classical particle mechanics was extended to rigid-body mechanics. Second, the older theory with its limited domain might be *absorbed* into another theory with a wider domain, as when Kepler's astronomical theory was derived from Newton's physics. In such a case, the original theory is understood to be true of the objects that are its concern, but only as a part of the greater story told by the reducing theory.

Diachronic ontological reductionism implies that the sequence of theories over time is both *progressive* and *cumulative*. It is progressive because successor theories constitute improvements on their predecessors (either because the successors expand the domain of their predecessors' application or because they are accurate in domains in which their predecessors were not). And it is cumulative because the successor theories endorse the previous theories as true within their domains but contribute further truths (in the domains outside the scope of the predecessor theory); the truth-content of the succession of theories is therefore steadily increasing.

According to diachronic methodological reductionism, there is no temporal change in scientific methodology; the scientific method that is employed by all scientists at a time is the same as that employed by all scientists across time, and therefore across theory-change. Of course certain methods might be improved, new instruments developed, and so on. But the logic of science itself – which underlies the explanation, prediction, confirmation, and interpretation relations between theory and observation – remains the same.

Finally, notice that the relations between theory and observation just mentioned do not include a procedure for the discovery of scientific theories. A logic of discovery would provide – ideally in algorithmic or recipe form – a procedure that would take a scientist from observations to the correct (or best) theory that those observations support. Having distinguished the project of formulating a method of theory-development (the *context of discovery*) from the method for the evaluation of theories (the *context of justification*), the positivists decided that the former is at best a subject for psychology, and abandoned it. The context of justification, they claimed, is the only context to which the philosopher of science is responsible.[24]

This ends our presentation of the various doctrines that make up the received view. Some, such as the partial interpretation account, were rejected fairly quickly; some, such as the D-N model of explanation, are no longer accepted, but survived much longer than positivism itself; and some, such as the syntactic approach and the attempt to develop a formal confirmation theory, still constitute viable positions in the philosophy of science. The influence of the received view, therefore, continues to be felt long after its demise.

9 "Empiricism, Semantics, and Ontology"

In the essay by this title, Carnap introduces a distinction between *internal* and *external* questions of existence. They are internal and external with respect to what he called a *linguistic framework*. Linguistic frameworks introduce the resources needed for discourse concerning certain kinds of entities, such as numbers, material objects, properties, and events. This requires both introducing a general term for the class of entity (e.g., "number") and allowing a particular class of expressions (e.g., numerals) to be substituted by variables that can be bound by quantifiers. Start, for example, with the use of numerals as adjectives, as in "There are three cats." The objects with which this sentence is concerned are obviously only the cats. But now allow substitution of the number by a variable n that comes within the scope of the *existential quantifier* which asserts that there is at least one n. The result in predicate logic is "($\exists$n)(There are n cats and n equals three)" which says that there is a number that is the number of cats, and that number is three. This sentence now speaks, not only of the cats, but of their number (which is identified with the number three). Numbers are thereby introduced as objects referred to by the numerals. Such frameworks also introduce rules that govern the assertion of the new sentences that they make available. To be proficient with the framework of numbers, for example, is not only to be able to form sentences that concern numbers; it is also to be able to determine the answers to such questions as what the value of two plus three is, or whether there is an even prime.

When a question of existence is answered in accordance with the rules of the framework, the question is *internal* to the framework. When doing your math homework, you know that there is an even prime, that three is the square root of nine, and so on. Notice that, since primes are numbers, if there is an even prime, then it trivially follows that there are numbers. But when philosophers question the existence of numbers, they are not so easily satisfied. They intend to ask a question, not within mathematics and answerable according to its rules, but *about* mathematics. They want to know whether there *really* are numbers, whether mathematics accurately represents some aspect of reality.

These are what Carnap called *external* questions of existence. And he simply denied that such questions are meaningful.[25] There is no way to answer them since the only procedures for their evaluation are specified by the framework, and the external questioner dismisses the answers provided by that evaluation as beside the point of the question.

However, that is not to say that there are no legitimate external questions whatsoever. We can consider the advantages and disadvantages of introducing a particular framework of entities, and propose that we do so (or not). But these are pragmatic questions concerning how the language is best engineered; they are not, Carnap insisted, questions of what to believe there is.

The distinction between internal and external questions also applies to questions of existence concerning theoretical entities within a system of science. The system provides rules for the evaluation of a particular existential question – for example, whether any molecules remain in an evacuated chamber – the answer to which is a matter of empirical investigation conducted by the scientist. But the philosopher wants to know whether there really are molecules and is not satisfied with the answer that the theory and its rules of application might deliver. After reflection (and argument) the *scientific realist* decides that indeed there are, whereas the *anti-realist* says that there are not (or else that we cannot tell whether there are or not).

Carnap, again, denies that such external questions make sense. There is the internal answer provided by empirical investigation conducted according to the rules of the scientific framework, and that is it. To demand more than this is to assume that there is more to the meaning of the term "molecule" than is provided by that term's role within the theory and its empirical consequences. But, as we saw above, Carnap insisted that such a role exhausts the term's legitimate interpretation.

It is not easy to say what Carnap's position on the realism debate in the philosophy of science amounts to. Traditionally, he is seen as an anti-realist. This is sometimes because he (and the positivists generally) are taken to be *reductionists*,[26] according to which assertions about theoretical entities can be completely translated into assertions about observable entities. This is not an accurate portrayal, however; Carnap explicitly rejected reductionism in this sense very early on. Certainly the method of partial interpretation is not a form of such reductionism.[27] So Carnap would have endorsed the claim that electrons exist, for example, since this follows from the application of the rules of the theory in conjunction with experimental results. And he would deny that the sentence "electrons exist" is translatable into a sentence about observables. This would suggest that he was a realist.

However, one might view the partial interpretation account as embodying anti-realism nevertheless. After all, the only significance that the theoretical terms receive is that which they absorb from their role in implying observation sentences; they, and the laws that embed them, are otherwise purely formal. This has struck many, including some positivists, to amount to a version of *instrumentalism* (which is a version of anti-realism) according to which theoretical terms and laws are really just instruments, or tools, mediating the prediction of observable phenomena. Such tools are useful or not for their purpose, but they are no more true (or false) than is a hammer.

Whether Carnap's "plague on both your houses" stance collapses into a form of either realism or anti-realism remains the subject of debate. Interestingly Carnap's stance foreshadows a similarly dismissive response to the realism debate formulated by Arthur Fine, which is otherwise very far removed from the received view. Fine calls his position the *Natural Ontological Attitude* (reading 9.4). A parallel debate has ensued concerning whether the Natural Ontological Attitude amounts to a form of either realism or anti-realism. We will discuss Fine's view, and that debate, in §11 of the Unit 9 commentary.

10 Reichenbach's Pragmatic Vindication

Hume presented a dilemma: induction can either be justified by "demonstrative reasoning" (what might be now called deductive proof) or by "moral reasoning" (reasoning on the basis of experience). But demonstrative reasoning guarantees its conclusions on the assumption that the premises are true, whereas inductive arguments do not. And to appeal to past and present experience of successful inductions in order to warrant future uses of induction is itself an inductive argument and so begs the question. There is therefore no justification for our belief that induction is a reliable method for the generation of a reasonably high proportion of true beliefs.

Hans Reichenbach agreed. But he claimed that we can nevertheless justify (or "vindicate") our use of induction, because we can argue that if there is *any* reliable method of predicting the future on the basis of the past, induction is it. The goal of an inductive argument, he suggested, is that of determining the limit of the frequency of "favorable" cases to all cases (e.g., of coin tosses that are heads, of ravens that are black, and so on; see §5 above). But if there *is* such a limit, then setting our prediction of that frequency to the frequency observed so far is guaranteed to eventually approach that limit. This is a consequence of the definition of a limit: to say that the limit of the frequency is f just is to say that, for any number ε (no matter how small), there is a number n of observed cases where the observed frequency will thereafter not deviate from f by more than ε. So, if there is a limit of the frequency, it just is that limit to which the observed proportion tends as those observations continue to be made.

There is, however, no guarantee that there is a limit. There is also no guarantee even that the number of observed cases is of a sufficient size as to provide us with any reason to believe that we are within any value ε of the limit, even assuming that one exists. These are, Reichenbach conceded, consequences of Hume's dilemma that must be accepted. However, it remains the case, he claimed, that if there is any method that will work, the above argument demonstrates that induction will do so. So, Reichenbach claims, it remains rational to use it, even in the face of Humean skepticism.

Consider an analogy. Suppose you are lost in the woods, and you discover a path. You have no idea, no justification for believing, that the path leads out of the woods at all. But you know this much: if there is any way out, then this path is it. In that case it remains rational for you to follow it, even in the face of no assurance of success. That is, Reichenbach suggests, our situation with respect to induction: we are not justified in believing that it will be successful, but we are nevertheless justified in using it to determine our expectations for the future.

There are various technical objections to Reichenbach's vindication, but the most forceful objection is simply that it provides far too little comfort. It might vindicate the *act* of our using induction, but it does not vindicate our *belief* in its reliability. It provides no support, for example, for the belief that the sun will very likely rise again tomorrow. One might reasonably consider Hume's critique to be directed, not at the act of performing an induction, but at the rationality of the belief that induction is reliable; and Reichenbach's vindication provides no protection against the critique so understood.

11 Strawson's "Ordinary Language" Response

Reichenbach took Hume's challenge seriously. We cannot, he conceded, provide such a justification as would meet that challenge, although we can nevertheless vindicate our use of induction. Peter Strawson, on the other hand, rejected the challenge itself, insisting that the question whether induction is rational is simply incoherent.

If someone routinely makes financial decisions on the basis of their horoscope, then we might challenge them, pointing out that the track record of those decisions has been disastrous. In so evaluating their behavior we implicitly (or explicitly) appeal to certain standards of rational belief, emphasizing accuracy in observation, adequacy of sample size, representativeness of the sample, avoidance of bias, and so on. To do all this is to appeal to the standards by which inductive reasoning is evaluated.

But if we are then asked to evaluate the inductive standard itself, Strawson asks, to what further standard are we to appeal? To demand that it meet the deductive standard, as Hume rightly noted, is to apply a standard applicable only to a distinct form of reasoning. But there is no other higher standard; induction just *is* the standard of rationality for beliefs formulated in response to experience. To ask whether such beliefs are rational is to judge them by the standards imposed by the inductive method. It is incoherent, therefore, to demand a justification for induction, when induction determines what it means to call a belief that is based on experience justified, just as it is incoherent to ask whether the laws of the land themselves are legal.

So attempting to respond to Hume's challenge is rather like attempting to answer the question what is north of the North Pole. Properly understood, the problem lies, not in the unavailability of satisfactory answers to a legitimate question, but in the posing of an incoherent question as a coherent one that demands an answer.

A problem with Strawson's solution is analogous to the problem with Reichenbach's vindication discussed above. Strawson understands Hume's challenge to concern the rationality of induction. But one might instead understand it to concern our belief in induction's reliability. We believe that the use of induction will generate a reasonably high proportion of true beliefs, and we want to know if that belief is correct. To be told that this belief is rational simply in virtue of the definition of "rational" seems to miss the point of the question. Suppose that we concede to Strawson that induction is rational. What we want to know is whether our judgment of its rationality is responsive to the reliability of the inductive method. And that still seems a coherent question to ask.

Notes

1 See reading 6.3 and §§4–6 of Unit 6 commentary.
2 See reading 3.7 and §5 of Unit 3 commentary.
3 See note 14 of Part II introduction.
4 Ayer 1946, *Language, Truth and Logic*, 2nd edn., London: Gollancz, p. 50.
5 The positivists distinguished pure geometry, which is purely mathematical, from applied or physical geometry, which purports to characterize the structure of physical space. The former has no empirical consequences of its own, while the latter amounts to an empirical scientific theory.
6 An advantage of the semantic account is that it allows the same theory to be expressed in two different languages, and also permits two different axiomatizations to count as expressions of the same theory.
7 The exception is Carnap's Ramsey-sentence approach, which employs second-order logic. See note 23 below.
8 Reading 5.1, p. 336. In a different work, Carnap characterized a property P as observable when a person "is able under suitable circumstances to come to a decision with the help of a few observations about a full sentence, say '~P(b)', i.e. to a confirmation of either 'P(b)' or 'P(b)' of such a high degree that he will either accept or reject 'P(b)'" (Carnap 1936–37, p. 455). We will discuss challenges to the observable/unobservable distinction in Unit 6.
9 The received view's observation vocabulary is often thought to be phenomenalistic, that is, as referring to sense-data concerning, for example, patches and colors in the visual field of the scientist's experience (rather than to features of the scientist's physical environment). But a phenomenalistic language was rejected early on as the observation basis in favor of a physicalistic language concerning ordinary physical objects, and was never the basis of the more sophisticated versions of the received view. See §8 of Part II introduction.
10 Reading 5.1, p. 337.
11 The rest are statistical generalizations; see below.
12 The claim that such decisive refutations of a theory are possible has, we will see, been challenged. See readings 4.13 and 6.3, as well as Unit 7 commentary, especially §§2–3.
13 Notice that the law so confirmed need not be an empirical law; it could concern unobservable theoretical entities. See §6 below.
14 See §3 of Part II introduction.
15 "Explananda" is the plural of "explanandum."
16 It is significant that Hempel did not insist that we use only the *most* general such class. In Unit 8 we will see Wesley Salmon suggest that Hempel's failing to do so is responsible for the counterexamples to Hempel's I-S model that are called *irrelevance* cases.
17 '~P' means "It is not the case that P."
18 As a result the positivist's account is sometimes called the "layer-cake" model of scientific theories.
19 See §5 of Part II introduction.
20 They were also called "operational definitions," "coordinating definitions," "dictionaries," and "interpretation rules."
21 The philosopher most closely associated with operationism is Percy Bridgman. See Bridgman 1927.
22 Consider the sentence "if you get an A, you will pass." This of course rules out your getting an A and failing (antecedent true, consequent false). But it does not rule out your not getting an A and

passing, since you could get a B, also a passing grade (antecedent false, consequent true). Nor does it rule out your not getting an A and failing, since you could get an F (antecedent false, consequent false). So logicians count the sentence true under those cases, which are the two cases when the antecedent is false.

23 It was not Carnap's last proposal, however. Very soon after formulating the partial interpretation approach, he abandoned it in favor of the "Ramsey-sentence" method (in fact he discusses both in the book from which selection 5.1 is excerpted). But that method faces serious difficulties, and, at any rate is not now associated with the standard version of the received view.

24 The distinction between the contexts of discovery and justification originates in Reichenbach, Hans, 1951, *The Rise of Scientific Philosophy*, Berkeley/Los Angeles: University of California Press.

25 It is not clear precisely how Carnap's charge of the meaninglessness of external questions was to be understood. Many assume it to constitute an expression of verificationism (for example Stroud, B. 1984, *Significance of Philosophical Scepticism*, Oxford: Oxford University Press). Others see in it an application of the analytic/synthetic distinction (Quine, "On Carnap's Views on Ontology" in *Philosophical Studies* 2, repr. in Quine, 1966, *Ways of Paradox*, Cambridge, MA: Harvard University Press, 2nd enlarged edn. 1976, pp. 203–11). See Alspector-Kelly, M., 2001, "On Quine on Carnap on Ontology", *Philosophical Studies*, Vol. 102 (January), pp. 93–122, and 2002, "Stroud's Carnap," *Philosophy and Phenomenological Research*, Vol. 64, No. 2 (March), pp. 276–302, for discussion.

26 The sense of this multiply abused term here is not that of ontological or methodological reductionism (either synchronic or diachronic), nor that in "reduction sentence," and should not be confused with those uses.

27 Ironically, the reduction sentences discussed in §7 above also do not provide reductions in the present sense of the term, since reduction sentences only constitute partial definitions.

Suggestions for Further Reading

Carnap 1956 contains a detailed statement of the received view and the role of the empiricist criterion of significance within it. His 1936–7 presents his earlier development of reduction sentences. The classic statement of operationism is Percy Bridgman 1927. Nagel 1979 is another detailed presentation of the received view that was highly influential in the United States. The D-N model of explanation was originally presented in Hempel and Oppenheim 1948. A fuller treatment of the D-N model, along with introduction of the D-S and I-S models, is given in Hempel 1965b (from which reading 5.2 is excerpted). An excellent, detailed discussion of the origins, development, critique, and downfall of the received view is Suppe 1977.

Bridgman, P. W., 1927, *The Logic of Modern Physics*. New York: Macmillan.

Carnap, R., 1936–7, "Testability and Meaning," in *Philosophy of Science* 3(4): 419–71 and *Philosophy of Science* 4(1): 1–40.

Carnap, R., 1956, "The Methodological Character of Theoretical Concepts," in Herbert Feigl and Michael Scriven (eds.), *The Foundations of Science and the Concepts of Science and Psychology*. Minnesota: University of Minneapolis Press, pp. 38–76.

Hempel, C., 1965a, *Aspects of Scientific Explanation and Other Essays in the Philosophy of Science*. New York: Free Press.

Hempel, C., 1965b, "Aspects of Scientific Explanation," in Hempel 1965a, pp. 331–496.

Hempel, C. and Oppenheim, P., 1948, "Studies in the Logic of Explanation," *Philosophy of Science* 15: 135–75. Reprinted in Hempel 1965a, pp. 245–90.

Nagel, E., 1979, *The Structure of Science: Problems in the Logic of Scientific Explanation*. Indianapolis: Hackett.

Suppe, F., 1977, "The Search for Philosophic Understanding of Scientific Theories," in Suppe, F., *The Structure of Scientific Theories*. Chicago: University of Illinois Press, pp. 3–232.

5.1

Theory and Observation

Rudolf Carnap

Rudolf Carnap (1891–1970) was a member of the Vienna Circle and the leading figure in the positivist philosophical movement that dominated the first half of the twentieth century. He was also the primary author of the received view of scientific theories. The following selection is an excerpt from his textbook in the philosophy of science, and it provides an overview of his views on the role of laws in scientific theories, the distinction between two concepts of probability, and the empirical significance of theoretical terms.

The Value of Laws: Explanation and Prediction

The observations we make in everyday life as well as the more systematic observations of science reveal certain repetitions or regularities in the world. Day always follows night; the seasons repeat themselves in the same order; fire always feels hot; objects fall when we drop them; and so on. The laws of science are nothing more than statements expressing these regularities as precisely as possible.

If a certain regularity is observed at all times and all places, without exception, then the regularity is expressed in the form of a "universal law". An example from daily life is, "All ice is cold." This statement asserts that any piece of ice – at any place in the universe, at any time, past, present, or future – is (was, or will be) cold. Not all laws of science are universal. Instead of asserting that a regularity occurs in *all* cases, some laws assert that it occurs in only a certain percentage of cases. If the percentage is specified or if in some other way a quantitative statement is made about the relation of one event to another, then the statement is called a "statistical law". For example: "Ripe apples are usually red", or "Approximately half the children born each year are boys." Both types of law – universal and statistical – are needed in science. The universal laws are logically simpler, and for this reason we shall consider them first. In the early part of this discussion "laws" will usually mean universal laws.

From Rudolf Carnap, *Philosophical Foundations of Physics: An Introduction to the Philosophy of Science*, ed. Martin Gardner (New York: Basic Books, 1966), pp. 3–8, 16–22, 32–9, 225–39.

Universal laws are expressed in the logical form of what is called in formal logic a "universal conditional statement". (In this book, we shall occasionally make use of symbolic logic, but only in a very elementary way.) For example, let us consider a law of the simplest possible type. It asserts that, whatever x may be, if x is P, then x is also Q. This is written symbolically as follows:

$$(x)(Px \supset Qx).$$

The expression "(x)" on the left is called a "universal quantifier." It tells us that the statement refers to *all* cases of x, rather than to just a certain percentage of cases. "Px" says that x is P, and "Qx" says that x is Q. The symbol "$\supset$" is a connective. It links the term on its left to the term on its right. In English, it corresponds roughly to the assertion, "If . . . then . . ."

If "x" stands for any material body, then the law states that, for any material body x, if x has the property P, it also has the property Q. For instance, in physics we might say: "For every body x, if that body is heated, that body will expand." This is the law of thermal expansion in its simplest, nonquantitative form. In physics, of course, one tries to obtain quantitative laws and to qualify them so as to exclude exceptions; but, if we forget about such refinements, then this universal conditional statement is the basic logical form of all universal laws. Sometimes we may say that, not only does Qx hold whenever Px holds, but the reverse is also true; whenever Qx holds, Px holds also. Logicians call this a biconditional statement – a statement that is conditional in both directions. But of course this does not contradict the fact that in all universal laws we deal with universal conditionals, because a biconditional may be regarded as the conjunction of two conditionals.

Not all statements made by scientists have this logical form. A scientist may say: "Yesterday in Brazil, Professor Smith discovered a new species of butterfly." This is not the statement of a law. It speaks about a specified single time and place; it states that something happened at that time and place. Because statements such as this are about single facts, they are called "singular" statements. Of course, all our knowledge has its origin in singular statements – the particular observations of particular individuals. One of the big, perplexing questions in the philosophy of science is how we are able to go from such singular statements to the assertion of universal laws.

[. . .]

When we use the word "fact", we will mean it in the singular sense in order to distinguish it clearly from universal statements. Such universal statements will be called "laws" even when they are as elementary as the law of thermal expansion or, still more elementary, the statement, "All ravens are black." I do not know whether this statement is true, but, assuming its truth, we will call such a statement a law of zoology. Zoologists may speak informally of such "facts" as "the raven is black" or "the octopus has eight arms", but, in our more precise terminology, statements of this sort will be called "laws".

Later we shall distinguish between two kinds of law – empirical and theoretical. Laws of the simple kind that I have just mentioned are sometimes called "empirical generalizations" or "empirical laws". They are simple because they speak of properties, like the color black or the magnetic properties of a piece of iron, that can be directly observed. The law of thermal expansion, for example, is a generalization based on many direct observations of bodies that expand when heated. In contrast, theoretical, nonobservable concepts, such as elementary particles and electromagnetic fields, must be dealt with by theoretical laws. We will discuss all this later. I mention it here because otherwise you might think that the examples I have given do not cover the kind of laws you have perhaps learned in theoretical physics.

To summarize, science begins with direct observations of single facts. Nothing else is observable. Certainly a regularity is not directly observable. It is only when many observations are compared with one another that regularities are discovered. These regularities are expressed by statements called "laws".

What good are such laws? What purposes do they serve in science and everyday life? The answer is twofold: they are used to *explain* facts already known, and they are used to *predict* facts not yet known.

First, let us see how laws of science are used for explanation. No explanation – that is, nothing that deserves the honorific title of "explanation" – can be given without referring to at least one law. (In simple cases, there is only one law, but in more complicated cases a set of many laws may be involved.) It is important to emphasize this point, because philosophers have often maintained that they could explain certain facts in history, nature, or human life in some other way. They usually do this by specifying some type of agent or force that is made responsible for the occurrence to be explained.

[. . .]

[F]act explanations are really law explanations in disguise. When we examine them more carefully, we find them to be abbreviated, incomplete statements that tacitly assume certain laws, but laws so familiar that it is unnecessary to express them. [. . .]

Consider one . . . example. We ask little Tommy why he is crying, and he answers with another fact: "Jimmy hit me on the nose." Why do we consider this a sufficient explanation? Because we know that a blow on the nose causes pain and that, when children feel pain, they cry. These are general psychological laws. They are so well known that they are assumed even by Tommy when he tells us why he is crying. If we were dealing with, say, a Martian child and knew very little about Martian psychological laws, then a simple statement of fact might not be considered an adequate explanation of the child's behavior. Unless facts can be connected with other facts by means of at least one law, explicitly stated or tacitly understood, they do not provide explanations.

The general schema involved in all explanation can be expressed symbolically as follows:

1 $(x)(Px \supset Qx)$
2 Pa
3 Qa

The first statement is the universal law that applies to any object x. The second statement asserts that a particular object a has the property P. These two statements taken together enable us to derive logically the third statement: object a has the property Q.

In science, as in everyday life, the universal law is not always explicitly stated. If you ask a physicist: "Why is it that this iron rod, which a moment ago fitted exactly into the apparatus, is now a trifle too long to fit?", he may reply by saying: "While you were out of the room, I heated the rod." He assumes, of course, that you know the law of thermal expansion; otherwise, in order to be understood, he would have added, "and, whenever a body is heated, it expands". The general law is essential to his explanation. If you know the law, however, and he knows that you know it, he may not feel it necessary to state the law. For this reason, explanations, especially in everyday life where common-sense laws are taken for granted, often seem quite different from the schema I have given.

At times, in giving an explanation, the only known laws that apply are statistical rather than universal. In such cases, we must be content with a statistical explanation. For example, we may know that a certain kind of mushroom is slightly poisonous and causes certain symptoms of illness in 90 per cent of those who eat it. If a doctor finds these symptoms when he examines a patient and the patient informs the doctor that yesterday he ate this particular kind of mushroom, the doctor will consider this an explanation of the symptoms even though the law involved is only a statistical one. And it is, indeed, an explanation.

[. . .]

In addition to providing *explanations* for observed facts, the laws of science also provide a means for *predicting* new facts not yet observed. The logical schema involved here is exactly the same as the schema underlying explanation. This, you recall, was expressed symbolically:

1 $(x)(Px \supset Qx)$
2 Pa
3 Qa

First we have a universal law: for any object x, if it has the property P, then it also has the property Q. Second, we have a statement saying that object a has the property P. Third, we deduce by elementary logic that object a has the property Q. This schema underlies both explanation and prediction; only the knowledge situation is different. In explanation, the fact Qa is already known. We explain Qa by showing how it can be deduced from statements 1 and 2. In prediction, *Qa* is a fact *not yet known*. We have a law, and we have the fact *Pa*. We conclude that *Qa* must also be a fact, even though it has not yet been observed. For example, I know the law of thermal expansion. I also know that I have heated a certain rod. By applying logic in the way shown in the schema, I infer that if I now measure the rod, I will find that it is longer than it was before.

In most cases, the unknown fact is actually a future event (for example, an astronomer predicts the time of the next eclipse of the sun); that is why I use the term "prediction" for this second use of laws. It need not, however, be prediction in the literal sense. In many cases the unknown fact is simultaneous with the known fact, as is the case in the example of the heated rod. The expansion of the rod occurs simultaneously with the heating. It is only our observation of the expansion that takes place after our observation of the heating.

In other cases, the unknown fact may even be in the past. On the basis of psychological laws, together with certain facts derived from historical documents, a historian infers certain unknown facts of history. An astronomer may infer that an eclipse of the moon must have taken place at a certain date in the past. A geologist may infer from striations on boulders that at one time in the past a region must have been covered by a glacier. I use the term "prediction" for all these examples because in every case we have the same logical schema and the same knowledge situation – a known fact and a known law from which an unknown fact is derived.

In many cases, the law involved may be statistical rather than universal. The prediction will then be only probable. A meteorologist, for instance, deals with a mixture of exact physical laws and various statistical laws. He cannot say that it will rain tomorrow; he can only say that rain is very likely.

This uncertainty is also characteristic of prediction about human behavior. On the basis of knowing certain psychological laws of a statistical nature and certain facts about a person, we can predict with varying degrees of probability how he will behave. Perhaps we ask a psychologist to tell us what effect a certain event will have on our child. He replies: "As I see the situation, your child will probably react in this way. Of course, the laws of psychology are not very exact. It is a young science, and as yet we know very little about its laws. But on the basis of what is known, I think it advisable that you plan to . . .". And so he gives us advice based on the best prediction he can make, with his probabilistic laws, about the future behavior of our child.

When the law is universal, then elementary deductive logic is involved in inferring unknown facts. If the law is statistical, we must use a different logic – the logic of probability. To give a simple example: a law states that 90 per cent of the residents of a certain region have black hair. I know that an individual is a resident of that region, but I do not know the color of his hair. I can infer, however, on the basis of the statistical law, that the probability his hair is black is $^{9}/_{10}$.

Prediction is, of course, as essential to everyday life as it is to science. Even the most trivial acts we perform during the day are based on predictions. You turn a doorknob. You do so because past observations of facts, together with universal laws, lead you to believe that turning the knob will open the door. You may not be conscious of the logical schema involved – no doubt you are thinking about other things – but all such deliberate actions presuppose the schema. There is a knowledge of specific facts, a knowledge of certain observed regularities that can be expressed as universal or statistical laws and provide a basis for the prediction of unknown facts. Prediction is involved in every act of human behavior that involves deliberate choice. Without it, both science and everyday life would be impossible.

Induction and Statistical Probability

[W]e [earlier] assumed that laws of science were available. We saw how such laws are used, in both science and everyday life, as explanations of known facts and as a means for predicting unknown facts. Let us now ask how we arrive at such laws. On what basis are we justified in believing that a law holds? We know, of course, that all laws are based on the observation of certain regularities. They constitute indirect knowledge, as opposed to direct knowledge of facts. What justifies us in going from the direct observation of facts to a law that expresses certain regularities of nature? This is what in traditional terminology is called "the problem of induction".

Induction is often contrasted with deduction by saying that deduction goes from the general to the specific or singular, whereas induction goes the other way, from the singular to the general. This is a misleading oversimplification. In deduction, there are kinds of inferences other than those from the general to the specific; in induction there are also many kinds of inference. The traditional distinction is also misleading because it suggests that deduction and induction are simply two branches of a single kind of logic. John Stuart Mill's famous work, *A System of Logic*, contains a lengthy description of what he called "inductive logic" and states various canons of inductive procedure. Today we are more reluctant to use the term "inductive inference". If it is used at all, we must realize that it refers to a kind of inference that differs fundamentally from deduction.

In deductive logic, inference leads from a set of premisses to a conclusion just as certain as the premisses. If you have reason to believe the premisses, you have equally valid reason to believe the conclusion that follows logically from the premisses. If the premisses are true, the conclusion cannot be false. With respect to induction, the situation is entirely different. The truth of an inductive conclusion is never certain. I do not mean only that the conclusion cannot be certain because it rests on premisses that cannot be known with certainty. Even if the premisses are assumed to be true and the inference is a valid inductive inference, the conclusion may be false. The most we can say is that, with respect to given premisses, the conclusion has a certain degree of probability. Inductive logic tells us how to calculate the value of this probability.

We know that singular statements of fact, obtained by observation, are never absolutely certain because we may make errors in our observations; but, in respect to laws, there is still greater uncertainty. A law about the world states that, in any particular case, at any place and any time, if one thing is true, another thing is true. Clearly, this speaks about an infinity of possible instances. The actual instances may not be infinite, but there is an infinity of possible instances. A physiological law says that, if you stick a dagger into the heart of any human being, he will die. Since no exception to this law has ever been observed, it is accepted as universal. It is true, of course, that the number of instances so far observed of daggers being thrust into human hearts is finite. It is possible that some day humanity may cease to exist; in that case, the number of human beings, both past and future, is finite. But we do not know that humanity will cease to exist. Therefore, we must say that there is an infinity of possible instances, all of which are covered by the law. And, if there is an infinity of instances, no number of finite observations, however large, can make the "universal" law certain.

Of course, we may go on and make more and more observations, making them in as careful and scientific a manner as we can, until eventually we may say: "This law has been tested so many times that we can have complete confidence in its truth. It is a well-established, well-founded law." If we think about it, however, we see that even the best-founded laws of physics must rest on only a finite number of observations. It is always possible that tomorrow a counterinstance may be found. At no time is it possible to arrive at *complete* verification of a law. In fact, we should not speak of "verification" at all – if by the word we mean a definitive establishment of truth – but only of confirmation.

Interestingly enough, although there is no way in which a law can be verified (in the strict sense), there is a simple way it can be falsified. One need find only a single counterinstance. The knowledge of a counterinstance may, in itself, be uncertain. You may have made an error of observation or have been deceived in some way. But, if we assume that the counterinstance is a fact, then the negation of the law follows immediately. If a law says that every object that is P is also Q and we find an object that is P and not Q, the law is refuted. A million positive instances are insufficient to verify the law; one counterinstance is sufficient to falsify it. The situation is strongly asymmetric. It is easy to refute a law; it is exceedingly difficult to find strong confirmation.

How do we find confirmation of a law? If we have observed a great many positive instances and no negative instance, we say that the confirmation is strong. How strong it is and whether the strength can be expressed numerically is still a controversial question in the philosophy of science.

We will return to this in a moment. Here we are concerned only with making clear that our first task in seeking confirmation of a law is to test instances to determine whether they are positive or negative. This is done by using our logical schema to make predictions. A law states that $(x)(Px \supset Qx)$; hence, for a given object a, $Pa \supset Qa$. We try to find as many objects as we can (here symbolized by "a") that have the property P. We then observe whether they also fulfill the condition Q. If we find a negative instance, the matter is settled. Otherwise, each positive instance is additional evidence adding to the strength of our confirmation.

There are, of course, various methodological rules for efficient testing. For example, instances should be diversified as much as possible. If you are testing the law of thermal expansion, you should not limit your tests to solid substances. If you are testing the law that all metals are good conductors of electricity, you should not confine your tests to specimens of copper. You should test as many metals as possible under various conditions – hot, cold, and so on. We will not go into the many methodological rules for testing; we will only point out that in all cases the law is tested by making predictions and then seeing whether those predictions hold. In some cases, we find in nature the objects that we wish to test. In other cases, we have to produce them. In testing the law of thermal expansion, for example, we do not look for objects that are hot; we take certain objects and heat them. Producing conditions for testing has the great advantage that we can more easily follow the methodological rule of diversification; but whether we create the situations to be tested or find them ready-made in nature, the underlying schema is the same.

A moment ago I raised the question of whether the degree of confirmation of a law (or a singular statement that we are predicting by means of the law) can be expressed in quantitative form. Instead of saying that one law is "well founded" and that another law "rests on flimsy evidence", we might say that the first law has a .8 degree of confirmation, whereas the degree of confirmation for the second law is only .2. This question has long been debated. My own view is that such a procedure is legitimate and that what I have called "degree of confirmation" is identical with logical probability.

Such a statement does not mean much until we know what is meant by "logical probability". Why do I add the adjective "logical"? It is not customary practice; most books on probability do not make a distinction between various kinds of probability, one of which is called "logical". It is my belief, however, that there are two fundamentally different kinds of probability, and I distinguish between them by calling one "statistical probability", and the other "logical probability". It is unfortunate that the same word, "probability", has been used in two such widely differing senses. Failing to make the distinction is a source of enormous confusion in books on the philosophy of science as well as in the discourse of scientists themselves.

Instead of "logical probability", I sometimes use the term "inductive probability", because in my conception this is the kind of probability that is meant whenever we make an inductive inference. By "inductive inference" I mean, not only inference from facts to laws, but also any inference that is "nondemonstrative"; that is, an inference such that the conclusion does not follow with logical necessity when the truth of the premisses is granted. Such inferences must be expressed in degrees of what I call "logical probability" or "inductive probability".

[...]

In my conception, logical probability is a logical relation somewhat similar to logical implication; indeed, I think probability may be regarded as a partial implication. If the evidence is so strong that the hypothesis follows logically from it – is logically implied by it – we have one extreme case in which the probability is 1. (Probability 1 also occurs in other cases, but this is one special case where is occurs.) Similarly, if the negation of a hypothesis is logically implied by the evidence, the logical probability of the hypothesis is 0. In between, there is a continuum of cases about which deductive logic tells us nothing beyond the negative assertion that neither the hypothesis nor its negation can be deduced from the evidence. On this continuum inductive logic must take over. But inductive logic is like deductive logic in being concerned solely with the statements involved, not with the facts of nature. By a logical analysis of a stated hypothesis h and stated evidence *e*, we conclude that *h* is not logically implied but is, so to speak, partially implied by *e* to the degree of so-and-so much.

At this point, we are justified, in my view, in assigning numerical value to the probability. If possible, we should like to construct a system of inductive logic of such a kind that for any pair of sentences, one asserting evidence *e* and the other stating a hypothesis *h*, we can assign a number giving the logical probability of *h* with respect to *e*. (We do not consider the trivial case in which the sentence *e* is contradictory; in such instances, no probability value can be assigned to *h*.) [. . .]

When I say I think it is possible to apply an inductive logic to the language of science, I do not mean that it is possible to formulate a set of rules, fixed once and for all, that will lead automatically, in any field, from facts to theories. It seems doubtful, for example, that rules can be formulated to enable a scientist to survey a hundred thousand sentences giving various observational reports and then find, by a mechanical application of those rules, a general theory (system of laws) that would explain the observed phenomena. This is usually not possible, because theories, especially the more abstract ones dealing with such non-observables as particles and fields, use a conceptual framework that goes far beyond the framework used for the description of observation material. One cannot simply follow a mechanical procedure based on fixed rules to devise a new system of theoretical concepts, and with its help a theory. Creative ingenuity is required. This point is sometimes expressed by saying that there cannot be an inductive machine – a computer into which we can put all the relevant observational sentences and get, as an output, a neat system of laws that will explain the observed phenomena.

I agree that there cannot be an inductive machine if the purpose of the machine is to invent new theories. I believe, however, that there can be an inductive machine with a much more modest aim. Given certain observations *e* and a hypothesis *h* (in the form, say, of a prediction or even of a set of laws), then I believe it is in many cases possible to determine, by mechanical procedures, the logical probability, or degree of confirmation, of *h* on the basis of *e*. For this concept of probability, I also use the term "inductive probability", because I am convinced that this is the basic concept involved in all inductive reasoning and that the chief task of inductive reasoning is the evaluation of this probability. [. . .]

Statements giving values of statistical probability are not purely logical; they are factual statements in the language of science. When a medical man says that the probability is "very good" (or perhaps he uses a numerical value and says .7) that a patient will react positively to a certain injection, he is making a statement in medical science. When a physicist says that the probability of a certain radioactive phenomenon is so-and-so much, he is making a statement in physics. Statistical probability is a scientific, empirical concept. Statements about statistical probability are "synthetic" statements, statements that cannot be decided by logic but which rest on empirical investigations . . .

[. . .]

On the other hand, we also need the concept of logical probability. It is especially useful in metascientific statements, that is, statements about science. We say to a scientist: "You tell me that I can rely on this law in making a certain prediction. How well established is the law? How trustworthy is the prediction?" The scientist today may or may not be willing to answer a metascientific question of this kind in quantitative terms. But I believe that, once inductive logic is sufficiently developed, he could reply: "This hypothesis is confirmed to degree .8 on the basis of the available evidence." A scientist who answers in this way is making a statement about a logical relation between the evidence and the hypothesis in question. The sort of probability he has in mind is logical probability, which I also call "degree of confirmation". His statement that the value of this probability is .8 is, in this context, not a synthetic (empirical) statement, but an analytic one. It is analytic because no empirical investigation is demanded. It expresses a logical relation between a sentence that states the evidence and a sentence that states the hypothesis.

Note that, in making an analytic statement of probability, it is always necessary to specify the evidence explicitly. The scientist must not say: "The hypothesis has a probability of .8." He must add, "with respect to such and such evidence." If this is not added, his statement might be taken as a statement of statistical probability. If he intends it to be a statement of logical probability, it is an elliptical statement in which an important component has been left out. In quantum theory, for instance, it is often difficult to know whether a physicist means statistical probability or logical probability. Physicists usually do not draw this distinction. They talk as though there were only one concept of probability with which they work. "We mean that kind of probability that fulfills the ordinary axioms of probability theory", they may say. But the ordinary axioms of probability theory are fulfilled by both concepts, so this remark does not clear up the question of exactly what type of probability they mean.

[. . .]

I will not go into greater detail here about my view of probability, because many technicalities are involved. But I will discuss the one inference in which the two concepts of probability may come together. This occurs when either the hypothesis or one of the premisses for the inductive inference contains a concept of statistical probability. We can see this easily by modifying the basic schema used in our discussion of universal laws. Instead of a universal law (1), we take as the first premiss a statistical law (1′), which says that the relative frequency (*rf*) of *Q* with respect to *P* is (say) .8. The second premiss (2) states, as before, that a certain individual *a* has the property *P*. The third statement (3) asserts that *a* has the property *Q*. This third statement, *Qa*, is the hypothesis we wish to consider on the basis of the two premisses.

In symbolic form:

(1′) $rf(Q,P) = .8$
(2) Pa
(3) Qa

What can we say about the logical relation of (3) to (1′) and (2)? In the previous case – the schema for a universal law – we could make the following logical statement:

(4) Statement (3) is logically implied by (1) and (2).

We cannot make such a statement about the schema given above because the new premiss (1′) is weaker than the former premiss (1); it states a relative frequency rather than a universal law. We *can*, however, make the following statement, which also asserts a logical relation, but in terms of logical probability or degree of confirmation, rather than in terms of implication:

(4′) Statement (3), on the basis of (1′) and (2), has a probability of .8.

Note that this statement, like statement (4), is not a logical inference from (1′) and (2). Both (4) and (4′) are statements in what is called a metalanguage; they are logical statements *about* three assertions: (1) [or (1′), respectively], (2), and (3).

[. . .]

The main points that I wish to stress here are these: Both types of probability – statistical and logical – may occur together in the same chain of reasoning. Statistical probability is part of the object language of science. To statements about statistical probability we can apply logical probability, which is part of the metalanguage of science. It is my conviction that this point of view gives a much clearer picture of statistical inference than is commonly found in books on statistics and that it provides an essential groundwork for the construction of an adequate inductive logic of science.

Theories and Nonobservables

One of the most important distinctions between two types of laws in science is the distinction between what may be called (there is no generally accepted terminology for them) empirical laws and theoretical laws. Empirical laws are laws that can be confirmed directly by empirical observations. The term "observable" is often used for any phenomenon that can be directly observed, so it can be said that empirical laws are laws about observables.

Here, a warning must be issued. Philosophers and scientists have quite different ways of using the terms "observable" and "nonobservable". To a philosopher, "observable" has a very narrow meaning. It applies to such properties as "blue", "hard", "hot". These are properties directly perceived by the senses. To the physicist, the word has a much broader meaning. It includes any quantitative magnitude that can be measured in a relatively simple, direct way. A philosopher would not consider a temperature of, perhaps, 80 degrees centigrade, or a weight of $93^1/2$ pounds, an observable because there is no direct sensory perception of such magnitudes. To a physicist, both are observables because they can be measured in an extremely simple way. The object to be weighed is placed on a balance scale. The temperature is measured with a thermometer. The physicist would not say that the mass of a molecule, let alone the mass of an electron, is something observable, because here the procedures of measurement are much more complicated and indirect. But magnitudes that can be established by relatively simple procedures – length with a ruler, time with a clock, or frequency of light waves with a spectrometer – are called observables.

A philosopher might object that the intensity of an electric current is not really observed. Only a pointer position was observed. An ammeter was attached to the circuit and it was noted that the pointer pointed to a mark labeled 5.3. Certainly the current's intensity was not observed. It was *inferred* from what was observed.

The physicist would reply that this was true enough, but the inference was not very complicated. The procedure of measurement is so simple, so well established, that it could not be doubted that the ammeter would give an accurate measurement of current intensity. Therefore, it is included among what are called observables.

There is no question here of who is using the term "observable" in a right or proper way. There is a continuum which starts with direct sensory observations and proceeds to enormously complex, indirect methods of observation. Obviously no sharp line can be drawn across this continuum; it is a matter of degree. A philosopher is sure that the sound of his wife's voice, coming from across the room, is an observable. But suppose he listens to her on the telephone. Is her voice an observable or isn't it? A physicist would certainly say that when he looks at something through an ordinary microscope, he is observing it directly. Is this also the case when he looks into an electron microscope? Does he observe the path of a particle when he sees the track it makes in a bubble chamber? In general, the physicist speaks of observables in a very wide sense compared with the narrow sense of the philosopher, but, in both cases, the line separating observable from nonobservable is highly arbitrary. It is well to keep this in mind whenever these terms are encountered in a book by a philosopher or scientist. Individual authors will draw the line where it is most convenient, depending on their points of view, and there is no reason why they should not have this privilege.

Empirical laws, in my terminology, are laws containing terms either directly observable by the senses or measurable by relatively simple techniques. Sometimes such laws are called empirical generalizations, as a reminder that they have been obtained by generalizing results found by observations and measurements. They include not only simple qualitative laws (such as, "All ravens are black") but also quantitative laws that arise from simple measurements. The laws relating pressure, volume, and temperature of gases are of this type. Ohm's law, connecting the electric potential difference, resistance, and intensity of current, is another familiar example. The scientist makes repeated measurements, finds certain regularities, and expresses them in a law. These are the empirical laws. As indicated in earlier chapters, they are used for explaining observed facts and for predicting future observable events.

There is no commonly accepted term for the second kind of laws, which I call *theoretical laws*. Sometimes they are called abstract or hypothetical laws. "Hypothetical" is perhaps not suitable because it suggests that the distinction between the two types of laws is based on the degree to which the laws are confirmed. But an empirical law, if it is a tentative hypothesis, confirmed only to a low degree, would still be an empirical law although it might be said that it was rather hypothetical. A theoretical law is not to be distinguished from an empirical law by the fact that it is not well established, but by the fact that it contains terms of a different kind. The terms of a theoretical law do not refer to observables even when the physicist's wide meaning for what can be observed is adopted. They are laws about such entities as molecules, atoms, electrons, protons, electromagnetic fields, and others that cannot be measured in simple, direct ways.

[. . .]

It is true, as shown earlier, that the concepts "observable" and "nonobservable" cannot be sharply defined because they lie on a continuum. In actual practice, however, the difference is usually great enough so there is not likely to be debate. All physicists would agree that the laws relating pressure, volume, and temperature of a gas, for example, are empirical laws. Here the amount of gas is large enough so that the magnitudes to be measured remain constant over a sufficiently large volume of space and period of time to permit direct, simple measurements which can then be generalized into laws. All physicists would agree that laws about the behavior of single molecules are theoretical. Such laws concern a microprocess about which generalizations cannot be based on simple, direct measurements.

Theoretical laws are, of course, more general than empirical laws. It is important to understand, however, that theoretical laws cannot be arrived at simply by taking the empirical laws, then generalizing a few steps further. How does a physicist arrive at an empirical law? He observes certain events in nature. He notices a certain regularity. He describes this regularity by making an inductive generalization. It might be supposed that he could now put together a group of empirical laws, observe some sort of pattern, make a wider inductive generalization, and arrive at a theoretical law. Such is not the case.

To make this clear, suppose it has been observed that a certain iron bar expands when heated. After the experiment has been repeated many times, always with the same result, the regularity is generalized by saying that this bar expands when heated. An empirical law has been stated, even though it has a narrow range and applies only to one particular iron bar. Now further tests are made of other iron objects with the ensuing discovery that every time an iron object is heated it expands. This permits a more general law to be formulated, namely that all bodies of iron expand when heated. In similar fashion, the still more general laws "All metals . . .", then "All solid bodies . . .", are developed. These are all simple generalizations, each a bit more general than the previous one, but they are all empirical laws. Why? Because in each case, the objects dealt with are observable (iron, copper, metal, solid bodies); in each case the increases in temperature and length are measurable by simple, direct techniques.

In contrast, a theoretical law relating to this process would refer to the behavior of molecules in the iron bar. In what way is the behavior of the molecules connected with the expansion of the bar when heated? You see at once that we are now speaking of nonobservables. We must introduce a theory – the atomic theory of matter – and we are quickly plunged into atomic laws involving concepts radically different from those we had before. It is true that these theoretical concepts differ from concepts of length and temperature only in the degree to which they are directly or indirectly observable, but the difference is so great that there is no debate about the radically different nature of the laws that must be formulated.

Theoretical laws are related to empirical laws in a way somewhat analogous to the way empirical laws are related to single facts. An empirical law helps to explain a fact that has been observed and to predict a fact not yet observed. In similar fashion, the theoretical law helps to explain empirical laws already formulated, and to permit the derivation of new empirical laws. Just as the single, separate facts fall into place in an orderly pattern when they are generalized in an empirical law, the single and separate empirical laws fit into the orderly pattern of a theoretical law. This raises one of the main problems in the methodology of science. How can the kind of knowledge that will justify the assertion of a theoretical law be obtained? An empirical law may be justified by making observations of single facts. But to justify a theoretical law, comparable observations cannot be made because the entities referred to in theoretical laws are nonobservables.

Before taking up this problem, some remarks made in an earlier chapter, about the use of the word "fact", should be repeated. It is important in the present context to be extremely careful in the use of this word because some authors, especially scientists, use "fact" or "empirical fact" for some propositions which I would call empirical laws. For example, many physicists will refer to the "fact" that the specific heat of copper is .090. I would call this a law because in its full formulation it is seen to be a universal conditional statement: "For any x, and any time t, if x is a solid body of copper, then the specific heat of x at t is .090." Some physicists may even speak of the law of thermal expansion, Ohm's law, and others, as facts. Of course, they can then say that theoretical laws help explain such facts. This sounds like my statement that empirical laws explain facts, but the word "fact" is being used here in two different ways. I restrict the word to particular, concrete facts that can be spatiotemporally specified, not thermal expansion in general, but *the* expansion of this iron bar observed this morning at ten o'clock when it was heated. It is important to bear in mind the restricted way in which I speak of facts. If the word "fact" is used in an ambiguous manner, the important difference between the ways in which empirical and theoretical laws serve for explanation will be entirely blurred.

How can theoretical laws be discovered? We cannot say: "Let's just collect more and more data, then generalize beyond the empirical laws until we reach theoretical ones." No theoretical law was ever found that way. We observe stones and trees and flowers, noting various regularities and describing them by empirical laws. But no matter how long or how carefully we observe such things, we never reach a point at which we observe a molecule. The term "molecule" never arises as a result of observations. For this reason, no amount of generalization from observations will ever produce a theory of molecular processes. Such a theory must arise in another way. It is stated not as a generalization of facts but as a hypothesis. The hypothesis is then tested in a manner analogous in certain ways to the testing of an empirical law. From the hypothesis, certain empirical laws are derived, and these empirical laws are tested in turn by observation of facts. Perhaps the empirical laws derived from the theory are already known and well confirmed. (Such laws may even have motivated the formulation of the theoretical law.) Regardless of whether the derived empirical laws are known and confirmed, or whether they are new laws confirmed by new observations, the confirmation of such derived laws provides indirect confirmation of the theoretical law.

The point to be made clear is this. A scientist does not start with one empirical law, perhaps Boyle's law for gases, and then seek a theory about molecules from which this law can be derived. The scientist tries to formulate a much more general theory from which a variety of empirical laws can be derived. The more such laws, the greater their variety and apparent lack of connection with one another, the stronger will be the theory that explains them. Some of these derived laws may have been known before, but the theory may also make it possible to derive new empirical laws which can be confirmed by new tests. If this is the case, it can be said that the theory made it possible to predict new empirical laws. The prediction is understood in a hypothetical way. If the theory holds, certain empirical laws will also hold. The predicted empirical law speaks about relations between observables, so it is now possible to make experiments to see if the empirical law holds. If the empirical law is confirmed, it provides indirect confirmation of the theory. Every confirmation of a law, empirical or theoretical, is, of course, only partial, never complete and absolute. But in the case of empirical laws, it is a more direct confirmation. The confirmation of a theoretical law is indirect, because it takes place only through the confirmation of empirical laws derived from the theory.

The supreme value of a new theory is its power to predict new empirical laws. It is true that it also has value in explaining known empirical laws, but this is a minor value. If a scientist proposes a new theoretical system, from which no new laws can be derived, then it is logically equivalent to the set of all known empirical laws. The theory may have a certain elegance, and it may simplify to some degree the set of all known laws, although it is not likely that there would be an essential simplification. On the other hand, every new theory in physics that has led to a great leap forward has been a theory from which new empirical laws could be derived. If Einstein had done no more than propose his theory of relativity as an elegant new theory that would embrace certain known laws – perhaps also simplify them to a certain degree – then his theory would not have had such a revolutionary effect.

Of course it was quite otherwise. The theory of relativity led to new empirical laws which explained for the first time such phenomena as the movement of the perihelion of Mercury, and the bending of light rays in the neighborhood of the sun. These predictions showed that relativity theory was more than just a new way of expressing the old laws. Indeed, it was a theory of great predictive power. The consequences that can be derived from Einstein's theory are far from being exhausted. These are consequences that could not have been derived from earlier theories. Usually a theory of such power does have an elegance, and a unifying effect on known laws. It is simpler than the total collection of known laws. But the great value of the theory lies in its power to suggest new laws that can be confirmed by empirical means.

Correspondence Rules

An important qualification must now be added to the discussion of theoretical laws and terms given in the last chapter. The statement that empirical laws are derived from theoretical laws is an oversimplification. It is not possible to derive them directly because a theoretical law contains theoretical terms, whereas an empirical law contains only observable terms. This prevents any direct deduction of an empirical law from a theoretical one.

To understand this, imagine that we are back in the nineteenth century, preparing to state for the first time some theoretical laws about molecules in a gas. These laws are to describe the number of molecules per unit volume of the gas, the molecular velocities, and so forth. To simplify matters, we assume that all the molecules have the same velocity. (This was indeed the original assumption; later it was abandoned in favor of a certain probability distribution of velocities.) Further assumptions must be made about what happens when molecules collide. We do not know the exact shape of molecules, so let us suppose that they are tiny spheres. How do spheres collide? There are laws about colliding spheres, but they concern large bodies. Since we cannot directly observe molecules, we assume their collisions are analogous to those of large bodies; perhaps they behave like perfect billiard balls on a frictionless table. These are, of course, only assumptions; guesses suggested by analogies with known macrolaws.

But now we come up against a difficult problem. Our theoretical laws deal exclusively with the behavior of molecules, which cannot be seen. How, therefore, can we deduce from such laws a law about observable properties such as the pressure or temperature of a gas or properties of sound waves that pass through the gas? The theoretical laws contain only theoretical terms. What we seek are empirical laws containing observable terms. Obviously, such laws cannot be derived without having something else given in addition to the theoretical laws.

The something else that must be given is this: a set of rules connecting the theoretical terms with the observable terms. Scientists and philosophers of science have long recognized the need for such a set of rules, and their nature has been often discussed. An example of such a rule is: "If there is an electromagnetic oscillation of a specified frequency, then there is a visible greenish-blue color of a certain hue." Here something observable is connected with a nonobservable microprocess.

Another example is: "The temperature (measured by a thermometer and, therefore, an observable in the wider sense explained earlier) of a gas is proportional to the mean kinetic energy of its molecules." This rule connects a nonobservable in molecular theory, the kinetic energy of molecules, with an observable, the temperature of the gas. If statements of this kind did not exist, there would be no way of deriving empirical laws about observables from theoretical laws about nonobservables.

Different writers have different names for these rules. I call them "correspondence rules". P. W. Bridgman calls them operational rules. Norman R. Campbell speaks of them as the "Dictionary".[1] Since the rules connect a term in one terminology with a term in another terminology, the use of the rules is analogous to the use of a French–English dictionary. What does the French word "cheval" mean? You look it up in the dictionary and find that it means "horse". It is not really that simple when a set of rules is used for connecting nonobservables with observables; nevertheless, there is an analogy here that makes Campbell's "Dictionary" a suggestive name for the set of rules.

There is a temptation at times to think that the set of rules provides a means for defining theoretical terms, whereas just the opposite is really true. A theoretical term can never be explicitly defined on the basis of observable terms, although sometimes an observable can be defined in theoretical terms. For example, "iron" can be defined as a substance consisting of small crystalline parts, each having a certain arrangement of atoms and each atom being a configuration of particles of a certain type. In theoretical terms then, it is possible to express what is meant by the observable term "iron", but the reverse is not true.

There is no answer to the question: "Exactly what is an electron?" Later we shall come back to this question, because it is the kind that philosophers are always asking scientists. They want the physicist to tell them just what he means by "electricity", "magnetism", "gravity", "a molecule". If the physicist explains them in theoretical terms, the philosopher may be disappointed. "That is not what I meant at all", he will say. "I want you to tell me, in ordinary language, what those terms mean." Sometimes the philosopher writes a book in which he talks about the great mysteries of nature. "No one", he writes, "has been able so far, and perhaps no one ever will be able, to give us a straightforward answer to the question: 'What is electricity?' And so electricity remains forever one of the great, unfathomable mysteries of the universe."

There is no special mystery here. There is only an improperly phrased question. Definitions that cannot, in the nature of the case, be given, should not be demanded. If a child does not know what an elephant is, we can tell him it is a huge animal with big ears and a long trunk. We can show him a picture of an elephant. It serves admirably to define an elephant in observable terms that a child can understand. By analogy, there is a temptation to believe that, when a scientist introduces theoretical terms, he should also be able to define them in familiar terms. But this is not possible. There is no way a physicist can show us a picture of electricity in the way he can show his child a picture of an elephant. Even the cell of an organism, although it cannot be seen with the unaided eye, can be represented by a picture because the cell can be seen when it is viewed through a microscope. But we do not possess a picture of the electron. We cannot say how it looks or how it feels, because it cannot be seen or touched. The best we can do is to say that it is an extremely small body that behaves in a certain manner. This may seem to be analogous to our description of an elephant. We can describe an elephant as a large animal that behaves in a certain manner. Why not do the same with an electron?

The answer is that a physicist can describe the behavior of an electron only by stating theoretical laws, and these laws contain only theoretical terms. They describe the field produced by an electron, the reaction of an electron to a field, and so on. If an electron is in an electrostatic field, its velocity will accelerate in a certain way. Unfortunately, the electron's acceleration is an unobservable. It is not like the acceleration of a billiard ball, which can be studied by direct observation. There is no way that a theoretical concept can be defined in terms of observables. We must, therefore, resign ourselves to the fact that definitions of the kind that can be supplied for observable terms cannot be formulated for theoretical terms.

It is true that some authors, including Bridgman, have spoken of the rules as "operational definitions". Bridgman had a certain justification, because he used his rules in a somewhat different way, I believe, than most physicists use them. He was a great physicist and was certainly aware of his departure from the usual use of rules, but he was willing to accept certain forms of speech that are not customary, and this explains his departure. It was pointed out in a previous chapter that Bridgman preferred to say that there is not just one concept of intensity of electric current, but a dozen concepts. Each procedure by which a magnitude can be measured provides an operational definition for that magnitude. Since there are different procedures for measuring current, there are different concepts. For the sake of convenience, the physicist speaks of just one concept of current. Strictly speaking, Bridgman believed, he should recognize many different concepts, each defined by a different operational procedure of measurement.

We are faced here with a choice between two different physical languages. If the customary procedure among physicists is followed, the various concepts of current will be replaced by one concept. This means, however, that you place the concept in your theoretical laws, because the operational rules are just correspondence rules, as I call them, which connect the theoretical terms with the empirical ones. Any claim to possessing a definition – that is, an operational definition – of the theoretical concept must be given up. Bridgman could speak of having operational definitions for his theoretical terms only because he was not speaking of a general concept. He was speaking of partial concepts, each defined by a different empirical procedure.

Even in Bridgman's terminology, the question of whether his partial concepts can be adequately defined by operational rules is problematic. Reichenbach speaks often of what he calls "correlative definitions". (In his German publications, he calls them *Zuordnungsdefinitionen*, from *zuordnen*, which means to correlate.) Perhaps correlation is a better term than definition for what Bridgman's rules actually do. In geometry, for instance, Reichenbach points out that the axiom system of geometry, as developed by David Hilbert, for example, is an uninterpreted axiom system. The basic concepts of point, line, and plane could just as well be called "class alpha", "class beta", and "class gamma". We must not be seduced by the sound of familiar words, such as "point" and "line", into thinking they must be taken in their ordinary meaning. In the axiom system, they are uninterpreted terms. But when geometry is applied to physics, these terms must be connected with something in the physical world. We can say, for example, that the lines of the geometry are exemplified by rays of light in a vacuum or by stretched cords. In order to connect the uninterpreted terms with observable physical phenomena, we must have rules for establishing the connection.

What we call these rules is, of course, only a terminological question; we should be cautious and not speak of them as definitions. They are not definitions in any strict sense. We cannot give a really adequate definition of the geometrical concept of "line" by referring to anything in nature. Light rays, stretched strings, and so on are only approximately straight; moreover, they are not lines, but only segments of lines. In geometry, a line is infinite in length and absolutely straight. Neither property is exhibited by any phenomenon in nature. For that reason, it is not possible to give an operational definition, in the strict sense of the word, of concepts in theoretical geometry. The same is true of all the other theoretical concepts of physics. Strictly speaking, there are no "definitions" of such concepts. I prefer not to speak of "operational definitions" or even to use Reichenbach's term "correlative definitions". In my publications (only in recent years have I written about this question), I have called them "rules of correspondence" or, more simply, "correspondence rules".

Campbell and other authors often speak of the entities in theoretical physics as mathematical entities. They mean by this that the entities are related to each other in ways that can be expressed by mathematical functions. But they are not mathematical entities of the sort that can be defined in pure mathematics. In pure mathematics, it is possible to define various kinds of numbers, the function of logarithm, the exponential function, and so forth. It is not possible, however, to define such terms as "electron" and "temperature" by pure mathematics. Physical terms can be introduced only with the help of nonlogical constants, based on observations of the actual world. Here we have an essential difference between an axiomatic system in mathematics and an axiomatic system in physics.

If we wish to give an interpretation to a term in a mathematical axiom system, we can do it by giving a definition in logic. Consider, for example, the term "number" as it is used in Peano's axiom system. We can define it in logical terms, by the Frege-Russell method, for example. In this way the concept of "number" acquires a complete, explicit definition on the basis of pure logic. There is no need to establish a connection between the number 5 and such observables as "blue" and "hot". The terms have only a logical interpretation; no connection with the actual world is needed. Sometimes an axiom system in mathematics is called a theory. Mathematicians speak of set theory, group theory, matrix theory, probability theory. Here the word "theory" is used in a purely analytic way. It denotes a deductive system that makes no reference to the actual world. We must always bear in mind that such a use of the word "theory" is entirely different from its use in reference to empirical theories such as relativity theory, quantum theory, psychoanalytical theory, and Keynesian economic theory.

A postulate system in physics cannot have, as mathematical theories have, a splendid isolation from the world. Its axiomatic terms – "electron", "field", and so on – must be interpreted by correspondence rules that connect the terms with observable phenomena. This interpretation is necessarily incomplete. Because it is always incomplete, the system is left open to make it possible to add new rules of correspondence. Indeed, this is what continually happens in the history of physics. I am not thinking now of a revolution in physics, in which an entirely new theory is developed, but of less radical changes that modify existing theories. Nineteenth-century physics provides a good example, because classical mechanics and electromagnetics had been established, and, for many decades, there was relatively little change in fundamental laws. The basic theories of physics remained unchanged. There was, however, a steady addition of new correspondence rules, because new procedures were continually being developed for measuring this or that magnitude.

Of course, physicists always face the danger that they may develop correspondence rules that will be incompatible with each other or with the theoretical laws. As long as such incompatibility does not occur, however, they are free to add new correspondence rules. The procedure is never-ending. There is always the possibility of adding new rules, thereby increasing the amount of interpretation specified for the theoretical terms; but no matter how much this is increased, the interpretation is never final. In a mathematical system, it is otherwise. There a logical interpretation of an axiomatic term *is* complete. Here we find another reason for reluctance in speaking of theoretical terms as "defined" by correspondence rules. It tends to blur the important distinction between the nature of an axiom system in pure mathematics and one in theoretical physics.

Is it not possible to interpret a theoretical term by correspondence rules so completely that no further interpretation would be possible? Perhaps the actual world is limited in its structure and laws. Eventually a point may be reached beyond which there will be no room for strengthening the interpretation of a term by new correspondence rules. Would not the rules then provide a final, explicit definition for the term? Yes, but then the term would no longer be theoretical. It would become part of the observation language. The history of physics has not yet indicated that physics will become complete; there has been only a steady addition of new correspondence rules and a continual modification in the interpretations of theoretical terms. There is no way of knowing whether this is an infinite process or whether it will eventually come to some sort of end.

It may be looked at this way. There is no prohibition in physics against making the correspondence rules for a term so strong that the term becomes explicitly defined and therefore ceases to be theoretical. Neither is there any basis for assuming that it will always be possible to add new correspondence rules. Because the history of physics has shown such a steady, unceasing modification of theoretical concepts, most physicists would advise against correspondence rules so strong that a theoretical term becomes explicitly defined. Moreover, it is a wholly unnecessary procedure. Nothing is gained by it. It may even have the adverse effect of blocking progress.

Of course, here again we must recognize that the distinction between observables and non-observables is a matter of degree. We might give an explicit definition, by empirical procedures, to a concept such as length, because it is so easily and directly measured, and is unlikely to be modified by new observations. But it would be rash to seek such strong correspondence rules that "electron" would be explicitly defined. The concept "electron" is so far removed from simple, direct observations that it is best to keep it theoretical, open to modifications by new observations.

Note

1 See Percy W. Bridgman, *The Logic of Modern Physics* (New York: Macmillan, 1927), and Norman R. Campbell, *Physics: The Elements* (Cambridge: Cambridge University Press, 1920). Rules of correspondence are discussed by Ernest Nagel, *The Structure of Science* (New York: Harcourt, Brace & World, 1961), pp. 97–105.

5.2

Scientific Explanation

Carl Hempel

Carl Hempel (1905–1997) was a member of the Berlin Society and the primary author of the received view on explanation. The following selection is an excerpt from his classic "Aspects of Scientific Explanation" and presents the deductive-nomological (D-N) and inductive-statistical (I-S) models of explanation. It also includes the requirement of maximal specificity (RMS) that constitutes his response to the problem of ambiguity of I-S explanations.

2 Deductive-Nomological Explanation

2.1 Fundamentals: D-N explanation and the concept of law

In his book, *How We Think*,[1] John Dewey describes a phenomenon he observed one day while washing dishes. Having removed some glass tumblers from the hot suds and placed them upside down on a plate, he noticed that soap bubbles emerged from under the tumbler's rims, grew for a while, came to a standstill and finally receded into the tumblers. Why did this happen? Dewey outlines an explanation to this effect: Transferring the tumblers to the plate, he had trapped cool air in them; that air was gradually warmed by the glass, which initially had the temperature of the hot suds. This led to an increase in the volume of the trapped air, and thus to an expansion of the soap film that had formed between the plate and the tumblers' rims. But gradually, the glass cooled off, and so did the air inside, and as a result, the soap bubbles receded.

The explanation here outlined may be regarded as an argument to the effect that the phenomenon to be explained, *the explanandum phenomenon*, was to be expected in virtue of certain explanatory facts. These fall into two groups: (i) particular facts and (ii) uniformities expressible by means of general laws. The first group includes facts such as these: the tumblers had been immersed in soap suds of a temperature considerably higher than that of the surrounding air; they were put, upside down, on a plate on which a puddle of soapy water had formed that provided a connecting soap film, and so on. The second group of explanatory facts would be expressed by the gas laws and by various other laws concerning the exchange of heat between bodies of different temperature, the elastic behavior of soap bubbles, and so on. While some of these laws are only hinted at by such phrasings as 'the warming of the trapped air led to an increase in its pressure', and others are not referred to even in this oblique fashion, they are clearly presupposed in the claim that certain stages in the process yielded others as their results. If we imagine the various explicit or tacit explanatory assumption to be fully stated, then the explanation may be conceived as a deductive argument of the form

$$\text{(D-N)}\quad \frac{\left.\begin{array}{l} C_1, C_2, \ldots, C_k \\ L_1, L_2, \ldots, L_r \end{array}\right\} \text{Explanans } S}{E \qquad \text{Explanandum-sentence}}$$

Here, $C_1, C_2, \ldots, C_k$ are sentences describing the particular facts invoked; $L_1, L_2, \ldots, L_r$ are the general laws on which the explanation rests. Jointly these sentences will be said to form the *explanans S*, where *S* may be thought of alternatively as the set of the explanatory sentences or as their conjunction. The conclusion *E* of the argument is a sentence describing the explanandum-phenomenon; I will call *E* the explanandum-sentence or explanandum-statement; the word 'explanandum' alone will be used to refer either to the explanandum-phenomenon or to the explanandum-sentence: the context will show which is meant.

The kind of explanation whose logical structure is suggested by the schema (D-N) will be called *deductive-nomological explanation* or D-N *explanation* for short; for it effects a deductive subsumption of the explanandum under principles that have the character of general laws. Thus a D-N explanation answers the question '*Why* did the explanandum-phenomenon occur?' by showing that the phenomenon resulted from certain particular circumstances, specified in $C_1, C_2, \ldots, C_k$, in accordance with the laws $L_1, L_2, \ldots, L_r$. By pointing this out, the argument shows that, given the particular circumstances and the laws in question, the occurrence of the phenomenon *was to be expected*; and it is in this sense that the explanation enables us to *understand why* the phenomenon occurred.[2]

In a D-N explanation, then, the explanandum is a logical consequence of the explanans. Furthermore, reliance on general laws is essential to a D-N explanation; it is in virtue of such laws that the particular facts cited in the explanans possess explanatory relevance to the explanandum phenomenon. Thus, in the case of Dewey's soap bubbles, the gradual warming of the cool air trapped under the hot tumblers would constitute a mere accidental antecedent rather than an explanatory factor for the growth of the bubbles, if it were not for the gas laws, which connect the two events. But what if the explanandum sentence *E* in an argument of the form (D-N) is a logical consequence of the sentences $C_1, C_2, \ldots, C_k$ alone? Then, surely, no empirical laws are *required* to deduce *E* from the explanans; and any laws included in the latter are gratuitous, dispensable premises. Quite so; but in this case, the argument would not count as an explanation. For example, the argument:

The soap bubbles first expanded
and then receded
―――――――――――――――
The soap bubbles first expanded

though deductively valid, clearly cannot qualify as an explanation of why the bubbles first expanded. The same remark applies to all other cases of this kind. A D-N explanation will have to contain, in its explanans, some general laws that are *required* for the deduction of the explanandum, i.e. whose deletion would make the argument invalid.

If the explanans of a given D-N explanation is true, i.e. if the conjunction of its constituent sentences is true, we will call the *explanation true*; a true explanation, of course, has a true explanandum as well. Next, let us call a *D-N explanation more or less strongly supported or confirmed* by a given body of evidence according as its explanans is more or less strongly confirmed by the given evidence. (One factor to be considered in appraising the empirical soundness of a given explanation will be the extent to which its explanans is supported by the total relevant evidence available.) Finally, by a *potential D-N explanation*, let us understand any argument that has the character of a D-N explanation except that the sentences constituting its explanans need not be true. In a potential D-N explanation, therefore, $L_1, L_2, \ldots, L_r$ will be what Goodman has called *lawlike sentences*, i.e. sentences that are like laws except for possibly being false. Sentences of this kind will also be referred to as *nomic* or *nomological.* We use the notion of a potential explanation, for example, when we ask whether a novel and as yet untested law or theory would provide an explanation for some empirical phenomenon; or when we say that the phlogiston theory, though now discarded, afforded an explanation for certain aspects of combustion.[3] Strictly speaking, only true lawlike statements can count as laws – one would hardly want to speak of false laws of nature. But for convenience I will occasionally use the term 'law' without implying that the sentence in question is true, as in fact, I have done already in the preceding sentence.

[. . .]

3.2 Deductive-statistical explanation

It is an instance of the so-called gambler's fallacy to assume that when several successive tossings of a fair coin have yielded heads, the next toss will more probably yield tails than heads. Why this is not the case can be explained by means of two hypotheses that have the form of statistical laws. The first is that the random experiment of flipping a fair coin yields heads with a statistical probability of 1/2. The second hypothesis is that the outcomes of different tossings of the coin are statistically independent, so that the probability of any specified sequence of outcomes – such as heads twice, then tails, then heads, then tails three times – equals the product of the probabilities of the constituent single outcomes. These two hypothesis in terms of statistical probabilities imply *deductively* that the probability for heads to come up after a long sequence of heads is still ½.

Certain statistical explanations offered in science are of the same deductive character, though often quite complex mathematically. Consider, for example, the hypothesis that for the atoms of every radioactive substance there is a characteristic probability of disintegrating during a given unit time interval. This complex statistical hypothesis explains, by deductive implication, various other statistical aspects of radioactive decay, among them, the following: Suppose that the decay of individual atoms of some radioactive substance is recorded by means of the scintillations produced upon a sensitive screen by the alpha particles emitted by the disintegrating atoms. Then the time intervals separating successive scintillations will vary considerably in length, but intervals of different lengths will occur with different statistical probabilities. Specifically, if the mean time interval between successive scintillations is s seconds, then the probability for two successive scintillations to be separated by more than $n{\cdot}s$ seconds is $(1/e)^n$, where e is the base of the natural logarithms.[4]

Explanations of the kind here illustrated will be called *deductive-statistical explanations*, or *D-S explanations*. They involve the deduction of a statement in the form of a statistical law from an explanans that contains indispensably at least one law or theoretical principle of statistical form. The deduction is effected by means of the mathematical theory of statistical probability, which makes it possible to calculate certain derivative probabilities (those referred to in the explanandum) on the basis of other probabilities (specified in the explanans) which have been empirically ascertained or hypothetically assumed. What a D-S explanation accounts for is thus always a general uniformity expressed by a presumptive law of statistical form.

Ultimately, however, statistical laws are meant to be applied to particular occurrences and to establish explanatory and predictive connections among them. In the next subsection, we will examine the statistical explanation of particular events. Our discussion will be limited to the case where the explanatory statistical laws are of basic form: this will suffice to exhibit the basic logical differences between the statistical and the deductive-nomological explanation of individual occurrences.

3.3 Inductive-statistical explanation

As an explanation of why patient John Jones recovered from a streptococcus infection, we might be told that Jones had been given penicillin. But if we try to amplify this explanatory claim by indicating a general connection between penicillin treatment and the subsiding of a streptococcus infection we cannot justifiably invoke a general law to the effect that in all cases of such infection, administration of penicillin will lead to recovery. What can be asserted, and what surely is taken for granted here, is only that penicillin will effect a cure in a high percentage of cases, or with a high statistical probability. This statement has the general character of a law of statistical form, and while the probability value is not specified, the statement indicates that it is high. But in contrast to the cases of deductive-nomological and deductive-statistical explanation, the explanans consisting of this statistical law together with the statement that the patient did receive penicillin obviously does not imply the explanandum statement, 'the patient recovered', with deductive certainty, but only, as we might say, with high likelihood, or near-certainty. Briefly, then, the explanation amounts to this argument:

(3a) The particular case of illness of John Jones – let us call it j – was an instance of severe streptococcal infection (Sj) which was treated with large doses of penicillin (Pj); and the statistical probability $p(R, S{\cdot}P)$ of recovery in cases where S and P are present is close to 1; hence, the case was practically certain to end in recovery (Rj).

This argument might invite the following schematization:

$$\text{(3b)}\quad \frac{\begin{array}{l} p(R, S{\cdot}P) \text{ is close to 1} \\ Sj \cdot Pj \end{array}}{\text{(Therefore:) It is practically certain (very likely) that } Rj}$$

In the literature on inductive inference, arguments thus based on statistical hypotheses have often been construed as having this form or a similar one. On this construal, the conclusion characteristically contains a modal qualifier such as 'almost certainly', 'with high probability', 'very likely', etc. But the conception of arguments having this character is untenable. For phrases of the form 'it is practically certain that p' or 'It is very likely that p', where the place of 'p' is taken by some statement, are not complete self-contained sentences that can be qualified as either true or false. The statement that takes the place of 'p' – for example, 'Rj' – is either true or false, quite independently of whatever relevant evidence may be available, but it can be qualified as more or less likely, probable, certain, or the like only *relative to some body of evidence*. One and the same statement, such as 'Rj', will be certain, very likely, not very likely, highly unlikely, and so forth, depending upon what evidence is considered. The phrase 'it is almost certain that Rj' taken by itself is therefore neither true nor false; and it cannot be inferred from the premises specified in (3b) nor from any other statements.

The confusion underlying the schematization (3b) might be further illuminated by considering its analogue for the case of deductive arguments. The force of a deductive inference, such as that from 'all F are G' and 'a is F' to 'a is G', is sometimes indicated by saying that if the premises are true, then the conclusion is necessarily true or is certain to be true – a phrasing that might suggest the schematization

$$\frac{\begin{array}{l} \text{All } F \text{ are } G \\ a \text{ is } F \end{array}}{\text{(Therefore:) It is necessary (certain) that } a \text{ is } G.}$$

But clearly the given premises – which might be, for example, 'all men are mortal' and 'Socrates is a man' – do not establish the sentence 'a is G' ('Socrates is mortal') as a necessary or certain truth. The certainty referred to in the informal paraphrase of the argument is relational: the statement 'a is G' is certain, or necessary, *relative to the specified premises*; i.e., their truth will guarantee its truth – which means nothing more than that 'a is G' is a logical consequence of those premises.

Analogously, to present our statistical explanation in the manner of schema (3b) is to misconstrue the function of the words 'almost certain' or 'very likely' as they occur in the formal wording of the explanation. Those words clearly must be taken to indicate that on the evidence provided by the explanans, or relative to that evidence, the explanandum is practically certain or very likely, i.e., that

(3c) 'Rj' is practically certain (very likely) relative to the explanans containing the sentences 'p (R, $S{\cdot}P$) is close to 1' and '$Sj \cdot Pj$'.[5]

The explanatory argument misrepresented by (3b) might therefore suitably be schematized as follows:

$$\text{(3d)}\quad \frac{\begin{array}{l} p(R, S{\cdot}P) \text{ is close to 1} \\ Sj{\cdot}Pj \end{array}}{Rj} \quad \text{[makes practically certain (very likely)]}$$

In this schema, the double line separating the "premises" from the "conclusion" is to signify that the relation of the former to the latter is not that of deductive implication but that of inductive support, the strength of which is indicated in square brackets.[6,7]

[. . .]

3.4 *The ambiguity of inductive-statistical explanation and the requirement of maximal specificity*

3.4.1 The problem of explanatory ambiguity

Consider once more the explanation (3d) of recovery in the particular case j of John Jones's illness. The statistical law there invoked claims recovery in response to penicillin only for a high percentage of streptococcal infections, but not for all of them; and in fact, certain streptococcus strains are resistant to penicillin. Let us say that an occurrence, e.g., a particular case of illness, has the property S^* (or belongs to the class S^*) if it is an instance of infection with a penicillin-resistant streptococcus strain. Then the probability of recovery among randomly chosen instances of S^* which are treated with penicillin will be quite small, i.e., $p(R, S^*{\cdot}P)$ will be close to 0 and the probability of non-recovery, $p(\bar{R}, S^*{\cdot}P)$ will be close to 1. But suppose now that Jones's illness is in fact a streptococcal infection of the penicillin-resistant variety, and consider the following argument:

$$\text{(3k)}\quad \frac{\begin{array}{l} p(\bar{R}, S^*{\cdot}P) \text{ is close to 1} \\ S^*j{\cdot}Pj \end{array}}{\bar{R}j} \quad \text{[makes practically certain]}$$

This "rival" argument has the same form as (3d), and on our assumptions, its premises are true, just like those of (3d). Yet its conclusion is the contradictory of the conclusion of (3d).

Or suppose that Jones is an octogenarian with a weak heart, and that in this group, S^{**}, the probability of recovery from a streptococcus infection in response to penicillin treatment, $p(R, S^{**}{\cdot}P)$, is quite small. Then, there is the following rival argument to (3d), which presents Jones's non-recovery as practically certain in the light of premises which are true:

$$\text{(3l)}\quad \frac{\begin{array}{l} p(\bar{R}, S^{**}{\cdot}P) \text{ is close to 1} \\ S^{**}j{\cdot}Pj \end{array}}{\bar{R}j} \quad \text{[makes practically certain]}$$

The peculiar logical phenomenon here illustrated will be called the *ambiguity of inductive-statistical explanation* or, briefly, of *statistical explanation*. This ambiguity derives from the fact that a given individual event (e.g., Jones's illness) will often be obtainable by random selection from any one of several "reference classes" (such as $S{\cdot}P$, $S^*{\cdot}P$, $S^{**}{\cdot}P$), with respect to which the kind of occurrence (e.g., R) instantiated by the given event has very different statistical probabilities. Hence, for a proposed probabilistic explanation with true explanans which confers near-certainty upon a particular event, there will often exist a rival argument of the same probabilistic form and with equally true premises which confers near-certainty upon the nonoccurrence of the same event. And any statistical explanation for the occurrence of an event must seem suspect if there is the possibility of a logically and empirically equally sound probabilistic account for its nonoccurrence. *This predicament has no analogue in the case of deductive explanation*; for if the premises of a proposed deductive explanation are true then so is its conclusion; and its contradictory, being false, cannot be a logical consequence of a rival set of premises that are equally true.

Here is another example of the ambiguity of I-S explanation: Upon expressing surprise at finding the weather in Stanford warm and sunny on a date as autumnal as November 27, I might be told, by way of explanation, that this was rather to be expected because the probability of warm and sunny weather (W) on a November day in Stanford (N) is, say, .95. Schematically, this account would take the following form, where 'n' stands for 'November 27':

$$p(W, N) = .95$$
$$(3m)\quad \frac{Nn}{Wn} \quad [.95]$$

But suppose it happens to be the case that the day before, November 26, was cold and rainy, and that the probability for the immediate successors (S) of cold and rainy days in Stanford to be warm and sunny is .2; then the account (3m) has a rival in the following argument which, by reference to equally true premises, presents it as fairly certain that November 27 is not warm and sunny:

$$p(\bar{W}, S) = .8$$
$$(3n)\quad \frac{Sn}{\bar{W}n} \quad [.8]$$

In this form, the problem of ambiguity concerns I-S arguments whose premises are in fact true, no matter whether we are aware of this or not. But, as will now be shown, the problem has a variant that concerns explanations whose explanans statements, no matter whether in fact true or not, are *asserted or accepted* by empirical science at the time when the explanation is proffered or contemplated. This variant will be called *the problem of the epistemic ambiguity of statistical explanation*, since it refers to what is presumed to be known in science rather than to what, perhaps unknown to anyone, is in fact the case.

Let K_t be the class of all statements asserted or accepted by empirical science at time t. This class then represents the total scientific information, or "scientific knowledge" at time t. The word 'knowledge' is here used in the sense in which we commonly speak of the scientific knowledge at a given time. It is not meant to convey the claim that the elements of K_t are true, and hence neither that they are definitely known to be true. No such claim can justifiably be made for any of the statements established by empirical science; and the basic standards of scientific inquiry demand that an empirical statement, however well supported, be accepted and thus admitted to membership in K_t only tentatively, i.e., with the understanding that the privilege may be withdrawn if unfavorable evidence should be discovered. The membership of K_t therefore changes in the course of time; for as a result of continuing research, new statements are admitted into that class; others may come to be discredited and dropped. Henceforth, the class of accepted statements will be referred to simply as K when specific reference to the time in question is not required. We will assume that K is logically consistent and that it is closed under logical implication, i.e., that it contains every statement that is logically implied by any of its subsets.

The *epistemic ambiguity of I-S explanation* can now be characterized as follows: The total set K of accepted scientific statements contains different subsets of statements which can be used as premises in arguments of the probabilistic form just considered, and which confer high probabilities on logically contradictory "conclusions." Our earlier examples (3k), (3l) and (3m), (3n) illustrate this point if we assume that the premises of those arguments all belong to K rather than that they are all true. If one of two such rival arguments with premises in K is proposed as an explanation of an event considered, or acknowledged, in science to have occurred, then the conclusion of the argument, i.e., the explanandum statement, will accordingly belong to K as well. And since K is consistent, the conclusion of the rival argument will not belong to K. Nonetheless it is disquieting that we should be able to say: No matter whether we are informed that the event in question (e.g., warm and sunny weather on November 27 in Stanford) did occur or that it did not occur, we can produce an explanation of the reported outcome in either case; and an explanation, moreover, whose premises are scientifically established statements that confer a high logical probability upon the reported outcome.

This epistemic ambiguity, again, has no analogue for deductive explanation; for since K is logically consistent, it cannot contain premise-sets that imply logically contradictory conclusions.

Epistemic ambiguity also bedevils the predictive use of statistical arguments. Here, it has the alarming aspect of presenting us with two rival arguments whose premises are scientifically well established, but one of which characterizes a contemplated future occurrence as practically certain, whereas the other characterizes it as practically impossible. Which of such conflicting arguments, if any, are rationally to be relied on for explanation or for prediction?

3.4.2 The requirement of maximal specificity and the epistemic relativity of inductive-statistical explanation

Our illustrations of explanatory ambiguity suggest that a decision on the acceptability of a proposed probabilistic explanation or prediction will have to be made in the light of all the relevant information at our disposal. This is indicated also by a general principle whose importance for inductive reasoning has been acknowledged, if not always very explicitly, by many writers, and which has recently been strongly emphasized by Carnap, who calls it *the requirement of total evidence.* Carnap formulates it as follows: "in the application of inductive logic to a given knowledge situation, the total evidence available must be taken as basis for determining the degree of confirmation."[8] Using only a part of the total evidence is permissible if the balance of the evidence is irrelevant to the inductive "conclusion," i.e., if on the partial evidence alone, the conclusion has the same confirmation, or logical probability, as on the total evidence.[9]

The requirement of total evidence is not a postulate nor a theorem of inductive logic; it is not concerned with the formal validity of inductive arguments. Rather, as Carnap has stressed, it is a maxim for the *application* of inductive logic; we might say that it states a necessary condition of rationality of any such application in a given "knowledge situation," which we will think of as represented by the set K of all statements accepted in the situation.

But in what manner should the basic idea of this requirement be brought to bear upon probabilistic explanation? Surely we should not insist that the explanans must contain all and only the empirical information available at the time. Not *all* the available information, because otherwise all probabilistic explanations acceptable at time t would have to have the same explanans, K_t; and not *only* the available information, because a proffered explanation may meet the intent of the requirement in not overlooking any relevant information available, and may nevertheless invoke some explanans statements which have not as yet been sufficiently tested to be included in K_t.

The extent to which the requirement of total evidence should be imposed upon statistical explanations is suggested by considerations such as the following. A proffered explanation of Jones's recovery based on the information that Jones had a streptococcal infection and was treated with penicillin, and that the statistical probability for recovery in such cases is very high is unacceptable if K includes the further information that Jones's streptococci were resistant to penicillin, or that Jones was an octogenarian with a weak heart, and that in these reference classes the probability of recovery is small. Indeed, one would want an acceptable explanation to be based on a statistical probability statement pertaining to the narrowest reference class of which, according to our total information, the particular occurrence under consideration is a member. Thus, if K tells us not only that Jones had a streptococcus infection and was treated with penicillin, but also that he was an octogenarian with a weak heart (and if K provides no information more specific than that) then we would require that an acceptable explanation of Jones's response to the treatment be based on a statistical law stating the probability of that response in the narrowest reference class to which our total information assigns Jones's illness, i.e., the class of streptococcal infections suffered by octogenarians with weak hearts.[10]

Let me amplify this suggestion by reference to our earlier example concerning the use of the law that the half-life of radon is 3.82 days in accounting for the fact that the residual amount of radon to which a sample of 10 milligrams was reduced in 7.64 days was within the range from 2.4 to 2.6 milligrams. According to present scientific knowledge, the rate of decay of a radioactive element depends solely upon its atomic structure as characterized by its atomic number and its mass number, and it is thus unaffected by the age of the sample and by such factors as temperature, pressure, magnetic and electric forces, and chemical interactions. Thus, by specifying the half-life of radon as well as the initial mass of the sample and the time interval in question, the explanans takes into account all the available information that is relevant to appraising the probability of the given outcome by means of statistical laws. To state the point somewhat differently: Under the circumstances here assumed, our total information K assigns the case under study first of all to the reference class say F_1, of cases where a 10 milligram sample of radon is allowed to decay for 7.64 days; and the half-life law for radon assigns a very high probability, within F_1, to the "outcome," say G, consisting in the fact that the residual mass of radon lies between 2.4 and 2.6 milligrams. Suppose now that K also contains information about the temperature of the given sample, the pressure and relative humidity under which it is kept, the surrounding electric and magnetic conditions, and so forth, so that K assigns the given case to a reference class much narrower than F_1, let us say, $F_1F_2F_3 \ldots F_n$. Now the theory of radioactive decay, which is equally included in K, tells us that the statistical probability of G within this narrower class is the same as within G. For this reason, it suffices in our explanation to rely on the probability $p(G, F_1)$.

Let us note, however, that "knowledge situations" are conceivable in which the same argument would not be an acceptable explanation. Suppose, for example, that in the case of the radon sample under study, the amount remaining one hour before the end of the 7.64 day period happens to have been measured and found to be 2.7 milligrams, and thus markedly in excess of 2.6 milligrams – an occurrence which, considering the decay law for radon, is highly improbable, but not impossible. That finding, which then forms part of the total evidence K, assigns the particular case at hand to a reference class, say F^*, within which, according to the decay law for radon, the outcome G is highly improbable since it would require a quite unusual spurt in the decay of the given sample to reduce the 2.7 milligrams, within the one final hour of the test, to an amount falling between 2.4 and 2.6 milligrams. Hence, the additional information here considered may not be disregarded, and an explanation of the observed outcome will be acceptable only if it takes account of the probability of G in the narrower reference class, i.e., $p(G, F_1F^*)$. (The theory of radioactive decay implies that this probability equals $p(G, F^*)$, so that as a consequence the membership of the given case in F_1 need not be explicitly taken into account.)

The requirement suggested by the preceding considerations can now be stated more explicitly; we will call it the *requirement of maximal specificity for inductive-statistical explanations*. Consider a proposed explanation of the basic statistical form

$$
\frac{p(G, F) = r \qquad Fb}{Gb} \; [r] \qquad (30)
$$

Let s be the conjunction of the premises, and, if K is the set of all statements accepted at the given time, let k be a sentence that is logically equivalent to K (in the sense that k is implied by K and in turn implies every sentence in K). Then, to be rationally acceptable in the knowledge situation represented by K, the proposed explanation (30) must meet the following condition (the requirement of maximal specificity): If $s{\cdot}k$ implies[11] that b belongs to a class F_1, and that F_1 is a subclass of F, then $s{\cdot}k$ must also imply a statement specifying the statistical probability of G in F_1, say

$$p(G, F_1) = r_1$$

Here, r_1 must equal r unless the probability statement just cited is simply a theorem of mathematical probability theory.

The qualifying unless-clause here appended is quite proper, and its omission would result in undesirable consequences. It is proper because theorems of pure mathematical probability theory cannot provide an explanation of empirical subject matter. They may therefore be discounted when we inquire whether $s{\cdot}k$ might not give us statistical laws specifying the probability of G in reference classes narrower than F. And the omission of the clause would prove troublesome, for if (30) is proffered as an explanation, then it is presumably accepted as a fact that Gb; hence 'Gb' belongs to K. Thus K assigns b to the narrower class $F{\cdot}G$, and concerning the probability of G in that class, $s{\cdot}k$ trivially implies the statement that $p(G, F{\cdot}G) = 1$, which is simply a consequence of the measure-theoretical postulates for statistical probability. Since $s{\cdot}k$ thus implies a more specific probability statement for G than that invoked in (30), the requirement of maximal specificity would be violated by (30) – and analogously by any proffered statistical explanation of an event that we take to have occurred – were it not for the unless-clause, which, in effect, disqualifies the notion that the statement '$p(G, F{\cdot}G) = 1$' affords a more appropriate law to account for the presumed fact that Gb.

The requirement of maximal specificity, then, is here tentatively put forward as characterizing the extent to which the requirement of total evidence properly applies to inductive-statistical explanations. The general idea thus suggested comes to this: In formulating or appraising an I-S explanation, we should take into account all that information provided by K which is of potential *explanatory* relevance to the explanandum event; i.e., all pertinent statistical laws, and such particular facts as might be connected, by the statistical laws, with the explanandum event.[12]

The requirement of maximal specificity disposes of the problem of epistemic ambiguity; for it is readily seen that of two rival statistical arguments with high associated probabilities and with premises that all belong to K, at least one violates the requirement of maximum specificity. Indeed, let

$$\frac{\begin{array}{c}p(G, F) = r_1 \\ F\,b\end{array}}{G\,b}\ [r_1] \quad \text{and} \quad \frac{\begin{array}{c}p(\bar{G}, H) = r_2 \\ H\,b\end{array}}{\bar{G}\,b}\ [r_2]$$

be the arguments in question, with r_1 and r_2 close to 1. Then, since K contains the premises of both arguments, it assigns b to both F and H and hence to $F{\cdot}H$. Hence if both arguments satisfy the requirement of maximal specificity, K must imply that

$$\begin{array}{ll} & p(G, F{\cdot}H) = p(G, F) = r_1 \\ & p(\bar{G}, F{\cdot}H) = p(\bar{G}, H) = r_2 \\ \text{But} & p(G, F{\cdot}H) + p(\bar{G}, F{\cdot}H) = 1 \\ \text{Hence} & r_1 + r_2 = 1 \end{array}$$

and this is an arithmetic falsehood, since r_1 and r_2 are both close to 1; hence it cannot be implied by the consistent class K.

Thus, for I-S explanations that meet the requirement of maximal specificity the problem of epistemic ambiguity no longer arises. We are *never* in a position to say: No matter whether this particular event did or did not occur, we can produce an acceptable explanation of either outcome; and an explanation, moreover, whose premises are scientifically accepted statements which confer a high logical probability upon the given outcome.

While the problem of epistemic ambiguity has thus been resolved, ambiguity in the first sense discussed in this section remains unaffected by our requirement; i.e., it remains the case that for a given statistical argument with true premises and a high associated probability, there may exist a rival one with equally true premises and with a high associated probability, whose conclusion contradicts that of the first argument. And though the set K of statements accepted at any time never includes all statements that are in fact true (and no doubt many that are false), it is perfectly possible that K should contain the premises of two such conflicting arguments; but as we have seen, at least one of the latter will fail to be rationally acceptable because it violates the requirement of maximal specificity.

The preceding considerations show that *the concept of statistical explanation for particular events is essentially relative to a given knowledge situation as represented by a class K of accepted statements.* Indeed, the requirement of maximal specificity makes explicit and unavoidable reference to such a class, and it thus serves to characterize the concept of "I-S explanation relative to the knowledge situation represented by K." We will refer to this characteristic as the *epistemic relativity of statistical explanation.*

It might seem that the concept of deductive explanation possesses the same kind of relativity, since whether a proposed D-N or D-S account is acceptable will depend not only on whether it is deductively valid and makes essential use of the proper type of general law, but also on whether its premises are well supported by the relevant evidence at hand. Quite so; and this condition of empirical confirmation applies equally to statistical explanations that are to be acceptable in a given knowledge situation. But the epistemic relativity that the requirement of maximal specificity implies for I-S explanations is of quite a different kind and has no analogue for D-N explanations. For the specificity requirement is not concerned with the evidential support that the total evidence K affords for the explanans statements: it does not demand that the latter be included in K, nor even that K supply supporting evidence for them. It rather concerns what may be called the concept of a *potential* statistical explanation. For it stipulates that no matter how much evidential support there may be for the explanans, a proposed I-S explanation is not acceptable if its potential explanatory force with respect to the specified explanandum is vitiated by statistical laws which are included in *K* but not in the explanans, and which might permit the production of rival statistical arguments. As we have seen, this danger never arises for deductive explanations. Hence, these are not subject to any such restrictive condition, and the notion of a potential deductive explanation (as contradistinguished from a deductive explanation with well-confirmed explanans) requires no relativization with respect to *K*.

As a consequence, we can significantly speak of true D-N and D-S explanations: they are those potential D-N and D-S explanations whose premises (and hence also conclusions) are true – no matter whether this happens to be known or believed, and thus no matter whether the premises are included in *K*. But this idea has no significant analogue for I-S explanation since, as we have seen, the concept of potential statistical explanation requires relativization with respect to *K*.

Notes

1 Dewey (1910), chap. VI.

2 A general conception of scientific explanation as involving a deductive subsumption under general laws was espoused, though not always clearly stated, by various thinkers in the past, and has been advocated by several recent or contemporary writers, among them N. R. Campbell [(1920), (1921)], who developed the idea in considerable detail. In a textbook published in 1934, the conception was concisely stated as follows: "Scientific explanation consists in subsuming under some rule or law which expresses an invariant character of a group of events, the particular events it is said to explain. Laws themselves may be explained, and in the same manner, by showing that they are consequences of more comprehensive theories." (Cohen and Nagel 1934, p. 397.) Popper has set forth this construal of explanation in several of his publications; cf. the note at the end of section 3 in Hempel and Oppenheim (1948) His earliest statement appears in section 12 of his book (1935), of which his work (1959) is an expanded English version. His book (1962) contains further observations on scientific explanation. For some additional references to other proponents of the general idea, see Donagan (1957), footnote 2; Scriven (1959), footnote 3. However, as will be shown in section 3, deductive subsumption under general laws does not constitute the only form of scientific explanation.

3 The explanatory role of the phlogiston theory is described in Conant (1951), pp. 164–71. The concept of potential explanation was introduced in Hempel and Oppenheim (1948), section 7. The concept of lawlike sentence, in the sense here indicated, is due to Goodman (1947).

4 Cf. Mises (1939), pp. 272–8, where both the empirical findings and the explanatory argument are presented. This book also contains many other illustrations of what is here called deductive-statistical explanation.

5 Phrases such as 'It is almost certain (very likely) that *j* recovers', even when given the relational construal here suggested, are ostensibly concerned with relations between propositions, such as those expressed by the sentences forming the conclusion and the premises of an argument. For the purpose of the present discussion, however, involvement with propositions can be avoided by construing the phrases in question as expressing logical relations between corresponding *sentences*, e.g., the conclusion-sentence and the premise-sentence of an argument. This construal, which underlies the formulation of (3c), will be adopted in this essay, though for the sake of convenience we may occasionally use a paraphrase.

6 In the familiar schematization of deductive arguments, with a single line separating the premises from the conclusion, no explicit distinction is made between a weaker and a stronger claim, either of which might be intended; namely (i) that the premises logically imply the conclusion and (ii) that, in addition, the premises are true. In the case of our probabilistic argument, (3c) expresses a weaker claim, analogous to (i), whereas (3d) may be taken to express a "proffered explanation" (the term is borrowed from Scheffler (1957), section 1) in which, in addition, the explanatory premises are – however tentatively – asserted as true.

7 The considerations here outlined concerning the use of terms like 'probably' and 'certainly' as modal qualifiers of individual statements seem to me to militate also against the notion of categorical probability statement that C. I. Lewis sets forth in the following passage (italics the author's):

> Just as 'If *D* then (certainly) *P*, and *D* is the fact,' leads to the categorical consequence, 'Therefore (certainly) *P*'; so too, 'If *D* then probably *P*, and *D* is the fact', leads to a categorical consequence expressed by 'It is probable that *P*'. And this conclusion is not merely the statement over again of the probability relation between '*P*' and '*D*'; any more than 'Therefore (certainly) *P*' is the

> statement over again of 'If *D* then (certainly) *P*'. 'If the barometer is high, tomorrow will probably be fair; and the barometer *is* high', categorically assures something expressed by 'Tomorrow will probably be fair'. This probability is still relative to the grounds of judgment; but if these grounds are actual, and contain all the available evidence which is pertinent, then it is not only categorical but may fairly be called *the* probability of the event in question. (1946, p. 319).

This position seems to me to be open to just those objections suggested in the main text. If '*P*' is a statement, then the expressions 'certainly *P*' and 'probably *P*' as envisaged in the quoted passage are not statements. If we ask how one would go about trying to ascertain whether they were true, we realize that we are entirely at a loss unless and until a reference set of statements or assumptions has been specified relative to which *P* may then be found to be certain, or to be highly probable, or neither. The expressions in question, then, are essentially incomplete; they are elliptic formulations of relational statements; neither of them can be the conclusion of an inference. However plausible Lewis's suggestion may seem, there is no analogue in inductive logic to *modus ponens*, or the "rule of detachment," of deductive logic, which, given the information that '*D*', and also 'if *D* then *P*', are true statements, authorizes us to detach the consequent '*P*' in the conditional premise and to assert it as a self-contained statement which must then be true as well.

At the end of the quoted passage, Lewis suggests the important idea that 'probably *P*' might be taken to mean that the total relevant evidence available at the time confers high probability upon *P*. But even this statement is relational in that it tacitly refers to some unspecified time, and, besides, his general notion of a categorical probability statement as a conclusion of an argument is not made dependent on the assumption that the premises of the argument include all the relevant evidence available.

It must be stressed, however, that elsewhere in his discussion, Lewis emphasizes the relativity of (logical) probability, and, thus, the very characteristic that rules out the conception of categorical probability statements.

Similar objections apply, I think, to Toulmin's construal of probabilistic arguments; cf. Toulmin (1958) and the discussion in Hempel (1960), sections 1–3.

8 Carnap (1950), p. 211.

The requirement is suggested, for example, in the passage from Lewis (1946) quoted in note 7 for this section. Similarly Williams speaks of "the most fundamental of all rules of probability logic, that 'the' probability of any proposition is its probability in relation to the known premises and them only." (Williams, 1947, p. 72).

I am greatly indebted to Professor Carnap for having pointed out to me in 1945, when I first noticed the ambiguity of probabilistic arguments, that this was but one of several apparent paradoxes of inductive logic that result from disregard of the requirement of total evidence.

Barker (1957), pp. 70–8, has given a lucid independent presentation of the basic ambiguity of probabilistic arguments, and a skeptical appraisal of the requirement of total evidence as a means of dealing with the problem. However, I will presently suggest a way of remedying the ambiguity of probabilistic explanation with the help of a rather severely modified version of the requirement of total evidence. It will be called the requirement of maximal specificity, and is not open to the same criticism.

9 Cf. Carnap (1950), p. 211 and p. 494.

10 This idea is closely related to one used by Reichenbach (cf. (1949), section 72) in an attempt to show that it is possible to assign probabilities to individual events within the framework of a strictly statistical conception of probability. Reichenbach proposed that the probability of a single event, such as the safe completion of a particular scheduled flight of a given commercial plane, be construed as the statistical probability which the *kind* of event considered (safe completion of a flight) possesses within the narrowest reference class to which the given case (the specified flight of the given plane) belongs, and for which reliable statistical information is available (for example, the class of scheduled flights undertaken so far by planes of the line to which the given plane belongs, and under weather conditions similar to those prevailing at the time of the flight in question).

11 Reference to $s{\cdot}k$ rather than to k is called for because, as was noted earlier, we do not construe the condition here under discussion as requiring that all the explanans statements invoked be scientifically accepted at the time in question, and thus be included in the corresponding class K.

12 By its reliance on this general idea, and specifically on the requirement of maximal specificity, the method here suggested for eliminating the epistemic ambiguity of statistical explanation differs substantially from the way in which I attempted in an earlier study (Hempel, 1962, especially section 10) to deal with the same problem. In that study, which did not distinguish explicitly between the

two types of explanatory ambiguity characterized earlier in this section, I applied the requirement of total evidence to statistical explanations in a manner which presupposed that the explanans of any acceptable explanation belongs to the class *K*, and which then demanded that the probability which the explanans confers upon the explanandum be equal to that which the total evidence, *K*, imparts to the explanandum. The reasons why this approach seems unsatisfactory to me are suggested by the arguments set forth in the present section. Note in particular that, if strictly enforced, the requirement of total evidence would preclude the possibility of any significant statistical explanation for events whose occurrence is regarded as an established fact in science; for any sentence describing such an occurrence is logically implied by *K* and thus trivially has the logical probability 1 relative to *K*.

References

Barker, S. F., 1957, *Induction and Hypothesis.* Ithaca, NY: Cornell University Press.

Campbell, N. R., 1920, *Physics: The Elements.* Cambridge: Cambridge University Press.

Campbell, N. R., 1952[1921], *What is Science?* New York: Dover.

Carnap, R., 1950, *Logical Foundations of Probability.* Chicago: University of Chicago Press.

Cohen, M. R. and Nagel, E., 1934, *An Introduction to Logic and Scientific Method.* New York: Harcourt, Brace and World.

Conant, James B., 1951, *Science and Common Sense.* New Haven: Yale University Press.

Dewey, John, 1910, *How We Think.* Boston: D. C. Heath.

Donagan, A., 1957, "Explanation in History." *Mind* 66: 145–64.

Goodman, N., 1947, "The Problem of Counterfactual Conditionals." *The Journal of Philosophy* 44: 113–28.

Hempel, C. G., 1960, "Inductive Inconsistencies." *Synthese* 12: 439–69.

Hempel, C. G., 1962, "Deductive-Nomological vs. Statistical Explanation," in H. Feigl and G. Maxwell (eds.), *Minnesota Studies in the Philosophy of Science.* Vol. III, Minneapolis: University of Minnesota Press.

Hempel, C. G. and Oppenheim, P., 1948, "Studies in the Logic of Explanation." *Philosophy of Science* 15: 135–75.

Lewis, C. I., 1946, *An Analysis of Knowledge and Valuation.* La Salle, Ill.: Open Court Publishing.

Mises, R. von, 1939, *Probability, Statistics and Truth.* London: William Hodge & Co.

Popper, K. R., 1935, *Logic der Forschung.* Vienna: Springer.

Popper, K. R., 1959, *The Logic of Scientific Discovery.* London: Hutchinson.

Popper, K. R., 1962, *Conjectures and Refutations.* New York: Basic Books.

Reichenbach, H., 1949, *The Theory of Probability.* Berkeley and Los Angeles: The University of California Press.

Scheffler, I., 1957, "Explanation, Prediction, and Abstraction." *The British Journal for the Philosophy of Science* 7: 293–309.

Scriven, M., 1959, "Truisms as the Grounds for Historical Explanations," in P. Gardiner (ed.), *Theories of History.* New York: The Free Press.

Toulmin, S., 1958, *The Uses of Argument.* Cambridge: Cambridge University Press.

Williams, D. C., 1947, *The Ground of Induction.* Cambridge, MA: Harvard University Press.

5.3

Empiricism, Semantics, and Ontology

Rudolf Carnap

In this selection, Carnap draws a distinction between "internal" and "external" questions concerning the existence of various kinds of entity (e.g., numbers). Internal questions are those answered by appeal to the rules of the "linguistic framework" (there is, for example, an even prime number according to the rules for mathematics). External questions are those posed by philosophers when they ask, for example, whether there really are numbers, notwithstanding what those rules imply. Carnap claims that only the internal questions of existence are meaningful; he therefore repudiates the external question of existence, and the philosophers' disputes concerning them, as incoherent.

1 The Problem of Abstract Entities

Empiricists are in general rather suspicious with respect to any kind of abstract entities like properties, classes, relations, numbers, propositions, etc. They usually feel much more in sympathy with nominalists than with realists (in the medieval sense). As far as possible they try to avoid any reference to abstract entities and to restrict themselves to what is sometimes called a nominalistic language, i.e., one not containing such references. However, within certain scientific contexts it seems hardly possible to avoid them. In the case of mathematics some empiricists try to find a way out by treating the whole of mathematics as a mere calculus, a formal system for which no interpretation is given, or can be given. Accordingly, the mathematician is said to speak not about numbers, functions and infinite classes but merely about meaningless symbols and formulas manipulated according to given formal rules. In physics it is more difficult to shun the suspected entities because the language of physics serves for the communication of reports and predictions and hence cannot be taken as a mere calculus. A physicist who is suspicious of abstract entities may perhaps try to declare a certain part of the language of physics as uninterpreted and uninterpretable,

From *Revue Internationale de Philosophie* 4 (1950): 20–40. As reprinted in the *Supplement to Meaning and Necessity: A Study in Semantics and Modal Logic*, enlarged edn. (University of Chicago Press, 1956).

that part which refers to real numbers as space-time coordinates or as values of physical magnitudes, to functions, limits, etc. More probably he will just speak about all these things like anybody else but with an uneasy conscience, like a man who in his everyday life does with qualms many things which are not in accord with the high moral principles he professes on Sundays. Recently the problem of abstract entities has arisen again in connection with semantics, the theory of meaning and truth. Some semanticists say that certain expressions designate certain entities, and among these designated entities they include not only concrete material things but also abstract entities e.g., properties as designated by predicates and propositions as designated by sentences.[1] Others object strongly to this procedure as violating the basic principles of empiricism and leading back to a metaphysical ontology of the Platonic kind.

It is the purpose of this article to clarify this controversial issue. The nature and implications of the acceptance of a language referring to abstract entities will first be discussed in general; it will be shown that using such a language does not imply embracing a Platonic ontology but is perfectly compatible with empiricism and strictly scientific thinking. Then the special question of the role of abstract entities in semantics will be discussed. It is hoped that the clarification of the issue will be useful to those who would like to accept abstract entities in their work in mathematics, physics, semantics, or any other field; it may help them to overcome nominalistic scruples.

2 Linguistic Frameworks

Are there properties classes, numbers, propositions? In order to understand more clearly the nature of these and related problems, it is above all necessary to recognize a fundamental distinction between two kinds of questions concerning the existence or reality of entities. If someone wishes to speak in his language about a new kind of entities, he has to introduce a system of new ways of speaking, subject to new rules; we shall call this procedure the construction of a linguistic *framework* for the new entities in question. And now we must distinguish two kinds of questions of existence: first, questions of the existence of certain entities of the new kind *within the framework*; we call them *internal questions*; and second, questions concerning the existence or reality *of the system of entities as a whole*, called *external questions*. Internal questions and possible answers to them are formulated with the help of the new forms of expressions. The answers may be found either by purely logical methods or by empirical methods, depending upon whether the framework is a logical or a factual one. An external question is of a problematic character which is in need of closer examination.

The world of things. Let us consider as an example the simplest kind of entities dealt with in the everyday language: the spatio-temporally ordered system of observable things and events. Once we have accepted the thing language with its framework for things, we can raise and answer internal questions, e.g., "Is there a white piece of paper on my desk?" "Did King Arthur actually live?", "Are unicorns and centaurs real or merely imaginary?" and the like. These questions are to be answered by empirical investigations. Results of observations are evaluated according to certain rules as confirming or disconfirming evidence for possible answers. (This evaluation is usually carried out, of course, as a matter of habit rather than a deliberate, rational procedure. But it is possible, in a rational reconstruction, to lay down explicit rules for the evaluation. This is one of the main tasks of a pure, as distinguished from a psychological, epistemology.) The concept of reality occurring in these internal questions is an empirical scientific non-metaphysical concept. To recognize something as a real thing or event means to succeed in incorporating it into the system of things at a particular space-time position so that it fits together with the other things as real, according to the rules of the framework.

From these questions we must distinguish the external question of the reality of the thing world itself. In contrast to the former questions, this question is raised neither by the man in the street nor by scientists, but only by philosophers. Realists give an affirmative answer, subjective idealists a negative one, and the controversy goes on for centuries without ever being solved. And it cannot be solved because it is framed in a wrong way. To be real in the scientific sense means to be an element of the system; hence this concept cannot be meaningfully applied to the system itself. Those who raise the question of the reality of the thing world itself have perhaps in mind not a theoretical question as their formulation seems to suggest, but rather a practical question, a matter of a practical decision concerning the structure of our language. We have to make the choice whether or not to accept and use the forms of expression in the framework in question.

In the case of this particular example, there is usually no deliberate choice because we all have accepted the thing language early in our lives as a matter of course. Nevertheless, we may regard it as a matter of decision in this sense: we are free to choose to continue using the thing language or not; in the latter case we could restrict ourselves to a language of sense data and other "phenomenal" entities, or construct an alternative to the customary thing language with another structure, or, finally, we could refrain from speaking. If someone decides to accept the thing language, there is no objection against saying that he has accepted the world of things. But this must not be interpreted as if it meant his acceptance of a *belief* in the reality of the thing world; there is no such belief or assertion or assumption, because it is not a theoretical question. To accept the thing world means nothing more than to accept a certain form of language, in other words, to accept rules for forming statements and for testing accepting or rejecting them. The acceptance of the thing language leads on the basis of observations made, also to the acceptance, belief, and assertion of certain statements. But the thesis of the reality of the thing world cannot be among these statements, because it cannot be formulated in the thing language or, it seems, in any other theoretical language.

The decision of accepting the thing language, although itself not of a cognitive nature, will nevertheless usually be influenced by theoretical knowledge, just like any other deliberate decision concerning the acceptance of linguistic or other rules. The purposes for which the language is intended to be used, for instance, the purpose of communicating factual knowledge, will determine which factors are relevant for the decision. The efficiency, fruitfulness, and simplicity of the use of the thing language may be among the decisive factors. And the questions concerning these qualities are indeed of a theoretical nature. But these questions cannot be identified with the question of realism. They are not yes-no questions but questions of degree. The thing language in the customary form works indeed with a high degree of efficiency for most purposes of everyday life. This is a matter of fact, based upon the content of our experiences. However, it would be wrong to describe this situation by saying: "The fact of the efficiency of the thing language is confirming evidence for the reality of the thing world; we should rather say instead: "This fact makes it advisable to accept the thing language."

The system of numbers. As an example of a system which is of a logical rather than a factual nature let us take the system of natural numbers. The framework for this system is constructed by introducing into the language new expressions with suitable rules: (1) numerals like "five" and sentence forms like "there are five books on the table"; (2) the general term "number" for the new entities, and sentence forms like "five is a number"; (3) expressions for properties of numbers (e.g. "odd," "prime"), relations (e.g., "greater than") and functions (e.g. "plus"), and sentence forms like "two plus three is five"; (4) numerical variables ("*m*," "*n*," etc.) and quantifiers for universal sentences ("for every *n* . . .") and existential sentences ("there is an *n* such that . . .") with the customary deductive rules.

Here again there are internal questions, e.g., "Is there a prime number greater than a hundred?" Here however the answers are found not by empirical investigation based on observations but by logical analysis based on the rules for the new expressions. Therefore the answers are here analytic, i.e., logically true.

What is now the nature of the philosophical question concerning the existence or reality of numbers? To begin with, there is the internal question which together with the affirmative answer, can be formulated in the new terms, say by "There are numbers" or, more explicitly, "There is an *n* such that *n* is a number." This statement follows from the analytic statement "five is a number" and is therefore itself analytic. Moreover, it is rather trivial (in contradistinction to a statement like "There is a prime number greater than a million which is likewise analytic but far from trivial), because it does not say more than that the new system is not empty; but this is immediately seen from the rule which states that words like "five" are substitutable for the new variables.

Therefore nobody who meant the question "Are there numbers?" in the internal sense would either assert or even seriously consider a negative answer. This makes it plausible to assume that those philosophers who treat the question of the existence of numbers as a serious philosophical problem and offer lengthy arguments on either side, do not have in mind the internal question. And indeed, if we were to ask them: "Do you mean the question as to whether the framework of numbers, *if* we were to accept it, would be found to be empty or not?" they would probably reply: "Not at all; we mean a question prior to the acceptance of the new framework." They might try to explain what they mean by saying that it is a question of the ontological status of numbers; the question whether or not numbers have a certain metaphysical characteristic called reality (but a kind of ideal reality, different from the material reality of the thing world) or subsistence or status of "independent entities." Unfortunately, these philosophers have so far not given a formulation of their question in terms of the common scientific language. Therefore our judgment must be that they have not succeeded in giving to the external question and to the possible answers any cognitive content. Unless and until they supply a clear cognitive interpretation, we are justified in our suspicion that their question is a pseudo-question, that is, one disguised in the form of a theoretical question while in fact it is a non-theoretical; in the present case it is the practical problem whether or not to incorporate into the language the new linguistic forms which constitute the framework of numbers.

The system of propositions. New variables, "p," "q," etc., are introduced with a role to the effect that any (declarative) sentence may be substituted for a variable of this kind; this includes, in addition to the sentences of the original thing language, also all general sentences with variables of any kind which may have been introduced into the language. Further, the general term "proposition" is introduced. "p is a proposition" may be defined by "p or not p" (or by any other sentence form yielding only analytic sentences). Therefore every sentence of the form "... is a proposition" (where any sentence may stand in the place of the dots) is analytic. This holds, for example, for the sentence:

(a) Chicago is large is a proposition.

(We disregard here the fact that the rules of English grammar require not a sentence but a that-clause as the subject of another sentence; accordingly instead of (a) we should have to say "That Chicago is large is a proposition.") Predicates may be admitted whose argument expressions are sentences; these predicates may be either extensional (e.g. the customary truth-functional connectives) or not (e.g. modal predicates like "possible," "necessary," etc.). With the help of the new variables, general sentences may be formed, e.g.,

(b) "For every *p*, either *p* or not-*p*."
(c) "There is a *p* such that *p* is not necessary and not-*p* is not necessary."
(d) "There is a *p* such that *p* is a proposition."

(c) and (d) are internal assertions of existence. The statement "There are propositions" may be meant in the sense of (d); in this case it is analytic (since it follows from (a)) and even trivial. If, however, the statement is meant in an external sense, then it is non-cognitive.

It is important to notice that the system of rules for the linguistic expressions of the propositional framework (of which only a few rules have here been briefly indicated) is sufficient for the introduction of the framework. Any further explanations as to the nature of the propositions (i.e., the elements of the system indicated, the values of the variables "*p*," "*q*," etc.) are theoretically unnecessary because, if correct, they follow from the rules. For example, are propositions mental events (as in Russell's theory)? A look at the rules shows us that they are not, because otherwise existential statements would be of the form: "If the mental state of the person in question fulfills such and such conditions, then there is a *p* such that...." The fact that no references to mental conditions occur in existential statements (like (c), (d), etc.) shows that propositions are not mental entities. Further, a statement of the existence of linguistic entities (e.g., expressions, classes of expressions, etc.) must contain a reference to a language. The fact that no such reference occurs in the existential statements here, shows that propositions are not linguistic entities. The fact that in these statements no reference to a subject (an observer or knower) occurs (nothing like: "There is a p which is necessary for Mr X."), shows that the propositions (and their properties, like necessity, etc.) are not subjective. Although characterizations of these or similar kinds are, strictly speaking, unnecessary, they may nevertheless be practically useful. If they are given, they should be understood, not as ingredient parts of the system, but merely as marginal notes with the purpose of supplying to the reader helpful hints or convenient pictorial associations which may make his learning of the use of the expressions easier than the bare system of the rules would do. Such a characterization is analogous to an extra-systematic explanation which a physicist sometimes gives to the beginner. He might, for example, tell him to imagine the atoms of a gas as small balls rushing around with great speed, or the electromagnetic field and its oscillations as quasi-elastic tensions and vibrations in an ether. In fact, however, all that can accurately be said about atoms or the field is implicitly contained in the physical laws of the theories in question.[2]

The system of thing properties The thing language contains words like "red," "hard," "stone," "house," etc., which we used for describing what things are like. Now we may introduce new variables, say "*f*," "*g*," etc., for which those words are substitutable and furthermore the general term "property." New rules are laid down which admit sentences like "Red is a property," "Red is a color," "These two pieces of paper have at least one color in common" (i.e., "There is an *f* such that *f* is a color, and . . ."). The last sentence is an internal assertion. It is an empirical, factual nature. However, the external statement, the philosophical statement of the reality of properties – a special case of the thesis of the reality of universals – is devoid of cognitive content.

The system of integers and rational numbers. Into a language containing the framework of natural numbers we may introduce first the (positive and negative) integers as relations among natural numbers and then the rational numbers as relations among integers. This involves introducing new types of variables, expressions substitutable for them, and the general terms "integer" and "rational number."

The system of real numbers. On the basis of the rational numbers, the real numbers may be introduced as classes of a special kind (segments) of rational numbers (according to the method developed by Dedekind and Frege). Here again a new type of variables is introduced, expressions substitutable for them (e.g., "$\sqrt{2}$"), and the general term "real number."

The spatio-temporal coordinate system for physics. The new entities are the space-time points. Each is an ordered quadruple of four real numbers, called its coordinates, consisting of three spatial and one temporal coordinates. The physical state of a spatio-temporal point or region is described either with the help of qualitative predicates (e.g., "hot") or by ascribing numbers as values of a physical magnitude (e.g., mass, temperature, and the like). The step from the system of things (which does not contain space-time points but only extended objects with spatial and temporal relations between them) to the physical coordinate system is again a matter of decision. Our choice of certain features, although itself not theoretical, is suggested by theoretical knowledge, either logical or factual. For example, the choice of real numbers rather than rational numbers or integers as coordinates is not much influenced by the facts of experience but mainly due to considerations of mathematical simplicity. The restriction to rational coordinates would not be in conflict with any experimental knowledge we have, because the result of any measurement is a rational number. However, it would prevent the use of ordinary geometry (which says, e.g., that the diagonal of a square with the side I has the irrational value $\sqrt{2}$) and thus lead to great complications. On the other hand, the decision to use three rather than two or four spatial coordinates is strongly suggested, but still not forced upon us, by the result of common observations. If certain events allegedly observed in spiritualistic seances, e.g., a ball moving out of a sealed box, were confirmed beyond any reasonable doubt, it might seem advisable to use four spatial coordinates. Internal questions are here, in general, empirical questions to be answered by empirical investigations. On the other hand, the external questions of the reality of physical space and physical time are pseudo-questions. A question like: "Are there (really) space-time points?" is ambiguous. It may be meant as an internal question; then the affirmative answer is, of course, analytic and trivial. Or it may be meant in the external sense: "Shall we introduce such and such forms into our language?"; in this case it is not a theoretical but a practical question, a matter of decision rather than assertion, and hence the proposed formulation would be misleading. Or finally, it may be meant in the following sense: "Are our experiences such that the use of the linguistic forms in question will be expedient and fruitful?" This is a theoretical question of a factual, empirical nature. But it concerns a matter of degree; therefore a formulation in the form "real or not?" would be inadequate.

3 What Does Acceptance of a Kind of Entities Mean?

Let us now summarize the essential characteristics of situations involving the introduction of a new kind of entities, characteristics which are common to the various examples outlined above.

The acceptance of a new kind of entities is represented in the language by the introduction of a framework of new forms of expressions to be used according to a new set of rules. There may be new names for particular entities of the kind in question; but some such names may already occur in the language before the introduction of the new framework. (Thus, for example, the thing language contains certainly words of the type of "blue" and "house" before the framework of properties is introduced; and it may contain words like "ten" in sentences of the form "I have ten fingers" before the framework of numbers is introduced.) The latter fact shows that the occurrence of constants of the type in question – regarded as names of entities of the new kind after the new framework is introduced – is not a sure sign of the acceptance of the new kind of entities. Therefore the introduction of such constants is not to be regarded as an essential step in the introduction of the framework. The two essential steps are rather the following. First, the introduction of a general term, a predicate of higher level, for the new kind of entities, permitting us to say for any particular entity that it belongs to this kind (e.g., "Red is a *property*," "Five is a *number*"). Second, the introduction of variables of the new type. The new entities are values of these variables; the constants (and the closed compound expressions, if any) are substitutable for the variables.[3] With the help of the variables, general sentences concerning the new entities can be formulated.

After the new forms are introduced into the language, it is possible to formulate with their help internal questions and possible answers to them. A question of this kind may be either empirical or logical; accordingly a true answer is either factually true or analytic.

From the internal questions we must clearly distinguish external questions, i.e., philosophical questions concerning the existence or reality of the total system of the new entities. Many philosophers regard a question of this kind as an ontological question which must be raised and answered before the introduction of the new language forms. The latter introduction, they believe, is legitimate only if it can be justified by an ontological insight supplying an affirmative answer to the question of reality. In contrast to this view, we take the position that the introduction of the new ways of speaking does not need any theoretical justification because it does not imply any assertion of reality. We may still speak (and have done so) of the "acceptance of the new entities" since this form of speech is customary; but one must keep in mind that this phrase does not mean for us anything more than acceptance of the new framework, i.e., of the new linguistic forms. Above all, it must not be interpreted as referring to an assumption, belief, or assertion of "the reality of the entities." There is no such assertion. An alleged statement of the reality of the system of entities is a pseudo-statement without cognitive content. To be sure, we have to face at this point an important question; but it is a practical, not a theoretical question; it is the question of whether or not to accept the new linguistic forms. The acceptance cannot be judged as being either true or false because it is not an assertion. It can only be judged as being more or less expedient, fruitful, conducive to the aim for which the language is intended. Judgments of this kind supply the motivation for the decision of accepting or rejecting the kind of entities.[4]

Thus it is clear that the acceptance of a linguistic framework must not be regarded as implying a metaphysical doctrine concerning the reality of the entities in question. It seems to me due to a neglect of this important distinction that some contemporary nominalists label the admission of variables of abstract types as "Platonism."[5] This is, to say the least, an extremely misleading terminology. It leads to the absurd consequence, that the position of everybody who accepts the language of physics with its real number variables (as a language of communication, not merely as a calculus) would be called Platonistic, even if he is a strict empiricist who rejects Platonic metaphysics.

A brief historical remark may here be inserted. The non-cognitive character of the questions which we have called here external questions was recognized and emphasized already by the Vienna Circle under the leadership of Moritz Schlick, the group from which the movement of logical empiricism originated. Influenced by ideas of Ludwig Wittgenstein, the Circle rejected both the thesis of the reality of the external world and the thesis of its irreality as pseudo-statements;[6] the same was the case for both the thesis of the reality of universals (abstract entities, in our present terminology) and the nominalistic thesis that they are not real and that their alleged names are not names of anything but merely *flatus vocis*. (It is obvious that the apparent negation of a pseudo-statement must also be a pseudo-statement.) It is therefore not correct to classify the members of the Vienna Circle as nominalists, as is sometimes done. However, if we look at the basic anti-metaphysical and pro-scientific attitude of most nominalists (and the same holds for many materialists and realists in the modern sense), disregarding their occasional pseudo-theoretical formulations, then it is, of course, true to say that the Vienna Circle was much closer to those philosophers than to their opponents.

[...]

... Generally speaking, if someone accepts a framework for a certain kind of entities, then he is bound to admit the entities as possible designata. Thus the question of the admissibility of entities of a certain type or of abstract entities in general as designata is reduced to the question of the acceptability of the linguistic framework for those entities. Both the nominalistic critics, who refuse the status of designators or names to expressions like "red," "five," etc., because they deny the existence of abstract entities, and the skeptics, who express doubts concerning the existence and demand evidence for it, treat the question of existence as a theoretical question. They do, of course, not mean the internal question; the affirmative answer to *this* question is analytic and trivial and too obvious for doubt or denial, as we have seen. Their doubts refer rather to the system of entities itself; hence they mean the external question. They believe that only after making sure that there really is a system of entities of the kind in question are we justified in accepting the framework by incorporating the linguistic forms into our language. However, we have seen that the external question is not a theoretical question but rather the practical question whether or not to accept those linguistic forms. This acceptance is not in need of a theoretical justification (except with respect to expediency and fruitfulness), because it does not imply a belief or assertion. Ryle says that the "Fido"–Fido principle is "a grotesque theory." Grotesque or not, Ryle is wrong in calling it a theory. It is rather the practical decision to accept certain frameworks. Maybe Ryle is historically right with respect to those whom he mentions as previous representatives of the principle, viz. John Stuart Mill, Frege, and Russell. If these philosophers regarded the acceptance of a system of entities as a theory, an assertion, they were victims of the same old, metaphysical confusion. But it is certainly wrong to regard my semantical method as involving a belief in the reality of abstract entities, since I reject a thesis of this kind as a metaphysical pseudo-statement.

The critics of the use of abstract entities in semantics overlook the fundamental difference between the acceptance of a system of entities and an internal assertion, e.g., an assertion that there are elephants or electrons or prime numbers greater than a million. Whoever makes an internal assertion is certainly obliged to justify it by providing evidence, empirical evidence in the case of electrons, logical proof in the case of the prime numbers. The demand for a theoretical justification, correct in the case of internal assertions, is sometimes wrongly applied to the acceptance of a system of entities. Thus, for example, Ernest Nagel in his review[7] asks for "evidence relevant for affirming with warrant that there are such entities as infinitesimals or propositions." He characterizes the evidence required in these cases – in distinction to the empirical evidence in the case of electrons – as "in the broad sense logical and dialectical." Beyond this no hint is given as to what might be regarded as relevant evidence. Some nominalists regard the acceptance of abstract entities as a kind of superstition or myth, populating the world with fictitious or at least dubious entities, analogous to the belief in centaurs or demons. This shows again the confusion mentioned, because a superstition or myth is a false (or dubious) internal statement.

Let us take as example the natural numbers as cardinal numbers, i.e., in contexts like "Here are three books." The linguistic forms of the framework of numbers, including variables and the general term "number," are generally used in our common language of communication; and it is easy to formulate explicit rules for their use. Thus the logical characteristics of this framework are sufficiently clear (while many internal questions, i.e., arithmetical questions, are, of course, still open). In spite of this, the controversy concerning the external question of the ontological reality of the system of numbers continues. Suppose that one philosopher says: "I believe that there are numbers as real entities. This gives me the right to use the linguistic forms of the numerical framework and to make semantical statements about numbers as designata of numerals." His nominalistic opponent replies: "You are wrong; there are no numbers. The numerals may still be used as meaningful expressions. But they are not names, there are no entities designated by them. Therefore the word "number" and numerical variables must not be used (unless a way were found to introduce them as merely abbreviating devices, a way of translating them into the nominalistic thing language)." I cannot think of any possible evidence that would be regarded as relevant by both philosophers, and therefore, if actually found, would decide the controversy or at least make one of the opposite theses more probable than the other. (To construe the numbers as classes or properties of the second level, according to the Frege–Russell method, does, of course, not solve the controversy, because the first philosopher would affirm and the second deny the existence of the system of classes or properties of the second level.) Therefore I feel compelled to regard the external question as a pseudo-question, until both parties to the controversy offer a common interpretation of the question as a cognitive question; this would involve an indication of possible evidence regarded as relevant by both sides.

There is a particular kind of misinterpretation of the acceptance of abstract entities in various fields of science and in semantics, that needs to be cleared up. Certain early British empiricists (e.g., Berkeley and Hume) denied the existence of abstract entities on the ground that immediate experience presents us only with particulars, not with universals, e.g., with this red patch, but not with Redness or Color-in-General; with this scalene triangle, but not with Scalene Triangularity or Triangularity-in-General. Only entities belonging to a type of which examples were to be found within immediate experience could be accepted as ultimate constituents of reality. Thus, according to this way of thinking, the existence of abstract entities could be asserted only if one could show either that some abstract entities fall within the given, or that abstract entities can be defined in terms of the types of entity which are given. Since these empiricists found no abstract entities within the realm of sense-data, they either denied their existence, or else made a futile attempt to define universals in terms of particulars. Some contemporary philosophers, especially English philosophers following Bertrand Russell, think in basically similar terms. They emphasize a distinction between the data (that which is immediately given in consciousness, e.g., sense-data, immediately past experiences, etc.) and the constructs based on the data. Existence or reality is ascribed only to the data; the constructs are not real entities; the corresponding linguistic expressions are merely ways of speech not actually designating anything (reminiscent of the nominalists' *flatus vocis*). We shall not criticize here this general conception. (As far as it is a principle of accepting certain entities and not accepting others, leaving aside any ontological, phenomenalistic and nominalistic pseudo-statements, there cannot be any theoretical objection to it.) But if this conception leads to the view that other philosophers or scientists who accept abstract entities thereby assert or imply their occurrence as immediate data, then such a view must be rejected as a misinterpretation. References to space-time points, the electromagnetic field, or electrons in physics, to real or complex numbers and their functions in mathematics, to the excitatory potential or unconscious complexes in psychology, to an inflationary trend in economics, and the like, do not imply the assertion that entities of these kinds occur as immediate data. And the same holds for references to abstract entities as designata in semantics. Some of the criticisms by English philosophers against such references give the impression that, probably due to the misinterpretation just indicated, they accuse the semanticist not so much of bad metaphysics (as some nominalists would do) but of bad psychology. The fact that they regard a semantical method involving abstract entities not merely as doubtful and perhaps wrong, but as manifestly absurd, preposterous and grotesque, and that they show a deep horror and indignation against this method, is perhaps to be explained by a misinterpretation of the kind described. In fact, of course, the semanticist does not in the least assert or imply that the abstract entities to which he refers can be experienced as immediately given either by sensation or by a kind of rational intuition. An assertion of this kind would indeed be very dubious psychology. The psychological question as to which kinds of entities do and which do not occur as immediate data is entirely irrelevant for semantics, just as it is for physics, mathematics, economics, etc., with respect to the examples mentioned above.[8]

5 Conclusion

For those who want to develop or use semantical methods, the decisive question is not the alleged ontological question of the existence of abstract entities but rather the question whether the rise of abstract linguistic foms or, in technical terms, the use of variables beyond those for things (or phenomenal data), is expedient and fruitful for the purposes for which semantical analyses are made, viz. the analysis, interpretation, clarification, or construction of languages of communication, especially languages of science. This question is here neither decided nor even discussed. It is not a question simply of yes or no, but a matter of degree. Among those philosophers who have carried out semantical analyses and thought about suitable tools for this work, beginning with Plato and Aristotle and, in a more technical way on the basis of modern logic, with C. S. Peirce and Frege, a great majority accepted abstract entities. This does, of course, not prove the case. After all, semantics in the technical sense is still in the initial phases of its development, and we must be prepared for possible fundamental changes in methods. Let us therefore admit that the nominalistic critics may possibly be right. But if so, they will have to offer better arguments than they have so far. Appeal to ontological insight will not carry much weight. The critics will have to show that it is possible to construct a semantical method which avoids all references to abstract entities and achieves by simpler means essentially the same results as the other methods.

The acceptance or rejection of abstract linguistic forms, just as the acceptance or rejection of any other linguistic forms in any branch of science, will finally be decided by their efficiency as instruments, the ratio of the results achieved to the amount and complexity of the efforts required. To decree dogmatic prohibitions of certain linguistic forms instead of testing them by their success or failure in practical use, is worse than futile; it is positively harmful because it may obstruct scientific progress. The history of science shows examples of such prohibitions based on prejudices deriving from religious, mythological, metaphysical, or other irrational sources, which slowed up the developments for shorter or longer periods of time. Let us learn from the lessons of history. Let us grant to those who work in any special field of investigation the freedom to use any form of expression which seems useful to them; the work in the field will sooner or later lead to the elimination of those forms which have no useful function. *Let us be cautious in making assertions and critical in examining them, but tolerant in permitting linguistic forms.*

Notes

1 The terms "sentence" and "statement" are here used synonymously for declarative (indicative propositional) sentences.

2 In my book *Meaning and Necessity* (Chicago, 1947) I have developed a semantical method which takes propositions as entities designated by sentences (more specifically, as intensions of sentences). In order to facilitate the understanding of the systematic development, I added some informal, extra-systematic explanations concerning the nature of propositions. I said that the term "proposition" "is

used neither for a linguistic expression nor for a subjective, mental occurrence, but rather for something objective that may or may not be exemplified in nature . . . We apply the term 'proposition' to any entities of a certain logical type, namely, those that may be expressed by (declarative) sentences in a language" (p. 27). After some more detailed discussions concerning the relation between propositions and facts, and the nature of false propositions, I added: "It has been the purpose of the preceding remarks to facilitate the understanding of our conception of propositions. If, however, a reader should find these explanations more puzzling than clarifying, or even unacceptable, he may disregard them" (p. 31) (that is, disregard these extra-systematic explanations, not the whole theory of the propositions as intensions of sentences, as one reviewer understood). In spite of this warning, it seems that some of those readers who were puzzled by the explanations, did not disregard them but thought that by raising objections against them they could refute the theory. This is analogous to the procedure of some laymen who by (correctly) criticizing the ether picture or other visualizations of physical theories, thought they had refuted those theories. Perhaps the discussions in the present paper will help in clarifying the role of the system of linguistic rules for the introduction of a framework for entities on the one hand, and that of extra-systematic explanations concerning the nature of the entities on the other.

3 W. V. O. Quine was the first to recognize the importance of the introduction of variables as indicating the acceptance of entities. "The ontology to which one's use of language commits him comprises simply the objects that he treats as falling . . . within the range of values of his variables." "Notes on Existence and Necessity," *Journal of Philosophy*, Vol. 40 (1943), pp. 113–127; compare also his "Designation and Existence," *Journal of Philosophy*, Vol. 36 (1939), pp. 702–9, and "On Universals," *The Journal of Symbolic Logic*, Vol. 12 (1947), pp. 74–84.

4 For a closely related point of view on these questions see the detailed discussions in Herbert Feigl, "Existential Hypotheses," *Philosophy of Science*, 17 (1950), pp. 35–62.

5 Paul Bernays, "Sur le platonisme dans les mathematiques" (*L'Enseignement math.*, 34 (1935), 52–69). W. V. O. Quine, see previous footnote and a recent paper ["On What There Is," *Review of Metaphysics*, Vol. 2 (1948), pp. 21–38]. Quine does not acknowledge the distinction which I emphasize above, because according to his general conception there are no sharp boundary lines between logical and factual truth, between questions of meaning and questions of fact, between the acceptance of a language structure and the acceptance of an assertion formulated in the language. This conception, which seems to deviate considerably from customary ways of thinking, is explained in his article "Semantics and Abstract Objects," *Proceedings of the American Academy of Arts and Sciences*, 80 (1951), 90–6. When Quine in the article ["On What There Is"] classifies my logistic conception of mathematics (derived from Frege and Russell) as "platonic realism" (p. 33), this is meant (according to a personal communication from him) not as ascribing to me agreement with Plato's metaphysical doctrine of universals, but merely as referring to the fact that I accept a language of mathematics containing variables of higher levels. With respect to the basic attitude to take in choosing a language form (an "ontology" in Quine's terminology, which seems to me misleading), there appears now to be agreement between us: "the obvious counsel is tolerance and an experimental spirit" (["On What There Is,"] p. 38).

6 See Carnap, *Scheinprobleme in der Philosophie; das Fremdpsychische und der Realismusstreit*, Berlin, 1928. Moritz Schlick, *Positivismus und Realismus*, reprinted in *Gesammelte Aufsatze*, Wien, 1938.

7 Ernest Nagel, "Review of Meaning and Necessity," *Journal of Philosophy* 45 (1948): 467–72.

8 Wilfrid Sellars ("Acquaintance and Description Again", in *Journal of Philosophy*, 46 (1949), 496–504; see pp. 502 f.) analyzes clearly the roots of the mistake "of taking the designation relation of semantic theory to be a reconstruction of being *present to an experience*."

5.4

The Pragmatic Vindication of Induction

Hans Reichenbach

Hans Reichenbach (1891–1953) was the founder of the Berlin Society, positivism's second point of origin (after the Vienna Circle). In this selection from his *Experience and Prediction*, he agrees with David Hume that we cannot justify belief in the reliability of the inductive method. Reichenbach argues that we can, nevertheless, justify our use of induction in guiding our expectations about the future. For it follows from the very concept of a limit of a frequency – the estimation of which is an induction's aim – that, if there *is* such a limit, the inductive method will succeed in uncovering it in the long run, and is the simplest rule that will do so.

§38 The Problem of Induction

So far we have only spoken of the useful qualities of the frequency interpretation. It also has dangerous qualities.

The frequency interpretation has two functions within the theory of probability. First, a frequency is used as a *substantiation* for the probability statement; it furnishes the reason why we believe in the statement. Second, a frequency is used for the *verification* of the probability statement; that is to say, it is to furnish the meaning of the statement. These two functions are not identical. The observed frequency from which we start is only the basis of the probability inference; we intend to state another frequency which concerns *future observations*. The probability inference proceeds from a known frequency to one unknown; it is from this function that its importance is derived. The probability statement sustains a prediction, and this is why we want it.

It is the problem of induction which appears with this formulation. The theory of probability involves the problem of induction, and a solution of the problem of probability cannot be given without an answer to the question of induction. The connection of both problems is well known; philosophers such as Peirce have expressed the idea

From Hans Reichenbach, *Experience and Prediction* (Chicago: University of Chicago Press, 1961), pp. 339–42, 348–57.

that a solution of the problem of induction is to be found in the theory of probability. The inverse relation, however, holds as well. Let us say, cautiously, that the solution of both problems is to be given within the same theory.

In uniting the problem of probability with that of induction, we decide unequivocally in favor of that determination of the degree of probability which mathematicians call the *determination a posteriori*. We refuse to acknowledge any so-called *determination a priori* such as some mathematicians introduce in the theory of the games of chance; on this point we refer to our remarks in §33, where we mentioned that the so-called determination a priori may be reduced to a determination a posteriori. It is, therefore, the latter procedure which we must now analyze.

By "determination a posteriori" we understand a procedure in which the relative frequency observed statistically is assumed to hold approximately for any future prolongation of the series. Let us express this idea in an exact formulation. We assume a series of events A and $\bar{A}$ (non-A); let n be the number of events, m the number of events of the type A among them. We have then the relative frequency

$$h^n = \frac{m}{n}$$

The assumption of the determination a posteriori may now be expressed:

For any further prolongation of the series as far as s *events* (s > n), *the relative frequency will remain within a small interval around* h^n *i.e., we assume the relation*

$$h^n - \varepsilon \leqq h^s \leqq h^n + \varepsilon$$

where ε *is a small number.*

This assumption formulates the *principle of induction*. We may add that our formulation states the principle in a form more general than that customary in traditional philosophy. The usual formulation is as follows: induction is the assumption that an event which occurred n times will occur at all following times. It is obvious that this formulation is a special case of our formulation, corresponding to the case $h^n = 1$. We cannot restrict our investigation to this special case because the general case occurs in a great many problems.

The reason for this is to be found in the fact that the theory of probability needs the definition of probability as the limit of the frequency. Our formulation is a necessary condition for the existence of a limit of the frequency near h^n; what is yet to be added is that there is an h^n of the kind postulated for every ε however small. If we include this idea in our assumption, our postulate of induction becomes the hypothesis that there is a limit to the relative frequency which does not differ greatly from the observed value.

If we enter now into a closer analysis of this assumption, one thing needs no further demonstration: the formula given is not a tautology. There is indeed no logical necessity that h^s remains within the interval $h^n \pm \varepsilon$; we may easily imagine that this does not take place.

The nontautological character of induction has been known a long time; Bacon had already emphasized that it is just this character to which the importance of induction is due. If inductive inference can teach us something new, in opposition to deductive inference, this is because it is not a tautology. This useful quality has, however, become the center of the epistemological difficulties of induction. It was David Hume who first attacked the principle from this side; he pointed out that the apparent constraint of the inductive inference, although submitted to by everybody, could not be justified. We believe in induction; we even cannot get rid of the belief when we know the impossibility of a logical demonstration of the validity of inductive inference; but as logicians we must admit that this belief is a deception – such is the result of Hume's criticism. We may summarize his objections in two statements:

1. We have no logical demonstration for the validity of inductive inference.
2. There is no demonstration a posteriori for the inductive inference; any such demonstration would presuppose the very principle which it is to demonstrate.

These two pillars of Hume's criticism of the principle of induction have stood unshaken for two centuries, and I think they will stand as long as there is a scientific philosophy.

[. . .]

39 The Justification of the Principle of Induction

We shall now begin to give the justification of induction which Hume thought impossible. In the pursuit of this inquiry, let us ask first what has been proved, strictly speaking, by Hume's objections.

Hume started with the assumption that a justification of inductive inference is only given if we can show that inductive inference must lead to success. In other words, Hume believed that any justified application of the inductive inference presupposes a demonstration that the conclusion is true. It is this assumption on which Hume's criticism is based. His two objections directly concern only the question of the truth of the conclusion; they prove that the truth of the conclusion cannot be demonstrated. The two objections, therefore, are valid only in so far as the Humean assumption is valid. It is this question to which we must turn: Is it necessary, for the justification of inductive inference, to show that its conclusion is true?

A rather simple analysis shows us that this assumption does not hold. Of course, if we were able to prove the truth of the conclusion, inductive inference would be justified; but the converse does not hold: a justification of the inductive inference does not imply a proof of the truth of the conclusion. The proof of the truth of the conclusion is only a sufficient condition for the justification of induction, not a necessary condition.

The inductive inference is a procedure which is to furnish us the best assumption concerning the future. If we do not know the truth about the future, there may be nonetheless a best assumption about it, i.e., a best assumption relative to what we know. We must ask whether such a characterization may be given for the principle of induction. If this turns out to be possible, the principle of induction will be justified.

An example will show the logical structure of our reasoning. A man may be suffering from a grave disease; the physician tells us: "I do not know whether an operation will save the man, but if there *is* any remedy, it is an operation." In such a case, the operation would be justified. Of course, it would be better to know that the operation will save the man; but, if we do not know this, the knowledge formulated in the statement of the physician is a sufficient justification. If we cannot realize the sufficient conditions of success, we shall at least realize the necessary conditions. If we were able to show that the inductive inference is a necessary condition of success, it would be justified; such a proof would satisfy any demands which may be raised about the justification of induction.

Now obviously there is a great difference between our example and induction. The reasoning of the physician presupposes inductions; his knowledge about an operation as the only possible means of saving a life is based on inductive generalizations, just as are all other statements of empirical character. But we wanted only to illustrate the logical structure of our reasoning. If we want to regard such a reasoning as a justification of the principle of induction, the character of induction as a necessary condition of success must be demonstrated in a way which does not presuppose induction. Such a proof, however, can be given.

If we want to construct this proof, we must begin with a determination of the aim of induction. It is usually said that we perform inductions with the aim of foreseeing the future. This determination is vague; let us replace it by a formulation more precise in character:

The aim of induction is to find series of events whose frequency of occurrence converges toward a limit.

We choose this formulation because we found that we need probabilities and that a probability is to be defined as the limit of a frequency; thus our determination of the aim of induction is given in such a way that it enables us to apply probability methods. If we compare this determination of the aim of induction with determinations usually given, it turns out to be not a confinement to a narrower aim but an expansion. What we usually call "foreseeing the future" is included in our formulation as a special case; the case of knowing with certainty for every event *A* the event *B* following it would correspond in our formulation to a case where the limit of the frequency is of the numerical value 1. Hume thought of this case only. Thus our inquiry differs from that of Hume in so far as it conceives the aim of induction in a generalized form. But we do not omit any possible applications if we determine the principle of induction as the means of obtaining the limit of a frequency. If we have limits of frequency, we have all we want, including the case considered by Hume; we have then the laws of nature in their most general form, including both statistical and so-called causal laws – the latter being nothing but a special case of statistical laws, corresponding to the numerical value 1 of the limit of the frequency. We are entitled, therefore, to consider the determination of the limit of a frequency as the aim of the inductive inference.

Now it is obvious that we have no guaranty that this aim is at all attainable. The world may be so disorderly that it is impossible for us to construct series with a limit. Let us introduce the term "predictable" for a world which is sufficiently ordered to enable us to construct series with a limit. We must admit, then, that we do not know whether the world is predictable.

But, if the world is predictable, let us ask what the logical function of the principle of induction will be. For this purpose, we must consider the definition of limit. The frequency h^n has a limit at p, if for any given ε there is an n such that h^n is within $p \pm \varepsilon$ and remains within this interval for all the rest of the series. Comparing our formulation of the principle of induction (§38) with this, we may infer from the definition of the limit that, if there is a limit, there is an element of the series from which the principle of induction leads to the true value of the limit. In this sense the principle of induction is a necessary condition for the determination of a limit.

It is true that, if we are faced with the value h^n for the frequency furnished by our statistics, we do not know whether this n is sufficiently large to be identical with, or beyond, the n of the "place of convergence" for ε. It may be that our n is not yet large enough, that after n there will be a deviation greater than ε from p. To this we may answer: We are not bound to stay at h^n; we may continue our procedure and shall always consider the last h^n obtained as our best value. This procedure must at sometime lead to the true value p, if there is a limit at all; the applicability of this procedure, as a whole, is a necessary condition of the existence of a limit at p.

To understand this, let us imagine a principle of a contrary sort. Imagine a man who, if h^n is reached, always makes the assumption that the limit of the frequency is at $h^n + a$, where a is a fixed constant. If this man continues his procedure for increasing n, he is sure to miss the limit; this procedure must at sometime become false, if there is a limit at all.

We have found now a better formulation of the necessary condition. We must not consider the individual assumption for an individual h^n; we must take account of the procedure of continued assumptions of the inductive type. The applicability of this procedure is the necessary condition sought.

If, however, it is only the whole procedure which constitutes the necessary condition, how may we apply this idea to the individual case which stands before us? We want to know whether the individual h^n observed by us differs less than ε from the limit of the convergence; this neither can be guaranteed nor can it be called a necessary condition of the existence of a limit. So what does our idea of the necessary condition imply for the individual case? It seems that for our individual case the idea turns out to be without any application.

This difficulty corresponds in a certain sense to the difficulty we found in the application of the frequency interpretation to the single case. It is to be eliminated by the introduction of a concept already used for the other problem: the concept of posit.

If we observe a frequency h^n and assume it to be the approximate value of the limit, this assumption is not maintained in the form of a true statement; it is a posit such as we perform in a wager. We posit h^n as the value of the limit, i.e., we wager on h^n, just as we wager on the side of a die. We know that h^n is our best wager, therefore we posit it. There is, however, a difference as to the type of posit occurring here and in the throw of the die.

In the case of the die, we know the weight belonging to the posit: it is given by the degree of probability. If we posit the case "side other than that numbered 1," the weight of this posit is 5/6. We speak in this case of a posit with appraised weight, or, in short, of an *appraised posit.*

In the case of our positing h^n, we do not know its weight. We call it, therefore, a *blind posit.* We know it is our best posit, but we do not know how good it is. Perhaps, although our best, it is a rather bad one.

The blind posit, however, may be corrected. By continuing our series, we obtain new values h^n; we always choose the last h^n. Thus the blind posit is of an approximative type; we know that the method of making and correcting such posits must in time lead to success, in case there is a limit of the frequency. It is this idea which furnishes the justification of the blind posit. The procedure described may be called the *method of anticipation*; in choosing h^n as our posit, we anticipate the case where n is the "place of convergence." It may be that by this anticipation we obtain a false value; we know, however, that a continued anticipation must lead to the true value, if there is a limit at all.

An objection may arise here. It is true that the principle of induction has the quality of leading to the limit, if there is a limit. But is it the only principle with such a property? There might be other methods which also would indicate to us the value of the limit.

Indeed, there might be. There might be even better methods, i.e., methods giving us the right value p of the limit, or at least a value better than ours, at a point in the series where h^n is still rather far from p. Imagine a clairvoyant who is able to foretell the value p of the limit in such an early stage of the series; of course we should be very glad to have such a man at our disposal. We may, however, without knowing anything about the predictions of the clairvoyant, make two general statements concerning them: (1) The indications of the clairvoyant can differ, if they are true, only in the beginning of the series, from those given by the inductive principle. In the end there must be an asymptotical convergence between the indications of the clairvoyant and those of the inductive principle. This follows from the definition of the limit. (2) The clairvoyant might be an imposter; his prophecies might be false and never lead to the true value p of the limit.

The second statement contains the reason why we cannot admit clairvoyance without control. How gain such control? It is obvious that the control is to consist in an application of the inductive principle: we demand the forecast of the clairvoyant and compare it with later observations; if then there is a good correspondence between the forecasts and the observations, we shall infer, by induction, that the man's prophecies will also be true in the future. Thus it is the principle of induction which is to decide whether the man is a good clairvoyant. This distinctive position of the principle of induction is due to the fact that we know about its function of finally leading to the true value of the limit, whereas we know nothing about the clairvoyant.

These considerations lead us to add a correction to our formulations. There are, of course, many necessary conditions for the existence of a limit; that one which we are to use however must be such that its character of being necessary must be known to us. This is why we must prefer the inductive principle to the indications of the clairvoyant and control the latter by the former: we control the unknown method by a known one.

Hence we must continue our analysis by restricting the search for other methods to those about which we may know that they must lead to the true value of the limit. Now it is easily seen not only that the inductive principle will lead to success but also that every method will do the same if it determines as our wager the value

$$h^n + c_n$$

where c_n is a number which is a function of n, or also of h^n, but bound to the condition

$$\lim_{n=\infty} c_n = 0$$

Because of this additional condition, the method must lead to the true value p of the limit; this condition indicates that all such methods, including the inductive principle, must converge asymptotically. The inductive principle is the special case where

$$c_n = 0$$

for all values of n.

Now it is obvious that a system of wagers of the more general type may have advantages. The "correction" c_n may be determined in such a way that the resulting wager furnishes even at an early stage of the series a good approximation of the limit p. The prophecies of a good clairvoyant would be of this type. On the other hand, it may happen also that c_n is badly determined, i.e., that the convergence is delayed by the correction. If the term c_n is arbitrarily formulated, we know nothing about the two possibilities. The value $c_n = 0$ – i.e., the inductive principle – is therefore the value of the smallest risk; any other determination may worsen the convergence. This is a practical reason for preferring the inductive principle.

These considerations lead, however, to a more precise formulation of the logical structure of the inductive inference. We must say that, if there is any method which leads to the limit of the frequency, the inductive principle will do the same; if there is a limit of the frequency, the inductive principle is a sufficient condition to find it. If we omit now the premise that there is a limit of the frequency, we cannot say that the inductive principle is the necessary condition of finding it because there are other methods using a correction c_n. There is a set of equivalent conditions such that the choice of one of the members of the set is necessary if we want to find the limit; and, if there is a limit, each of the members of the set is an appropriate method for finding it. We may say, therefore, that the *applicability* of the inductive principle is a necessary condition of the existence of a limit of the frequency.

The decision in favor of the inductive principle among the members of the set of equivalent means may be substantiated by pointing out its quality of embodying the smallest risk; after all, this decision is not of a great relevance, as all these methods must lead to the same value of the limit if they are sufficiently continued. It must not be forgotten, however, that the method of clairvoyance is not, without further ado, a member of the set because we do not know whether the correction c_n occurring here is submitted to the condition of convergence to zero. This must be proved first, and it can only be proved by using the inductive principle, viz., a method known to be a member of the set: this is why clairvoyance, in spite of all occult pretensions, is to be submitted to the control of scientific methods, i.e., by the principle of induction.

It is in the analysis expounded that we see the solution of Hume's problem.[1] Hume demanded too much when he wanted for a justification of the inductive inference a proof that its conclusion is true. What his objections demonstrate is only that such a proof cannot be given. We do not perform, however, an inductive inference with the pretension of obtaining a true statement. What we obtain is a wager; and it is the best wager we can lay because it corresponds to a procedure the applicability of which is the necessary condition of the possibility of predictions. To fulfil the conditions sufficient for the attainment of true predictions does not lie in our power; let us be glad that we are able to fulfil at least the conditions necessary for the realization of this intrinsic aim of science.

Note

1 This theory of induction was first published by the author in *Erkenntnis*, III (1933), 421–5. A more detailed exposition was given in the author's *Wehrscheinlichkeitslehre*, §80.

5.5

Dissolving the Problem of Induction

Peter Strawson

Peter Strawson (1919–2006) was a prominent British philosopher in the "ordinary language" tradition, according to which philosophical disputes arise from conceptual misunderstandings that can be resolved by careful analysis of ordinary language. In this selection, he argues that David Hume's problem of induction rests on such a misunderstanding. Hume insists that we provide a reason to believe that induction is rational. But, Strawson responds, the inductive method precisely is the standard of rationality when it comes to reasoning from experience; to say that such reasoning is rational just is to say that it is in accord with the inductive method. The question whether the inductive method itself is rational therefore manifests a misunderstanding of the concept of rationality and of its range of application.

[. . .]

II The 'Justification' of Induction

7 We have seen something, then, of the nature of inductive reasoning; of how one statement or set of statements may support another statement, S, which they do not entail, with varying degrees of strength, ranging from being conclusive evidence for S to being only slender evidence for it; from making S as certain as the supporting statements, to giving it some slight probability. We have seen, too, how the question of degree of support is complicated by consideration of relative frequencies and numerical chances.

There is, however, a residual philosophical question which enters so largely into discussion of the subject that it must be discussed. It can be raised, roughly, in the following forms. What reason have we to place reliance on inductive procedures? Why should we suppose that the accumulation of instances of *As* which are *Bs*, however various the conditions in which they are

From P. F. Strawson, *Introduction to Logical Theory* (New York: John Wiley & Sons, 1952), pp. 248–52, 256–63.

observed, gives any good reason for expecting the next *A* we encounter to be a *B*? It is our habit to form expectations in this way; but can the habit be rationally justified? When this doubt has entered our minds it may be difficult to free ourselves from it. For the doubt has its source in a confusion; and some attempts to resolve the doubt preserve the confusion; and other attempts to show that the doubt is senseless seem altogether too facile. The root-confusion is easily described; but simply to describe it seems an inadequate remedy against it. So the doubt must be examined again and again, in the light of different attempts to remove it.

If someone asked what grounds there were for supposing that deductive reasoning was valid, we might answer that there were in fact no grounds for supposing that deductive reasoning was always valid; sometimes people made valid inferences, and sometimes they were guilty of logical fallacies. If he said that we had misunderstood his question, and that what he wanted to know was what grounds there were for regarding deduction *in general* as a valid method of argument, we should have to answer that his question was without sense, for to say that an argument, or a form or method of argument, was valid or invalid would *imply* that it was deductive; the concepts of validity and invalidity had application only to individual deductive arguments or forms of deductive argument. Similarly, if a man asked what grounds there were for thinking it reasonable to hold beliefs arrived at inductively, one might at first answer that there were good and bad inductive arguments, that sometimes it was reasonable to hold a belief arrived at inductively and sometimes it was not. If he, too, said that his question had been misunderstood, that he wanted to know whether induction in general was a reasonable method of inference, then we might well think his question senseless in the same way as the question whether deduction is in general valid; for to call a particular belief reasonable or unreasonable is to apply inductive standards, just as to call a particular argument valid or invalid is to apply deductive standards. The parallel is not wholly convincing; for words like 'reasonable' and 'rational' have not so precise and technical a sense as the word 'valid'. Yet it is sufficiently powerful to make us wonder how the second question could be raised at all, to wonder why, in contrast with the corresponding question about deduction, it should have seemed to constitute a genuine problem.

Suppose that a man is brought up to regard formal logic as the study of the science and art of reasoning. He observes that all inductive processes are, by deductive standards, invalid; the premises never entail the conclusions. Now inductive processes are notoriously important in the formation of beliefs and expectations about everything which lies beyond the observation of available witnesses. But an *invalid* argument is an *unsound* argument; an *unsound* argument is one in which *no good reason* is produced for accepting the conclusion. So if inductive processes are invalid, if all the arguments we should produce, if challenged, in support of our beliefs about what lies beyond the observation of available witnesses are unsound, then we have no good reason for any of these beliefs. This conclusion is repugnant. So there arises the demand for a justification, not of this or that particular belief which goes beyond what is entailed by our evidence, but a justification of induction in general. And when the demand arises in this way it is, in effect, the demand that induction shall be shown to be really a kind of deduction; for nothing less will satisfy the doubter when this is the route to his doubts.

Tracing this, the most common route to the general doubt about the reasonableness of induction, shows how the doubt seems to escape the absurdity of a demand that induction in general shall be justified by inductive standards. The demand is that induction should be shown to be a rational process; and this turns out to be the demand that one kind of reasoning should be shown to be another and different kind. Put thus crudely, the demand seems to escape one absurdity only to fall into another. Of course, inductive arguments are not deductively valid; if they were, they would be deductive arguments. Inductive reasoning must be assessed, for soundness, by inductive standards. Nevertheless, fantastic as the wish for induction to be deduction may seem, it is only in terms of it that we can understand some of the attempts that have been made to justify induction.

8 The first kind of attempt I shall consider might be called the search for the supreme premise of inductions. In its primitive form it is quite a crude attempt; and I shall make it cruder by

caricature. We have already seen that for a particular inductive step, such as 'The kettle has been on the fire for ten minutes, so it will be boiling by now', we can substitute a deductive argument by introducing a generalization (e.g., 'A kettle always boils within ten minutes of being put on the fire') as an additional premise. This manœuvre shifted the emphasis of the problem of inductive support on to the question of how we established such generalizations as these, which rested on grounds by which they were not entailed. But suppose the manœuvre could be repeated. Suppose we could find one supremely general proposition, which taken in conjunction with the evidence for any accepted generalization of science or daily life (or at least of science) would entail that generalization. Then, so long as the status of the supreme generalization could be satisfactorily explained, we could regard all sound inductions to unqualified general conclusions as, at bottom, valid deductions. The justification would be found, for at least these cases. The most obvious difficulty in this suggestion is that of formulating the supreme general proposition in such a way that it shall be precise enough to yield the desired entailments, and yet not obviously false or arbitrary. Consider, for example, the formula: 'For all *f*, *g*, wherever *n* cases of *f*. *g*, and no cases of *f*. ~ *g*, are observed, then all cases of *f* are cases of *g*.' To turn it into a sentence, we have only to replace '*n*' by some number. But what number? If we take the value of '*n*' to be 1 or 20 or 500, the resulting statement is obviously false. Moreover, the choice of any number would seem quite arbitrary; there is no privileged number of favourable instances which we take as decisive in establishing a generalization. If, on the other hand, we phrase the proposition vaguely enough to escape these objections – if, for example, we phrase it as 'Nature is uniform' – then it becomes too vague to provide the desired entailments. It should be noticed that the impossibility of framing a general proposition of the kind required is really a special case of the impossibility of framing precise rules for the assessment of evidence. If we could frame a rule which would tell us precisely when we had *conclusive* evidence for a generalization, then it would yield just the proposition required as the supreme premise.

Even if these difficulties could be met, the question of the status of the supreme premise would remain. How, if a non-necessary proposition, could it be established? The appeal to experience, to inductive support, is clearly barred on pain of circularity. If, on the other hand, it were a necessary truth and possessed, in conjunction with the evidence for a generalization, the required logical power to entail the generalization (e.g., if the latter were the conclusion of a hypothetical syllogism, of which the hypothetical premise was the necessary truth in question), then the evidence would entail the generalization independently, and the problem would not arise: a conclusion unbearably paradoxical. In practice, the extreme vagueness with which candidates for the role of supreme premise are expressed prevents their acquiring such logical power, and at the same time renders it very difficult to classify them as analytic or synthetic: under pressure they may tend to tautology; and, when the pressure is removed, assume an expansively synthetic air.

In theories of the kind which I have here caricatured the ideal of deduction is not usually so blatantly manifest as I have made it. One finds the 'Law of the Uniformity of Nature' presented less as the suppressed premise of crypto-deductive inferences than as, say, the 'presupposition of the validity of inductive reasoning'. I shall have more to say about this in my last section.

[. . .]

10 Let us turn from attempts to justify induction to attempts to show that the demand for a justification is mistaken. We have seen already that what lies behind such a demand is often the absurd wish that induction should be shown to be some kind of deduction – and this wish is clearly traceable in the two attempts at justification which we have examined. What other sense could we give to the demand? Sometimes it is expressed in the form of a request for proof that induction is a *reasonable* or *rational* procedure, that we have *good grounds* for placing reliance upon it. Consider the uses of the phrases 'good grounds', 'justification', 'reasonable', &c. Often we say such things as 'He has *every justification* for believing that *p*'; 'I have *very good reasons* for believing it'; 'There are *good grounds* for the view that *q*'; 'There is *good evidence* that *r*'. We often talk, in such ways as these, of justification, good grounds or reasons or evidence for certain

beliefs. Suppose such a belief were one expressible in the form 'Every case of *f* is a case of *g*'. And suppose someone were asked what he meant by saying that he had good grounds or reasons for holding it. I think it would be felt to be a satisfactory answer if he replied: 'Well, in all my wide and varied experience I've come across innumerable cases of *f* and never a case of *f* which wasn't a case of *g*.' In saying this, he is clearly claiming to have *inductive* support, *inductive* evidence, of a certain kind, for his belief; and he is also giving a perfectly proper answer to the question, what he meant by saying that he had ample justification, good grounds, good reasons for his belief. It is an analytic proposition that it is reasonable to have a degree of belief in a statement which is proportional to the strength of the evidence in its favour; and it is an analytic proposition, though not a proposition of mathematics, that, other things being equal, the evidence for a generalization is strong in proportion as the number of favourable instances, and the variety of circumstances in which they have been found, is great. So to ask whether it is reasonable to place reliance on inductive procedures is like asking whether it is reasonable to proportion the degree of one's convictions to the strength of the evidence. Doing this is what 'being reasonable' *means* in such a context.

As for the other form in which the doubt may be expressed, viz., 'Is induction a justified, or justifiable, procedure?', it emerges in a still less favourable light. No sense has been given to it, though it is easy to see why it seems to have a sense. For it is generally proper to inquire *of a particular belief*, whether its adoption is justified; and, in asking this, we are asking whether there is good, bad, or any, evidence for it. In applying or withholding the epithets 'justified', 'well founded', &c., in the case of specific beliefs, we are appealing to, and applying, inductive standards. But to what standards are we appealing when we ask whether the application of inductive standards is justified or well grounded? If we cannot answer, then no sense has been given to the question. Compare it with the question: Is the law legal? It makes perfectly good sense to inquire of a particular action, of an administrative regulation, or even, in the case of some states, of a particular enactment of the legislature, whether or not it is legal. The question is answered by an appeal to a legal system, by the application of a set of legal (or constitutional) rules or standards. But it makes no sense to inquire in general whether the law of the land, the legal system as a whole, is or is not legal. For to what legal standards are we appealing?

The only way in which a sense might be given to the question, whether induction is in general a justified or justifiable procedure, is a trival one which we have already noticed. We might interpret it to mean 'Are all conclusions, arrived at inductively, justified?', i.e., 'Do people always have adequate evidence for the conclusions they draw?' The answer to this question is easy, but uninteresting: it is that sometimes people have adequate evidence, and sometimes they do not.

11 It seems, however, that this way of showing the request for a general justification of induction to be absurd is sometimes insufficient to allay the worry that produces it. And to point out that 'forming rational opinions about the unobserved on the evidence available' and 'assessing the evidence by inductive standards' are phrases which describe the same thing, is more apt to produce irritation than relief. The point is felt to be 'merely a verbal' one; and though the point of this protest is itself hard to see, it is clear that something more is required. So the question must be pursued further. First, I want to point out that there is something a little odd about talking of 'the inductive method', or even 'the inductive policy', as if it were just one possible method among others of arguing from the observed to the unobserved, from the available evidence to the facts in question. If one asked a meteorologist what method or methods he used to forecast the weather, one would be surprised if he answered: 'Oh, just the inductive method.' If one asked a doctor by what means he diagnosed a certain disease, the answer 'By induction' would be felt as an impatient evasion, a joke, or a rebuke. The answer one hopes for is an account of the tests made, the signs taken account of, the rules and recipes and general laws applied. When such a specific method of prediction or diagnosis is in question, one can ask whether the method is justified in practice; and here again one is asking whether its employment is inductively justified, whether it commonly gives correct results. This question would normally seem an admissible one. One might be tempted to conclude

that, while there are many different specific methods of prediction, diagnosis, &c., appropriate to different subjects of inquiry, all such methods could properly be called 'inductive' in the sense that their employment rested on inductive support; and that, hence, the phrase 'non-inductive method of finding out about what lies deductively beyond the evidence' was a description without meaning, a phrase to which no sense had been given; so that there could be no question of justifying our selection of one method, called 'the inductive', of doing this.

However, someone might object: 'Surely it is possible, though it might be foolish, to use methods utterly different from accredited scientific ones. Suppose a man, whenever he wanted to form an opinion about what lay beyond his observation or the observation of available witnesses, simply shut his eyes, asked himself the appropriate question, and accepted the first answer that came into his head. Wouldn't this be a non-inductive method?' Well, let us suppose this. The man is asked: 'Do you usually get the right answer by your method?' He might answer: 'You've mentioned one of its drawbacks; I never do get the right answer; but it's an extremely easy method.' One might then be inclined to think that it was not a method of finding things out at all. But suppose he answered: Yes, it's usually (always) the right answer. Then we might be willing to call it a method of finding out, though a strange one. But, then, by the very fact of its success, it would be an inductively supported method. For each application of the method would be an application of the general rule, 'The first answer that comes into my head is generally (always) the right one'; and for the truth of this generalization there would be the inductive evidence of a long run of favourable instances with no unfavourable ones (if it were 'always'), or of a sustained high proportion of successes to trials (if it were 'generally').

So every successful method or recipe for finding out about the unobserved must be one which has inductive support; for to say that a recipe is successful is to say that it has been repeatedly applied with success; and repeated successful application of a recipe constitutes just what we mean by inductive evidence in its favour. Pointing out this fact must not be confused with saying that 'the inductive method' is justified by its success, justified because it works. This is a mistake, and an important one. I am not seeking to 'justify the inductive method', for no meaning has been given to this phrase. *A fortiori*, I am not saying that induction is justified by its success in finding out about the unobserved. I am saying, rather, that any successsful method of finding out about the unobserved is necessarily justified by induction. This is an analytic proposition. The phrase 'successful method of finding things out which has no inductive support' is self-contradictory. Having, or acquiring, inductive support is a necessary condition of the success of a method.

Why point this out at all? First, it may have a certain therapeutic force, a power to reassure. Second, it may counteract the tendency to think of 'the inductive method' as something on a par with specific methods of diagnosis or prediction and therefore, like them, standing in need of (inductive) justification.

12 There is one further confusion, perhaps the most powerful of all in producing the doubts, questions, and spurious solutions discussed in this Part. We may approach it by considering the claim that induction is justified by its success in practice. The phrase 'success of induction' is by no means clear and perhaps embodies the confusion of induction with some specific method of prediction, &c., appropriate to some particular line of inquiry. But, whatever the phrase may mean, the claim has an obviously circular look. Presumably the suggestion is that we should argue from the past 'successes of induction' to the continuance of those successes in the future; from the fact that it has worked hitherto to the conclusion that it will continue to work. Since an argument of this kind is plainly inductive, it will not serve as a justification of induction. One cannot establish a principle of argument by an argument which uses that principle. But let us go a little deeper. The argument rests the justification of induction on a matter of fact (its 'past successes'). This is characteristic of nearly all attempts to find a justification. The desired premise of Section 8 was to be some fact about the constitution of the universe which, even if it could not be used as a suppressed premise to give inductive arguments a deductive turn, was at any rate a 'presupposition of the validity of induction'. Even the mathematical argument of Section 9 required

buttressing with some large assumption about the make-up of the world. I think the source of this general desire to find out some fact about the constitution of the universe which will 'justify induction' or 'show it to be a rational policy' is the confusion, the running together, of two fundamentally different questions: to one of which the answer is a matter of non-linguistic fact, while to the other it is a matter of meanings.

There is nothing self-contradictory in supposing that all the uniformities in the course of things that we have hitherto observed and come to count on should cease to operate to-morrow; that all our familiar recipes should let us down, and that we should be unable to frame new ones because such regularities as there were were too complex for us to make out. (We may assume that even the expectation that all of us, in such circumstances, would perish, were falsified by someone surviving to observe the new chaos in which, roughly speaking, nothing foreseeable happens.) Of course, we do not believe that this will happen. We believe, on the contrary, that our inductively supported expectation-rules, though some of them will have, no doubt, to be dropped or modified, will continue, on the whole, to serve us fairly well; and that we shall generally be able to replace the rules we abandon with others similarly arrived at. We might give a sense to the phrase 'success of induction' by calling this vague belief the belief that induction will continue to be successful. It is certainly a factual belief, not a necessary truth; a belief, one may say, about the constitution of the universe. We might express it as follows, choosing a phraseology which will serve the better to expose the confusion I wish to expose:

I. (The universe is such that) induction will continue to be successful.

I is very vague: it amounts to saying that there are, and will continue to be, natural uniformities and regularities which exhibit a humanly manageable degree of simplicity. But, though it is vague, certain definite things can be said about it. (1) It is not a necessary, but a contingent, statement; for chaos is not a self-contradictory concept. (2) We have good inductive reasons for believing it, good inductive evidence for it. We believe that some of our recipes will continue to hold good because they have held good for so long. We believe that we shall be able to frame new and useful ones, because we have been able to do so repeatedly in the past. Of course, it would be absurd to try to use I to 'justify induction', to show that it is a reasonable policy; because I is a conclusion inductively supported.

Consider now the fundamentally different statement:

II. Induction is rational (reasonable).

We have already seen that the rationality of induction, unlike its 'successfulness', is not a fact about the constitution of the world. It is a matter of what we mean by the word 'rational' in its application to any procedure for forming opinions about what lies outside our observations or that of available witnesses. For to have good reasons for any such opinion is to have good inductive support for it. The chaotic universe just envisaged, therefore, is not one in which induction would cease to be rational; it is simply one in which it would be impossible to form rational expectations to the effect that specific things would happen. It might be said that in such a universe it would at least be rational to refrain from forming specific expectations, to expect nothing but irregularities. Just so. But this is itself a higher-order induction: where irregularity is the rule, expect further irregularities. Learning not to count on things is as much learning an inductive lesson as learning what things to count on.

So it is a contingent, factual matter that it is sometimes possible to form rational opinions concerning what specifically happened or will happen in given circumstances (I); it is a non-contingent, *a priori* matter that the only ways of doing this must be inductive ways (II). What people have done is to run together, to conflate, the question to which I is answer and the quite different question to which II is an answer, producing the muddled and senseless questions: 'Is the universe such that inductive procedures are rational?' or 'What must the universe be like in order for inductive procedures to be rational?' It is the attempt to answer these confused questions which leads to statements like 'The uniformity of nature is a presupposition of the validity of induction'. The statement that nature is uniform might be taken to be a vague way of expressing what we expressed by I; and certainly this fact is a condition of, for it is identical with, the likewise contingent fact that we are, and shall continue to be, able to form rational opinions, of the kind we are most anxious to form, about the unobserved. But neither this fact about the world, nor any other, is a condition of the necessary truth that, if it is possible to form rational opinions of this kind, these will be inductively supported opinions. The discordance of the conflated questions manifests itself in an uncertainty about the status to be accorded to the alleged presupposition of the 'validity' of induction. For it was dimly, and correctly, felt that the reasonableness of inductive procedures was not merely a contingent, but a necessary, matter; so any necessary condition of their reasonableness had likewise to be a necessary matter. On the other hand, it was uncomfortably clear that chaos is not a self-contradictory concept; that the fact that some phenomena do exhibit a tolerable degree of simplicity and repetitiveness is not guaranteed by logic, but is a contingent affair. So the presupposition of induction had to be both contingent and necessary: which is absurd. And the absurdity is only lightly veiled by the use of the phrase 'synthetic a priori' instead of 'contingent necessary'.

Unit 6

After the Received View: Confirmation and Observation

In this unit, we will review a number of criticisms of positivism in general and of the received view in particular. They are organized around two themes: the first concerns the confirmation of scientific theories (§§1–6), and the second concerns the observable/unobservable (or observation language/theoretical language) distinction (§§7–9).

1 Hempel's Critique of the Empiricist Criterion of Cognitive Significance

Carl Hempel was an advocate of the received view of scientific theories and the author of the received view on scientific explanation. But he was nevertheless skeptical of two fundamental tenets of the version of that view that we reviewed in the Unit 5 commentary, namely, verificationism (that is, the empiricist criterion of cognitive significance) and the analytic/synthetic distinction. In "Empiricist Criteria of Cognitive Significance: Problems and Changes" (reading 6.1), he challenges verificationism.

Hempel does not criticize the idea that there is an observation vocabulary or that sentences devoted to the description of observations are constructed from it.[1] Instead, he is concerned with the prospects for a solution to the "problem of theoretical terms," which is the attempt to certify the empirical significance of theoretical terms in science by relating them to the language of observation.[2] He argues that those prospects are very dim.

The first proposal that he considers is the (conclusive) verifiability requirement, which requires of a meaningful sentence that it can be conclusively established by observation. We have already seen the essential problem with this.[3] Laws, such as "all ravens are black," concern a potential infinity of cases, and therefore can never be conclusively established by any finite number of observations of black ravens.

In *Language, Truth, and Logic*, A. J. Ayer tried to formulate a criterion that requires only confirmability rather than complete verifiability. His first attempt required that we can deduce observation sentences from the sentence in question, perhaps with the help of additional sentences. A problem with this is that from "The absolute is lazy" (a paradigmatically metaphysical claim) we can, with the help of "If the absolute is lazy, then it is sunny out" deduce "It is sunny out" (an observation sentence). We would then have to count "The absolute is lazy" as significant by this criterion. Aware of this problem, Ayer then suggested that we require that the additional sentences that we use in the deduction pass the significance test independently. But Hempel points out that any consequence of "It is sunny out" is also a consequence of "It is sunny out and the absolute is lazy." So no matter what other sentences are used to derive empirical consequences from the first, they can also be used to derive the same consequences from the second.

Hempel then considers operational definitions and reduction sentences. We considered these options, as well as objections of the sort that Hempel raises, in §7 of the Unit 5 commentary. We will not repeat those discussions here. Instead, we will focus on the primary difficulty that Hempel raises for Carnap's 'partial interpretation' formulation of the criterion (also §7 of Unit 5 commentary).

Recall that the method of partial interpretation consists first in setting out the scientific system as a formal, uninterpreted calculus consisting in a set of primitive terms and axioms; the primitive terms are used to define other expressions and the axioms are used to derive theorems. We then formulate correspondence rules (typically reduction sentences) that assign empirical meaning, not to the primitive terms, but instead to more complex expressions defined by those primitive terms. Empirical significance is then drawn up to the primitive theoretical terms; their meaningfulness is due to their role in a theoretical network that has empirical consequences. So understood, a sentence will be insignificant (relative to that network) if its removal does not reduce the collection of empirical consequences. Such a sentence is *isolated*. We then require that a theoretical system contain no isolated sentences.

Hempel presents a devastating problem for this account. To understand it, first consider the following collection of sentences:

1 The absolute is lazy;
2 Sarah is going to the bar; and
3 If Sarah is going to the bar, then so is Andre.

(1)–(3) imply:

4 Andre is going to the bar.

But (1) is unnecessary for this inference and is therefore isolated; since it is isolated, we can safely remove it. So far, so good.

But now replace (1) and (2) with:

5 The absolute is lazy and Sarah is going to the bar.

(5) and (3) also imply (4). But this time (5) is *not* isolated since we need it to infer (4).

The obvious response is that, while (5) is not isolated, it is equivalent to the conjunction of (1) and (2); and (1), we saw, *is* isolated. In general, we might require of a theoretical system that it neither have any isolated sentences nor be logically equivalent to any system that has such sentences.

But now consider a system that has theoretical sentences T_1 and T_2 from which we can derive observation sentence O. Suppose that both T_1 and T_2 are required in order to derive O, so that neither are isolated. However, precisely because the collection of sentences (T_1, T_2) implies O, it is logically equivalent to the collection (T_1, T_2, O).[4] But now, since O is listed separately, T_1 and T_2 are no longer required to derive it; T_1 and T_2 are now isolated. But then the original set (T_1, T_2) is logically equivalent to one that contains isolated sentences and therefore violates the criterion.

The problem is that this will be true of *any* system in which theoretical sentences are used to derive observations: all such systems will be logically equivalent to ones in which those observations are listed separately, which will then isolate the theoretical sentences. So the only legitimate systems will be ones that only list the observations; no theoretical assertions will ever be legitimate. But the point of the criterion is to ensure the legitimacy of the theoretical terms.

Hempel does not conclude that theoretical terms should be dispensed with. He points out that the "history of the scientific endeavor shows that if we wish to arrive at precise, comprehensive, and well-confirmed general laws, we have to rise above the level of direct observation" (reading 6.1, p. 399). We would therefore "deprive ourselves of the tremendous fertility of theoretical constructs, and we would often render the formal structure of the expurgated theory clumsy and inefficient" (reading 6.1, pp. 399–400) if we were to ban theoretical constructs in the formulation of scientific theories. If the empiricist criterion of cognitive significance requires that we do this, then so much the worse for that criterion.

2 The Raven Paradox

Hempel also presented a notorious paradox for the received view's conception of the relation between a scientific law and its confirming instances. According to the received view, laws are universal generalizations of the form "All P are

Q" (written "(x)(Px ⊃ Qx)" in logical notation). We identify cases in which the property specified in the antecedent is realized and determine whether the consequent is realized as well; if it is, the generalization is confirmed, and if not, it is disconfirmed. So, a black raven confirms the (putative) law "All ravens are black", and an orange raven disconfirms it. This is now called *Nicod's criterion* (after Jean Nicod who formulated it).[5]

But "All P are Q" is logically equivalent to "All non-Q are non-P" by the logical rule of contraposition. "All ravens are black," for example, is equivalent to "All non-black things are non-ravens" (more colloquially: "Anything that is not black is not a raven"). By Nicod's criterion, a black raven will not confirm the latter law, since it is not a non-black non-raven, and so does not realize the antecedent and consequent conditions.

This violates what Hempel calls the *equivalence condition*, which requires that anything that confirms (or disconfirms) a sentence also confirms (or disconfirms) any sentence logically equivalent to it. He then argues that we should accept this condition, for two reasons. First, it is intuitively odd to think that which formulation of a sentence we choose from among logically equivalent formations should make a difference to its confirmation. Second, the fundamental role of laws, according to the received view, is to predict and explain phenomena. And laws do this (on Hempel's own account) by being premises in arguments from which sentences describing the phenomena are derived.[6] But it does not matter for that purpose how the law is formulated; any logically equivalent formulation will permit the derivation (if the original formulation did).

One way out is to characterize Nicod's criterion as only a sufficient condition for confirmation (so that anything that is a P and a Q confirms "All P are Q"), but not a necessary one (there are confirming instances of "All P are Q" that are not both P and Q). Then we could count the black raven as confirming "All non-black things are non-ravens" as well as "All ravens are black" in accordance with the equivalence condition.

But now consider a red pencil. That is a non-black non-raven; it therefore confirms "All non-black things are non-ravens" by Nicod's criterion. But "All non-black things are non-ravens" is equivalent to "All ravens are black." So by the equivalence condition, it also confirms "all ravens are black," and so do green leaves, yellow bananas, and so on.[7] But the idea that yellow bananas confirm the law that all ravens are black is very unintuitive.

Philosophers who attempt to resolve the raven paradox fall generally into two camps: those who think that the criterion can be modified so as to rule out these odd cases, and those who think that, despite their oddity, these cases really do confirm the law. After considering some versions of the first option, Hempel himself settled on the second, suggesting that the oddity is an illusion generated by the influence of our background knowledge. Unfortunately we do not have the space to discuss the voluminous literature on the paradox and the different proposed solutions that have been produced. Suffice it to say that it is far from clear that a decisive solution is on the table.

3 Goodman's New Riddle of Induction

We have seen the problem of induction that Hume raised (reading 3.7) and some responses to it (readings 4.4, 4.10, 4.11, 5.4, and 5.5). Suppose that some such solution (or vindication, or dissolution) works. Nelson Goodman suggests that, even then, we are not out of the woods, because there is another problem to contend with: his *new riddle of induction*.

Suppose that we have observed a large number of emeralds under a variety of circumstances, and they have all been green. We draw the conclusion that all emeralds are green. We thus expect emeralds that we inspect in the future – after January 1, 2040, say – will be green. This is, of course, the only color of future emeralds our evidence supports; to infer that future emeralds will be blue, for example, on that evidence is to not use induction at all. But now consider the property "grue". An object is grue if and only if it is green when it is examined before January 1, 2040 and blue afterwards. All the emeralds we have examined so far have been grue as well as green (since they were green and examined before that date). So the same evidence we have for the hypothesis that all emeralds are green appears to equally support the hypothesis that all emeralds are grue.

We thus should also expect that emeralds examined after that date will be grue. But, given the definition of "grue", that means that we should expect that emeralds examined after that date should be blue. We should also expect that they will be red, because they are also "gred" (green before January 1, 2040 and red afterward), and yellow, because they are also "grellow", and so on. So the prediction that future emeralds will be green is not favored by our observations any more than is any other color; nothing indeed follows about the future at all, since we can construct a similarly "gruesome" predicate for any property whatsoever. We obviously need to bar the use of gruesome properties, distinguishing those that are *projectable* – legitimately used in inductive arguments – from those that are not. But just what determines whether a property is projectable? What exactly is wrong with "grue"?

One might focus on the reference to January 1, 2040. The word "grue" itself does not refer to a time, but it is logically equivalent to an expression that does (according to its given definition). So we might suggest that no predicates that involve reference to particular times (or places, or people, or whatever else we wished to exclude), or that are logically equivalent to ones that do, are projectable. The problem for this is that "all emeralds are green" is logically equivalent to "all emeralds before January 1, 2040 are green and all emeralds after that date are green," which includes a reference to a specific time as well.

But, one might protest, we do not *have* to refer to that date to understand "green", but we do in order to understand "grue" (since its definition necessarily includes such a reference). Goodman argues that this reflects nothing more than the historical accident that we started with "green" and "blue" and defined "grue" in terms of them. But we could have just as easily started with "grue". To see this, we need another predicate, "bleen": something is bleen if it is blue when examined before January 1, 2040 and green otherwise. Now we can define "green" as "grue if examined before January 1, 2040 and bleen otherwise" (and correspondingly for "blue"). So we could have started with grue and bleen as our primitive terms, and defined green and blue in terms of them, instead of the other way around.

Goodman's own proposal is simply that the projectable predicates are those that are *entrenched*, that is, those that we have already been using. But other philosophers have found this to be no more attractive than Hume's appeal to habit in his response to his original problem of induction. Many other approaches to the new riddle have been proposed. A popular response is that the predicate must be a *natural kind* instead of the artificial kinds that grue and bleen seem to be. But this requires an account of the difference between natural and artificial kinds, which is itself the subject of considerable dispute.

Whether or not the natural kind response or some other approach is successful, one crucial consequence of the new riddle appears to be unavoidable. The positivists hoped that it is possible to formulate a purely formal theory of confirmation analogous to the formal theories of deductive validity.[8] But the inter-translatability of grue and bleen with blue and green suggests that no purely formal theory can work; we have to attend to the content as well as the form, that is, to the character of the properties themselves that are referred to by the predicates in addition to (or perhaps instead of) the syntactical structure of the predicates. The natural kind response exemplifies this since it requires that we determine whether the property referred to by the predicate is a natural kind, something that cannot be determined by examination of the predicate alone. To at least this extent, the conception of inductive confirmation as a probabilistic extension of the formal theory of deductive validity is seriously threatened.

4 Quine and the Closed-Circle Argument Against Analyticity

In §§6–7 of the Part II introduction and in §7 of the Unit 5 commentary, we saw how important the concept of analytic truth was for the positivists. First, it allows them to endorse the ubiquitous use of logic and mathematics in the sciences without violating their scruples against the synthetic a priori. Second, it grounded their conception of the philosophical endeavor as part conceptual analysis and part language engineering. Third, it allowed them to recognize multiple systems, or frameworks (e.g., of geometry or even logic), without having to settle the question which is really correct. (Construed as systems of analytic propositions, they all are correct, but do not conflict since the meanings are different.) And fourth, it underlay the many versions of the empiricist criterion of cognitive significance since the various correspondence rules they invoke confer meaning on the theoretical terms and are, therefore, analytic.

During his career, Carnap developed a variety of different formulations of the analytic/synthetic distinction; each had its difficulties, and debate is ongoing as to whether they are surmountable. But, not surprisingly in light of its significance for the received view in particular and the positivists' philosophical outlook in general, he never gave up. However, a different kind of criticism of the analytic/synthetic distinction began to emerge. It concerned, not whether this or that formulation of the distinction was tenable, but rather whether we really need it at all. The *locus classicus* of this critique is W. V. O. Quine's landmark essay, "Two Dogmas of Empiricism."

The first part of the paper lays out what might be called the *closed circle argument*. This argument presupposes that the concept of analyticity needs a definition, and is illegitimate unless such a definition can be provided. But the concept of analyticity belongs to a group of concepts, each of which is definable in terms of the others but not by any concepts outside of the group.[9] And each of those other concepts needs defining as badly as does the target concept. So no definition of the concept of analyticity can be provided. The concept of analyticity is illegitimate.

Our tour around the circle begins with Gottlob Frege's approach: an analytic proposition is either a logical truth or can be transformed into a logical truth by replacing synonyms for synonyms.[10] For example, 'bachelor' in "All bachelors are unmarried" can be replaced by 'unmarried adult males'; and "Unmarried adult males are unmarried" is a logical truth.

There are two problems with this definition of analyticity. First, it just pushes the issue back to the status of logical truth. To say that a logical truth is analytic is, on this analysis, just to say that a logical truth is a logical truth. But that, of course, does not tell us *why* logical truths are true; it is, for example, compatible with J. S. Mill's contention that logical truths are highly confirmed empirical claims rather than true in virtue of meaning. Second, it relies on the concept of synonymy. But synonymy, Quine insists, stands in as much need of clarification as does analyticity.

At one point Carnap suggested that analyticities are sentences that are true in all *state descriptions*. The rough idea is that a state description assigns a truth-value (that is, assigns "true" or "false") to every statement that can be formulated in the language.[11] Thus, for example, one state description might declare that the sentence "The Eiffel Tower is made of iron" is true and that "The Eiffel Tower is in England" is false, whereas another state description might declare that the first is false and second true, a third that they are both true, and a fourth that they are both false. As it happens, the first state description describes the actual world; the others describe non-actual but possible worlds. To be true in all state descriptions implies that there is no possible world (at least as describable by this language) in which the proposition is false.

But then it looks like one state description could list the sentence "Quine is a bachelor" as true and "Quine is unmarried" as false, and thus imply that there is a married bachelor. But this is supposed to be analytically false and therefore should be false in all state descriptions. Ruling it out requires taking into account the meaning equivalence between 'bachelor' and 'unmarried adult male'; and we are back to synonymy.

Another approach identifies analytic sentences as those that are true by definition. But a definition just is a report of synonymy. Admittedly, there are some definitions that are not obviously so. Explicative definitions, for example, redefine a vague concept more precisely (e.g., defining "adult" as "someone of at least eighteen years of age") for some purpose (in this case, a legal one). But insofar as the new definition resembles the old, it concerns sameness of meaning or synonymy again. And in so far as it assigns a new meaning, it does not tell us how a sentence already established in the language can be analytic. Sometimes we do just assign a new notation to abbreviate an existing expression (e.g., "let x equal any real number between five and twenty-five"). At the time the assignment is made, Quine is prepared to grant, the old and new expressions are indeed synonymous. But this assignment is an event, and it tells us nothing of how the trait of synonymy, or analyticity, is supposed to linger on after the event has passed.

Approaching synonymy more directly, we might define it as interchangeability *salva veritate* (i.e., saving truth). One concept is synonymous with another if, for every sentence in which the former occurs, replacing it with the latter will not change the sentence's truth-value. The problem is that any two expressions which refer to the same things will satisfy this condition even if they are not synonymous. For example, anything that has a heart has a kidney, not by definition, but simply as a fact about how such creatures are put together. So any sentence with 'creature with a heart', such as "There is a creature with a heart looking in the window!" can be replaced with "There is a creature with a kidney looking in the window!" without changing the sentence's truth-value. But these two expressions are not synonymous.

The situation changes if we allow ourselves use of the word "necessarily", since "Necessarily, creatures with a heart are creatures with a heart" is true, whereas "Necessarily, creatures with a heart are creatures with a kidney" is false. But the necessity of the first sentence is just a reflection of the fact that it is logical truth, and therefore analytic; in general, a sentence is necessary simply because it is analytic.[12] But now we are back to analyticity.

Another of Carnap's attempts (in addition to the appeal to state descriptions) was to explicitly indicate the analytic sentences as those that followed from the meaning-rules, or *semantical rules* of the language. Quine objects that this just selects certain truths and puts them on a page with the title "semantical rule"; it does not tell us what a semantical rule in general (for all languages) is.

5 Against the Second Dogma

Perhaps we can make better progress in our attempt to define the concept of analyticity if we appeal to the verification theory of meaning. According to this theory (according to Quine), the meaning of a statement is the method of confirming (or falsifying) it; an analytic sentence would then be one that is confirmed come what may. The problem, Quine claims, is that this presupposes *reductionism* according to which a meaningful statement is translatable into a statement about experience or, more generally, is associated with a set of experiences that confirms it and another set that disconfirms it.

But that assumption is false, as was long ago indicated by Pierre Duhem.[13] It is only with the aid of a host of *auxiliary hypotheses*, that is, initial and boundary conditions, theories borrowed from other sciences, background theories concerning the instruments used, assumptions about the experimental setup, and so on, that we can extract an empirical prediction from an hypothesis. And all those auxiliary hypotheses themselves presuppose others within our "web of belief" in the same way. The end result is that "our statements about the external world face the tribunal of sense experience not individually but only as a corporate body" (reading 6.3, p. 420).

This is now called the *Duhem-Quine thesis*. Hypothesis H does not, by itself, imply observation O, as the hypothetico-deductive model would have it (see §2 of Unit 5 commentary). Rather, only H & A_1 & A_2 & . . . & A_n does (where A_1–A_n are auxiliary assumptions). So, if the observed result is that O is false, the fault may lie with any of the auxiliary assumptions rather than with the hypothesis; the hypothesis itself can always be protected from falsification, come what may, so long as we are prepared to make other adjustments among one or more auxiliary hypotheses. So the fact that a proposition can be saved come what may is no mark of analyticity, but rather applies to virtually every hypothesis in the web of belief.

The Duhem-Quine thesis grounds another, called the *underdetermination thesis*. Since we could respond to a failed observation by revising different auxiliary hypotheses, there are a variety of different resolutions of the tension between observation and theory. The evidence therefore underdetermines (i.e., does not itself settle) which set of theory-plus-auxiliaries should be accepted.

But some adjustments to the web are more sensible than others. Rejection of a belief on the "periphery" – that is, relatively close to the boundary of experience, such as a low-level empirical law in biology – will not have a particularly dramatic impact on the rest of the web. (It will have no impact on quantum physics, for example.) But rejection of a belief in the "center" – that is, one that is more abstract and theoretical – might well have a devastating impact. The laws of physics, for example, are invoked in everything from cosmology to archeology to geology to sports medicine; change those laws, and many other changes will also have to be made in order to recover stability. Logic and mathematics are even more central than the laws of physics. There are very few regions of human inquiry that do not employ mathematics, and none at all that do not employ logic. To revise those would therefore be deeply disruptive, and so is best avoided.

We have just explained the apparent resistance to empirical refutation demonstrated by logic and math that led the positivists to characterize them as analytic. But we did it without appeal to the concept of analyticity. Logical and mathematical truth is, as Mill said, fundamentally just another kind of empirical truth. But, contrary to Mill, it does not consist in highly confirmed empirical generalizations vulnerable to a falsifying counter-instance. Instead, they are deeply embedded theoretical hypotheses, protected by the auxiliary hypotheses in the way that the Duhem-Quine thesis allows, because of their centrality to our web of belief.[14] This permits the possibility that even our current systems of logic and mathematics can themselves be overthrown. And indeed Einstein employed non-Euclidean geometry in his formulation of general relativity, and "quantum logic" has been proposed as a solution to certain problems in quantum mechanics. Not only can any statement be maintained come what may; all – even logic and mathematics – are susceptible to revision if the occasion calls for it. The difference between logical and mathematical truth on the one hand and empirical truth on the other is, Quine insisted, a difference of degree rather than of kind.

6 Empiricism without the Dogmas

The closed circle argument is of questionable force. After all, any philosophically significant concept will seem to be only interdefinable within a circle of related concepts, so long as the circle is big enough. But that argument is really only stage-setting. Quine's fundamental attack on the concept of analyticity is that, once we absorb the impact of the Duhem-Quine thesis, we simply do not need the concept; we can explain everything we need to explain about the character of logic and mathematics without it (and without returning to the rationalists' synthetic a priori). That is the core of Quine's attack on the analytic/synthetic distinction.

Quine did not, however, reject verificationism. Viewed as the general doctrine that meaning is a function of empirical consequence, he endorsed it. But in the face of the Duhem-Quine thesis, meaning does not accrue to the theory sentence-by-sentence; instead, the "unit of empirical significance is the whole of science" (reading 6.3, p. 420). *Semantic holism* – that meaning accrues to words and sentences in virtue of their role in a system – thus results from the combination of the Duhem-Quine thesis and verificationism.

The web of belief must always square with experience at its boundary. But, because of the Duhem-Quine thesis, that leaves considerable latitude concerning how the web should be adjusted in response. The web is, however, subject to two other pressures. First, we should change as little as possible; to respond to one unexpected experimental result by revising the laws of logic would produce disastrous consequences for the rest of the web, and should be avoided if at all possible. This is the virtue of *conservatism* (or as Quine once put it, the "maxim of minimum mutilation").

But we also want a tidy system of belief, rather than one that is complex, disorganized, and unwieldy. And sometimes even fairly drastic changes are required to keep the overall system as manageable as possible. (Quantum mechanics, for example, requires overthrow of the conviction that physical laws are deterministic.) This is the virtue of *simplicity*. The trajectory of the web of belief is a product of these three pressures – conformity with experience, conservatism, and simplicity – in operation.

Conservatism and simplicity are likely to strike one as merely pragmatic; they both seem to constitute a kind of intellectual laziness, after all. But moral laziness is a vice; why should intellectual laziness count as a virtue? Surely we should determine that these are indeed virtues by investigating whether they will lead us closer to the truth.

Quine's answer is essentially that the question presupposes a standpoint from outside the web of belief from which we can conduct such an inquiry. But there is, he insists, no such 'cosmic exile'; we are like mariners on Neurath's boat at sea, forced to keep the boat afloat while making repairs. We can only operate from within the scientific world view and make adjustments as we go. This is Quine's *naturalism*: the scientific standpoint is our intellectual standpoint generally, since there is no real alternative.

The positivists were aware of these consequences to some extent. The Duhem-Quine thesis was recognized as far back as, well, Duhem (1906), and was widely endorsed by the positivists.[15] Carnap repudiated reductionism long before the development of the received view, and the method of partial interpretation embodies a holistic account of the meaning of theoretical terms. And his "Empiricism, Semantics, and Ontology" (reading 5.3) recognizes that external framework-questions are fundamentally pragmatic questions concerning how the language is most conveniently structured.

But the distinction between those questions and the internal questions – between questions of language and questions of fact – is, Quine suggested, itself a manifestation of the analytic/synthetic distinction, and equally untenable. Quine's "more thorough pragmatism"[16] is, Quine might suggest, the destination to which the path of the positivist's increasingly holistic account of scientific methodology leads.

7 Putnam against the Observational-Theoretical Dichotomy

Readings 6.5–6.7 by Hilary Putnam, N. R. Hanson, and Grover Maxwell all attack the observable/unobservable (or observation-vocabulary/theoretical-vocabulary) distinction, but in very different ways. Putnam's argument is directed against the assumption underlying the received view that the language of science can be divided into two vocabularies, the observation vocabulary and the theoretical vocabulary.[17]

This distinction, we saw in §7 of the Unit 5 commentary, underlies the problem of theoretical terms; it was assumed that observation terms are unproblematic, and that the empirical significance of the theoretical terms needs to be accounted for by characterizing their relation to the observation terms (by means of operational definitions, reduction sentences, or partial interpretation).[18] Assuming that we learn observation terms first, theoretical terms are then introduced later, on the basis, somehow, of the observation terms; hence the problem of accounting for their meaning. Putnam denied that this distinction is coherent, and therefore claimed that there is no such problem of theoretical terms.

Putnam's argument is straightforward. The distinction between observation terms and theoretical terms is a linguistic distinction, whereas the distinction between observable and unobservable entities is an ontological distinction, that is, a distinction between objects. The positivists assumed that these two distinctions run together. Putnam claims that they do not, and supports his claim with examples. A children's story about people who are too little to see concerns (putative) unobservable objects, but it employs only non-theoretical terms (at least in the sense that "people", "little", and so on are not a part of any systematic scientific theory). And to describe an object passing overhead as a "satellite" (or one in the kitchen as a "microwave", or one at the grocery store checkout as a "laser scanner") is to employ theoretical terminology in the description of observable entities.

The positivist might respond by saying that these terms are observable in one context and theoretical in another. But if any putative theoretical term can be used in observation reports (as in "we observed the creation of two electron-positron pairs"), and any non-theoretical term can be used to characterize unobservables (as in an early explanation of red light as a stream of red particles), then there is no longer a point to the attempt to separate a vocabulary exclusively used to describe observations from one used solely within the theory.

8 Hanson and the Theory-Ladenness of Observation

Reading 6.6 by Hanson concerns, not the language used to describe observations, but the experience of observing itself. He argues that this experience is *theory-laden*, that is, that what we observe is influenced by our background theoretical commitments. Intuitively, our sensory perception of the world seems to be transparent: when our perceptual system is working properly (e.g., when we are not under the influence of hallucinogenic drugs), what we see just is what is in front of us. If we (clearly) see an apple on the table, for example, then there is indeed an apple there that we see, with the characteristics that we see it as having (e.g., being red, having a stem). What we see is a function solely of our physical environment; if what we see changes (e.g., we see the apple roll off the table), then there must be a corresponding change in that environment (e.g., the apple indeed does roll off the table). In this way our perceptions are an accurate and objective indication of the character of our physical environment, which simply "floods in" when we open our eyes. This explains their role in science as the theory-neutral testing ground against which theories can be assessed and competing theories compared.

Hanson denies this *theory-neutrality* of perception. He discusses some examples drawn from Gestalt psychology. Consider the image, called the *Necker cube* (see p. 435). If you are like most people shown this image, you will first see a cube with its front face below its back face (as if you were viewing a box on a table). But you will soon see it change to a cube with its front face above its back face (as if you were viewing a cube-shaped speaker attached high on a wall). The transition, called a *Gestalt shift*, is instantaneous and complete. You do not slowly see a cube coming into view one way, and then slowly shift into another; you just see one, and then immediately see another. And you cannot see both at the same time.

Of course in one sense of "what you see", what you see are flat lines on a page, since that is, in fact, what is in front of you. And you see those lines, in this sense, through the Gestalt shift. But this sense of "what you see" just amounts to "what is physically in front of you that is causing your experience." It does not characterize the nature of your perceptual experience itself, which is far better described as "a cube in a certain orientation." Most people cannot in fact see flat lines on a page in the latter sense; the perception of it *as* a three-dimensional cube in one or the other orientation is almost inevitable. There is no intermediary stage in your perceptual experience where you first see flat lines on a page, and then interpret it as a cube. Interpretation, at least in the usual sense, is an intellectual act, a process of thought, rather than a perceptual act. And no such conscious intellectual act intervenes in your perception of the cube. You do not see lines and then think about it; instead, you just see a cube.

This second sense is, however, the sense relevant for scientific observation. After all, scientists report (or intend to report) what is in front of them by reporting what they *observe*, that is, what their perceptual experience presents their environment (e.g., the lab) as being like. So the sense of seeing that is relevant to scientific observation, and reports of it, is the sense in which you see a cube in one orientation, and then in another.

You might object that the image projected on your retina is always the same, and indeed it is. But Hanson points out that there are, in fact, two such images (one in each eye), and both are very small, and upside down. But, of course, you do not see two small, inverted images. Those images are part of what make it possible for you to see, but they are not themselves what you do see.

When you see the cube in one orientation or another, you know perfectly well that nothing in front of your eyes is in fact changing. What you see in the relevant sense is, then, not just a function of what is in front of your eyes; it also depends how your perceptual system "makes sense of," or organizes your experience. If you had never encountered box-shaped objects in your life, you would almost certainly not see what you do in fact see; you would instead see only lines on a page. What you see, then, is in part a result of training (although very early training).

Such training of one's perceptual system continues, Hanson argues, in science education. What, to a novice, looks like light and dark spots under a microscope, for example, looks like a cell wall, mitochondria, and so on to a trained cell biologist. When most of us look at an X-ray of an injured knee, we see only vague shapes and contours; your doctor, however, sees torn ligaments. As before, she does not *see* vague shapes and contours and then *infer* the torn ligaments; she just sees the torn ligaments themselves.

Suppose that Johannes Kepler, who said that the earth is revolving on its axis while the sun is stationary, and Tycho Brahe, who said that the earth is still and the sun revolves around it, are looking at the sunset. Kepler sees the earth fall back away from the sun, bringing the obscuring earth between them, whereas Brahe sees the sun sink below the stationary horizon. As before, in one sense both obviously see the same thing – the physical sun is in front of both. But, as before, that is only the "what-is-physically-in-front-of-them" sense, which is not the sense that concerns the character of their distinct perceptual experiences.

Observation, within and without science, Hanson concludes, is therefore theory-laden: what you see is a function of both what is in front of you and your background theoretical commitments. This seems to pose a threat to the objectivity of science: if observation is influenced by theory, then it does not constitute a neutral, objective basis with reference to which theories can be tested and compared. Kepler's and Brahe's reports of their sunset experiences, for example, cannot sensibly be appealed to in way of deciding between their respective astronomical views, since those views are influencing the experiences themselves.

One initial response to this threat is that 'theory-laden' does not mean 'theory-determined'; however much your perceptual experience is influenced by observation, what you perceive nevertheless does depend to a great extent on the physical environment in front of you. You cannot therefore see whatever you like, however strong your theoretical commitments might be. Second – and in contrast to the Kepler and Brahe example – the theory with which the observation is laden may not be the theory that is being tested. So the same theory might underlie the observations of two scientists witnessing an experiment, resulting in agreement concerning the experimental result, even while conflict between their other theoretical commitments leads only one of them to be surprised by that result. The experiment can thereby provide an independent test of those commitments, notwithstanding the theory-ladenness of the scientists' perceptions of it. Whether these two responses are sufficient is an open question.[19]

9 Maxwell on the Distinction between Observable and Unobservable Objects

Notice that Putnam's and Hanson's critiques do not undermine a distinction between observable and unobservable objects per se; Putnam's critique is directed instead against a claim about language, and Hanson's is directed against a claim about perceptual experience. Maxwell, however, argues that there is not even a legitimate distinction between the objects themselves, however experienced or described.

Maxwell presents three arguments. The first points out that there is a continuum running from observation through a vacuum, through air (which involves slight distortions resulting from temperature gradients), through a window, through glasses, through binoculars, through low-power microscopes, through high-power microscopes and through electron microscopes. Advocates of the distinction are bound to call the objects seen by the naked eye through a vacuum observable, and those "seen" through an electron microscope (e.g., the surface of a molecule) unobservable. But, Maxwell points out, there is no obvious cutoff point; any observable/unobservable line through this continuum will look arbitrary. We could, of course, consider observability to be a matter of degree. But advocates of the distinction often intend to affirm the existence of observables and deny that of the unobservables. There is, however, no such thing as degrees of existence to correspond to degrees of observability.

Maxwell's second argument invokes another continuum running from very small molecules (like hydrogen) through medium-sized ones (like proteins and viruses) to very large ones (like crystals of salt, diamond, and lumps of polymeric plastic). Advocates of the distinction are surely going to count the hydrogen molecule as unobservable, and the lump of plastic as observable (notwithstanding its being a single molecule). But again, where is the line to be drawn? And even if we could legitimately draw it somewhere, we would nevertheless no longer be able to insist that all molecules are unobservable; that would depend on the molecule. In general, there may be no category of object whose instances are exclusively observable (or unobservable).

Maxwell's third argument concerns the fact that what is observable or not is determined by certain biological characteristics of the human body, and these characteristics could change. Suppose, for example, that a mutant is born who is able to observe ultraviolet radiation, or X-rays; these would then be observable. There is no obvious reason why such a mutation could not take place (given enough time and the right selection pressures). This suggests that there is no such thing as an entity that is impossible to observe.[20]

Maxwell's arguments, if successful, are troublesome for the scientific anti-realist who denies that science provides us with knowledge of unobservable reality. Such an anti-realist does concede that we possess knowledge of observable reality. That distinction in her attitudes toward observable and unobservable reality cannot obviously be maintained without a clear distinction between those realities, which is the distinction that Maxwell's arguments are designed to undermine. The scientific realist, however, who endorses knowledge of both realities, appears not to require such a clear distinction. So Maxwell's arguments are standardly taken to constitute an argument in favor of scientific realism.

In reading 9.3 Bas van Fraassen, an anti-realist, exploits the fact that Maxwell's arguments do not concern either a distinction in vocabulary (as challenged by Putnam) or the characterization of experience as theory-independent (as challenged by Hanson). Van Fraassen can therefore cheerfully endorse the rejection of the positivist's two-vocabulary view as well as the theory-ladenness of observation, without undermining the observable/unobservable object distinction (which is presupposed by his anti-realism). He therefore needs only to contend with Maxwell's arguments. We will discuss his attempt to do this in §§7–10 of the Unit 9 commentary.

Notes

1 This is why his essay is grouped with the first theme rather than the second.
2 See §6 of Unit 5 commentary for discussion of the problem of theoretical terms.
3 See §5 of the Part II introduction.
4 A collection of sentences A is always logically equivalent to a collection B that contains the sentences of A as well as some (or all) of A's logical consequences.
5 Nicod, 1930, *Foundations of Geometry & Induction, Containing Geometry in a Sensible World and the Logical Problem of Induction*, London: Paul, Trench, Trubner; New York: Harcourt, Brace. Reprinted London: Routledge, 2000.
6 See §§2–3 of Unit 5 commentary.
7 In fact, anything that is either not a raven or is black (or both) will confirm a logically equivalent formulation of "All ravens are black." See reading 6.2, pages 406–407.
8 Recall from §5 of the Unit 5 commentary that Carnap conceived of the confirmation relation as a "logical probability" akin to deductive consequence.
9 Ironically, the concept "definition" itself is a member of the group.
10 In fact it begins with Kant's definition, according to which an analytic proposition is one where the predicate is contained in the subject. The indictment of this definition is not its interdefinability with the others, but the fact that it is a vague metaphor with a too-limited range of application. For example, it will not account for the seeming analyticity of: "If Sarah is shorter than Bob, then Bob is taller than Sarah."
11 Either directly to the atomic sentences or indirectly, by construction, using the truth-functions and quantifiers.
12 Quine agreed with the positivists that there is no such thing as non-linguistic necessity.
13 See reading 4.13 and §5 of Unit 4 commentary.
14 The Duhem-Quine thesis is really the extension of the Duhem thesis about empirical hypotheses to logic and mathematics, and so to the entire web of belief.
15 This includes the otherwise very conservative A. J. Ayer; see Ayer, Alfred J., 1946, *Language, Truth and Logic*, 2nd edn., London: Gollancz, pp. 90–5.
16 Reading 6.3, p. 422.
17 Hanson's attack is the topic of §8 and Maxwell's of §9.
18 See §7 of the Unit 5 commentary.
19 In Unit 7, the theory-ladenness of perception will become an integral component in Thomas Kuhn's argument for the conclusion that paradigms are incommensurable.
20 Another possibility is that we might find ourselves expanding our community of observers to include aliens that happen to have evolved eyes that operate in a manner akin to electron microscopes.

Suggestions for Further Reading

Hempel discusses the problem of the seeming dispensability of theoretical terms in Hempel 1958 as well as in reading 6.1. Scheffler 1963 contains a detailed discussion of the raven paradox. Quine's critique of the positivists' views on logic and mathematics continues in Quine 1966b and Quine 1966c. In Quine 1969, he reviews what he takes to be the downfall of the traditional epistemological approach advocated by Carnap and advocates a naturalized alternative. The definitive statement of his views is Quine 1960. Larry Laudan in

Laudan 1990 denies that the Duhem-Quine thesis is as significant as it has been taken to be. A debate between Paul Churchland and Jerry Fodor concerning the doctrine that observation is theory-laden is in Churchland 1979, Fodor 1984, Churchland 1988, and Fodor 1988. Bas van Fraassen 1980 defends the observable/unobservable distinction against Maxwell's arguments in §2 of Chapter 2. Hacking 1985 defends the claim (against van Fraassen) that we do indeed observe through microscopes. A discussion of the difference in the uses of the concept of observability in science and philosophy is in Shapere 1982.

Churchland, P., 1979, *Scientific Realism and the Plasticity of Mind*. Cambridge: Cambridge University Press.

Churchland, P., 1988, "Perceptual Plasticity and Theory Neutrality: A Reply to Jerry Fodor," *Philosophy of Science* 55: 167–87.

Fodor, J., 1984, "Observation Reconsidered," *Philosophy of Science*, 51(1) (March): 23–43.

Fodor, J., 1988, "A Reply to Churchland's Perceptual Plasticity and Theory Neutrality," *Philosophy of Science* 55(2) (June): 188–98.

Hacking, I., 1985, "Do We See Through a Microscope?" in P. Churchland and C. Hooker (eds.), *Images of Science*. Chicago: University of Chicago Press, pp. 132–52.

Hempel, C., 1958, "The Theoretician's Dilemma," in H. Feigl, M. Scriven, and G. Maxwell (eds.), *Minnesota Studies in the Philosophy of Science Vol. 2*. Minneapolis: University of Minnesota Press, pp. 37–98.

Laudan, L., 1990, "De-Mystifying Underdetermination," in W. Savage (ed.), *Minnesota Studies in the Philosophy of Science Vol. 14: Scientific Theories*. Minneapolis: University of Minnesota Press, pp. 267–97.

Quine, W. V. O., 1960, *Word and Object*. Cambridge: MIT Press.

Quine, W. V. O., 1966a, *The Ways of Paradox and Other Essays*. New York: Random House.

Quine, W. V. O., 1966b, "Truth by Convention," in Quine 1966a, pp. 70–99.

Quine, W. V. O., 1966c, "Carnap and Logical Truth," in Quine 1966a, pp. 100–25.

Quine, W. V. O., 1969, "Epistemology Naturalized," in *Ontological Relativity and Other Essays*. New York: Columbia University Press, pp. 69–90.

Scheffler, I., 1963, *The Anatomy of Inquiry*. New York: Knopf.

Shapere, D., 1982, "The Concept of Observation in Science and Philosophy," *Philosophy of Science* 49: 485–525.

van Fraassen, B., 1980, *The Scientific Image*. Oxford: Oxford University Press.

6.1

Empiricist Criteria of Cognitive Significance: Problems and Changes

Carl Hempel

In this selection, Carl Hempel surveys a variety of attempts to formulate an empiricist or "verificationist" criterion of meaning, culminating in a devastating critique of the "partial interpretation" approach contained in the received view developed by Rudolf Carnap. Hempel argues that the partial interpretation version of the criterion is incompatible with the use of the very theoretical concepts that this version of the criterion was designed to legitimize.

1 The General Empiricist Conception of Cognitive and Empirical Significance

It is a basic principle of contemporary empiricism that a sentence makes a cognitively significant assertion, and thus can be said to be either true or false, if and only if either (1) it is analytic or contradictory – in which case it is said to have purely logical meaning or significance – or else (2) it is capable, at least potentially, of test by experiential evidence – in which case it is said to have empirical meaning or significance. The basic tenet of this principle, and especially of its second part, the so-called testability criterion of empirical meaning (or better: meaningfulness), is not peculiar to empiricism alone: it is characteristic also of contemporary operationism, and in a sense of pragmatism as well; for the pragmatist maxim that a difference must make a difference to be a difference may well be construed as insisting that a verbal difference between two sentences must make a difference in experiential implications if it is to reflect a difference in meaning.

From Carl G. Hempel, *Aspects of Scientific Explanation and other Essays in the Philosophy of Science* (New York: The Free Press, 1965), pp. 101–19. This essay combines, with certain omissions and some changes, the contents of two articles: "Problems and Changes in the Empiricist Criterion of Meaning," *Revue Internationale de Philosophie* No. 11, pp. 41–63 (January, 1950); and "The Concept of Cognitive Significance: A Reconsideration," *Proceedings of the American Academy of Arts and Sciences* 80, No. 1, pp. 61–77 (1951). This material is reprinted with kind permission of *Revue Internationale de Philosophie* and of the American Academy of Arts and Sciences.

How this general conception of cognitively significant discourse led to the rejection, as devoid of logical and empirical meaning, of various formulations in speculative metaphysics, and even of certain hypotheses offered within empirical science, is too well known to require recounting. I think that the general intent of the empiricist criterion of meaning is basically sound, and that notwithstanding much oversimplification in its use, its critical application has been, on the whole, enlightening and salutary. I feel less confident, however, about the possibility of restating the general idea in the form of precise and general criteria which establish sharp dividing lines (a) between statements of purely logical and statements of empirical significance, and (b) between those sentences which do have cognitive significance and those which do not.

In the present paper, I propose to reconsider these distinctions as conceived in recent empiricism, and to point out some of the difficulties they present. The discussion will concern mainly the second of the two distinctions; in regard to the first, I shall limit myself to a few brief remarks.

2 The Earlier Testability Criteria of Meaning and Their Shortcomings

Let us note first that any general criterion of cognitive significance will have to meet certain requirements if it is to be at all acceptable. Of these, we note one, which we shall consider here as expressing a necessary, though by no means sufficient, *condition of adequacy* for criteria of cognitive significance.

(A) If under a given criterion of cognitive significance, a sentence *N* is nonsignificant, then so must be all truth-functional compound sentences in which *N* occurs nonvacuously as a component. For if *N* cannot be significantly assigned a truth value, then it is impossible to assign truth values to the compound sentences containing *N*; hence, they should be qualified as nonsignificant as well.

We note two corollaries of requirement (A):

(A1) If under a given criterion of cognitive significance, a sentence *S* is nonsignificant, then so must be its negation, ~ *S*.

(A2) If under a given criterion of cognitive significance, a sentence *N* is nonsignificant, then so must be any conjunction *N·S* and any disjunction *NvS*, no matter whether *S* is significant under the given criterion or not.

We now turn to the initial attempts made in recent empiricism to establish general criteria of cognitive significance. Those attempts were governed by the consideration that a sentence, to make an empirical assertion must be capable of being borne out by, or conflicting with, phenomena which are potentially capable of being directly observed. Sentences describing such potentially observable phenomena – no matter whether the latter do actually occur or not – may be called observation sentences. More specifically, an *observation sentence* might be construed as a sentence – no matter whether true or false – which asserts or denies that a specified object, or group of objects, of macroscopic size has a particular *observable characteristic*, i.e., a characteristic whose presence or absence can, under favorable circumstances, be ascertained by direct observation.[1]

The task of setting up criteria of empirical significance is thus transformed into the problem of characterizing in a precise manner the relationship which obtains between a hypothesis and one or more observation sentences whenever the phenomena described by the latter either confirm or disconfirm the hypothesis in question. The ability of a given sentence to enter into that relationship to some set of observation sentences would then characterize its testability-in-principle, and thus its empirical significance. Let us now briefly examine the major attempts that have been made to obtain criteria of significance in this manner.

One of the earliest criteria is expressed in the so-called *verifiability requirement*. According to it, a sentence is empirically significant if and only if it is not analytic and is capable, at least in principle, of complete verification by observational evidence; i.e., if observational evidence can be described which, if actually obtained, would conclusively establish the truth of the sentence.[2] With the help of the concept of observation sentence, we can restate this requirement as follows: A sentence *S* has empirical meaning if and only if it is possible to indicate a finite set of observation sentences, $O_1, O_2, \ldots, O_n$, such that if these are true, then *S* is necessarily true, too. As stated, however, this condition is satisfied also if *S* is an analytic sentence or if the given observation sentences are logically incompatible with each other. By the following formulation, we rule these cases out and at the same time express the intended criterion more precisely:

(2.1) REQUIREMENT OF COMPLETE VERIFIABILITY IN PRINCIPLE. A sentence has empirical meaning if and only if it is not analytic and follows logically from some finite and logically consistent class of observation sentences.[3] These observation sentences need not be true, for what the criterion is to explicate is testability by "potentially observable phenomena," or testability "in principle."

In accordance with the general conception of cognitive significance outlined earlier, a sentence will now be classified as cognitively significant if either it is analytic or contradictory, or it satisfies the verifiability requirement.

This criterion, however, has several serious defects. One of them has been noted by several writers:

a. Let us assume that the properties of being a stork and of being red-legged are both observable characteristics, and that the former does not logically entail the latter. Then the sentence

(*S*1) All storks are red-legged

is neither analytic nor contradictory; and clearly, it is not deducible from a finite set of observation sentences. Hence, under the contemplated criterion, *S*1 is devoid of empirical significance; and so are all other sentences purporting to express universal regularities or general laws. And since sentences of this type constitute an integral part of scientific theories, the verifiability requirement must be regarded as overly restrictive in this respect.

Similarly, the criterion disqualifies all sentences such as 'For any substance there exists some solvent', which contain both universal and existential quantifiers (i.e., occurrences of the terms 'all' and 'Some' or their equivalents); for no sentences of this kind can be logically deduced from any finite set of observation sentences.

Two further defects of the verifiability requirement do not seem to have been widely noticed:

b. As is readily seen, the negation of *S*1

(~*S*1) There exists at least one stork that is not red-legged

is deducible from any two observation sentences of the type 'a is a stork' and 'a is not red-legged'. Hence, ~*S*1 is cognitively significant under our criterion, but *S*1 is not, and this constitutes a violation of condition (A1).

c. Let *S* be a sentence which does, and *N* a sentence which does not satisfy the verifiability requirement. Then *S* is deducible from some set of observation sentences; hence, by a familiar rule of logic, *S*v*N* is deducible from the same set, and therefore cognitively significant according to our criterion. This violates condition (A2) above.[4]

Strictly analogous considerations apply to an alternative criterion, which makes complete falsifiability in principle the defining characteristic of empirical significance. Let us formulate this criterion as follows:

(2.2) REQUIREMENT OF COMPLETE FALSIFIABILITY IN PRINCIPLE. A sentence has empirical meaning if and only if its negation is not analytic and follows logically from some finite logically consistent class of observation sentences.

This criterion qualifies a sentence as empirically meaningful if its negation satisfies the requirement of complete verifiability; as it is to be expected, it is therefore inadequate on similar grounds as the latter:

(a) It denies cognitive significance to purely existential hypotheses, such as 'There exists at least one unicorn', and all sentences whose formulation calls for mixed – i.e., universal and existential – quantification, such as 'For every compound there exists some solvent', for none of these can possibly be conclusively falsified by a finite number of observation sentences.

(b) If '*P*' is an observation predicate, then the assertion that all things have the property *P* is qualified as significant, but its negation, being equivalent to a purely existential hypothesis, is disqualified [cf. (*a*)]. Hence, criterion (2.2) gives rise to the same dilemma as (2.1).

(c) If a sentence *S* is completely falsifiable whereas *N* is a sentence which is not, then their conjunction, *S*·*N* (i.e., the expression obtained by connecting the two sentences by the word 'and') is completely falsifiable; for if the negation of *S* is entailed by a class of observation sentences, then the negation of *S*·*N* is, *a fortiori*, entailed by the same class. Thus, the criterion allows empirical significance to many sentences which an adequate empiricist criterion should rule out, such as 'All swans are white and the absolute is perfect.'

In sum, then, interpretations of the testability criterion in terms of complete verifiability or of complete falsifiability are inadequate because they are overly restrictive in one direction and overly inclusive in another, and because both of them violate the fundamental requirement A.

Several attempts have been made to avoid these difficulties by construing the testability criterion as demanding merely a partial and possibly indirect confirmability of empirical hypotheses by observational evidence.

A formulation suggested by Ayer[5] is characteristic of these attempts to set up a clear and sufficiently comprehensive criterion of confirmability. It states, in effect, that a sentence *S* has empirical import if from *S* in conjunction with suitable subsidiary hypotheses it is possible to derive observation sentences which are not derivable from the subsidiary hypotheses alone.

This condition is suggested by a closer consideration of the logical structure of scientific testing; but it is much too liberal as it stands. Indeed, as Ayer himself has pointed out in the second edition of his book, *Language, Truth, and Logic*,[6] his criterion allows empirical import to any sentence whatever. Thus, e.g., if *S* is the sentence 'The absolute is perfect', it suffices to choose as a subsidiary hypothesis the sentence 'If the absolute is perfect then this apple is red' in order to make possible the deduction of the observation sentence 'This apple is red', which clearly does not follow from the subsidiary hypothesis alone.

To meet this objection, Ayer proposed a modified version of his testability criterion. In effect, the modification restricts the subsidiary hypotheses mentioned in the previous version to sentences which either are analytic or can independently be shown to be testable in the sense of the modified criterion.[7]

But it can readily be shown that this new criterion, like the requirement of complete falsifiability, allows empirical significance to any conjunction *S·N*, where *S* satisfies Ayer's criterion while *N* is a sentence such as 'The absolute is perfect', which is to be disqualified by that criterion. Indeed, whatever consequences can be deduced from *S* with the help of permissible subsidiary hypotheses can also be deduced from *S·N* by means of the same subsidiary hypotheses; and as Ayer's new criterion is formulated essentially in terms of the deducibility of a certain type of consequence from the given sentence, it countenances *S·N* together with *S*. Another difficulty has been pointed out by Church, who has shown[8] that if there are any three observation sentences none of which alone entails any of the others, then it follows for any sentence *S* whatsoever that either it or its denial has empirical import according to Ayer's revised criterion.

All the criteria considered so far attempt to explicate the concept of empirical significance by specifying certain logical connections which must obtain between a significant sentence and suitable observation sentences. It seems now that this type of approach offers little hope for the attainment of precise criteria of meaningfulness: this conclusion is suggested by the preceding survey of some representative attempts, and it receives additional support from certain further considerations, some of which will be presented in the following sections.

3 Characterization of Significant Sentences by Criteria for Their Constituent Terms

An alternative procedure suggests itself which again seems to reflect well the general viewpoint of empiricism: It might be possible to characterize cognitively significant sentences by certain conditions which their constituent terms have to satisfy. Specifically, it would seem reasonable to say that all extralogical terms[9] in a significant sentence must have experiential reference, and that therefore their meanings must be capable of explication by reference to observables exclusively.[10] In order to exhibit certain analogies between this approach and the previous one, we adopt the following terminological conventions:

Any term that may occur in a cognitively significant sentence will be called a *cognitively significant term*. Furthermore, we shall understand by an *observation term* any term which either (a) is an *observation predicate*, i.e., signifies some observable characteristic (as do the terms 'blue', 'warm', 'soft', 'coincident with', 'of greater apparent brightness than') or (b) names some physical object of macroscopic size (as do the terms 'the needle of this instrument', 'the Moon', 'Krakatoa Volcano', 'Greenwich, England', 'Julius Caesar').

Now while the testability criteria of meaning aimed at characterizing the cognitively significant sentences by means of certain inferential connections in which they must stand to some observation sentences, the alternative approach under consideration would instead try to specify the vocabulary that may be used in forming significant sentences. This vocabulary, the class of significant terms, would be characterized by the condition that each of its elements is either a logical term or else a term with empirical significance; in the latter case, it has to stand in certain definitional or explicative connections to some observation terms. This approach certainly avoids any violations of our earlier conditions of adequacy. Thus, e.g., if S is a significant sentence, i.e., contains cognitively significant terms only, then so is its denial, since the denial sign, and its verbal equivalents, belong to the vocabulary of logic and are thus significant. Again, if N is a sentence containing a non-significant term, then so is any compound sentence which contains N.

But this is not sufficient, of course. Rather, we shall now have to consider a crucial question analogous to that raised by the previous approach: Precisely how are the logical connections between empirically significant terms and observation terms to be construed if an adequate criterion of cognitive significance is to result? Let us consider some possibilities.

(3.1) The simplest criterion that suggests itself might be called the *requirement of definability*. It would demand that any term with empirical significance must be explicitly definable by means of observation terms.

This criterion would seem to accord well with the maxim of operationism that all significant terms of empirical science must be introduced by operational definitions. However, the requirement of definability is vastly too restrictive, for many important terms of scientific and even pre-scientific discourse cannot be explicitly defined by means of observation terms.

In fact, as Carnap[11] has pointed out, an attempt to provide explicit definitions in terms of observables encounters serious difficulties as soon as disposition terms, such as 'soluble', 'malleable', 'electric conductor', etc., have to be accounted for; and many of these occur even on the pre-scientific level of discourse.

Consider, for example, the word 'fragile'. One might try to define it by saying that an object x is fragile if and only if it satisfies the following condition: If at any time t the object is sharply struck, then it breaks at that time. But if the statement connectives in this phrasing are construed truth-functionally, so that the definition can be symbolized by

(D) $Fx \equiv (t)(Sxt \supset Bxt)$

then the predicate 'F' thus defined does not have the intended meaning. For let a be any object which is not fragile (e.g., a raindrop or a rubber band), but which happens not to be sharply struck at any time throughout its existence. Then 'Sat' is false and hence '$Sat \supset Bat$' is true for all values of 't'; consequently, 'Fa' is true though a is not fragile.

To remedy this defect, one might construe the phrase 'if . . . then . . .' in the original definiens as having a more restrictive meaning than the truth-functional conditional. This meaning might be suggested by the subjunctive phrasing 'If x were to be sharply struck at any time t, then x would break at t.' But a satisfactory elaboration of this construal would require a clarification of the meaning and the logic of counterfactual and subjunctive conditionals, which is a thorny problem.[12]

An alternative procedure was suggested by Carnap in his theory of reduction sentences.[13] These are sentences which, unlike definitions, specify the meaning of a term only conditionally or partially. The term 'fragile', for example, might be introduced by the following reduction sentence:

(R) $(x)(t)[Sxt \supset (Fx \equiv Bxt)]$

which specifies that if x is sharply struck at any time t, then x is fragile if and only if x breaks at t.

Our earlier difficulty is now avoided, for if a is a nonfragile object that is never sharply struck, then that expression in R which follows the quantifiers is true of a; but this does not imply that 'Fa' is true. But the reduction sentence R specifies the meaning of 'F' only for application to those objects which meet the "test condition" of being sharply struck at some time; for these it states that fragility then amounts to breaking. For objects that fail to meet the test condition, the meaning of '*F*' is left undetermined. In this sense, reduction sentences have the character of partial or conditional definitions.

Reduction sentences provide a satisfactory interpretation of the experiential import of a large class of disposition terms and permit a more adequate formulation of so-called operational definitions, which, in general, are not complete definitions at all. These considerations suggest a greatly liberalized alternative to the requirement of definability:

(3.2) *The requirement of reducibility*. Every term with empirical significance must be capable of introduction, on the basis of observation terms, through chains of reduction sentences.

This requirement is characteristic of the liberalized versions of positivism and physicalism which, since about 1936, have superseded the older, overly narrow conception of a full definability of all terms of empirical science by means of observables,[14] and it avoids many of the shortcomings of the latter. Yet, reduction sentences do not seem to offer an adequate means for the introduction of the central terms of advanced scientific theories, often referred to as theoretical constructs. This is indicated by the following considerations: A chain of reduction sentences provides a necessary and a sufficient condition for the applicability of the term it introduces. (When the two conditions coincide, the chain is tantamount to an explicit definition.) But now take, for example, the concept of length as used in classical physical theory. Here, the length in centimeters of the distance between two points may assume any positive real number as its value; yet it is clearly impossible to formulate, by means of observation terms, a sufficient condition for the applicability of such expressions as 'having a length of $\sqrt{2}$ cm' and 'having a length of $\sqrt{2} + 10^{-100}$ cm'; for such conditions would provide a possibility for discrimination, in observational terms, between two lengths which differ by only 10^{-100} cm.[15]

It would be ill-advised to argue that for this reason, we ought to permit only such values of the magnitude, length, as permit the statement of sufficient conditions in terms of observables. For this would rule out, among others, all irrational numbers and would prevent us from assigning, to the diagonal of a square with sides of length 1, the length $\sqrt{2}$, which is required by Euclidean geometry. Hence, the principles of Euclidean geometry would not be universally applicable in physics. Similarly, the principles of the calculus would become inapplicable, and the system of scientific theory as we know it today would be reduced to a clumsy, unmanageable torso. This, then, is no way of meeting the difficulty. Rather, we shall have to analyze more closely the function of constructs in scientific theories, with a view to obtaining through such an analysis a more adequate characterization of cognitively significant terms.

Theoretical constructs occur in the formulation of scientific theories. These may be conceived of, in their advanced stages, as being stated in the form of deductively developed axiomatized systems. Classical mechanics, or Euclidean or some Non-Euclidean form of geometry in physical interpretation, present examples of such systems. The extralogical terms used in a theory of this kind may be divided, in familiar manner, into primitive or basic terms, which are not defined within the theory, and defined terms, which are explicitly defined by means of the primitives. Thus, e.g., in Hilbert's axiomatization of Euclidean geometry, the terms 'point', 'straight line', 'between' are among the primitives, while 'line segment', 'angle', 'triangle', 'length' are among the defined terms. The basic and the defined terms together with the terms of logic constitute the vocabulary out of which all the sentences of the theory are constructed. The latter are divided, in an axiomatic presentation, into primitive statements (also called postulates or basic statements) which, in the theory, are not derived from any other statements, and derived ones, which are obtained by logical deduction from the primitive statements.

From its primitive terms and sentences, an axiomatized theory can be developed by means of purely formal principles of definition and deduction, without any consideration of the empirical significance of its extralogical terms. Indeed, this is the standard procedure employed in the axiomatic development of uninterpreted mathematical theories such as those of abstract groups or rings or lattices, or any form of pure (i.e., noninterpreted) geometry.

However, a deductively developed system of this sort can constitute a scientific theory only if it has received an empirical interpretation[16] which renders it relevant to the phenomena of our experience. Such interpretation is given by assigning a meaning, in terms of observables, to certain terms or sentences of the formalized theory. Frequently, an interpretation is given not for the primitive terms or statements but rather for some of the terms definable by means of the primitives, or for some of the sentences deducible from the postulates.[17] Furthermore, interpretation may amount to only a partial assignment of meaning. Thus, e.g., the rules for the measurement of length by means of a standard rod may be considered as providing a *partial* empirical interpretation for the term 'the length, in centimeters, of interval *i*', or alternatively, for some sentences of the form 'the length of interval *i* is *r* centimeters'. For the method is applicable only to intervals of a certain medium size, and even for the latter it does not constitute a full interpretation since the use of a standard rod does not constitute the only way of determining length: various alternative procedures are available involving the measurement of other magnitudes which are connected, by general laws, with the length that is to be determined.

This last observation, concerning the possibility of an indirect measurement of length by virtue of certain laws, suggests an important reminder. It is not correct to speak, as is often done, of "the experiential meaning" of a term or a sentence in isolation. In the language of science, and for similar reasons even in pre-scientific discourse, a single statement usually has no experiential implications. A single sentence in a scientific theory does not, as a rule, entail any observation sentences; consequences asserting the occurrence of certain observable phenomena can be derived from it only by conjoining it with a set of other, subsidiary, hypotheses. Of the latter, some will usually be observation sentences, others will be previously accepted theoretical statements. Thus, e.g., the relativistic theory of the deflection of light rays in the gravitational field of the sun entails assertions about observable phenomena only if it is conjoined with a considerable body of astronomical and optical theory as well as a large number of specific statements about the instruments used in those observations of solar eclipses which serve to test the hypothesis in question.

Hence, the phrase, 'the experiential meaning of expression *E*' is elliptical: What a given expression "means" in regard to potential empirical data is relative to two factors, namely:

I. *the linguistic framework L* to which the expression belongs. Its rules determine, in particular, what sentences – observational or otherwise – may be inferred from a given statement or class of statements;
II. the theoretical context in which the expression occurs, i.e., the class of those statements in *L* which are available as subsidiary hypotheses.

Thus, the sentence formulating Newton's law of gravitation has no experiential meaning by itself; but when used in a language whose logical apparatus permits the development of the calculus, and when combined with a suitable system of other hypotheses – including sentences which connect some of the theoretical terms with observation terms and thus establish a partial interpretation – then it has a bearing on observable phenomena in a large variety of fields. Analogous considerations are applicable to the term 'gravitational field', for example. It can be considered as having experiential meaning only within the context of a theory, which must be at least partially interpreted; and the experiential meaning of the term – as expressed, say, in the form of operational criteria for its application – will depend again on the theoretical system at hand, and on the logical characteristics of the language within which it is formulated.

4 Cognitive Significance as a Characteristic of Interpreted Systems

The preceding considerations point to the conclusion that a satisfactory criterion of cognitive significance cannot be reached through the second avenue of approach here considered, namely by means of specific requirements for the terms which make up significant sentences. This result accords with a general characteristic of scientific (and, in principle, even pre-scientific) theorizing: Theory formation and concept formation go hand in hand; neither can be carried on successfully in isolation from the other.

If, therefore, cognitive significance can be attributed to anything, then only to entire theoretical systems formulated in a language with a well-determined structure. And the decisive mark of cognitive significance in such a system appears to be the existence of an interpretation for it in terms of observables. Such an interpretation might be formulated, for example, by means of conditional or biconditional sentences connecting nonobservational terms of the system with observation terms in the given language; the latter as well as the connecting sentences may or may not belong to the theoretical system.

But the requirement of partial interpretation is extremely liberal; it is satisfied, for example, by the system consisting of contemporary physical theory combined with some set of principles of speculative metaphysics, even if the latter have no empirical interpretation at all. Within the total system, these metaphysical principles play the role of what K. Reach and also O. Neurath liked to call *isolated sentences*: They are neither purely formal truths or falsehoods, demonstrable or refutable by means of the logical rules of the given language system; nor do they have any experiential bearing; i.e., their omission from the theoretical system would have no effect on its explanatory and predictive power in regard to potentially observable phenomena (i.e., the kind of phenomena described by observation sentences). Should we not, therefore, require that a cognitively significant system contain no isolated sentences? The following criterion suggests itself:

(4.1) A theoretical system is cognitively significant if and only if it is partially interpreted to at least such an extent that none of its primitive sentences is isolated.

But this requirement may bar from a theoretical system certain sentences which might well be viewed as permissible and indeed desirable. By way of a simple illustration, let us assume that our theoretical system T contains the primitive sentence

$$(S1) \quad (x)[P_1x \supset (Qx \equiv P_2x)]$$

where 'P_1' and 'P_2' are observation predicates in the given language L, while 'Q' functions in T somewhat in the manner of a theoretical construct and occurs in only one primitive sentence of T, namely $S1$. Now $S1$ is not a truth or falsehood of formal logic; and furthermore, if $S1$ is omitted from the set of primitive sentences of T, then the resulting system, T', possesses exactly the same systematic, i.e., explanatory and predictive, power as T. Our contemplated criterion would therefore qualify $S1$ as an isolated sentence which has to be eliminated – excised by means of Occam's razor, as it were – if the theoretical system at hand is to be cognitively significant.

But it is possible to take a much more liberal view of $S1$ by treating it as a partial definition for the theoretical term 'Q'. Thus conceived, $S1$ specifies that in all cases where the observable characteristic P_1 is present, 'Q' is applicable if and only if the observable characteristic P_2 is present as well. In fact, $S1$ is an instance of those partial, or conditional, definitions which Carnap calls bilateral reduction sentences. These sentences are explicitly qualified by Carnap as analytic (though not, of course, as truths of formal logic), essentially on the ground that all their consequences which are expressible by means of observation predicates (and logical terms) alone are truths of formal logic.[18]

Let us pursue this line of thought a little further. This will lead us to some observations on analytic sentences and then back to the question of the adequacy of (4.1).

Suppose that we add to our system T the further sentence

$$(S2) \quad (x)[P_3x \supset (Qx \equiv P_4x)]$$

where 'P_3', 'P_4' are additional observation predicates. Then, on the view that "every bilateral reduction sentence is analytic",[19] $S2$ would be analytic as well as $S1$. Yet, the two sentences jointly entail non-analytic consequences which are expressible in terms of observation predicates alone, such as[20]

$$(O) \quad (x)[\sim (P_1x \cdot P_2x \cdot Px_3 \cdot \sim P_4x) \cdot \sim (P_1x \cdot \sim P_2x \cdot P_3x \cdot P_4x)]$$

But one would hardly want to admit the consequence that the conjunction of two analytic sentences may be synthetic. Hence if the concept of analyticity can be applied at all to the sentences of interpreted deductive systems, then it will have to be relativized with respect to the theoretical context at hand. Thus, e.g., *S1* might be qualified as analytic relative to the system *T*, whose remaining postulates do not contain the term '*Q*', but as synthetic relative to the system *T* enriched by *S2*. Strictly speaking, the concept of analyticity has to be relativized also in regard to the rules of the language at hand, for the latter determine what observational or other consequences are entailed by a given sentence. This need for at least a twofold relativization of the concept of analyticity was almost to be expected in view of those considerations which required the same twofold relativization for the concept of experiential meaning of a sentence.

If, on the other hand, we decide not to permit *S1* in the role of a partial definition and instead reject it as an isolated sentence, then we are led to an analogous conclusion: Whether a sentence is isolated or not will depend on the linguistic frame and on the theoretical context at hand: While *S1* is isolated relative to *T* (and the language in which both are formulated), it acquires definite experiential implications when *T* is enlarged by *S2*.

Thus we find, on the level of interpreted theoretical systems, a peculiar rapprochement, and partial fusion, of some of the problems pertaining to the concepts of cognitive significance and of analyticity: Both concepts need to be relativized; and a large class of sentences may be viewed, apparently with equal right, as analytic in a given context, or as isolated, or nonsignificant, in respect to it.

In addition to barring, as isolated in a given context, certain sentences which could just as well be construed as partial definitions, the criterion (4.1) has another serious defect. Of two logically equivalent formulations of a theoretical system it may qualify one as significant while barring the other as containing an isolated sentence among its primitives. For assume that a certain theoretical system *T1* contains among its primitive sentences *S′*, *S″*, . . . exactly one, *S′*, which is isolated. Then *T1* is not significant under (4.1). But now consider the theoretical system *T2* obtained from *T1* by replacing the two first primitive sentences, *S′*, *S″*, by one, namely their conjunction. Then, under our assumptions, none of the primitive sentences of *T2* is isolated, and *T2*, though equivalent to *T1*, is qualified as significant by (4.1). In order to do justice to the intent of (4.1), we would therefore have to lay down the following stricter requirement:

(4.2) A theoretical system is cognitively significant if and only if it is partially interpreted to such an extent that in no system equivalent to it at least one primitive sentence is isolated.

Let us apply this requirement to some theoretical system whose postulates include the two sentences *S1* and *S2* considered before, and whose other postulates do not contain '*Q*' at all. Since the sentences *S1* and *S2* together entail the sentence *O*, the set consisting of *S1* and *S2* is logically equivalent to the set consisting of *S1*, *S2* and *O*. Hence, if we replace the former set by the latter, we obtain a theoretical system equivalent to the given one. In this new system, both *S1* and *S2* are isolated since, as can be shown, their removal does not affect the explanatory and predictive power of the system in reference to observable phenomena. To put it intuitively, the systematic power of *S1* and *S2* is the same as that of *O*. Hence, the original system is disqualified by (4.2). From the viewpoint of a strictly sensationalist positivism as perhaps envisaged by Mach, this result might be hailed as a sound repudiation of theories making reference to fictitious entities, and as a strict insistence on theories couched exclusively in terms of observables. But from a contemporary vantage point, we shall have to say that such a procedure overlooks or misjudges the important function of constructs in scientific theory: The history of scientific endeavor shows that if we wish to arrive at precise, comprehensive, and well-confirmed general laws, we have to rise above the level of direct observation. The phenomena directly accessible to our experience are not connected by general laws of great scope and rigor. Theoretical constructs are needed for the formulation of such higher-level laws. One of the most important functions of a well-chosen construct is its potential ability to serve as a constituent in ever new general connections that may be discovered; and to such connections we would blind ourselves if we insisted on banning from scientific theories all those terms and sentences which could be "dispensed with" in the sense indicated in (4.2). In following such a narrowly phenomenalistic or positivistic course, we would deprive ourselves of the tremendous fertility of theoretical constructs, and we would often render the formal structure of the expurgated theory clumsy and inefficient.

Criterion (4.2), then, must be abandoned, and considerations such as those outlined in this paper seem to lend strong support to the conjecture that no adequate alternative to it can be found; i.e., that it is not possible to formulate general and precise criteria which would separate those partially interpreted systems whose isolated sentences might be said to have a significant function from those in which the isolated sentences are, so to speak, mere useless appendages.

We concluded earlier that cognitive significance in the sense intended by recent empiricism and operationism can at best be attributed to sentences forming a theoretical system, and perhaps rather to such systems as wholes. Now, rather than try to replace (4.2) by some alternative, we will have to recognize further that cognitive significance in a system is a matter of degree: Significant systems range from those whose entire extralogical vocabulary consists of observation terms, through theories whose formulation relies heavily on theoretical constructs, on to systems with hardly any bearing on potential empirical findings. Instead of dichotomizing this array into significant and non-significant systems it would seem less arbitrary and more promising to appraise or compare different theoretical systems in regard to such characteristics as these:

a. the clarity and precision with which the theories are formulated, and with which the logical relationships of their elements to each other and to expressions couched in observational terms have been made explicit;
b. the systematic, i.e., explanatory and predictive, power of the systems in regard to observable phenomena;
c. the formal simplicity of the theoretical system with which a certain systematic power is attained;
d. the extent to which the theories have been confirmed by experiential evidence.

Many of the speculative philosophical approaches to cosmology, biology, or history, for example, would make a poor showing on practically all of these counts and would thus prove no matches to available rival theories, or would be recognized as so unpromising as not to warrant further study or development.

If the procedure here suggested is to be carried out in detail, so as to become applicable also in less obvious cases, then it will be necessary, of course, to develop general standards, and theories pertaining to them, for the appraisal and comparison of theoretical systems in the various respects just mentioned. To what extent this can be done with rigor and precision cannot well be judged in advance. In recent years, a considerable amount of work has been done towards a definition and theory of the concept of degree of confirmation, or logical probability, of a theoretical system;[21] and several contributions have been made towards the clarification of some of the other ideas referred to above.[22] The continuation of this research represents a challenge for further constructive work in the logical and methodological analysis of scientific knowledge.

Notes

1 Observation sentences of this kind belong to what Carnap has called the thing-language, cf., e.g., (1938), pp. 52–53. That they are adequate to formulate the data which serve as the basis for empirical tests is clear in particular for the intersubjective testing procedures used in science as well as in large areas of empirical inquiry on the common-sense level. In epistemological discussions, it is frequently assumed that the ultimate evidence for beliefs about empirical matters consists in perceptions and sensations whose description calls for a phenomenalistic type of language. The specific problems connected with the phenomenalistic approach cannot be discussed here; but it should be mentioned that at any rate all the critical considerations presented in this article in regard to the testability criterion are applicable, *mutatis mutandis*, to the case of a phenomenalistic basis as well.

2 Originally, the permissible evidence was meant to be restricted to what is observable by the speaker and perhaps his fellow beings during their life times. Thus construed, the criterion rules out, as cognitively meaningless, all statements about the distant future or the remote past, as has been pointed out, among others, by Ayer (1946), chapter I; by Pap (1949), chapter 13, esp. pp. 333 ff.; and by Russell (1948), pp. 445–7. This difficulty is avoided, however, if we permit the evidence to

consist of any finite set of "logically possible observation data", each of them formulated in an observation sentence. Thus, e.g., the sentence S_1, "The tongue of the largest dinosaur in New York's Museum of Natural History was blue or black" is completely verifiable in our sense; for it is a logical consequence of the sentence S_2, "The tongue of the largest dinosaur in New York's Museum of Natural History was blue"; and this is an observation sentence, in the sense just indicated.

And if the concept of *verifiability in principle* and the more general concept of *confirmability in principle*, which will be considered later, are construed as referring to *logically possible evidence* as expressed by observation sentences, then it follows similarly that the class of statements which are verifiable, or at least confirmable, in principle include such assertions as that the planet Neptune and the Antarctic Continent existed before they were discovered, and that atomic warfare, if not checked, will lead to the extermination of this planet. The objections which Russell (1948), pp. 445 and 447, raises against the verifiability criterion by reference to those examples do not apply therefore if the criterion is understood in the manner here suggested. Incidentally, statements of the kind mentioned by Russell, which are not actually verifiable by any human being, were explicitly recognized as cognitively significant already by Schlick (1936), Part V, who argued that the impossibility of verifying them was "merely empirical." The characterization of verifiability with the help of the concept of observation sentence as suggested here might serve as a more explicit and rigorous statement of that conception.

3 As has frequently been emphasized in the empiricist literature, the term "verifiability" is to indicate, of course, the conceivability, or better, the logical possibility, of evidence of an observational kind which, if actually encountered, would constitute conclusive evidence for the given sentence; it is not intended to mean the technical possibility of performing the tests needed to obtain such evidence, and even less the possibility of actually finding directly observable phenomena which constitute conclusive evidence for that sentence – which would be tantamount to the actual existence of such evidence and would thus imply the truth of the given sentence. Analogous remarks apply to the terms "falsifiability" and "confirmability". This point has clearly been disregarded in some critical discussions of the verifiability criterion. Thus, e.g., Russell (1948), p. 448 construes verifiability as the actual existence of a set of conclusively verifying occurrences. This conception, which has never been advocated by any logical empiricist, must naturally turn out to be inadequate since according to it the empirical meaningfulness of a sentence could not be established without gathering empirical evidence, and moreover enough of it to permit a conclusive proof of the sentence in question! It is not surprising, therefore, that his extraordinary interpretation of verifiability leads Russell to the conclusion: "In fact, that a proposition is verifiable is itself not verifiable" (*l.c.*). Actually, under the empiricist interpretation of complete verifiability, any statement asserting the verifiability of some sentence *S* whose text is quoted, is either analytic or contradictory; for the decision whether there exists a class of observation sentences which entail *S*, i.e., whether such observation sentences can be formulated, no matter whether they are true or false – that decision is a purely logical matter.

4 The arguments here adduced against the verifiability criterion also prove the inadequacy of a view closely related to it, namely that two sentences have the same cognitive significance if any set of observation sentences which would verify one of them would also verify the other, and conversely. Thus, e.g., under this criterion, any two general laws would have to be assigned the same cognitive significance, for no general law is verified by any set of observation sentences. The view just referred to must be clearly distinguished from a position which Russell examines in his critical discussion of the positivistic meaning criterion. It is "the theory that two propositions whose verified consequences are identical have the same significance" (1948), p. 448. This view is untenable indeed, for what consequences of a statement have actually been verified at a given time is obviously a matter of historical accident which cannot possibly serve to establish identity of cognitive significance. But I am not aware that any logical empiricist ever subscribed to that "theory."

5 (1936, 1946), Chap. I. The case against the requirements of verifiability and of falsifiability, and in favor of a requirement of partial confirmability and disconfirmability, is very clearly presented also by Pap (1949), chapter 13.

6 (1946), 2d ed., pp. 11–12.

7 This restriction is expressed in recursive form and involves no vicious circle. For the full statement of Ayer's criterion, see Ayer (1946), p. 13.

8 Church (1949). An alternative criterion recently suggested by O'Connor (1950) as a revision of Ayer's formulation is subject to a slight variant of Church's stricture: It can be shown that if there are three observation sentences none of which entails any of the others, and if *S* is any

noncompound sentence, then either *S* or ~S is significant under O'Connor's criterion.

9 An extralogical term is one that does not belong to the specific vocabulary of logic. The following phrases, and those definable by means of them, are typical examples of logical terms: 'not', 'or', 'if . . . then', 'all', 'some', '. . . is an element of class . . .'. Whether it is possible to make a sharp theoretical distinction between logical and extra-logical terms is a controversial issue related to the problem of discriminating between analytic and synthetic sentences. For the purpose at hand, we may simply assume that the logical vocabulary is given by enumeration.

10 For a detailed exposition and critical discussion of this idea, see H. Feigl's stimulating and enlightening article (1950).

11 Cf. (1936–37), especially section 7.

12 On this subject, see for example Langford (1941); Lewis (1946), pp. 210–30; Chisholm (1946); Goodman (1947); Reichenbach (1947), Chapter VIII; Hempel and Oppenheim (1948), Part III; Popper (1949); and especially Goodman's further analysis (1955).

13 Cf. Carnap, *loc. cit.* note 11. For a brief elementary presentation of the main idea, see Carnap (1938), Part III. The sentence *R* here formulated for the predicate '*F*' illustrates only the simplest type of reduction sentence, the so-called bilateral reduction sentence.

14 Cf. the analysis in Carnap (1936–7), especially section 15; also see the briefer presentation of the liberalized point of view in Carnap (1938).

15 (Added in 1964.) This is not strictly correct. For a more circumspect statement, see note 12 in "A Logical Appraisal of Operationism" and the fuller discussion in section 7 of the essay "The Theoretician's Dilemma."

16 The interpretation of formal theories has been studied extensively by Reichenbach, especially in his pioneer analyses of space and time in classical and in relativistic physics. He describes such interpretation as the establishment of *coordinating definitions* (Zuordnungsdefinitionen) for certain terms of the formal theory. See, for example, Reichenbach (1928). More recently, Northrop [cf. (1947), Chap. VII, and also the detailed study of the use of deductively formulated theories in science, ibid., Chaps. IV, V, VI] and H. Margenau [cf., for example, (1935)] have discussed certain aspects of this process under the title of *epistemic correlation.*

17 A somewhat fuller account of this type of interpretation may be found in Carnap (1939), §24. The articles by Spence (1944) and by MacCorquodale and Meehl (1948) provide enlightening illustrations of the use of theoretical constructs in a field outside that of the physical sciences, and of the difficulties encountered in an attempt to analyze in detail their function and interpretation.

18 Cf. Carnap (1936–37), especially sections 8 and 10.

19 Carnap (1936–7), p. 452.

20 The sentence *O* is what Carnap calls the *representative sentence* of the couple consisting of the sentences *S*1 and *S*2; see (1936–7), pp. 450–3.

21 Cf., for example, Carnap (1945)1 and (1945)2, and especially (1950). Also see Helmer and Oppenheim (1945).

22 On simplicity, cf. especially Popper (1935), Chap. V; Reichenbach (1938), §42; Goodman (1949)1, (1949)2, (1950); on explanatory and predictive power, cf. Hempel and Oppenheim (1948), Part IV.

References

Ayer, A. J., *Language, Truth and Logic*, London, 1936; 2nd edn. 1946.

Carnap, R., "Testability and Meaning," *Philosophy of Science*, 3 (1936) and 4 (1937).

Carnap, R., "Logical Foundations of the Unity of Science," in: *International Encyclopedia of Unified Science*, I, 1; Chicago, 1938.

Carnap, R., *Foundations of Logic and Mathematics*, Chicago, 1939.

Carnap, R., "On Inductive Logic," *Philosophy of Science*, 12 (1945). Referred to as (1945)1 in this article.

Carnap, R., "The Two Concepts of Probability," *Philosophy and Phenomenological Research*, 5 (1945). Referred to as (1945)2 in this article.

Carnap, R., *Logical Foundations of Probability*, Chicago, 1950.

Chisholm, R. M., "The Contrary-to-Fact Conditional," *Mind*, 55 (1946).

Church, A., Review of Ayer (1946), *The Journal of Symbolic Logic*, 14 (1949), 52–3.

Feigl, H., "Existential Hypotheses: Realistic vs. Phenomenalistic Interpretations," *Philosophy of Science*, 17 (1950).

Goodman, N., "The Problem of Counterfactual Conditionals," *The Journal of Philosophy*, 44 (1947).

Goodman, N., "The Logical Simplicity of Predicates," *The Journal of Symbolic Logic*, 14 (1949). Referred to as (1949)1 in this article.

Goodman, N., "Some Reflections on the Theory of Systems," *Philosophy and Phenomenological Research*, 9 (1949). Referred to as (1949)2 in this article.

Goodman, N., "An Improvement in the Theory of Simplicity," *The Journal of Symbolic Logic*, 15 (1950).

Goodman, N., *Fact, Fiction, and Forecast*, Cambridge, Massachusetts, 1955.

Helmer, O. and P. Oppenheim, "A Syntactical Definition of Probability and of Degree of Confirmation." *The Journal of Symbolic Logic*, 10 (1945).

Hempel, C. G. and P. Oppenheim, "Studies in the Logic of Explanation," *Philosophy of Science*, 15 (1948). (Reprinted in this volume.)

Langford, C. H., Review in *The Journal of Symbolic Logic*, 6(1941), 67–8.

Lewis, C. I., *An Analysis of Knowledge and Valuation*, La Salle, Ill., 1946.

MacCorquodale, K. and P. E. Meehl, "On a Distinction Between Hypothetical Constructs and Intervening Variables," *Psychological Review*, 55 (1948).

Margenau, H., "Methodology of Modern Physics," *Philosophy of Science*, 2 (1935).

Northrop, F. S. O., *The Logic of the Sciences and the Humanities*, New York, 1947.

O'Connor, D. J., "Some Consequences of Professor A. J. Ayer's Verification Principle," *Analysis*, 10 (1950).

Pap, A., *Elements of Analytic Philosophy*, New York, 1949.

Popper, K., *Logik der Forschung*, Wien, 1935.

Popper, K., "A Note on Natural Laws and So-Called 'Contrary-to-Fact Conditionals'," *Mind*, 58 (1949).

Reichenbach, H., *Philosophic der Raum-Zeit-Lehre*, Berlin, 1928.

Reichenbach, H., *Elements of Symbolic Logic*, New York, 1947.

Russell, B., *Human Knowledge*, New York, 1948.

Schlick, M., "Meaning and Verification," *Philosophical Review*, 45 (1936). Also reprinted in Feigl, H. and W. Sellars, (eds.) *Readings in Philosophical Analysis*, New York, 1949.

Spence, Kenneth W., "The Nature of Theory Construction in Contemporary Psychology," *Psychological Review*, 51 (1944).

6.2

The Raven Paradox

Carl Hempel

According to the received view, scientific laws are universal or statistical generalizations of the form "all (or most) As are Bs," such as "all ravens are (creatures that are) black." The sighting of a black raven would intuitively confirm this law. But, Hempel points out in this selection, if one assumes a very intuitive principle – that an observation that confirms a generalization also confirms any other statement that is logically equivalent to that generalization – it appears that a red pencil also confirms the law that all ravens are black, which is highly unintuitive. Hempel's paradox presents a serious obstacle to the attempt to formulate a purely formal theory of confirmation.

1 Nicod's Criterion of Confirmation and Its Shortcomings

We consider first a conception of confirmation which underlies many recent studies of induction and of scientific method. A very explicit statement of this conception has been given by Jean Nicod in the following passage: "Consider the formula or the law: *A entails B*. How can a particular proposition, or more briefly, a fact, affect its probability? If this fact consists of the presence of B in a case of A, it is favorable to the law: '*A entails B*'; on the contrary, if it consists of the absence of B in a case of A, it is unfavorable to this law. It is conceivable that we have here the only two direct modes in which a fact can influence the probability of a law. . . . Thus, the entire influence of particular truths or facts on the probability of universal propositions or laws would operate by means of these two elementary relations which we shall call *confirmation* and *invalidation*."[1] Note that the applicability of this criterion is restricted to hypotheses of the form '*A entails B*'. Any hypothesis H of this kind may be expressed in the notation of symbolic logic[2] by means of a universal conditional sentence, such as, in the simplest case,

$$(x)[P(x) \supset Q(x)]$$

i.e. 'For any object x: if x is a P, then x is a Q,' or also 'Occurrence of the quality P entails occurrence of the quality Q.' According to the above

From Carl G. Hempel, "Studies in the Logic of Confirmation (I)," *Mind*, Vol. LIV, No. 213 (1945): 1–26.

criterion this hypothesis is confirmed by an object *a* if *a* is *P* and *Q*; and the hypothesis is disconfirmed by *a* if *a* is *P*, but not *Q*.[3] In other words, an object confirms a universal conditional hypothesis if and only if it satisfies both the antecedent (here: '$P(x)$') and the consequent (here: '$Q(x)$') of the conditional; it disconfirms the hypothesis if and only if it satisfies the antecedent, but not the consequent of the conditional; and (we add this to Nicod's statement) it is neutral, or irrelevant, with respect to the hypothesis if it does not satisfy the antecedent.

This criterion can readily be extended so as to be applicable also to universal conditionals containing more than one quantifier, such as 'Twins always resemble each other', or, in symbolic notation, '$(x)(y)(\text{Twins}(x, y) \supset \text{Rsbl}(x, y))$'. In these cases, a confirming instance consists of an ordered couple, or triple, etc., of objects satisfying the antecedent and the consequent of the conditional. (In the case of the last illustration, any two persons who are twins and resemble each other would confirm the hypothesis; twins who do not resemble each other would disconfirm it; and any two persons not twins – no matter whether they resemble each other or not – would constitute irrelevant evidence.)

We shall refer to this criterion as Nicod's criterion.[4] It states explicitly what is perhaps the most common tacit interpretation of the concept of confirmation. While seemingly quite adequate, it suffers from serious shortcomings, as will now be shown.

(*a*) First, the applicability of this criterion is restricted to hypotheses of universal conditional form; it provides no standards of confirmation for existential hypotheses (such as 'There exists organic life on other stars', or 'Poliomyelitis is caused by some virus') or for hypotheses whose explicit formulation calls for the use of both universal and existential quantifiers (such as 'Every human being dies some finite number of years after his birth', or the psychological hypothesis, 'You can fool all of the people some of the time and some of the people all of the time, but you cannot fool all of the people all of the time', which may be symbolized by '$(x)(Et)\text{Fl}(x, t) \cdot (Ex)(t)\text{Fl}(x, t) \cdot \sim (x)(t)\text{Fl}(x, t)$', (where '$\text{Fl}(x, t)$' stands for 'You can fool person *x* at time *t*'). We note, therefore, the desideratum of establishing a criterion of confirmation which is applicable to hypotheses of *any* form.[5]

(*b*) We now turn to a second shortcoming of Nicod's criterion. Consider the two sentences

S_1: '$(x)[\text{Raven}(x) \supset \text{Black}(x)]$';
S_2: '$(x)[\sim\text{Black}(x) \supset \sim \text{Raven}(x)]$'

(i.e. 'All ravens are black' and 'Whatever is not black is not a raven'), and let *a*, *b*, *c*, *d* be four objects such that *a* is a raven and black, *b* a raven but not black, *c* not a raven but black, and *d* neither a raven nor black. Then according to Nicod's criterion, *a* would confirm S_1, but be neutral with respect to S_2; *b* would disconfirm both S_1 and S_2; *c* would be neutral with respect to both S_1 and S_2, and *d* would confirm S_2, but be neutral with respect to S_1.

But S_1 and S_2 are logically equivalent; they have the same content, they are different formulations of the same hypothesis. And yet, by Nicod's criterion, either of the objects *a* and *d* would be confirming for one of the two sentences, but neutral with respect to the other. This means that Nicod's criterion makes confirmation depend not only on the content of the hypothesis, but also on its formulation.[6]

One remarkable consequence of this situation is that every hypothesis to which the criterion is applicable – i.e. every universal conditional – can be stated in a form for which there cannot possibly exist any confirming instances. Thus, e.g. the sentence

$$(x)[(\text{Raven}(x) \cdot \sim \text{Black}(x)) \supset (\text{Raven}(x) \cdot \sim \text{Raven}(x)]$$

is readily recognized as equivalent to both S_1 and S_2 above; yet no object whatever can confirm this sentence, i.e. satisfy both its antecedent and its consequent; for the consequent is contradictory. An analogous transformation is, of course, applicable to any other sentence of universal conditional form.

4 The Equivalence Condition

The results just obtained call attention to the following condition which an adequately defined concept of confirmation should satisfy, and in the light of which Nicod's criterion has to be rejected as inadequate:

> *Equivalence condition*: Whatever confirms (disconfirms) one of two equivalent sentences, also confirms (disconfirms) the other.

Fulfillment of this condition makes the confirmation of a hypothesis independent of the way in which it is formulated; and no doubt it will be conceded that this is a necessary condition for the adequacy of any proposed criterion of confirmation. Otherwise, the question as to whether certain data confirm a given hypothesis would have to be answered by saying: "That depends on which of the different equivalent formulations of the hypothesis is considered" – which appears absurd. Furthermore – and this is a more important point than an appeal to a feeling of absurdity – an adequate definition of confirmation will have to do justice to the way in which empirical hypotheses function in theoretical scientific contexts such as explanations and predictions; but when hypotheses are used for purposes of explanation or prediction,[7] they serve as premises in a deductive argument whose conclusion is a description of the event to be explained or predicted. The deduction is governed by the principles of formal logic, and according to the latter, a deduction which is valid will remain so if some or all of the premises are replaced by different but equivalent statements; and indeed, a scientist will feel free, in any theoretical reasoning involving certain hypotheses, to use the latter in whichever of their equivalent formulations are most convenient for the development of his conclusions. But if we adopted a concept of confirmation which did not satisfy the equivalence condition, then it would be possible, and indeed necessary, to argue in certain cases that it was sound scientific procedure to base a prediction on a given hypothesis if formulated in a sentence S_1, because a good deal of confirming evidence had been found for S_1; but that it was altogether inadmissible to base the prediction (say, for convenience of deduction) on an equivalent formulation S_2, because no confirming evidence for S_2 was available. Thus, the equivalence condition has to be regarded as a necessary condition for the adequacy of any definition of confirmation.

5 The Paradoxes of Confirmation

Perhaps we seem to have been laboring the obvious in stressing the necessity of satisfying the equivalence condition. This impression is likely to vanish upon consideration of certain consequences which derive from a combination of the equivalence condition with a most natural and plausible assumption concerning a sufficient condition of confirmation.

The essence of the criticism we have leveled so far against Nicod's criterion is that it certainly cannot serve as a necessary condition of confirmation; thus, in the illustration given in the beginning of section 3, object *a* confirms S_1 and should therefore also be considered as confirming S_2, while according to Nicod's criterion it is not. Satisfaction of the latter is therefore not a necessary condition for confirming evidence.

On the other hand, Nicod's criterion might still be considered as stating a particularly obvious and important sufficient condition of confirmation. And indeed, if we restrict ourselves to universal conditional hypotheses in one variable[8] – such as S_1 and S_2 in the above illustration – then it seems perfectly reasonable to qualify an object as confirming such a hypothesis if it satisfies both its antecedent and its consequent. The plausibility of this view will be further corroborated in the course of our subsequent analyses.

Thus, we shall agree that if *a* is both a raven and black, then *a* certainly confirms S_1: '$(x)(\text{Raven}(x) \supset \text{Black}(x))$', and if *d* is neither black nor a raven, *d* certainly confirms S_2: $(x)[\sim \text{Black}(x) \supset \sim \text{Raven}(x)]$'.

Let us now combine this simple stipulation with the equivalence condition. Since S_1 and S_2 are equivalent, *d* is confirming also for S_1; and thus, we have to recognize as confirming for S_1 any object which is neither black nor a raven. Consequently, any red pencil, any green leaf, any yellow cow, etc., becomes confirming evidence for the hypothesis that all ravens are black. This surprising consequence of two very adequate assumptions (the equivalence condition and the above sufficient condition of confirmation) can be further expanded: The sentence S_1 can readily be shown to be equivalent to S_3: '$(x)[(\text{Raven}(x) \vee \sim \text{Raven}(x)) \supset (\sim \text{Raven}(x) \vee \text{Black}(x))]$', i.e. 'Anything which is or is not a raven is either no raven or black'. According to the above sufficient condition, S_3 is certainly confirmed by any object, say *e*, such that (1) *e* is or is not a raven and, in addition (2) *e* is not a raven or is also black. Since (1) is analytic, these conditions reduce to (2). By virtue of the equivalence condition, we have therefore to consider as confirming for S_1 any object which is either no raven or also black (in other words: any object which is no raven at all, or a black raven).

Of the four objects characterized in section 3, *a*, *c* and *d* would therefore constitute confirming evidence for S_1, while *b* would be disconforming for S_1. This implies that any nonraven represents confirming evidence for the hypothesis that all ravens are black.[9]

We shall refer to these implications of the equivalence condition and of the above sufficient condition of confirmation as the *paradoxes of confirmation*.

How are these paradoxes to be dealt with? Renouncing the equivalence condition would not represent an acceptable solution, as it is shown by the considerations presented in section 4. Nor does it seem possible to dispense with the stipulation that an object satisfying two conditions, C_1 and C_2, should be considered as confirming a general hypothesis to the effect that any object which satisfies C_1 also satisfies C_2.

But the deduction of the above paradoxical results rests on one other assumption which is usually taken for granted, namely, that the meaning of general empirical hypotheses, such as that all ravens are black, or that all sodium salts burn yellow, can be adequately expressed by means of sentences of universal conditional form, such as '$(x)[\text{Raven}(x) \supset \text{Black}(x)]$' and '$(x)(\text{Sod. Salt}(x) \supset \text{Burn Yellow}(x))$', etc. Perhaps this customary mode of presentation has to be modified; and perhaps such a modification would automatically remove the paradoxes of confirmation? If this is not so, there seems to be only one alternative left, namely to show that the impression of the paradoxical character of those consequences is due to misunderstanding and can be dispelled, so that no theoretical difficulty remains. We shall now consider these two possibilities in turn: Subsections 5.11 and 5.12 are devoted to a discussion of two different proposals for a modified representation of general hypotheses; in subsection 5.2, we shall discuss the second alternative, i.e. the possibility of tracing the impression of paradoxicality back to a misunderstanding.

5.11. It has often been pointed out that while Aristotelian logic, in agreement with prevalent everyday usage, confers existential import upon sentences of the form 'All *P*'s are *Q*'s', a universal conditional sentence, in the sense of modern logic, has no existential import; thus, the sentence

$$`(x)[\text{Mermaid}(x) \supset \text{Green}(x)]'$$

does not imply the existence of mermaids; it merely asserts that any object either is not a mermaid at all, or a green mermaid; and it is true simply because of the fact that there are no mermaids. General laws and hypotheses in science, however – so it might be argued – are meant to have existential import; and one might attempt to express the latter by supplementing the customary universal conditional by an existential clause. Thus, the hypothesis that all ravens are black would be expressed by means of the sentence S_1: '$[(x)(\text{Raven}(x) \supset \text{Black}(x)] \cdot (Ex)\text{Raven}(x)$'; and the hypothesis that no nonblack things are ravens by S_2: '$(x)[\sim\text{Black}(x) \supset \sim \text{Raven}(x)] \cdot (Ex) \sim \text{Black}(x)$'. Clearly, these sentences are not equivalent, and of the four objects *a*, *b*, *c*, *d* characterized in section 3, part (*b*), only *a* might reasonably be said to confirm S_1, and only *d* to confirm S_2. Yet this method of avoiding the paradoxes of confirmation is open to serious objections:

(*a*) First of all, the representation of every general hypothesis by a conjunction of a universal conditional and an existential sentence would invalidate many logical inferences which are generally accepted as permissible in a theoretical argument. Thus, for example, the assertions that all sodium salts burn yellow, and that whatever does not burn yellow is no sodium salt are logically equivalent according to customary understanding and usage, and their representation by universal conditionals preserves this equivalence; but if existential clauses are added, the two assertions are no longer equivalent, as is illustrated above by the analogous case of S_1 and S_2.

(*b*) Second, the customary formulation of general hypotheses in empirical science clearly does not contain an existential clause, nor does it, as a rule, even indirectly determine such a clause unambiguously. Thus, consider the hypothesis that if a person after receiving an injection of a certain test substance has a positive skin reaction, he has diphtheria. Should we construe the existential clause here as referring to persons, to persons receiving the injection, or to persons who, upon receiving the injection, show a positive skin reaction? A more or less arbitrary decision has to be made; each of the possible decisions gives a different interpretation to the hypothesis, and none of them seems to be really implied by the latter.

(*c*) Finally, many universal hypotheses cannot be said to imply an existential clause at all. Thus, it may happen that from a certain astrophysical theory a universal hypothesis is deduced concerning the character of the phenomena which would take place under certain specified extreme conditions. A hypothesis of this kind need not (and, as a rule, does not) imply that such extreme conditions ever were or will be realized; it has no existential import. Or consider a biological hypothesis to the effect that whenever man and ape are crossed, the offspring will have such and such characteristics. This is a general hypothesis; it might be contemplated as a mere conjecture, or as a consequence of a broader genetic theory, other implications of which may already have been tested with positive results; but unquestionably the hypothesis does not imply an existential clause asserting that the contemplated kind of cross-breeding referred to will, at some time, actually take place.

5.12 Perhaps the impression of the paradoxical character of the cases discussed in the beginning of section 5 may be said to grow out of the feeling that the hypothesis that all ravens are black is about ravens, and not about non-black things, nor about all things. The use of an existential clause was one attempt at exhibiting this presumed peculiarity of the hypothesis. The attempt has failed, and if we wish to express the point in question, we shall have to look for a stronger device. The idea suggests itself of representing a general hypothesis by the customary universal conditional, supplemented by the indication of the specific "field of application" of the hypothesis; thus, we might represent the hypothesis that all ravens are black by the sentence '$(x)[\text{Raven}(x) \supset \text{Black}(x)]$' or any one of its equivalents, plus the indication 'Class of ravens', characterizing the field of application; and we might then require that every confirming instance should belong to the field of application. This procedure would exclude the objects *c* and *d* from those constituting confirming evidence and would thus avoid those undesirable consequences of the existential-clause device which were pointed out in 5.11 (*c*). But apart from this advantage, the second method is open to objections similar to those which apply to the first: (*a*) The way in which general hypotheses are used in science never involves the statement of a field of application; and the choice of the latter in a symbolic formulation of a given hypothesis thus introduces again a considerable measure of arbitrariness. In particular, for a scientific hypothesis to the effect that all *P*'s are *Q*'s, the field of application cannot simply be said to be the class of all *P*'s; for a hypothesis such as that all sodium salts burn yellow finds important application in tests with negative results; e.g., it may be applied to a substance of which it is not known whether it contains sodium salts, nor whether it burns yellow; and if the flame does not turn yellow, the hypothesis serves to establish the absence of sodium salts. The same is true of all other hypotheses used for tests of this type. (*b*) Again, the consistent use of a field of application in the formulation of general hypotheses would involve considerable logical complications, and yet would have no counterpart in the theoretical procedure of science, where hypotheses are subjected to various kinds of logical transformation and inference without any consideration that might be regarded as referring to changes in the fields of application. This method of meeting the paradoxes would therefore amount to dodging the problem by means of an ad hoc device which cannot be justified by reference to actual scientific procedure.

5.2 We have examined two alternatives to the customary method of representing general hypotheses by means of universal conditionals; neither of them proved an adequate means of precluding the paradoxes of confirmation. We shall now try to show that what is wrong does not lie in the customary way of construing and representing general hypotheses, but rather in our reliance on a misleading intuition in the matter: The impression of a paradoxical situation is not objectively founded; it is a psychological illusion.

(*a*) One source of misunderstanding is the view, referred to before, that a hypothesis of the simple form 'Every *P* is a *Q*', such as 'All sodium salts burn yellow', asserts something about a certain limited class of objects only, namely, the class of all *P*'s. This idea involves a confusion of logical and practical considerations: Our interest in the hypothesis may be focussed upon its applicability to that particular class of objects, but the hypothesis nevertheless asserts something about, and indeed imposes restrictions upon, *all* objects (within the logical type of the variable occurring in the hypothesis, which in the case of our last illustration might be the class of all physical objects). Indeed, a hypothesis of the form 'Every *P* is a *Q*' forbids the occurrence of any objects having the property *P* but lacking the property *Q*; i.e. it restricts all objects whatsoever to the class of those which either lack the property *P* or also have the property *Q*. Now, every object either belongs to this class or falls outside it, and thus, every object – and not only the *P*'s – either conforms to the hypothesis or violates it; there is no object which is not implicitly referred to by a hypothesis of this type. In particular, every object which either is no sodium salt or burns yellow conforms to, and thus bears out, the hypothesis that all sodium salts burn yellow; every other object violates that hypothesis.

The weakness of the idea under consideration is evidenced also by the observation that the class of objects about which a hypothesis is supposed to assert something is in no way clearly determined, and that it changes with the context, as was shown in 5.12 (*a*).

(*b*) A second important source of the appearance of paradoxicality in certain cases of confirmation is exhibited by the following consideration.

Suppose that in support of the assertion 'All sodium salts burn yellow' somebody were to adduce an experiment in which a piece of pure ice was held into a colorless flame and did not turn the flame yellow. This result would confirm the assertion, 'Whatever does not burn yellow is no sodium salt' and consequently, by virtue of the equivalence condition, it would confirm the original formulation. Why does this impress us as paradoxical? The reason becomes clear when we compare the previous situation with the case where an object whose chemical constitution is as yet unknown to us is held into a flame and fails to turn it yellow, and where subsequent analysis reveals it to contain no sodium salt. This outcome, we should no doubt agree, is what was to be expected on the basis of the hypothesis that all sodium salts burn yellow – no matter in which of its various equivalent formulations it may be expressed; thus, the data here obtained constitute confirming evidence for the hypothesis. Now the only difference between the two situations here considered is that in the first case we are told beforehand the test substance is ice, and we happen to "know anyhow" that ice contains no sodium salt; this has the consequence that the outcome of the flame-color test becomes entirely irrelevant for the confirmation of the hypothesis and thus can yield no new evidence for us. Indeed, if the flame should not turn yellow, the hypothesis requires that the substance contain no sodium salt – and we know beforehand that ice does not; and if the flame should turn yellow, the hypothesis would impose no further restrictions on the substance: hence, either of the possible outcomes of the experiment would be in accord with the hypothesis.

The analysis of this example illustrates a general point: In the seemingly paradoxical cases of confirmation, we are often not actually judging the relation of the given evidence *E* alone to the hypothesis *H* (we fail to observe the methodological fiction, characteristic of every case of confirmation, that we have no relevant evidence for *H* other than that included in *E*); instead, we tacitly introduce a comparison of *H* with a body of evidence which consists of *E* in conjunction with additional information that we happen to have at our disposal; in our illustration, this information includes the knowledge (1) that the substance used in the experiment is ice, and (2) that ice contains no sodium salt. If we assume this additional information as given, then, of course, the outcome of the experiment can add no strength to the hypothesis under consideration. But if we are careful to avoid this tacit reference to additional knowledge (which entirely changes the character of the problem), and if we formulate the question as to the confirming character of the evidence in a manner adequate to the concept of confirmation as used in this paper, we have to ask: Given some object *a* (it happens to be a piece of ice, but this fact is not included in the evidence), and given the fact that *a* does not turn the flame yellow and is no sodium salt: does *a* then constitute confirming evidence for the hypothesis? And now – no matter whether *a* is ice or some other substance – it is clear that the answer has to be in the affirmative; and the paradoxes vanish.

So far, in section (*b*), we have considered mainly that type of paradoxical case which is illustrated by the assertion that any nonblack nonraven constitutes confirming evidence for the hypothesis, 'All ravens are black.' However, the general idea just outlined applies as well to the even more extreme cases exemplified by the assertion that any nonraven as well as any black object confirms the hypothesis in question. Let us illustrate this by reference to the latter case. If the given evidence *E* – i.e. in the sense of the required methodological fiction, all data relevant for the hypothesis – consists only of one object which, in addition, is black, then *E* may reasonably be said to support even the hypothesis that all objects are black, and a fortiori *E* supports the weaker assertion that all ravens are black. In this case, again, our factual knowledge that not all objects are black tends to create an impression of paradoxicality which is not justified on logical grounds. Other paradoxical cases of confirmation may be dealt with analogously. Thus it turns out that the paradoxes of confirmation, as formulated above, are due to a misguided intuition in the matter rather than to a logical flaw in the two stipulations from which they were derived.[10,11]

Notes

1 Jean Nicod, *Foundations of Geometry and Induction* (trans. by P. P. Wiener), London, 1930; 219; *cf.* also R. M. Eaton's discussion of "Confirmation and Infirmation," which is based on Nicod's views; it is included in Chap. III of his *General Logic* (New York, 1931).

2 In this essay, only the most elementary devices of this notation are used; the symbolism is essentially that of *Principia Mathematica*, except that parentheses are used instead of dots, and that existential quantification is symbolized by '(E)' instead of by the inverted 'E.'

3 (Added in 1964). More precisely we would have to say, in Nicod's parlance, that the hypothesis is confirmed by the *proposition* that *a* is both *P* and *Q*, and is disconfirmed by the *proposition* that *a* is *P* but not *Q*.

4 This term is chosen for convenience, and in view of the above explicit formulation given by Nicod; it is not, of course, intended to imply that this conception of confirmation originated with Nicod.

5 For a rigorous formulation of the problem, it is necessary first to lay down assumptions as to the means of expression and the logical structure of the language in which the hypotheses are supposed to be formulated; the desideratum then calls for a definition of confirmation applicable to any hypothesis which can be expressed in the given language. Generally speaking, the problem becomes increasingly difficult with increasing richness and complexity of the assumed language of science.

6 This difficulty was pointed out, in substance, in my article "Le problème de la vérité," *Theoria* (Göteborg), vol. 3 (1937), esp. p. 222.

7 For a more detailed account of the logical structure of scientific explanation and prediction, *cf.* C. G. Hempel, "The Function of General Laws in History," *The Journal of Philosophy*, vol. 39 (1942), esp. sections 2, 3, 4. The characterization, given in that paper as well as in the above text, of explanations and predictions as arguments of a deductive logical structure, embodies an oversimplification: as will be shown in section 7 of the present essay, explanations and predictions often involve "quasi-inductive" steps besides deductive ones. This point, however, does not affect the validity of the above argument.

8 This restriction is essential: In its general form which applies to universal conditionals in any number of variables, Nicod's criterion cannot even be construed as expressing a sufficient condition of confirmation. This is shown by the following rather surprising example: Consider the hypothesis:

$$S_1: (x)(y)[\sim (R(x, y) \cdot R(y, x)) \supset (R(x, y) \cdot \sim R(y, x))]$$

Let *a*, *b* be two objects such that $R(a, b)$ and $\sim R(b, a)$. Then clearly, the couple (a, b) satisfies both the antecedent and the consequent of the universal conditional S_1; hence, if Nicod's criterion in its general form is accepted as stating a sufficient condition of confirmation, (a, b) constitutes confirming evidence for S_1. But S_1 can be shown to be equivalent to

S_2: $(x)(y)R(x, y)$

Now, by hypothesis, we have $\sim R(b, a)$; and this flatly contradicts S_2 and thus S_1. Thus, the couple (a, b), although satisfying both the antecedent and the consequent of the universal conditional S_1, actually constitutes disconfirming evidence of the strongest kind (conclusively disconfirming evidence, as we shall say later) for that sentence. This illustration reveals a striking and – as far as I am aware – hitherto unnoticed weakness of that conception of confirmation which underlies Nicod's criterion. In order to realize the bearing of our illustration upon Nicod's original formulation, let A and B be $\sim (R(x, y) \cdot R(y, x))$ and $R(x, y) \cdot \sim (R(y, x))$, respectively. Then S_1 asserts that A entails B, and the couple (a, b) is a case of the presence of B in the presence of A; this should, according to Nicod, be favorable to S_1.

9 (Added in 1964). The following further "paradoxical" consequence of our two conditions might be noted: Any hypothesis of universal conditional form can be equivalently rewritten as another hypothesis of the same form which, even if true, can have no confirming instances in Nicod's sense at all, since the proposition that a given object satisfies the antecedent and the consequent of the second hypothesis is self-contradictory. For example, '$(x)[P(x) \supset Q(x)]$' is equivalent to the sentence '$(x)[(P(x) \cdot \sim Q(x)) \supset (P(x) \cdot \sim P(x))]$', whose consequent is true of nothing.

10 The basic idea of section (*b*) in the above analysis is due to Dr Nelson Goodman, to whom I wish to reiterate my thanks for the help he rendered me, through many discussions, in clarifying my ideas on this point.

11 The considerations presented in section (*b*) above are also influenced by, though not identical in content with, the very illuminating discussion of the paradoxes by the Polish methodologist and logician Janina Hosiasson-Lindenbaum; *cf.* her article "On Confirmation," *The Journal of Symbolic Logic*, vol. 5 (1940), especially section 4. Dr Hosiasson's attention had been called to the paradoxes by my article "Le problème de la vérité" (*cf.* note 20) and by discussions with me. To my knowledge, hers has so far been the only publication which presents an explicit attempt to solve the problem. Her solution is based on a theory of degrees of confirmation, which is developed in the form of an uninterpreted axiomtaic system, and most of her arguments presuppose that theoretical framework. I have profited, however, by some of Miss Hosiasson's more general observations which proved relevant for the analysis of the paradoxes of the nongraduated or qualitative concept of confirmation which forms the object of the present study.

One point in those of Miss Hosiasson's comments which rest on her theory of degrees of confirmation is of particular interest, and I should like to discuss it briefly. Stated in reference to the raven hypothesis, it consists in the suggestion that the finding of one nonblack object which is no raven, while constituting confirming evidence for the hypothesis, would increase the degree of confirmation of the hypothesis by a smaller amount than the finding of one raven which is black. This is said to be so because the class of all ravens is much less numerous than that of all nonblack objects, so that – to put the idea in suggestive though somewhat misleading terms – the finding of one black raven confirms a larger portion of the total content of the hypothesis than the finding of one nonblack nonraven. In fact, from the basic assumptions of her theory, Miss Hosiasson is able to derive a theorem according to which the above statement about the relative increase in degree of confirmation will hold provided that actually the number of all ravens is small compared with the number of all nonblack objects. But is this last numerical assumption actually warranted in the present case and analogously in all other "paradoxical" cases? The answer depends in part upon the logical structure of the language of science. If a "coordinate language" is used, in which, say, finite space–time regions figure as individuals, then the raven hypothesis assumes some such form as 'Every space–time region which contains a raven contains something black'; and even if the total number of ravens ever to exist is finite, the class of space–time regions containing a raven has the power of the continuum, and so does the class of space–time regions containing something nonblack; thus, for a coordinate language of the type under consideration, the above numerical assumption is not warranted. Now the use of a coordinate language may appear quite artificial in this particular illustration; but it will seem very appropriate in many other contexts, such as, e.g., that of physical field theories. On the other hand, Miss Hosiasson's numerical assumption may well be justified on the basis of a "thing language," in which physical objects of finite size function as individuals. Of course, even on this basis, it remains an empirical question, for every hypothesis of the form 'All P's are Q's', whether actually the class of non-Q's is much more numerous than the class of P's; and in many cases this question will be very difficult to decide.

6.3

Two Dogmas of Empiricism

W. V. O. Quine

W. V. O. Quine (1908–2000) was one of the most prominent early critics of positivism and one of the most influential philosophers of the second half of the twentieth century. His famous "Two Dogmas of Empiricism" excerpted below is likely the most widely cited single essay in philosophy. In it, Quine challenges the concept of analytic truth, or truth in virtue of meaning alone, which was a crucial element in the positivists' philosophical program (including the received view of scientific theories). First, he argues that the concept admits of no definition that does not appeal to concepts that are equally suspect. Second, he argues that it is unnecessary; an adequate empiricist philosophy can be formulated without it. The result is a shift away from reductionism (the second dogma) and toward confirmational and semantic holism and a more thoroughgoing pragmatism.

[. . .]

1. Background for Analyticity

Kant's cleavage between analytic and synthetic truths was foreshadowed in Hume's distinction between relations of ideas and matters of fact, and in Leibniz's distinction between truths of reason and truths of fact. Leibniz spoke of the truths of reason as true in all possible worlds. Picturesqueness aside, this is to say that the truths of reason are those which could not possibly be false. In the same vein we hear analytic statements defined as statements whose denials are self-contradictory. But this definition has small explanatory value; for the notion of self-contradictoriness, in the quite broad sense needed for this definition of analyticity, stands in exactly the same need of clarification as does the notion of analyticity itself. The

From *From a Logical Point of View,* Harvard University Press, 1961, pp. 20–46. Original version appears in *The Philosophical Review*, Vol. 69 (1951): 20–43 (excerpts).

two notions are the two sides of a single dubious coin.

Kant conceived of an analytic statement as one that attributes to its subject no more than is already conceptually contained in the subject. This formulation has two shortcomings: it limits itself to statements of subject-predicate form, and it appeals to a notion of containment which is left at a metaphorical level. But Kant's intent, evident more from the use he makes of the notion of analyticity than from his definition of it, can be restated thus: a statement is analytic when it is true by virtue of meanings and independently of fact.

[. . .]

The Aristotelian notion of essence was the forerunner, no doubt, of the modern notion of intension or meaning. For Aristotle it was essential in men to be rational, accidental to be two-legged. But there is an important difference between this attitude and the doctrine of meaning. From the latter point of view it may indeed be conceded (if only for the sake of argument) that rationality is involved in the meaning of the word 'man' while two-leggedness is not; but two-leggedness may at the same time be viewed as involved in the meaning of 'biped' while rationality is not. Thus from the point of view of the doctrine of meaning it makes no sense to say of the actual individual, who is at once a man and a biped, that his rationality is essential and his two-leggedness accidental or vice versa. Things had essences, for Aristotle, but only linguistic forms have meanings. Meaning is what essence becomes when it is divorced from the object of reference and wedded to the word.

For the theory of meaning the most conspicuous question is as to the nature of its objects: what sort of things are meanings? A felt need for meant entities may derive from an earlier failure to appreciate that meaning and reference are distinct. Once the theory of meaning is sharply separated from the theory of reference, it is a short step to recognizing as the business of the theory of meaning simply the synonymy of linguistic forms and the analyticity of statements; meanings themselves, as obscure intermediary entities, may well be abandoned.[1]

The problem of analyticity confronts us anew. Statements which are analytic by general philosophical acclaim are not, indeed, far to seek. They fall into two classes. Those of the first class, which may be called logically true, are typified by:

(1) No unmarried man is married.

The relevant feature of this example is that it is not merely true as it stands, but remains true under any and all reinterpretations of 'man' and 'married.' If we suppose a prior inventory of *logical* particles, comprising 'no,' 'un-' 'if,' 'then,' 'and,' etc., then in general a logical truth is a statement which is true and remains true under all reinterpretations of its components other than the logical particles.

But there is also a second class of analytic statements, typified by:

(2) No bachelor is married.

The characteristic of such a statement is that it can be turned into a logical truth by putting synonyms for synonyms; thus (2) can be turned into (1) by putting 'unmarried man' for its synonym 'bachelor.' We still lack a proper characterization of this second class of analytic statements, and therewith of analyticity generally, inasmuch as we have had in the above description to lean on a notion of 'synonymy' which is no less in need of clarification than analyticity itself.

In recent years Carnap has tended to explain analyticity by appeal to what he calls state-descriptions.[2] A state-description is any exhaustive assignment of truth values to the atomic, or noncompound, statements of the language. All other statements of the language are, Carnap assumes, built up of their component clauses by means of the familiar logical devices, in such a way that the truth value of any complex statement is fixed for each state-description by specifiable logical laws. A statement is then explained as analytic when it comes out true under every state-description. This account is an adaptation of Leibniz's "true in all possible worlds." But note that this version of analyticity serves its purpose only if the atomic statements of the language are, unlike 'John is a bachelor' and 'John is married,' mutually independent. Otherwise there would be a state-description which assigned truth to 'John is a bachelor' and falsity to 'John is married,' and consequently 'All bachelors are married' would turn out synthetic rather than analytic under the proposed criterion. Thus the criterion of analyticity in terms of state-descriptions serves only for languages devoid of extralogical synonym-pairs, such as 'bachelor' and 'unmarried man': synonym-pairs of the type which give rise to the "second class" of analytic statements. The criterion in terms of state-descriptions is a reconstruction at best of logical truth, not of analyticity.

I do not mean to suggest that Carnap is under any illusions on this point. His simplified model language with its state-descriptions is aimed primarily not at the general problem of analyticity but at another purpose, the clarification of probability and induction. Our problem, however, is analyticity; and here the major difficulty lies not in the first class of analytic statements, the logical truths, but rather in the second class, which depends on the notion of synonymy.

2. Definition

There are those who find it soothing to say that the analytic statements of the second class reduce to those of the first class, the logical truths, by *definition*; 'bachelor,' for example, is *defined* as 'unmarried man.' But how do we find that 'bachelor' is defined as 'unmarried man'? Who defined it thus, and when? Are we to appeal to the nearest dictionary, and accept the lexicographer's formulation as law? Clearly this would be to put the cart before the horse. The lexicographer is an empirical scientist, whose business is the recording of antecedent facts; and if he glosses 'bachelor' as 'unmarried man' it is because of his belief that there is a relation of synonymy between these forms, implicit in general or preferred usage prior to his own work. The notion of synonymy presupposed here has still to be clarified, presumably in terms relating to linguistic behavior. Certainly the "definition" which is the lexicographer's report of an observed synonymy cannot be taken as the ground of the synonymy.

Definition is not, indeed, an activity exclusively of philologists. Philosophers and scientists frequently have occasions to "define" a recondite term by paraphrasing it into terms of a more familiar vocabulary. But ordinarily such a definition, like the philologist's, is pure lexicography, affirming a relationship of synonymy antecedent to the exposition in hand.

Just what it means to affirm synonymy, just what the interconnections may be which are necessary and sufficient in order that two linguistic forms be properly describable as synonymous, is far from clear; but, whatever these interconnections may be, ordinarily they are grounded in usage. Definitions reporting selected instances of synonymy come then as reports upon usage.

There is also, however, a variant type of definitional activity which does not limit itself to the reporting of pre-existing synonymies. I have in mind what Carnap calls *explication* – an activity to which philosophers are given, and scientists also in their more philosophical moments. In explication the purpose is not merely to paraphrase the definiendum into an outright synonym, but actually to improve upon the definiendum by refining or supplementing its meaning. But even explication, though not merely reporting a pre-existing synonymy between definiendum and definiens, does rest nevertheless on *other* pre-existing synonymies. The matter may be viewed as follows. Any word worth explicating has some contexts which, as wholes, are clear and precise enough to be useful; and the purpose of explication is to preserve the usage of these favored contexts while sharpening the usage of other contexts. In order that a given definition be suitable for purposes of explication, therefore, what is required is not that the definiendum in its antecedent usage be synonymous with the definiens, but just that each of these favored contexts of the definiendum taken as a whole in its antecedent usage, be synonymous with the corresponding context of the definiens.

Two alternative definientia may be equally appropriate for the purposes of a given task of explication and yet not be synonymous with each other, for they may serve interchangeably within the favored contexts but diverge elsewhere. By cleaving to one of these definientia rather than the other, a definition of explicative kind generates, by fiat, a relationship of synonymy between definiendum and definiens which did not hold before. But such a definition still owes its explicative function, as seen, to pre-existing synonymies.

There does, however, remain still an extreme sort of definition which does not hark back to prior synonymies at all; namely, the explicitly conventional introduction of novel notations for purposes of sheer abbreviation. Here the definiendum becomes synonymous with the definiens simply because it has been created expressly for the purpose of being synonymous with the definiens. Here we have a really transparent case of synonymy created by definition; would that all species of synonymy were as intelligible. For the rest, definition rests on synonymy rather than explaining it.

The word "definition" has come to have a dangerously reassuring sound, due no doubt to its frequent occurrence in logical and mathematical writings. We shall do well to digress now into a brief appraisal of the role of definition in formal work.

In logical and mathematical systems either of two mutually antagonistic types of economy may be striven for, and each has its peculiar practical utility. On the one hand we may seek economy of practical expression: ease and brevity in the statement of multifarious relationships. This sort of economy calls usually for distinctive concise notations for a wealth of concepts. Second, however, and oppositely, we may seek economy in grammar and vocabulary; we may try to find a minimum of basic concepts such that, once a distinctive notation has been appropriated to each of them, it becomes possible to express any desired further concept by mere combination and iteration of our basic notations. This second sort of economy is impractical in one way, since a poverty in basic idioms tends to a necessary lengthening of discourse. But it is practical in another way: it greatly simplifies theoretical discourse *about* the language, through minimizing the terms and the forms of construction wherein the language consists.

Both sorts of economy, though prima facie incompatible, are valuable in their separate ways. The custom has consequently arisen of combining both sorts of economy by forging in effect two languages, the one a part of the other. The inclusive language, though redundant in grammar and vocabulary, is economical in message lengths, while the part, called *primitive notation*, is economical in grammar and vocabulary. Whole and part are correlated by rules of translation whereby each idiom not in primitive notation is equated to some complex built up of primitive notation. These rules of translation are the so-called *definitions* which appear in formalized systems. They are best viewed not as adjuncts to one language but as correlations between two languages, the one a part of the other.

But these correlations are not arbitrary. They are supposed to show how the primitive notations can accomplish all purposes, save brevity and convenience, of the redundant language. Hence the definiendum and its definiens may be expected, in each case, to be related in one or another of the three ways lately noted. The definiens may be a faithful paraphrase of the definiendum into the narrower notation, preserving a direct synonymy[3] as of antecedent usage; or the definiens may, in the spirit of explication, improve upon the antecedent usage of the definiendum; or finally, the definiendum may be a newly created notation, newly endowed with meaning here and now.

In formal and informal work alike, thus, we find that definition – except in the extreme case of the explicitly conventional introduction of new notation – hinges on prior relationships of synonymy. Recognizing then that the notation of definition does not hold the key to synonymy and analyticity, let us look further into synonymy and say no more of definition.

3. Interchangeability

A natural suggestion, deserving close examination, is that the synonymy of two linguistic forms consists simply in their interchangeability in all contexts without change of truth value; interchangeability, in Leibniz's phrase, *salva veritate*.[4] Note that synonyms so conceived need not even be free from vagueness, as long as the vaguenesses match.

But it is not quite true that the synonyms 'bachelor' and 'unmarried man' are everywhere interchangeable *salva veritate*. Truths which become false under substitution of 'unmarried man' for 'bachelor' are easily constructed with help of 'bachelor of arts' or 'bachelor's buttons.' Also with help of quotation, thus:

'Bachelor' has less than ten letters.

Such counterinstances can, however, perhaps be set aside by treating the phrases 'bachelor of arts' and 'bachelor's buttons' and the quotation "bachelor" each as a single indivisible word and then stipulating that the interchangeability *salva veritate* which is to be the touchstone of synonymy is not supposed to apply to fragmentary occurrences inside of a word. This account of synonymy, supposing it acceptable on other counts, has indeed the drawback of appealing to a prior conception of "word" which can be counted on to present difficulties of formulation in its turn. Nevertheless some progress might be claimed in having reduced the problem of synonymy to a problem of wordhood. Let us pursue this line a bit, taking "word" for granted.

The question remains whether interchangeability *salva veritate* (apart from occurrences within words) is a strong enough condition for synonymy, or whether, on the contrary, some non-synonymous expressions might be thus interchangeable. Now let us be clear that we are not concerned here with synonymy in the sense of complete identity in psychological associations or poetic quality; indeed no two expressions are synonymous in such a sense. We are concerned only with what may be called *cognitive synonymy*. Just what this is cannot be said without successfully finishing the present study; but we know something about it from the need which arose for it in connection with analyticity in §1. The sort of synonymy needed there was merely such that any analytic statement could be turned into a logical truth by putting synonyms for synonyms. Turning the tables and assuming analyticity, indeed, we could explain cognitive synonymy of terms as follows (keeping to the familiar example): to say that 'bachelor' and 'unmarried man' are cognitively synonymous is to say no more nor less than that the statement:

(3) All and only bachelors are unmarried men

is analytic.[5]

What we need is an account of cognitive synonymy not presupposing analyticity – if we are to explain analyticity conversely with help of cognitive synonymy as undertaken in Section 1. And indeed such an independent account of cognitive synonymy is at present up for consideration, namely, interchangeability *salva veritate* everywhere except within words. The question before us, to resume the thread at last, is whether such interchangeability is a sufficient condition for cognitive synonymy. We can quickly assure ourselves that it is, by examples of the following sort. The statement:

(4) Necessarily all and only bachelors are bachelors

is evidently true, even supposing 'necessarily' so narrowly construed as to be truly applicable only to analytic statements. Then, *if* 'bachelor' and 'unmarried man' are interchangeable *salva veritate*, the result

(5) Necessarily, all and only bachelors are unmarried men

of putting 'unmarried man' for an occurrence of 'bachelor' in (4) must, like (4), be true. But to say that (5) is true is to say that (3) is analytic, and hence that 'bachelor' and 'unmarried man' are cognitively synonymous.

Let us see what there is about the above argument that gives it its air of hocus-pocus. The condition of interchangeability *salva veritate* varies in its force with variations in the richness of the language at hand. The above argument supposes we are working with a language rich enough to contain the adverb 'necessarily,' this adverb being so construed as to yield truth when and only when applied to an analytic statement. But can we condone a language which contains such an adverb? Does the adverb really make sense? To suppose that it does is to suppose that we have already made satisfactory sense of 'analytic.' Then what are we so hard at work on right now?

[. . .]

Now a language . . . is *extensional*, in this sense: any two predicates which *agree extensionally* (i.e., are true of the same objects) are interchangeable *salva veritate*.[6]

In an extensional language, therefore, interchangeability *salva veritate* is no assurance of cognitive synonymy of the desired type. That 'bachelor' and 'unmarried man' are interchangeable *salva veritate* in an extensional language assures us of no more than that (3) is true. There is no assurance here that the extensional agreement of 'bachelor' and 'unmarried man' rests on meaning rather than merely on accidental matters of fact, as does extensional agreement of 'creature with a heart' and 'creature with kidneys.'

For most purposes extensional agreement is the nearest approximation to synonymy we need care about. But the fact remains that extensional agreement falls far short of cognitive synonymy of the type required for explaining analyticity in the manner of §1. The type of cognitive synonymy required there is such as to equate the synonymy of 'bachelor' and 'unmarried man' with the analyticity of (3), not merely with the truth of (3).

So we must recognize that interchangeability *salva veritate*, if construed in relation to an extensional language, is not a sufficient condition of cognitive synonymy in the sense needed for deriving analyticity in the manner of §1. If a language contains an intensional adverb 'necessarily' in the sense lately noted, or other particles to the same effect, then interchangeability *salva veritate* in such a language does afford a sufficient condition of cognitive synonymy; but such a language is intelligible only if the notion of analyticity is already clearly understood in advance.

The effort to explain cognitive synonymy first, for the sake of deriving analyticity from it afterward as in §1, is perhaps the wrong approach. Instead we might try explaining analyticity somehow without appeal to cognitive synonymy. Afterward we could doubtless derive cognitive synonymy from analyticity satisfactorily enough if desired. We have seen that cognitive synonymy of 'bachelor' and 'unmarried man' can be explained as analyticity of (3). The same explanation works for any pair of one-place predicates, of course, and it can be extended in obvious fashion to many-place predicates. Other syntactical categories can also be accommodated in fairly parallel fashion. Singular terms may be said to be cognitively synonymous when the statement of identity formed by putting ' = ' between them is analytic. Statements may be said simply to be cognitively synonymous when their biconditional (the result of joining them by 'if and only if') is analytic.[7] If we care to lump all categories into a single formulation, at the expense of assuming again the notion of "word" which was appealed to early in this section, we can describe any two linguistic forms as cognitively synonymous when the two forms are interchangeable (apart from occurrences within "words") *salva* (no longer *veritate* but) *analyticitate*. Certain technical questions arise, indeed, over cases of ambiguity or homonymy; let us not pause for them, however, for we are already digressing. Let us rather turn our backs on the problem of synonymy and address ourselves anew to that of analyticity.

4. Semantical Rules

Analyticity at first seemed most naturally definable by appeal to a realm of meanings. On refinement, the appeal to meanings gave way to an appeal to synonymy or definition. But definition turned out to be a will-o'-the-wisp, and synonymy turned out to be best understood only by dint of a prior appeal to analyticity itself. So we are back at the problem of analyticity.

I do not know whether the statement 'Everything green is extended' is analytic. Now does my indecision over this example really betray an incomplete understanding, an incomplete grasp of the "meanings," of 'green' and 'extended'? I think not. The trouble is not with 'green' or 'extended,' but with 'analytic.'

It is often hinted that the difficulty in separating analytic statements from synthetic ones in ordinary language is due to the vagueness of ordinary language and that the distinction is clear when we have a precise artificial language with explicit "semantical rules." This, however, as I shall now attempt to show, is a confusion.

The notion of analyticity about which we are worrying is a purported relation between statements and languages: a statement S is said to be *analytic for* a language L, and the problem is to make sense of this relation generally, for example, for variable 'S' and 'L.' The point that I want to make is that the gravity of this problem is not perceptibly less for artificial languages than for natural ones. The problem of making sense of the idiom 'S *is analytic for* L,' with variable 'S' and 'L,' retains its stubbornness even if we limit the range of the variable 'L' to artificial languages. Let me now try to make this point evident.

[. . .]

... Instead of appealing to an unexplained word 'analytic,' we are now appealing to an unexplained phrase 'semantical rule.' Not every true statement which says that the statements of some class are true can count as a semantical rule – otherwise all truths would be "analytic" in the sense of being true according to semantical rules. Semantical rules are distinguishable, apparently, only by the fact of appearing on a page under the heading 'Semantical Rules'; and this heading is itself then meaningless.

We can say indeed that a statement is *analytic-for-*L_0 if and only if it is true according to such and such specifically appended "semantical rules," but then we find ourselves back at essentially the same case which was originally discussed: 'S is analytic-for-L_0 if and only if. . . .' Once we seek to explain 'S is analytic for L' generally for variable 'L' (even allowing limitation of 'L' to artificial languages), the explanation 'true according to the semantical rules of L' is unavailing; for the relative term 'semantical rule of' is as much in need of clarification, at least, as 'analytic for.'

[. . .]

Not all the explanations of analyticity known to Carnap and his readers have been covered explicitly in the above considerations, but the extension to other forms is not hard to see. Just one additional factor should be mentioned which sometimes enters: sometimes the semantical rules are in effect rules of translation into ordinary language, in which case the analytic statements of the artificial language are in effect recognized as such from the analyticity of their specified translations in ordinary language. Here certainly there can be no thought of an illumination of the problem of analyticity from the side of the artificial language.

From the point of view of the problem of analyticity the notion of an artificial language with semantical rules is *a feu follet par excellence.* Semantical rules determining the analytic statements of an artificial language are of interest only in so far as we already understand the notion of analyticity; they are of no help in gaining this understanding.

Appeal to hypothetical languages of an artificially simple kind could conceivably be useful in clarifying analyticity, if the mental or behavioral or cultural factors relevant to analyticity – whatever they may be – were somehow sketched into the simplified model. But a model which takes analyticity merely as an irreducible character is unlikely to throw light on the problem of explicating analyticity.

It is obvious that truth in general depends on both language and extra-linguistic fact. The statement 'Brutus killed Caesar' would be false if the world had been different in certain ways, but it would also be false if the word 'killed' happened rather to have the sense of 'begat.' Hence the temptation to suppose in general that the truth of a statement is somehow analyzable into a linguistic component and a factual component. Given this supposition, it next seems reasonable that in some statements the factual component should be null; and these are the analytic statements. But, for all its a priori reasonableness, a boundary between analytic and synthetic statement simply has not been drawn. That there is such a distinction to be drawn at all is an unempirical dogma of empiricists, a metaphysical article of faith.

5. The Verification Theory and Reductionism

In the course of these somber reflections we have taken a dim view first of the notion of meaning, then of the notion of cognitive synonymy: and finally of the notion of analyticity. But what, it may be asked, of the verification theory of meaning? This phrase has established itself so firmly as a catchword of empiricism that we should be very unscientific indeed not to look beneath it for a possible key to the problem of meaning and the associated problems.

The verification theory of meaning, which has been conspicuous in the literature from Peirce onward, is that the meaning of a statement is the method of empirically confirming or infirming it. An analytic statement is that limiting case which is confirmed no matter what.

As urged in §1, we can as well pass over the question of meanings as entities and move straight to sameness of meaning, or synonymy. Then what the verification theory says is that statements are synonymous if and only if they are alike in point of method of empirical confirmation or infirmation.

This is an account of cognitive synonymy not of linguistic forms generally, but of statements.[8] However, from the concept of synonymy of statements we could derive the concept of synonymy for other linguistic forms, by considerations somewhat similar to those at the end of §3. Assuming the notion of "word," indeed, we could explain any two forms as synonymous when the putting of the one form for an occurrence of the other in any statement (apart from occurrences within "words") yields a synonymous statement. Finally, given the concept of synonymy thus for linguistic forms generally, we could define analyticity in terms of synonymy and logical truth as in §1. For that matter, we could define analyticity more simply in terms of just synonymy of statements together with logical truth; it is not necessary to appeal to synonymy of linguistic forms other than statements. For a statement may be described as analytic simply when it is synonymous with a logically true statement.

So, if the verification theory can be accepted as an adequate account of statement synonymy, the notion of analyticity is saved after all. However, let us reflect. Statement synonymy is said to be likeness of method of empirical confirmation or infirmation. Just what are these methods which are to be compared for likeness? What, in other words, is the nature of the relationship between a statement and the experiences which contribute to or detract from its confirmation?

The most naive view of the relationship is that it is one of direct report. This is *radical reductionism.* Every meaningful statement is held to be translatable into a statement (true or false) about immediate experience. Radical reductionism, in one form or another, well antedates the verification theory of meaning explicitly so called. Thus Locke and Hume held that every idea must either originate directly in sense experience or else be compounded of ideas thus originating; and taking a hint from Tooke we might rephrase this doctrine in semantical jargon by saying that a term, to be significant at all, must be either a name of a sense datum or a compound of such names or an abbreviation of such a compound. So stated, the doctrine remains ambiguous as between sense data as sensory events and sense data as sensory qualities; and it remains vague as to the admissible ways of compounding. Moreover, the doctrine is unnecessarily and intolerably restrictive in the term-by-term critique which it imposes. More reasonably, and without yet exceeding the limits of what I have called radical reductionism, we may take full statements as our significant units – thus demanding that our statements as wholes be translatable into sense-datum language, but not that they be translatable term by term.

[...]

Radical reductionism, conceived now with statements as units, sets itself the task of specifying a sense-datum language and showing how to translate the rest of significant discourse, statement by statement, into it. Carnap embarked on this project in the *Aufbau.*

The language which Carnap adopted as his starting point was not a sense-datum language in the narrowest conceivable sense, for it included also the notations of logic, up through higher set theory. In effect it included the whole language of pure mathematics. The ontology implicit in it (i.e., the range of values of its variables) embraced not only sensory events but classes, classes of classes, and so on. Empiricists there are who would boggle at such prodigality. Carnap's starting point is very parsimonious, however, in its extralogical or sensory part. In a series of constructions in which he exploits the resources of modern logic with much ingenuity, Carnap succeeds in defining a wide array of important additional sensory concepts which, but for his constructions, one would not have dreamed were definable on so slender a basis. Carnap was the first empiricist who, not content with asserting the reducibility of science to terms of immediate experience, took serious steps toward carrying out the reduction.

If Carnap's starting point is satisfactory, still his constructions were, as he himself stressed, only a fragment of the full program. The construction of even the simplest statements about the physical world was left in a sketchy state. Carnap's suggestions on this subject were, despite their sketchiness, very suggestive. He explained spatio-temporal point-instants as quadruples of real numbers and envisaged assignment of sense qualities to point-instants according to certain canons. Roughly summarized, the plan was that qualities should be assigned to point-instants in such a way as to achieve the laziest world compatible with our experience. The principle of least action was to be our guide in constructing a world from experience.

Carnap did not seem to recognize, however, that his treatment of physical objects fell short of reduction not merely through sketchiness, but in principle. Statements of the form 'Quality *q* is at point-instant *x; y; z; t*' were, according to his canons, to be apportioned truth values in such a way as to maximize and minimize certain over-all features, and with growth of experience the truth values were to be progressively revised in the same spirit. I think this is a good schematization (deliberately oversimplified, to be sure) of what science really does; but it provides no indication, not even the sketchiest, of how a statement of the form 'Quality *q* is at *x; y; z; t*' could ever be translated into Carnap's initial language of sense data and logic. The connective 'is at' remains an added undefined connective; the canons counsel us in its use but not in its elimination.

Carnap seems to have appreciated this point afterward; for in his later writings he abandoned all notion of the translatability of statements about the physical world into statements about immediate experience. Reductionism in its radical form has long since ceased to figure in Carnap's philosophy.

But the dogma of reductionism has, in a subtler and more tenuous form, continued to influence the thought of empiricists. The notion lingers that to each statement, or each synthetic statement, there is associated a unique range of possible sensory events such that the occurrence of any of them would add to the likelihood of truth of the statement, and that there is associated also another unique range of possible sensory events whose occurrence would detract from that likelihood. This notion is of course implicit in the verification theory of meaning.

The dogma of reductionism survives in the supposition that each statement, taken in isolation from its fellows, can admit of confirmation or infirmation at all. My countersuggestion, issuing essentially from Carnap's doctrine of the physical world in the *Aufbau*, is that our statements about the external world face the tribunal of sense experience not individually but only as a corporate body.'

The dogma of reductionism, even in its attenuated form, is intimately connected with the other dogma: that there is a cleavage between the analytic and the synthetic. We have found ourselves led, indeed, from the latter problem to the former through the verification theory of meaning. More directly, the one dogma clearly supports the other in this way: as long as it is taken to be significant in general to speak of the confirmation and infirmation of a statement, it seems significant to speak also of a limiting kind of statement which is vacuously confirmed, *ipso facto*, come what may; and such a statement is analytic.

The two dogmas are, indeed, at root identical. We lately reflected that in general the truth of statements does obviously depend both upon extra-linguistic fact; and we noted that this obvious circumstance carries in its train, not logically but all too naturally, a feeling that the truth of a statement is somehow analyzable into a linguistic component and a factual component. The factual component must, if we are empiricists, boil down to a range of confirmatory experiences. In the extreme case where the linguistic component is all that matters, a true statement is analytic. But I hope we are now impressed with how stubbornly the distinction between analytic and synthetic has resisted any straightforward drawing. I am impressed also, apart from prefabricated examples of black and white balls in an urn, with how baffling the problem has always been of arriving at any explicit theory of the empirical confirmation of a synthetic statement. My present suggestion is that it is nonsense, and the root of much nonsense, to speak of a linguistic component and a factual component in the truth of any individual statement. Taken collectively, science has its double dependence upon language and experience; but this duality is not significantly traceable into the statements of science taken one by one.

The idea of defining a symbol in use was, as remarked, an advance over the impossible term-by-term empiricism of Locke and Hume. The statement, rather than the term, came with Frege to be recognized as the unit accountable to an empiricist critique. But what I am now urging is that even in taking the statement as unit we have drawn our grid too finely. The unit of empirical significance is the whole of science.

6. Empiricism Without the Dogmas

The totality of our so-called knowledge or beliefs, from the most casual matters of geography and history to the profoundest laws of atomic physics or even of pure mathematics and logic, is a man-made fabric which impinges on experience only along the edges. Or, to change the figure, total science is like a field of force whose boundary conditions are experience. A conflict with experience at the periphery occasions readjustments in the interior of the field. Truth values have to be redistributed over some of our statements. Re-evaluation of some statements entails re-evaluation of others, because of their logical interconnections – the logical laws being in turn simply certain further statements of the system, certain further elements of the field. Having re-evaluated one statement we must re-evaluate some others, whether they be statements logically connected with the first or whether they be the statements of logical connections themselves. But the total field is so undetermined by its boundary conditions, experience, that there is much latitude of choice as to what statements to re-evaluate in the light of any single contrary experience. No particular experiences are linked with any particular statements in the interior of the field, except indirectly through considerations of equilibrium affecting the field as a whole.

If this view is right, it is misleading to speak of the empirical content of an individual statement – especially if it be a statement at all remote from the experiential periphery of the field. Furthermore it becomes folly to seek a boundary between synthetic statements, which hold contingently on experience, and analytic statements which hold come what may. Any statement can be held true come what may, if we make drastic enough adjustments elsewhere in the system. Even a statement very close to the periphery can be held true in the face of recalcitrant experience by pleading hallucination or by amending certain statements of the kind called logical laws. Conversely, by the same token, no statement is immune to revision. Revision even of the logical law of the excluded middle has been proposed as a means of simplifying quantum mechanics; and what difference is there in principle between such a shift and the shift whereby Kepler superseded Ptolemy, or Einstein Newton, or Darwin Aristotle?

For vividness I have been speaking in terms of varying distances from a sensory periphery. Let me try now to clarify this notion without metaphor. Certain statements, though about physical objects and not sense experience, seem peculiarly germane to sense experience – and in a selective way: some statements to some experiences, others to others. Such statements, especially germane to particular experiences, I picture as near the periphery. But in this relation of "germaneness" I envisage nothing more than a loose association reflecting the relative likelihood, in practice, of our choosing one statement rather than another for revision in the event of recalcitrant experience. For example, we can imagine recalcitrant experiences to which we would surely be inclined to accommodate our system by re-evaluating just the statement that there are brick houses on Elm Street, together with related statements on the same topic. We can imagine other recalcitrant experiences to which we would be inclined to accommodate our system by re-evaluating just the statement that there are no centaurs, along with kindred statements. A recalcitrant experience can, I have already urged, be accommodated by any of various alternative re-evaluations in various alternative quarters of the total system; but, in the cases which we are now imagining, our natural tendency to disturb the total system as little as possible would lead us to focus our revisions upon these specific statements concerning brick houses or centaurs. These statements are felt, therefore, to have a sharper empirical reference than highly theoretical statements of physics or logic or ontology. The latter statements may be thought of as relatively centrally located within the total network, meaning merely that little preferential connection with any particular sense data obtrudes itself.

As an empiricist I continue to think of the conceptual scheme of science as a tool, ultimately, for predicting future experience in the light of past experience. Physical objects are conceptually imported into the situation as convenient intermediaries – not by definition in terms of experience, but simply as irreducible posits comparable, epistemologically, to the gods of Homer. Let me interject that for my part I do, qua lay physicist, believe in physical objects and not in Homer's gods; and I consider it a scientific error to believe otherwise. But in point of epistemological footing the physical objects and the gods differ only in degree and not in kind. Both sorts of entities enter our conception only as cultural posits. The myth of physical objects is epistemologically superior to most in that it has proved more efficacious than other myths as a device for working a manageable structure into the flux of experience.

Positing does not stop with macroscopic physical objects. Objects at the atomic level and beyond are posited to make the laws of macroscopic objects, and ultimately the laws of experience, simpler and more manageable; and we need not expect or demand full definition of atomic and subatomic entities in terms of macroscopic ones, any more than definition of macroscopic things in terms of sense data. Science is a continuation of common sense, and it continues the common-sense expedient of swelling ontology to simplify theory.

Physical objects, small and large, are not the only posits. Forces are another example; and indeed we are told nowadays that the boundary between energy and matter is obsolete. Moreover, the abstract entities which are the substance of mathematics – ultimately classes and classes of classes and so on up – are another posit in the same spirit. Epistemologically these are myths on the same footing with physical objects and gods, neither better nor worse except for differences in the degree to which they expedite our dealings with sense experiences.

The over-all algebra of rational and irrational numbers is underdetermined by the algebra of rational numbers, but is smoother and more convenient; and it includes the algebra of rational numbers as a jagged or gerrymandered part.[10] Total science, mathematical and natural and human, is similarly but more extremely underdetermined by experience. The edge of the system must be kept squared with experience; the rest, with all its elaborate myths or fictions, has as its objective the simplicity of laws.

Ontological questions, under this view, are on a par with questions of natural science.[11] Consider the question whether to countenance classes as entities. This, as I have argued elsewhere,[12] is the question whether to quantify with respect to variables which take classes as values. Now Carnap ["Empiricism, Semantics, and Ontology," *Revue internationale de philosophie* 4 (1950), 20–40.] has maintained that this is a question not of matters of fact but of choosing a convenient language form, a convenient conceptual scheme or framework for science. With this I agree, but only on the proviso that the same be conceded regarding scientific hypotheses generally. Carnap ["Empiricism, Semantics, and Ontology", p. 32n.] has recognized that he is able to preserve a double standard for ontological questions and scientific hypotheses only by assuming an absolute distinction between the analytic and the synthetic; and I need not say again that this is a distinction which I reject.[13]

The issue over there being classes seems more a question of convenient conceptual scheme; the issue over there being centaurs, or brick houses on Elm Street, seems more a question of fact. But I have been urging that this difference is only one of degree, and that it turns upon our vaguely pragmatic inclination to adjust one strand of the fabric of science rather than another in accommodating some particular recalcitrant experience. Conservatism figures in such choices, and so does the quest for simplicity.

Carnap, Lewis, and others take a pragmatic stand on the question of choosing between language forms, scientific frameworks; but their pragmatism leaves off at the imagined boundary between the analytic and the synthetic. In repudiating such a boundary I espouse a more thorough pragmatism. Each man is given a scientific heritage plus a continuing barrage of sensory stimulation; and the considerations which guide him in warping his scientific heritage to fit his continuing sensory promptings are, where rational, pragmatic.

Notes

1 See "On What There Is", p. 11f, and "The Problem of Meaning in Linguistics," p. 48f.

2 R. Carnap, *Meaning and Necessity* (Chicago, 1947), pp. 9 ff.; *Logical Foundations of Probability* (Chicago, 1950), pp. 70 ff.

3 According to an important variant sense of 'definition', the relation preserved may be the weaker relation of mere agreement in reference; see "Notes on the Theory of Reference," p. 132. But, definition in this sense is better ignored in the present connection, being irrelevant to the question of synonymy.

4 Cf. C. I. Lewis, *A Survey of Symbolic Logic* (Berkeley, 1918), p. 373.

5 This is cognitive synonymy in a primary, broad sense. Carnap (*Meaning and Necessity*, pp. 56 ff.) and C. I. Lewis (*Analysis of Knowledge and Valuation* [La Salle, Ill., 1946], pp. 83 ff.) have suggested how, once this notion is at hand, a narrower sense of cognitive synonymy which is preferable for some purposes can in turn be derived. But this special ramification of concept-building lies aside from the present purposes and must not be confused with the broad sort of cognitive synonymy here concerned.

6 This is the substance of Quine, *Mathematical Logic* (New York: Morton 1940; rev. edn., 1951, *121.)

7 The 'if and only if' itself is intended in the truth functional sense. See Carnap, *Meaning and Necessity*, p. 14.

8 The doctrine can indeed be formulated with terms rather than statements as the units. Thus Lewis describes the meaning of a term as "a criterion in mind, by reference to which one is able to apply or refuse to apply the expression in question in the case of presented, or imagined things or situations" (Lewis, *An Analysis of Knowledge and Valuation*, p. 133). – For an instructive account of the vicissitudes of the verification theory of meaning, centered however on the question of meaning*fulness* rather than synonymy and analyticity see C. G. Hempel, "Problems and Changes in the Empiricist Criterion of Meaning," *Revue Internationale de Philosophie 4* (1950), 41–43.

9 This doctrine was well argued by Pierre Duhem, *La Theorie physique: Son object et sa structure* (Paris, 1906): 303–28. Or See Armand Lowinger, *The Methodology of Pierre Duhem* (New York: Columbia University Press, 1941), pp. 132–40.

10 Cf. "On What There Is," p. 18.

11 "L'ontologie fait corps avec la science elle-même et ne peut en être separée." Emile Meyerson, *Identité et realité* (Paris, 1908 4th edn., 1932), p. 439.

12 "On What There Is," pp. 12 f.; Logic and the Reification of Universals," 102 ff.

13 For an effective expression of further misgivings over this distinction, see M. White, "The Analytic and the Synthetic: an Untenable Dualism," in Sidney Hook (ed.), *John Dewey: Philosopher of Science and Freedom* (New York: Dial Press, 1950), pp. 316–30.

6.4

The New Riddle of Induction

Nelson Goodman

Nelson Goodman's (1906–1998) work has had a significant impact in diverse philosophical areas ranging from the philosophy of mathematics to aesthetics. In this selection, Goodman presents his new riddle of induction. Intuitively (and putting aside David Hume's "old" problem of induction) the sighting of a green emerald confirms the law that all emeralds are green, and therefore that an emerald examined after the year 2040 will be green. But by construction of a distinct predicate called "grue", Goodman argues that the very same sighting also seems to confirm that emeralds will be blue after 2040 (or have any other property). An inductive argument can therefore be formulated to support any prediction whatsoever, a disastrous result for the attempt to formulate a logic of induction or confirmation.

Confirmation of a hypothesis by an instance depends rather heavily upon features of the hypothesis other than its syntactical form. That a given piece of copper conducts electricity increases the credibility of statements asserting that other pieces of copper conduct electricity, and thus confirms the hypothesis that all copper conducts electricity. But the fact that a given man now in this room is a third son does not increase the credibility of statements asserting that other men now in this room are third sons, and so does not confirm the hypothesis that all men now in this room are third sons. Yet in both cases our hypothesis is a generalization of the evidence statement. The difference is that in the former case the hypothesis is a *lawlike* statement; while in the latter case, the hypothesis is a merely contingent or accidental generality. Only a statement that is *lawlike* – regardless of its truth or falsity or its scientific importance – is capable of receiving confirmation from an instance of it; accidental statements are not. Plainly, then, we must look for a way of distinguishing lawlike from accidental statements.

So long as what seems to be needed is merely a way of excluding a few odd and unwanted

From Goodman's *Fact, Fiction, and Forecast*, 4th edn. (Cambridge, Mass.: Harvard University Press, 1983), pp. 72–81.

cases that are inadvertently admitted by our definition of confirmation, the problem may not seem very hard or very pressing. We fully expect that minor defects will be found in our definition and that the necessary refinements will have to be worked out patiently one after another. But some further examples will show that our present difficulty is of a much graver kind.

Suppose that all emeralds examined before a certain time *t* are green.[1] At time *t*, then, our observations support the hypothesis that all emeralds are green; and this is in accord with our definition of confirmation. Our evidence statements assert that emerald *a* is green, that emerald *b* is green, and so on; and each confirms the general hypothesis that all emeralds are green. So far, so good.

Now let me introduce another predicate less familiar than "green". It is the predicate "grue" and it applies to all things examined before *t* just in case they are green but to other things just in case they are blue. Then at time *t* we have, for each evidence statement asserting that a given emerald is green, a parallel evidence statement asserting that that emerald is grue. And the statements that emerald *a* is grue, that emerald *b* is grue, and so on, will each confirm the general hypothesis that all emeralds are grue. Thus according to our definition, the prediction that all emeralds subsequently examined will be green and the prediction that all will be grue are alike confirmed by evidence statements describing the same observations. But if an emerald subsequently examined is grue, it is blue and hence not green. Thus although we are well aware which of the two incompatible predictions is genuinely confirmed, they are equally well confirmed according to our present definition. Moreover, it is clear that if we simply choose an appropriate predicate, then on the basis of these same observations we shall have equal confirmation, by our definition, for any prediction whatever about other emeralds – or indeed about anything else.[2] As in our earlier example, only the predictions subsumed under lawlike hypotheses are genuinely confirmed; but we have no criterion as yet for determining lawlikeness. And now we see that without some such criterion, our definition not merely includes a few unwanted cases, but is so completely ineffectual that it virtually excludes nothing. We are left once again with the intolerable result that anything confirms anything. This difficulty cannot be set aside as an annoying detail to be taken care of in due course. It has to be met before our definition will work at all.

Nevertheless, the difficulty is often slighted because on the surface there seem to be easy ways of dealing with it. Sometimes, for example, the problem is thought to be much like the paradox of the ravens. We are here again, it is pointed out, making tacit and illegitimate use of information outside the stated evidence: the information, for example, that different samples of one material are usually alike in conductivity, and the information that different men in a lecture audience are usually not alike in the number of their older brothers. But while it is true that such information is being smuggled in, this does not by itself settle the matter as it settles the matter of the ravens. There the point was that when the smuggled information is forthrightly declared, its effect upon the confirmation of the hypothesis in question is immediately and properly registered by the definition we are using. On the other hand, if to our initial evidence we add statements concerning the conductivity of pieces of other materials or concerning the number of older brothers of members of other lecture audiences, this will not in the least affect the confirmation, according to our definition, of the hypothesis concerning copper or of that concerning this lecture audience. Since our definition is insensitive to the bearing upon hypotheses of evidence so related to them, even when the evidence is fully declared, the difficulty about accidental hypotheses cannot be explained away on the ground that such evidence is being surreptitiously taken into account.

A more promising suggestion is to explain the matter in terms of the effect of this other evidence not directly upon the hypothesis in question but *in*directly through other hypotheses that *are* confirmed, according to our definition, by such evidence. Our information about other materials does by our definition confirm such hypotheses as that all pieces of iron conduct electricity, that no pieces of rubber do, and so on; and these hypotheses, the explanation runs, impart to the hypothesis that all pieces of copper conduct electricity (and also to the hypothesis that none do) the character of lawlikeness – that is, amenability to confirmation by direct positive instances when found. On the other hand, our information about other lecture audiences *dis*confirms many hypotheses to the effect that all the men in one audience are third sons, or that none are; and this strips any character of lawlikeness from the hypothesis that all (or the hypothesis that none) of the men in *this* audience are third sons. But clearly if this course is to be followed, the circumstances under which hypotheses are thus related to one another will have to be precisely articulated.

The problem, then, is to define the relevant way in which such hypotheses must be alike. Evidence for the hypothesis that all iron conducts electricity enhances the lawlikeness of the hypothesis that all zirconium conducts electricity, but does not similarly affect the hypothesis that all the objects on my desk conduct electricity. Wherein lies the difference? The first two hypotheses fall under the broader hypothesis – call it "*H*" – that every class of things of the same material is uniform in conductivity; the first and third fall only under some such hypothesis as – call it "*K*" – that every class of things that are either all of the same material or all on a desk is uniform in conductivity. Clearly the important difference here is that evidence for a statement affirming that one of the classes covered by *H* has the property in question increases the credibility of any statement affirming that another such class has this property; while nothing of the sort holds true with respect to *K*. But this is only to say that *H* is lawlike and *K* is not. We are faced anew with the very problem we are trying to solve: the problem of distinguishing between lawlike and accidental hypotheses.

The most popular way of attacking the problem takes its cue from the fact that accidental hypotheses seem typically to involve some spatial or temporal restriction, or reference to some particular individual. They seem to concern the people in some particular room, or the objects on some particular person's desk; while lawlike hypotheses characteristically concern all ravens or all pieces of copper whatsoever. Complete generality is thus very often supposed to be a sufficient condition of lawlikeness; but to define this complete generality is by no means easy. Merely to require that the hypothesis contain no term naming, describing, or indicating a particular thing or location will obviously not be enough. The troublesome hypothesis that all emeralds are grue contains no such term; and where such a term does occur, as in hypotheses about men in *this room*, it can be suppressed in favor of some predicate (short or long, new or old) that contains no such term but applies only to exactly the same things. One might think, then, of excluding not only hypotheses that actually contain terms for specific individuals but also all hypotheses that are equivalent to others that do contain such terms. But, as we have just seen, to exclude only hypotheses of which *all* equivalents contain such terms is to exclude nothing. On the other hand, to exclude all hypotheses that have *some* equivalent containing such a term is to exclude everything; for even the hypothesis

All grass is green

has as an equivalent

All grass in London or elsewhere is green.

The next step, therefore, has been to consider ruling out predicates of certain kinds. A syntactically universal hypothesis is lawlike, the proposal runs, if its predicates are 'purely qualitative' or 'non-positional'.[3] This will obviously accomplish nothing if a purely qualitative predicate is then conceived either as one that is equivalent to some expression free of terms for specific individuals, or as one that is equivalent to no expression that contains such a term; for this only raises again the difficulties just pointed out. The claim appears to be rather that at least in the case of a simple enough predicate we can readily determine by direct inspection of its meaning whether or not it is purely qualitative. But even aside from obscurities in the notion of 'the meaning' of a predicate, this claim seems to me wrong. I simply do not know how to tell whether a predicate is qualitative or positional, except perhaps by completely begging the question at issue and asking whether the predicate is 'well-behaved' – that is, whether simple syntactically universal hypotheses applying it are lawlike.

This statement will not go unprotested. "Consider", it will be argued, "the predicates 'blue' and 'green' and the predicate 'grue' introduced earlier, and also the predicate 'bleen' that applies to emeralds examined before time *t* just in case they are blue and to other emeralds just in case they are green. Surely it is clear", the argument runs, "that the first two are purely qualitative and the second two are not; for the meaning of each of the latter two plainly involves reference to a specific temporal position." To this I reply that indeed I do recognize the first two as well-behaved predicates admissible in lawlike hypotheses, and the second two as ill-behaved predicates. But the argument that the former but not the latter are purely qualitative seems to me quite unsound. True enough, if we start with "blue" and "green", then "grue" and "bleen" will be explained in terms of "blue" and "green" and a temporal term. But equally truly, if we start with "grue" and "bleen", then "blue" and "green" will be explained in terms of "grue" and "bleen" and a temporal term; "green", for example, applies to emeralds examined before time *t* just in case they are grue, and to other emeralds just in case they are bleen. Thus qualitativeness is an entirely relative matter and does not by itself establish any dichotomy of predicates. This relativity seems to be completely overlooked by those who contend that the qualitative character of a predicate is a criterion for its good behavior.

Of course, one may ask why we need worry about such unfamiliar predicates as "grue" or about accidental hypotheses in general, since we are unlikely to use them in making predictions. If our definition works for such hypotheses as are normally employed, isn't that all we need? In a sense, yes; but only in the sense that we need no definition, no theory of induction, and no philosophy of knowledge at all. We get along well enough without them in daily life and in scientific research. But if we seek a theory at all, we cannot excuse gross anomalies resulting from a proposed theory by pleading that we can avoid them in practice. The odd cases we have been considering are clinically pure cases that, though seldom encountered in practice, nevertheless display to best advantage the symptoms of a widespread and destructive malady.

We have so far neither any answer nor any promising clue to an answer to the question what distinguishes lawlike or confirmable hypotheses from accidental or non-confirmable ones; and what may at first have seemed a minor technical difficulty has taken on the stature of a major obstacle to the development of a satisfactory theory of confirmation. It is this problem that I call the new riddle of induction.

Notes

1 Although the example used is different, the argument to follow is substantially the same as that set forth in my note 'A Query on Confirmation', *Journal of Philosophy* XLIII (1946): 383–5.

2 For instance, we shall have equal confirmation, by our present definition, for the prediction that roses subsequently examined will be blue. Let "emerose" apply just to emeralds examined before time *t*, and to roses examined later. Then all emeroses so far examined are grue, and this confirms the hypothesis that all emeroses are grue and hence the prediction that roses subsequently examined will be blue. The problem raised by such antecedents has been little noticed, but is no easier to meet than that raised by similarly perverse consequents.

3 Carnap took this course in his paper 'On the Application of Inductive Logic', *Philosophy and Phenomenological Research*, vol. 8 (1947), pp. 133–47, which is in part a reply to my 'A Query on Confirmation', cited in note 1. The discussion was continued in my note 'On Infirmities of Confirmation Theory', *Philosophy and Phenomenological Research*, vol. 8 (1947), pp. 149–51; and in Carnap's 'Reply to Nelson Goodman', same journal, same volume, pp. 461–2.

6.5

What Theories Are Not

Hilary Putnam

Hilary Putnam (b. 1926) occupies a unique position in philosophy as one who has not only advocated a wide variety of competing positions at different stages of his career, but who also presented some of the strongest arguments for those positions while endorsing them, as well as some of the strongest arguments against them when advocating their competitors. In this short essay, he argues that the distinction between observation and theoretical vocabularies underlying the received view is untenable.

In this paper I consider the role of *theories* in empirical science, and attack what may be called the "received view" – that theories are to be thought of as "partially interpreted calculi" in which only the "observation terms" are "directly interpreted" (the theoretical terms being only "partially interpreted," or, some people even say, "partially understood").

To begin, let us review this received view. The view divides the nonlogical vocabulary of science into two parts:

Observation Terms	**Theoretical Terms**
such terms as	such terms as
"red"	"electron"
"touches"	"dream"
"stick," etc.	"gene," etc.

The basis for the division appears to be as follows: the observation terms apply to what may be called publicly observable things and signify observable qualities of these things, while the theoretical terms correspond to the remaining unobservable qualities and things.

This division of terms into two classes is then allowed to generate a division of statements into two[1] classes as follows:

From Putnam's *Mathematics, Matter, and Method, Philosophical Papers Vol. 1* (Second Edition) (Cambridge: Cambridge University Press, 1979, pp. 215–20. First published in E. Nagel, P. Suppes, and A. Tarski (eds.), *Logic, Methodology and Philosophy of Science* (Stanford: Stanford University Press, 1962).

Observational Statements	**Theoretical Statements**
statements containing only observation terms and logical vocabulary	statements containing theoretical terms

Lastly, a scientific theory is conceived of as an axiomatic system which may be thought of as initially uninterpreted, and which gains "empirical meaning" as a result of a specification of meaning *for the observation terms alone.* A kind of partial meaning is then thought of as drawn up to the theoretical terms, by osmosis, as it were.

The Observational–Theoretical Dichotomy

One can think of many distinctions that are crying out to be made ("new" terms vs. "old" terms, technical terms vs. non-technical ones, terms more-or-less peculiar to one science vs. terms common to many, for a start). My contention here is simply:

(1) The *problem* for which this dichotomy was invented ("how is it possible to interpret theoretical terms?") does not exist.

(2) A basic reason some people have given for introducing the dichotomy is false: namely, justification in science does *not* proceed "down" in the direction of observation terms. In fact, justification in science proceeds in any direction that may be handy – more observational assertions sometimes being justified with the aid of more theoretical ones, and vice versa. Moreover, as we shall see, while the notion of an *observation report* has some importance in the philosophy of science, such reports cannot be identified on the basis of the vocabulary they do or do not contain.

(3) In any case, whether the reasons for introducing the dichotomy were good ones or bad ones, the double distinction (observation terms–theoretical terms, observation statements–theoretical statements) presented above is, in fact, completely broken-backed. This I shall try to establish now.

In the first place, it should be noted that the dichotomy under discussion was intended as an explicative and not merely a stipulative one. That is, the words "observational" and "theoretical" are not having arbitrary new meanings bestowed upon them; rather, pre-existing uses of these words (especially in the philosophy of science) are presumably being sharpened and made clear. And, in the second place, it should be recalled that we are dealing with a double, not just a single, distinction. That is to say, part of the contention I am criticizing is that, once the distinction between observational and theoretical *terms* has been drawn as above, the distinction between theoretical statements and observational reports or assertions (in something like the sense usual in methodological discussions) can be drawn in terms of it. What I mean when I say that the dichotomy is "completely broken-backed" is this:

(A) If an "observation term" is one that cannot apply to an unobservable, then there are no observation terms.[2]

(B) Many terms that refer primarily to what Carnap would class as "unobservables" are not theoretical terms; and at least some theoretical terms refer primarily to observables.

(C) Observational reports can and frequently do contain theoretical terms.

(D) A scientific theory, properly so-called, may refer only to observables. (Darwin's theory of evolution, as originally put forward, is one example.)

To start with the notion of an "observation term": Carnap's formulation in *Testability and Meaning* (Carnap, 1955) was that for a term to be an observation term not only must it correspond to an observable quality, but the determination whether the quality is present or not must be able to be made by the observer in a relatively short time, and with a high degree of confirmation. In his most recent authoritative publication (Carnap, 1956), Carnap is rather brief. He writes, "the terms of V_O [the 'observation vocabulary'] are predicates designating observable properties of events or things (e.g. 'blue', 'hot', 'large', etc.) or observable relations between them (e.g. 'x is warmer than y', 'x is contiguous to y', etc.)" (Carnap, 1956, p. 41). The only other clarifying remarks I could find are the following: "The name 'observation language' may be understood in a narrower or in a wider sense; the observation language in the wider sense includes the disposition terms. In this article I take the observation language L_O in the narrower sense" (Carnap, 1956, p. 63). "An observable property may be regarded as a simple special case of a testable disposition: for example, the operation for finding out whether a thing is blue or hissing or cold, consists simply in looking or listening or touching the thing, respectively. Nevertheless, *in the reconstruction of the language* [italics mine] it seems convenient to take some properties for which the test procedure is extremely simple (as in the examples given) as directly observable, and use them as primitives L_O" (Carnap, 1956, p. 63).

These paragraphs reveal that Carnap, at least, thinks of observation terms as corresponding to qualities that can be detected without the aid of instruments. But always so detected? Or can an observation term refer sometimes to an observable thing and sometimes to an unobservable? While I have not been able to find any explicit statement on this point, it seems to me that writers like Carnap must be *neglecting* the fact that *all* terms – including the "observation terms" – have at least the possibility of applying to unobservables. Thus their problem has sometimes been formulated in quasi-historical terms – "How could theoretical terms have been introduced into the language?" And the usual discussion strongly suggests that the following puzzle is meant; if we imagine a time at which people could only talk about observables (had not available any theoretical terms), how did they ever manage to *start* talking about unobservables?

It is possible that I am here doing Carnap and his followers an injustice. However, polemics aside, the following points must be emphasized.

(1) Terms referring to unobservables are *invariably* explained, in the actual history of science, with the aid of already present locutions referring to unobservables. There never was a stage of language at which it was impossible to talk about unobservables. Even a three-year-old child can understand a story about "people too little to see"[3] and not a single "theoretical term" occurs in this phrase.

(2) There is not even a single *term* of which it is true to say that it *could not* (without changing or extending its meaning) be used to refer to unobservables. "Red," for example, was so used by Newton when he postulated that red light consists of *red corpuscles*.[4]

In short: if an "observation term" is a term which *can*, in principle, only be used to refer to observable things, then *there are no observation terms*. If, on the other hand, it is granted that locutions consisting of just observation terms can refer to unobservables, there is no longer any reason to maintain *either* that theories and speculations about the unobservable parts of the world must contain "theoretical (= non-observation) terms" *or* that there is any general problem as to how one can introduce terms referring to unobservables. Those philosophers who find difficulty in how we understand theoretical terms should find an equal difficulty in how we understand "red" and "smaller than."

So much for the notion of an "observation term." Of course, one may recognize the point just made – that the "observation terms" also apply, in some contexts, to unobservables – and retain the class (with a suitable warning as to how the label "observational term" is to be understood). But can we agree that the complementary class – what should be called the "nonobservable terms" – is to be labelled "theoretical terms"? No, for the identification of "theoretical term" with "term (other than the 'disposition terms,' which are given a special place in Carnap's scheme) designating an unobservable quality" is unnatural and misleading. On the one hand, it is clearly an enormous (and, I believe, insufficiently motivated) extension of common usage to classify such terms as "angry," "loves," and so forth, as "theoretical terms" simply because they allegedly do not refer to public observables. A theoretical term, properly so-called, is one which comes from a scientific *theory* (and the almost untouched problem, in thirty years of writing about "theoretical terms" is what is *really* distinctive about such terms). In this sense (and I think it the sense important for discussions of science) "satellite" is, for example, a theoretical term (although the things it refers to are quite observable[5]) and "dislikes" clearly is not.

Our criticisms so far might be met by relabelling the first dichotomy (the dichotomy of terms) "observation vs. non-observation," and suitably "hedging" the notion of "observation." But more serious difficulties are connected with the identification upon which the second dichotomy is based – the identification of "theoretical statements" with statements containing non-observation ("theoretical" terms) and "observation statements" with "statements in the observational vocabulary."

That observation statements may contain theoretical terms is easy to establish. For example, it is easy to imagine a situation in which the following sentence might occur: "We also *observed* the creation of two electron-positron pairs."

This objection is sometimes dealt with by proposing to "relativize" the observation-theoretical dichotomy to the context. (Carnap, however, rejects this way out in the article we have been citing.) This proposal to "relativize" the dichotomy does not seem to me to be very helpful. In the first place, one can easily imagine a context in which "electron" would occur, in the same text, in *both* observational reports and in theoretical conclusions from those reports. (So that one would have distortions if one tried to put the term in either the "observational term" box or in the "theoretical term" box.) In the second place, for what philosophical problem or point does one require even the relativized dichotomy?

The usual answer is that sometimes a statement A (observational) is offered in support of a statement B (theoretical). Then, in order to explain why A is not itself questioned in the context, we need to be able to say that A is functioning, in that context, as an observation report. But this misses the point I have been making! I do not deny the need for some such notion as "observation report." What I deny is that the distinction between observation reports and, among other things, theoretical statements, can or should be drawn on the basis of vocabulary. In addition, a relativized dichotomy will not serve Carnap's purposes. One can hardly maintain that theoretical terms are only partially interpreted, whereas observation terms are completely interpreted, if no sharp line exists between the two classes. (Recall that Carnap takes his problem to be "reconstruction of the language," not of some isolated scientific context.)

Notes

1 Sometimes a *tripartite* division is used: observation statements, theoretical statements (containing *only* theoretical terms), and "mixed" statements (containing both kinds of terms). This refinement is not considered here, because it avoids none of the objections presented below.

2 I neglect the possibility of trivially constructing terms that refer only to observables: namely, by conjoining "and is an observable thing" to terms that would otherwise apply to some unobservables. "Being an observable thing" is, in a sense, highly theoretical and yet applies only to observables!

3 Von Wright has suggested (in conversation) that this is an *extended* use of language (because we first learn words like "people" in connection with people we *can* see). This argument from "the way we learn to use the word" appears to be unsound, however (cf. Fodor, 1961).

4 Some authors (although not Carnap) explain the intelligibility of such discourse in terms of logically possible submicroscopic observers. But (a) such observers could not see single photons (or light corpuscles) even on Newton's theory; and (b) once such physically impossible (though logically possible) "observers" are introduced, why not go further and have observers with sense organs for electric charge, or the curvature of space, etc.! Presumably because *we* can see *red*, but not *charge*. But then, this just makes the point that we *understand* "red" even when applied outside our normal "range," even though we learn it ostensively, without *explaining* that fact. (The explanation lies in this: that understanding any term – even "red" – involves at least two elements: internalizing the syntax of a natural language, and acquiring a background of ideas. Overemphasis on the way "red" is *taught* has led some philosophers to misunderstand how it is *learned.*)

5 Carnap might exclude "satellite" as an observation term, on the ground that it takes a comparatively long time to verify that something is a satellite with the naked eye, even if the satellite is close to the parent body (although this could be debated). However, "satellite" cannot be excluded on the quite different ground that many satellites are too far away to see (which is the ground that first comes to mind) since the same is true of the huge majority of all *red* things.

References

Carnap, R. 1955. "Testability and Meaning" in H. Feigl and M. Brodbeck (eds.), *Readings in the Philosophy of Science*, Appleton-Century-Crofts. Reprinted from *Philosophy of Science 3* (1939) and *4* (1937).

Carnap, R. 1956. "The Methodological Character of Theoretical Concepts" in H. Feigl *et al.* (eds.), *Minnesota Studies in the Philosophy of Science*, Minneapolis, 1–74.

Fodor, J. A. 1961. "Of Words and Uses," *Inquiry* 4:3 (Autumn), 190–208.

6.6

On Observation

N. R. Hanson

In the selection below, Norwood Russell Hanson (1924–1967) predates Thomas Kuhn in formulating the doctrine that observation is "theory-laden." Intuitively, the nature of a perceiver's experience is a function only of the nature of the physical environment in front of her. Hanson argues that this is mistaken: what one sees is a function, not only of that environment, but also of one's background theoretical commitments, so that two people with different such commitments might end up seeing different things even when confronted by the same physical environment.

Were the eye not attuned to the Sun,
The Sun could never be seen by it.

Goethe[1]

A

Consider two microbiologists. They look at a prepared slide; when asked what they see, they may give different answers. One sees in the cell before him a cluster of foreign matter: it is an artefact, a coagulum resulting from inadequate staining techniques. This clot has no more to do with the cell, *in vivo*, than the scars left on it by the archaeologists spade have to do with the original shape of some Grecian urn. The other biologist identifies the clot as a cell organ, a 'Golgi body'. As for techniques, he argues: 'The standard way of detecting a cell organ is by fixing and staining. Why single out this one technique as producing artefacts, while others disclose genuine organs?'

The controversy continues.[2] It involves the whole theory of microscopical technique; nor is it an obviously experimental issue. Yet it affects what scientists say they see. Perhaps there is a sense in which two such observers do not see the same thing, do not begin from the same data, though their eyesight is normal and they are visually aware of the same object.

From Norwood Russell Hanson, *Patterns of Discovery: An Inquiry into the Conceptual Foundations of Science* (Cambridge: Cambridge University Press, 1965), pp. 4–25.

Imagine these two observing a Protozoon – *Amoeba*. One sees a one-celled animal, the other a non-celled animal. The first sees *Amoeba* in all its analogies with different types of single cells: liver cells, nerve cells, epithelium cells. These have a wall, nucleus, cytoplasm, etc. Within this class *Amoeba* is distinguished only by its independence. The other, however, sees *Amoeba's* homology not with single cells, but with whole animals. Like all animals *Amoeba* ingests its food, digests and assimilates it. It excretes, reproduces and is mobile – more like a complete animal than an individual tissue cell.

This is not an experimental issue, yet it can affect experiment. What either man regards as significant questions or relevant data can be determined by whether he stresses the first or the last term in 'unicellular animal'.[3]

Some philosophers have a formula ready for such situations: 'Of course they see the same thing. They make the same observation since they begin from the same visual data. But they interpret what they see differently. They construe the evidence in different ways.'[4] The task is then to show how these data are moulded by different theories or interpretations or intellectual constructions.

Considerable philosophers have wrestled with this task. But in fact the formula they start from is too simple to allow a grasp of the nature of observation within physics. Perhaps the scientists cited above do not begin their inquiries from the same data, do not make the same observations, do not even see the same thing? Here many concepts run together. We must proceed carefully, for wherever it makes sense to say that two scientists looking at *x* do not see the same thing, there must always be a prior sense in which they do see the same thing. The issue is, then, 'Which of these senses is most illuminating for the understanding of observational physics?'

These biological examples are too complex. Let us consider Johannes Kepler: imagine him on a hill watching the dawn. With him is Tycho Brahe. Kepler regarded the sun as fixed: it was the earth that moved. But Tycho followed Ptolemy and Aristotle in this much at least: the earth was fixed and all other celestial bodies moved around it. *Do Kepler and Tycho see the same thing in the east at dawn?*

We might think this an experimental or observational question, unlike the questions 'Are there Golgi bodies?' and 'Are Protozoa one-celled or non-celled?'. Not so in the sixteenth and seventeenth centuries. Thus Galileo said to the Ptolemaist '. . . neither Aristotle nor you can prove that the earth is *de facto* the centre of the universe . . .'.[5] 'Do Kepler and Tycho see the same thing in the east at dawn?' is perhaps not a *de facto* question either, but rather the beginning of an examination of the concepts of seeing and observation.

The resultant discussion might run:

'Yes, they do.'

'No, they don't.'

'Yes, they do!'

'No, they don't!' . . .

That this is possible suggests that there may be reasons for both contentions.[6] Let us consider some points in support of the affirmative answer.

The physical processes involved when Kepler and Tycho watch the dawn are worth noting. Identical photons are emitted from the sun; these traverse solar space, and our atmosphere. The two astronomers have normal vision; hence these photons pass through the cornea, aqueous humour, iris, lens and vitreous body of their eyes in the same way. Finally their retinas are affected. Similar electro-chemical changes occur in their selenium cells. The same configuration is etched on Kepler's retina as on Tycho's. So they see the same thing.

Locke sometimes spoke of seeing in this way: a man sees the sun if his is a normally-formed retinal picture of the sun. Dr Sir W. Russell Brain speaks of our retinal sensations as indicators and signals. Everything taking place behind the retina is, as he says, 'an intellectual operation based largely on non-visual experience . . .'.[7] What we *see* are the changes in the *tunica retina*. Dr Ida Mann regards the macula of the eye as itself 'seeing details in bright light', and the rods as 'seeing approaching motor-cars'. Dr Agnes Arber speaks of the eye as itself seeing.[8] Often, talk of seeing can direct attention to the retina. Normal people are distinguished from those for whom no retinal pictures can form: we may say of the former that they can see whilst the latter cannot see. Reporting when a certain red dot can be seen may supply the occulist with direct information about the condition of one's retina.[9]

This need not be pursued, however. These writers speak carelessly: seeing the sun is not seeing retinal pictures of the sun. The retinal images which Kepler and Tycho have are four in number, inverted and quite tiny.[10] Astronomers cannot be referring to these when they say they see the sun. If they are hypnotized, drugged, drunk or distracted they may not see the sun, even though their retinas register its image in exactly the same way as usual.

Seeing is an experience. A retinal reaction is only a physical state – a photochemical excitation. Physiologists have not always appreciated the differences between experiences and physical states.[11] People, not their eyes, see. Cameras, and eye-balls, are blind. Attempts to locate within the organs of sight (or within the neurological reticulum behind the eyes) some nameable called 'seeing' may be dismissed. That Kepler and Tycho do, or do not, see the same thing cannot be supported by reference to the physical states of their retinas, optic nerves or visual cortices: there is more to seeing than meets the eyeball.

Naturally, Tycho and Kepler see the same physical object. They are both visually aware of the sun. If they are put into a dark room and asked to report when they see something – anything at all – they may both report the same object at the same time. Suppose that the only object to be seen is a certain lead cylinder. Both men see the same thing: namely this object – whatever it is. It is just here, however, that the difficulty arises, for while Tycho sees a mere pipe, Kepler will see a telescope, the instrument about which Galileo has written to him.

Unless both are visually aware of the same object there can be nothing of philosophical interest in the question whether or not they see the same thing. Unless they both see the sun in this prior sense our question cannot even strike a spark.

Nonetheless, both Tycho and Kepler have a common visual experience of some sort. This experience perhaps constitutes their seeing the same thing. Indeed, this may be a seeing logically more basic than anything expressed in the pronouncement 'I see the sun' (where each means something different by 'sun'). If what they meant by the word 'sun' were the only clue, then Tycho and Kepler could not be seeing the same thing, even though they were gazing at the same object.

If, however, we ask, not 'Do they see the same thing?' but rather 'What is it that they both see?', an unambiguous answer may be forthcoming. Tycho and Kepler are both aware of a brilliant yellow-white disc in a blue expanse over a green one. Such a 'sense-datum' picture is single and uninverted. To be unaware of it is not to have it. Either it dominates one's visual attention completely or it does not exist.

If Tycho and Kepler are aware of anything visual, it must be of some pattern of colours. What else could it be? We do not touch or hear with our eyes, we only take in light.[12] This private pattern is the same for both observers. Surely if asked to sketch the contents of their visual fields they would both draw a kind of semi-circle on a horizon-line.[13] They say they see the sun. But they do not see every side of the sun at once; so what they really see is discoid to begin with. It is but a visual aspect of the sun. In any single observation the sun is a brilliantly luminescent disc, a penny painted with radium.

So something about their visual experiences at dawn is the same for both: a brilliant yellow-white disc centred between green and blue colour patches. Sketches of what they both see could be identical – congruent. In this sense Tycho and Kepler see the same thing at dawn. The sun appears to them in the same way. The same view, or scene, is presented to them both.

In fact, we often speak in this way. Thus the account of a recent solar eclipse:[14] 'Only a thin crescent remains; white light is now completely obscured; the sky appears a deep blue, almost purple, and the landscape is a monochromatic green . . . there are the flashes of light on the disc's circumference and now the brilliant crescent to the left. . . .' Newton writes in a similar way in the *Opticks*: 'These Arcs at their first appearance were of a violet and blue Colour, and between them were white Arcs of Circles, which . . . became a little tinged in their inward Limbs with red and yellow. . . .'[15] Every physicist employs the language of lines, colour patches, appearances, shadows. In so far as two normal observers use this language of the same event, they begin from the same data: they are making the same observation. Differences between them must arise in the interpretations they put on these data.

Thus, to summarize, saying that Kepler and Tycho see the same thing at dawn just because their eyes are similarly affected is an elementary mistake. There is a difference between a physical state and a visual experience. Suppose, however, that it is argued as above – that they see the same thing because they have the same sense-datum experience. Disparities in their accounts arise in *ex post facto* interpretations of what is seen, not in the fundamental visual data. If this is argued, further difficulties soon obtrude.

B

Normal retinas and cameras are impressed similarly by Fig. 8.[16] Our visual sense-data will be the same too. If asked to draw what we see, most of us will set out a configuration like Fig. 8.

Do we all see the same thing?[17] Some will see a perspex cube viewed from below. Others will see it from above. Still others will see it as a kind of polygonally cut gem. Some people see only criss-crossed lines in a plane. It may be seen as a block of ice, an aquarium, a wire frame for a kite – or any of a number of other things.

Do we, then, all see the same thing? If we do, how can these differences be accounted for?

Here the 'formula' re-enters: 'These are different *interpretations* of what all observers see in common. Retinal reactions to Fig. 8 are virtually identical; so too are our visual sense-data, since our drawings of what we see will have the same content. There is no place in the seeing for these differences, so they must lie in the interpretations put on what we see.'

This sounds as if I do two things, not one, when I see boxes and bicycles. Do I put different interpretations on Fig. 8 when I see it now as a box from below, and now as a cube from above? I am aware of no such thing. I mean no such thing when I report that the box's perspective has snapped back into the page.[18] If I do not mean this, then the concept of seeing which is natural in this connexion does not designate two diaphanous components, one optical, the other interpretative. Fig. 8 is simply seen now as a box from below, now as a cube from above; one does not first soak up an optical pattern and then clamp an interpretation on it. Kepler and Tycho just see the sun. That is all. That is the way the concept of seeing works in this connexion.

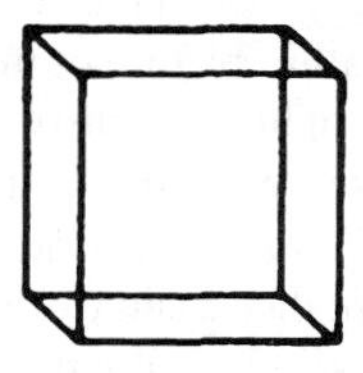

Figure 8 Necker cube

'But', you say, 'seeing Fig. 8 first as a box from below, then as a cube from above, involves interpreting the lines differently in each case.' Then for you and me to have a different interpretation of Fig. 8 just *is* for us to see something different. This does not mean we see the same thing and then interpret it differently. When I suddenly exclaim 'Eureka – a box from above', I do not refer simply to a different interpretation. (Again, there is a logically prior sense in which seeing Fig. 8 as from above and then as from below is seeing the same thing differently, i.e. being aware of the same diagram in different ways. We can refer just to this, but we need not. In this case we do not.)

Besides, the word 'interpretation' is occasionally useful. We know where it applies and where it does not. Thucydides presented the facts objectively; Herodotus put an interpretation on them. The word does not apply to everything – it has a meaning. Can interpreting always be going on when we see? Sometimes, perhaps, as when the hazy outline of an agricultural machine looms up on a foggy morning and, with effort, we finally identify it. Is this the 'interpretation' which is active when bicycles and boxes are clearly seen? Is it active when the perspective of Fig. 8 snaps into reverse? There was a time when Herodotus was half-through with his interpretation of the Graeco-Persian wars. Could there be a time when one is half-through interpreting Fig. 8 as a box from above, or as anything else?

'But the interpretation takes very little time – it is instantaneous.' Instantaneous interpretation hails from the Limbo that produced unsensed sensibilia, unconscious inference, incorrigible statements, negative facts and *Objektive*. These are ideas which philosophers force on the world to preserve some pet epistemological or metaphysical theory.

Only in contrast to 'Eureka' situations (like perspective reversals, where one cannot interpret the data) is it clear what is meant by saying that though Thucydides could have put an interpretation on history, he did not. Moreover, whether or not an historian is advancing an interpretation is an empirical question: we know what would count as evidence one way or the other. But whether we are employing an interpretation when we see Fig. 8 in a certain way is not empirical. What could count as evidence? In no ordinary sense of 'interpret' do I interpret Fig. 8 differently when its perspective reverses for me. If there is some extraordinary sense of word it is not clear, either in ordinary language, or in extraordinary (philosophical) language. To insist that different reactions to Fig. 8 *must* lie in the interpretations put on a common visual experience is just to reiterate (without reasons) that the seeing of x *must* be the same for all observers looking at x.

'But "I see the figure as a box" means: I am having a particular visual experience which I always have when I interpret the figure as a box, or when I look at a box. . . .' '. . . if I meant this, I ought to know it. I ought to be able to refer to the experience directly and not only indirectly. . . .'[19]

Ordinary accounts of the experiences appropriate to Fig. 8 do not require visual grist going into an intellectual mill: theories and interpretations are 'there' in the seeing from the outset. How can interpretations 'be there' in the seeing? How is it possible to see an object according to an interpretation? 'The question represents it as a queer fact; as if something were being forced into a form it did not really fit. But no squeezing, no forcing took place here.'[20]

Consider now the reversible perspective figures which appear in textbooks on Gestalt psychology: the tea-tray, the shifting (Schröder) staircase, the tunnel. Each of these can be seen as concave, as convex, or as a flat drawing.[21] Do I really see something different each time, or do I only interpret what I see in a different way? To interpret is to think, to do something; seeing is an experiential state.[22] The different ways in which these figures are seen are not due to different thoughts lying behind the visual reactions. What could 'spontaneous' mean if these reactions are not spontaneous? When the staircase 'goes into reverse' it does so spontaneously. One does not think of anything special; one does not think at all. Nor does one interpret. One just sees, now a staircase as from above, now a staircase as from below.

The sun, however, is not an entity with such variable perspective. What has all this to do with suggesting that Tycho and Kepler may see different things in the east at dawn? Certainly the cases are

Figure 9 Old/young woman

different. But these reversible perspective figures are examples of different things being seen in the same configuration, where this difference is due neither to differing visual pictures, nor to any 'interpretation' superimposed on the sensation.

Some will see in Fig. 9 an old Parisienne, others a young woman (à la Toulouse-Lautrec).[23] All normal retinas 'take' the same picture; and our sense-datum pictures must be the same, for even if you see an old lady and I a young lady, the pictures we draw of what we see may turn out to be geometrically indistinguishable. (Some can see this *only* in one way, not both. This is like the difficulty we have after finding a face in a tree-puzzle; we cannot thereafter see the tree without the face.)

When what is observed is characterized so differently as 'young woman' or 'old woman', is it not natural to say that the observers see different things? Or must 'see different things' mean only 'see different objects'? This is a primary sense of the expression, to be sure. But is there not also a sense in which one who cannot see the young lady in Fig. 9 sees something different from me, who sees the young lady? Of course there is.

Similarly, in Köhler's famous drawing of the Goblet-and-Faces[24] we 'take' the same retinal/cortical/sense-datum picture of the configuration; our drawings might be indistinguishable. I see a goblet, however, and you see two men staring at one another. Do we see the same thing? Of course we do. But then again we do not. (The sense in which we *do* see the same thing begins to lose its philosophical interest.)

I draw my goblet. You say 'That's just what I saw, two men in a staring contest'. What steps must be taken to get you to see what I see? When attention shifts from the cup to the faces does one's visual picture change? How? What is it that changes? What could change? Nothing optical or sensational is modified. Yet one sees different things. The organization of what one sees changes.[25]

How does one describe the difference between the *jeune fille* and the *vieille femme* in Fig. 9? Perhaps the difference is not describable: it may just show itself.[26] That two observers have not seen the same things in Fig. 9 could show itself in their behaviour. What is the difference between us when you see the zebra as black with white stripes and I see it as white with black stripes? Nothing optical. Yet there might be a context (for instance, in the genetics of animal pigmentation), where such a difference could be important.

A third group of figures will stress further this organizational element of seeing and observing. They will hint at how much more is involved when Tycho and Kepler witness the dawn than 'the formula' suggests.

What is portrayed in Fig. 10? Your retinas and visual cortices are affected much as mine are; our sense-datum pictures would not differ. Surely we could all produce an accurate sketch of Fig. 10. Do we see the same thing?

I see a bear climbing up the other side of a tree. Did the elements 'pull together'/cohere/organize, when you learned this?[27] You might even say with Wittgenstein 'it has not changed, and yet I see it differently . . .'.[28] Now, does it not have '. . . a quite particular "organization" '?

Organization is not itself seen as are the lines and colours of a drawing. It is not itself a line, shape, or a colour. It is not an element in the visual field, but rather the way in which elements are appreciated. Again, the plot is not another detail in the story. Nor is the tune just one more note. Yet without plots and tunes details and notes would not hang together. Similarly the organization of Fig. 10 is nothing that registers on the retina along with other details. Yet it gives the lines and shapes a pattern. Were this lacking we would be left with nothing but an unintelligible configuration of lines.

How do visual experiences become organized? How is seeing possible?

Consider Fig. 11 in the context of Fig. 12. The context gives us the clue. Here, some people could not see the figure as an antelope. Could people who had never seen an antelope, but only birds, see an antelope in Fig. 11?

In the context of Fig. 13 the figure may indeed stand out as an antelope. It might even be urged that the figure seen in Fig. 12 has no similarity to the one in Fig. 13 although the two are congruent. Could anything be more opposed to a sense-datum account of seeing?

Of a figure similar to the Necker cube (Fig. 8) Wittgenstein writes, 'You could imagine [this] appearing in several places in a text-book. In the relevant text something different is in question every time: here a glass cube, there an inverted open box, there a wire frame of that shape, there three boards forming a solid angle. Each time the text supplies the interpretation of the illustration. But we can also see the illustration now as one thing, now as another. So we interpret it, and see it as we interpret it.'[29]

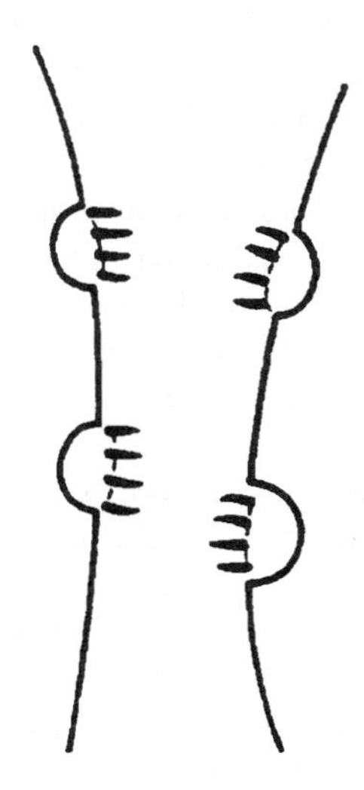

Figure 10 Bear

Figure 11 Bird/antelope

Figure 12 Birds

Figure 13 Antelopes

Consider now the head-and-shoulders in Fig. 14. The upper margin of the picture cuts the brow, thus the top of the head is not shown. The point of the jaw, clean shaven and brightly illuminated, is just above the geometric center of the picture. A white mantle . . . covers the right shoulder. The right upper sleeve is exposed as the rather black area at the lower left. The hair and beard are after the manner of a late mediaeval representation of Christ.[30]

The appropriate aspect of the illustration is brought out by the verbal context in which it appears. It is not an illustration of anything determinate unless it appears in some such context. In the same way, I must talk and gesture around Fig. 11 to get you to see the antelope when only the bird has revealed itself. I must provide a context. The context is part of the illustration itself.

Such a context, however, need not be set out explicitly. Often it is 'built into' thinking, imagining and picturing. We are set[31] to appreciate the visual aspect of things in certain ways. Elements in our experience do not cluster at random.

Figure 14 Man

A trained physicist could see one thing in Fig. 15: an X-ray tube viewed from the cathode. Would Sir Lawrence Bragg and an Eskimo baby see the same thing when looking at an X-ray tube? Yes, and no. Yes – they are visually aware of the same object. No – the *ways* in which they are visually aware are profoundly different. Seeing is not only the having of a visual experience; it is also the way in which the visual experience is had.

At school the physicist had gazed at this glass-and-metal instrument. Returning now, after years in University and research, his eye lights upon the same object once again. Does he see the same thing now as he did then? Now he sees the instrument in terms of electrical circuit theory, thermodynamic theory, the theories of metal and glass structure, thermionic emission, optical transmission, refraction, diffraction, atomic theory, quantum theory and special relativity.

Contrast the freshman's view of college with that of his ancient tutor. Compare a man's first glance at the motor of his car with a similar glance ten exasperating years later.

'Granted, one learns all these things', it may be countered, 'but it all figures in the interpretation the physicist puts on what he sees. Though the layman sees exactly what the physicist sees, he cannot interpret it in the same way because he has not learned so much.'

Is the physicist doing more than just seeing? No; he does nothing over and above what the layman does when he sees an X-ray tube. What are you doing over and above reading these words? Are you interpreting marks on a page? When would this ever be a natural way of speaking? Would an infant see what you see here, when you see words and sentences and he sees but marks and

Figure 15 X-ray tube

lines? One does nothing beyond looking and seeing when one dodges bicycles, glances at a friend, or notices a cat in the garden.

'The physicist and the layman see the same thing', it is objected, 'but they do not make the same thing of it.' The layman can make nothing of it. Nor is that just a figure of speech. I can make nothing of the Arab word for *cat*, though my purely visual impressions may be indistinguishable from those of the Arab who can. I must learn Arabic before I can see what he sees. The layman must learn physics before he can see what the physicist sees.

If one must find a paradigm case of seeing it would be better to regard as such not the visual apprehension of colour patches but things like seeing what time it is, seeing what key a piece of music is written in, and seeing whether a wound is septic.[32]

Pierre Duhem writes:

> Enter a laboratory; approach the table crowded with an assortment of apparatus, an electric cell, silk-covered copper wire, small cups of mercury, spools, a mirror mounted on an iron bar; the experimenter is inserting into small openings the metal ends of ebony-headed pins; the iron oscillates, and the mirror attached to it throws a luminous band upon a celluloid scale; the forward-backward motion of this spot enables the physicist to observe the minute oscillations of the iron bar. But ask him what he is doing. Will he answer 'I am studying the oscillations of an iron bar which carries a mirror'? No, he will say that he is measuring the electric resistance of the spools. If you are astonished, if you ask him what his words mean, what relation they have with the phenomena he has been observing and which you have noted at the same time as he, he will answer that your question requires a long explanation and that you should take a course in electricity.[33]

The visitor must learn some physics before he can see what the physicist sees. Only then will the context throw into relief those features of the objects before him which the physicist sees as indicating resistance.

This obtains in all seeing. Attention is rarely directed to the space between the leaves of a tree, save when a Keats brings it to our notice.[34] (Consider also what was involved in Crusoe's seeing a vacant space in the sand as a footprint.) Our attention most naturally rests on objects and events which dominate the visual field. What a blooming, buzzing, undifferentiated confusion visual life would be if we all arose tomorrow without attention capable of dwelling only on what had heretofore been overlooked.[35]

The infant and the layman can see: they are not blind. But they cannot see what the physicist sees; they are blind to what he sees.[36] We may not hear that the oboe is out of tune, though this will be painfully obvious to the trained musician. (Who, incidentally, will not hear the tones and *interpret* them as being out of tune, but will simply hear the oboe to be out of tune.[37] We simply see what time it is; the surgeon simply sees a wound to be septic; the physicist sees the X-ray tube's anode overheating.) The elements of the visitor's visual field, though identical with those of the physicist, are not organized for him as for the physicist; the same lines, colours, shapes are apprehended by both, but not in the same way. There are indefinitely many ways in which a constellation of lines, shapes, patches, may be seen. *Why* a visual pattern is seen differently is a question for psychology, but *that* it may be seen differently is important in any examination of the concepts of seeing and observation. Here, as Wittgenstein might have said, the psychological is a symbol of the logical.

You see a bird, I see an antelope; the physicist sees an X-ray tube, the child a complicated lamp bulb; the microscopist sees coelenterate mesoglea, his new student sees only a gooey, formless stuff. Tycho and Simplicius see a mobile sun, Kepler and Galileo see a static sun.[38]

It may be objected, 'Everyone, whatever his state of knowledge, will see Fig. 8 as a box or cube, viewed as from above or as from below'. True; almost everyone, child, layman, physicist, will see the figure as box-like one way or another. But could such observations be made by people ignorant of the construction of box-like objects? No. This objection only shows that most of us – the blind, babies, and dimwits excluded – have learned enough to be able to see this figure as a three-dimensional box. This reveals something about the sense in which Simplicius and Galileo do see the same thing (which I have never denied): they both see a brilliant heavenly body. The schoolboy and the physicist both see that the X-ray tube will smash if dropped. Examining how observers see different things in *x* marks something important about their seeing the same thing when looking at *x*. If seeing different things involves having different knowledge and theories about *x*, then perhaps the sense in which they see the same thing involves their sharing knowledge and theories about *x*. Bragg and the baby share no knowledge of X-ray tubes. They see the same thing only in that if they are looking at *x* they are both having some visual experience of it. Kepler and Tycho agree on more: they see the same thing in a stronger sense. Their visual fields are organized in much the same way. Neither sees the sun about to break out in a grin, or about to crack into ice cubes. (The baby is not 'set' even against these eventualities.) Most people today see the same thing at dawn in an even stronger sense: we share much knowledge of the sun. Hence Tycho and Kepler see different things, and yet they see the same thing. That these things can be said depends on their knowledge, experience, and theories.

Kepler and Tycho are to the sun as we are to Fig. 11, when I see the bird and you see only the antelope. The elements of their experiences are identical; but their conceptual organization is vastly different. Can their visual fields have a different organization? Then they can see different things in the east at dawn.

It is the sense in which Tycho and Kepler do not observe the same thing which must be grasped if one is to understand disagreements within microphysics. Fundamental physics is primarily a search for intelligibility – it is philosophy of matter. Only secondarily is it a search for objects and facts (though the two endeavours are as hand and glove). Microphysicists seek new modes of conceptual organization. If that can be done the finding of new entities will follow. Gold is rarely discovered by one who has not got the lay of the land.

To say that Tycho and Kepler, Simplicius and Galileo, Hooke and Newton, Priestley and Lavoisier, Soddy and Einstein, De Broglie and Born, Heisenberg and Bohm all make the same observations but use them differently is too easy.[39] It does not explain controversy in research science. Were there no sense in which they were different observations they could not be used differently. This may perplex some: that researchers sometimes do not appreciate data in the same way is a serious matter. It is important to realize, however, that sorting out differences about data, evidence, observation, may require more than simply gesturing at observable objects. It may require a comprehensive reappraisal of one's subject matter. This may be difficult, but it should not obscure the fact that nothing less than this may do.

C

There is a sense, then, in which seeing is a 'theory-laden' undertaking. Observation of *x* is shaped by prior knowledge of *x*. Another influence on observations rests in the language or notation used to express what we know, and without which there would be little we could recognize as knowledge. This will be examined.[40]

I do not mean to identify seeing with *seeing as*. Seeing an X-ray tube is not seeing a glass-and-metal object as an X-ray tube.[41] However, seeing an antelope and seeing an object as an antelope have much in common. Something of the concept of seeing can be discerned from tracing uses of 'seeing . . . as . . .'. Wittgenstein is reluctant[42] to concede this, but his reasons are not clear to me. On the contrary, the logic of 'seeing as' seems to illuminate the general perceptual case.[43] Consider again the footprint in the sand. Here all the organizational features of *seeing as* stand out clearly, in the absence of an '*object*'. One can even imagine cases where 'He sees it as a footprint' would be a way of referring to another's apprehension of what actually is a footprint. So, while I do not identify, for example, Hamlet's seeing of a camel in the clouds with his seeing of Yorick's skull, there is still something to be learned about the latter from noting what is at work in the former.

There is, however, a further element in seeing and observation. If the label 'seeing as' has drawn out certain features of these concepts, 'seeing that . . .' may bring out more. Seeing a bear in Fig. 10 was to see that were the 'tree' circled we should come up behind the beast. Seeing the dawn was for Tycho and Simplicius to see that the earth's brilliant satellite was beginning its diurnal circuit around us, while for Kepler and Galileo it was to see that the earth was spinning them back into the light of our local star. Let us examine 'seeing that' in these examples. It may be the logical element which connects observing with our knowledge, and with our language.

Of course there are cases where the data are confused and where we may have no clue to guide us. In microscopy one often reports sensations in a phenomenal, lustreless way: 'it is green in this light; darkened areas mark the broad end. . . .' So too the physicist may say: 'the needle oscillates, and there is a faint streak near the neon parabola. Scintillations appear on the periphery of the cathode-scope. . . .' To deny that these are genuine cases of seeing, even observing, would be unsound, just as is the suggestion that they are the *only* genuine cases of seeing.

These examples are, however, overstressed. The language of shapes, colour patches, oscillations and pointer-readings is appropriate to the unsettled experimental situation, where confusion and even conceptual muddle may dominate. The observer may not know what he is seeing: he aims only to get his observations to cohere against a background of established knowledge. This seeing is the goal of observation. It is in these terms, and not in terms of 'phenomenal' seeing, that new inquiry proceeds. Every physicist forced to observe his data as in an oculist's office finds himself in a special, unusual situation. He is obliged to forget what he knows and to watch events like a child. These are non-typical cases, however spectacular they may sometimes be.

First registering observations and then casting about for knowledge of them gives a simple model of how the mind and the eye fit together. The relationship between seeing and the corpus of our knowledge, however, is not a simple one.

What is it to see boxes, staircases, birds, antelopes, bears, goblets, X-ray tubes? It is (at least) to have knowledge of certain sorts. (Robots and electric eyes are blind, however efficiently they react to light. Cameras cannot see.) It is to see that, were certain things done to objects before our eyes, other things would result. How should we regard a man's report that he sees x if we know him to be ignorant of all x-ish things? Precisely as we would regard a four-year-old's report that he sees a meson shower. 'Smith sees x' suggests that Smith could specify some things pertinent to x. To see an X-ray tube is at least to see that, were it dropped on stone, it would smash. To see a goblet is to see something with concave interior. We may be wrong, but not always – not even usually. Besides, deceptions proceed in terms of what is normal, ordinary. Because the world is not a cluster of conjurer's tricks, conjurers can exist. Because the logic of 'seeing that' is an intimate part of the concept of seeing, we sometimes rub our eyes at illusions.

'Seeing as' and 'seeing that' are not components of seeing, as rods and bearings are parts of motors: seeing is not composite. Still, one *can* ask logical questions. What must have occurred, for instance, for us to describe a man as having found a collar stud, or as having seen a bacillus? Unless he had had a visual sensation and knew what a bacillus was (and looked like) we would not say that he had seen a bacillus, except in the sense in which an infant could see a bacillus. 'Seeing as' and 'seeing that', then, are not psychological components of seeing. They are logically distinguishable elements in seeing-talk, in our concept of seeing.

To see Fig. 8 as a transparent box, an ice-cube, or a block of glass is to see that it is six-faced, twelve-edged, eight-cornered. Its corners are solid right angles; if constructed it would be of rigid, or semi-rigid material, not of liquescent or gaseous stuff like oil, vapour or flames. It would be tangible. It would take up space in an exclusive way, being locatable here, there, but at least somewhere. Nor would it cease to exist when we blinked. Seeing it as a cube is just to see that all these things would obtain.

This is knowledge: it is knowing what kind of a thing 'box' or 'cube' denotes and something about what materials can make up such an entity. 'Transparent box' or 'glass cube' would not express what was seen were any of these further considerations denied. Seeing a bird in the sky involves seeing that it will not suddenly do vertical snap rolls; and this is more than marks the retina. We could be wrong. But to see a bird, even momentarily, is to see it in all these connexions. As Wisdom would say, every perception involves an aetiology and a prognosis.[44]

Sense-datum theorists stress how we can go wrong in our observations, as when we call aeroplanes 'birds'. Thus they seek what we are right about, even in these cases. Preoccupation with this problem obscures another one, namely, that of describing what is involved when we are right about what we say we see; and after all this happens very often. His preoccupation with mistakes leads the phenomenalist to portray a world in which we are usually deceived; but the world of physics is not like that. Were a physicist in an ordinary laboratory situation to react to his visual environment with purely sense-datum responses – as does the infant or the idiot – we would think him out of his mind. We would think him *not* to be seeing what was around him.

'Seeing that' threads knowledge into our seeing; it saves us from re-identifying everything that meets our eye; it allows physicists to observe new data as physicists, and not as cameras. We do not ask 'What's that?' of every passing bicycle. The knowledge is there in the seeing and not an adjunct of it. (The pattern of threads is there in the cloth and not tacked on to it by ancillary operations.) We rarely catch ourselves tacking knowledge on to what meets the eye. Seeing this page as having an opposite side requires no squeezing or forcing, yet nothing optical guarantees that when you turn the sheet it will not cease to exist. This is but another way of saying that ordinary seeing is corrigible, which everybody would happily concede. The search for incorrigible seeing has sometimes led some philosophers to deny that anything less than the incorrigible is seeing at all.

Seeing an object *x* is to see that it may behave in the ways we know *x*'s do behave: if the object's behaviour does not accord with what we expect of *x*'s we may be blocked from seeing it as a straightforward *x* any longer. Now we rarely see dolphin as fish, the earth as flat, the heavens as an inverted bowl or the sun as our satellite. '. . . what I perceive as the dawning of an aspect is not a property of the object, but an internal relation between it and other objects.'[45] To see in Fig. 15 an X-ray tube is to see that a photo-sensitive plate placed below it will be irradiated. It is to see that the target will get extremely hot, and as it has no water-jacket it must be made of metal with a high melting-point – molybdenum or tungsten. It is to see that at high voltages green fluorescence will appear at the anode. Could a physicist see an X-ray tube without seeing that these other things would obtain? Could one see something as an incandescent light bulb and fail to see that it is the wire filament which 'lights up' to a white heat? The answer may sometimes be 'yes', but this only indicates that different things can be meant by 'X-ray tube' and 'incandescent bulb'. Two people confronted with an *x* may mean different things by *x*. Must their saying 'I see *x*' mean that they see the same thing? A child could parrot 'X-ray tube', or 'Kentucky' or 'Winston', when confronted with the figure above, but he would not see that these other things followed. And this is what the physicist does see.

If in the brilliant disc of which he is visually aware Tycho sees only the sun, then he cannot but see that it is a body which will behave in characteristically 'Tychonic' ways. These serve as the foundation for Tycho's general geocentric-geostatic theories about the sun. They are not imposed on his visual impressions as a tandem interpretation: they are 'there' in the seeing. (So too the interpretation of a piece of music is there in the music. Where else could it be? It is not something superimposed upon pure, unadulterated sound.)

Similarly we see Fig. 8 as from underneath, as from above, or as a diagram of a rat maze or a gem-cutting project. However construed, the construing is there in the seeing. One is tempted to say 'the construing *is* the seeing'. The thread and its arrangement *is* the fabric, the sound and its composition *is* the music, the colour and its disposition *is* the painting. There are not two operations involved in my seeing Fig. 8 as an ice-cube; I simply see it as an ice-cube. Analogously, the physicist sees an X-ray tube, not by first soaking up reflected light and then clamping on interpretations, but just as you see this page before you.

Tycho sees the sun beginning its journey from horizon to horizon. He sees that from some celestial vantage point the sun (carrying with it the moon and planets) could be watched circling our fixed earth. Watching the sun at dawn through Tychonic spectacles would be to see it in something like this way.

Kepler's visual field, however, has a different conceptual organization. Yet a drawing of what he sees at dawn could be a drawing of exactly what Tycho saw,[46] and could be recognized as such by Tycho. But Kepler will see the horizon dipping, or turning away, from our fixed local star. The shift from sunrise to horizon-turn is analogous to the shift-of-aspect phenomena already considered; it is occasioned by differences between what Tycho and Kepler think they know.

These logical features of the concept of seeing are inextricable and indispensable to observation in research physics. Why indispensable? That men do see in a way that permits analysis into 'seeing as' and 'seeing that' factors is one thing; 'indispensable', however, suggests that the world must be seen thus. This is a stronger claim, requiring a stronger argument. Let us put it differently: that observation in physics is not an encounter with unfamiliar and unconnected flashes, sounds and bumps, but rather a calculated meeting with these as flashes, sounds and bumps of a particular kind – this might figure in an account of what observation is. It would not secure the point that observation could not be otherwise. This latter type of argument is now required: it must establish that an alternative account would be not merely false, but absurd. To this I now turn.

D

Fortunately, we do not see the sun and the moon as we see the points of colour and light in the oculist's office; nor does the physicist see his laboratory equipment, his desk, or his hands in the baffled way that he may view a cloud-chamber photograph or an oscillograph pattern. In most cases we could give further information about what sort of thing we see. This might be expressed in a list: for instance, that *x* would break if dropped, that *x* is hollow, and so on.

To see Fig. 10 as a bear on a tree is to see that further observations are possible; we can imagine the bear as viewed from the side or from behind. Indeed, seeing Fig. 10 as a bear is just to have seen that these other views could all be simultaneous. It is also to see that certain observations are not possible: for example, the bear cannot be waving one paw in the air, nor be dangling one foot. This too is 'there' in the seeing.

'Is it a question of both seeing and thinking? or an amalgam of the two, as I should almost like to say?'[47] Whatever one would like to say, there is more to seeing Fig. 10 as a bear, than optics, photochemistry or phenomenalism can explain.[48]

Notice a logical feature: 'see that' and 'seeing that' are always followed by 'sentential' clauses. The addition of but an initial capital letter and a full stop sets them up as independent sentences. One can see an ice-cube, or see a kite as a bird. One cannot see that an ice-cube, nor see that a bird. Nor is this due to limitations of vision. Rather, one may see that *ice-cubes can melt*; that *birds have 'hollow' bones*. Tycho and Simplicius see that *the universe is geocentric*; Kepler and Galileo see that *it is heliocentric*. The physicist sees that *anode-fluorescence will appear in an X-ray tube at high voltages*. The phrases in italics are complete sentential units.

Pictures and statements differ in logical type, and the steps between visual pictures and the statements of what is seen are many and intricate. Our visual consciousness is dominated by pictures; scientific knowledge, however, is primarily linguistic. Seeing is, as I should almost like to say, an amalgam of the two – pictures and language. At the least, the concept of seeing embraces the concepts of visual sensation and of knowledge.[49]

The gap between pictures and language locates the logical function of 'seeing that'. For vision is essentially pictorial, knowledge fundamentally linguistic. Both vision and knowledge are indispensable elements in seeing; but differences between pictorial and linguistic representation may mark differences between the optical and conceptual features of seeing. This may illuminate what 'seeing that' consists in.

Not all the elements of statement correspond to the elements of pictures: only someone who misunderstood the uses of language would expect otherwise.[50] There is a 'linguistic' factor in seeing, although there is nothing linguistic about what forms in the eye, or in the mind's eye. Unless there were this linguistic element, nothing we ever observed could have relevance for our knowledge. We could not speak of significant observations: nothing seen would make sense, and microscopy would only be a kind of kaleidoscopy. For what is it for things to make sense other than for descriptions of them to be composed of meaningful sentences?

We must explore the gulf between pictures and language, between sketching and describing, drawing and reporting. Only by showing how picturing and speaking are different can one suggest how 'seeing that' may bring them together; and brought together they must be if observations are to be *significant* or *noteworthy*.

Knowledge here is of what there is, as factually expressed in books, reports, and essays. How to do things is not our concern. I know how to whistle; but could I express that knowledge in language? Could I describe the taste of salt, even though I know perfectly well how salt tastes? I know how to control a parachute – much of that knowledge was imparted in lectures and drills, but an essential part of it was not *imparted* at all; it was 'got on the spot'. Physicists rely on 'know-how', on the 'feel' of things and the 'look' of a situation, for these control the direction of research. Such imponderables, however, rarely affect the corpus of physical truths. It is not Galileo's insight, Newton's genius and Einstein's imagination which have per se changed our knowledge of what there is: it is the true things they have said. 'Physical knowledge', therefore, will mean 'what is reportable in the texts, reports and discussions of physics.' We are concerned with *savoir*, not *savoir faire*.[51]

The 'foundation' of the language of physics, the part closest to mere sensation, is a series of statements. Statements are true or false. Pictures are not at all like statements: they are neither true nor false; retinal, cortical, or sense-datum pictures are neither true nor false. Yet what we see can determine whether statements like 'The sun is above the horizon' and 'The cube is transparent', are true or false. Our visual sensations may be 'set' by language forms; how else could they be appreciated in terms of what we know? Until they *are* so appreciated they do not constitute observation: they are more like the buzzing confusion of fainting or the vacant vista of aimless staring through a railway window.[52] Knowledge of the world is not a *montage* of sticks, stones, colour patches and noises, but a system of propositions.

Fig. 15, p. 439 asserts nothing. It could be inaccurate, but it could not be a lie. This is the wedge between pictures and language.

Significance, relevance – these notions depend on what we already know. Objects, events, pictures, are not intrinsically significant or relevant. If seeing were just an optical-chemical process, then nothing we saw would ever be relevant to what we know, and nothing known could have significance for what we see. Visual life would be unintelligible; intellectual life would lack a visual aspect. Man would be a blind computer harnessed to a brainless photoplate.[53]

Pictures often copy originals. All the elements of a copy, however, have the same kind of function. The lines depict elements in the original. The arrangement of the copy's elements shows the disposition of elements in the original. Copy and original are of the same logical type; you and your reflexion are of the same type. Similarly, language might copy what it describes.[54]

Consider Fig. 10 alongside 'The bear is on the tree'. The picture contains a bear-element and a tree-element. If it is true to life, then in the original there is a bear and a tree. If the sentence is true *of* life, then (just as it contains 'bear' and 'tree') the situation it describes contains a bear and a tree. The picture combines its elements, it mirrors the actual relation of the bear and the tree. The sentence likewise conjoins 'bear' and 'tree' in the schema 'The —is on the —'. This verbal relation signifies the actual relation of the real bear and the real tree. Both picture and sentence are true copies: they contain nothing the original lacks, and lack nothing the original contains. The elements of the picture stand for (represent) elements of the original: so do 'bear' and 'tree'. This is more apparent when expressed symbolically as $b\ R\ t$, where b = bear, t = tree and R = the relation of being on.

By the arrangement of their elements these copies show the arrangement in the original situation. Thus Fig. 10, 'The bear is on the tree', and 'bRt' show what obtains with the real bear and the real tree; while 'The tree is on the bear', and 'tRb', and a certain obvious cluster of lines do not show what actually obtains.

The copy is of the same type as the original. We can sketch the bear's teeth, but not his growl, any more than we could see the growl of the original bear. Leonardo could draw Mona Lisa's smile, but not her laugh. Language, however, is more versatile. Here is a dissimilarity between picturing and asserting which will grow to fracture the account once tendered by Wittgenstein, Russell and Wisdom. Language can encapsulate scenes and sounds, teeth and growls, smiles and laughs; a picture, or a gramophone, can do one or the other, but not both. Pictures and recordings stand for things by possessing certain properties of the original itself. Images, reflections, pictures and maps duplicate the spatial properties of what they image, reflect, picture or map; gramophone recordings duplicate audio-temporal properties. Sentences are not like this. They do not stand for things in virtue of possessing properties of the original; they do not *stand for* anything. They can state what is, or could be, the case. They can be used to make assertions, convey descriptions, supply narratives, accounts, etc., none of which depend on the possession of some property in common with what the statement is about. We need not write 'THE BEAR is bigger than ITS CUB' to show our meaning.

Images, reflexions, pictures and maps in fact copy originals with different degrees of strictness. A reflexion of King's Parade does not copy in the same sense that a charcoal sketch does, and both differ from the representation of 'K.P.' on a map of Cambridge and from a town-planner's drawing. The more like a reflection a map becomes, the less useful it is as a map. Drawings are less like copies of originals than are photographs. Of a roughly sketched ursoid shape one says either 'That's a bear' or 'That's supposed to be a bear'. Similarly with maps. Of a certain dot on the map one says either 'This is Cambridge' or 'This stands for Cambridge'.

Language copies least of all. There are exceptional words like 'buzz', 'tinkle' and 'toot', but they only demonstrate how conventional our languages and notations are. Nothing about 'bear' looks like a bear; nothing in the sound of 'bear' resembles a growl. That b-e-a-r can refer to bears is due to a convention which co-ordinates the word with the object. There is nothing dangerous about a red flag, yet it is a signal for danger. Of Fig. 10 we might say 'There is a bear'. We would never say this of the word 'bear'. At the cinema we say 'It's a bear', or 'There's K.P.' – not 'That stands for a bear', or 'That denotes K.P.' It is words that denote; but they are rarely like what they denote.

Sentences do not show, for example, bears climbing trees, but they can state that bears climb trees. Showing the sun climbing into the sky consists in representing sun and sky and arranging them appropriately. Stating that the sun is climbing into the sky consists in referring to the sun and then characterizing it as climbing into the sky. The differences between representing and referring, between arranging and characterizing – these are the differences between picturing and language-using.

These differences exist undiminished between visual sense-data and basic sentences. Early logical constructionists were inattentive to the difficulties in fitting visual sense-data to basic sentences. Had they heeded the differences between pictures and maps, they might have detected greater differences still between pictures and language. One's visual awareness of a brown ursoid patch is logically just as remote from the utterance '(I am aware of a) brown, ursoid patch now', as with any of the pictures and sentences we have considered. The picture is of *x*; the statement is to the effect that *x*. The picture shows *x*; the statement refers to and describes *x*. The gap between pictures and language is not closed one millimetre by focusing on sense-data and basic sentences.

The prehistory of languages need not detain us. The issue concerns differences between *our* languages and *our* pictures, and not the smallness of those differences at certain historical times. Wittgenstein is misleading about this: '. . . and from [hieroglyphic writing] came the alphabet without the essence of representation being lost.'[55] This strengthened the picture theory of meaning, a truth-functional account of language and a theory of atomic sentences. But unless the essence of representation had been lost, languages could not be used in speaking the truth, telling lies, referring and characterizing.

Not all elements of a sentence do the same work. All the elements of pictures, however, just represent.[56] A picture of the dawn could be cut into small pictures, but sentences like 'The sun is on the horizon' and 'I perceive a solaroid patch now' cannot be cut into small sentences. All the elements of the picture show something; none of the elements of the sentences state anything. 'Bear!' may serve as a statement, as may 'Tree!' from the woodcutter, or 'Sun!' in eclipse-observations. But 'the', 'is' and 'on' are not likely ever to behave as statements.

Pictures are of the picturable. Recordings are of the recordable. You cannot play a smile or a wink on the gramophone. But language is more versatile: we can describe odours, sounds, feels, looks, smiles and winks. This freedom makes type-mistakes possible: for example, 'They found his pituitary but not his mind', 'We surveyed his retina but could not find his sight'. Only when we are free from the natural limitations of pictures and recordings can such errors occur. They are just possible in maps; of the hammer and sickle which signifies Russia on a school map, for instance, a child might ask 'How many miles long is the sickle?'. Maps with their partially conventional characters must be read (as pictures and photographs need not be); yet they must copy.

There is a corresponding gap between visual pictures and what we know. Seeing bridges this, for while seeing is at the least a 'visual copying' of objects, it is also more than that. It is a certain sort of seeing of objects: seeing that if x were done to them y would follow. This fact got lost in all the talk about knowledge arising from sense experience, memory, association and correlation. Memorizing, associating, correlating and comparing mental pictures may be undertaken *ad indefinitum* without one step having been taken towards scientific knowledge, that is, propositions known to be true. How long must one shuffle photographs, diagrams and sketches of antelopes before the statement 'antelopes are ungulates' springs forth?

When language and notation are ignored in studies of observation, physics is represented as resting on sensation and low-grade experiment. It is described as repetitious, monotonous concatenation of spectacular sensations, and of school-laboratory experiments. But physical science is not just a systematic exposure of the senses to the world; it is also a way of thinking about the world, a way of forming conceptions. The paradigm observer is not the man who sees and reports what all normal observers see and report, but the man who sees in familiar objects what no one else has seen before.[57]

Notes

1 Wär' nicht das Auge sonnenhaft,
Die Sonne könnt' es nie erblicken;
Goethe, *Zahme Xenien* (Werke, Weimar, 1887–1918), Bk. 3, 1805.

2 Cf. the papers by Baker and Gatonby in *Nature*, 1949–present.

3 This is not a *merely* conceptual matter, of course. Cf. Wittgenstein, *Philosophical Investigations* (Blackwell, Oxford, 1953), p. 196.

4 (1) G. Berkeley, *Essay Towards a New Theory of Vision* (in *Works*, vol. I (London, T. Nelson, 1948–56)), pp. 51 ff.
(2) James Mill, *Analysis of the Phenomena of the Human Mind* (Longmans, London, 1869), vol. I, p. 97.
(3) J. Sully, *Outlines of Psychology* (Appleton, New York, 1885).
(4) William James, *The Principles of Psychology* (Holt, New York, 1890–1905), vol. II, pp. 4, 78, 80 and 81; vol. I, p. 221.
(5) A. Schopenhauer, *Satz vom Grunde* (in *Sämmtliche Werke*, Leipzig, 1888), ch. IV.
(6) H. Spencer, *The Principles of Psychology* (Appleton, New York, 1897), vol. IV, chs. IX, IX.
(7) E. von Hartmann, *Philosophy of the Unconscious* (K. Paul, London, 1931), B, chs. VII, VIII.
(8) W. M. Wundt, *Vorlesungen über die Menschen und Thierseele* (Voss, Hamburg, 1892), IV, XIII.
(9) H. L. F. von Helmholtz, *Handbuch der Physiologischen Optik* (Leipzig, 1867), pp. 430, 447.
(10) A. Binet, *La psychologie du raisonnement, recherches expérimentales par l'hypnotisme* (Alcan, Paris, 1886), chs. III, V.
(11) J. Grote, *Exploratio Philosophica* (Cambridge, 1900), vol. II, pp. 201 ff.
(12) B. Russell, in *Mind* (1913), p. 76. *Mysticism and Logic* (Longmans, New York, 1918), p. 209. *The Problems of Philosophy* (Holt, New York, 1912), pp. 73, 92, 179, 203.
(13) Dawes Hicks, *Arist. Soc. Sup.* vol. II (1919), pp. 176–8.
(14) G. F. Stout, *A Manual of Psychology* (Clive, London, 1907, 2nd ed.), vol. II, 1 and 2, pp. 324, 561–4.
(15) A. C. Ewing, *Fundamental Questions of Philosophy* (New York, 1951), pp. 45 ff.
(16) G. W. Cunningham, *Problems of Philosophy* (Holt, New York, 1924), pp. 96–7.

5 Galileo, *Dialogue Concerning the Two Chief World Systems* (California, 1953), 'The First Day', p. 33.

6 '"Das ist doch kein Sehen!" – "Das ist doch ein Sehen!" Beide müssen sich begrifflich rechtfertigen lassen' (Wittgenstein, *Phil. Inv.* p. 203).

7 Brain, *Recent Advances in Neurology* (with Strauss) (London, 1929), p. 88. Compare Helmholtz: 'The sensations are signs to our consciousness, and it is the task of our intelligence to learn to understand their meaning' (*Handbuch der Physiologischen Optik* (Leipzig, 1867), vol. III, p. 433).

See also Husserl, 'Ideen zu einer Reinen Phaenomenologie', in *Jahrbuch für Philosophice*, vol. I (1913), pp. 75, 79, and Wagner's *Handwörterbuch der Physiologie*, vol. III, section 1 (1846), p. 183.

8 Mann, *The Science of Seeing* (London, 1949), pp. 48–9. Arber, *The Mind and the Eye* (Cambridge, 1954). Compare Müller: 'In any field of vision, the retina sees only itself in its spatial extension during a state of affection. It perceives itself as . . . etc.' (*Zur vergleichenden Physiologie des Gesichtesinnes des Menschen und der Thiere* (Leipzig, 1826), p. 54).

9 Kolin: 'An astigmatic eye when looking at millimeter paper can accommodate to see sharply either the vertical lines or the horizontal lines' (*Physics* (New York, 1950), pp. 570 ff.).

10 Cf. Whewell, *Philosophy of Discovery* (London, 1860), 'The Paradoxes of Vision'.

11 Cf. e.g. J. Z. Young, *Doubt and Certainty in Science* (Oxford, 1951, The Reith Lectures), and Gray Walter's article in *Aspects of Form*, ed. by L. L. Whyte (London, 1953). Compare Newton: 'Do not the Rays of Light in falling upon the bottom of the Eye excite Vibrations in the Tunica Retina? Which Vibrations, being propagated along the solid Fibres of the Nerves into the Brain, cause the Sense of seeing' (*Opticks* (London, 1769), Bk. III, part I).

12 'Rot und grün kann ich nur sehen, aber nicht hören' (Wittgenstein, *Phil. Inv.* p. 209).

13 Cf. 'An appearance is the same whenever the same eye is affected in the same way' (Lambert, *Photometria* (Berlin, 1760)); 'We are justified, when different perceptions offer themselves to us, to infer that the underlying real conditions are different' (Helmholtz, *Wissenschaftliche Abhandlungen* (Leipzig, 1882), vol. II, p. 656), and Hertz: 'We form for ourselves images or symbols of the external objects; the manner in which we form them is such that the logically necessary (*denknotwendigen*) consequences of the images in thought are invariably the images of materially necessary (*naturnotwendigen*) consequences of the corresponding objects' (*Principles of Mechanics* (London, 1889), p. 1).

Broad and Price make depth a feature of the private visual pattern. However, Weyl (*Philosophy of Mathematics and Natural Science* (Princeton, 1949), p. 125) notes that a single eye perceives qualities spread out in a *two*-dimensional field, since the latter is dissected by any one-dimensional line running through it. But our conceptual difficulties remain even when Kepler and Tycho keep one eye closed.

Whether or not two observers are having the same visual sense-data reduces directly to the question of whether accurate pictures of the contents of their visual fields are identical, or differ in some detail. We can then discuss the publicly observable pictures which Tycho and Kepler draw of what they see, instead of those private, mysterious entities locked in their visual consciousness. The accurate picture and the sense-datum must be identical; how could they differ?

14 From the BBC report, 30 June 1954.

15 Newton, *Opticks*, Bk. II, part I. The writings of Claudius Ptolemy sometimes read like a phenomenalist's textbook. Cf. e.g. *The Almagest* (Venice, 1515), VI, section 11, 'On the Directions in the Eclipses', 'When it touches the shadow's circle from within', 'When the circles touch each other from without'. Cf. also VII and VIII, IX (section 4). Ptolemy continually seeks to chart and predict 'the appearances' – the points of light on the celestial globe. *The Almagest* abandons any attempt to explain the machinery behind these appearances.

Cf. Pappus: 'The (circle) dividing the milk-white portion which owes its colour to the sun, and the portion which has the ashen colour natural to the moon itself is indistinguishable from a great circle' (*Mathematical Collection* (Hultsch, Berlin and Leipzig, 1864), pp. 554–60).

16 This famous illusion dates from 1832, when L. A. Necker, the Swiss naturalist, wrote a letter to Sir David Brewster describing how when certain rhomboidal crystals were viewed on end the perspective could shift in the way now familiar to us. Cf. *Phil. Mag.* III, no. 1 (1832), 329–37, especially p. 336. It is important to the present argument to note that this observational phenomenon began life not as a psychologist's trick, but at the very frontiers of observational science.

17 Wittgenstein answers: 'Denn wir sehen eben wirklich zwei verschiedene Tatsachen' (*Tractatus*, 5. 5423).

18 'Auf welche Vorgänge spiele ich an?' (Wittgenstein, *Phil. Inv.* p. 214).

19 *Ibid.* p. 194 (top).

20 *Ibid.* p. 200.

21 This is *not* due to eye movements, or to local retinal fatigue. Cf. Flugel, *Brit. J. Psychol.* VI (1913), 60; *Brit. J. Psychol.* V (1913), 357. Cf. Donahue and Griffiths, *Amer. J. Psychol.* (1931), and Luckiesh,

Visual Illusions and their Applications (London, 1922). Cf. also Peirce, *Collected Papers* (Harvard, 1931), 5, 183. References to psychology should not be misunderstood; but as one's acquaintance with the psychology of perception deepens, the character of the conceptual problems one regards as significant will deepen accordingly. Cf. Wittgenstein, *Phil. Inv.* p. 206 (top). Again, p. 193: 'Its causes are of interest to psychologists. We are interested in the concept and its place among the concepts of experience.'

22 Wittgenstein, *Phil. Inv.* p. 212.

23 From Boring, *Amer. J. Psychol.* XLII (1930), 444 and cf. Allport, *Brit. J. Psychol*, XXI (1930), 133; Leeper, *J. Genet. Psychol.* XLVI (1935), 41; Street, *Gestalt Completion Test* (Columbia Univ., 1931); Dees and Grindley, *Brit. J. Psychol.* (1947).

24 Köhler, *Gestalt Psychology* (London, 1929). Cf. his *Dynamics in Psychology* (London, 1939).

25 'Mein Gesichteseindruck hat sich geändert; – wie war er früher; wie ist er jetzt? – Stelle ich ihn durch eine genaue Kopie dar – und ist das keine gute Darstellung? – so zeigt sich keine Änderung' (Wittgenstein, *Phil. Inv.* p. 196).

26 'Was gezeigt werden kann, kann nicht gesagt werden' (Wittgenstein, *Tractatus*, 4. 1212).

27 This case is different from fig. 8. Now I can help a 'slow' percipient by tracing in the outline of the bear. In fig. 8 a percipient either gets the perspectival arrangement, or he does not, though even here Wittgenstein makes some suggestions as to how one might help; cf. *Tractatus*, 5. 5423, last line,

28 Wittgenstein, *Phil. Inv.* p. 193. Helmholtz speaks of the 'integrating' function which converts the figure into the appearance of an object hit by a visual ray (*Phys. Optik*, vol. III, p. 239). This is reminiscent of Aristotle, for whom seeing consisted in emanations from our eyes. They reach out, tentacle-fashion, and touch objects whose shapes are 'felt' in the eye. (Cf. *De Caelo* (Oxford, 1928), 290a, 18; and *Meteorologica* (Oxford, 1928), III, iv, 373 b, 2. (Also Plato, *Meno*, London, 1869), 76C-D.) But he controverts this in *Topica* (Oxford, 1928), 105 b, 6.) Theophrastus argues that 'Vision is due to the gleaming . . . which [in the eye] reflects to the object' (*On the Senses*, 26, trans. G. M. Stratton). Hero writes: 'Rays proceeding from our eyes are reflected by mirrors . . . that our sight is directed in straight lines proceeding from the organ of vision may be substantiated as follows' (*Catoptrics*, 1–5, trans. Schmidt in *Heronis Alexandrini Opera* (Leipzig, 1899–1919)). Galen is of the same opinion. So too is Leonardo: 'The eye sends its image to the object . . . the power of vision extends by means of the visual rays . . .' (*Notebooks*, C.A. 135 v.b. and 270 v.c). Similarly Donne in *The Ecstasy* writes: 'Our eyebeams twisted and . . . pictures in our eyes to get was all *our* propagation.'

This is the view that all perception is really a species of touching, e.g. Descartes' *impressions*, and the analogy of the wax. Compare: '[Democritus] explains [vision] by the air between the eye and the object [being] compressed . . . [it] thus becomes imprinted . . . "as if one were to take a mould in wax" . . .' Theophrastus (*op. cit.* 50–3). Though it lacks physical and physiological support, the view is attractive in cases where lines seem suddenly to be forced into an intelligible pattern–by us.

29 *Ibid.* p. 193. Cf. Helmholtz, *Phys. Optik*, vol. III, pp. 4, 18 and Fichte (*Bestimmung des Menschen*, ed. Medicus (Bonn, 1834), vol. III, p. 326). Cf. also Wittgenstein, *Tractatus*, 2. 0123.

30 P. B. Porter, *Amer. J. Psychol.* LXVII (1954), 550.

31 Writings by Gestalt psychologists on 'set' and 'Aufgabe' are many. Yet they are overlooked by most philosophers. A few fundamental papers are: Külpe, *Ber. I Kongress Exp. Psychol., Giessen* (1904); Bartlett, *Brit. J. Psychol.* VIII (1916), 222; George, *Amer. J. Psychol.* XXVIII (1917), 1; Fernberger, *Psychol. Monogr.* XXVI (1919), 6; Zigler, *Amer. J. Psychol*, XXXI (1920), 273; Boring, *Amer. J. Psychol*, XXXV (1924), 301; Wilcocks, *Amer. J. Psychol*, XXXVI (1925), 324; Gilliland, *Psychol. Bull*, XXIV (1927), 622; Gottschaldt, *Psychol. Forsch.* XII (1929), 1; Boring, *Amer. J. Psychol*, XLII (1930), 444; Street, *Gestalt Completion Test* (Columbia University, 1931); Ross and Schilder, *J. Gen. Psychol.* X (1934), 152; Hunt, *Amer. J. Psychol*, XLVII (1935), 1; Süpola, *Psychol. Monogr.* XLVI (1935), 210, 27; Gibson, *Psychol. Bull.* XXXVIII (1941), 781; Henle, *J. Exp. Psychol.* XXX (1942), 1; Luchins, *J. Soc. Psychol*, XXI (1945), 257; Wertheimer, *Productive Thinking* (1945); Russell Davis and Sinha, *Quart, J. Exp. Psychol.* (1950); Hall, *Quart. J. Exp. Psychol.* II (1950), 153.

Philosophy has no concern with fact, only with conceptual matters (cf. Wittgenstein, *Tractatus*, 4. 111); but discussions of perception could not but be improved by the reading of these twenty papers.

32 Often 'What do you see?' only poses the question 'Can you identify the object before you?'. This is calculated more to test one's knowledge than one's eyesight.

33 Duhem, *La théorie physique* (Paris, 1914), p. 218.

34 Chinese poets felt the significance of 'negative features' like the hollow of a clay vessel or the central vacancy of the hub of a wheel (cf. Waley, *Three Ways of Thought in Ancient China* (London, 1939), p. 155).

35 Infants are indiscriminate; they take in spaces, relations, objects and events as being of equal value. They still must learn to organize their visual attention. The camera-clarity of their visual reactions is not by itself sufficient to differentiate elements in their visual fields. Contrast Mr W. H. Auden who recently said of the poet that he is 'bombarded by a stream of varied sensations which would drive him mad if he took them all in. It is impossible to guess how much energy we have to spend every day in not-seeing, not-hearing, not-smelling, not-reacting.'

36 Cf. 'He was blind to the *expression* of a face. Would his eyesight on that account be defective?' (Wittgenstein, *Phil. Inv.* p. 210) and 'Because they seeing see not; and hearing they hear not, neither do they understand' (Matt. xiii. 10–13).

37 'Es hört doch jeder nur, was er versteht' (Goethe, *Maxims* (*Werke*, Weimar, 1887–1918)).

38 Against this Professor H. H. Price has argued: 'Surely it appears to both of them to be rising, to be moving upwards, across the horizon . . . they both see a moving sun: they both see a round bright body which appears to be rising.' Philip Frank retorts: 'Our sense observation shows only that in the morning the distance between horizon and sun is increasing, but it does not tell us whether the sun is ascending or the horizon is descending . . .' (*Modern Science and its Philosophy* (Harvard, 1949), p. 231). Precisely. For Galileo and Kepler the horizon drops; for Simplicius and Tycho the sun rises. This is the difference Price misses, and which is central to this essay.

39 This parallels the too-easy epistemological doctrine that all normal observers see the same things in *x*, but interpret them differently.

40 Cf. the important paper by Carmichael, Hogan and Walter, 'An Experimental Study of the Effect of Language on the Reproduction of Visually Perceived Form', *J. Exp. Psychol.* XV (1932), 73–86. (Cf. also Wulf, *Beiträge zur Psychologic der Gestalt.* VI. 'Über die Veränderung von Vorstellungen (Gedächtnis und Gestalt).' *Psychol. Forsch.* I (1921), 333–73.) Cf. also Wittgenstein, *Tractatus*, 5. 6; 5. 61.

41 Wittgenstein, *Phil. Inv.* p. 206.

42 '"Seeing as . . ." is not part of perception. And for that reason it is like seeing and again not like' (*ibid.* p. 197).

43 'All seeing is seeing as . . . if a person sees something at all it must look like something to him . . .' (G. N. A. Vesey, 'Seeing and Seeing As', *Proc. Aristotelian Soc.* (1956), p. 114.)

44 'Is the pinning on of a medal merely the pinning on of a bit of metal?' (Wisdom, 'Gods', *Proc. Aristotelian Soc.* (1944–5)).

45 Wittgenstein, *Phil. Inv.* p. 212. Cf. *Tractatus* 2. 0121. Cf. Also Helmholtz, *Phys. Optik*, vol. III, p. 18.

46 Drawn on grid paper the two visual pictures could be geometrically identical. Cf. 'If the two different 'appearances' of a reversible figure were indeed things ('pictures') we could conceive of them projected out from our minds, on to a screen, side by side, and distinguishable. But the only images on a screen which could serve as projections of the two different 'appearances' would be identical' (G. N. A. Vesey, 'Seeing and Seeing As', *Proc. Aristotelian Soc.* (1956)).

47 Wittgenstein, *Phil. Inv.* p. 197.

48 'Of the sense-data we cannot know more . . . than that they are in agreement with one another' (Leibniz, *Die Philosophische Schriften* (Berlin, 1875–90), vol. IV, p. 356).

49 We speak of 'phototropism' in flatworms, but not seeing. (If dogs talked, Descartes might not have regarded them as blind machines.)

50 In their logical-construction, 'picture-theory' periods, Russell (*Logical Atomism* (Minnesota, 1950)), Wittgenstein (*Tractatus*) and Wisdom ('Logical Constructions' (*Mind*, 1931–4)) must fall into this class.

51 '"Knowing" it means only: being able to describe it' (Wittgenstein, *Phil. Inv.* p. 185).

52 'I looked at the flower, but was thinking of something else and was not conscious of its colour . . . [I] looked at it without seeing it . . .' (*ibid.* p. 211). The history of physics supplies further examples, cf. chs. II, IV and VI.

53 Cf. Kant: 'Intuition without concepts is blind. . . . Concepts without intuition are empty.' Indeed how is 'interpretation' of a *pure* visual sense-datum possible?

54 Cf. Wittgenstein, *Tractatus*, 2. 1–2, 2 and 3–3. 1.

55 *Ibid.* 4. 016.

56 Thus the pattern of a picture is not another element of the picture. The difference between the bird and the antelope is like that between *bRt* and *tRb*. We may see different things in the same visual elements; just as when you say '*bRt*' and I say '*tRb*' we have said different things with the same words. Cf. Wittgenstein, *Tractatus*, 3. 141: 'Der Satz ist kein Wörtengemisch. – (Wie das musikalische Thema kein Gemisch von Tönen.)'

57 '"Natural Philosophy" . . . lies not in discovering facts, but in discovering new ways of thinking about them. The test which we apply to these ideas is this – do they enable us to fit the facts to each other?' Bragg, 'The Atom', in *The History of Science* (London, 1948), p. 167.

'Orderly arrangement is the task of the scientist. A science is built out of facts just as a house is built out of bricks. But a mere collection of facts cannot be called a science any more than a pile of bricks can be called a house' (Poincaré, *Foundations of Science* (Science Press, Lancaster, Pa., 1946), p. 127). 'An object is frequently not *seen from not knowing how to see it*, rather than from any defect in the organ of vision . . . [Herschel said] "I will prepare the apparatus, and put you in such a position that [Fraunhofer's dark lines] shall be visible, and yet you shall look for them and not find them: after which, while you remain in the same position, I will instruct you *how to see them*, and you shall see them, and not merely wonder you did not see them before, but you shall find it impossible to look at the spectrum without seeing them"' (Babbage, *The Decline of Science in England* (R. Clay, London, 1830)).

Figure acknowledgments

Figures 8, 9, 10, 11, 12, 13, 14, and 15: from Hanson, selections from "On Observation," *Patterns of Discovery: An Inquiry into the Conceptual Foundations of Science* (Cambridge University Press, 1965), pp. 4–25; pp. 9, 11, 12, 13, 14, 15.

6.7

The Ontological Status of Theoretical Entities

Grover Maxwell

The positivists (and many anti-realists, both historical and contemporary) assume that a clear distinction can be made between those entities that are observable (chairs, the Eiffel Tower, a thermometer) from those that are not (electrons, electromagnetic waves, tectonic plates). In the selection below, Grover Maxwell (1918–1981) challenges this claim by presenting three arguments that purport to show that the observable/unobservable object distinction is vague and arbitrary.

That anyone today should seriously contend that the entities referred to by scientific theories are only convenient fictions, or that talk about such entities is translatable without remainder into talk about sense contents or everyday physical objects, or that such talk should be regarded as belonging to a mere calculating device and, thus, without cognitive content – such contentions strike me as so incongruous with the scientific and rational attitude and practice that I feel this paper *should* turn out to be a demolition of straw men. But the instrumentalist views of outstanding physicists such as Bohr and Heisenberg are too well known to be cited, and in a recent book of great competence, Professor Ernest Nagel concludes that "the opposition between [the realist and the instrumentalist] views [of theories] is a conflict over preferred modes of speech" and "the question as to which of them is the 'correct position' has only terminological interest."[1] The phoenix, it seems, will not be laid to rest.

The literature on the subject is, of course, voluminous, and a comprehensive treatment of the problem is far beyond the scope of one essay. I shall limit myself to a small number of constructive arguments (for a radically realistic interpretation of theories) and to a critical examination of some of the more crucial assumptions (sometimes

From Maxwell, "The Ontological Status of Theoretical Entities," from *Minnesota Studies in The Philosophy of Science*, vol. III; *Scientific Explanation, Space, and Time*, ed. by Herbert Feigl and Grover Maxwell (University of Minnesota Press 1962), pp. 3–27.

tacit, sometimes explicit) that seem to have generated most of the problems in this area.[2]

The Problem

Although this essay is not comprehensive, it aspires to be fairly self-contained. Let me, therefore, give a pseudohistorical introduction to the problem with a piece of science fiction (or fictional science).

In the days before the advent of microscopes, there lived a Pasteur-like scientist whom, following the usual custom, I shall call Jones. Reflecting on the fact that certain diseases seemed to be transmitted from one person to another by means of bodily contact or by contact with articles handled previously by an afflicted person, Jones began to speculate about the mechanism of the transmission. As a "heuristic crutch," he recalled that there is an obvious *observable* mechanism for transmission of certain afflictions (such as body lice), and he postulated that all, or most, infectious diseases were spread in a similar manner but that in most cases the corresponding "bugs" were too small to be seen and, possibly, that some of them lived inside the bodies of their hosts. Jones proceeded to develop his theory and to examine its testable consequences. Some of these seemed to be of great importance for preventing the spread of disease.

After years of struggle with incredulous recalcitrance, Jones managed to get some of his preventative measures adopted. Contact with or proximity to diseased persons was avoided when possible, and articles which they handled were "disinfected" (a word coined by Jones) either by means of high temperatures or by treating them with certain toxic preparations which Jones termed "disinfectants." The results were spectacular: within ten years the death rate had declined 40 per cent. Jones and his theory received their well-deserved recognition.

However, the "crobes" (the theoretical term coined by Jones to refer to the disease-producing organisms) aroused considerable anxiety among many of the philosophers and philosophically inclined scientists of the day. The expression of this anxiety usually began something like this: "In order to account for the facts, Jones must assume that his crobes are too small to be seen. Thus the very postulates of his theory preclude their being observed; they are *unobservable in principle*." (Recall that no one had envisaged such a thing as a microscope.) This common prefatory remark was then followed by a number of different "analyses" and "interpretations" of Jones' theory. According to one of these, the tiny organisms were merely convenient fictions – *façons de parler* – extremely useful as heuristic devices for facilitating (in the "context of discovery") the thinking of scientists but not to be taken seriously in the sphere of cognitive knowledge (in the "context of justification"). A closely related view was that Jones' theory was merely an instrument, useful for organizing observation statements and (thus) for producing desired results, and that, therefore, it made no more sense to ask what was the nature of the entities to which it referred than it did to ask what was the nature of the entities to which a hammer or any other tool referred.[3] "Yes," a philosopher might have said, "Jones' theoretical expressions are just meaningless sounds or marks on paper which, when correlated with observation sentences by appropriate syntactical rules, enable us to predict successfully and otherwise organize data in a convenient fashion." These philosophers called themselves "instrumentalists."

According to another view (which, however, soon became unfashionable), although expressions containing Jones' theoretical terms were genuine sentences, they were translatable without remainder into a set (perhaps infinite) of observation sentences. For example. 'There are crobes of disease X on this article' was said to translate into something like this: 'If a person handles this article without taking certain precautions, he will (probably) contract disease X; and if this article is first raised to a high temperature, then if a person handles it at any time afterward, before it comes into contact with another person with disease X, he will (probably) not contract disease X; and . . .'

Now virtually all who held any of the views so far noted granted, even insisted, that theories played a useful and legitimate role in the scientific enterprise. Their concern was the elimination of "pseudo problems" which might arise, say, when one began wondering about the "reality of supra-empirical entities," etc. However, there was also a school of thought, founded by a psychologist named Pelter, which differed in an interesting manner from such positions as these. Its members held that while Jones' crobes might very well exist and enjoy "full-blown reality," they should not be the concern of medical research at all. They insisted that if Jones had employed the correct methodology, he would have discovered, even sooner and with much less effort, all of the observation laws relating to disease contraction, transmission, etc. without introducing superfluous links (the crobes) into the causal chain.

Now, lest any reader find himself waxing impatient, let me hasten to emphasize that this crude parody is not intended to convince anyone, or even to cast serious doubt upon sophisticated varieties of any of the reductionistic positions caricatured (some of them not too severely, I would contend) above. I am well aware that there are theoretical entities and theoretical entities, some of whose conceptual and theoretical statuses differ in important respects from Jones' crobes. (I shall discuss some of these later.) Allow me, then, to bring the Jonesean prelude to our examination of observability to a hasty conclusion.

Now Jones had the good fortune to live to see the invention of the compound microscope. His crobes were "observed" in great detail, and it became possible to identify the specific kind of *microbe* (for so they began to be called) which was responsible for each different disease. Some philosophers freely admitted error and were converted to realist positions concerning theories. Others resorted to subjective idealism or to a thoroughgoing phenomenalism, of which there were two principal varieties. According to one, the one "legitimate" observation language had for its descriptive terms only those which referred to sense data. The other maintained the stronger thesis that *all* "factual" statements were *translatable* without remainder into the sense-datum language. In either case, any two non-sense data (e.g., a theoretical entity and what would ordinarily be called an "observable physical object") had virtually the same status. Others contrived means of modifying their views much less drastically. One group maintained that Jones' crobes actually never had been unobservable in principle, for, they said, the theory did not imply the impossibility of finding a means (e.g., the microscope) of observing them. A more radical contention was that the crobes were not observed at all; it was argued that what was seen by means of the microscope was just a shadow or an image rather than a corporeal organism.

The Observational-Theoretical Dichotomy

Let us turn from these fictional philosophical positions and consider some of the actual ones to which they roughly correspond. Taking the last one first, it is interesting to note the following passage from Bergmann: "But it is only fair to point out that if this . . . methodological and terminological analysis [for the thesis that there are no atoms] . . . is strictly adhered to, even stars and microscopic objects are not physical things in a literal sense, but merely by courtesy of language and pictorial imagination. This might seem awkward. But when I look through a microscope, all I see is a patch of color which creeps through the field like a shadow over a wall. And a shadow, though real, is certainly not a physical thing."[4]

I should like to point out that it is also the case that if this analysis is strictly adhered to, we cannot observe physical things through opera glasses, or even through ordinary spectacles, and one begins to wonder about the status of what we see through an ordinary windowpane. And what about distortions due to temperature gradients – however small and, thus, always present – in the ambient air? It really *does* "seem awkward" to say that when people who wear glasses describe what they see they are talking about shadows, while those who employ unaided vision talk about physical things – or that when we look through a windowpane, we can only *infer* that it is raining, while if we raise the window, we may "observe directly" that it is. The point I am making is that there is, in principle, a continuous series beginning with looking through a vacuum and containing these as members: looking through a windowpane, looking through glasses, looking through binoculars, looking through a low-power microscope, looking through a high-power microscope, etc., in the order given. The important consequence is that, so far, we are left without criteria which would enable us to draw a non-arbitrary line between "observation" and "theory". Certainly, we will often find it convenient to draw such a to-some-extent-arbitrary line; but its position will vary widely from context to context. (For example, if we are determining the resolving characteristics of a certain microscope, we would certainly draw the line beyond ordinary spectacles, probably beyond simple magnifying glasses, and possibly beyond another microscope with a lower power of resolution.) But what ontological ice does a mere methodologically convenient observational-theoretical dichotomy cut? Does an entity attain physical thinghood and/or "real existence" in one context only to lose it in another? Or, we may ask, recalling the continuity from observable to unobservable, is what is seen through spectacles a "little bit less real" or does it "exist to a slightly less extent" than what is observed by unaided vision?[5]

However, it might be argued that things seen through spectacles and binoculars look like ordinary physical objects, while those seen through microscopes and telescopes look like shadows and patches of light. I can only reply that this does not seem to me to be the case, particularly when looking at the moon, or even Saturn, through a telescope or when looking at a small, though "directly observable," physical object through a low-power microscope. Thus, again, a continuity appears.

"But," it might be objected, "theory tells us that what we see by means of a microscope is a real image, which is certainly distinct from the object on the stage." Now first of all, it should be remarked that it seems odd that one who is espousing an austere empiricism which requires a sharp observational-language/theoretical-language distinction (and one in which the former language has a privileged status) should need a theory in order to tell him what is observable. But, letting this pass, what is to prevent us from saying that we still observe the object on the stage, even though a "real image" may be involved? Otherwise, we shall be strongly tempted by phenomenalistic demons, and at this point we are considering a physical-object observation language rather than a sense-datum one. (Compare the traditional puzzles: Do I see one physical object or two when I punch my eyeball? Does one object split into two? Or do I see one object and one image? Etc.)

Another argument for the continuous transition from the observable to the unobservable (theoretical) may be adduced from theoretical considerations themselves. For example, contemporary valency theory tells us that there is a virtually continuous transition from very small molecules (such as those of hydrogen) through "medium-sized" ones (such as those of the fatty acids, polypeptides, proteins, and viruses) to extremely large ones (such as crystals of the salts, diamonds, and lumps of polymeric plastic). The molecules in the last-mentioned group are macro, "directly observable" physical objects but are, nevertheless, genuine, single molecules; on the other hand, those in the first mentioned group have the same perplexing properties as subatomic particles (de Broglie waves, Heisenberg indeterminacy, etc.). Are we to say that a large protein molecule (e.g., a virus) which can be "seen" only with an electron microscope is a little less real or exists to somewhat less an extent than does a molecule of a polymer which can be seen with an optical microscope? And does a hydrogen molecule partake of only an infinitesimal portion of existence or reality? Although there certainly *is* a continuous transition from observability to unobservability, any talk of such a continuity from full-blown existence to nonexistence is, clearly, nonsense.

Let us now consider the next to last modified position which was adopted by our fictional philosophers. According to them, it is only those entities which are *in principle* impossible to observe that present special problems. What kind of impossibility is meant here? Without going into a detailed discussion of the various types of impossibility, about which there is abundant literature with which the reader is no doubt familiar, I shall assume what usually seems to be granted by most philosophers who talk of entities which are unobservable in principle – i.e., that the theory(s) itself (coupled with a physiological theory of perception, I would add) entails that such entities are unobservable.

We should immediately note that if this analysis of the notion of unobservability (and, hence, of observability) is accepted, then its use as a means of delimiting the observation language seems to be precluded for those philosophers who regard theoretical expressions as elements of a calculating device – as meaningless strings of symbols. For suppose they wished to determine whether or not 'electron' was a theoretical term. First, they must see whether the theory entails the sentence 'Electrons are unobservable.' So far, so good, for their calculating devices are said to be able to select genuine sentences, provided they contain no theoretical terms. But what about the selected "sentence" itself? Suppose that 'electron' is an observation term. It follows that the expression is a genuine sentence and asserts that electrons are unobservable. But this entails that 'electron' is *not* an observation term. Thus if 'electron' is an observation term, then it is *not* an observation term. Therefore it is not an observation term. But then it follows that 'Electrons are unobservable' is not a genuine sentence and does not assert that electrons are unobservable, since it is a meaningless string of marks and does not assert anything whatever. Of course, it could be stipulated that when a theory "selects" a meaningless expression of the form 'Xs are unobservable,' then 'X' is to be taken as a theoretical term. But this seems rather arbitrary.

But, assuming that well-formed theoretical expressions are genuine sentences, what shall we say about unobservability in principle? I shall begin by putting my head on the block and argue that the present-day status of, say, electrons is in many ways similar to that of Jones' crobes before microscopes were invented. I am well aware of the numerous theoretical arguments for the impossibility of observing electrons. But suppose new entities are discovered which interact with electrons in such a mild manner that if an electron is, say, in an eigenstate of position, then, in certain circumstances, the interaction does not disturb it. Suppose also that a drug is discovered which vastly alters the human perceptual apparatus – perhaps even activates latent capacities so that a new sense modality emerges. Finally, suppose that in our altered state we are able to perceive (not necessarily visually) by means of these new entities in a manner roughly analogous to that by which we now see by means of photons. To make this a little more plausible, suppose that the energy eigenstates of the electrons in some of the compounds present in the relevant perceptual organ are such that even the weak interaction with the new entities alters them and also that the cross sections, relative to the new entities, of the electrons and other particles of the gases of the air are so small that the chance of any interaction here is negligible. Then we might be able to "observe directly" the position and possibly the approximate diameter and other properties of some electrons. It would follow, of course, that quantum theory would have to be altered in some respects, since the new entities do not conform to all its principles. But however improbable this may be, it does not, I maintain, involve any logical or conceptual absurdity. Furthermore, the modification necessary for the inclusion of the new entities would not necessarily change the meaning of the term 'electron.'[6]

Consider a somewhat less fantastic example, and one which does not involve any change in physical theory. Suppose a human mutant is born who is able to "observe" ultraviolet radiation, or even X rays, in the same way we "observe" visible light.

Now I think that it is extremely improbable that we will ever observe electrons directly (i.e., that it will ever be reasonable to assert that we have so observed them). But this is neither here nor there; it is not the purpose of this essay to predict the future development of scientific theories, and, hence, it is not its business to decide what actually is observable or what will become observable (in the more or less intuitive sense of 'observable' with which we are now working). After all, we are operating, here, under the assumption that it is theory, and thus science itself, which tells us what is or is not, in this sense, observable (the 'in principle' seems to have become superfluous). And this is the heart of the matter; for it follows that, at least for this sense of 'observable', there are no a priori or philosophical criteria for separating the observable from the unobservable. By trying to show that we can talk about the *possibility* of observing electrons without committing logical or conceptual blunders, I have been trying to support the thesis that any (nonlogical) term is a *possible* candidate for an observation term.

There is another line which may be taken in regard to delimitation of the observation language. According to it, the proper term with which to work is not 'observ*able*' but, rather 'observ*ed*.' There immediately comes to mind the tradition beginning with Locke and Hume (No idea without a preceding impression!), running through Logical Atomism and the Principle of Acquaintance, and ending (perhaps) in contemporary positivism. Since the numerous facets of this tradition have been extensively examined and criticized in the literature, I shall limit myself here to a few summary remarks.

Again, let us consider at this point only observation languages which contain ordinary physical-object terms (along with observation predicates, etc., of course). Now, according to this view, all descriptive terms of the observation language must refer to that which has been observed. How is this to be interpreted? Not too narrowly, presumably, otherwise each language user would have a different observation language. The name of my Aunt Mamie, of California, whom I have never seen, would not be in my observation language, nor would 'snow' be an observation term for many Floridians. One could, of course, set off the observation language by means of this awkward restriction, but then, obviously, not being the referent of an observation term would have no bearing on the ontological status of Aunt Mamie or that of snow.

Perhaps it is intended that the referents of observation terms must be members of a *kind* some of whose members have been observed or instances of a *property* some of whose instances have been observed. But there are familiar difficulties here. For example, given any entity, we can always find a kind whose only member is the entity in question; and surely expressions such as 'men over 14 feet tall' should be counted as observational even though no instances of the "property" of being a man over 14 feet tall have been observed. It would seem that this approach must soon fall back upon some notion of simples or determinables vs. determinates. But is it thereby saved? If it is held that only those terms which refer to observed simples or observed determinates are observation terms, we need only remind ourselves of such instances as Hume's notorious missing shade of blue. And if it is contended that in order to be an observation term an expression must at least refer to an observed determinable, then we can always find such a determinable which is broad enough in scope to embrace any entity whatever. But even if these difficulties can be circumvented, we see (as we knew all along) that this approach leads inevitably into phenomenalism, which is a view with which we have not been concerning ourselves.

Now it is not the purpose of this essay to give a detailed critique of phenomenalism. For the most part, I simply assume that it is untenable, at least in any of its translatability varieties.[7] However, if there are any unreconstructed phenomenalists among the readers, my purpose, insofar as they are concerned, will have been largely achieved if they will grant what I suppose most of them would stoutly maintain anyway, i.e., that theoretical entities are no worse off than so-called observable physical objects.

Nevertheless, a few considerations concerning phenomenalism and related matters may cast some light upon the observational-theoretical dichotomy and, perhaps, upon the nature of the "observation language." As a preface, allow me some overdue remarks on the latter. Although I have contended that the line between the observable and the unobservable is diffuse, that it shifts from one scientific problem to another, and that it is constantly being pushed toward the "unobservable" end of the spectrum as we develop better means of observation – better instruments – it would, nevertheless, be fatuous to minimize the importance of the observation base, for it is absolutely necessary as a confirmation base for statements which do refer to entities which are unobservable at a given time. But we should take as its basis and its unit not the "observational term" but, rather, the quickly decidable sentence. (I am indebted to Feyerabend, *loc. cit.*, for this terminology.) A quickly decidable sentence (in the technical sense employed here) may be defined as a singular, nonanalytic sentence such that a reliable, reasonably sophisticated language user can very quickly decide[8] whether to assert it or deny it when he is reporting on an occurrent situation. 'Observation term' may now be defined as a 'descriptive (nonlogical) term which may occur in a quickly decidable sentence,' and 'observation sentence' as a 'sentence whose only descriptive terms are observation terms.'

Returning to phenomenalism, let me emphasize that I am not among those philosophers who hold that there are no such things as sense contents (even sense data), nor do I believe that they play no important role in our peception of "reality." But the fact remains that the referents of most (not all) of the statements of the linguistic framework used in everyday life and in science are *not* sense contents but, rather, physical objects and other publicly observable entities. Except for pains, odors, "inner states," etc., *we do not usually observe sense contents*; and although there is good reason to believe that they play an indispensable role in observation, *we are usually not aware of them when we* (visually or tactilely) *observe physical objects.* For example, when I observe a distorted, obliquely reflected image in a mirror, I may seem to be seeing a baby elephant standing on its head; later I discover it is an image of Uncle Charles taking a nap with his mouth open and his hand in a peculiar position. Or, passing my neighbor's home at a high rate of speed, I observe that he is washing a car. If asked to report these observations I could quickly and easily report a baby elephant and a washing of a car; I probably would not, without subsequent observations, be able to report what colors, shapes, etc. (i.e., what sense data) were involved.

Two questions naturally arise at this point. How is it that we can (sometimes) quickly decide the truth or falsity of a pertinent observation sentence? and, What role do sense contents play in the appropriate tokening of such sentences? The heart of the matter is that these are primarily scientific-theoretical questions rather than "purely logical," "purely conceptual," or "purely epistemological." If theoretical physics, psychology, neurophysiology, etc., were sufficiently advanced, we could give satisfactory answers to these questions, using, in all likelihood, the physical-thing language as our observation language and *treating sensations, sense contents, sense data, and "inner states" as theoretical* (yes, theoretical!) *entities.*[9]

It is interesting and important to note that, even before we give completely satisfactory answers to the two questions considered above, we can, with due effort and reflection, train ourselves to "observe directly" what were once theoretical entities – the sense contents (color sensations, etc.) – involved in our perception of physical things. As has been pointed out before, we can also come to observe other kinds of entities which were once theoretical. Those which most readily come to mind involve the use of instruments as aids to observation. Indeed, using our painfully acquired theoretical knowledge of the world, we come to see that we "directly observe" many kinds of so-called theoretical things. After listening to a dull speech while sitting on a hard bench, we begin to become poignantly aware of the presence of a considerably strong gravitational field, and as Professor Feyerabend is fond of pointing out, if we were carrying a heavy suitcase in a changing gravitational field, we could observe the changes of the $G\mu\nu$ of the metric tensor.

I conclude that our drawing of the observational-theoretical line at any given point is an accident and a function of our physiological make-up, our current state of knowledge, and the instruments we happen to have available and, therefore, that it has no ontological significance whatever.

Notes

1 E. Nagel, *The Structure of Science* (New York: Harcourt, Brace, and World, 1961), Ch. 6.

2 For the genesis and part of the content of some of the ideas expressed herein, I am indebted to a number of sources; some of the more influential are H. Feigl, "Existential Hypothesis," *Philosophy of Science*, 17: 35–62 (1950); P. K. Feyerabend, "An Attempt at a Realistic Interpretation of Experience," *Proceedings of the Aristotelian Society*, 58: 144–70 (1958); N. R. Hanson, *Patterns of Discovery* (Cambridge: Cambridge University Press, 1958); E. Nagel, loc. cit.; Karl Popper, *The Logic of Scientific Discovery* (London: Hutchinson, 1959); M. Scriven, "Definitions, Explanations, and Theories," in *Minnesota Studies in the Philosophy of Science*, vol. II, H. Feigl, M. Scriven, and G. Maxwell, eds. (Minneapolis: University of Minnesota Press, 1958); Wilfrid Sellars, "Empiricism and the Philosophy of Mind," in *Minnesota Studies in the Philosophy of Science*, vol. I, H. Feigl and M. Scriven, eds. (Minneapolis: University of Minnesota Press, 1956), and "The Language of Theories," in *Current Issues in the Philosophy of Science*, H. Feigl and G. Maxwell, eds. (New York: Holt, Rinehart, and Winston, 1961).

3 I have borrowed the hammer analogy from E. Nagel, "Science and [Feigl's] Semantic Realism," *Philosophy of Science*, 17: 174–81 (1950), but it should be pointed out that Professor Nagel makes it clear that he does not necessarily subscribe to the view which he is explaining.

4 G. Bergmann, "Outline of an Empiricist Philosophy of Physics," *American Journal of Physics*, 11: 248–258; 335–342 (1943), reprinted in *Readings in the Philosophy of Science*, H. Feigl and M. Brodbeck, eds. (New York: Appleton-Century-Crofts, 1953), pp. 262–87.

5 I am not attributing to Professor Bergmann the absurd views suggested by these questions. He seems to take a sense-datum language as his observation language (the base of what he called "the empirical hierarchy"), and, in some ways, such a position is more difficult to refute than one which purports to take an "observable-physical-object"

view. However, I believe that demolishing the straw men with which I am now dealing amounts to desirable preliminary "therapy." Some nonrealist interpretations of theories which embody the presupposition that the observable-theoretical distinction is sharp and ontologically crucial seem to me to entail positions which correspond to such straw men rather closely.

6 For arguments that it is possible to alter a theory without altering the meanings of its terms, see my "Meaning Postulates in Scientific Theories," in *Current Issues in the Philosophy of Science*, Feigl and Maxwell, eds.

7 The reader is no doubt familiar with the abundant literature concerned with this issue. See, for example, Sellars' "Empiricism and the Philosophy of Mind," which also contains references to other pertinent works.

8 We may say "noninferentially" decide, provided this is interpreted liberally enough to avoid starting the entire controversy about observability all over again.

9 Cf. Sellars, "Empiricism and the Philosophy of Mind." As Professor Sellars points out, this is the crux of the "other minds" problem. Sensations and inner states (relative to an intersubjective observation language, I would add) are theoretical entities (and they "really exist") and *not* merely actual and/or possible behavior. Surely it is the unwillingness to countenance theoretical entities – the hope that every sentence is translatable not only into some observation language but into the physical-thing language – which is responsible for the "logical behaviorism" of the neo-Wittgensteinians.

Unit 7

After the Received View: Methodology

1 Popper and the Demarcation Criterion

The positivists' verification criterion was designed to distinguish sensible empirical doctrine from senseless metaphysics. The metaphysics they opposed (e.g., transcendental idealism, or theological doctrine) purports to concern matters beyond the reach of empirical scientific inquiry. But some non-scientific doctrines – astrology, phrenology, creation science – do present themselves as scientific, hoping thereby to garner the respect that accrues to scientific doctrines without earning that respect by subjecting themselves to empirical scrutiny. Such *pseudo-science* received the same treatment by the positivists as did those doctrines that are more blatantly metaphysical: legitimate scientific doctrines admit of empirical confirmation; but pseudo-scientific doctrines admit of no such confirmation; according to the verification criterion they are therefore meaningless.[1]

Although Karl Popper interacted with members of the Vienna Circle, he never joined them (he was identified by Otto Neurath as the circle's "official opposition," a title Popper enthusiastically endorsed).[2] He did share the Circle's deep respect for the scientific endeavor, and also considered the formulation of a distinction between science and non-science (of both the metaphysical and pseudo-scientific varieties) to be a fundamental task for the philosophy of science.

However, he vehemently denied that appeal to the verification criterion is the right way to distinguish between science and non-science for two reasons. First, he denied that the *demarcation criterion* by which science and pseudo-science are distinguished is equivalent to that between the cognitively meaningful and meaningless. Both metaphysical and pseudo-scientific doctrines might well be meaningful, he thought, but meeting that minimal qualification is no guarantee that they constitute legitimate scientific claims. Second, he denied that a theory's capacity to be confirmed is any indication of its scientific character.

Indeed, Popper argued, a theory can be *too* confirmable. He noticed that advocates of both Marxism and Freudian psychoanalytic theory celebrated their theories' ability to explain virtually any phenomenon. Confronted by behavior (political or personal) that might appear to conflict with the theory, some manner of adjustment of that theory seemed always available that would permit the theory's advocates to explain that behavior (e.g., as a manifestation of false consciousness, or Oedipal conflict) and announce that they had achieved another confirmation of the theory. Popper compared this with Arthur Eddington's recent experimental test of Albert Einstein's theory of relativity, in which Eddington checked the theory's predictions regarding the bending of starlight by the sun's gravitational field against observations of star position near the sun during an eclipse. That prediction was spectacularly successful, and so Einstein's theory was also confirmed. But Popper was struck by the *risk* that Einstein's theory took; if there was no apparent shift in the star's position in the sky, or if it varied substantially from that predicted by Einstein's theory, then this one observation would (Popper thought) decide the matter against that theory. Marxism and psychoanalysis took no such risks: the theories were so structured that no disconfirmation of the theory was possible.

This apparent strength of the latter two theories was instead, Popper insisted, their crucial weakness. Confirmation is easy to come by, so long as the theory is so formulated as to render it immune to refutation. But then the victory of confirmation, where no disconfirmation is possible, is empty: if a theory takes no risks, then the fact that it suffers no defeat can command no respect. Einstein's theory, by contrast, took a great risk in predicting a result that was entirely unexpected, and which could easily have been empirically disconfirmed; its success in the face of such risk is spectacular.

Falsification, therefore, rather than verification (or confirmation) is the key to the demarcation criterion. A theory is scientific insofar as it presents opportunities to be proven wrong; and the success of its predictions is of interest in proportion to the risk it takes.

It should be noted that Popper's demarcation criterion is not intended to distinguish those scientific theories we should accept from those we should not. Newtonian mechanics is (Popper thought) highly falsifiable, and therefore a legitimate scientific theory; but it is in fact false. And a horoscope prediction that a "valuable financial opportunity may soon present itself" is bound to be true (it can remain true, after all, even if such an opportunity does not in fact present itself, since it still "may" have). Nevertheless, according to the falsification criterion, the former is a paradigmatic scientific theory and the latter is not scientific at all. The criterion demarcates the scientific from the unscientific, not correct from incorrect science.

2 Conjecture and Refutation

Falsifiability corresponds nicely to some intuitively valuable characteristics of a scientific theory. First, a theory is more attractive the more informative it is; a theory that tells us that the planets periodically traverse the sky is of less interest than one that tells us where they will be in the sky at a particular day and time. But a theory that is more informative is also more falsifiable, since the additional information may be incorrect (e.g., the planet might be in the wrong place). Second, scientific theories are more attractive if they are more precise, and the more precise a prediction the more falsifiable it is, since smaller discrepancies between predicted and observed values will count as errors. And third, a theory is more attractive the wider the range of phenomena that it attempts to accommodate, since the additional phenomena might end up conflicting with the theory. For example, Isaac Newton's theory provided a mechanics for both the celestial and terrestrial realms, whereas Johannes Kepler's theory concerned only the former. So Newton's theory was open to refutation by possible observations on earth to which Kepler's theory was immune. Popper therefore came to see falsifiability as the key, not only to the demarcation criterion, but to scientific methodology generally.[3]

Popper was one of the very few philosophers of science to fully endorse David Hume's skepticism regarding induction. But far from despairing of the rationality of scientific method in the face of that seemingly devastating result, he proceeded to formulate an account that did not depend on induction at all. In §2 of the Unit 5 commentary we saw the positivist Rudolf Carnap suggest that, while a theory cannot be proven, it can be refuted. No finite number of ravens, for example, decisively demonstrates the truth of the claim that all ravens are black, since the domain of that law is potentially infinite. However, a single non-black raven will refute that proposition. Incredibly, Popper proposed an account of scientific methodology that depends *solely* on this fact and which therefore entirely dispenses with the attempt to develop a logic of confirmation. It is a testimony to his genius that he was able to do so much, and convince so many, with so little.

He proposed that science proceeds, not from careful observation to induced theory, but by *conjecture* and *refutation*. A conjecture (or hypothesis) is formulated first, he insisted, before the collection of data. There is otherwise nothing to determine what should be observed and, without such guidance, the data would amount to a random collection of useless trivia.[4] The theory formulated must be *bold*, that is, highly falsifiable. The scientist then proceeds, not to attempt to confirm it, but instead to ruthlessly subject it to tests that will demonstrate that it is wrong. The only successes that count are those when a theory survives such rigorous attempts at falsification, which constitute what Popper called *corroboration* of the theory.

But corroboration is, he insisted, not a reason for belief. An argument to the effect that multiple corroborations make it likely that a theory is true would itself be an inductive argument; but Popper rejected inductive reasoning. Instead, we *accept* the corroborated theory, which means that we use it for our present purposes but not that we believe it to be true, or even likely to be so. To the question why it is reasonable to accept the theory that has not been falsified over ones that have, Popper responds that "the only correct answer is the straightforward one: because we search for the truth (even though we can never be sure we have found it), and because the falsified theories are known or believed to be false, while the not-falsified theories may still be true" (reading 7.1, p. 484).

Popper correspondingly rejected the inductivist conception of scientific progress.[5] Progress, according to the inductivist, consists in either the steady accumulation of confirmations of the current theory or the replacement of the current theory with a successor that covers a wider range of phenomena (according to which successor the current theory is correct within its more restricted domain). The result is expected to consist in a sequence of increasingly broad and more accurate representations of empirical reality.

Popper's alternative conception of progress is comparable to an evolutionary process. The current theory, unfalsified to date, has not suffered the fate of its refuted predecessors, as it has survived the rigorous tests that brought them down (although this provides no assurance that it will continue to do so). The new theory is also more informative, more precise, and covers a broader range than the old. So it is, in these ways, better adapted to the harshly critical scientific environment than the old theories were, as will be its successor in comparison with it. This "backward-looking" account of progress corresponds to the succession of increasingly successful adaptations generated by environmental selection pressure. And it requires no more of a guiding hand directing the process toward an already determined goal than does evolution. This contrasts with the "forward-looking" account of the inductivist who sees the sequence of theories as explicitly guided toward the truth.

3 Criticism of Popper

As Pierre Duhem and W. V. O. Quine pointed out, no single theory generates empirical consequences on its own, but only with the help of various initial and boundary conditions, assumptions about the experimental setup and the instruments used within it, and a host of other *auxiliary assumptions*. When a prediction fails, the responsibility might lie, not with the theory itself, but rather with one of these auxiliary assumptions that were employed in extracting the prediction from the theory.[6] There can, therefore, be no such decisive refutation of a theory as Popper suggests.

Another concern with Popper's account is captured by the slogan that all theories "are born refuted."[7] That is, every non-trivial scientific theory that has actually been developed made predictions that did not perfectly conform to observation. If Popperian methodology had been scrupulously followed, then every such theory would have been rejected, and science would have come to a standstill. Instead, scientists proceeded to attempt to resolve those failed predictions, which tenacity has paid off in the development of spectacularly empirically successful theories.

A third criticism concerns Popper's account of progress. He endorsed truth as the aim of science; we would, otherwise, have no motive for preferring unfalsified theories to falsified ones. He was, therefore, a realist, in at least that sense (a "conjectural realist" as he described his position). But he steadfastly denied that we can ever legitimately take our currently accepted theories to be true, or even likely to be so. He also denied that progress is progress toward that aim. But an aim, surely, is that to which we strive. It is far from clear that it is rational to strive for something while convinced that we can never legitimately take ourselves to be anywhere near achieving it.[8]

Finally, a fourth criticism concerns Popper's attitude toward observation sentences. The positivists considered observation sentences to be the theory-neutral touchstones against which theories are evaluated. But Popper denied this, maintaining instead that the "basic sentences" against which theories are tested are themselves influenced by the observer's theoretical commitments; they are, as the terminology goes, theory-laden.[9] But then, when a theory seems to have been falsified by a contrary basic sentence, it is open to the scientist to counter that it is the theory with which the observation is laden, rather than the theory that supposedly failed the test, that has been shown up. This again undermines the idea of a decisive refutation of a theory by observation.

Popper was well aware of these criticisms and attempted to modify his account of scientific methodology in order to accommodate them. While we do not have the space to explore the details of those modifications here, the fundamental criticism has been made: the image of science as a history of bold conjecture and decisive refutation, governed straightforwardly by one deductive rule of inference, is untenable.

4 Kuhn: Attending to History

Ideally, an account of scientific methodology would meet both a normative and a descriptive constraint. The normative constraint requires that the account should vindicate our intuitive conviction that science is a rational endeavor. (There is, however, as much latitude in this as there is in the concept of rationality itself.) The descriptive constraint requires that the account should identify as scientists those who are unquestionably so (e.g., Darwin, Newton, Einstein), and as scientific theories those that are unquestionably such (e.g., evolution by natural selection, Newtonian mechanics, relativity theory). If an account implies, for example, that Newton was not a scientist, this will not count as a remarkable discovery about Newton but instead as a decisive strike against the account.[10]

The positivists took the normative constraint very seriously, and the project of developing a logic of confirmation was the positivists' attempt to satisfy it. They expended considerably less effort, however, to ensure that their characterization of the scientific endeavor satisfies the descriptive constraint, paying only scant attention to the history of science and attending only to its finished theories rather than to the processes that produced them.

Such a dismissive attitude toward the history of science is, to some extent at least, understandable. After all, their work was grounded in the assumption that science is a rational endeavor. If they then uncover the conditions of rationality, then, of course, science will conform to those conditions. Any departures from those conditions that are discovered by the historian must amount to social and psychological bias interfering with objective scientific inquiry. Explaining those departures is, therefore, the business of the psychologist and sociologist rather than the philosopher of science.

Thomas Kuhn's views in his tremendously influential *Structure of Scientific Revolutions* (excerpted in reading 7.2) are notoriously difficult to characterize, but the underlying theme of his work is, arguably, that careful reflection on the actual history of science severely undermines the idea that the normative constraint can be met by a set of rules of the sort that the positivists had in mind or, for that matter, by a set of rules of any sort. This is not to say that science is irrational; although Kuhn has been frequently accused of saying as much, he insisted that this was not his intent. Rather, he argued that the conceptual resources provided by the received view were simply inadequate to the task of making sense of science as it is actually practiced and that much of the problem is due to the positivists' assumption that scientific methodology can be codified in a set of rules that provide every scientist at every time with the same methodological algorithm.[11] (As we will see, however, it is far from clear that Kuhn's own account is any more successful at meeting these two constraints.)

The history represented in the textbook of a particular science, Kuhn granted, conforms closely to the account of progress enshrined in the received view. The contemporary scientific account is presented as the culmination of a slow but nevertheless inexorable accumulation of empirically certified knowledge, the accumulation of which sometimes required the heroic repudiation of obscuring myth and religious bias. This patient and determined accumulation is occasionally accelerated by the efforts of a scientific genius; but for the most part the working scientist stands on the shoulders of such giants, and adds her own modest contribution to the store of knowledge that is now very plentiful indeed.

But that textbook history is, Kuhn insisted, a lie. The image of slow but steady cumulative progress that it presents entirely obscures from view the scientific revolution that periodically, and catastrophically, disrupts that progress. Kuhn chose "revolution" consciously, trading on an analogy with politics. In politics, it is distinguished from a reformation, which indicates change from within, that is, the exploitation of mechanisms for change built into the current system (and so which mechanisms survive the resulting change they make possible). A revolution, however, is change from without and implies alteration in those mechanisms themselves. Scientific revolutions are, Kuhn argued, akin to the political in at least this respect since they involve alteration in the very mechanisms of scientific inquiry.

Indeed science resembles politics in another important respect, one typically ignored by advocates of the received view. Both science and politics are social institutions, and their social character is one of their intrinsic features. For Kuhn, there is really no such thing as a scientist who works alone; science is conducted in and by a community of scientists with a common research agenda.

5 Pre-Science, Exemplar, and Paradigm

Before such an agenda is established, there is *pre-science*. Humans have always attempted to make sense of certain phenomena by appeal to stories that posit an underlying reality. But in the pre-scientific period, different individuals do not just advocate different such theories. They also take different phenomena to demand accommodation, and they rank the significance of those phenomena in different ways. They also differ in the background assumptions that they take for granted in formulating their theory, what adequacy in a theoretical account amounts to, and so on. With such ongoing disagreement over fundamentals, it is no surprise that nothing resembling progress is made.

Eventually, though, a group of practitioners manages to formulate a reasonably coherent outlook that, while not accommodating everything that everyone wants accommodated, nevertheless demonstrates some success in dealing with a number of phenomena that are widely recognized as important. That view gains a sufficient number of adherents to constitute a community of practitioners working largely in concert on mutually recognized problems in order to formulate mutually recognizable solutions. This is the rise of what Kuhn called *normal science*. Normal science is the nature of scientific activity most of the time.

Notice that the distinction between pre-science and normal science can only apply to communities of practitioners. This exemplifies Kuhn's conception of science as fundamentally a social endeavor, so that one cannot really be a scientist in isolation but only when acting in concert with fellow normal scientists to solve a number of shared problems in mutually recognized ways.

Normal science is governed by a *paradigm*. The difficulty in pinning this concept down is evidenced by Kuhn's own later distinction between two senses, the *exemplar* and the *disciplinary matrix*, where the former is a component of the latter. In what follows, we will reserve the term 'paradigm' for the disciplinary matrix, and will (attempt to) use 'exemplar' as Kuhn intended.

Those problem solutions around which the normal scientific community coalesces become the theory's exemplars. They are the prototypical applications of the theory, illustrating the theory's ability to solve important problems. But their significance is far more than merely illustrative. The training of new scientists into the theory is primarily organized around them: far from learning a set of formulas and rules, the student learns how the theory is applied in these particular cases. Then, on the basis of her familiarity with them, she learns how to recognize new problems as similar to the exemplars, allowing her to apply the techniques she learned to apply to the exemplars to these new cases. Eventually her repertoire of successful applications is sufficiently broad that she can attempt to formulate solutions to problems for which there is no current solution; she then becomes a professional member of the scientific community.

The training that the budding scientist receives through exemplar applications of the theory, Kuhn insists, cannot be characterized as the learning of an algorithm that applies in the same way to all problems. Recognizing that a new problem is similar to an exemplar does not require that the practitioner must be able to list a set of characteristics in common, any more than recognizing that family members look alike requires a list of common facial features. Kuhn inherited from Ludwig Wittgenstein[12] the idea that family resemblance does not require a set of necessary and sufficient conditions of application. Similarly, scientific training built around exemplary applications of the theory does not require abstraction of a single set of rules applicable to all problems the normal scientist wants to solve. Such training inculcates a skill; and possession of a skill does not inevitably reduce to a grasp of a set of abstract principles.

Exemplars are thus the most important aspect of the paradigm: they illustrate the problem-solving power of the theory, they are the fundamental pedagogical tool in the instruction of new scientists, and they guide the normal scientist's attempts to solve new problems. They do not, however, exhaust the paradigm. The normal scientist must also take much for granted if she is to make any progress in her detailed work. These assumptions include certain basic metaphysical principles, such as that the motion appropriate to the heavenly bodies is uniform rotation, that the terrestrial and celestial realms are fundamentally different, that there is no action at a distance, that all natural laws are deterministic, and so on. These principles have no specific empirical implications since nothing observable immediately follows from, for example, the separation of the heavens from the earth. But they act as constraints on what empirical theories are acceptable. The Copernican proposal that the earth is a planet, for example, was seen at the time not just as false, but as absurd, since it violated that separation of the celestial and terrestrial realms. Some of these principles may be codified as laws, as, for example, in Newtonian mechanics, whose laws embody the principle that change of motion, rather than motion itself, requires explanation by appeal to a force (in contrast with Aristotle's account). Paradigm change brings with it a change in the metaphysical principles and, therefore, in the boundaries that delimit legitimate solutions to problems.

The paradigm does not only equip the scientist with certain beliefs (both broadly metaphysical and more specifically empirical), it also determines the *conceptual scheme* in which those beliefs are couched. For all his differences with the positivists, Kuhn shared their holistic account of theoretical terms: the meaning of a theoretical concept is a function of its place within the theoretical system in which it is embedded. Unlike the positivists, however, he drew the consequence that, when the theory changes, the conceptual scheme inevitably changes with it. "Mass", for example, refers to a fixed intrinsic property of an object in Newtonian mechanics, but, in Einstein's theory, it identifies a variable property that is relative to reference frame. The meaning of theoretical terms is not just different in different theories; since they owe their content to their role in the theory, and the theories are distinct, no scheme of translation can be provided to compare them. This is Kuhn's doctrine of the *incommensurability of meaning.*[13]

The paradigm also determines the values that guide the scientist in theory evaluation. Must a theory be compatible with religious doctrine? Can a theory legitimately posit unobservables? Is naked-eye observation the final arbiter, or can it be superseded by the view through the telescope? There is, Kuhn insisted, no paradigm-independent answer to such questions, and so no paradigm-independent standards of evaluation to which theorists operating in different paradigms can appeal in order to compare their views. The scientist's own theory is bound to come out the winner in light of the values endorsed by that paradigm, and similarly for that of the scientist within the opposing paradigm. The arguments that they offer to one another by appeal to those values are therefore circular, presupposing as they do values that are intrinsic to the paradigm they mean to defend. Their debate is therefore inevitably at cross-purposes. This is Kuhn's doctrine of the *incommensurability of standards.* Kuhn also accepted the theory-ladenness of observation (for which he was, at least to some extent, indebted to N. R. Hanson[14]). This is the doctrine that what the scientist observes, and how they describe what they observe, is determined, not only by what is in front of their eyes, but also by their background theoretical commitments. In conjunction with the incommensurability of meaning, this implies that the reports of observation, and even the observations themselves, are inexorably tied to the paradigm, and are therefore also incommensurable across theories. The objective touchstone of observation, appeal to which provided the positivists with a theory-neutral basis for the assessment and comparison of theories, is gone. Kuhn notoriously went so far as to say that scientists operating within different paradigms "live in different worlds" (reading 7.2, p. 510) (although he left it frustratingly unsettled just how literally he intended this claim to be interpreted).

6 Normal Science, Crisis, and Revolution

In its early life, a paradigm is far from fully formed. There are a few exemplars, and some idea of the domain to which it applies, but many details remain to be worked out. Normal scientific work consists in the working out of those details, and is fundamentally a *puzzle-solving* endeavor. Consider crossword puzzles. The individual who works on a crossword puzzle knows that some solutions are acceptable and some not: words cannot be made up; they must fit in the spaces provided with one letter to a space; and so on. And the idea of challenging those rules in way of solving the puzzle is nonsensical; solving the crossword by putting three letters in one square is no solution at all. The puzzle-solver is assured that there is a solution and, if she has not found it, that is her fault rather than somehow that of the puzzle itself. And while she has clues to which to appeal, she has no algorithm or recipe that will automatically take her to a solution; rather, it requires insight, lateral thinking, recognition of similarities, and the mobilization of much background experience. And that background experience itself does not consist in the acquisition of instructions for puzzle-solving, but instead simply in much practice with previous puzzles. Puzzle-solving is a skill to be honed rather than a method to be memorized.

Normal scientific activity is much the same. The paradigm determines what problems are there to solve. It also determines what counts as a legitimate solution; a solution that violates the paradigm's standards is no solution at all. The scientist is assured by the paradigm that the solution she seeks does indeed exist; failure to find it is therefore only the fault of the scientist rather than the paradigm. And the scientist has no algorithm to which to appeal; she must also use insight, lateral thinking, recognition of similarities (with the exemplars in particular), association, and much background experience. That background, finally, does not consist in the memorization of an abstract scientific method, but instead much practice with previous problems, beginning with the exemplars.

The problems that normal scientists attempt to solve include (among other things): extension of the theory to areas that it was not originally designed to accommodate; derivation and application of low-level empirical laws and constants; determination of the values of various parameters; derivation and observation of phenomena predicted by the theory that were not known to be a consequence of it when originally proposed; and resolution of *anomalies*, observations that contradict the predictions derived from the theory.

This last task must be carefully understood. The paradigm inevitably comes into being with anomalies, which are typically recognized by the scientific community. Far from challenging the paradigm, they are merely opportunities for refinement or further articulation of the paradigm. It is taken for granted by the normal scientist that a solution exists, although discovering it might take some ingenuity. The paradigm remains unquestioned throughout.

The paradigm determines what experiments need to be conducted, what facts there are to be uncovered, what new domains are legitimate extensions of the theory, what conditions must be satisfied for a solution to be legitimate, how the observations are described, what the theoretical terms mean, and what values are to be realized. It is only because all this is settled that the normal scientist can get on with her work without continually returning to fundamentals.[15] But it is also because of this that her work gets increasingly detailed, precise, and narrow. And that, in turn, ensures that anomalies that resist solution are bound to come into view and multiply, much as close inspection of an apparently smooth table's surface will reveal bumps and ridges that become more and more noticeable with every increase in magnification.

As the community continues to struggle with these recalcitrant and multiplying anomalies, its failure to resolve them becomes increasingly acute and unsettling. Having tried every trick in the paradigm's book without success, there is the sense that the paradigm has let them down. The paradigm itself finally becomes the object of critical scrutiny. Assumptions integral to it that have long gone unchallenged become questioned, but scientists disagree as to which of those assumptions should be reconsidered. They try to formulate explicit rules in way of resolving their disagreement – which rules, notwithstanding the positivists' conception, show up for the first time – to no avail, and the dispute becomes increasingly abstract and philosophical. The community splinters, reverting to a state akin to that of pre-science. This is what Kuhn calls *crisis*.

Eventually, a subset of the community – often a particular young not-yet-indoctrinated scientist – formulates a view that neatly resolves at least a number of the particularly acute anomalies that generated crisis, around which a new community forms. Although some members of the original community convert to the new view, they more often remain committed to the old paradigm until retirement or death. A new paradigm, and new community of scientists which accept and proceed to operate under it, is subsequently launched. This replacement of one paradigm by another is what Kuhn calls a *scientific revolution*; the result of such a revolution is a *paradigm shift*.

Such revolutions are, apparently, total. Exemplars, metaphysical assumptions, conceptual scheme, observation-content and methodology are all paradigm-dependent, and shift with the paradigm. A transfer of allegiance from one paradigm to another as a result is, Kuhn said, more akin to a religious "conversion experience" than a response to carefully considered argument.

There is, as a result, little room in Kuhn's view for the idea that the history of science has been progressive. There is recognizable progress during the periods of normal science in the slow but steady accumulation of solutions to the various puzzles with which the normal scientist is occupied. But the idea of comparing the puzzle-solving capacity of one paradigm with another across a revolution is inherently problematic. The incommensurability of meaning implies that the ability to comprehend both a problem and its solution is limited to the paradigm's own scientists, so that neither problem nor solution can be recognized across paradigms. And even if they could be so recognized, the incommensurability of standards implies that the question whether a particular puzzle is worth solving will be subject to irresolvable disagreement between scientists within different paradigms, as will the question whether a particular proposed solution counts as a legitimate solution.

So scientists within the pre-revolution paradigm will inevitably characterize the paradigm shift as regressive and those within the post-revolution paradigm will inevitably characterize it as progressive. And there appears to be no paradigm-independent way to determine which is correct. Indeed, the idea that such a correct answer exists appears to be a casualty of Kuhn's account. As mentioned in §4 above, this is not how the science textbooks present their disciplines' histories, which instead describe slow but steady progress from ancient times to the present day. But, Kuhn points out, the textbooks are written by the victors, that is, by the post-revolution scientists. The fact that they characterize the revolution that led to their standpoint (and the revolutions that came before that one) as progressive is hardly a surprise.

7 Criticism of Kuhn

Some criticisms of Kuhn's view concern its internal consistency. The resolution of crisis, for example, requires that the same anomalies that proved so resistant to resolution within the old paradigm are resolved in the new. But then the very same anomaly must be recognizable from the standpoint of both paradigms, which seems to contradict the incommensurability of meaning thesis as applied to observation.

Other criticisms concern the historical accuracy of his account of long periods of normal science governed by a monolithic paradigm punctuated by the occasional relatively brief revolution. His own favorite illustration of a paradigm shift, for example, was the Copernican revolution. And that revolution, initiated by Copernicus and finally completed by Newton, was a drawn-out, piecemeal affair, involving the iconoclastic work of a number of scientists, which took over one hundred years to complete. It therefore hardly resembles the brief, dramatic, wholesale shift that is supposed to characterize a Kuhnian revolution.

Moreover, it is not always easy to discern in the historical record the sharp distinction between normal and revolutionary science that Kuhn's account requires. Consider the Copernican revolution again. Copernicus was intent on restoring uniform circular motion to the heavens (and was in this way as ultra-conservative as he was revolutionary). Kepler, however, made the planets travel with continuously varying speed along elliptical orbits. We call Kepler a Copernican, but it is far from clear that Copernicus would himself have done so. Was Kepler a normal scientist in the Copernican paradigm, or did he introduce a new paradigm entirely? There does not appear to be any way to answer such questions.

These historical considerations have led some philosophers to deny that the components of a paradigm are inextricably bound together, suggesting instead that different components can change while others persist through the change, as when Kepler rejected Copernicus' circular orbits while endorsing his placement of the sun at the center – or, more accurately, at one focus – of the system.[16]

But the most vociferous criticisms of Kuhn's view concerned the apparent absence of a paradigm-neutral standpoint from which paradigms can be objectively assessed and compared and from which a paradigm shift can be judged as progressive (or not). Critics charged that this implies either that science is fundamentally irrational (at least through paradigm shifts) or it suggests an unpalatable relativism according to which the world itself that the paradigm purports to represent is itself paradigm-dependent.

Kuhn maintained that this misrepresents his intended position, and he later attempted to mitigate the pronouncements in *Structure* that seemed to lay him open to this critique (as he did in the 1969 postscript, excerpts from which are included in reading 7.2). There are, he affirmed, indeed cross-paradigm scientific values. These include problem-solving ability, empirical accuracy and precision, consistency (both internally and with other theories accepted at the time), breadth of scope, simplicity, and fecundity in generating further research.

But these do not, he insisted, amount to a set of rules for two reasons. First, they are vague and subject to multiple interpretations. Which of these two theories, for example, is the simplest: one introducing many different entities but few laws, or another introducing many laws but few entities? Second, different scientists can legitimately disagree about how these values should be ranked as to their significance. An advocate of the Copernican view, for example, might tout its mathematical simplicity, whereas an advocate of the Ptolemaic account might cite its consistency with the then-contemporary Aristotelian physics and cosmology. Thus, notwithstanding the existence of a common set of values to which every scientist is answerable, it remains possible for scientists to rationally disagree as to which of two theories is best.

This seems to raise once again the specter of irrationalism. But Kuhn insisted that the possibility of such disagreement, far from being detrimental to a conception of science as rational, is necessary. The alternative is that no two scientists could rationally disagree as to which theoretical approach is worth pursuing; only one approach can be so judged by the community of scientists. But paradigms are, in their early stages, fragile and underdeveloped; they require time and attention before their articulation is far enough along to fairly assess their worth. This requires that scientists be encouraged to pursue them, and to be counted as scientists when doing so, even when the rest of the community considers the fledgling paradigm to score badly with respect to the cross-paradigm values.

Kuhn also appeals to these virtues in order to counter the objection that his account makes no room for scientific progress across revolutions. We can (although only with hindsight) judge the succession of theories over time as manifesting an improvement in the realization of these virtues overall. In this sense, Kuhn is a "convinced believer of scientific progress" (reading 7.2, p. 512) But he nevertheless repudiates the idea that the succession of theories is guided in any way. In particular, he rejects the idea that this succession is one that increasingly approximates the truth: "the notion of a match between the ontology of a theory and its 'real' counterpart in nature now seems to me illusive in principle . . . I can see [in the succession of theories] no coherent direction of ontological development" (reading 7.2, p. 512).

Whether these replies to his critics are adequate is an open question. But it is not open to doubt the permanent and dramatic impact that Kuhn's work has had on the subsequent philosophy of science.

8 Lakatos and the Methodology of Research Programs

Imre Lakatos hoped to reconcile Popper's falsificationism with the reflections on the history of science to be found in Kuhn's *Structure*. He agreed with Popper that the scientist is obliged to respond to a failed prediction; anomalies cannot simply be ignored. But he agreed with Kuhn that not only are anomalies tolerated (which Popper did recognize, as Lakatos pointed out), they are not even perceived as threats except under the extreme circumstances that generate crisis. He agreed also that the fundamental unit of scientific inquiry is larger than the individual theory. Kuhn's units were paradigms, whereas Lakatos called them *research programs*; but they shared much in common.

A research program consists in a *hard core* of central doctrine and an *auxiliary belt* of less central assumptions, initial and boundary conditions, methodological rules, experimental techniques, and so on. The hard core generates no empirical consequences on its own, but only in conjunction with various elements drawn from the auxiliary belt. Blame for an anomaly, therefore, can always be placed on the elements of the auxiliary belt that were employed in the derivation of the failed prediction instead of on the hard core.[17] The research program includes a *negative heuristic* and a *positive heuristic*. The negative heuristic forbids certain lines of research (namely, those that invade the hard core). The positive heuristic provides the scientist with guidance in deciding which elements of the auxiliary belt she should adjust. Conformity with the positive and negative heuristics protects the hard core from refutation; the result is a sequence of *problem shifts* that constitute the research program.

So far, the analogy with Kuhn's paradigm and normal science is obvious. But Lakatos goes on to distinguish *progressive* from *degenerating* research programs. A progressive research program is one that both successfully resolves anomalies and does so by introducing problem shifts that generate novel and successful predictions. (The second condition ensures that anomalies are not resolved in an ad hoc fashion.) A degenerating program is one which either fails to resolve anomalies or resolves them only by recourse to ad hoc problem shifts, that is, changes to the auxiliary belt that generate no novel predictions (or that generates novel predictions that themselves turn out to be anomalies).

There is no decisive point at which the scientist must concede that her research program is degenerating. It might, after all, have fallen into a degenerating slump, but then proceed to turn itself around and resolve its anomalies, producing a host of successful novel predictions. With the benefit of hindsight we can, however, determine whether a transition was progressive by noting whether the shift was from a degenerating to a progressive program. The incapacity of Kuhn's account to characterize progress across program transitions (or paradigm shifts) is, as a result, corrected by appeal to the distinction between progressive and degenerating research programs.

9 Criticism of Lakatos

The similarity between Lakatos's and Kuhn's positions opens the former to some of the same criticisms that afflict the latter. Was Kepler's use of elliptical orbits a violation of the Copernican hard core or is the shape of the orbit a part of the auxiliary belt? Did Ptolemy's use of the equant (which implied that the planet actually changes speed on its path around the earth) violate the hard core of mathematical astronomy formulated around Plato's challenge to represent the motions of the heavens as composed of uniform circular motion, or was the equant just a novel way to answer that challenge (by separating "uniform" and "circular")? Again, it is hard to see to what one can appeal in formulating an answer.

Lakatos's claim that a program can be deemed degenerating only in hindsight also limits the attractions of his view, since it implies that his methodology cannot provide the working scientist with any practical advice as to how to proceed; she can only hope that history will see her as making the right decision at the right time.

Finally, although the methodology of research programs permits cross-program comparison, it remains difficult to see what the sequence of programs – one replacing the other apparently without end – achieves, since no discernible goal seems to be satisfied by the process overall. (This is essentially the same criticism raised against Popper's evolutionary account of scientific progress.)

Notes

1. The term 'verification criterion' is used in the wide sense here that requires of a meaningful claim that it generates – or at least plays an indispensable role in generating – empirical consequences, rather than in the more specific sense, rejected early on, that requires conclusive verification. See discussion in §5 of the Part II introduction.
2. See §2 of Unit II commentary for discussion of the Vienna Circle.
3. Alan Chalmers nicely describes the relation between falsification on the one hand and informativeness, precision, and range (although not in precisely those terms) in his *What Is This Thing Called Science, Third Edition* (Indianapolis: Hackett, 1999), pp. 61–8.
4. The instruction to "observe!" as Popper points out (reading 7.1, p. 478) is absurd unless it is clear *what* it is that one should be observing.
5. Equivalently, the 'confirmationist', 'verificationist', or 'positivist' conception.
6. This is the "Duhem-Quine thesis," whose implications are explored in §5 of the Unit 6 commentary.
7. Reading 7.3, p. 517. Lakatos's account of scientific methodology is the subject of §§8–9.
8. In response to this, Popper later attempted to formulate an account of progress toward the truth compatible with his overall falsificationism that relied on the concept of increase in *verisimilitude*, or truth-content. The idea was that the verisimilitude of theory A can be greater than theory B – and so A closer to the truth than B – even if both are false (as are all past theories according to the falsificationist). But Popper's account of verisimilitude fell prey to devastating critique (Pavel Tichý, 1974, "On Popper's Definitions of Verisimilitude," *The British Journal for the Philosophy of Science* 25; and David Miller, 1972, "The Truth-likeness of Truthlikeness," *Analysis* 33(2): 50–5) and it remains an open question whether a coherent version of the concept can be formulated.
9. See reading 6.6 and §8 of the Unit 8 commentary.
10. The second criticism of Popper above constitutes an application of the descriptive constraint.
11. An algorithm is a step-by-step procedure that will accomplish a certain task. A good recipe, for example, is an algorithm.
12. Ludwig Wittgenstein, 1953/2001, *Philosophical Investigations*, Oxford: Blackwell Publishing.
13. The thesis of incommensurability was formulated simultaneously and independently by Kuhn and Paul Feyerabend. Feyerabend's philosophy of science is similar to that of Kuhn in other respects, particularly in the critique of an algorithmic, ahistorical conception of scientific methodology. Feyerabend's critique is more radical, however, in that he repudiates the very idea of a methodology specific to science that distinguishes it from other human endeavors. See Feyerabend, *Against Method.*
14. See reading 6.6 and §8 of the Unit 6 commentary.
15. This contrasts sharply with Popper, who encourages skepticism toward one's scientific beliefs.
16. See Laudan, 1985, "Dissecting the Holist Picture of Scientific Change," in J. Kourany, *Scientific Knowledge*, Belmont, CA: Wadsworth, pp. 276–95.
17. In this way Lakatos hopes to turn the impact of the Duhem-Quine thesis to his advantage. See reading 6.3 and §6 of the Unit 6 commentary.

Suggestions for Further Reading

The classic statement of Popper's falsificationism is his 1959[1935]. Further elaborations and alterations of his view can be found in Popper 1963 and Popper 1983. Kuhn's views were presented in his monumental *The Structure of Scientific Revolutions* (Kuhn 1970[1962]) from which reading 7.2 is excerpted. He reconsiders those views in the essays collected in Kuhn 1977. Lakatos 1977 contains some of Lakatos's most important essays. Paul Feyerabend, who independently formulated the incommensurability of meaning thesis advocated by Kuhn, argues against the very idea of a methodology that distinguishes science from other human endeavors in Feyerabend 1975. In Laudan 1977, Larry Laudan develops an account of progress which consists in a research program's accumulation of solved empirical and conceptual problems. And in Laudan 1984 he criticizes Kuhn's "holistic" model of scientific change and offers in its stead his "reticulational" model that permits piecemeal changes. Both

Lakatos and Musgrave 1970 and Hacking 1981 contain a number of classic papers on the issue, including pieces by Popper, Kuhn, Feyerabend, and Laudan.

Feyerabend, P., 1975, *Against Method*. London: Verso.

Hacking, I. (ed.), 1981, *Scientific Revolutions*. Oxford: Oxford University Press.

Kuhn, T., 1970[1962], *The Structure of Scientific Revolutions*. Chicago: University of Chicago Press. 2nd edn., with postscript, 1970.

Kuhn, T., 1977, *The Essential Tension: Selected Studies in Scientific Tradition and Change*. Chicago: University of Chicago Press.

Lakatos, I., 1977, *The Methodology of Scientific Research Programmes: Philosophical Papers Vol. 1*. Cambridge: Cambridge University Press.

Lakatos, I. and Musgrave, A. (eds.), 1970, *Criticism and the Growth of Knowledge*. Cambridge: Cambridge University Press.

Laudan, L., 1977, *Progress and Its Problems: Toward a Theory of Scientific Growth*. London: Routledge and Kegan Paul.

Laudan, L., 1984, *Science and Values: The Aims of Science and Their Role in Scientific Debate*. Berkeley: University of California Press.

Popper, K., 1959[1935], *The Logic of Scientific Discovery*. London: Hutchinson. Translation of *Logik der Forschung*, Vienna: Julius Springer Verlag.

Popper, K., 1963, *Conjectures and Refutations: The Growth of Scientific Knowledge*. London: Routledge.

Popper, K., 1983, *Realism and the Aim of Science*. London: Hutchinson.

7.1

Science: Conjectures and Refutations

Karl R. Popper

Together with Rudolf Carnap and Thomas Kuhn, Karl Popper (1902–1994) is one of the three most influential philosophers of science of the twentieth century. Schooled in Vienna when the Vienna Circle was dominant, Popper rejected both their preoccupation with a meaning criterion and their attempt to formulate a logic of confirmation, substituting instead a criterion of demarcation between science and non-science. His criterion concerns, not a theory's confirmability, but instead its falsifiability, that is, its susceptibility to be proven incorrect. That criterion, and his falsificationist account of scientific methodology – the method of conjecture and refutation – is sketched in the selection below.

> *Mr Turnbull had predicted evil consequences, . . . and was now doing the best in his power to bring about the verification of his own prophecies.*
>
> Anthony Trollope

I

When I received the list of participants in this course and realized that I had been asked to speak to philosophical colleagues I thought, after some hesitation and consultation, that you would probably prefer me to speak about those problems which interest me most, and about those developments with which I am most intimately acquainted. I therefore decided to do what I have never done before: to give you a report on my own work in the philosophy of science, since the autumn of 1919 when I first began to grapple with the problem, "*When should a theory be ranked as scientific?*" or "*Is there a criterion for the scientific character or status of a theory?*"

The problem which troubled me at the time was neither, "When is a theory true?" nor, "When is a theory acceptable?" My problem was different. I *wished to distinguish between science and*

A lecture given at Peterhouse, Cambridge, in Summer 1953, originally published by the British Council under the title "Philosophy of Science: a Personal Report" in *British Philosophy in Mid-Century*, ed. C. A. Mace, 1957 (George Allen & Unwin, 1957).

pseudo-science; knowing very well that science often errs, and that pseudo-science may happen to stumble on the truth.

I knew, of course, the most widely accepted answer to my problem: that science is distinguished from pseudo-science – or from "metaphysics" – by its *empirical method*, which is essentially *inductive*, proceeding from observation or experiment. But this did not satisfy me. On the contrary, I often formulated my problem as one of distinguishing between a genuinely empirical method and a non-empirical or even a pseudo-empirical method – that is to say, a method which, although it appeals to observation and experiment, nevertheless does not come up to scientific standards. The latter method may be exemplified by astrology, with its stupendous mass of empirical evidence based on observation – on horoscopes and on biographies.

But as it was not the example of astrology which led me to my problem I should perhaps briefly describe the atmosphere in which my problem arose and the examples by which it was stimulated. After the collapse of the Austrian Empire there had been a revolution in Austria: the air was full of revolutionary slogans and ideas, and new and often wild theories. Among the theories which interested me Einstein's theory of relativity was no doubt by far the most important. Three others were Marx's theory of history, Freud's psycho-analysis, and Alfred Adler's so-called "individual psychology."

There was a lot of popular nonsense talked about these theories, and especially about relativity (as still happens even today), but I was fortunate in those who introduced me to the study of this theory. We all – the small circle of students to which I belonged – were thrilled with the result of Eddington's eclipse observations which in 1919 brought the first important confirmation of Einstein's theory of gravitation. It was a great experience for us, and one which had a lasting influence on my intellectual development.

The three other theories I have mentioned were also widely discussed among students at that time. I myself happened to come into personal contact with Alfred Adler, and even to co-operate with him in his social work among the children and young people in the working-class districts of Vienna where he had established social guidance clinics.

It was during the summer of 1919 that I began to feel more and more dissatisfied with these three theories – the Marxist theory of history, psycho-analysis, and individual psychology; and I began to feel dubious about their claims to scientific status. My problem perhaps first took the simple form, "What is wrong with Marxism, psycho-analysis, and individual psychology? Why are they so different from physical theories, from Newton's theory, and especially from the theory of relativity?"

To make this contrast clear I should explain that few of us at the time would have said that we believed in the *truth* of Einstein's theory of gravitation. This shows that it was not my doubting the *truth* of those other three theories which bothered me, but something else. Yet neither was it that I merely felt mathematical physics to be more *exact* than the sociological or psychological type of theory. Thus what worried me was neither the problem of truth, at that stage at least, nor the problem of exactness or measurability. It was rather that I felt that these other three theories, though posing as sciences, had in fact more in common with primitive myths than with science; that they resembled astrology rather than astronomy.

I found that those of my friends who were admirers of Marx, Freud, and Adler, were impressed by a number of points common to these theories, and especially by their apparent *explanatory power*. These theories appeared to be able to explain practically everything that happened within the fields to which they referred. The study of any of them seemed to have the effect of an intellectual conversion or revelation, opening your eyes to a new truth hidden from those not yet initiated. Once your eyes were thus opened you saw confirming instances everywhere: the world was full of *verifications* of the theory. Whatever happened always confirmed it. Thus its truth appeared manifest; and unbelievers were clearly people who did not want to see the manifest truth; who refused to see it, either because it was against their class interest, or because of their repressions which were still "un-analysed" and crying aloud for treatment.

The most characteristic element in this situation seemed to me the incessant stream of confirmations, of observations which "verified" the theories in question; and this point was constantly emphasized by their adherents. A Marxist could not open a newspaper without finding on every page confirming evidence for his interpretation of history; not only in the news, but also in its presentation – which revealed the class bias of the paper – and especially of course in what the paper did *not* say. The Freudian analysts emphasized that their theories were constantly verified by their "clinical observations." As for Adler, I was much impressed by a personal experience. Once, in 1919, I reported to him a case which to me did not seem particularly Adlerian, but which he found no difficulty in analysing in terms of his theory of inferiority feelings, although he had not even seen the child. Slightly shocked, I asked him how he could be so sure. "Because of my thousandfold experience," he replied; whereupon I could not help saying: "And with this new case, I suppose, your experience has become thousand-and-one-fold."

What I had in mind was that his previous observations may not have been much sounder than this new one; that each in its turn had been interpreted in the light of "previous experience," and at the same time counted as additional confirmation. What, I asked myself, did it confirm? No more than that a case could be interpreted in the light of the theory. But this meant very little, I reflected, since every conceivable case could be interpreted in the light of Adler's theory, or equally of Freud's. I may illustrate this by two very different examples of human behaviour: that of a man who pushes a child into the water with the intention of drowning it; and that of a man who sacrifices his life in an attempt to save the child. Each of these two cases can be explained with equal ease in Freudian and in Adlerian terms. According to Freud the first man suffered from repression (say, of some component of his Oedipus complex), while the second man had achieved sublimation. According to Adler the first man suffered from feelings of inferiority (producing perhaps the need to prove to himself that he dared to commit some crime), and so did the second man (whose need was to prove to himself that he dared to rescue the child). I could not think of any human behaviour which could not be interpreted in terms of either theory. It was precisely this fact – that they always fitted, that they were always confirmed – which in the eyes of their admirers constituted the strongest argument in favour of these theories. It began to dawn on me that this apparent strength was in fact their weakness.

With Einstein's theory the situation was strikingly different. Take one typical instance – Einstein's prediction, just then confirmed by the findings of Eddington's expedition. Einstein's gravitational theory had led to the result that light must be attracted by heavy bodies (such as the sun), precisely as material bodies were attracted. As a consequence it could be calculated that light from a distant fixed star whose apparent position was close to the sun would reach the earth from such a direction that the star would seem to be slightly shifted away from the sun; or, in other words, that stars close to the sun would look as if they had moved a little away from the sun, and from one another. This is a thing which cannot normally be observed since such stars are rendered invisible in daytime by the sun's overwhelming brightness; but during an eclipse it is possible to take photographs of them. If the same constellation is photographed at night one can measure the distances on the two photographs, and check the predicted effect.

Now the impressive thing about this case is the *risk* involved in a prediction of this kind. If observation shows that the predicted effect is definitely absent, then the theory is simply refuted. The theory is *incompatible with certain possible results of observation* – in fact with results which everybody before Einstein would have expected.[1] This is quite different from the situation I have previously described, when it turned out that the theories in question were compatible with the most divergent human behaviour, so that it was practically impossible to describe any human behaviour that might not be claimed to be a verification of these theories.

These considerations led me in the winter of 1919–20 to conclusions which I may now reformulate as follows.

(1) It is easy to obtain confirmations, or verifications, for nearly every theory – if we look for confirmations.

(2) Confirmations should count only if they are the result of *risky predictions*; that is to say, if, unenlightened by the theory in question, we should have expected an event which was incompatible with the theory –

an event which would have refuted the theory.

(3) Every "good" scientific theory is a prohibition: it forbids certain things to happen. The more a theory forbids, the better it is.

(4) A theory which is not refutable by any conceivable event is nonscientific. Irrefutability is not a virtue of a theory (as people often think) but a vice.

(5) Every genuine *test* of a theory is an attempt to falsify it, or to refute it. Testability is falsifiability; but there are degrees of testability: some theories are more testable, more exposed to refutation, than others; they take, as it were, greater risks.

(6) Confirming evidence should not count *except when it is the result of a genuine test of the theory*; and this means that it can be presented as a serious but unsuccessful attempt to falsify the theory. (I now speak in such cases of "corroborating evidence.")

(7) Some genuinely testable theories, when found to be false, are still upheld by their admirers – for example by introducing *ad hoc* some auxiliary assumption, or by reinterpreting the theory *ad hoc* in such a way that it escapes refutation. Such a procedure is always possible, but it rescues the theory from refutation only at the price of destroying, or at least lowering, its scientific status. (I later described such a rescuing operation as a "*conventionalist twist*" or a "*conventionalist stratagem*.")

One can sum up all this by saying that *the criterion of the scientific status of a theory is its falsifiability, or refutability, or testability.*

II

I may perhaps exemplify this with the help of the various theories so far mentioned. Einstein's theory of gravitation clearly satisfied the criterion of falsifiability. Even if our measuring instruments at the time did not allow us to pronounce on the results of the tests with complete assurance, there was clearly a possibility of refuting the theory.

Astrology did not pass the test. Astrologers were greatly impressed, and misled, by what they believed to be confirming evidence – so much so that they were quite unimpressed by any unfavourable evidence. Moreover, by making their interpretations and prophecies sufficiently vague they were able to explain away anything that might have been a refutation of the theory had the theory and the prophecies been more precise. In order to escape falsification they destroyed the testability of their theory. It is a typical soothsayer's trick to predict things so vaguely that the predictions can hardly fail: that they become irrefutable.

The Marxist theory of history, in spite of the serious efforts of some of its founders and followers, ultimately adopted this soothsaying practice. In some of its earlier formulations (for example in Marx's analysis of the character of the "coming social revolution") their predictions were testable, and in fact falsified.[2] Yet instead of accepting the refutations the followers of Marx re-interpreted both the theory and the evidence in order to make them agree. In this way they rescued the theory from refutation; but they did so at the price of adopting a device which made it irrefutable. They thus gave a "conventionalist twist" to the theory; and by this stratagem they destroyed its much advertised claim to scientific status.

The two psycho-analytic theories were in a different class. They were simply non-testable, irrefutable. There was no conceivable human behaviour which could contradict them. This does not mean that Freud and Adler were not seeing certain things correctly: I personally do not doubt that much of what they say is of considerable importance, and may well play its part one day in a psychological science which is testable. But it does mean that those 'clinical observations' which analysts naïvely believe confirm their theory cannot do this any more than the daily confirmations which astrologers find in their practice.[3] And as for Freud's epic of the Ego, the Super-ego, and the Id, no substantially stronger claim to scientific status can be made for it than for Homer's collected stories from Olympus. These theories describe some facts, but in the manner of myths. They contain most interesting psychological suggestions, but not in a testable form.

At the same time I realized that such myths may be developed, and become testable; that historically speaking all – or very nearly all – scientific theories originate from myths, and that a myth may contain important anticipations of scientific theories. Examples are Empedocles' theory of evolution by trial and error, or Parmenides' myth of the unchanging block universe in which nothing ever happens and which, if we add another dimension, becomes Einstein's block universe (in which, too, nothing ever happens, since everything is, four-dimensionally speaking, determined and laid down from the beginning). I thus felt that if a theory is found to be non-scientific, or "metaphysical" (as we might say), it is not thereby found to be unimportant, or insignificant, or "meaningless," or "nonsensical."[4] But it cannot claim to be backed by empirical evidence in the scientific sense – although it may easily be, in some genetic sense, the "result of observation."

(There were a great many other theories of this pre-scientific or pseudo-scientific character, some of them, unfortunately, as influential as the Marxist interpretation of history; for example, the racialist interpretation of history – another of those impressive and all-explanatory theories which act upon weak minds like revelations.)

Thus the problem which I tried to solve by proposing the criterion of falsifiability was neither a problem of meaningfulness or significance, nor a problem of truth or acceptability. It was the problem of drawing a line (as well as this can be done) between the statements, or systems of statements, of the empirical sciences, and all other statements – whether they are of a religious or of a metaphysical character, or simply pseudo-scientific. Years later – it must have been in 1928 or 1929 – I called this first problem of mine the "*problem of demarcation.*" The criterion of falsifiability is a solution to this problem of demarcation, for it says that statements or systems of statements, in order to be ranked as scientific, must be capable of conflicting with possible, or conceivable, observations.

III

Today I know, of course, that this *criterion of demarcation* – the criterion of testability, or falsifiability, or refutability – is far from obvious; for even now its significance is seldom realized. At that time, in 1920, it seemed to me almost trivial, although it solved for me an intellectual problem which had worried me deeply, and one which also had obvious practical consequences (for example, political ones). But I did not yet realize its full implications, or its philosophical significance. When I explained it to a fellow student of the Mathematics Department (now a distinguished mathematician in Great Britain), he suggested that I should publish it. At the time I thought this absurd for I was convinced that my problem, since it was so important for me, must have agitated many scientists and philosophers who would surely have reached my rather obvious solution. That this was not the case I learnt from Wittgenstein's work, and from its reception; and so I published my results thirteen years later in the form of a criticism of Wittgenstein's *criterion of meaningfulness.*

Wittgenstein, as you all know, tried to show in the *Tractatus* (see for example his propositions 6.53; 6.54; and 5) that all so-called philosophical or metaphysical propositions were actually non-propositions or pseudo-propositions: that they were senseless or meaningless. All genuine (or meaningful) propositions were truth functions of the elementary or atomic propositions which described "atomic facts," i.e. – facts which can in principle be ascertained by observation. In other words, meaningful propositions were fully reducible to elementary or atomic propositions which were simple statements describing possible states of affairs, and which could in principle be established or rejected by observation. If we call a statement an "observation statement" not only if it states an actual observation but also if it states anything that *may* be observed, we shall have to say (according to the *Tractatus*, 5 and 4.52) that every genuine proposition must be a truth-function of, and therefore deducible from, observation statements. All other apparent propositions will be meaningless pseudo-propositions; in fact they will be nothing but nonsensical gibberish.

This idea was used by Wittgenstein for a characterization of science, as opposed to philosophy. We read (for example in 4.11, where natural science is taken to stand in opposition to philosophy): "The totality of true propositions is the total natural science (or the totality of the natural sciences)." This means that the propositions which belong to science are those deducible from *true* observation statements; they are those propositions which can be *verified* by true observation statements. Could we know all true observation statements, we should also know all that may be asserted by natural science.

This amounts to a crude verifiability criterion of demarcation. To make it slightly less crude, it could be amended thus: "The statements which may possibly fall within the province of science are those which may possibly be verified by observation statements; and these statements, again, coincide with the class of *all* genuine or meaningful statements." For this approach, then, *verifiability, meaningfulness, and scientific character all coincide*.

I personally was never interested in the so-called problem of meaning; on the contrary, it appeared to me a verbal problem, a typical pseudo-problem. I was interested only in the problem of demarcation, i.e. in finding a criterion of the scientific character of theories. It was just this interest which made me see at once that Wittgenstein's verifiability criterion of meaning was intended to play the part of a criterion of demarcation as well; and which made me see that, as such, it was totally inadequate, even if all misgivings about the dubious concept of meaning were set aside. For Wittgenstein's criterion of demarcation – to use my own terminology in this context – is verifiability, or deducibility from observation statements. But this criterion is too narrow (*and* too wide): it excludes from science practically everything that is, in fact, characteristic of it (while failing in effect to exclude astrology). No scientific theory can ever be deduced from observation statements, or be described as a truth-function of observation statements.

All this I pointed out on various occasions to Wittgensteinians and members of the Vienna Circle. In 1931–2 I summarized my ideas in a largish book (read by several members of the Circle but never published; although part of it was incorporated in my *Logic of Scientific Discovery*); and in 1933 I published a letter to the Editor of *Erkenntnis* in which I tried to compress into two pages my ideas on the problems of demarcation and induction.[5] In this letter and elsewhere I described the problem of meaning as a pseudo-problem, in contrast to the problem of demarcation. But my contribution was classified by members of the Circle as a proposal to replace the verifiability criterion of *meaning* by a falsifiability criterion of *meaning* – which effectively made nonsense of my views.[6] My protests that I was trying to solve, not their pseudo-problem of meaning, but the problem of demarcation, were of no avail.

My attacks upon verification had some effect, however. They soon led to complete confusion in the camp of the verificationist philosophers of sense and nonsense. The original proposal of verifiability as the criterion of meaning was at least clear, simple, and forceful. The modifications and shifts which were now introduced were the very opposite.[7] This, I should say, is now seen even by the participants. But since I am usually quoted as one of them I wish to repeat that although I created this confusion I never participated in it. Neither falsifiability nor testability were proposed by me as criteria of meaning; and although I may plead guilty to having introduced both terms into the discussion, it was not I who introduced them into the theory of meaning.

Criticism of my alleged views was widespread and highly successful. I have yet to meet a criticism of my views.[8] Meanwhile, testability is being widely accepted as a criterion of demarcation.

IV

I have discussed the problem of demarcation in some detail because I believe that its solution is the key to most of the fundamental problems of the philosophy of science. I am going to give you later a list of some of these other problems, but only one of them – the *problem of induction* – can be discussed here at any length.

I had become interested in the problem of induction in 1923. Although this problem is very closely connected with the problem of demarcation, I did not fully appreciate the connection for about five years.

I approached the problem of induction through Hume. Hume, I felt, was perfectly right in pointing out that induction cannot be logically justified. He held that there can be no valid logical[9] arguments allowing us to establish "*that those instances, of which we have had no experience, resemble those, of which we have had experience.*" Consequently "*even after the observation of the frequent or constant conjunction of objects, we have no reason to draw any inference concerning any object beyond those of which we have had experience.*" For "shou'd it be said that we have experience"[10] – experience teaching us that objects constantly conjoined with certain other objects continue to be so conjoined – then, Hume says, "I wou'd renew my question, *why from this experience we form any conclusion beyond those past instances, of which we have had experience.*" In other words, an attempt to justify the practice of induction by an appeal to experience must lead to an *infinite regress*. As a result we can say that theories can never be inferred from observation statements, or rationally justified by them.

I found Hume's refutation of inductive inference clear and conclusive. But I felt completely dissatisfied with his psychological explanation of induction in terms of custom or habit.

[. . .]

We must thus replace, for the purposes of a psychological theory of the origin of our beliefs, the naïve idea of events which *are* similar by the idea of events to which we react by *interpreting* them as being similar. But if this is so (and I can see no escape from it) then Hume's psychological theory of induction leads to an infinite regress, precisely analogous to that other infinite regress which was discovered by Hume himself, and used by him to explode the logical theory of induction. For what do we wish to explain? In the example of the puppies we wish to explain behaviour which may be described as *recognizing or interpreting* a situation as a repetition of another. Clearly, we cannot hope to explain this by an appeal to earlier repetitions, once we realize that the earlier repetitions must also have been repetitions-for-them, so that precisely the same problem arises again: that of *recognizing or interpreting* a situation as a repetition of another.

To put it more concisely, similarity-for-us is the product of a response involving interpretations (which may be inadequate) and anticipations or expectations (which may never be fulfilled). It is therefore impossible to explain anticipations, or expectations, as resulting from many repetitions, as suggested by Hume. For even the first repetition-for-us must be based upon similarity-for-us, and therefore upon expectations – precisely the kind of thing we wished to explain.

This shows that there is an infinite regress involved in Hume's psychological theory.

Hume, I felt, had never accepted the full force of his own logical analysis. Having refuted the logical idea of induction he was faced with the following problem: how do we actually obtain our knowledge, as a matter of psychological fact, if induction is a procedure which is logically invalid and rationally unjustifiable? There are two possible answers: (1) We obtain our knowledge by a non-inductive procedure. This answer would have allowed Hume to retain a form of rationalism. (2) We obtain our knowledge by repetition and induction, and therefore by a logically invalid and rationally unjustifiable procedure, so that all apparent knowledge is merely a kind of belief – belief based on habit. This answer would imply that even scientific knowledge is irrational, so that rationalism is absurd, and must be given up. (I shall not discuss here the age-old attempts, now again fashionable, to get out of the difficulty by asserting that though induction is of course logically invalid if we mean by "logic" the same as "deductive logic," it is not irrational by its own standards, as may be seen from the fact that every reasonable man applies it *as a matter of fact*: it was Hume's great achievement to break this uncritical identification of the question of fact – *quid facti?* – and the question of justification or validity – *quid juris?* . . .

It seems that Hume never seriously considered the first alternative. Having cast out the logical theory of induction by repetition he struck a bargain with common sense, meekly allowing the re-entry of induction by repetition, in the guise of a psychological theory. I proposed to turn the tables upon this theory of Hume's. Instead of explaining our propensity to expect regularities as the result of repetition, I proposed to explain repetition-for-us as the result of our propensity to expect regularities and to search for them.

Thus I was led by purely logical considerations to replace the psychological theory of induction by the following view. Without waiting, passively, for repetitions to impress or impose regularities upon us, we actively try to impose regularities upon the world. We try to discover similarities in it, and to interpret it in terms of laws invented by us. Without waiting for premises we jump to conclusions. These may have to be discarded later, should observation show that they are wrong.

This was a theory of trial and error – of *conjectures and refutations*. It made it possible to understand why our attempts to force interpretations upon the world were logically prior to the observation of similarities. Since there were logical reasons behind this procedure, I thought that it would apply in the field of science also; that scientific theories were not the digest of observations, but that they were inventions – conjectures boldly put forward for trial, to be eliminated if they clashed with observations; with observations which were rarely accidental but as a rule undertaken with the definite intention of testing a theory by obtaining, if possible, a decisive refutation.

V

The belief that science proceeds from observation to theory is still so widely and so firmly held that my denial of it is often met with incredulity. I have even been suspected of being insincere – of denying what nobody in his senses can doubt.

But in fact the belief that we can start with pure observations alone, without anything in the nature of a theory, is absurd; as may be illustrated by the story of the man who dedicated his life to natural science, wrote down everything he could observe, and bequeathed his priceless collection of observations to the Royal Society to be used as inductive evidence. This story should show us that though beetles may profitably be collected, observations may not.

Twenty-five years ago I tried to bring home the same point to a group of physics students in Vienna by beginning a lecture with the following instructions: "Take pencil and paper, carefully observe, and write down what you have observed" They asked, of course, *what* I wanted them to observe. Clearly the instruction, "Observe!" is absurd.[11] (It is not even idiomatic, unless the object of the transitive verb can be taken as understood.) Observation is always selective. It needs a chosen object, a definite task, an interest, a point of view, a problem. And its description presupposes a descriptive language, with property words; it presupposes similarity and classification, which in its turn presupposes interests, points of view, and problems. "A hungry animal," writes Katz,[12] "divides the environment into edible and inedible things. An animal in flight sees roads to escape and hiding places. . . . Generally speaking, objects change . . . according to the needs of the animal." We may add that objects can be classified, and can become similar or dissimilar, *only* in this way – by being related to needs and interests. This rule applies not only to animals but also to scientists. For the animal a point of view is provided by its needs, the task of the moment, and its expectations; for the scientist by his theoretical interests, the special problem under investigation, his conjectures and anticipations, and the theories which he accepts as a kind of background: his frame of reference, his "horizon of expectations."

The problem "Which comes first, the hypothesis (*H*) or the observation (*O*)," is soluble; as is the problem, "Which comes first, the hen (*H*) or the egg (*O*)." The reply to the latter is, "An earlier kind of egg"; to the former, "An earlier kind of hypothesis." It is quite true that any particular hypothesis we choose will have been preceded by observations – the observations, for example, which it is designed to explain. But these observations, in their turn, presupposed the adoption of a frame of reference: a frame of expectations: a frame of theories. If they were significant, if they created a need for explanation and thus gave rise to the invention of a hypothesis, it was because they could not be explained within the old theoretical framework, the old horizon of expectations. There is no danger here of an infinite regress. Going back to more and more primitive theories and myths we shall in the end find unconscious, *inborn* expectations.

The theory of inborn *ideas* is absurd, I think; but every organism has inborn *reactions* or *responses*; and among them, responses adapted to impending events. These responses we may describe as "expectations" without implying that these "expectations" are conscious. The newborn baby "expects," in this sense, to be fed (and, one could even argue, to be protected and loved). In view of the close relation between expectation and knowledge we may even speak in quite a reasonable sense of "inborn knowledge." This "knowledge" is not, however, *valid a priori*; an inborn expectation, no matter how strong and specific, may be mistaken. (The newborn child may be abandoned, and starve.)

Thus we are born with expectations; with "knowledge" which, although not *valid* a priori, is *psychologically or genetically* a priori, i.e. prior to all observational experience. One of the most important of these expectations is the expectation of finding a regularity. It is connected with an inborn propensity to look out for regularities, or with a *need* to *find* regularities, as we may see from the pleasure of the child who satisfies this need.

This "instinctive" expectation of finding regularities, which is psychologically a priori, corresponds very closely to the "law of causality" which Kant believed to be part of our mental outfit and to be a priori valid. One might thus be inclined to say that Kant failed to distinguish between psychologically a priori ways of thinking or responding and a priori valid beliefs. But I do not think that his mistake was quite as crude as that. For the expectation of finding regularities is not only psychologically a priori, but also logically a priori: it is logically prior to all observational experience, for it is prior to any recognition of similarities, as we have seen; and all observation involves the recognition of similarities (or dissimilarities). But in spite of being logically a priori in this sense the expectation is not valid a priori. For it may fail: we can easily construct an environment (it would be a lethal one) which, compared with our ordinary environment, is so chaotic that we completely fail to find regularities. (All natural laws could remain valid: environments of this kind have been used in the animal experiments mentioned in the next section.)

Thus Kant's reply to Hume came near to being right; for the distinction between an a priori valid expectation and one which is both genetically *and* logically prior to observation, but not a priori valid, is really somewhat subtle. But Kant proved too much. In trying to show how knowledge is possible, he proposed a theory which had the unavoidable consequence that our quest for knowledge must necessarily succeed, which is clearly mistaken. When Kant said, "Our intellect does not draw its laws from nature but imposes its laws upon nature," he was right. But in thinking that these laws are necessarily true, or that we necessarily succeed in imposing them upon nature, he was wrong.[13] Nature very often resists quite successfully, forcing us to discard our laws as refuted; but if we live we may try again.

To sum up this logical criticism of Hume's psychology of induction we may consider the idea of building an induction machine. Placed in a simplified "world" (for example, one of sequences of coloured counters) such a machine may through repetition "learn," or even "formulate," laws of succession which hold in its "world." If such a machine can be constructed (and I have no doubt that it can) then, it might be argued, my theory must be wrong; for if a machine is capable of performing inductions on the basis of repetition, there can be no logical reasons preventing us from doing the same.

The argument sounds convincing, but it is mistaken. In constructing an induction machine we, the architects of the machine, must decide *a priori* what constitutes its "world"; what things are to be taken as similar or equal; and what *kind* of "laws" we wish the machine to be able to "discover" in its "world." In other words we must build into the machine a framework determining what is relevant or interesting in its world: the machine will have its "inborn" selection principles. The problems of similarity will have been solved for it by its makers who thus have interpreted the "world" for the machine.

VI

Our propensity to look out for regularities, and to impose laws upon nature, leads to the psychological phenomenon of *dogmatic thinking* or, more generally, dogmatic behaviour: we expect regularities everywhere and attempt to find them even where there are none; events which do not yield to these attempts we are inclined to treat as a kind of "background noise"; and we stick to our expectations even when they are inadequate and we ought to accept defeat. This dogmatism is to some extent necessary. It is demanded by a situation which can only be dealt with by forcing our conjectures upon the world. Moreover, this dogmatism allows us to approach a good theory in stages, by way of approximations: if we accept defeat too easily, we may prevent ourselves from finding that we were very nearly right.

It is clear that this *dogmatic attitude*, which makes us stick to our first impressions, is indicative of a strong belief; while a *critical attitude*, which is ready to modify its tenets, which admits doubt and demands tests, is indicative of a weaker belief. Now according to Hume's theory, and to the popular theory, the strength of a belief should be a product of repetition; thus it should always grow with experience, and always be greater in less primitive persons. But dogmatic thinking, an uncontrolled wish to impose regularities, a manifest pleasure in rites and in repetition as such, are characteristic of primitives and children; and increasing experience and maturity sometimes create an attitude of caution and criticism rather than of dogmatism.

I may perhaps mention here a point of agreement with psycho-analysis. Psycho-analysts assert that neurotics and others interpret the world in accordance with a personal set pattern which is not easily given up, and which can often be traced back to early childhood. A pattern or scheme which was adopted very early in life is maintained throughout, and every new experience is interpreted in terms of it; verifying it, as it were, and contributing to its rigidity. This is a description of what I have called the dogmatic attitude, as distinct from the critical attitude, which shares with the dogmatic attitude the quick adoption of a schema of expectations – a myth, perhaps, or a conjecture or hypothesis – but which is ready to modify it, to correct it, and even to give it up. I am inclined to suggest that most neuroses may be due to a partially arrested development of the critical attitude; to an arrested rather than a natural dogmatism; to resistance to demands for the modification and adjustment of certain schematic interpretations and responses. This resistance in its turn may perhaps be explained, in some cases, as due to an injury or shock, resulting in fear and in an increased need for assurance or certainty analogous to the way in which an injury to a limb makes us afraid to move it, so that it becomes stiff. (It might even be argued that the case of the limb is not merely analogous to the dogmatic response, but an instance of it.) The explanation of any concrete case will have to take into account the weight of the difficulties involved in making the necessary adjustments – difficulties which may be considerable, especially in a complex and changing world: we know from experiments on animals that varying degrees of neurotic behaviour may be produced at will by correspondingly varying difficulties.

I found many other links between the psychology of knowledge and psychological fields which are often considered remote from it – for example the psychology of art and music; in fact, my ideas about induction originated in a conjecture about the evolution of Western polyphony. But you will be spared this story.

VII

My logical criticism of Hume's psychological theory, and the considerations connected with it (most of which I elaborated in 1926–7, in a thesis entitled "On Habit and Belief in Laws"[14]) may seem a little removed from the field of the philosophy of science. But the distinction between dogmatic and critical thinking, or the dogmatic and the critical attitude, brings us right back to our central problem. For the dogmatic attitude is clearly related to the tendency to *verify* our laws and schemata by seeking to apply them and to confirm them, even to the point of neglecting refutations, whereas the critical attitude is one of readiness to change them – to test them; to refute them; to *falsify* them, if possible. This suggests that we may identify the critical attitude with the scientific attitude, and the dogmatic attitude with the one which we have described as pseudo-scientific.

It further suggests that genetically speaking the pseudo-scientific attitude is more primitive than, and prior to, the scientific attitude: that it is a prescientific attitude. And this primitivity or priority also has its logical aspect. For the critical attitude is not so much opposed to the dogmatic attitude as super-imposed upon it: criticism must be directed against existing and influential beliefs in need of critical revision – in other words, dogmatic beliefs. A critical attitude needs for its raw material, as it were, theories or beliefs which are held more or less dogmatically.

Thus science must begin with myths, and with the criticism of myths; neither with the collection of observations, nor with the invention of experiments, but with the critical discussion of myths, and of magical techniques and practices. The scientific tradition is distinguished from the prescientific tradition in having two layers. Like the latter, it passes on its theories; but it also passes on a critical attitude towards them. The theories are passed on, not as dogmas, but rather with the challenge to discuss them and improve upon them. This tradition is Hellenic: it may be traced back to Thales, founder of the first *school* (I do not mean "of the first *philosophical* school," but simply "of the first school") which was not mainly concerned with the preservation of a dogma.

The critical attitude, the tradition of free discussion of theories with the aim of discovering their weak spots so that they may be improved upon, is the attitude of reasonableness, of rationality. It makes far-reaching use of both verbal argument and observation – of observation in the interest of argument, however. The Greeks' discovery of the critical method gave rise at first to the mistaken hope that it would lead to the solution of all the great old problems; that it would establish certainty; that it would help to *prove* our theories, to *justify* them. But this hope was a residue of the dogmatic way of thinking; in fact nothing can be justified or proved (outside of mathematics and logic). The demand for rational proofs in science indicates a failure to keep distinct the broad realm of rationality and the narrow realm of rational certainty: it is an untenable, an unreasonable demand.

Nevertheless, the role of logical argument, of deductive logical reasoning, remains all-important for the critical approach; not because it allows us to prove our theories, or to infer them from observation statements, but because only by purely deductive reasoning is it possible for us to discover what our theories imply, and thus to criticize them effectively. Criticism, I said, is an attempt to find the weak spots in a theory, and these, as a rule, can be found only in the more remote logical consequences which can be derived from it. It is here that purely logical reasoning plays an important part in science.

Hume was right in stressing that our theories cannot be validly inferred from what we can know to be true – neither from observations nor from anything else. He concluded from this that our belief in them was irrational. If "belief" means here our inability to doubt our natural laws, and the constancy of natural regularities, then Hume is again right: this kind of dogmatic belief has, one might say, a physiological rather than a rational basis. If, however, the term "belief" is taken to cover our critical acceptance of scientific theories – a *tentative* acceptance combined with an eagerness to revise the theory if we succeed in designing a test which it cannot pass – then Hume was wrong. In such an acceptance of theories there is nothing irrational. There is not even anything irrational in relying for practical purposes upon well-tested theories, for no more rational course of action is open to us.

Assume that we have deliberately made it our task to live in this unknown world of ours; to adjust ourselves to it as well as we can; to take advantage of the opportunities we can find in it; and to explain it, *if* possible (we need not assume that it is), and as far as possible, with the help of laws and explanatory theories. *If we have made this our task, then there is no more rational procedure than the method of trial and error – of conjecture and refutation*: of boldly proposing theories; of trying our best to show that these are erroneous; and of accepting them tentatively if our critical efforts are unsuccessful.

From the point of view here developed all laws, all theories, remain essentially tentative, or conjectural, or hypothetical, even when we feel unable to doubt them any longer. Before a theory has been refuted we can never know in what way it may have to be modified. That the sun will always rise and set within twenty-four hours is still proverbial as a law "established by induction beyond reasonable doubt." It is odd that this example is still in use, though it may have served well enough in the days of Aristotle and Pytheas of Massalia – the great traveller who for centuries was called a liar because of his tales of Thule, the land of the frozen sea and the *midnight sun*.

The method of trial and error is not, of course, simply identical with the scientific or critical approach – with the method of conjecture and refutation. The method of trial and error is applied not only by Einstein but, in a more dogmatic fashion, by the amoeba also. The difference lies not so much in the trials as in a critical and constructive attitude towards errors; errors which the scientist consciously and cautiously tries to uncover in order to refute his theories with searching arguments, including appeals to the most severe experimental tests which his theories and his ingenuity permit him to design.

The critical attitude may be described as the conscious attempt to make our theories, our conjectures, suffer in our stead in the struggle for the survival of the fittest. It gives us a chance to survive the elimination of an inadequate hypothesis – when a more dogmatic attitude would eliminate it by eliminating us. (There is a touching story of an Indian community which disappeared because of its belief in the holiness of life, including that of tigers.) We thus obtain the fittest theory within our reach by the elimination of those which are less fit. (By "fitness" I do not mean merely "usefulness" but truth; . . . .) I do not think that this procedure is irrational or in need of any further rational justification.

VIII

Let us now turn from our logical criticism of the *psychology of experience* to our real problem – the problem of *the logic of science*. Although some of the things I have said may help us here, in so far as they may have eliminated certain psychological prejudices in favour of induction, my treatment of the *logical problem of induction* is completely independent of this criticism, and of all psychological considerations. Provided you do not dogmatically believe in the alleged psychological fact that we make inductions, you may now forget my whole story with the exception of two logical points: my logical remarks on testability or falsifiability as the criterion of demarcation; and Hume's logical criticism of induction.

From what I have said it is obvious that there was a close link between the two problems which interested me at that time: demarcation, and induction or scientific method. It was easy to see that the method of science is criticism, i.e. attempted falsifications. Yet it took me a few years to notice that the two problems – of demarcation and of induction – were in a sense one.

Why, I asked, do so many scientists believe in induction? I found they did so because they believed natural science to be characterized by the inductive method – by a method starting from, and relying upon, long sequences of observations and experiments. They believed that the difference between genuine science and metaphysical or pseudo-scientific speculation depended solely upon whether or not the inductive method was employed. They believed (to put it in my own terminology) that only the inductive method could provide a satisfactory *criterion of demarcation*.

I recently came across an interesting formulation of this belief in a remarkable philosophical book by a great physicist – Max Born's *Natural Philosophy of Cause and Chance*.[15] He writes: "Induction allows us to generalize a number of observations into a general rule: that night follows day and day follows night . . . But while everyday life has no definite criterion for the validity of an induction, . . . science has worked out a code, or rule of craft, for its application." Born nowhere reveals the contents of this inductive code (which, as his wording shows, contains a "definite criterion for the validity of an induction"); but he stresses that "there is no logical argument" for its acceptance: "it is a question of faith"; and he is therefore "willing to call induction a metaphysical principle." But why does he believe that such a code of valid inductive rules must exist? This becomes clear when he speaks of the "vast communities of people ignorant of, or rejecting, the rule of science, among them the members of anti-vaccination societies and believers in astrology. It is useless to argue with them; I cannot compel them to accept the same criteria of valid induction in which I believe: the code of scientific rules." This makes it quite clear that "*valid induction*" *was here meant to serve as a criterion of demarcation between science and pseudo-science.*

But it is obvious that this rule or craft of "valid induction" is not even metaphysical: it simply does not exist. No rule can ever guarantee that a generalization inferred from true observations, however often repeated, is true. (Born himself does not believe in the truth of Newtonian physics, in spite of its success, although he believes that it is based on induction.) And the success of science is not based upon rules of induction, but depends upon luck, ingenuity, and the purely deductive rules of critical argument.

I may summarize some of my conclusions as follows:

(1) Induction, i.e. inference based on many observations, is a myth. It is neither a psychological fact, nor a fact of ordinary life, nor one of scientific procedure.
(2) The actual procedure of science is to operate with conjectures: to jump to conclusions – often after one single observation (as noticed for example by Hume and Born).

(3) Repeated observations and experiments function in science as *tests* of our conjectures or hypotheses, i.e. as attempted refutations.
(4) The mistaken belief in induction is fortified by the need for a criterion of demarcation which, it is traditionally but wrongly believed, only the inductive method can provide.
(5) The conception of such an inductive method, like the criterion of verifiability, implies a faulty demarcation.
(6) None of this is altered in the least if we say that induction makes theories only probable rather than certain.

IX

If, as I have suggested, the problem of induction is only an instance or facet of the problem of demarcation, then the solution to the problem of demarcation must provide us with a solution to the problem of induction. This is indeed the case, I believe, although it is perhaps not immediately obvious.

For a brief formulation of the problem of induction we can turn again to Born, who writes: "... no observation or experiment, however extended, can give more than a finite number of repetitions"; therefore, "the statement of a law – B depends on A – always transcends experience. Yet this kind of statement is made everywhere and all the time, and sometimes from scanty material."[16]

In other words, the logical problem of induction arises from (*a*) Hume's discovery (so well expressed by Born) that it is impossible to justify a law by observation or experiment, since it "transcends experience"; (*b*) the fact that science proposes and uses laws "everywhere and all the time." (Like Hume, Born is struck by the "scanty material," i.e. the few observed instances upon which the law may be based.) To this we have to add (*c*) *the principle of empiricism* which asserts that in science, only observation and experiment may decide upon the *acceptance or rejection* of scientific statements, including laws and theories.

These three principles, (*a*), (*b*), and (*c*), appear at first sight to clash; and this apparent clash constitutes the *logical problem of induction*.

Faced with this clash, Born gives us (*c*), the principle of empiricism (as Kant and many others, including Bertrand Russell, have done before him), in favour of what he calls a "metaphysical principle"; a metaphysical principle which he does not even attempt to formulate; which he vaguely describes as a "code or rule of craft"; and of which I have never seen any formulation which even looked promising and was not clearly untenable.

But in fact the principles (*a*) to (*c*) do not clash. We can see this the moment we realize that the acceptance by science of a law or of a theory is *tentative only*; which is to say that all laws and theories are conjectures, or tentative *hypotheses* (a position which I have sometimes called "hypotheticism"); and that we may reject a law or theory on the basis of new evidence, without necessarily discarding the old evidence which originally led us to accept it.[17]

The principle of empiricism (*c*) can be fully preserved, since the fate of a theory, its acceptance or rejection, is decided by observation and experiment – by the result of tests. So long as a theory stands up to the severest tests we can design, it is accepted; if it does not, it is rejected. But it is never inferred, in any sense, from the empirical evidence. There is neither a psychological nor a logical induction. *Only the falsity of the theory can be inferred from empirical evidence, and this inference is a purely deductive one.*

Hume showed that it is not possible to infer a theory from observation statements; but this does not affect the possibility of refuting a theory by observation statements. The full appreciation of this possibility makes the relation between theories and observations perfectly clear.

This solves the problem of the alleged clash between the principles (*a*), (*b*), and (*c*), and with it Hume's problem of induction.

X

Thus the problem of induction is solved. But nothing seems less wanted than a simple solution to an age-old philosophical problem. Wittgenstein and his school hold that genuine philosophical problems do not exist;[18] from which it clearly follows that they cannot be solved. Others among my contemporaries do believe that there are philosophical problems, and respect them; but they seem to respect them too much; they seem to believe that they are insoluble, if not taboo; and they are shocked and horrified by the claim that there is a simple, neat, and lucid, solution to any of them. If there is a solution it must be deep, they feel, or at least complicated.

However this may be, I am still waiting for a simple, neat and lucid criticism of the solution which I published first in 1933 in my letter to the Editor of *Erkenntnis*,[19] and later in *The Logic of Scientific Discovery*.

Of course, one can invent new problems of induction, different from the one I have formulated and solved. (Its formulation was half its solution.) But I have yet to see any reformulation of the problem whose solution cannot be easily obtained from my old solution. I am now going to discuss some of these re-formulations.

One question which may be asked is this: how do we really jump from an observation statement to a theory?

Although this question appears to be psychological rather than philosophical, one can say something positive about it without invoking psychology. One can say first that the jump is not from an observation statement, but from a problem-situation, and that the theory must allow us *to explain* the observations which created the problem (that is, *to deduce* them from the theory strengthened by other accepted theories and by other observation statements, the so-called initial conditions). This leaves, of course, an immense number of possible theories, good and bad; and it thus appears that our question has not been answered.

But this makes it fairly clear that when we asked our question we had more in mind than, "How do we jump from an observation statement to a theory?" The question we had in mind was, it now appears, "How do we jump from an observation statement to a *good* theory?" But to this the answer is: by jumping first to *any* theory and then testing it, to find whether it is good or not; i.e. by repeatedly applying the critical method, eliminating many bad theories, and inventing many new ones. Not everybody is able to do this; but there is no other way.

Other questions have sometimes been asked. The original problem of induction, it was said, is the problem of *justifying* induction, i.e. of justifying inductive inference. If you answer this problem by saying that what is called an "inductive inference" is always invalid and therefore clearly not justifiable, the following new problem must arise: how do you justify your method of trial and error? Reply: the method of trial and error is a *method of eliminating false theories* by observation statements; and the justification for this is the purely logical relationship of deducibility which allows us to assert the falsity of universal statements if we accept the truth of singular ones.

Another question sometimes asked is this: why is it reasonable to prefer non-falsified statements to falsified ones? To this question some involved answers have been produced, for example pragmatic answers. But from a pragmatic point of view the question does not arise, since false theories often serve well enough: most formulae used in engineering or navigation are known to be false, although they may be excellent approximations and easy to handle; and they are used with confidence by people who know them to be false.

The only correct answer is the straightforward one: because we search for truth (even though we can never be sure we have found it), and because the falsified theories are known or believed to be false, while the non-falsified theories may still be true. Besides, we do not prefer *every* non-falsified theory – only one which, in the light of criticism, appears to be better than its competitors: which solves our problems, which is well tested, and of which we think, or rather conjecture or hope (considering other provisionally accepted theories), that it will stand up to further tests.

It has also been said that the problem of induction is, "Why is it *reasonable* to believe that the future will be like the past?," and that a satisfactory answer to this question should make it plain that such a belief is, in fact, reasonable. My reply is that it is reasonable to believe that the future will be very different from the past in many vitally important respects. Admittedly it is perfectly reasonable to *act* on the assumption that it will, in many respects, be like the past, and that well-tested laws will continue to hold (since we can have no better assumption to act upon); but it is also reasonable to believe that such a course of action will lead us at times into severe trouble, since some of the laws upon which we now heavily rely may easily prove unreliable. (Remember the midnight sun!) One might even say that to judge from past experience, and from our general scientific knowledge, the future will *not* be like the past, in perhaps most of the ways which those have in mind who say that it will. Water will sometimes not quench thirst, and air will choke those who breathe it. An apparent way out is to say that the future will be like the past *in the sense that the laws of nature will not change*, but this is begging the question. We speak of a "law of nature" only if we think that we have before us a regularity which does not change; and if we find that it changes then we shall not continue to call it a "law of nature." Of course our search for natural laws indicates that we hope to find them, and that we believe that there are natural laws; but our belief in any particular natural law cannot have a safer basis than our unsuccessful critical attempts to refute it.

I think that those who put the problem of induction in terms of the *reasonableness* of our beliefs are perfectly right if they are dissatisfied with a Humean, or post-Humean, sceptical despair of reason. We must indeed reject the view that a belief in science is as irrational as a belief in primitive magical practices – that both are a matter of accepting a "total ideology," a convention or a tradition based on faith. But we must be cautious if we formulate our problem, with Hume, as one of the reasonableness of our *beliefs*. We should split this problem into three – our old problem of demarcation, or of how to *distinguish* between science and primitive magic; the problem of the rationality of the scientific or critical *procedure*, and of the role of observation within it; and lastly the problem of the rationality of our *acceptance* of theories for scientific and for practical purposes. To all these three problems solutions have been offered here.

One should also be careful not to confuse the problem of the reasonableness of the scientific procedure and the (tentative) acceptance of the results of this procedure – i.e. the scientific theories – with the problem of the rationality or otherwise *of the belief that this procedure will succeed*. In practice, in practical scientific research, this belief is no doubt unavoidable and reasonable, there being no better alternative. But the belief is certainly unjustifiable in a theoretical sense, as I have argued. Moreover, if we could show, on general logical grounds, that the scientific quest is likely to succeed, one could not understand why anything like success has been so rare in the long history of human endeavours to know more about our world.

Yet another way of putting the problem of induction is in terms of probability. Let t be the theory and e the evidence: we can ask for $P(t,e)$, that is to say, the probability of t, given e. The problem of induction, it is often believed, can then be put thus: construct a *calculus of probability* which allows us to work out for any theory t what its probability is, relative to any given empirical evidence e; and show that $P(t,e)$ increases with the accumulation of supporting evidence, and reaches high values – at any rate values greater than $^1/_2$.

In *The Logic of Scientific Discovery* I explained why I think that this approach to the problem is fundamentally mistaken.[20] To make this clear, I introduced there the distinction between *probability* and *degree of corroboration or confirmation*. (The term "confirmation" has lately been so much used and misused that I have decided to surrender it to the verificationists and to use for my own purposes "corroboration" only. The term "probability" is best used in some of the many senses which satisfy the well-known calculus of probability, axiomatized, for example, by Keynes, Jeffreys, and myself; but nothing of course depends on the choice of words, as long as we do not *assume*, uncritically, that degree of corroboration must also be a probability – that is to say, that it must satisfy the calculus of probability.)

I explained in my book why we are interested in theories with a *high degree of corroboration*, And I explained why it is a mistake to conclude from this that we are interested in *highly probable* theories. I pointed out that the probability of a statement (or set of statements) is always the greater the less the statement says: it is inverse to the content or the deductive power of the statement, and thus to its explanatory power. Accordingly every interesting and powerful statement must have a low probability; and *vice versa*: a statement with a high probability will be scientifically uninteresting, because it says little and has no explanatory power. Although we seek theories with a high degree of corroboration, *as scientists we do not seek highly probable theories but explanations; that is to say, powerful and improbable theories.*[21] The opposite view – that science aims at high probability – is a characteristic development of verificationism: if you find that you cannot verify a theory, or make it certain by induction, you may turn to probability as a kind of "*Ersatz*" for certainty, in the hope that induction may yield at least that much.

I have discussed the two problems of demarcation and induction at some length. Yet since I set out to give you in this lecture a kind of report on the work I have done in this field I shall have to add, in the form of an Appendix, a few words about some other problems on which I have been working, between 1934 and 1953. I was led to most of these problems by trying to think out the consequences of the solutions to the two problems of demarcation and induction. But time does not allow me to continue my narrative, and to tell you how my new problems arose out of my old ones. Since I cannot even start a discussion of these further problems now, I shall have to confine myself to giving you a bare list of them, with a few explanatory words here and there. But even a bare list may be useful, I think. It may serve to give an idea of the fertility of the approach. It may help to illustrate what our problems look like; and it may show how many there are, and so convince you that there is no need whatever to worry over the question whether philosophical problems exist, or what philosophy is really about. So this list contains, by implication, an apology for my unwillingness to break with the old tradition of trying to solve problems with the help of rational argument, and thus for my unwillingness to participate wholeheartedly in the developments, trends, and drifts, of contemporary philosophy.

Notes

1 This is a slight oversimplification, for about half of the Einstein effect may be derived from the classical theory, provided we assume a ballistic theory of light.

2 See, for example, my *Open Society and Its Enemies*, ch. 15, section iii, and notes 13–14.

3 "Clinical observations," like all other observations, are *interpretations in the light of theories* . . . ; and for this reason alone they are apt to seem to support those theories in the light of which they were interpreted. But real support can be obtained only from observations undertaken as tests (by "attempted refutations"); and for this purpose *criteria of refutation* have to be laid down beforehand: it must be agreed which observable situations, if actually observed, mean that the theory is refuted. But what kind of clinical responses would refute to the satisfaction of the analyst not merely a particular analytic diagnosis but psychoanalysis itself? And have such criteria ever been discussed or agreed upon by analysts? Is there not, on the contrary, a whole family of analytic concepts, such as "ambivalence" (I do not suggest that there is no such thing as ambivalence), which would make it difficult, if not impossible, to agree upon such criteria? Moreover, how much headway has been made in investigating the question of the extent to which the (conscious or unconscious) expectations and theories held by the analyst influence the "clinical responses" of the patient? (To say nothing about the conscious attempts to influence the patient by proposing interpretations to him, etc.) Years ago I introduced the term "*Oedipus effect*" to describe the influence of a theory or expectation or prediction *upon the event which it predicts* or describes: it will be remembered that the causal chain leading to Oedipus' parricide was started by the oracle's prediction of this event. This is a characteristic and recurrent theme of such myths, but one which seems to have failed to attract the interest of the analysts, perhaps not accidentally. (The problem of confirmatory dreams suggested by the analyst is discussed by Freud, for example in *Gesammelte Schriften*, III, 1925, where he says on p. 314: "If anybody asserts that most of the dreams which can be utilized in an analysis . . . owe their origin to [the analyst's] suggestion, then no objection can be made from the point of view of analytic theory. Yet there is nothing in this fact," he surprisingly adds, "which would detract from the reliability of our results.")

4 The case of astrology, nowadays a typical pseudoscience, may illustrate this point. It was attacked by Aristotelians and other rationalists, down to Newton's day, for the wrong reason – for its now accepted assertion that the planets had an "influence" upon terrestrial ("sublunar") events. In fact Newton's theory of gravity, and especially the lunar theory of the tides, was historically speaking an offspring of astrological lore. Newton, it seems, was most reluctant to adopt a theory which came from the same stable as for example the theory that "influenza" epidemics are due to an astral "influence." And Galileo, no doubt for the same reason, actually rejected the lunar theory of the tides; and his misgivings about Kepler may easily be explained by his misgivings about astrology.

5 *My Logic of Scientific Discovery* (1959, 1960, 1961), here usually referred to as *L.Sc.D.*, is the translation of *Logik der Forschung* (1934), with a number of

additional notes and appendices, including (on pp. 312–14) the letter to the Editor of *Erkenntnis* mentioned here in the text which was first published in *Erkenntnis*, 3, 1933, pp. 426 f.

Concerning my never published book mentioned here in the text, see R. Carnap's paper "*Ueber Protokollstäze*" (On Protocol-Sentences), *Erkenntnis*, 3, 1932, pp. 215–28 where he gives an outline of my theory on pp. 223–8, and accepts it. He calls my theory "procedure B," and says (p. 224, top): "Starting from a point of view different from Neurath's" (who developed what Carnap calls on p. 223 "procedure A"), "Popper developed procedure B as part of his system." And after describing in detail my theory of tests, Carnap sums up his views as follows (p. 228): "After weighing the various arguments here discussed, it appears to me that the second language form with procedure B – that is in the form here described – is the most adequate among the forms of scientific language at present advocated . . . in the . . . theory of knowledge." This paper of Carnap's contained the first published report of my theory of critical testing. (See also my critical remarks in *L.Sc.D.*, note 1 to section 29, p. 104, where the date "1933" should read "1932".)

6 Wittgenstein's example of a nonsensical pseudo-proposition is: "Socrates is identical." Obviously, "Socrates is not identical" must also be nonsense. Thus the negation of any nonsense will be nonsense, and that of a meaningful statement will be meaningful. *But the negation of a testable (or falsifiable) statement need not be testable*, as was pointed out, first in my *L.Sc.D.*, (e.g. pp. 38 f.) and later by my critics. The confusion caused by taking testability as a criterion of *meaning* rather than of *demarcation* can easily be imagined.

7 The most recent example of the way in which the history of this problem is misunderstood is A. R. White's "Note on Meaning and Verification," *Mind*, 63, 1954, pp. 66 ff. J. L. Evans's article, *Mind*, 62, 1953, pp. 1 ff., which Mr White criticizes, is excellent in my opinion, and unusually perceptive. Understandably enough, neither of the authors can quite reconstruct the story. (Some hints may be found in my *Open Society*, notes 46, 51 and 52 to ch. 11.)

8 In *L.Sc.D.* I discussed, and replied to, some likely objections which afterwards were indeed raised without reference to my replies. One of them is the contention that the falsification of a natural law is just as impossible as its verification. The answer is that this objection mixes two entirely different levels of analysis (like the objection that mathematical demonstrations are impossible since checking, no matter how often repeated, can never make it quite certain that we have not overlooked a mistake). On the first level, there is a logical asymmetry: one singular statement – say about the perihelion of Mercury – can formally falsify Kepler's laws; but these cannot be formally verified by any number of singular statements. The attempt to minimize this asymmetry can only lead to confusion. On another level, we may hesitate to accept any statement, even the simplest observation statement; and we may point out that every statement involves *interpretation in the light of theories*, and that it is therefore uncertain. This does not affect the fundamental asymmetry, but it is important: most dissectors of the heart before Harvey observed the wrong things – those, which they expected to see. There can never be anything like a completely safe observation, free from the dangers of misinterpretation. (This is one of the reasons why the theory of induction does not work.) The "empirical basis" consists largely of a mixture of *theories* of lower degree of universality (of "reproducible effects"). But the fact remains that, relative to whatever basis the investigator may accept (at his peril), he can test his theory only by trying to refute it.

9 Hume does not say "logical" but "demonstrative," a terminology which, I think, is a little misleading. The following two quotations are from the *Treatise of Human Nature*, Book I, Part III, sections vi and xii. (The italics are all Hume's.)

10 This and the next quotation are from *loc. cit.*, section vi. See also Hume's *Enquiry Concerning Human Understanding*, section IV, Part II, and his *Abstract*, edited 1938 by J. M. Keynes and P. Sraffa, p. 15, and quoted in *L.Sc.D.*, new appendix *VII, text to note 6.

11 See section 30 of *L.Sc.D.*

12 Katz, *loc. cit.*

13 Kant believed that Newton's dynamics was *a priori* valid. (See his *Metaphysical Foundations of Natural Science*, published between the first and the second editions of the *Critique of Pure Reason*.) But if, as he thought, we can explain the validity of Newton's theory by the fact that our intellect imposes its laws upon nature, it follows, I think, that our intellect *must succeed* in this; which makes it hard to understand why a priori knowledge such as Newton's should be so hard to come by.

14 A thesis submitted under the title "*Gewohnheit und Gesetzerlebnis*" to the Institute of Education of the City of Vienna in 1927. (Unpublished.)

15 Max Born, *Natural Philosophy of Cause and Chance*, Oxford, 1949, p. 7.

16 *Natural Philosophy of Cause and Chance*, p. 6.

17 I do not doubt that Born and many others would agree that theories are accepted only tentatively.

But the widespread belief in induction shows that the far-reaching implications of this view are rarely seen.

18 Wittgenstein still held this belief in 1946.

19 See note 5, above.

20 *L.Sc.D.* (see note 5 above), ch. x, especially sections 80 to 83, also section 34 ff. See also my note "A Set of Independent Axioms for Probability," *Mind*, N.S. 47, 1938, p. 275. (This note has since been reprinted, with corrections, in the new appendix *ii of *L.Sc.D.*)

21 A definition, in terms of probabilities (see the next note), of $C(t,e)$, i.e. of the degree of corroboration (of a theory t relative to the evidence e) satisfying the demands indicated in my *L.Sc.D.*, sections 82 to 83, is the following:

$$C(t,e) = E(t,e)(1 + P(t)P(t,e)),$$

where $E(t,e) = (P(e,t) - P(e))/(P(e,t) + P(e))$ is a (non-additive) measure of the explanatory power of t with respect to e. Note that $C(t,e)$ is not a probability: it may have values between -1 (refutation of t by e) and $C(t,t) \leq +1$. Statements t which are lawlike and thus non-verifiable cannot even reach $C(t,e) = C(t,t)$ upon empirical evidence e. $C(t,t)$ is the *degree of corroborability* of t, and is equal to the *degree of testability* of t, or to the *content* of t. Because of the demands implied in point (6) at the end of section I above, I do not think, however, that it is possible to give a complete formalization of the idea of corroboration (or, as I previously used to say, of confirmation).

(Added 1955 to the first proofs of this paper:)

See also my note "Degree of Confirmation," *British Journal for the Philosophy of Science*, 5, 1954, pp. 143 ff. (See also 5, p. 334.) I have since simplified this definition as follows (*B.J.P.S.*, 1955, 5, p. 359:)

$$C(t,e) = (P(e,t) - P(e))/(P(e,t) - P(et) + P(e))$$

For a further improvement, see *B.J.P.S.* 6, 1955, p. 56.

7.2

The Structure of Scientific Revolutions

Thomas Kuhn

It is difficult to overemphasize the impact of Thomas Kuhn's (1922–1996) monumental work, *The Structure of Scientific Revolutions*, excerpted in the following selection. Notwithstanding its origin as a volume in the positivist's series *The International Encyclopedia of Unified Science*, it is arguably the work that, more than any other, undermined the received view of scientific theories. Its account of the history of science as normal science governed by a paradigm punctuated by occasional but cataclysmic revolutions or paradigm shifts challenged the traditional conception of science as an objective and progressive endeavor governed by a universal inductive logic or scientific method. Kuhn was widely understood to argue in *Structure* that the shift from one theory to another across revolutions is irrational and that the realities that scientists study are constructed by and so relative to the paradigm. Whether or not these are correct interpretations of the work (which is notoriously difficult to interpret), *Structure* made the history of science relevant to the philosophy of science in a way it had not been before.

I Introduction: A Role for History

History, if viewed as a repository for more than anecdote or chronology, could produce a decisive transformation in the image of science by which we are now possessed. That image has previously been drawn, even by scientists themselves, mainly from the study of finished scientific achievements as these are recorded in the classics and, more recently, in the textbooks from which each new scientific generation learns to practice its trade. Inevitably, however, the aim of such books is persuasive and pedagogic; a concept of science drawn from them is no more likely to fit the enterprise that produced them than an image of a national culture drawn from a tourist brochure

Selections from Thomas S. Kuhn, *The Structure of Scientific Revolutions*, 3rd edn. (Chicago: University of Chicago Press, 1996).

or a language text. This essay attempts to show that we have been misled by them in fundamental ways. Its aim is a sketch of the quite different concept of science that can emerge from the historical record of the research activity itself.

[. . .]

If science is the constellation of facts, theories, and methods collected in current texts, then scientists are the men who, successfully or not, have striven to contribute one or another element to that particular constellation. Scientific development becomes the piecemeal process by which these items have been added, singly and in combination, to the ever growing stockpile that constitutes scientific technique and knowledge. And history of science becomes the discipline that chronicles both these successive increments and the obstacles that have inhibited their accumulation. . . .

In recent years, however, a few historians of science have been finding it more and more difficult to fulfil the functions that the concept of development-by-accumulation assigns to them. . . .

The result of all these doubts and difficulties is a historiographic revolution in the study of science, though one that is still in its early stages. . . .

[. . .]

. . . Normal science, the activity in which most scientists inevitably spend almost all their time, is predicated on the assumption that the scientific community knows what the world is like. Much of the success of the enterprise derives from the community's willingness to defend that assumption, if necessary at considerable cost. Normal science, for example, often suppresses fundamental novelties because they are necessarily subversive of its basic commitments. Nevertheless, so long as those commitments retain an element of the arbitrary, the very nature of normal research ensures that novelty shall not be suppressed for very long. Sometimes a normal problem, one that ought to be solvable by known rules and procedures, resists the reiterated onslaught of the ablest members of the group within whose competence it falls. On other occasions a piece of equipment designed and constructed for the purpose of normal research fails to perform in the anticipated manner, revealing an anomaly that cannot, despite repeated effort, be aligned with professional expectation. In these and other ways besides, normal science repeatedly goes astray. And when it does – when, that is, the profession can no longer evade anomalies that subvert the existing tradition of scientific practice – then begin the extraordinary investigations that lead the profession at last to a new set of commitments, a new basis for the practice of science. The extraordinary episodes in which that shift of professional commitments occurs are the ones known in this essay as scientific revolutions. They are the tradition-shattering complements to the tradition-bound activity of normal science.

[. . .]

II The Route to Normal Science

In this essay, 'normal science' means research firmly based upon one or more past scientific achievements, achievements that some particular scientific community acknowledges for a time as supplying the foundation for its further practice. . . . Aristotle's *Physica*, Ptolemy's *Almagest*, Newton's *Principia* and *Opticks*, Franklin's *Electricity*, Lavoisier's *Chemistry*, and Lyell's *Geology* – these and many other works served for a time implicitly to define the legitimate problems and methods of a research field for succeeding generations of practitioners. They were able to do so because they shared two essential characteristics. Their achievement was sufficiently unprecedented to attract an enduring group of adherents away from competing modes of scientific activity. Simultaneously, it was sufficiently open-ended to leave all sorts of problems for the redefined group of practitioners to resolve.

Achievements that share these two characteristics I shall henceforth refer to as 'paradigms,' a term that relates closely to 'normal science.' By choosing it, I mean to suggest that some accepted examples of actual scientific practice – examples which include law, theory, application, and instrumentation together – provide models from which spring particular coherent traditions of scientific research. These are the traditions which the historian describes under such rubrics as 'Ptolemaic astronomy' (or 'Copernican'), 'Aristotelian dynamics' (or 'Newtonian'), 'corpuscular optics' (or 'wave optics'), and so on. . . . Men whose research is based on shared paradigms are committed to the same rules and standards for scientific practice. That commitment and the apparent consensus it produces are prerequisites for normal science, i.e., for the genesis and continuation of a particular research tradition.

[. . .]

. . . No period between remote antiquity and the end of the seventeenth century exhibited a single generally accepted view about the nature of light. Instead there were a number of competing schools and sub-schools, most of them espousing one variant or another of Epicurean, Aristotelian, or Platonic theory. One group took light to be particles emanating from material bodies; for another it was a modification of the medium that intervened between the body and the eye; still another explained light in terms of an interaction of the medium with an emanation from the eye; and there were other combinations and modifications besides. Each of the corresponding schools derived strength from its relation to some particular metaphysic, and each emphasized, as paradigmatic observations, the particular cluster of optical phenomena that its own theory could do most to explain. Other observations were dealt with by ad hoc elaborations, or they remained as outstanding problems for further research.[1]

[. . .]

Excluding those fields, like mathematics and astronomy, in which the first firm paradigms date from prehistory and also those, like biochemistry, that arose by division and recombination of specialties already matured, the [situation] outlined above [is] historically typical. . . . History suggests that the road to a firm research consensus is extraordinarily arduous.

History also suggests, however, some reasons for the difficulties encountered on that road. In the absence of a paradigm or some candidate for paradigm, all of the facts that could possibly pertain to the development of a given science are likely to seem equally relevant. As a result, early fact-gathering is a far more nearly random activity than the one that subsequent scientific development makes familiar. Furthermore, in the absence of a reason for seeking some particular form of more recondite information, early fact-gathering is usually restricted to the wealth of data that lie ready to hand. The resulting pool of facts contains those accessible to casual observation and experiment together with some of the more esoteric data retrievable from established crafts like medicine, calendar making, and metallurgy. Because the crafts are one readily accessible source of facts that could not have been casually discovered, technology has often played a vital role in the emergence of new sciences.

[. . .]

This is the situation that creates the schools characteristic of the early stages of a science's development. No natural history can be interpreted in the absence of at least some implicit body of intertwined theoretical and methodological belief that permits selection, evaluation, and criticism. If that body of belief is not already implicit in the collection of facts – in which case more than "mere facts" are at hand – it must be externally supplied, perhaps by a current metaphysic, by another science, or by personal and historical accident. No wonder, then, that in the early stages of the development of any science different men confronting the same range of phenomena, but not usually all the same particular phenomena, describe and interpret them in different ways. What is surprising, and perhaps also unique in its degree to the fields we call science, is that such initial divergences should ever largely disappear.

For they do disappear to a very considerable extent and then apparently once and for all. Furthermore, their disappearance is usually caused by the triumph of one of the pre-paradigm schools, which, because of its own characteristic beliefs and preconceptions, emphasized only some special part of the too sizable and inchoate pool of information. . . . To be accepted as a paradigm, a theory must seem better than its competitors, but it need not, and in fact never does, explain all the facts with which it can be confronted.

[. . .]

. . . When, in the development of at natural science, an individual or group first produces a synthesis able to attract most of the next generation's practitioners, the older schools gradually disappear. In part their disappearance is caused by their members' conversion to the new paradigm. But there are always some men who cling to one or another of the older views, and they are simply read out of the profession, which thereafter ignores their work. The new paradigm implies a new and more rigid definition of the field. Those unwilling or unable to accommodate their work to it must proceed in isolation or attach themselves to some other group.[2] Historically, they have often simply stayed in the departments of philosophy from which so many of the special sciences have been spawned. As these indications hint, it is sometimes just its reception of a paradigm that transforms a group previously interested merely in the study of nature into a profession or, at least, a discipline. In the sciences (though not in fields like medicine, technology, and law, of which the principal *raison d'être* is an external social need), the formation of specialized journals, the foundation of specialists' societies, and the claim for a special place in the curriculum have usually been associated with a group's first reception of a single paradigm. At least this was the case between the time, a century and a half ago, when the institutional pattern of scientific specialization first developed and the very recent time when the paraphernalia of specialization acquired a prestige of their own.

The more rigid definition of the scientific group has other consequences. When the individual scientist can take a paradigm for granted, he need no longer, in his major works, attempt to build his field anew, starting from first principles and justifying the use of each concept introduced. That can be left to the writer of textbooks. Given a textbook, however, the creative scientist can begin his research where it leaves off and thus concentrate exclusively upon the subtlest and most esoteric aspects of the natural phenomena that concern his group. And as he does this, his research communiqués will begin to change in ways whose evolution has been too little studied but whose modern end products are obvious to all and oppressive to many. No longer will his researches usually be embodied in books addressed, like Franklin's *Experiments . . . on Electricity* or Darwin's *Origin of Species*, to anyone who might be interested in the subject matter of the field. Instead they will usually appear as brief articles addressed only to professional colleagues, the men whose knowledge of a shared paradigm can be assumed and who prove to be the only ones able to read the papers addressed to them.

[. . .]

III The Nature of Normal Science

[. . .]

. . . [W]e must recognize how very limited in both scope and precision a paradigm can be at the time of its first appearance. Paradigms gain their status because they are more successful than their competitors in solving a few problems that the group of practitioners has come to recognize is acute. To be more successful is not, however, to be either completely successful with a single problem or notably successful with any large number. The success of a paradigm – whether Aristotle's analysis of motion, Ptolemy's computations of planetary position, Lavoisier's application of the balance, or Maxwell's mathematization of the electromagnetic field – is at the start largely a promise of success discoverable in selected and still incomplete examples. Normal science consists in the actualization of that promise, an actualization achieved by extending the knowledge of those facts that the paradigm displays as particularly revealing, by increasing the extent of the match between those facts and the paradigm's predictions, and by further articulation of the paradigm itself.

Few people who are not actually practitioners of a mature science realize how much mop-up work of this sort a paradigm leaves to be done or quite how fascinating such work can prove in the execution. And these points need to be understood. Mopping-up operations are what engage most scientists throughout their careers. They constitute what I am here calling normal science. Closely examined, whether historically or in the contemporary laboratory, that enterprise seems an attempt to force nature into the preformed and relatively inflexible box that the paradigm supplies. No part of the aim of normal science is to call forth new sorts of phenomena; indeed those that will not fit the box are often not seen at all. Nor do scientists normally aim to invent new theories, and they are often intolerant of those invented by others.[3] Instead, normal-scientific research is directed to the articulation of those phenomena and theories that the paradigm already supplies.

Perhaps these are defects. The areas investigated by normal science are, of course, minuscule; the enterprise now under discussion has drastically restricted vision. But those restrictions, born from confidence in a paradigm, turn out to be essential to the development of science. By focusing attention upon a small range of relatively esoteric problems, the paradigm forces scientists to investigate some part of nature in a detail and depth that would otherwise be unimaginable. . . .

[. . .]

There are, I think, only three normal foci for factual scientific investigation, and they are neither always nor permanently distinct. First is that class of facts that the paradigm has shown to be particularly revealing of the nature of things. By employing them in solving problems, the paradigm has made them worth determining both with more precision and in a larger variety of situations. At one time or another, these significant factual determinations have included: in astronomy – stellar position and magnitude, the periods of eclipsing binaries and of planets; in physics – the specific gravities and compressibilities of materials, wave lengths and spectral intensities, electrical conductivities and contact potentials; and in chemistry – composition and combining weights, boiling points and acidity of solutions, structural formulas and optical activities. Attempts to increase the accuracy and scope with which facts like these are known occupy a significant fraction of the literature of experimental and observational science. . . .

A second usual but smaller class of factual determinations is directed to those facts that, though often without much intrinsic interest, can be compared directly with predictions from the paradigm theory. As we shall see shortly, when I turn from the experimental to the theoretical problems of normal science, there are seldom many areas in which a scientific theory, particularly if it is cast in a predominantly mathematical form, can be directly compared with nature. No more than three such areas are even yet accessible to Einstein's general theory of relativity.[4] Furthermore, even in those areas where application is possible, it often demands theoretical and instrumental approximations that severely limit the agreement to be expected. Improving that agreement or finding new areas in which agreement can be demonstrated at all presents a constant challenge to the skill and imagination of the experimentalist and observer. . . .

A third class of experiments and observations exhausts, I think, the fact-gathering activities of normal science. It consists of empirical work undertaken to articulate the paradigm theory, resolving some of its residual ambiguities and permitting the solution of problems to which it had previously only drawn attention. This class proves to be the most important of all, and its description demands its subdivision. In the more mathematical sciences, some of the experiments aimed at articulation are directed to the determination of physical constants. . . . examples . . . include determinations of the astronomical unit, Avogadro's number, Joule's coefficient, the electronic charge, and so on. . . .

Efforts to articulate a paradigm are not, however, restricted to the determination of universal constants. They may, for example, also aim at quantitative laws: Boyle's Law relating gas pressure to volume, Coulomb's Law of electrical attraction, and Joule's formula relating heat generated to electrical resistance and current are all in this category. . . .

Finally, there is a third sort of experiment which aims to articulate a paradigm. More than the others this one can resemble exploration, and it is particularly prevalent in those periods and sciences that deal more with the qualitative than with the quantitative aspects of nature's regularity. Often a paradigm developed for one set of phenomena is ambiguous in its application to other closely related ones. Then experiments are necessary to choose among the alternative ways of applying the paradigm to the new area of interest. For example, the paradigm applications of the caloric theory were to heating and cooling by mixtures and by change of state. But heat could be released or absorbed in many other ways – e.g., by chemical combination, by friction, and by compression or absorption of a gas – and to each of these other phenomena the theory could be applied in several ways. If the vacuum had a heat capacity, for example, heating by compression could be explained as the result of mixing gas with void. Or it might be due to a change in the specific heat of gases with changing pressure. And there were several other explanations besides. Many experiments were undertaken to elaborate these various possibilities and to distinguish between them; all these experiments arose from the caloric theory as paradigm, and all exploited it in the design of experiments and in the interpretation of results.[5]...

Turn now to the theoretical problems of normal science, which fall into very nearly the same classes as the experimental and observational. A part of normal theoretical work, though only a small part, consists simply in the use of existing theory to predict factual information of intrinsic value. The manufacture of astronomical ephemerides, the computation of lens characteristics, and the production of radio propagation curves are examples of problems of this sort. Scientists, however, generally regard them as hack work to be relegated to engineers or technicians. At no time do very many of them appear in significant scientific journals. But these journals do contain a great many theoretical discussions of problems that, to the non-scientist, must seem almost identical. These are the manipulations of theory undertaken, not because the predictions in which they result are intrinsically valuable, but because they can be confronted directly with experiment. Their purpose is to display a new application of the paradigm or to increase the precision of an application that has already been made.

The need for work of this sort arises from the immense difficulties often encountered in developing points of contact between a theory and nature.... For the heavens Newton had derived Kepler's Laws of planetary motion and also explained certain of the observed respects in which the moon failed to obey them. For the earth he had derived the results of some scattered observations on pendulums and the tides. With the aid of additional but ad hoc assumptions, he had also been able to derive Boyle's Law and an important formula for the speed of sound in air. Given the state of science at the time, the success of the demonstrations was extremely impressive. Yet given the presumptive generality of Newton's Laws, the number of these applications was not great, and Newton developed almost no others.... How to adapt it for terrestrial applications, particularly for those of motion under constraint, was by no means clear....

[...]

But it is not all of this sort. Even in the mathematical sciences there are also theoretical problems of paradigm articulation; and during periods when scientific development is predominantly qualitative, these problems dominate. Some of the problems, in both the more quantitative and more qualitative sciences, aim simply at clarification by reformulation. The *Principia*, for example, did not always prove an easy work to apply, partly because it retained some of the clumsiness inevitable in a first venture and partly because so much of its meaning was only implicit in its applications....

[...]

These three classes of problems – determination of significant fact, matching of facts with theory, and articulation of theory – exhaust, I think, the literature of normal science, both empirical and theoretical....

IV Normal Science as Puzzle-solving

Perhaps the most striking feature of the normal research problems we have just encountered is how little they aim to produce major novelties, conceptual or phenomenal. Sometimes, as in a wave-length measurement, everything but the most esoteric detail of the result is known in advance, and the typical latitude of expectation is only somewhat wider.... And the project whose outcome does not fall in that narrower range is usually just a research failure, one which reflects not on nature but on the scientist.

[. . .]

But if the aim of normal science is not major substantive novelties – if failure to come near the anticipated result is usually failure as a scientist – then why are these problems undertaken at all? . . . Bringing a normal research problem to a conclusion is achieving the anticipated in a new way, and it requires the solution of all sorts of complex instrumental, conceptual, and mathematical puzzles. The man who succeeds proves himself an expert puzzle-solver, and the challenge of the puzzle is an important part of what usually drives him on.

[. . .]

. . . [O]ne of the things a scientific community acquires with a paradigm is a criterion for choosing problems that, while the paradigm is taken for granted, can be assumed to have solutions. To a great extent these are the only problems that the community will admit as scientific or encourage its members to undertake. Other problems, including many that had previously been standard, are rejected as metaphysical, as the concern of another discipline, or sometimes as just too problematic to be worth the time. . . .

If, however, the problems of normal science are puzzles in this sense, we need no longer ask why scientists attack them with such passion and devotion. A man may be attracted to science for all sorts of reasons. Among them are the desire to be useful, the excitement of exploring new territory, the hope of finding order, and the drive to test established knowledge. These motives and others besides also help to determine the particular problems that will later engage him. Furthermore, though the result is occasional frustration, there is good reason why motives like these should first attract him and then lead him on.[6] The scientific enterprise as a whole does from time to time prove useful, open up new territory, display order, and test long-accepted belief. Nevertheless, *the individual* engaged on a normal research problem *is almost never doing any one of these things.* Once engaged, his motivation is of a rather different sort. What then challenges him is the conviction that, if only he is skilful enough, he will succeed in solving a puzzle that no one before has solved or solved so well. . . .

. . . If it is to classify as a puzzle, a problem must be characterized by more than an assured solution. There must also be rules that limit both the nature of acceptable solutions and the steps by which they are to he obtained. . . .

. . . What can we say are the main categories into which these rules fall?[7] The most obvious and probably the most binding is exemplified by the sorts of generalizations we have just noted. These are explicit statements of scientific law and about scientific concepts and theories. While they continue to be honored, such statements help to set puzzles and to limit acceptable solutions. Newton's Laws, for example, performed those functions during the eighteenth and nineteenth centuries. . . .

Rules like these are, however, neither the only nor even the most interesting variety displayed by historical study. At a level lower or more concrete than that of laws and theories, there is, for example, a multitude of commitments to preferred types of instrumentation and to the ways in which accepted instruments may legitimately be employed. . . .

[. . .]

Finally, at a still higher level, there is another set of commitments without which no man is a scientist. The scientist must, for example, be concerned to understand the world and to extend the precision and scope with which it has been ordered. That commitment must, in turn, lead him to scrutinize, either for himself or through colleagues, some aspect of nature in great empirical detail. And, if that scrutiny displays pockets of apparent disorder, then these must challenge him to a new refinement of his observational techniques or to a further articulation of his theories. . . .

V The Priority of Paradigms

. . . Close historical investigation of a given specialty at a given time discloses a set of recurrent and quasi-standard illustrations of various theories in their conceptual, observational, and instrumental applications. These are the community's paradigms, revealed in its textbooks, lectures, and laboratory exercises. By studying them and by practicing with them, the members of the corresponding community learn their trade. . . .

[. . .]

. . . Scientists can agree that a Newton, Lavoisier, Maxwell, or Einstein has produced an apparently permanent solution to a group of outstanding problems and still disagree, sometimes without being aware of it, about the particular abstract characteristics that make those solutions permanent. They can, that is, agree in their *identification* of a paradigm without agreeing on, or even attempting to produce, a full *interpretation* or *rationalization* of it. Lack of a standard interpretation or of an agreed reduction to rules will not prevent a paradigm from guiding research. Normal science can be determined in part by the direct inspection of paradigms, a process that is often aided by but does not depend upon the formulation of rules and assumptions. Indeed, the existence of a paradigm need not even imply that any full set of rules exists.[8]

Inevitably, the first effect of those statements is to raise problems. In the absence of a competent body of rules, what restricts the scientist to a particular normal-scientific tradition? What can the phrase 'direct inspection of paradigms' mean? Partial answers to questions like these were developed by the the late Ludwig Wittgenstein, though in a very different context. . . .

[C]onfronted with a previously unobserved activity, we apply the term 'game' because what we are seeing bears a close "family resemblance" to a number of the activities that we have previously learned to call by that name. For Wittgenstein, in short, games, and chairs, and leaves are natural families, each constituted by a network of overlapping and crisscross resemblances. . . .

Something of the same sort may very well hold for the various research problems and techniques that arise within a single normal-scientific tradition. What these have in common is not that they satisfy some explicit or even some fully discoverable set of rules and assumptions that gives the tradition its character and its hold upon the scientific mind. Instead, they may relate by resemblance and by modeling to one or another part of the scientific corpus which the community in question already recognizes as among its established achievements. . . .

So far this point has been entirely theoretical: paradigms *could* determine normal science without the intervention of discoverable rules. Let me now try to increase both its clarity and urgency by indicating some of the reasons for believing that paradigms actually do operate in this manner. The first, which has already been discussed quite fully, is the severe difficulty of discovering the rules that have guided particular normal-scientific traditions. That difficulty is very nearly the same as the one the philosopher encounters when he tries to say what all games have in common. The second, to which the first is really a corollary, is rooted in the nature of scientific education. Scientists, it should already be clear, never learn concepts, laws, and theories in the abstract and by themselves. Instead, these intellectual tools are from the start encountered in a historically and pedagogically prior unit that displays them with and through their applications. A new theory is always announced together with applications to some concrete range of natural phenomena; without them it would not be even a candidate for acceptance. After it has been accepted, those same applications or others accompany the theory into the textbooks from which the future practitioner will learn his trade. They are not there merely as embroidery or even as documentation. On the contrary, the process of learning a theory depends upon the study of applications, including practice problem-solving both with a pencil and paper and with instruments in the laboratory. If, for example, the student of Newtonian dynamics ever discovers the meaning of terms like 'force,' 'mass,' 'space,' and 'time,' he does so less from the incomplete though sometimes helpful definitions in his text than by observing and participating in the application of these concepts to problem-solution.

[. . .]

. . . Normal science can proceed without rules only so long as the relevant scientific community accepts without question the particular problem-solutions already achieved. Rules should therefore become important and the characteristic unconcern about them should vanish whenever paradigms or models are felt to be insecure. That is, moreover, exactly what does occur. The preparadigm period, in particular, is regularly marked by frequent and deep debates over legitimate methods, problems, and standards of solution, though these serve rather to define schools than to produce agreement. . . . Though almost non-existent during periods of normal science, they recur regularly just before and during scientific revolutions, the periods when paradigms are first under attack and then subject to change. The transition from Newtonian to quantum mechanics evoked many debates about both the nature and the standards of physics, some of which still continue.[9] . . .

A fourth reason for granting paradigms a status prior to that of shared rules and assumptions can conclude this section. The introduction to this essay suggested that there can be small revolutions as well as large ones, that some revolutions affect only the members of a professional subspecialty, and that for such groups even the discovery of a new and unexpected phenomenon may be revolutionary. . . . If normal science is so rigid and if scientific communities are so close-knit as the preceding discussion has implied, how can a change of paradigm ever affect only a small subgroup? What has been said so far may have seemed to imply that normal science is a single monolithic and unified enterprise that must stand or fall with any one of its paradigms as well as with all of them together. But science is obviously seldom or never like that. . . . [S]ubstituting paradigms for rules should make the diversity of scientific fields and specialties easier to understand. Explicit rules, when they exist, are usually common to a very broad scientific group, but paradigms need not be. . . .

[. . .]

VI Anomaly and the Emergence of Scientific Discoveries

Normal science, the puzzle-solving activity we have just examined, is a highly cumulative enterprise, eminently successful in its aim, the steady extension of the scope and precision of scientific knowledge. In all these respects it fits with great precision the most usual image of scientific work. Yet one standard product of the scientific enterprise is missing. Normal science does not aim at novelties of fact or theory and, when successful, finds none. New and unsuspected phenomena are, however, repeatedly uncovered by scientific research, and radical new theories have again and again been invented by scientists. History even suggests that the scientific enterprise has developed a uniquely powerful technique for producing surprises of this sort. If this characteristic of science is to be reconciled with what has already been said, then research under a paradigm must be a particularly effective way of inducing paradigm change. That is what fundamental novelties of fact and theory do. Produced inadvertently by a game played under one set of rules, their assimilation requires the elaboration of another set. After they have become parts of science, the enterprise, at least of those specialists in whose particular field the novelties lie, is never quite the same again.

. . . Discovery commences with the awareness of anomaly, i.e., with the recognition that nature has somehow violated the paradigm-induced expectations that govern normal science. It then continues with a more or less extended exploration of the area of anomaly. And it closes only when the paradigm theory has been adjusted so that the anomalous has become the expected. . . .

[. . .]

. . . [N]ovelty emerges only with difficulty, manifested by resistance, against a background provided by expectation. Initially, only the anticipated and usual are experienced even under circumstances where anomaly is later to be observed. Further acquaintance, however, does result in awareness of something wrong or does relate the effect to something that has gone wrong before. That awareness of anomaly opens a period in which conceptual categories are adjusted until the initially anomalous has become the anticipated. At this point the discovery has been completed. I have already urged that that process or one very much like it is involved in the emergence of all fundamental scientific novelties. Let me now point out that, recognizing the process, we can at last begin to see why normal science, a pursuit not directed to novelties and tending at first to suppress them, should nevertheless be so effective in causing them to arise.

In the development of any science, the first received paradigm is usually felt to account quite successfully for most of the observations and experiments easily accessible to that science's practitioners. Further development, therefore, ordinarily calls for the construction of elaborate equipment, the development of an esoteric vocabulary and skills, and a refinement of concepts that increasingly lessens their resemblance to their usual common-sense prototypes. That professionalization leads, on the one hand, to an immense restriction of the scientist's vision and to a considerable resistance to paradigm change. The science has become increasingly rigid. On the other hand, within those areas to which the paradigm directs the attention of the group, normal science leads to a detail of information and to a precision of the observation-theory match that could be achieved in no other way. . . . [N]ovelty ordinarily emerges only for the man who, knowing *with precision* what he should expect, is able to recognize that something has gone wrong. Anomaly appears only against the background provided by the paradigm. The more precise and far-reaching that paradigm is, the more sensitive an indicator it provides of anomaly and hence of an occasion for paradigm change. . . . By ensuring that the paradigm will not be too easily surrendered, resistance guarantees that scientists will not be lightly distracted and that the anomalies that lead to paradigm change will penetrate existing knowledge to the core. The very fact that a significant scientific novelty so often emerges simultaneously from several laboratories is an index both to the strongly traditional nature of normal science and to the completeness with which that traditional pursuit prepares the way for its own change.

VII Crisis and the Emergence of Scientific Theories

[. . .]

. . . [T]he awareness of anomaly had lasted so long and penetrated so deep that one can appropriately describe the fields affected by it as in a state of growing crisis. Because it demands large-scale paradigm destruction and major shifts in the problems and techniques of normal science, the emergence of new theories is generally preceded by a period of pronounced professional insecurity. As one might expect, that insecurity is generated by the persistent failure of the puzzles of normal science to come out as they should. Failure of existing rules is the prelude to a search for new ones.

Look first at a particularly famous case of paradigm change, the emergence of Copernican astronomy. When its predecessor, the Ptolemaic system, was first developed during the last two centuries before Christ and the first two after, it was admirably successful in predicting the changing positions of both stars and planets. No other ancient system had performed so well; for the stars, Ptolemaic astronomy is still widely used today as an engineering approximation; for the planets, Ptolemy's predictions were as good as Copernicus'. But to be admirably successful is never, for a scientific theory, to be completely successful. With respect both to planetary position and to precession of the equinoxes, predictions made with Ptolemy's system never quite conformed with the best available observations. Further reduction of those minor discrepancies constituted many of the principal problems of normal astronomical research for many of Ptolemy's successors, just as a similar attempt to bring celestial observation and Newtonian theory together provided normal research problems for Newton's eighteenth-century successors. For some time astronomers had every reason to suppose that these attempts would be as successful as those that had led to Ptolemy's system. Given a particular discrepancy, astronomers were invariably able to eliminate it by making some particular adjustment in Ptolemy's system of compounded circles. But as time went on, a man looking at the net result of the normal research effort of many astronomers could observe that astronomy's complexity was increasing far more rapidly than its accuracy and that a discrepancy corrected in one place was likely to show up in another.[10]

[. . .]

. . . [A] novel theory emerged only after a pronounced failure in the normal problem-solving activity . . . The novel theory seems a direct response to crisis. Note also, though this may not be quite so typical, that the problems with respect to which breakdown occurred were all of a type that had long been recognized. Previous practice of normal science had given every reason to consider them solved or all but solved, which helps to explain why the sense of failure, when it came, could be so acute. Failure with a new sort of problem is often disappointing but never surprising. Neither problems nor puzzles yield often to the first attack. Finally, these examples share another characteristic that may help to make the case for the role of crisis impressive: the solution to each of them had been at least partially anticipated during a period when there was no crisis in the corresponding science; and in the absence of crisis those anticipations had been ignored.

[...]

VIII The Response to Crisis

Let us then assume that crises are a necessary precondition for the emergence of novel theories and ask next how scientists respond to their existence. Part of the answer, as obvious as it is important, can be discovered by noting first what scientists never do when confronted by even severe and prolonged anomalies. Though they may begin to lose faith and then to consider alternatives, they do not renounce the paradigm that has led them into crisis. They do not, that is, treat anomalies as counterinstances, though in the vocabulary of philosophy of science that is what they are.... [O]nce it has achieved the status of paradigm, a scientific theory is declared invalid only if an alternate candidate is available to take its place. No process yet disclosed by the historical study of scientific development at all resembles the methodological stereotype of falsification by direct comparison with nature.... The decision to reject one paradigm is always simultaneously the decision to accept another, and the judgment leading to that decision involves the comparison of both paradigms with nature *and* with each other.

... By themselves they cannot and will not falsify that philosophical theory, for its defenders will do what we have already seen scientists doing when confronted by anomaly. They will devise numerous articulations and *ad hoc* modifications of their theory in order to eliminate any apparent conflict. Many of the relevant modifications and qualifications are, in fact, already in the literature. If, therefore, these epistemological counterinstances are to constitute more than a minor irritant, that will be because they help to permit the emergence of a new and different analysis of science within which they are no longer a source of trouble. Furthermore, if a typical pattern, which we shall later observe in scientific revolutions, is applicable here, these anomalies will then no longer seem to be simply facts. From within a new theory of scientific knowledge, they may instead seem very much like tautologies, statements of situations that could not conceivably have been otherwise.

It has often been observed, for example, that Newton's second law of motion, though it took centuries of difficult factual and theoretical research to achieve, behaves for those committed to Newton's theory very much like a purely logical statement that no amount of observation could refute.[11]...

[...]

The same point can be made at least equally effectively in reverse: there is no such thing as research without counterinstances. For what is it that differentiates normal science from science in a crisis state? Not, surely, that the former confronts no counterinstances. On the contrary, what we previously called the puzzles that constitute normal science exist only because no paradigm that provides a basis for scientific research ever completely resolves all its problems. The very few that have ever seemed to do so (e.g., geometric optics) have shortly ceased to yield research problems at all and have instead become tools for engineering. Excepting those that are exclusively instrumental, every problem that normal science sees as a puzzle can be seen, from another viewpoint, as a counterinstance and thus as a source of crisis. Copernicus saw as counterinstances what most of Ptolemy's other successors had seen as puzzles in the match between observation and theory. Lavoisier saw as a counterinstance what Priestley had seen as a successfully solved puzzle in the articulation of the phlogiston theory. And Einstein saw as counterinstances what Lorentz, Fitzgerald, and others had seen as puzzles in the articulation of Newton's and Maxwell's theories....

How can the situation have seemed otherwise? That question necessarily leads to the historical and critical elucidation of philosophy, and those topics are here barred. But we can at least note two reasons why science has seemed to provide so apt an illustration of the generalization that truth and falsity are uniquely and unequivocally determined by the confrontation of statement with fact. Normal science does and must continually strive to bring theory and fact into closer agreement, and that activity can easily be seen as testing or as a search for confirmation or falsification. Instead, its object is to solve a puzzle for whose very existence the validity of the paradigm must be assumed. Failure to achieve a solution discredits only the scientist and not the theory. Here, even more than above, the proverb applies: "It is a poor carpenter who blames his tools." In addition, the manner in which science pedagogy entangles discussion of a theory with remarks on its exemplary applications has helped to reinforce a confirmation-theory drawn predominantly from other sources. Given the slightest reason for doing so, the man who reads a science text can easily take the applications to be the evidence for the theory, the reasons why it ought to be believed. But science students accept theories on the authority of teacher and text, not because of evidence. What alternatives have they, or what competence? The applications given in texts are not there as evidence but because learning them is part of learning the paradigm at the base of current practice. . . .

It follows that if an anomaly is to evoke crisis, it must usually be more than just an anomaly. . . . Sometimes an anomaly will clearly call into question explicit and fundamental generalizations of the paradigm, as the problem of ether drag did for those who accepted Maxwell's theory. Or, as in the Copernican revolution, an anomaly without apparent fundamental import may evoke crisis if the applications that it inhibits have a particular practical importance, in this case for calendar design and astrology. . . .

When, for these reasons or others like them, an anomaly comes to seem more than just another puzzle of normal science, the transition to crisis and to extraordinary science has begun. The anomaly itself now comes to be more generally recognized as such by the profession. . . .

[. . .]

. . . All crises begin with the blurring of a paradigm and the consequent loosening of the rules for normal research. In this respect research during crisis very much resembles research during the pre-paradigm period, except that in the former the locus of difference is both smaller and more clearly defined. And all crises close in one of three ways. Sometimes normal science ultimately proves able to handle the crisis-provoking problem despite the despair of those who have seen it as the end of an existing paradigm. On other occasions the problem resists even apparently radical new approaches. Then scientists may conclude that no solution will be forthcoming in the present state of their field. The problem is labelled and set aside for a future generation with more developed tools. Or, finally, the case that will most concern us here, a crisis may end with the emergence of a new candidate for paradigm and with the ensuing battle over its acceptance. . . .

[. . .]

. . . Faced with an admittedly fundamental anomaly in theory, the scientist's first effort will often be to isolate it more precisely and to give it structure. Though now aware that they cannot be quite right, he will push the rules of normal science harder than ever to see, in the area of difficulty, just where and how far they can be made to work. Simultaneously he will seek for ways of magnifying the breakdown, of making it more striking and perhaps also more suggestive than it had been when displayed in experiments the outcome of which was thought to be known in advance. And in the latter effort, more than in any other part of the post-paradigm development of science, he will look almost like our most prevalent image of the scientist. He will, in the first place, often seem a man searching at random, trying experiments just to see what will happen, looking for an effect whose nature he cannot quite guess. Simultaneously, since no experiment can be conceived without some sort of theory, the scientist in crisis will constantly try to generate speculative theories that, if successful, may disclose the road to a new paradigm and, if unsuccessful, can be surrendered with relative ease.

[. . .]

This sort of extraordinary research is often, though by no means generally, accompanied by another. It is, I think, particularly in periods of acknowledged crisis that scientists have turned to philosophical analysis as a device for unlocking the riddles of their field. Scientists have not generally needed or wanted to be philosophers. Indeed, normal science usually holds creative philosophy at arm's length, and probably for good reasons. To the extent that normal research work can be conducted by using the paradigm as a model, rules and assumptions need not be made explicit. In Section V we noted that the full set of rules sought by philosophical analysis need not even exist. But that is not to say that the search for assumptions (even for non-existent ones) cannot be an effective way to weaken the grip of a tradition upon the mind and to suggest the basis for a new one. It is no accident that the emergence of Newtonian physics in the seventeenth century and of relativity and quantum mechanics in the twentieth should have been both preceded and accompanied by fundamental philosophical analyses of the contemporary research tradition.[12] . . .

[. . .]

. . . Almost always the men who achieve these fundamental inventions of a new paradigm have been either very young or very new to the field whose paradigm they change.[13] And perhaps that point need not have been made explicit, for obviously these are the men who, being little committed by prior practice to the traditional rules of normal science, are particularly likely to see that those rules no longer define a playable game and to conceive another set that can replace them.

[. . .]

. . . Confronted with anomaly or with crisis, scientists take a different attitude toward existing paradigms, and the nature of their research changes accordingly. The proliferation of competing articulations, the willingness to try anything, the expression of explicit discontent, the recourse to philosophy and to debate over fundamentals, all these are symptoms of a transition from normal to extraordinary research. It is upon their existence more than upon that of revolutions that the notion of normal science depends.

IX The Nature and Necessity of Scientific Revolutions

. . . [S]cientific revolutions are here taken to be those non-cumulative developmental episodes in which an older paradigm is replaced in whole or in part by an incompatible new one. There is more to be said, however, and an essential part of it can be introduced by asking one further question. Why should a change of paradigm be called a revolution? In the face of the vast and essential differences between political and scientific development, what parallelism can justify the metaphor that finds revolutions in both?

One aspect of the parallelism must already be apparent. Political revolutions are inaugurated by a growing sense, often restricted to a segment of the political community, that existing institutions have ceased adequately to meet the problems posed by an environment that they have in part created. In much the same way, scientific revolutions are inaugurated by a growing sense, again often restricted to a narrow subdivision of the scientific community, that an existing paradigm has ceased to function adequately in the exploration of an aspect of nature to which that paradigm itself had previously led the way. In both political and scientific development the sense of malfunction that can lead to crisis is prerequisite to revolution. . . .

. . . Like the choice between competing political institutions, that between competing paradigms proves to be a choice between incompatible modes of community life. Because it has that character, the choice is not and cannot be determined merely by the evaluative procedures characteristic of normal science, for these depend in part upon a particular paradigm, and that paradigm is at issue. When paradigms enter, as they must, into a debate about paradigm choice, their role is necessarily circular. Each group uses its own paradigm to argue in that paradigm's defense.

The resulting circularity does not, of course, make the arguments wrong or even ineffectual. The man who premises a paradigm when arguing in its defense can nonetheless provide a clear exhibit of what scientific practice will be like for those who adopt the new view of nature. That exhibit can be immensely persuasive, often compellingly so. Yet, whatever its force, the status of the circular argument is only that of persuasion. It cannot be made logically or even probabilistically compelling for those who refuse to step into the circle. . . .

[. . .]

. . . Cumulative acquisition of unanticipated novelties proves to be an almost non-existent exception to the rule of scientific development. The man who takes historic fact seriously must suspect that science does not tend toward the ideal that our image of its cumulativeness has suggested. Perhaps it is another sort of enterprise.

If, however, resistant facts can carry us that far, then a second look at the ground we have already covered may suggest that cumulative acquisition of novelty is not only rare in fact but improbable in principle. Normal research, which *is* cumulative, owes its success to the ability of scientists regularly to select problems that can be solved with conceptual and instrumental techniques close to those already in existence. (That is why an excessive concern with useful problems, regardless of their relation to existing knowledge and technique, can so easily inhibit scientific development.) The man who is striving to solve a problem defined by existing knowledge and technique is not, however, just looking around. He knows what he wants to achieve, and he designs his instruments and directs his thoughts accordingly. Unanticipated novelty, the new discovery, can emerge only to the extent that his anticipations about nature and his instruments prove wrong. Often the importance of the resulting discovery will itself be proportional to the extent and stubbornness of the anomaly that foreshadowed it. Obviously, then, there must be a conflict between the paradigm that discloses anomaly and the one that later renders the anomaly lawlike. . . .

[. . .]

A century ago it would, I think, have been possible to let the case for the necessity of revolutions rest at this point. But today, unfortunately, that cannot be done because the view of the subject developed above cannot be maintained if the most prevalent contemporary interpretation of the nature and function of scientific theory is accepted. That interpretation, closely associated with early logical positivism and not categorically rejected by its successors, would restrict the range and meaning of an accepted theory so that it could not possibly conflict with any later theory that made predictions about some of the same natural phenomena. The best-known and the strongest case for this restricted conception of a scientific theory emerges in discussions of the relation between contemporary Einsteinian dynamics and the older dynamical equations that descend from Newton's *Principia*. From the viewpoint of this essay these two theories are fundamentally incompatible in the sense illustrated by the relation of Copernican to Ptolemaic astronomy: Einstein's theory can be accepted only with the recognition that Newton's was wrong. Today this remains a minority view.[14] We must therefore examine the most prevalent objections to it.

The gist of these objections can be developed as follows. Relativistic dynamics cannot have shown Newtonian dynamics to be wrong, for Newtonian dynamics is still used with great success by most engineers and, in selected applications, by many physicists. Furthermore, the propriety of this use of the older theory can be proved from the very theory that has, in other applications, replaced it. Einstein's theory can be used to show that predictions from Newton's equations will be as good as our measuring instruments in all applications that satisfy a small number of restrictive conditions. For example, if Newtonian theory is to provide a good approximate solution, the relative velocities of the bodies considered must be small compared with the velocity of light. Subject to this condition and a few others, Newtonian theory seems to be derivable from Einsteinian, of which it is therefore a special case.

But, the objection continues, no theory can possibly conflict with one of its special cases. If Einsteinian science seems to make Newtonian dynamics wrong, that is only because some Newtonians were so incautious as to claim that Newtonian theory yielded entirely precise results or that it was valid at very high relative velocities. Since they could not have had any evidence for such claims, they betrayed the standards of science when they made them. In so far as Newtonian theory was ever a truly scientific theory supported by valid evidence, it still is. Only extravagant claims for the theory – claims that were never properly parts of science – can have been shown by Einstein to be wrong. Purged of these merely human extravagances, Newtonian theory has never been challenged and cannot be.

[. . .]

[But] [c]an Newtonian dynamics really be *derived* from relativistic dynamics? What would such a derivation look like? Imagine a set of statements, $E_1, E_2, \ldots, E_n$, which together embody the laws of relativity theory. These statements contain variables and parameters representing spatial position, time, rest mass, etc. From them, together with the apparatus of logic and mathematics, is deducible a whole set of further statements including some that can be checked by observation. To prove the adequacy of Newtonian dynamics as a special case, we must add to the E_i's additional statements, like $(v/c)^2 << 1$, restricting the range of the parameters and variables. This enlarged set of statements is then manipulated to yield a new set, $N_1, N_2, \ldots, N_m$, which is identical in form with Newton's laws of motion, the law of gravity, and so on. Apparently Newtonian dynamics has been derived from Einsteinian, subject to a few limiting conditions.

Yet the derivation is spurious, at least to this point. Though the N_i's are a special case of the laws of relativistic mechanics, they are not Newton's Laws. Or at least they are not unless those laws are reinterpreted in a way that would have been impossible until after Einstein's work. The variables and parameters that in the Einsteinian E_i's represented spatial position, time, mass, etc., still occur in the N_i's; and they there still represent Einsteinian space, time, and mass. But the physical referents of these Einsteinian concepts are by no means identical with those of the Newtonian concepts that bear the same name. (Newtonian mass is conserved; Einsteinian is convertible with energy. Only at low relative velocities may the two be measured in the same way, and even then they must not be conceived to be the same.) Unless we change the definitions of the variables in the N_i's, the statements we have derived are not Newtonian. If we do change them, we cannot properly be said to have *derived* Newton's Laws, at least not in any sense of "derive" now generally recognized. . . .

This need to change the meaning of established and familiar concepts is central to the revolutionary impact of Einstein's theory. . . . Just because it did not involve the introduction of additional objects or concepts, the transition from Newtonian to Einsteinian mechanics illustrates with particular clarity the scientific revolution as a displacement of the conceptual network through which scientists view the world.

[. . .]

. . . Successive paradigms tell us different things about the population of the universe and about that population's behavior. They differ, that is, about such questions as the existence of subatomic particles, the materiality of light, and the conservation of heat or of energy. These are the substantive differences between successive paradigms, and they require no further illustration. But paradigms differ in more than substance, for they are directed not only to nature but also back upon the science that produced them. They are the source of the methods, problem-field, and standards of solution accepted by any mature scientific community at any given time. As a result, the reception of a new paradigm often necessitates a redefinition of the corresponding science. Some old problems may be relegated to another science or declared entirely "unscientific." Others that were previously non-existent or trivial may, with a new paradigm, become the very archetypes of significant scientific achievement. And as the problems change, so, often, does the standard that distinguishes a real scientific solution from a mere metaphysical speculation, word game, or mathematical play. The normal-scientific tradition that emerges from a scientific revolution is not only incompatible but often actually incommensurable with that which has gone before.

The impact of Newton's work upon the normal seventeenth-century tradition of scientific practice provides a striking example of these subtler effects of paradigm shift. Before Newton was born the "new science" of the century had at last succeeded in rejecting Aristotelian and scholastic explanations expressed in terms of the essences of material bodies. To say that a stone fell because its "nature" drove it toward the center of the universe had been made to look a mere tautological word-play, something it had not previously been. Henceforth the entire flux of sensory appearances, including color, taste, and even weight, was to be explained in terms of the size, shape, position, and motion of the elementary corpuscles of base matter. The attribution of other qualities to the elementary atoms was a resort to the occult and therefore out of bounds for science. Molière caught the new spirit precisely when he ridiculed the doctor who explained opium's efficacy as a soporific by attributing to it a dormitive potency. During the last half of the seventeenth century many scientists preferred to say that the round shape of the opium particles enabled them to sooth the nerves about which they move.[15]

[. . .]

. . . In learning a paradigm the scientist acquires theory, methods, and standards together, usually in an inextricable mixture. Therefore, when paradigms change, there are usually significant shifts in the criteria determining the legitimacy both of problems and of proposed solutions.

[. . .]

X Revolutions as Changes of World View

Examining the record of past research from the vantage of contemporary historiography, the historian of science may be tempted to exclaim that when paradigms change, the world itself changes with them. Led by a new paradigm, scientists adopt new instruments and look in new places. Even more important, during revolutions scientists see new and different things when looking with familiar instruments in places they have looked before. It is rather as if the professional community had been suddenly transported to another planet where familiar objects are seen in a different light and are joined by unfamiliar ones as well. Of course, nothing of quite that sort does occur: there is no geographical transplantation; outside the laboratory everyday affairs usually continue as before. Nevertheless, paradigm changes do cause scientists to see the world of their research-engagement differently. In so far as their only recourse to that world is through what they see and do, we may want to say that after a revolution scientists are responding to a different world.

It is as elementary prototypes for these transformations of the scientist's world that the familiar demonstrations of a switch in visual gestalt prove so suggestive. What were ducks in the scientist's world before the revolution are rabbits afterwards. The man who first saw the exterior of the box from above later sees its interior from below. Transformations like these, though usually more gradual and almost always irreversible, are common concomitants of scientific training. Looking at a contour map, the student sees lines on paper, the cartographer a picture of a terrain. Looking at a bubble-chamber photograph, the student sees confused and broken lines, the physicist a record of familiar subnuclear events. Only after a number of such transformations of vision does the student become an inhabitant of the scientist's world, seeing what the scientist sees and responding as the scientist does. The world that the student then enters is not, however, fixed once and for all by the nature of the environment, on the one hand, and of science, on the other. Rather, it is determined jointly by the environment and the particular normal-scientific tradition that the student has been trained to pursue. Therefore, at times of revolution, when the normal-scientific tradition changes, the scientist's perception of his environment must be re-educated – in some familiar situations he must learn to see a new gestalt. After he has done so the world of his research will seem, here and there, incommensurable with the one he had inhabited before. That is another reason why schools guided by different paradigms are always slightly at cross-purposes.

[. . .]

. . . Surveying the rich experimental literature from which these examples are drawn makes one suspect that something like a paradigm is prerequisite to perception itself. What a man sees depends both upon what he looks at and also upon what his previous visual-conceptual experience has taught him to see. In the absence of such training there can only be, in William James's phrase, "a bloomin' buzzin' confusion."

In recent years several of those concerned with the history of science have found the sorts of experiments described above immensely suggestive. N. R. Hanson, in particular, has used gestalt demonstrations to elaborate some of the same consequences of scientific belief that concern me here.[16]. . .

[. . .]

. . . The history of astronomy provides many examples of paradigm-induced changes in scientific perception, some of them even less equivocal. Can it conceivably be an accident, for example, that Western astronomers first saw change in the previously immutable heavens during the half-century after Copernicus' new paradigm was first proposed? The Chinese, whose cosmological beliefs did not preclude celestial change, had recorded the appearance of many new stars in the heavens at a much earlier date. Also, even without the aid of a telescope, the Chinese had systematically recorded the appearance of sunspots centuries before these were seen by Galileo and his contemporaries.[17] Nor were sunspots and a new star the only examples of celestial change to emerge in the heavens of Western astronomy immediately after Copernicus. Using traditional instruments, some as simple as a piece of thread, late sixteenth-century astronomers repeatedly discovered that comets wandered at will through the space previously reserved for the immutable planets and stars.[18] The very ease and rapidity with which astronomers saw new things when looking at old objects with old instruments may make us wish to say that, after Copernicus, astronomers lived in a different world. In any case, their research responded as though that were the case.

[. . .]

Do we, however, really need to describe what separates Galileo from Aristotle, or Lavoisier from Priestley, as a transformation of vision? Did these men really *see* different things when *looking at* the same sorts of objects? Is there any legitimate sense in which we can say that they pursued their research in different worlds? Those questions can no longer be postponed, for there is obviously another and far more usual way to describe all of the historical examples outlined above. Many readers will surely want to say that what changes with a paradigm is only the scientist's interpretation of observations that themselves are fixed once and for all by the nature of the environment and of the perceptual apparatus. . . .

. . . [T]he process by which either the individual or the community makes the transition from constrained fall to the pendulum or from dephlogisticated air to oxygen is not one that resembles interpretation. How could it do so in the absence of fixed data for the scientist to interpret? Rather than being an interpreter, the scientist who embraces a new paradigm is like the man wearing inverting lenses. Confronting the same constellation of objects as before and knowing that he does so, he nevertheless finds them transformed through and through in many of their details.

[. . .]

Under these circumstances we may at least suspect that scientists are right in principle as well as in practice when they treat oxygen and pendulums (and perhaps also atoms and electrons) as the fundamental ingredients of their immediate experience. As a result of the paradigm-embodied experience of the race, the culture, and, finally, the profession, the world of the scientist has come to be populated with planets and pendulums, condensers and compound ores, and other such bodies besides. Compared with these objects of perception, both meter stick readings and retinal imprints are elaborate constructs to which experience has direct access only when the scientist, for the special purposes of his research, arranges that one or the other should do so. . . .

. . . The child who transfers the word 'mama' from all humans to all females and then to his mother is not just learning what 'mama' means or who his mother is. Simultaneously he is learning some of the differences between males and females as well as something about the ways in which all but one female will behave toward him. His reactions, expectations, and beliefs – indeed, much of his perceived world – change accordingly. By the same token, the Copernicans who denied its traditional title 'planet' to the sun were not only learning what 'planet' meant or what the sun was. Instead, they were changing the meaning of 'planet' so that it could continue to make useful distinctions in a world where all celestial bodies, not just the sun, were seen differently from the way they had been seen before. . . .

[. . .]

XI The Invisibility of Revolutions

. . . I suggest that there are excellent reasons why revolutions have proved to be so nearly invisible. Both scientists and laymen take much of their image of creative scientific activity from an authoritative source that systematically disguises – partly for important functional reasons – the existence and significance of scientific revolutions. . . .

As the source of authority, I have in mind principally text books of science together with both the popularizations and the philosophical works modeled on them. . . .

[. . .]

. . . both the layman's and the practitioner's knowledge of science is based on textbooks and a few other types of literature derived from them. Textbooks, however, being pedagogic vehicles for the perpetuation of normal science, have to be rewritten in whole or in part whenever the language, problem-structure, or standards of normal science change. In short, they have to be rewritten in the aftermath of each scientific revolution, and, once rewritten, they inevitably disguise not only the role but the very existence of the revolutions that produced them. . . .

[. . .]

XII The Resolution of Revolutions

[. . .]

. . . All historically significant theories have agreed with the facts, but only more or less. There is no more precise answer to the question whether or how well an individual theory fits the facts. But questions much like that can be asked when theories are taken collectively or even in pairs. It makes a great deal of sense to ask which of two actual and competing theories fits the facts *better*. Though neither Priestley's nor Lavoisier's theory, for example, agreed precisely with existing observations, few contemporaries hesitated more than a decade in concluding that Lavoisier's theory provided the better fit of the two.

This formulation, however, makes the task of choosing between paradigms look both easier and more familiar than it is. If there were but one set of scientific problems, one world within which to work on them, and one set of standards for their solution, paradigm competition might be settled more or less routinely by some process like counting the number of problems solved by each. But, in fact, these conditions are never met completely. The proponents of competing paradigms are always at least slightly at cross-purposes. Neither side will grant all the non-empirical assumptions that the other needs in order to make its case. Like Proust and Berthollet arguing about the composition of chemical compounds, they are bound partly to talk through each other. Though each may hope to convert the other to his way of seeing his science and its problems, neither may hope to prove his case. The competition between paradigms is not the sort of battle that can be resolved by proofs.

We have already seen several reasons why the proponents of competing paradigms must fail to make complete contact with each other's viewpoints. Collectively these reasons have been described as the incommensurability of the pre- and postrevolutionary normal-scientific traditions, and we need only recapitulate them briefly here. In the first place, the proponents of competing paradigms will often disagree about the list of problems that any candidate for paradigm must resolve. Their standards or their definitions of science are not the same. Must a theory of motion explain the cause of the attractive forces between particles of matter or may it simply note the existence of such forces? Newton's dynamics was widely rejected because, unlike both Aristotle's and Descartes's theories, it implied the latter answer to the question. When Newton's theory had been accepted, a question was therefore banished from science. That question, however, was one that general relativity may proudly claim to have solved. . . .

[. . .]

How, then, are scientists brought to make this transposition? Part of the answer is that they are very often not. Copernicanism made few converts for almost a century after Copernicus' death. Newton's work was not generally accepted, particularly on the Continent, for more than half a century after the *Principia* appeared.[19] Priestley never accepted the oxygen theory, nor Lord Kelvin the electromagnetic theory, and so on. The difficulties of conversion have often been noted by scientists themselves. Darwin, in a particularly perceptive passage at the end of his *Origin of Species*, wrote: "Although I am fully convinced of the truth of the views given in this volume . . . , I by no means expect to convince experienced naturalists whose minds are stocked with a multitude of facts all viewed, during a long course of years, from a point of view directly opposite to mine. . . . [B]ut I look with confidence to the future, – to young and rising naturalists, who will be able to view both sides of the question with impartiality."[20] And Max Planck, surveying his own career in his *Scientific Autobiography*, sadly remarked that "a new scientific truth does not triumph by convincing its opponents and making them see the light, but rather because its opponents eventually die, and a new generation grows up that is familiar with it."[21]

These facts and others like them are too commonly known to need further emphasis. But they do need re-evaluation. In the past they have most often been taken to indicate that scientists, being only human, cannot always admit their errors, even when confronted with strict proof. I would argue, rather, that in these matters neither proof nor error is at issue. The transfer of allegiance from paradigm to paradigm is a conversion experience that cannot be forced. . . .

[. . .]

Probably the single most prevalent claim advanced by the proponents of a new paradigm is that they can solve the problems that have led the old one to a crisis. . . .

[. . .]

All the arguments for a new paradigm discussed so far have been based upon the competitors' comparative ability to solve problems. To scientists those arguments are ordinarily the most significant and persuasive. The preceding examples should leave no doubt about the source of their immense appeal. But, for reasons to which we shall shortly revert, they are neither individually nor collectively compelling. Fortunately, there is also another sort of consideration that can lead scientists to reject an old paradigm in favor of a new. These are the arguments, rarely made entirely explicit, that appeal to the individual's sense of the appropriate or the aesthetic – the new theory is said to be "neater," "more suitable," or "simpler" than the old. . . .

[. . .]

But paradigm debates are not really about relative problem-solving ability, though for good reasons they are usually couched in those terms. Instead, the issue is which paradigm should in the future guide research on problems many of which neither competitor can yet claim to resolve completely. A decision between alternate ways of practicing science is called for, and in the circumstances that decision must be based less on past achievement than on future promise. The man who embraces a new paradigm at an early stage must often do so in defiance of the evidence provided by problem-solving. He must, that is, have faith that the new paradigm will succeed with the many large problems that confront it, knowing only that the older paradigm has failed with a few. A decision of that kind can only be made on faith. . . .

[. . .]

At the start a new candidate for paradigm may have few supporters, and on occasions the supporters' motives may be suspect. Nevertheless, if they are competent, they will improve it, explore its possibilities, and show what it would be like to belong to the community guided by it. And as that goes on, if the paradigm is one destined to win its fight, the number and strength of the persuasive arguments in its favor will increase. More scientists will then be converted, and the exploration of the new paradigm will go on. Gradually the number of experiments, instruments, articles, and books based upon the paradigm will multiply. Still more men, convinced of the new view's fruitfulness, will adopt the new mode of practicing normal science, until at last only a few elderly hold-outs remain. And even they, we cannot say, are wrong. Though the historian can always find men – Priestley, for instance – who were unreasonable to resist for as long as they did, he will not find a point at which resistance becomes illogical or unscientific. At most he may wish to say that the man who continues to resist after his whole profession has been converted has ipso facto ceased to be a scientist.

XIII Progress through Revolutions

... Why is progress a perquisite reserved almost exclusively for the activities we call science? The most usual answers to that question have been denied in the body of this essay....

[...]

... Throughout the pre-paradigm period when there is a multiplicity of competing schools, evidence of progress, except within schools, is very hard to find. This is the period described in Section II as one during which individuals practice science, but in which the results of their enterprise do not add up to science as we know it. And again, during periods of revolution when the fundamental tenets of a field are once more at issue, doubts are repeatedly expressed about the very possibility of continued progress if one or another of the opposed paradigms is adopted. ... In short, it is only during periods of normal science that progress seems both obvious and assured. During those periods, however, the scientific community could view the fruits of its work in no other way.

[...]

... Why should progress also be the apparently universal concomitant of scientific revolutions? Once again, there is much to be learned by asking what else the result of a revolution could be. Revolutions close with a total victory for one of the two opposing camps. Will that group ever say that the result of its victory has been something less than progress? That would be rather like admitting that they had been wrong and their opponents right. To them, at least, the outcome of revolution must be progress, and they are in an excellent position to make certain that future members of their community will see past history in the same way....

[...]

Inevitably those remarks will suggest that the member of a mature scientific community is, like the typical character of Orwell's *1984*, the victim of a history rewritten by the powers that be.... [N]o explanation of progress through revolutions may stop at this point. To do so would be to imply that in the sciences might makes right, a formulation which would again not be entirely wrong if it did not suppress the nature of the process and of the authority by which the choice between paradigms is made....

[...]

... First, the new candidate must seem to resolve some outstanding and generally recognized problem that can be met in no other way. Second, the new paradigm must promise to preserve a relatively large part of the concrete problem-solving ability that has accrued to science through its predecessors....

[...]

... In the sciences there need not be progress of another sort. We may, to be more precise, have to relinquish the notion, explicit or implicit, that changes of paradigm carry scientists and those who learn from them closer and closer to the truth.

... The developmental process described in this essay has been a process of evolution *from* primitive beginnings – a process whose successive stages are characterized by an increasingly detailed and refined understanding of nature. But nothing that has been or will be said makes it a process of evolution *toward* anything. Inevitably that lacuna will have disturbed many readers. We are all deeply accustomed to seeing science as the one enterprise that draws constantly nearer to some goal set by nature in advance.

But need there be any such goal? Can we not account for both science's existence and its success in terms of evolution from the community's state of knowledge at any given time? Does it really help to imagine that there is some one full, objective, true account of nature and that the proper measure of scientific achievement is the extent to which it brings us closer to that ultimate goal? If we can learn to substitute evolution-from-what-we-do-know for evolution-toward-what-we-wish-to-know, a number of vexing problems may vanish in the process. Somewhere in this maze, for example, must lie the problem of induction.

[. . .]

. . . The net result of a sequence of such revolutionary selections, separated by periods of normal research, is the wonderfully adapted set of instruments we call modern scientific knowledge. Successive stages in that developmental process are marked by an increase in articulation and specialization. And the entire process may have occurred, as we now suppose biological evolution did, without benefit of a set goal, a permanent fixed scientific truth, of which each stage in the development of scientific knowledge is a better exemplar.

[. . .]

Postscript – 1969

[. . .]

. . . [I]n much of the book the term 'paradigm' is used in two different senses. On the one hand, it stands for the entire constellation of beliefs, values, techniques, and so on shared by the members of a given community. On the other, it denotes one sort of element in that constellation, the concrete puzzle-solutions which, employed as models or examples, can replace explicit rules as a basis for the solution of the remaining puzzles of normal science. . . .

[. . .]

. . . [A] few readers of this book have concluded that my concern is primarily or exclusively with major revolutions such as those associated with Copernicus, Newton, Darwin, or Einstein. A clearer delineation of community structure should, however, help to enforce the rather different impression I have tried to create. A revolution is for me a special sort of change involving a certain sort of reconstruction of group commitments. But it need not be a large change, nor need it seem revolutionary to those outside a single community, consisting perhaps of fewer than twenty-five people. . . .

[. . .]

Having isolated a particular community of specialists . . . one may usefully ask: What do its members share that accounts for the relative fulness of their professional communication and the relative unanimity of their professional judgments? To that question my original text licenses the answer, a paradigm or set of paradigms. But for this use, unlike the one to be discussed below, the term is inappropriate. . . . For present purposes I suggest 'disciplinary matrix': 'disciplinary' because it refers to the common possession of the practitioners of a particular discipline; 'matrix' because it is composed of ordered elements of various sorts, each requiring further specification. . . .

One important sort of component I shall label 'symbolic generalizations,' having in mind those expressions, deployed without question or dissent by group members, which can readily be cast in a logical form like $(x)(y)(z)\phi(x, y, z)$. They are the formal or the readily formalizable components of the disciplinary matrix. . . .

[. . .]

Consider next a second type of component of the disciplinary matrix, one about which a good deal has been said in my original text under such rubrics as 'metaphysical paradigms' or 'the metaphysical parts of paradigms.' . . . [N]ow I would describe such commitments as beliefs in particular models, and I would expand the category models to include also the relatively heuristic variety . . . [A]ll models have similar functions. Among other things they supply the group with preferred or permissible analogies and metaphors. By doing so they help to determine what will be accepted as an explanation and as a puzzle-solution; conversely, they assist in the determination of the roster of unsolved puzzles and in the evaluation of the importance of each. . . .

A third sort of element in the disciplinary matrix I shall here describe as values. . . . Probably the most deeply held values concern predictions:

they should be accurate; quantitative predictions are preferable to qualitative ones; whatever the margin of permissible error, it should be consistently satisfied in a given field; and so on. There are also, however, values to be used in judging whole theories: they must, first and foremost, permit puzzle-formulation and solution; where possible they should be simple, self-consistent, and plausible, compatible, that is, with other theories currently deployed. (I now think it a weakness of my original text that so little attention is given to such values as internal and external consistency in considering sources of crisis and factors in theory choice.) . . .

. . . But judgments of simplicity, consistency, plausibility, and so on often vary greatly from individual to individual. What was for Einstein an insupportable inconsistency in the old quantum theory, one that rendered the pursuit of normal science impossible, was for Bohr and others a difficulty that could be expected to work itself out by normal means. Even more important, in those situations where values must be applied, different values, taken alone, would often dictate different choices. One theory may be more accurate but less consistent or plausible than another; again the old quantum theory provides an example. In short, though values are widely shared by scientists and though commitment to them is both deep and constitutive of science, the application of values is sometimes considerably affected by the features of individual personality and biography that differentiate the members of the group.

To many readers of the preceding chapters, this characteristic of the operation of shared values has seemed a major weakness of my position. Because I insist that what scientists share is not sufficient to command uniform assent about such matters as the choice between competing theories or the distinction between an ordinary anomaly and a crisis-provoking one, I am occasionally accused of glorifying subjectivity and even irrationality.[22] But that reaction ignores two characteristics displayed by value judgments in any field. First, shared values can be important determinants of group behavior even though the members of the group do not all apply them in the same way. . . . Second, individual variability in the application of shared values may serve functions essential to science. The points at which values must be applied are invariably also those at which risks must be taken. . . . In matters like these the resort to shared values rather than to shared rules governing individual choice may be the community's way of distributing risk and assuring the long-term success of its enterprise.

Turn now to a fourth sort of element in the disciplinary matrix, not the only other kind but the last I shall discuss here. For it the term 'paradigm' would be entirely appropriate, both philologically and autobiographically; this is the component of a group's shared commitments which first led me to the choice of that word. Because the term has assumed a life of its own, however, I shall here substitute 'exemplars.' By it I mean, initially, the concrete problem-solutions that students encounter from the start of their scientific education, whether in laboratories, on examinations, or at the ends of chapters in science texts. To these shared examples should, however, be added at least some of the technical problem-solutions found in the periodical literature that scientists encounter during their post-educational research careers and that also show them by example how their job is to be done. . . .

[. . .]

. . . That sort of learning is not acquired by exclusively verbal means. Rather it comes as one is given words together with concrete examples of how they function in use; nature and words are learned together. To borrow once more Michael Polanyi's useful phrase, what results from this process is "tacit knowledge" which is learned by doing science rather than by acquiring rules for doing it.

[. . .]

Notice now that two groups, the members of which have systematically different sensations on receipt of the same stimuli, do *in some sense* live in different worlds. We posit the existence of stimuli to explain our perceptions of the world, and we posit their immutability to avoid both individual and social solipsism. About neither posit have I the slightest reservation. But our world is populated in the first instance not by stimuli but by the objects of our sensations, and these need not be the same, individual to individual or group to group. . . .

Returning now to exemplars and rules . . . One of the fundamental techniques by which the members of a group, whether an entire culture or a specialists' sub-community within it, learn to see the same things when confronted with the same stimuli is by being shown examples of situations that their predecessors in the group have already learned to see as like each other and as different from other sorts of situations. . . .

[. . .]

. . . A number of [philosophers], however, have reported that I believe the following:[23] the proponents of incommensurable theories cannot communicate with each other at all; as a result, in a debate over theory-choice there can be no recourse to *good* reasons; instead theory must be chosen for reasons that are ultimately personal and subjective; some sort of mystical apperception is responsible for the decision actually reached. . . .

. . . Debates over theory-choice cannot be cast in a form that fully resembles logical or mathematical proof. . . .

Nothing about that relatively familiar thesis implies either that there are no good reasons for being persuaded or that those reasons are not ultimately decisive for the group. Nor does it even imply that the reasons for choice are different from those usually listed by philosophers of science: accuracy, simplicity, fruitfulness, and the like. What it should suggest, however, is that such reasons function as values and that they can thus be differently applied, individually and collectively, by men who concur in honoring them. . . .

[. . .]

. . . Two men who perceive the same situation differently but nevertheless employ the same vocabulary in its discussion must be using words differently. They speak, that is, from what I have called incommensurable viewpoints. . . .

[. . .]

The men who experience such communication breakdowns must, however, have some recourse. The stimuli that impinge upon them are the same. So is their general neural apparatus, however differently programmed. Furthermore, except in a small, if all-important, area of experience even their neural programming must be very nearly the same, for they share a history, except the immediate past. As a result, both their everyday and most of their scientific world and language are shared. Given that much in common, they should be able to find out a great deal about how they differ. . . .

Briefly put, what the participants in a communication breakdown can do is recognize each other as members of different language communities and then become translators.[24] . . .

[. . .]

The conversion experience that I have likened to a gestalt switch remains, therefore, at the heart of the revolutionary process. Good reasons for choice provide motives for conversion and a climate in which it is more likely to occur. Translation may, in addition, provide points of entry for the neural reprogramming that, however inscrutable at this time, must underlie conversion. But neither good reasons nor translation constitute conversion, and it is that process we must explicate in order to understand an essential sort of scientific change.

[. . .]

One consequence of the position just outlined has particularly bothered a number of my critics.[25] They find my viewpoint relativistic, particularly as it is developed in the last section of this book. . . .

[. . .]

Imagine an evolutionary tree representing the development of the modern scientific specialties from their common origins in, say, primitive natural philosophy and the crafts. A line drawn up that tree, never doubling back, from the trunk to the tip of some branch would trace a succession of theories related by descent. Considering any two such theories, chosen from points not too near their origin, it should be easy to design a list of criteria that would enable an uncommitted observer to distinguish the earlier from the more recent theory time after time. Among the most useful would be: accuracy of prediction, particularly of quantitative prediction; the balance between esoteric and everyday subject matter; and the number of different problems solved. Less useful for this purpose, though also important determinants of scientific life, would be such values as simplicity, scope, and compatibility with other specialties. Those lists are not yet the ones required, but I have no doubt that they can be completed. If they can, then scientific development is, like biological, a unidirectional and irreversible process. Later scientific theories are better than earlier ones for solving puzzles in the often quite different environments to which they are applied. That is not a relativist's position, and it displays the sense in which I am a convinced believer in scientific progress.

Compared with the notion of progress most prevalent among both philosophers of science and laymen, however, this position lacks an essential element. . . . One often hears that successive theories grow ever closer to, or approximate more and more closely to, the truth. . . .

Perhaps there is some other way of salvaging the notion of 'truth' for application to whole theories, but this one will not do. There is, I think, no theory-independent way to reconstruct phrases like 'really there'; the notion of a match between the ontology of a theory and its "real" counterpart in nature now seems to me illusive in principle. Besides, as a historian, I am impressed with the implausability of the view. I do not doubt, for example, that Newton's mechanics improves on Aristotle's and that Einstein's improves on Newton's as instruments for puzzle-solving. But I can see in their succession no coherent direction of ontological development.

Notes

1. Vasco Ronchi, *Histoire de la lumière*, trans. Jean Taton (Paris, 1956), chaps. i–iv.
2. The history of electricity provides an excellent example which could be duplicated from the careers of Priestley, Kelvin, and others. Franklin reports that Nollet, who at mid-century was the most influential of the Continental electricians, "lived to see himself the last of his Sect, except Mr B. – his Eleve and immediate Disciple" (Max Farrand [ed.], *Benjamin Franklin's Memoirs* [Berkeley, Calif., 1949], pp. 384–86). More interesting, however, is the endurance of whole schools in increasing isolation from professional science. Consider, for example, the case of astrology, which was once an integral part of astronomy. Or consider the continuation in the late eighteenth and early nineteenth centuries of a previously respected tradition of "romantic" chemistry. This is the tradition discussed by Charles C. Gillispie in 'The *Encyclopédie* and the Jacobin Philosophy of Science: A Study in Ideas and Consequences," *Critical Problems in the History of Science*, ed. Marshall Clagett (Madison, Wis., 1959), pp. 255–89; and "The Formation of Lamarck's Evolutionary Theory," *Archives internationales d'histoire des sciences*, XXXVII (1956), 323–38.
3. Bernard Barber, "Resistance by Scientists to Scientific Discovery," *Science*, CXXXIV (1961), 596–602.
4. The only long-standing check point still generally recognized is the precession of Mercury's perihelion. The red shift in the spectrum of light from distant stars can be derived from considerations more elementary than general relativity, and the same may be possible for the bending of light around the sun, a point now in some dispute. In any case, measurements of the latter phenomenon remain equivocal. One additional check point may have been established very recently: the gravitational shift of Mossbauer radiation. Perhaps there will soon be others in this now active but long dormant field. For an up-to-date capsule account of the problem, see L. I. Schiff, "A Report on the NASA Conference on Experimental Tests of Theories of Relativity," *Physics Today*, XIV (1961), 42–8.
5. T. S. Kuhn, "The Caloric Theory of Adiabatic Compression," *Isis*, XLIX (1958), 132–40.
6. The frustrations induced by the conflict between the individual's role and the over-all pattern of scientific development can, however, occasionally be quite serious. On this subject, see Lawrence S. Kubie, "Some Unsolved Problems of the Scientific Career," *American Scientist*, XLI (1953), 596–613; and XLII (1954), 104–12.
7. I owe this question to W. O. Hagstrom, whose work in the sociology of science sometimes overlaps my own.
8. Michael Polanyi has brilliantly developed a very similar theme, arguing that much of the scientist's success depends upon "tacit knowledge," i.e., upon knowledge that is acquired through practice and that cannot be articulated explicitly. See his *Personal Knowledge* (Chicago, 1958), particularly chaps. v and vi.

9 For controversies over quantum mechanics, see Jean Ullmo, *La crise de la physique quantique* (Paris, 1950), chap. ii.

10 J. L. E. Dreyer, *A History of Astronomy from Thales to Kepler* (2d ed.; New York, 1953), chaps, xi–xii.

11 See particularly the discussion in N. R. Hanson, *Patterns of Discovery* (Cambridge, 1958), pp. 99–105.

12 For the philosophical counterpoint that accompanied seventeenth-century mechanics, see René Dugas, *La mécanique au XVIIe siècle* (Neuchatel, 1954), particularly chap. xi. For the similar nineteenth-century episode, see the same author's earlier book, *Histoire de la mécanique* (Neuchatel, 1950), pp. 419–43.

13 This generalization about the role of youth in fundamental scientific research is so common as to be a cliché. Furthermore, a glance at almost any list of fundamental contributions to scientific theory will provide impressionistic confirmation. Nevertheless, the generalization badly needs systematic investigation. Harvey C. Lehman (*Age and Achievement* [Princeton, 1953]) provides many useful data; but his studies make no attempt to single out contributions that involve fundamental reconceptualization. Nor do they inquire about the special circumstances, if any, that may accompany relatively late productivity in the sciences.

14 See, for example, the remarks by P. P. Wiener in *Philosophy of Science*, XXV (1958), 298.

15 For corpuscularism in general, see Marie Boas, "The Establishment of the Mechanical Philosophy," *Osiris*, X (1952), 412–541. For the effect of particle-shape on taste, see ibid., p. 483.

16 N. R. Hanson, *Patterns of Discovery* (Cambridge, 1958), chap. i.

17 Joseph Needham, *Science and Civilization in China*, III (Cambridge, 1959), 423–29, 434–36.

18 T. S. Kuhn, *The Copernican Revolution* (Cambridge, Mass., 1957), pp. 206–9.

19 I. B. Cohen, *Franklin and Newton: An Inquiry into Speculative Newtonian Experimental Science and Franklin's Work in Electricity as an Example Thereof* (Philadelphia, 1956), pp. 93–4.

20 Charles Darwin, *On the Origin of Species*... (authorized edition from 6th English ed.; New York, 1889), II, 295–6.

21 Max Planck, *Scientific Autobiography and Other Papers*, trans. F. Gaynor (New York, 1949), pp. 33–4.

22 See particularly: Dudley Shapere, "Meaning and Scientific Change," in *Mind and Cosmos: Essays in Contemporary Science and Philosophy*, The University of Pittsburgh Series in the Philosophy of Science, III (Pittsburgh, 1966), 41–85; Israel Scheffler, *Science and Subjectivity* (New York, 1967); and the essays of Sir Karl Popper and Imre Lakatos in *Growth of Knowledge*.

23 See the works cited in note 22, above, and also the essay by Stephen Toulmin in *Growth of Knowledge*.

24 The already classic source for most of the relevant aspects of translation is W. V. O. Quine, *Word and Object* (Cambridge, Mass., and New York, 1960), chaps, i and ii. But Quine seems to assume that two men receiving the same stimulus must have the same sensation and therefore has little to say about the extent to which a translator must be able to *describe* the world to which the language being translated applies. For the latter point see, E. A. Nida, "Linguistics and Ethnology in Translation Problems," in Del Hymes (ed.), *Language and Culture in Society* (New York, 1964), pp. 90–7.

25 Shapere, "Structure of Scientific Revolutions," and Popper in *Growth of Knowledge*.

7.3

Science and Pseudoscience

Imre Lakatos

Sympathetic with Popper's conviction that falsifications are crucial to the proper understanding of scientific methodology, Imre Lakatos (1922–1974) was also familiar with Kuhn's seemingly contrary suggestion that scientists never take an anomalous observation to refute or even challenge the paradigm under which they work. In an attempt to resolve this apparent conflict, Lakatos formulated the *methodology of research programs*, described in the reading below.

Man's respect for knowledge is one of his most peculiar characteristics. Knowledge in Latin is *scientia*, and science came to be the name of the most respectable kind of knowledge. But what distinguishes knowledge from superstition, ideology or pseudoscience? The Catholic Church ex-communicated Copernicans, the Communist Party persecuted Mendelians on the ground that their doctrines were pseudoscientific. The demarcation between science and pseudoscience is not merely a problem of armchair philosophy: it is of vital social and political relevance.

Many philosophers have tried to solve the problem of demarcation in the following terms: a statement constitutes knowledge if sufficiently many people believe it sufficiently strongly. But the history of thought shows us that many people were totally committed to absurd beliefs. If the strength of beliefs were a hallmark of knowledge, we should have to rank some tales about demons, angels, devils, and of heaven and hell as knowledge. Scientists, on the other hand, are very sceptical even of their best theories. Newton's is the most powerful theory science has yet produced, but Newton himself never believed that bodies attract each other at a distance. So no degree of commitment to beliefs makes them knowledge. Indeed, the hallmark of scientific behaviour is a certain scepticism even towards one's most cherished theories. Blind commitment to a theory is not an intellectual virtue: it is an intellectual crime.

Thus a statement may be pseudoscientific even if it is eminently 'plausible' and everybody

From Imre Lakatos, *The Methodology of Scientific Research Programmes: Vol. 1: Philosophical Papers* (Cambridge: Cambridge University Press, 1977), pp. 1–7.

believes in it, and it may be scientifically valuable even if it is unbelievable and nobody believes in it. A theory may even be of supreme scientific value even if no one understands it, let alone believes it.

The cognitive value of a theory has nothing to do with its psychological influence on people's minds. Belief, commitment, understanding are states of the human mind. But the objective, scientific value of a theory is independent of the human mind which creates it or understands it. Its scientific value depends only on what objective support these conjectures have in facts. As Hume said:

> If we take in our hand any volume; of divinity, or school metaphysics, for instance; let us ask, does it contain any abstract reasoning concerning quantity or number? No. Does it contain any experimental reasoning concerning matter of fact and existence? No. Commit it then to the flames. For it can contain nothing but sophistry and illusion.

But what is 'experimental' reasoning? If we look at the vast seventeenth-century literature on witchcraft, it is full of reports of careful observations and sworn evidence – even of experiments. Glanvill, the house philosopher of the early Royal Society, regarded witchcraft as the paradigm of experimental reasoning. We have to define experimental reasoning before we start Humean book burning.

In scientific reasoning, theories are confronted with facts; and one of the central conditions of scientific reasoning is that theories must be supported by facts. Now how exactly can facts support theory?

Several different answers have been proposed. Newton himself thought that he proved his laws from facts. He was proud of not uttering mere hypotheses: he only published theories proven from facts. In particular, he claimed that he deduced his laws from the 'phenomena' provided by Kepler. But his boast was nonsense, since according to Kepler, planets move in ellipses, but according to Newton's theory, planets would move in ellipses only if the planets did not disturb each other in their motion. But they do. This is why Newton had to devise a perturbation theory from which it follows that no planet moves in an ellipse.

One can today easily demonstrate that there can be no valid derivation of a law of nature from any finite number of facts; but we still keep reading about scientific theories being proved from facts. Why this stubborn resistance to elementary logic?

There is a very plausible explanation. Scientists want to make their theories respectable, deserving of the title 'science', that is, genuine knowledge. Now the most relevant knowledge in the seventeenth century, when science was born, concerned God, the Devil, Heaven and Hell. If one got one's conjectures about matters of divinity wrong, the consequence of one's mistake was eternal damnation. Theological knowledge cannot be fallible: it must be beyond doubt. Now the Enlightenment thought that we were fallible and ignorant about matters theological. There is no scientific theology and, therefore, no theological knowledge. Knowledge can only be about Nature, but this new type of knowledge had to be judged by the standards they took over straight from theology: it had to be proven beyond doubt. Science had to achieve the very certainty which had escaped theology. A scientist, worthy of the name, was not allowed to guess: he had to prove each sentence he uttered from facts. This was the criterion of scientific honesty. Theories unproven from facts were regarded as sinful pseudoscience, heresy in the scientific community.

It was only the downfall of Newtonian theory in this century which made scientists realize that their standards of honesty had been utopian. Before Einstein most scientists thought that Newton had deciphered God's ultimate laws by proving them from the facts. Ampère, in the early nineteenth century, felt he had to call his book on his speculations concerning electromagnetism: *Mathematical Theory of Electrodynamic Phenomena Unequivocally Deduced from Experiment.* But at the end of the volume he casually confesses that some of the experiments were never performed and even that the necessary instruments had not been constructed!

If all scientific theories are equally unprovable, what distinguishes scientific knowledge from ignorance, science from pseudoscience?

One answer to this question was provided in the twentieth century by 'inductive logicians'. Inductive logic set out to define the probabilities of different theories according to the available total evidence. If the mathematical probability of a theory is high, it qualifies as scientific; if it is low or even zero, it is not scientific. Thus the hallmark of scientific honesty would be never to say anything that is not at least highly probable. Probabilism has an attractive feature: instead of simply providing a black-and-white distinction between science and pseudoscience, it provides a continuous scale from poor theories with low probability, to good theories with high probability. But, in 1934, Karl Popper, one of the most influential philosophers of our time, argued that the mathematical probability of all theories, scientific or pseudoscientific, given *any* amount of evidence is zero. If Popper is right, scientific theories are not only equally unprovable but also equally improbable. A new demarcation criterion was needed and Popper proposed a rather stunning one. A theory may be scientific even if there is not a shred of evidence in its favour, and it may be pseudoscientific even if all the available evidence is in its favour. That is, the scientific or non-scientific character of a theory can be determined independently of the facts. A theory is 'scientific' if one is prepared to specify in advance a crucial experiment (or observation) which can falsify it, and it is pseudoscientific if one refuses to specify such a 'potential falsifier'. But if so, we do not demarcate scientific theories from pseudo-scientific ones, but rather scientific method from non-scientific method. Marxism, for a Popperian, is scientific if the Marxists are prepared to specify facts which, if observed, make them give up Marxism. If they refuse to do so, Marxism becomes a pseudoscience. It is always interesting to ask a Marxist, what conceivable event would make him abandon his Marxism. If he is committed to Marxism, he is bound to find it immoral to specify a state of affairs which can falsify it. Thus a proposition may petrify into pseudoscientific dogma or become genuine knowledge, depending on whether we are prepared to state observable conditions which would refute it.

Is, then, Popper's falsifiability criterion the solution to the problem of demarcating science from pseudoscience? No. For Popper's criterion ignores the remarkable tenacity of scientific theories. Scientists have thick skins. They do not abandon a theory merely because facts contradict it. They normally either invent some rescue hypothesis to explain what they then call a mere anomaly or, if they cannot explain the anomaly, they ignore it, and direct their attention to other problems. Note that scientists talk about anomalies, recalcitrant instances, not refutations. History of science, of course, is full of accounts of how crucial experiments allegedly killed theories. But such accounts are fabricated long after the theory had been abandoned. Had Popper ever asked a Newtonian scientist under what experimental conditions he would abandon Newtonian theory, some Newtonian scientists would have been exactly as nonplussed as are some Marxists.

What, then, is the hallmark of science? Do we have to capitulate and agree that a scientific revolution is just an irrational change in commitment, that it is a religious conversion? Tom Kuhn, a distinguished American philosopher of science, arrived at this conclusion after discovering the naïvety of Popper's falsificationism. But if Kuhn is right, then there is no explicit demarcation between science and pseudoscience, no distinction between scientific progress and intellectual decay, there is no objective standard of honesty. But what criteria can he then offer to demarcate scientific progress from intellectual degeneration?

In the last few years I have been advocating a methodology of scientific research programmes, which solves some of the problems which both Popper and Kuhn failed to solve.

First, I claim that the typical descriptive unit of great scientific achievements is not an isolated hypothesis but rather a research programme. Science is not simply trial and error, a series of conjectures and refutations. 'All swans are white' may be falsified by the discovery of one black swan. But such trivial trial and error does not rank as science. Newtonian science, for instance, is not simply a set of four conjectures – the three laws of mechanics and the law of gravitation. These four laws constitute only the 'hard core' of the Newtonian programme. But this hard core is tenaciously protected from refutation by a vast 'protective belt' of auxiliary hypotheses. And, even more importantly, the research programme also has a 'heuristic', that is, a powerful problem-solving machinery, which, with the help of sophisticated mathematical techniques, digests anomalies and even turns them into positive evidence. For instance, if a planet does not move exactly as it should, the Newtonian scientist checks his conjectures concerning atmospheric refraction, concerning propagation of light in magnetic storms, and hundreds of other conjectures which are all part of the programme. He may even invent a hitherto unknown planet and calculate its position, mass and velocity in order to explain the anomaly.

Now, Newton's theory of gravitation, Einstein's relativity theory, quantum mechanics, Marxism, Freudianism, are all research programmes, each with a characteristic hard core stubbornly defended, each with its more flexible protective belt and each with its elaborate problem-solving machinery. Each of them, at any stage of its development, has unsolved problems and undigested anomalies. All theories, in this sense, are born refuted and die refuted. But are they equally good? Until now I have been describing what research programmes are like. But how can one distinguish a scientific or progressive programme from a pseudoscientific or degenerating one?

Contrary to Popper, the difference cannot be that some are still unrefuted, while others are already refuted. When Newton published his *Principia*, it was common knowledge that it could not properly explain even the motion of the moon; in fact, lunar motion refuted Newton. Kaufmann, a distinguished physicist, refuted Einstein's relativity theory in the very year it was published. But all the research programmes I admire have one characteristic in common. They all predict novel facts, facts which had been either undreamt of, or have indeed been contradicted by previous or rival programmes. In 1686, when Newton published his theory of gravitation, there were, for instance, two current theories concerning comets. The more popular one regarded comets as a signal from an angry God warning that He will strike and bring disaster. A little known theory of Kepler's held that comets were celestial bodies moving along straight lines. Now according to Newtonian theory, some of them moved in hyperbolas or parabolas never to return; others moved in ordinary ellipses. Halley, working in Newton's programme, calculated on the basis of observing a brief stretch of a comet's path that it would return in seventy-two years' time; he calculated to the minute when it would be seen again at a well-defined point of the sky. This was incredible. But seventy-two years later, when both Newton and Halley were long dead, Halley's comet returned exactly as Halley predicted. Similarly, Newtonian scientists predicted the existence and exact motion of small planets which had never been observed before. Or let us take Einstein's programme. This programme made the stunning prediction that if one measures the distance between two stars in the night and if one measures the distance between them during the day (when they are visible during an eclipse of the sun), the two measurements will be different. Nobody had thought to make such an observation before Einstein's programme. Thus, in a progressive research programme, theory leads to the discovery of hitherto unknown novel facts. In degenerating programmes, however, theories are fabricated only in order to accommodate known facts. Has, for instance, Marxism ever predicted a stunning novel fact successfully? Never! It has some famous unsuccessful predictions. It predicted the absolute impoverishment of the working class. It predicted that the first socialist revolution would take place in the industrially most developed society. It predicted that socialist societies would be free of revolutions. It predicted that there will be no conflict of interests between socialist countries. Thus the early predictions of Marxism were bold and stunning but they failed. Marxists explained all their failures: they explained the rising living standards of the working class by devising a theory of imperialism; they even explained why the first socialist revolution occurred in industrially backward Russia. They 'explained' Berlin 1953, Budapest 1956, Prague 1968. They 'explained' the Russian–Chinese conflict. But their auxiliary hypotheses were all cooked up after the event to protect Marxian theory from the facts. The Newtonian programme led to novel facts; the Marxian lagged behind the facts and has been running fast to catch up with them.

To sum up. The hallmark of empirical progress is not trivial verifications: Popper is right that there are millions of them. It is no success for Newtonian theory that stones, when dropped, fall towards the earth, no matter how often this is repeated. But so-called 'refutations' are not the hallmark of empirical failure, as Popper has preached, since all programmes grow in a permanent ocean of anomalies. What really count are dramatic, unexpected, stunning predictions: a few of them are enough to tilt the balance; where theory lags behind the facts, we are dealing with miserable degenerating research programmes.

Now, how do scientific revolutions come about? If we have two rival research programmes, and one is progressing while the other is degenerating, scientists tend to join the progressive programme. This is the rationale of scientific revolutions. But while it is a matter of intellectual honesty to keep the record public, it is not dishonest to stick to a degenerating programme and try to turn it into a progressive one.

As opposed to Popper the methodology of scientific research programmes does not offer instant rationality. One must treat budding programmes leniently: programmes may take decades before they get off the ground and become empirically progressive. Criticism is not a Popperian quick kill, by refutation. Important criticism is always constructive: there is no refutation without a better theory. Kuhn is wrong in thinking that scientific revolutions are sudden, irrational changes in vision. The history of science refutes both Popper and Kuhn: on close inspection both Popperian crucial experiments and Kuhnian revolutions turn out to be myths: what normally happens is that progressive research programmes replace degenerating ones.

The problem of demarcation between science and pseudoscience has grave implications also for the institutionalization of criticism. Copernicus's theory was banned by the Catholic Church in 1616 because it was said to be pseudoscientific. It was taken off the index in 1820 because by that time the Church deemed that facts had proved it and therefore it became scientific. The Central Committee of the Soviet Communist Party in 1949 declared Mendelian genetics pseudoscientific and had its advocates, like Academician Vavilov, killed in concentration camps; after Vavilov's murder Mendelian genetics was rehabilitated; but the Party's right to decide what is science and publishable and what is pseudoscience and punishable was upheld. The new liberal Establishment of the West also exercises the right to deny freedom of speech to what it regards as pseudoscience, as we have seen in the case of the debate concerning race and intelligence. All these judgments were inevitably based on some sort of demarcation criterion. This is why the problem of demarcation between science and pseudoscience is not a pseudoproblem of armchair philosophers: it has grave ethical and political implications.

Unit 8

After the Received View: Explanation

The model of explanation advocated by the positivists (and particularly by Carl Hempel, see reading 5.2) has been subject to a barrage of criticisms, to the extent that it is no longer tenable. But, as with other aspects of the received view, there is little consensus as to what is fundamentally wrong with it, and therefore no single successor.

In this unit, we will briefly review a number of counterexamples to Hempel's model, many of which are presented in the first reading by Wesley Salmon. Then we will discuss three prominent contenders to Hempel's model: the *statistical-relevance* (S-R) model, developed by Salmon and presented in reading 8.2; the *causal-mechanical* (C-M) model, also developed by Salmon (who has renounced his earlier S-R model) and presented in reading 8.3; and the *unificationist* model developed initially by Michael Friedman and then by Philip Kitcher (Kitcher's version is presented in reading 8.4 and discussed below). This is certainly not an exhaustive list of alternative models that have been developed; but these three are of great historical as well as theoretical significance, and they represent very different themes that continue to be heard.

1 The Counterexamples

You might remember that Hempel offered not one but two models of explanation: deductive-nomological (D-N) and inductive-statistical (I-S).[1] But common to both are the following four principles:

1. An explanation must essentially involve at least one law in the explanans.
2. The explanandum must be expectable given the explanans. D-N explanations satisfy this requirement by implying the explanans; I-S explanations satisfy it by rendering the explanandum probable given the explanans.[2]
3. The explanans must be empirically testable.
4. The explanans must be true.

There has been no sustained critique of (3), and little of (4).[3] The most damaging criticisms concern (1) and (2), and it is counterexamples to these two conditions that we will consider below in detail. They fall into two groups: those that purport to show that the conditions that Hempel specified for an adequate explanation are not sufficient (so that, at minimum, more than (1)–(4) is required), and those that purport to show that they are not necessary (that, in particular, either (1) or (2) is not required).

Examples that demonstrate that (1)–(4) are insufficient are arguments in which phenomena are derived from laws but that are intuitively not explanations. Some of these are called *asymmetry* cases. Given the laws of geometry and the linear path of sunlight we can derive – and explain – the length of a flagpole's shadow from its height and the position of the sun. However, we can equally well derive the flagpole's height from the length of its shadow given those same laws. Although both of these derivations conform to Hempel's conditions, only the first is explanatory. Such cases demonstrate that explanation is asymmetric in ways that predictive arguments are not.[4]

Another example related to asymmetry cases is that of the barometer and storm. A barometer's rapid fall is reliably correlated with the approach of a storm. But, of course, the barometer's fall does not explain the storm's arrival. Intuitively, the drop in atmospheric pressure is the explanation (or at least part of the explanation) for both the storm's arrival and the barometer reading (and, therefore, for the correlation between them).[5]

Other counterexamples that attack the sufficiency of (1)–(4) are cases of *irrelevance*. Men who take birth control pills do not become pregnant. This seems lawful; after all, it is not as though such men could become pregnant but just have not to date.[6] Suppose John Jones takes his wife's pills. Thanks to the law, we can predict with complete assurance that he will not become pregnant. But of course this is only because he is male; the fact that he took the pills is explanatorily irrelevant. The example nevertheless conforms to Hempel's conditions.[7]

Another example related to irrelevance cases is that of Jones's recovery from a cold within a week. We might try to explain that recovery by appeal to the fact that he took a large dose of vitamin C and the fact that most people who take such doses recover from a cold within a week. The results conform to Hempel's requirements for an I-S explanation. But, in fact, most colds clear up within a week, whether or not the sufferer takes vitamin C; taking it makes no statistical difference to time to recovery. This suggests that Jones's having taken the vitamin C is irrelevant to the explanation of his recovery.[8]

A similar example applies against the D-N model. Jones eats a pound of arsenic which, given the laws of biochemistry, ensures his death within twenty-four hours. And indeed he does die. But it turns out that he died only because he was hit by a bus just after his arsenic meal, which meal is, therefore, explanatorily irrelevant.

Examples that purport to show that Hempel's conditions are not necessary attack either the first condition (that laws must be essentially involved in the explanation) or the second (that the explanation must make the explanandum expectable). Here is an example against (1): you ask why there is a puddle of ink on the floor, and your friend tells you that she knocked over the inkwell. We would normally take that as a perfectly good explanation; but neither you nor your friend is likely aware of the various physical laws (and detailed initial conditions) that would have to be cited in order for this to qualify as a D-N explanation.

Aware of such examples, Hempel claimed that these are not truly explanations, but only "elliptical" or "partial" explanations, or only "explanation sketches," to be filled in with the relevant laws and all the requisite details that would be required to give a full explanation. A concern about this response is that a great many everyday explanations are like the inkwell case in that we are clueless as to the relevant laws and incapable of drawing the required inference, but remain confident that we have indeed provided an explanation.

Counterexamples to the necessity of (2) (that explanations must render the explanandum expectable) include two sorts of cases. The first are those where an explanation seems to provide little to no grounds for any prediction whatsoever. We can appeal to evolutionary theory, for example, to explain why a particular organism displays a certain trait by appeal to natural selection and the survival advantage conferred on the organism's predecessors in virtue of their possessing that trait. But no predictions (of new species, of new traits of existing species, etc.) have been derived from the theory; it appears to have only explanatory, but little predictive, capacity.[9]

The second sort of case involves explanations that, far from rendering the explanandum likely, render it positively unlikely. Only people with advanced syphilis develop paresis (a degenerative neurological condition). We might therefore explain why Jones developed paresis by pointing out that he has advanced syphilis. But even among those with advanced syphilis, paresis is rare; so we could not have predicted that Jones would develop it on the basis of his advanced syphilis.

2 Salmon's S-R Model

In the paresis example just discussed, having advanced syphilis does not provide grounds to predict the onset of paresis. But it is nevertheless relevant to it: since people who do not have syphilis do not get paresis, having syphilis makes it more likely that one will have paresis as compared to the general population. Syphilis is therelfore *statistically relevant* to paresis in the sense that having the one will make a difference to the probability that one will have the other; and it is this statistical relevance that seems to provide the basis for the explanation.

In the earlier case of John Jones who takes his wife's birth-control pills and avoids pregnancy, statistical relevance also seems crucial. Jones's pill consumption does not change the probability that he will become pregnant, and that seems to be why we would not cite it in explaining his avoidance of pregnancy.

With these examples in mind, Salmon proposed statistical relevance as the fundamental explanatory concept: to provide an explanation of Xs having property P is essentially to provide all, and only, the statistically relevant information concerning whether X will have P.

In order to do this, we divide the population into various subclasses depending on whether the members share (or share the absence of) various characteristics. We keep dividing the population in this way so long as the resulting subclasses differ in the probability that they will have P; when they no longer differ, we stop dividing. We have then divided (or "partitioned") the population into a number of subclasses, each with a different probability that its members will have P, where the members of each subclass all have the same probability of having P. We then complete the explanation simply by indicating to which of these subclasses X belongs.

Consider the pregnancy example. The probability of a woman being pregnant is lower if she takes birth-control pills than it is if she does not (that being the point of such pills). Also, the probability of a woman who took such pills' being pregnant is higher than that of men (since those pills are not perfectly effective). And the probability of a woman who did not take such pills' being pregnant is obviously higher than that of men. However, the probability that men who take such pills become pregnant is equal to that for men who do not take such pills (that is, zero).

So we should divide the population of humans into the following subclasses:

1 Females taking birth-control pills;
2 Females not taking birth-control pills; and
3 Males.

For each of these subclasses, the probability that members of that subclass become pregnant is different from that of members of the other classes. And within each subclass, each member has the same probability of being pregnant; that is why subclass (3) is not divided between males who take such pills and those who do not. When a class admits of no further statistically relevant subdivisions in this way (relative to the probability of having the property of interest, which in this case is being pregnant), it is called a *homogenous* class. Salmon required that each class be divided into further subclasses whenever the class is not homogenous.[10] But when the subclass is homogenous, he insisted that it must be divided no further. The problem with all of the irrelevance cases is that they involve divisions among homogenous subclasses.[11]

Since we know that, in fact, the class of females can be divided further into statistically relevant subclasses – those that are sexually active and those that are not, for example – subclasses (1) and (2) would in fact require further subdivision in order to be homogenous. And in addition to those characteristics that we know about, there may be others (consumption of certain foods, for example) that would also result in statistically relevant subclasses of which we are not aware. According to Salmon, a correct and complete explanation is provided only when the classes are in fact or "objectively" homogenous, whether or not we know, or are able to know, that they are homogenous.[12] It is therefore possible that what looks like a complete explanation given the information we have may not in fact be so. But we do our best, and hope that the divisions that we have identified are the only ones that are statistically relevant, so that the resulting subclasses are truly homogenous.

We are now finally able to explain Jones's failure to become pregnant. We first subdivide the population into homogenous subclasses. We then indicate, for each subclass, the probability that members of that subclass will be pregnant. Finally, we simply state to which of the subclasses Jones belongs (subclass (3), with probability zero, in this case). We have then provided all of the statistically relevant information we have concerning the explanandum. "What more," Salmon asked, "could one ask of an explanation?" (reading 8.2, p. 549).

3 S-R and Low-Probability Explanations

We will explain Jones's paresis in the same way. We divide the population between those with advanced syphilis and those without, indicate the probability that members in each subclass get paresis (a low probability for those with, zero for those without), and indicate that Jones is a member of the first subclass.

Notice that in this case the explanation does not render the explanandum expectable; members of the subclass to which Jones belongs only have a low probability of getting paresis. Explanations, on the S-R model, are therefore not arguments; an explanation just gathers statistically relevant information pertaining to the explanandum, but need not provide a reason to believe that the explanandum took place. It therefore need not provide the basis for a prediction as required by the explanation/prediction symmetry thesis. That thesis and the claim that explanations are arguments go hand in hand, and the S-R model rejects both.

Suppose that a coin is tossed that is biased toward heads, so that the probability of heads is 0.9 and the probability of tails is 0.1. Suppose also that the class of coin tosses is homogenous (there is no statistically relevant way to divide it further).[13] If it lands on heads, then both Hempel's I-S and Salmon's S-R models will explain this event in much the same way. But if it lands on tails, then Hempel will have to say that no explanation is possible (since it was unlikely given the coin's bias). On the S-R model, however, the same explanation is available as when the coin lands on heads (except that the toss is a member of the other subclass).

Intuitions (and theories of explanation) divide on this point. Some think that even statistical explanations with high probability were bad enough, and that only when an explanation guarantees the explanandum is an explanation provided.[14] Others (including Hempel) allow explanations that confer a high probability on the explanandum, but not ones that render the explanandum unlikely. Finally, others (such as Salmon's S-R) countenance the latter kinds of explanations as well.

Salmon's argument for such explanations is that it is arbitrary to admit only the high-probability cases as explanatory. Take the coin example again. We toss the coin; according to Hempel, if it lands heads we have an explanation for why it does so but if it lands tails we do not. And yet, there is a positive chance that either will happen. And we have no more information to appeal to in distinguishing the two cases; we understand them equally well. In light of this it seems odd to say that we can explain a result of heads but cannot explain a result of tails.

4 S-R and the Counterexamples

It might seem as though the S-R model fails to accommodate the barometer and storm example. After all, the fact that the barometer falls *is* statistically relevant to the storm's arrival: the probability of such a storm is higher when the barometer falls. Nevertheless, the barometer's falling does not explain the storm's arrival. Salmon's method of handling such cases is by appeal to *screening off*: the statistical relevance of the barometer's fall to the storm's arrival (and vice versa) disappears – it is screened off – if we take into account the fact that the air pressure has dropped. More precisely: the probability that the storm will come if the air pressure drops is the same as the probability that the storm will come if the air pressure drops *and* the barometer falls. But the reverse is not true: the probability that the storm will come if the barometer falls is not the same as the probability that the storm will come if the barometer falls *and* the air pressure drops. (This is because the barometer might malfunction, so that the barometer's fall would not indicate the onset of a storm.) So the proper subclasses are "air pressure drops" and "air pressure does not drop" rather than "barometer falls" and "barometer does not fall," in line with our intuition that it is the air pressure drop that explains the storm's arrival rather than the barometer's fall. In general, the property that screens off, instead of the property that is screened off, is used to partition the class. Salmon called this the *screening-off rule*.

Salmon also appeals to the screening-off rule in order to handle asymmetry cases, such as the flagpole and shadow example. It is true that the length of the shadow is statistically relevant to the height of the flagpole (although the shadow does not explain the height). But whatever the explanation for the flagpole's height really is, that explanation will screen off the shadow explanation. Suppose, for example, that the flagpole is 10 feet high because its owner instructed the flag's manufacturer to make it that high. The probability that the flagpole is 10 feet high given that the manufacturer was so instructed is the same as the probability that it is that high given that the manufacturer was so instructed and that an 8 foot shadow appears. But the probability that it is 10 feet high if an 8 foot shadow appears is not the same as the probability that it is that high if an 8 foot shadow appears and the manufacturer was so instructed. This is because the shadow might be shortened, for example, by the beam of an inconveniently placed lamp, without changing the height of the flagpole. So we should use the screening-off property – flagpole's manufacture according to instructions – rather than the property screened off – length of shadow – to partition the class when explaining the flagpole's height, again in line with intuition.

5 Problems with S-R

One problem concerns the objective homogeneity condition. Consider the pregnancy example again. A great many characteristics (age, health, genetic factors, and so on) influence the probability that a female will become pregnant; so many, in fact, that the probability may well differ from one individual female to the next. But then we would have to partition the class so finely that only one female is a member of each subclass. And that intuitively is just too much information; we hardly need to know such an unmanageably huge mass of information, for example, to know why one female who is taking birth control pills did not get pregnant.

Another problem concerns the relation between statistical relevance and causality. The S-R model was motivated, in part, by Salmon's intention to capture the causal relationships (between the flagpole and the shadow, for example) that seem important to explanations. His thought was that causal relations could be analyzed as statistical relevance relationships. At the end of reading 8.2, however, he concedes that he would be "inclined to harbor serious misgivings about the adequacy of my view of statistical explanation if the statistical analysis of causation cannot be carried through successfully, for the relation between causation and explanation seems extremely intimate" (reading 8.2, p. 551).

Unfortunately, there are good reasons for thinking that such an analysis is unsuccessful. For example, consider again Jones's failure to become pregnant. Suppose that birth-control pills are in fact perfectly effective, so that every female who takes them avoids becoming pregnant. Then the probability that a female who takes such pills will become pregnant is the same as that of a man (whether or not the latter takes such pills). So we will have only two subclasses:

1 Women who do not take birth-control pills; and
2 People who are either men or birth-control-taking women.

The explanation for Jones's failure to become pregnant now cites his membership in the second group. But precisely the same explanation will be given for a female who does not become pregnant because she took such pills. The explanations in these two cases, however, are entirely different.

In general, while statistical relations may provide significant clues that indicate the underlying causal relationships, they do not perfectly mimic them. Some can indeed be realized by a number of different causal relations. (Notice, for example, that both the barometer-and-storm and flagpole-and-shadow cases exhibit the same screening-off relation.) Causal relationships therefore cannot be analyzed as statistical relevance relations. If explanation hinges fundamentally on causal relationships, then the S-R model is inadequate as a model of explanation.

Salmon eventually decided that this was correct and, therefore, that the theory of explanation had to be more directly characterized in terms of explanation relationships; we must, he said, put "cause" back into "because."[15]

But causation has been at least as mysterious to philosophers as explanation, particularly in light of Hume's critique of the idea of causation as a "mysterious power" (see §5 of the Unit 3 commentary). So we cannot just leave the issue with the slogan that to explain a phenomenon is to reveal its cause; we need a viable account of causation that does not run afoul of Hume's critique. The attempt to provide such an account led to Salmon's C-M model, which is the subject of the next section.

6 The C-M Model

Philosophers have long understood causal relations to hold between momentary events: Sarah threw the baseball, which broke the window, which frightened the cat, which knocked over the vase, and so on. The resulting image is a chain of discrete events, each caused by the previous one and causing the next. Salmon suggests that we should reject this image. In the place of events – momentary entities with very short durations – he substitutes causal *processes* with considerably longer duration. Sally's arm movement as she throws the baseball, the ball's traveling through the air toward the window, the window's breaking and falling to the ground, the cat's running off, the vase's fall to the ground, etc. are all examples of processes. Others (that are somewhat simpler in certain respects) include a car traveling along the road, a beam of starlight traveling across empty space, and even a mountain's immobility.

Such processes are identified by the continuity of certain characteristics or structure; the car retains its shape as it moves along the road, the starlight retains its spectral characteristics as it passes through space, the mountain retains its location, etc. Some characteristics do change; the car's gas tank is continually emptying, for example. But to be a process there must be continuity in at least some characteristics.

Causal processes are capable of transmitting information. They do so in virtue of their capacity to be *marked*. A rock thrown up by a passing truck hits the car and dents it; a red filter is placed in the path of starlight, turning it red. In both cases, the process is altered in some way (the shape of the car, the color of the light), which alteration persists unless and until the process is interfered with again. A process need not actually be marked to constitute a bona fide causal process; but it must be possible to mark it in some way.

Processes that cannot be marked are *pseudo-processes*. The shadow of the car as the car speeds along the highway appears to be a process; it exhibits a continuity of structure – of shape, for example – as does the car itself. But the shadow cannot itself be marked. Suppose, for example, that it passes over a boulder on the roadside, deforming its shape. Unlike the car's dent, the shadow's deformed shape is not retained without further interaction; as soon as it passes the boulder, it returns to its original shape.

Sometimes independent processes emanate from a central source. After a nuclear blast, witnesses near ground zero have a higher incidence of leukemia. This is due to the radiation emanating from the blast, which is a common cause of the various victims' illness.

Salmon understands the momentary events that philosophers have traditionally identified as cause and effect – the baseball's hitting the window, for example – to constitute *interactions* between processes (between the baseball's trip to the window and the window itself in this case).[16] The interaction marks both processes: the ball's velocity and trajectory change, and the glass shatters and begins to fall to the floor.

"Interaction" and "mark-transmission" both sound like causal phenomena; since the goal is to give an analysis of such phenomena, Salmon must characterize these in non-causal terms. To say that a process is marked is just to say that an alteration appears in a particular characteristic of that process (a dent appears on the car), which continues to appear as a characteristic of that process until the process is marked again. An interaction occurs when two processes spatiotemporally intersect, after which intersection both processes are marked.

The result is a vision of the world as a network of spatiotemporally extended causal processes, some of which interact, which interaction produces marks in those processes. Those processes then transmit the mark until further interactions, and so on. On occasion, one event will generate a number of processes as their common cause. When we explain an event we display at least some of the causal network leading up to the phenomenon, and describe the phenomenon itself in terms of those processes and their interactions.

This is not to say that Salmon has abandoned statistical relevance relations. When we construct an explanation our first task is as it was in the S-R model: we collect together the various statistical relevance relations surrounding the event we wish to explain. But instead of taking those statistical relations to constitute the causal network, Salmon now sees those relations as indicators of the underlying causal relations (characterized in terms of processes, interactions, and common causes) which must then be identified, and which then explain those statistical relations.

The statistical indicators of common causes are *conjunctive forks*. The probability that one person randomly chosen from the population will contract leukemia is statistically independent of that of another person similarly chosen; but they are not independent when both are near the blast of a nuclear bomb. When that proximity is taken into account, however – when we ask what the probability is of both contracting the disease given that they are near the blast – they are probabilistically independent of one another once again. In this way the statistical coincidence in the incidence of leukemia among people near the blast of a nuclear bomb disappears – is screened off – when that proximity is taken into account.

Interactions between processes are indicated by a different statistical relation that Salmon calls an *interactive fork*, in which the statistical dependence is not screened off. For example, consider a cue ball shot by a relative novice and striking the 8-ball. Suppose the probability that the 8-ball will end up in the corner pocket given that the novice attempts the shot is about 0.5. Suppose also that the probability that the cue ball ends up in one of the corner pockets (resulting in a scratch) given that the novice attempts the shot is also 0.5. But suppose also that the two balls are so positioned that if novice manages to get the 8-ball in the corner pocket, it is very likely (0.95, say) that the cue ball will go into the other corner pocket. Then the probability that the cue ball is pocketed is greater given that the novice attempts the shot *and* the 8-ball is pocketed than it is given only that the novice attempts the shot. This, Salmon suggests, is the statistical manifestation of an interaction. The interaction is between, in this case, the process of the cue ball moving across the table toward the 8-ball and colliding with it and the process of the 8-ball stationary on the table which is then hit by the cue ball, resulting in a change in the velocity of both.[17]

A complete scientific explanation, then, requires two steps: we first identify all statistical relevance relations (including conjunctive and interactive forks) surrounding the event we wish to explain. We then explain those relations by appeal to underlying causal processes, common causes, and interactions leading up to and constituting the event.

7 Problems with the C-M Model

A great virtue of the C-M approach – indeed, of any causal theory of explanation – is that it appears to easily resolve the standard counterexamples. The flagpole's shadow does not explain its height because it is the height that is (part of) the cause of the former rather than vice versa; Jones's having taken birth control pills does not explain his failure to become pregnant because it does not cause it (his being male does instead); the drop in air pressure caused the storm, not the barometer's fall; the syphilis caused the paresis; and so on.

But there are problems. One problem (which also applies to any causal approach) concerns how much of the causal network must be displayed in giving an explanation. For example, one's birth is an indispensable part of that causal network leading to one's death; but we would hardly cite one's birth as part of the explanation for one's death. We need some method to identify some aspects of that network as more relevant than others.[18]

Another problem concerns explanations that do not seem to involve the sort of causal processes to which Salmon appeals. Neither explanations within pure mathematics nor explanations of laws by appeal to other laws, for example, seem to involve causal processes. Other explanations appeal to laws that constrain global properties of a system (such as the gas laws that limit simultaneous values of pressure, temperature and volume of a gas) rather than describing the immensely complex underlying causal processes. Finally, some theories countenance action at a distance (such as Newton's theory of gravity), which are not spatiotemporally continuous in the manner of Salmon's causal processes.

A further problem concerns Salmon's appeal to counterfactuals in describing the various concepts. For example, processes are distinguished from pseudo-processes, not by being marked but instead by the fact that they *could* be marked. The concern is that the status of such counterfactuals has always been murky (and indeed difficulties in understanding them have standardly been taken to be part of what makes understanding causality in general problematic). Salmon suggests that these counterfactuals are easily tested empirically; but the adequacy of this response is uncertain.

Another related concern is a threat of circularity. An alteration in a process is a mark if it continues without any further interactions. But there is no independent characterization of an interaction than that it leaves a mark on the processes involved. Consider the car and shadow again. The shadow, Salmon says, cannot be marked; when it is deformed by a boulder it returns to its original shape once the boulder has passed. But why does the boulder's passage itself not constitute an interaction with the shadow? The shadow's return to its original shape seems, after all, to have the characteristics of a mark; it remains until further interaction. Such "pseudo-interactions" will need to be distinguished from real causal interactions; but it is not clear how this can be done in a non-circular fashion.

Finally, it is not clear why Salmon insists that statistical relevance relations must be part of the explanation. They are, after all, themselves *explained* by the causal relations; why would they themselves be part of what those causal relations explain?

Some of these problems apply specifically to Salmon's version of a causal theory and some to causal theories of explanation generally. A very different approach, one that reverses the relation between causation and explanation, is Kitcher's unificationist model, which is the subject of the next section.

8 The Unificationist Model

Like Hempel and unlike Salmon (of both S-R and C-M models), Kitcher believes that explanations are arguments.[19] But he denies that there is a set of conditions that can be applied to arguments one by one to determine whether they count as explanations. In contrast to all the models of explanation we have considered so far, he claims that putative explanations must be assessed *en masse* rather than individually. In particular, we have to determine whether the collection of arguments of which the candidate is a member unifies our body of knowledge; it only counts as a bona fide explanation if that collection does the best job of unifying that knowledge.

The basic idea is fairly simple (although its expression, as we will see, is rather complex). Scientific theories are impressive when they can explain a lot with a little. Newtonian mechanics, for example, could account for a wide variety of phenomena, ranging from the orbit of the planets around the sun, to the tides and their relation to the moon, to the path of a cannonball. In so doing it reduces the number of facts that we have to accept as "brute" and shows how others can be derived from them. If explanations provide understanding, and understanding is a matter of showing how superficially distinct phenomena are really manifestations of the same few underlying principles, then unification is the key to explanation.

The question is how this intuitive relation between explanation and unification is to be cashed out. Call the collection of sentences that constitutes the current body of scientific doctrine K. We are after the collection of arguments that best unifies K; Kitcher calls this the *Explanatory Store*, E(K), over K. The arguments in the explanatory store (and only those) will count as explanations in K. The task is to identify E(K).

Unification, as Kitcher understands it, is realized in the similarities among the arguments in E(K). Newton's theory, for example, repeatedly uses the same patterns of argument, substituting appropriate terms for the variables within the laws to derive specific values. So Kitcher appeals to the concept of an *argument pattern* that has many similar arguments as instances.

An argument is composed of sentences, some of which are the premises and one of which is the conclusion. Correspondingly, an argument pattern is composed of *schematic sentences*, in which some of the terms (indicating the masses of the bodies involved, for example) are substituted with dummy letters. A *schematic argument* is a sequence of such sentences. A set of *filling instructions* indicates how the dummy letters are to be filled in, with one such instruction for each dummy letter. A *classification* describes the general properties of the argument: which sentences are the premises and which the conclusion, which sentences are inferred from which, what the rules of inference are, and so on. An argument pattern, finally, consists of a schematic argument, a set of filling instructions, and a classification.

An argument pattern can be more *stringent* by having fewer dummy letters to fill, by imposing greater restrictions on what can fill them, or by imposing more constraints on the logical structure. The least stringent pattern is one that will impose no such constraints whatsoever, counting any argument as an instance. The most stringent pattern will impose enough constraints so that only one argument can count as an instance. Between these two extremes are patterns that are more or less stringent.

Consider the total set of accepted scientific sentences K. There will be various ways of deriving some of these sentences from others, that is, different sets of arguments where some of these sentences are premises and some are conclusions. Kitcher calls these different sets of arguments *systematizations*.

Now consider one of those systematizations. The arguments in it will be instances of various argument patterns; and some will be instances of the same argument pattern. Suppose that we construct a collection of argument patterns that has every argument in the systematization as an instance. There will be a number of ways of doing this, in part because the argument patterns involved can be more or less stringent. Some will have more argument patterns, some will have less. Each set of such argument patterns is called a *generating set*. Each systematization will have a number of generating sets.

The generating sets for a particular systematization can be evaluated in terms of two conditions: (a) employing a small number of argument patterns; and (b) employing stringent argument patterns. The more a generating set satisfies both of these conditions, Kitcher suggests, the more unifying it is. These criteria are bound to conflict, since the more stringent a pattern is the more difficult it is for an argument to count as an instance of it and, therefore, the more argument patterns will be required. Nevertheless, the goal is to maximize both; Kitcher provides no method for the resolution of conflict, suggesting only that the best choice will be clear in particular cases. The generating set that is most unifying for a particular systematization is called that systematization's *basis*.

So, for the total set of sentences K there are a number of systematizations; for each systematization there are a number of generating sets; and for each systematization one of those generating sets wins as most unifying and is called that systematization's basis. The last step is to compare the bases for the different systematizations. To do this we apply the two criteria (a) and (b) as before, but now we also consider (c) the number of sentences in K that end up being conclusions of the arguments in each basis's systematization.

For example, suppose there are three systematizations (I, II, and III), and we have identified the basis for each (1, 2, and 3) in accordance with criteria (a) and (b). In addition to comparing bases 1, 2, and 3 in terms of criteria (a) and (b), we can also compare how many sentences in K are among the conclusions of I, II, and III.

Again, the criteria (a), (b), and (c) may pull in different directions; it is, for example, possible that basis 1 scores best on (a), 2 on (b), and 3 on (c). But, as before, we attempt to identify the basis that maximizes these three criteria together (so, for example, 3 might do so well with criterion 3 that it overshadows (a)'s and (b)'s performances on 1 and 2).

Finally, the basis that wins is the most unifying generating set for K. The arguments within its systematization, and only those, count as the explanations. In this way explanations are selected in light of the overall unifying capacity of the systematization of which they are members.

9 Unificationism and the Counterexamples

Fortunately, with Kitcher's view in hand, his solution to the standard counterexamples is fairly simple. Consider again Jones and the birth-control pills. Suppose we explain Jones's failure to become pregnant by appeal to his having taken the pills. That will presumably be an instance of a pattern that, for any man who takes such pills, will derive his non-pregnancy from pill consumption. But how do we explain the fact that males who do not take such pills also do not become pregnant? If we do not explain these at all, then we will score poorly on criterion (c), since there will be many sentences like "Bob is not pregnant" that will not be the conclusions of explanatory arguments. Suppose we explain these cases in the usual way (by pointing out that men do not possess the necessary biological equipment). Then we will have two argument patterns, one for men who take such pills (appealing to the pill consumption) and one for men who do not (appealing to biology). But the latter pattern can *also* explain the non-pregnancy of the male pill-consumers. So if we keep both patterns where one would do, we will not satisfy criterion (a). Therefore, we will clearly satisfy these criteria best by employing only the biological explanation for all men, as intuition suggests.

Precisely the same technique is applied to asymmetry cases. If we explain the flagpole's height by appeal to the length of its shadow (realizing a height-derived-from-shadow pattern) we will have to deal with the height of flagpoles (and other objects) that does not happen to be casting shadows. If we do not provide explanations for these, then we will again score poorly on criterion (c). If we do provide explanations, they will presumably be what Kitcher calls *origin and development* explanations, which appeal to whatever process it was that produced the object (and maintains its existence now), such as the manufacturing process that produced the flagpole. And we will be faced with the same options as before: origin-and-development explanations will also explain the height of flagpoles with shadows, so we do not need the shadow-based explanations. The most unifying system will therefore countenance only origin-and-development explanations, as intuition suggests. In general, Kitcher suggests that arguments that intuitively do not count as explanations will be members of systematizations that are demonstrably not the most unifying.

10 Problems with the Unificationist Model

One problem is that causal relations, that are apparently so important to explanation, seem to have been left out of the picture. In a complete reversal of the C-M model, according to which explanations are grounded in causal relations, Kitcher suggests that causal relations are ultimately based on explanations. Causal relations are determined, he claims, by the most unifying systematization of scientific knowledge once scientific inquiry is complete. The causes are those events cited in the explanans of the arguments in that systematization, their effects those events cited in the explananda of those arguments.

But many find this kind of anti-realism about causation – according to which what causes what is ontologically dependent on the structure of ideal scientific knowledge – very difficult to swallow. Whether a particular rock broke a particular window, for example, hardly seems to depend on complex structural features of the system of knowledge we will end up with far down the road of scientific inquiry.

A related problem concerns the sheer complexity of Kitcher's system. Upon seeing Sally throw the rock, thereby breaking the window, we would reasonably view ourselves as knowing the explanation for the window's breaking. On Kitcher's view, however, we will not know that it really is the right explanation unless we have completed scientific inquiry and then performed the complex operations upon it that Kitcher requires in order to determine the most unifying systematization of the body of scientific knowledge. For all we know now, the concepts 'rock' and 'window' might not even show up in the final scientific view. But surely we are now able to correctly identify this and many other such explanations, often (as in this case) just by witnessing the relevant events, both within science and in ordinary life.

Finally, it is not clear that his attempt to avoid the counterexamples will ultimately work. Consider, for example, the flagpole and shadow again. If flagpoles (and other objects) *always* cast shadows, then there would be no loss in the number of conclusions (criterion (c)) if we use shadow-explanations rather than origin-and-development explanations. Of course not all objects do in fact cast shadows. But in some systems such backward or "retrodictive" arguments are always available. Given Kepler's laws and the position and velocity of Mars yesterday at noon, we can derive – and, we might say, explain – the position and velocity of Mars today at noon. But using those same laws we can always do the reverse, deriving yesterday's position and velocity from those of today. Kepler's laws are in this sense time-symmetric; and when the laws of a system are time-symmetric in this way, there will always be a retrodictive derivation for every predictive one, and therefore no reduction in the number of conclusions will result by selecting the predictive derivations as the explanations.

11 Summary

We have considered three models of explanation in this unit commentary, and Hempel's model in §§3–4 of the Unit 5 commentary. Abstracting from the details, they each voice very different themes. The theme of the Hempelian model is that explanations are arguments that individually meet certain criteria. The theme of the S-R model is that they are not arguments but instead collections of all information relevant to the probability of the explanandum. The theme of the C-M model is that explanations identify the causes of the explanandum. And the theme of the unificationist account is that explanatory capacity is fundamentally a global characteristic of scientific theories.[20] All four themes (and others) continue to be heard in contemporary philosophical explorations of the issue; whether they can be made harmonious with one another remains an open question.

Notes

1. In fact he offered the D-S model as well. But D-S explanations are essentially D-N explanations where the explanandum is a statistical law, and so need not be explicitly discussed here.
2. This condition is a consequence of the explanation/prediction symmetry thesis.
3. Some insist that false theories (e.g., phlogiston theory) can explain, although of course they cannot provide the *correct* explanation. It is not clear that there is more at stake here than terminology: what Hempel calls explanations they call correct explanations, and what they call false explanations Hempel calls putative explanations.
4. Salmon's example CE-1 ("The Eclipse", p. 531) is another asymmetry case. A difference is that the flagpole's being the height it is and shadow are simultaneous, whereas in the eclipse case the symmetry is between inferences from past to present (which seem explanatory) and "retrodictions" or inferences from present to past (which do not). The flagpole case indicates that the symmetry problem cannot be solved simply by requiring that the event cited in the explanandum be later than the event cited in the explanans.
5. Salmon's example CE-4 ("The Moon and the Tides", p. 532) is also closely related to the symmetry cases.
6. More precisely, this is an exceptionless, non-accidental, counterfactually supporting regularity; and these are the characteristics of (universal) laws on the received view.
7. Another famous irrelevance example is Salmon's CE-6 ("The Hexed Salt", p. 534).
8. Salmon's example CE-9 (p. 534) is another along the same lines.
9. At least so claimed Michael Scriven (1959), "Explanation and Prediction in Evolutionary Theory," *Science* 130: 477–82. This claim is, however, contentious.
10. This corresponds to Hempel's requirement of maximal specificity (RMS), which also requires that all statistically relevant information be taken into account in constructing the explanans. See §4 of the Unit 5 commentary.
11. Hempel's RMS did not impose a requirement corresponding to this widest-homogenous-class condition, which is why his view is susceptible to the irrelevance counterexamples.
12. This also differs from Hempel's RMS, which is relative to our current scientific understanding.
13. This is, in fact, bound to be false. There is information about the precise angular momentum of the coin, the rigidity of the surface it lands on, the height to which it is thrown, etc. that could be used to distinguish subclasses.
14. This includes Kitcher, whose unificationist model is discussed in §8 below.
15. Salmon 1977, "At-At Theory of Causal Influence", *Philosophy of Science* 44(2), (June), p. 215.
16. The continued persistence through time of such a physical object as the window constitutes a process for Salmon.
17. See note 16.
18. A prominent view suggests that the selection of certain factors over others as explanatorily relevant is a pragmatic matter that can vary from context to context (e.g., van Fraassen 1980).
19. Indeed – and unlike Hempel – Kitcher recognizes only deductive arguments as explanations. The unificationist model that he develops, however, does not itself impose this limitation (he does not insist on it in reading 8.4).
20. The discussion here of the S-R, C-M, and unificationist models has benefited greatly from James

Woodward's excellent discussion of these same three views in his *Stanford Encyclopedia of Philosophy* entry on "Scientific Explanation." (http://plato.stanford.edu/entries/scientific-explanation/). We recommend that article for a lucid discussion of further details concerning these views.

Suggestions for Further Reading

Salmon's C-M model is presented in detail in Salmon 1984. The unificationist approach was originally developed in Friedman 1974. Kitcher's version of this approach presented in reading 8.4 is more thoroughly developed in Kitcher 1989. Peter Railton develops an alternative approach to statistical explanation in Railton 1978. Van Fraassen 1980, ch. 5, emphasizes the pragmatic and contextual aspects of explanation. Railton develops the important concept of an "ideal explanatory text" (which would contain all causal and nomological information relevant to an explanandum) in Railton 1981. An excellent review of the history of the theory of explanation is given in Salmon 1989.

Friedman, M., 1974, "Explanation and Scientific Understanding," *Journal of Philosophy* 71: 5–19.

Kitcher, P., 1989, "Explanatory Unification and the Causal Structure of the World," in P. Kitcher and W. Salmon, *Scientific Explanation*, Minneapolis: University of Minnesota Press, pp. 410–505.

Railton, P., 1978, "A Deductive-Nomological Model of Probabilistic Explanation," *Philosophy of Science* 45: 206–26.

Railton, P., 1981, "Probability, Explanation, and Information," *Synthese* 48: 233–5.

Salmon, W., 1984, *Scientific Explanation and the Causal Structure of the World*. Princeton: Princeton University Press.

Salmon, W., 1989, *Four Decades of Scientific Explanation*. Minneapolis: University of Minnesota Press.

van Fraassen, B., 1980, "The Pragmatics of Explanation," chapter 5 of *The Scientific Image*. Oxford: Oxford University Press.

8.1

Counterexamples to the D-N and I-S Models of Explanation

Wesley Salmon

Wesley Salmon (1925–2001) is the most prominent twentieth-century philosopher of explanation after Carl Hempel and the author of two of the most influential models of explanation (presented in the next two readings). The reading below is excerpted from his review of the history of the philosophy of explanation, *Four Decades of Scientific Explanation*, and provides a succinct summary of the main counterexamples to Hempel's D-N and I-S models.

One rather obvious problem has to do with the temporal relations between the explanatory facts (as expressed by the singular sentences in the explanans) and the fact-to-be-explained (as expressed by the explanandum-sentence). In the schema reproduced above (p. 13; Hempel-Oppenheim 1948, 249) the Cs are labeled as "antecedent conditions," but in the formal explication no temporal constraints are given. Indeed, no such temporal constraints are mentioned even in the informal conditions of adequacy. This issue has been posed in terms of the explanation of an eclipse.

(CE-1) The eclipse. Going along with the D-N model, we might, for example, regard a total lunar eclipse as satisfactorily explained by deducing its occurrence from the relative positions of the earth, sun, and moon at some prior time in conjunction with the laws of celestial mechanics that govern their motions. It is equally possible, however, to deduce the occurrence of the eclipse from the relative positions of the earth, sun, and moon at some time after the eclipse in conjunction with the very same laws. Yet, hardly anyone would admit that the latter deduction qualifies as an explanation.[1] One might suppose that the failure to impose temporal restrictions was merely an oversight that could be corrected later, but Hempel (1965a, 353) raises this question explicitly and declines to add any temporal constraint.[2]

Another issue, closely related to the matter of temporal priority, has to do with the role of causality in scientific explanation. Our commonsense notion of explanation seems to take it

From Wesley C. Salmon, *Four Decades of Scientific Explanation* (Pittsburgh, PA: University of Pittsburgh Press, 2006). "2.3 Famous Counterexamples to the Deductive-Nomological Model" is from. *Four Decades of Scientific Explanation*, by Wesley C. Salmon, © 2006. Reprinted with permission of the University of Pittsburgh Press.

for granted that to explain some particular event is to identify its cause and, possibly, point out the causal connection. Hempel and Oppenheim seem to share this intuition, for they remark, "The type of explanation which has been considered here so far is often referred to as causal explanation" (Hempel-Oppenheim 1948, 250). In "Aspects of Scientific Explanation," while admitting that some D-N explanations are causal, Hempel explicitly denies that all are (1965a, 352–4). The problems that arise in this connection can readily be seen by considering several additional well-known examples.

(CE-2) Bromberger's flagpole example. A vertical flagpole of a certain height stands on a flat level piece ground.[3] The sun is at a certain elevation and is shining brightly. The flagpole casts a shadow of a certain length. Given the foregoing facts about the height of the flagpole and the position of the sun, along with the law of rectilinear propagation of light, we can deduce the length of the shadow. This deduction may be accepted as a legitimate D-N explanation of the length of the shadow. Similarly, given the foregoing facts about the position of the sun and the length of the shadow, we can invoke the same law to deduce the height of the flagpole. Nevertheless, few people would be willing to concede that the height of the flagpole is explained by the length of its shadow.[4] The reason for this asymmetry seems to lie in the fact that a flagpole of a certain height causes a shadow of a given length, and thereby explains the length of the shadow, whereas the shadow does not cause the flagpole, and consequently cannot explain its height.

(CE-3) The barometer. If a sharp drop in the reading on a properly functioning barometer occurs, we can infer that there will be a storm – for the sake of argument, let us assume that there is a law that whenever the barometric pressure drops sharply a storm will occur. Nevertheless, we do not want to say that the barometric reading explains the storm, since both the drop in barometric reading and the occurrence of the storm are caused by atmospheric conditions in that region. When two different occurrences are effects of a common cause, we do not allow that either one of the effects explains the other. However, the explanation of the storm on the basis of the barometric reading fits the D-N model.

(CE-4) The moon and the tides. Long before the time of Newton, mariners were fully aware of the correlation between the position and phase of the moon and the rising and falling of the tides. They had no knowledge of the causal connection between the moon and the tides, so they had no explanation for the rising and falling of the tides, and they made no claim to any scientific explanation. To whatever extent they thought they had an explanation, it was probably that God in his goodness put the moon in the sky as a sign for the benefit of mariners. Nevertheless, given the strict law correlating the position and phase of the moon with the ebb and flow of the tides,[5] it was obviously within their power to construct D-N explanations of the behavior of the tides. It was not until Newton furnished the causal connection, however, that the tides could actually be explained.

One of the most controversial theses propounded by Hempel and Oppenheim is *the symmetry of explanation and prediction.* According to this view, the very same logical schema fits scientific explanation and scientific prediction; the sole difference between them is pragmatic. If the event described by E has already occurred, we may ask why. A D-N explanation consisting of a derivation of E from laws and antecedent conditions provides a suitable response. If, however, we are in possession of the same laws and antecedent conditions before the occurrence of E, then that same argument provides a prediction of E. Any D-N explanation is an argument that, were we in possession of it early enough, would enable us to anticipate, on a sound scientific basis, the occurrence of E. Since every D-N explanation involves laws, a hallmark of explanations of this type is that they provide *nomic expectability.*[6]

In discussing the symmetry of explanation and prediction in the preceding paragraph, I was tacitly assuming that the so-called antecedent conditions in the explanans are, in fact, earlier than the explanandum event. However, in view of Hempel's rejection of any requirement of temporal priority, the symmetry thesis must be construed a bit more broadly. Suppose, for example, that the explanandum-event E occurs before the conditions C in the explanans. Then, as I construe the symmetry thesis, we would be committed to the view that the D-N explanation is an argument that could be used subsequent to the occurrence of the explanatory conditions C to retrodict E. It is quite possible, of course, that E has occurred, but that we are ignorant of that fact. With knowledge of the appropriate laws, our subsequent knowledge of conditions C would enable us to learn that E did, in fact, obtain. Parallel remarks could be made about the case in which C and E are simultaneous. Thus, in its full generality, the symmetry thesis should be interpreted in such a way that "prediction" is construed as "inference from the known to the unknown."[7]

As Hempel later pointed out in "Aspects of Scientific Explanation," the symmetry thesis can be separated into two parts: (i) Every D-N explanation is a prediction – in the sense explained in the preceding paragraph – and (ii) every (nonstatistical) scientific prediction is a D-N explanation. It is worthwhile, I think, to distinguish a *narrower symmetry thesis*, which applies only to D-N explanations of particular facts, and a *broader symmetry thesis*, which applies to both D-N and I-S explanations of particular facts. According to the narrower thesis, every *nonstatistical* prediction is a D-N explanation; according to the broader thesis, every prediction is an explanation of either the D-N or I-S variety. Given the fact that statistical explanation is not explicated in the Hempel-Oppenheim article, only the narrower symmetry thesis is asserted there. The broader thesis, as we shall see, was advocated (with certain limitations) in "Aspects."

Nevertheless, various critics of the Hempel-Oppenheim article failed to take sufficient notice of the explicit assertion that not all legitimate scientific explanations are D-N – that some are statistical. Scriven (1959) strongly attacked subthesis (i) – that all explanations could serve as predictions under suitable pragmatic circumstances – by citing evolutionary biology and asserting that it furnishes explanations (of what has evolved) but not predictions (of what will evolve). If, as I believe, evolutionary biology is a statistical theory, then Scriven's argument applies at best to the broader, not the narrower symmetry thesis. Although this argument was published in 1959,[8] it does, I believe, pose a serious problem for the theory of statistical explanation Hempel published three years later. In the same article Scriven set forth a widely cited counterexample:

(CE-5) Syphilis and paresis. Paresis is one form of tertiary syphilis, and it can occur only in individuals who go through the primary, secondary, and latent stages of the disease without treatment with penicillin. If a subject falls victim to paresis, the explanation is that it was due to latent untreated syphilis. However, only a relatively small percentage – about 25% – of victims of latent untreated syphilis develop paresis. Hence, if a person has latent untreated syphilis, the correct prediction is that he or she will not develop paresis. This counterexample, like the argument from evolutionary biology, applies only to the broader symmetry thesis.

When the narrower symmetry thesis is spelled out carefully, it seems impossible to provide a counterexample for subthesis (i) – that every explanation is a prediction (given the right pragmatic situation). That subthesis amounts only to the assertion that the conclusion of a D-N argument follows from its premises. Against subthesis (i) of the broader symmetry thesis the syphilis/paresis counterexample is, I think, quite telling.

When we turn to subthesis (ii) of the narrower symmetry thesis – i.e., that every (nonstatistical) prediction is an explanation – the situation is quite different. Here (CE-3) and (CE-4) provide important counterexamples. From the barometric reading, the storm can be predicted, but the barometric reading does not explain the storm. From the position and phase of the moon, pre-Newtonians could predict the behavior of the tides, but they had no explanation of them. Various kinds of correlations exist that provide excellent bases for prediction, but because no suitable causal relations exist (or are known), these correlations do not furnish explanations.

There is another basis for doubting that every scientific prediction can serve, in appropriate pragmatic circumstances, as an explanation. Hempel and Oppenheim insist strongly upon the covering law character of explanations. However, it seems plausible to suppose that some respectable scientific predictions can be made without benefit of laws – i.e., some predictions are inferences from particular instances to particular instances. Suppose, for instance, that I have tried quite a number of figs from a particular tree, and have found each of them tasteless. A friend picks a fig from this tree and is about to eat it. I warn the friend, "Don't eat it; it will be tasteless." This is, to be sure, low-level science, but I do not consider it an unscientific prediction. Moreover,

I do not think any genuine laws are involved in the prediction. In (1965, 376) Hempel considers the acceptability of subthesis (ii) of the symmetry thesis an open question.

There is another fundamental difficulty with Hempel and Oppenheim's explication of D-N explanation; this one has to do with explanatory relevance. It can be illustrated by a few well-known counterexamples.

(CE-6) The hexed salt. A sample of table salt has been placed in water and it has dissolved. Why? Because a person wearing a funny hat mumbled some nonsense syllables and waved a wand over it – i.e., cast a dissolving spell upon it. The explanation offered for the fact that it dissolved is that it was hexed, and all hexed samples of table salt dissolve when placed in water. In this example it is *not* being supposed that any actual magic occurs. All hexed table salt is water-soluble because all table salt is water-soluble. This example fulfills the requirements for D-N explanation, but it manifestly fails to be a bona fide explanation.[9]

(CE-7) Birth-control pills. John Jones (a male) has not become pregnant during the past year because he has faithfully consumed his wife's birth-control pills, and any male who regularly takes oral contraceptives will avoid becoming pregnant. Like (CE-6), this example conforms to the requirements for D-N explanation.

[. . .]

2.5 Early Objections to the Inductive-Statistical Model

My own particular break with the received view occurred shortly after the incident with Smart that I related at the beginning of this essay. In a paper (W. Salmon 1965) presented at the 1963 meeting of the American Association for the Advancement of Science, Section L, organized by Adolf Grünbaum, I argued that Hempel's I-S model (as formulated in his (1962)), with its high probability requirement and its demand for expectability, is fundamentally mistaken. Hempel's example of John Jones's rapid recovery from his strep infection immediately called to mind such issues as the alleged efficacy of vitamin C in preventing, shortening, or mitigating the severity of common colds,[10] and the alleged efficacy of various types of psychotherapy. I offered the following examples:

(CE-8) John Jones was almost certain to recover from his cold within a week because he took vitamin C, and almost all colds clear up within a week after administration of vitamin C.

(CE-9) John Jones experienced significant remission of his neurotic symptoms because he underwent psychotherapy, and a sizable percentage of people who undergo psychotherapy experience significant remission of neurotic symptoms.

Because almost all colds clear up within a week whether or not the patient takes vitamin C, I suggested, the first example is not a bona fide explanation. Because many sorts of psychological problems have fairly large spontaneous remission rates, I called into question the legitimacy of the explanation proffered in the second example. What is crucial for statistical explanation, I claimed, is not how probable the explanans renders the explanandum, but rather, whether the facts cited in the explanans *make a difference* to the probability of the explanandum.

To test the efficacy of any sort of therapy, physical or psychological, controlled experiments are required. By comparing the outcomes in an experimental group (the members of which receive the treatment in question) with those of a control group (the members of which do not receive that treatment), we procure evidence concerning the effectiveness of the treatment. This determines whether we are justified in claiming explanatory import for the treatment vis-à-vis the remission of the disease. If, for example, the rate of remission of a certain type of neurotic symptom during or shortly after psychotherapy is high, but no higher than the spontaneous remission rate, it would be illegitimate to cite the treatment as the explanation (or even part of the explanation) of the disappearance of that symptom. Moreover, if the rate of remission of a symptom in the presence of psychotherapy is not very high, but is nevertheless significantly higher than the spontaneous remission rate, the therapy can legitimately be offered as at least part of the explanation of the patient's recovery. It follows from these considerations that high probability is neither necessary nor sufficient for bona fide statistical explanation.

Statistical relevance, not high probability, I argued, is the key desideratum in statistical explanation.

Notes

1 However, see Grünbaum (1963), Fetzer (1974), and Rescher (1963).
2 . . . Hempel did not base his qualms about explanations of these sorts on the problem of the temporal relation between the function and its goal.
3 I do not believe Bromberger ever published this precise example; his actual examples, which have the same import, are the height of a tower and the height of a utility pole to which a guy wire is attached.
4 Bas van Fraassen is an exception. In his 1980 work (pp. 132–4) he suggests that there are possible contexts in which such an explanation would be legitimate. We shall discuss his theory of explanation in §4.4 below.
5 This 'law' does, of course, make reference to particular entitites – earth, sun, and moon – but that, in itself, is not too damaging to the example. After all, in this respect it is just like Kepler's laws of planetary motion and Galileo's law of falling bodies. Like these, it qualifies as a derivative law, though not a fundamental law.
6 Scheffler (1957) subjected the symmetry thesis to searching criticism. To the best of my knowledge, this article is the first significant published critique of the Hemel–Oppenheim article.
7 See Grünbaum (1963) for a fuller discussion of this point.
8 As Philip Kitcher pointed out to me in a personal communication, Hempel could be defended against this example by arguing the "a *natural selection* explanation of the presence of a trait is really a deduction that the probability that the trait becomes fixed in a finite population is high."
9 This example is due to Henry Kybury (1965). A variant of this example (due to Noretta Koertge) has unhexed table salt placed in holy water; and the 'explanation' of the dissolving is that the water was blessed and whenever table salt is placed in holy water it dissolves.
10 Linus Pauling's claims in this regard were receiving a good deal of publicity in the early 1960s (see Pauling, 1970).

References

Fetzer, James H., 1974, "Grünbaum's 'Defense' of the Symmetry Thesis," *Philosophical Studies* 25: 173–87.

Grünbaum, Adolf, 1963, "Temporally Asymmetric Principles, Parity Between Explanation and Prediction, and Mechanism vs. Teleology," in Bernard H. Baumrin (ed.), *Philosophy of Science: The Delaware Seminar*, vols. 1–2, New York: John Wiley, pp. 57–96.

Hempel, Carl G., *Aspects of Scientific Explanation and Other Essays in the Philosophy of Science*. New York: The Free Press, 1965.

Hempel, Carl G., 1965a, "Aspects of Scientific Explanation," in Hempel 1965.

Hempel, Carl G. and Paul Oppenheim, 1948, "Studies in the Logic of Explanation," *Philosophy of Science* 15, pp. 135–75. Reprinted in Hempel (1965).

Kyburg, Henry E., Jr., 1965, "Comment," *Philosophy of Science* 32: 147–51.

Rescher, Nicholas, 1963, "Discrete State Systems, Markov Chains, and Problems in the Theory of Scientific Explanation and Prediction," *Philosophy of Science* 30: 325–45

Salmon, Wesley C., "The Status of Prior Probabilities in Statistical Explanation," *Philosophy of Science* 32, pp. 137–46.

Scheffler, Israel, 1957, "Explanation, Prediction, and Abstraction," *British Journal for the Philosophy of Science* 7: 293–309.

Scriven, Michael, "Explanation and Prediction in Evolutionary Theory," *Science* 30, pp. 477–82.

van Fraassen, Bas C., 1980, *The Scientific Image*. Oxford: Clarendon Press.

8.2

The Statistical Relevance Model of Explanation

Wesley Salmon

In this selection Salmon presents his first model of explanation, the *statistical-relevance* (S-R) model. According to this model, an explanation is not an argument whose premises are the explanans, and conclusion the explanandum, as in the D-N and I-S models. An explanation is instead an assembly of all the information that is relevant to the probability of the explanandum. That information need not make the explanandum probable given the explanans, as Hempel required.

Preliminary Analysis

The obvious trouble with our horrible examples is that the "explanatory" argument is not needed to make us see that the explanandum event was to be expected. There are other, more satisfactory, grounds for this expectation. The "explanatory facts" adduced are irrelevant to the explanandum event despite the fact that the explanandum follows (deductively or inductively) from the explanans. Table salt dissolves in water regardless of hexing, almost all colds clear up within a week regardless of treatment, males do not get pregnant regardless of pills, the moon reappears regardless of the amount of Chinese din, and there are no wild tigers in Times Square regardless of our friend's moans. Each of these explanandum events has a high prior probability independent of the explanatory facts, and the probability of the explanandum event relative to the explanatory facts is the same as this prior probability. In this sense the explanatory facts are irrelevant to the explanandum event. The explanatory facts do nothing to enhance the probability of the explanandum event or to make us more certain of its occurrence than we would otherwise have been. This is not because we know that the fact to be explained has occurred; it is because we had other grounds for expecting it to occur, *even if we had not already witnessed it.*

From Wesley C. Salmon, "Statistical Explanation and Statistical Relevance," from *Nature and Function of Scientific Theories*, ed. Robert G. Colodny (Pittsburgh: University of Pittsburgh Press, 1970), pp. 173–231. Reprinted in Wesley C. Salmon, Richard Jeffrey, and James Greeno (eds.), *Statistical Explanation and Statistical Relevance* (Pittsburgh: University of Pittsburgh Press, 1971), extracts from pp. 29–87.

Our examples thus show that it is not correct, even in a preliminary and inexact way, to characterize explanatory accounts as arguments showing that the explanandum event was to be expected. It is more accurate to say that an explanatory argument shows that the probability of the explanandum event relative to the explanatory facts is substantially greater than its prior probability.[1] An explanatory account, on this view, increases the degree to which the explanandum event was to be expected. As will emerge later in this paper, I do not regard such a statement as fully accurate; in fact, the increase in probability is merely a pleasant by-product which often accompanies a much more fundamental characteristic. Nevertheless, it makes a useful starting point for further analysis.

[...]

The Single Case

Let A be an unending sequence of draws of balls from an urn, and let B be the class of red things. A is known as the *reference class*, and B the *attribute class*. The probability of red draws from this urn, $P(A,B)$, is the limit of the relative frequency with which members of the reference class belong to the attribute class, that is, the limit of the relative frequency with which draws from the urn result in a red ball as the number of draws increases without any bound.[2]

Frequentists like John Venn and Hans Reichenbach have dealt with the problem of the single case by assigning each single event to a reference class and by transferring the probability value from that reference class to the single event in question.[3] Thus, if the limit of the relative frequency of red among draws from our urn is one-third, then we say that the probability of getting red on *the next draw* is one-third. To this way the meaning of the probability concept has been extended so that it applies to single events as well as to large aggregates.

The fundamental difficulty arises because a given event can be referred to any of a large number of reference classes, and the probability of the attribute in question may vary considerably from one of these to another. For instance, we could place two urns on a table, the one on the left containing only red balls, the one on the right containing equal numbers of red, white, and blue balls. The reference class A might consist of blind drawings from the right-hand urn, the ball being replaced and the urn thoroughly shaken after each draw. Another reference class A' might consist of draws made alternately from the left- and the right-hand urns. Infinitely many other reference classes are easily devised to which the next draw – the draw with which we are concerned – belongs. From which reference class shall we transfer our probability of value to this single case? A method must be established for choosing the appropriate reference class. Notice, however, that there is no difficulty in selecting an attribute class. The question we ask determines the attribute class. We want to know the probability of getting red, so there is no further problem about the attribute class.

[...]

Statistical relevance is the essential notion here. It is desirable to narrow the reference class in statistically relevant ways, but not in statistically irrelevant ways. When we choose a reference class to which to refer a given single case, we must ask whether there is any statistically relevant way to subdivide that class. If so, we may choose the narrower subclass that results from the subdivision; if no statistically relevant way is known, we must avoid making the reference class any narrower. Consider, for example, the probability that a particular individual, John Smith, will still be alive ten years hence. To determine this probability, we take account of his age, sex, occupation, and health; we ignore his eye color, his automobile license number, and his last initial. We expect the relative frequency of survival for ten more years to vary among the following reference classes: humans, Americans, American males, forty-two-year-old American males, forty-two-year-old American male steeplejacks, and forty-two-year-old American male steeplejacks suffering from advanced cases of lung cancer. We believe that the relative frequency of survival for another ten years is the same in the following classes: forty-two-year-old American male steeplejacks with advanced cases of lung cancer, forty-two-year-old blue-eyed American male steeplejacks with advanced cases of lung cancer, and forty-two-year-old blue-eyed American male steeplejacks with even automobile license plate numbers who suffer from advanced cases of lung cancer.

Suppose we are dealing with some particular object or event x, and we seek to determine the probability (weight) that it has attribute B. Let x be assigned to a reference class A, of which it is a member. $P(A,B)$ is the probability of this attribute within this reference class. A set of mutually exclusive and exhaustive subclasses of a class is a *partition* of that class. We shall often be concerned with partitions of reference classes into two subclasses; such partitions can be effected by a property C which divides the class A into two subclasses, $A.C$ and $A.\bar{C}$. A property C is said to be *statistically relevant* to B within A if and only if $P(A.C,B) \neq P(A,B)$. This notion of statistical relevance is the fundamental concept upon which I hope to build an explication of inductive explanation.

In his development of a frequency theory based essentially upon the concept of randomness, Richard von Mises introduced the notion of a *place selection*: "By a place selection we mean the selection of a partial sequence in such a way that we decide whether an element should or should not be included without making use of the attribute of the element."[4] A place selection effects a partition of a reference class into two subclasses, elements of the place selection and elements not included in the place selection. In the reference class of draws from our urn, every third draw starting with the second, every kth draw where k is prime, every draw following a red result, every draw made with the left hand, and every draw made while the sky is cloudy all would be place selections. "Every draw of a red ball" and "every draw of a ball whose color is at the opposite end of the spectrum from violet" do not define place selections, for membership in these classes cannot be determined without reference to the attribute in question.

A place selection may or may not be statistically relevant to a given attribute in a given reference class. If the place selection is statistically irrelevant to an attribute within a reference class, the probability of that attribute within the subclass determined by the place selection is equal to the probability of that attribute within the entire original reference class. If every place selection is irrelevant to a given attribute in a given sequence, von Mises called the sequence *random*. If every property that determines a place selection is statistically irrelevant to B in A, I shall say that A is a *homogeneous reference class* for B. A reference class is homogeneous if there is no way, even in principle, to effect a statistically relevant partition without already knowing which elements have the attribute in question and which do not. Roughly speaking, each member of a homogeneous reference class is a random member.

The aim in selecting a reference class to which to assign a single case is not to select the narrowest, but the widest, available class. However, the reference class should be homogeneous, and achieving homogeneity requires making the reference class narrower if it was not already homogeneous. I would reformulate Reichenbach's method of selection of a reference class as follows: choose the broadest homogeneous reference class to which the single event belongs. I shall call this the *reference class rule*.

Let me make it clear immediately that, although I regard the above formulation as an improvement over Reichenbach's, I do not suppose that it removes all ambiguities about the selection of reference classes either in principle or in practice. In principle it is possible for an event to belong to two equally wide homogeneous reference classes, and the probabilities of the attribute in these two classes need not be the same. For instance, suppose that the drawing from the urn is not random and that the limit of the relative frequency of red for every kth draw (k prime) is $1/4$, whereas the limit of the relative frequency of red for every even draw is $3/4$. Each of these subsequences may be perfectly random; each of the foregoing place selections may, therefore, determine a homogeneous reference class. Since the intersection of these two place selections is finite, it does not determine a reference class for a probability. The second draw, however, belongs to both place selections; in this fictitious case there is a genuine ambiguity concerning the probability to be taken as the weight of red on the second draw.

In practice we often lack full knowledge of the properties relevant to a given attribute, so we do not know whether our reference class is homogeneous or not. Sometimes we have strong reason to believe that our reference class is not homogeneous, but we do not know what property will effect a statistically relevant partition. For instance, we may believe that there are causal factors that determine which streptococcus infections will respond to penicillin and which ones will not, but we may not yet know what these causal factors are. When we know or suspect that a reference class is not homogeneous, but we do not know how to make any statistically relevant partition, we may say that the reference class is *epistemically homogeneous*. In other cases, we know that a reference class is inhomogeneous and we know what attributes would effect a statistically relevant partition, but it is too much trouble to find out which elements belong to each subclass of the partition. For instance, we believe that a sufficiently detailed knowledge of the initial conditions under which a coin is tossed would enable us to predict (perfectly or very reliably) whether the outcome will be heads or tails, but practically speaking we are in no position to determine these initial conditions or make the elaborate calculations required to predict the outcome. In such cases we may say that the reference class is *practically homogeneous*.[5]

The reference class rule remains, then, a methodological rule for the application of probability knowledge to single events. In practice we attempt to refer our single cases to classes that are practically or epistemically homogeneous. When something important is at stake, we may try to extend our knowledge in order to improve the degree of homogeneity we can achieve. Strictly speaking, we cannot meaningfully refer to degrees of homogeneity until a quantitative concept of homogeneity has been provided....

It would, of course, be a serious methodological error to assign a single case to an inhomogeneous reference class if neither epistemic nor practical considerations prevent partitioning to achieve homogeneity. This fact constitutes another basis for regarding the reference class rule as the counterpart of the requirement of total evidence. The requirement of total evidence demands that we use all available relevant evidence; the reference class rule demands that we partition whenever we have available a statistically relevant place selection by means of which to effect the partition.

Although we require homogeneity, we must also prohibit partitioning of the reference class by means of statistically irrelevant place selections. The reason is obvious. Irrelevant partitioning reduces, for no good reason, the inductive evidence available for ascertaining the limiting frequency of our attribute in a reference class that is as homogeneous as we can make it. Another important reason for prohibiting irrelevant partitions will emerge below when we discuss the importance of multiple homogeneous reference classes.

A couple of fairly obvious facts about homogeneous reference classes should be noted at this point. If all A's are B, A is a homogeneous reference class for B. (Somewhat counterintuitively, perhaps, B occurs perfectly randomly in A.) In this case, $P(A,B) = 1$ and $P(A.C,B) = 1$ for any C whatever; consequently, no place selection can yield a probability for B different from that in the reference class A. Analogously, A is homogeneous for B if no A's are B. In the frequency interpretation, of course, $P(A,B)$ can equal one even though not all A's are B. It follows that a probability of one does not entail that the reference class is homogeneous.

Some people maintain, often on a priori grounds, that A is homogeneous (not merely practically or epistemically homogeneous) for B only if all A's are B or no A's are B; such people are determinists. They hold that causal factors always determine which A's are B and which A's are not B; these causal factors can, in principle, be discovered and used to construct a place selection for making a statistically relevant partition of A. I do not believe in this particular form of determinism. It seems to me that there are cases in which A is a homogeneous reference class for B even though not all A's are B. In a sample of radioactive material a certain percentage of atoms disintegrate in a given length of time; no place selection can give us a partition of the atoms for which the frequency of disintegration differs from that in the whole sample. A beam of electrons is shot at a potential barrier and some pass through while others are reflected; no place selection will enable us to make a statistically relevant partition in the class of electrons in the beam. A beam of silver atoms is sent through a strongly inhomogeneous magnetic field (Stern–Gerlach experiment); some atoms are deflected upward and some are deflected downward, but there is no way of partitioning the beam in a statistically relevant manner. Some theorists maintain, of course, that further investigation will yield information that will enable us to make statistically relevant partitions in these cases, but this is, at present, no more than a declaration of faith in determinism. Whatever the final resolution of this controversy, the homogeneity of *A* for *B* does not logically entail that all *A's* are *B*. The truth or falsity of determinism cannot be settled a priori.

The purpose of the foregoing excursus on the frequency treatment of the problem of the single case has been to set the stage for a discussion of the explanation of particular events. Let us reconsider some of our examples in the light of this theory. The relative frequency with which we encounter instances of water-soluble substances in the normal course of things is noticeably less than one; therefore, the probability of water solubility in the reference class of samples of unspecified substances is significantly less than one. If we ask why a particular sample of unspecified material has dissolved in water, the prior weight of this explanandum event is less than one as referred to the class of samples of unspecified substances. This broad reference class is obviously inhomogeneous with respect to water solubility. If we partition it into the subclass of samples of table salt and samples of substances other than table salt, it turns out that every member of the former subclass is water-soluble. The reference class of samples of table salt is homogeneous with respect to water solubility. The weight for the single case, referred to this homogeneous reference class, is much greater than its prior weight. By referring the explanandum event to a homogeneous reference class and substantially increasing its weight, we have provided an inductive explanation of its occurrence. As the discussion develops, we shall see that the homogeneity of the reference class is the key to the explanation. The increase in weight is a fortunate dividend in many cases.

If we begin with the reference class of samples of table salt, asking why this sample of table salt dissolves in water, we already have a homogeneous reference class. If, however, we subdivide that reference class into hexed and unhexed samples, we have added nothing to the explanation of dissolving, for no new probability value results and we have not made the already homogeneous reference class any more homogeneous. Indeed, we have made matters worse by introducing a statistically irrelevant partition.

The original reference class of samples of unspecified substances can be partitioned into hexed and unhexed samples. If this partition is accomplished by means of a place selection – that is, if the hexing is done without reference to previous knowledge about solubility – the probabilities of water solubility in the subclasses will be no different from the probability in the original reference class. The reference class of hexed samples of unspecified substances is no more homogeneous than the reference class of samples of unspecified substances; moreover, it is narrower. The casting of a dissolving spell is statistically irrelevant to water solubility, so it cannot contribute to the homogeneity of the reference class, and it must not be used in assigning a weight to the single case. For this reason it contributes nothing to the explanation of the fact that this substance dissolves in water.

The vitamin C example involves the same sort of consideration. In the class of colds in general, there is a rather high frequency of recovery within a week. In the narrower reference class of colds for which the victim has taken vitamin C, the frequency of recovery within a week is no different. Vitamin C is not efficacious, and that fact is reflected in the statistical irrelevance of administration of vitamin C to recovery from a cold within a week. Subdivision of the reference class in terms of administration of vitamin C does not yield a more homogeneous reference class and, consequently, does not yield a higher weight for the explanandum event. [. . .]

[. . .]

Causal and Statistical Relevance

The attempt to explicate explanation in terms of probability, statistical relevance, and homogeneity is almost certain to give rise to a standard objection. Consider the barometer example introduced by Michael Scriven in a discussion of the thesis of symmetry between explanation and prediction[6] – a thesis whose discussion I shall postpone until a later section. If the barometer in my house shows a sudden drop, a storm may be predicted with high reliability. But the barometric reading is only an indicator; it does not cause the storm and, according to Scriven, it does not explain its occurrence. The storm is caused by certain widespread atmospheric conditions, and the behavior of the barometer is merely symptomatic of them. "In explanation we are looking for a *cause*, an event that not only occurred earlier but stands in a *special relation* to the other event. Roughly speaking, the prediction requires only a correlation, the explanation more."[7]

The objection takes the following form. There is a correlation between the behavior of the barometer and the occurrence of storms. If we take the general reference class of days in the vicinity of my house and ask for the probability of a storm, we get a rather low prior probability. If we partition that reference class into two subclasses, namely, days on which there is a sudden drop in the barometer and days on which there is not, we have a posterior probability of a storm in the former class much higher than the prior probability. The new reference class is far more homogeneous than the old one. Thus, according to the view I am suggesting, the drop in barometric reading would seem to explain the storm.

I am willing to admit that symptomatic explanations seem to have genuine explanatory value in the absence of knowledge of causal relations, that is, as long as we do not know that we are dealing only with symptoms. Causal explanations supersede symptomatic ones when they can be given, and when we suspect we are dealing with symptoms, we look hard for a causal explanation. The reason is that a causal explanation provides a more homogeneous reference class than does a symptomatic explanation. Causal proximity increases homogeneity. The reference class of days on which there is a local drop in barometric pressure inside my house, for instance, is more homogeneous than the reference class of days on which my barometer shows a sudden drop, for my barometer may be malfunctioning. Similarly, the reference class of days on which there is a widespread sudden drop in atmospheric pressure is more homogeneous than the days on which there is a local drop, for the house may be tightly sealed or the graduate students may be playing a joke on me.[8] It is not that we obtain a large increase in the probability of a storm as we move from one of these reference classes to another; rather, each progressively better partitioning makes the preceding partitioning *statistically irrelevant*.

It will be recalled that the property C is statistically irrelevant to the attribute B in the reference class A iff $P(A,B) = P(A.C,B)$. The probability of a storm on a day when there is a sudden drop in atmospheric pressure and when my barometer executes a sudden drop is precisely the same as the probability of a storm on a day when there is a sudden widespread drop in atmospheric pressure. To borrow a useful notion from Reichenbach, we may say that the sudden widespread drop in atmospheric pressure *screens off* the drop in barometer reading from the occurrence of the storm.[9] The converse relation does not hold. The probability of a storm on a day when the reading on my barometer makes a sudden drop is not equal to the probability of a storm on a day when the reading on my barometer makes a sudden drop and there is a sudden widespread drop in the atmospheric pressure. The sudden drop in barometric reading does not screen off the sudden widespread drop in atmospheric pressure from the occurrence of the storm.

More formally, we may say that D screens off C from B in reference class A iff (if and only if)

$$P(A.C.D,B) = P(A.D,B) \neq P(A.C,B).$$

For purposes of the foregoing example, let A = the class of days in the vicinity of my house, let B = the class of days on which there is an occurrence of a storm, let C = the class of days on which there is a sudden drop in reading on my barometer, and let D = the class of days on which there is a widespread drop in atmospheric pressure in the area in which my house is located. By means of this formal definition, we see that D screens off C from B, but C does not screen off D from B. The screening-off relation is, therefore, not symmetrical, although the relation of statistical relevance is symmetrical.[10]

When one property in terms of which a statistically relevant partition in a reference class can be effected screens off another property in terms of which another statistically relevant partition of that same reference class can be effected, then the screened-off property must give way to the property which screens it off. This is the *screening-off rule*. The screened-off property then becomes irrelevant and no longer has explanatory value. This consideration shows how we can handle the barometer example and a host of others, such as the explanation of measles in terms of spots, in terms of exposure to someone who has the disease, and in terms of the presence of the virus. The unwanted "symptomatic explanations" can be blocked by use of the screening-off concept, which is defined in terms of statistical irrelevance alone. We have not found it necessary to introduce an independent concept of causal relation in order to handle this problem. Reichenbach believed it was possible to define causal relatedness in terms of screening-off relations; but whether his program can be carried through or not, it seems that many causal relations exhibit the desired screening-off relations.[11]

Explanations with Low Weight

According to Hempel, the basic requirement for an inductive explanation is that the posterior weight (as I have been describing it) must be high whereas I have been suggesting that the important characteristic is the increase of the posterior weight over the prior weight as a result of incorporating the event into a homogeneous reference class. The examples discussed thus far satisfy both of these desiderata, so they do not serve well to discriminate between the two views. I would maintain, however, that when the prior weight of an event is very low, it is not necessary that its posterior weight be made high in order to have an inductive explanation. This point is illustrated by the well-known paresis example.[12]

No one ever contracts paresis unless he has had latent syphilis which has gone untreated, but only a small percentage of victims of untreated latent syphilis develop paresis. Still, it has been claimed, the occurrence of paresis is explained by the fact that the individual has had syphilis. This example has been hotly debated because of its pertinence to the issue of symmetry between explanation and prediction – an issue I still wish to postpone. Nevertheless, the following observations are in order. The prior probability of a person contracting paresis is very low, and the reference class of people in general is inhomogeneous. We can make a statistically relevant partition into people who have untreated latent syphilis and those who do not. (Note that latent syphilis screens off primary syphilis and secondary syphilis.) The probability that a person with untreated latent syphilis will contract paresis is still low, but it is considerably higher than the prior probability of paresis among people in general. To cite untreated latent syphilis as an explanation of paresis is correct, for it does provide a partition of the general reference class which yields a more homogeneous reference class and a higher posterior weight for the explanandum event.

When the posterior weight of an event is low, it is tempting to think that we have not fully explained it. We are apt to feel that we have not yet found a completely homogeneous reference class. If only we had fuller understanding, we often believe, we could sort out causal antecedents in order to be able to say which cases of untreated latent syphilis will become paretic and which ones will not. With this knowledge we would be able to partition the reference class of victims of untreated latent syphilis into two subclasses in terms of these causal antecedents so that all (or an overwhelming majority of) members of one subclass will develop paresis whereas none (or very few) of the members of the other will become paretic. This conviction may be solidly based upon experience with medical explanation, and it may provide a sound empirical basis for the search for additional explanatory facts – more relevant properties in terms of which to improve the homogeneity of the reference class. Nevertheless, the reference class of untreated latent syphilitics is (as I understand it) epistemically homogeneous in terms of our present knowledge, so we have provided the most adequate explanation possible in view of the knowledge we possess.[13]

A parallel example could be constructed in physics where we have much greater confidence in the actual homogeneity of the reference class. Suppose we had a metallic substance in which one of the atoms experienced radioactive decay within a particular, small time period, say one minute. For purposes of the example, let us suppose that only one such decay occurred within that time period. When asked why that particular atom decayed, we might reply that the substance is actually an alloy of two metals, one radioactive (for example, uranium 238) and one stable (for example, lead 206). Since the half-life of U^{238} is 4.5×10^9 years, the probability of a given uranium atom's decaying in an interval of one minute is not large, yet there is explanatory relevance in pointing out that the atom that did decay was a

U^{238} atom.[14] According to the best theoretical knowledge now available, the class of U^{238} atoms is homogeneous with respect to radioactive decay, and there is in principle no further relevant partition that can be made. Thus, it is not necessary to suppose that examples such as the paresis case derive their explanatory value solely from the conviction that they are partial explanations which can someday be made into full explanations by means of further knowledge.[15]

There is one further way to maintain that explanations of improbable events involve, nevertheless, high probabilities. If an outcome is improbable (though not impossible) on a given trial, its occurrence at least once in a sufficiently large number of trials can be highly probable. For example, the probability of getting a double six on a toss of a pair of standard dice is $1/36$; in twenty-five tosses the probability of at least one double six is over one-half. No matter how small the probability p of an event on a single trial, provided $p > 0$, and no matter how large r, provided $r < 1$, there is some n such that the probability of at least one occurrence in n trials is greater than r.[16] On this basis, it might be claimed, we explain improbable events by saying, in effect, that given enough trials the event is probable. This is a satisfactory explanation of the fact that certain types of events occur occasionally, but it still leaves open the question of how to explain the fact that this particular improbable event happened on this particular occasion.

To take a somewhat more dramatic example, each time an alpha particle bombards the potential barrier of a uranium nucleus, it has a chance of about 10^{-38} of tunneling through and escaping from the nucleus; one can appreciate the magnitude of this number by noting that whereas the alpha particle bombards the potential barrier about 10^{21} times per second, the half-life of uranium is of the order of a billion years.[17] For any given uranium atom, if we wait long enough, there is an overwhelmingly large probability that it will decay, but if we ask, "Why did the alpha particle tunnel out on this particular bombardment of the potential barrier?" the only answer is that in the homogeneous reference class of approaches to the barrier, it has a 1-in-10^{38} chance of getting through. I do not regard the fact that it gets through on a particular trial inexplicable, but certainly anyone who takes explanations to be arguments showing that the event was to be expected must conclude that this fact defies all explanation.

Multiple Homogeneity

The paresis example illustrates another important methodological point. I have spoken so far as if one homogeneous reference class were sufficient for the explanation of a particular event. This, I think, is incorrect. When a general reference class is partitioned, we can meaningfully ask about the homogeneity of each subclass in the partition (as we did in defining degree of inhomogeneity above). To be sure, when we are attempting to provide an explanation of a particular explanandum event x, we focus primary attention upon the subclass to which x belongs. Nevertheless, I think we properly raise the question of the homogeneity of other subclasses when we evaluate our explanation. In the paresis example we may be convinced that the reference class of untreated latent syphilitics is inhomogeneous, but the complementary class is perfectly homogeneous. Since no individuals who do not have untreated latent syphilis develop paresis, no partition statistically relevant to the development of paresis can be made in the reference class of people who do not have untreated latent syphilis.

Consider the table salt example from the standpoint of the homogeneity of more than one reference class. Although the reference class of samples of table salt is completely homogeneous for water solubility, the complementary class certainly is not. It is possible to make further partitions in the class of samples of substances other than table salt which are statistically relevant to water solubility: samples of sand, wood, and gold are never water soluble; samples of baking soda, sugar, and rock salt always are.

If we explain the fact that this sample dissolves in water by observing that it is table salt and all table salt is water soluble, we may feel that the explanation is somewhat inadequate. Some theorists would say that it is an adequate explanation, but that we can equally legitimately ask for an explanation of the general fact that table salt is water soluble. Although I have great sympathy with the idea that general facts need explanation and are amenable to explanation, I think it is important to recognize the desideratum of homogeneity of the complementary reference class. I think we may rightly claim fully adequate explanation of a particular fact when (but not necessarily only when) the original reference class A, with respect to which its prior probability is assessed, can be partitioned into two subclasses $A.C$ and $A.\bar{C}$, each of which is homogeneous for the attribute in question. In the ideal case all $A.C$'s are B and no $A.\bar{C}$'s are B – that is, if x is A, then x is B if and only if x is C. However, there is no reason to believe that Nature is so accommodating as to provide in all cases even the possibility in principle of such fully deterministic explanations.

We now have further reason to reject the dissolving spell explanation for the fact that a sample of table salt dissolves in water. If we partition the general reference class of samples of unspecified substances into hexed samples of table salt and all other samples of substances, this latter reference class is less homogeneous than the class of all samples of substances other than table salt, for we know that all unhexed samples of table salt are water soluble. To make a statistically irrelevant partition not only reduces the available statistics; it also reduces the homogeneity of the complementary reference class. This consideration also applies to such other examples as John Jones and his wife's birth control pills.

It would be a mistake to suppose that it must always be possible in principle to partition the original reference class A into two homogeneous subclasses. It may be that the best we can hope for is a partition into k subclasses $A.C_k$, each completely homogeneous, and such that $P(A.C_i,B) \neq P(A.C_j,B)$ if $i \neq j$. This is the *multiple homogeneity rule*. It expresses the fundamental condition for adequate explanation of particular events, and it will serve as a basis for the general characterization of deductive and inductive explanation.

[. . .]

Explanation Without Increase of Weight

It is tempting to suppose, as I have been doing so far and as all the examples have suggested, that explanation of an explanandum event somehow confers upon it a posterior weight that is greater than its prior weight. When I first enunciated this principle in section 3, however, I indicated that, though heuristically beneficial, it should not be considered fully accurate. Although most explanations may conform to this general principle, I think there may be some that do not. It is now time to consider some apparent exceptions and to see whether they are genuine exceptions.

Suppose, for instance, that a game of heads and tails is being played with two crooked pennies, and that these pennies are brought in and out of play in some irregular manner. Let one penny be biased for heads to the extent that 90 percent of the tosses with it yield heads; let the other be similarly biased for tails. Furthermore, let the two pennies be used with equal frequency in this game, so that the overall probability of heads is one-half. (Perhaps a third penny, which is fair, is tossed to decide which of the two biased pennies is to be used for any given play.) Suppose a play of this game results in a head; the prior weight of this event is one-half. The general reference class of plays can, however, be partitioned in a statistically relevant way into two homogeneous reference classes. If the toss was made with the penny biased for heads, the result is explained by that fact, and the weight of the explanandum event is raised from 0.5 to 0.9.

Suppose, however, that the toss were made with the penny biased for tails; the explanandum event is now referred to the other subclass of the original reference class, and its weight is decreased from 0.5 to 0.1. Do we want to say in this case that the event is thereby explained? Many people would want to deny it, for such cases conflict with their intuitions (which, in many cases, have been significantly conditioned by Hempel's persuasive treatment) about what an explanation ought to be. I am inclined, on the contrary, to claim that this is genuine explanation. There are, after all, improbable occurrences – such events are not explained by making them probable. Is it not a peculiar prejudice to maintain that only those events which are highly probable are capable of being explained – that improbable events are in principle inexplicable? Any event, regardless of its probability, is amenable to explanation, I believe; in the case of improbable events, the correct explanation is that they are highly improbable occurrences which happen, nevertheless, with a certain definite frequency. If the reference class is actually homogeneous, there are no other circumstances with respect to which they are probable. No further explanation can be required or can be given.[18]

There are various reasons for which my view might be rejected. In the first place, I am inclined to think that the deterministic prejudice may often be operative – namely, that x is B is not explained until x is incorporated within a reference class all of whose members are B. This is the feeling that seemed compelling in connection with the paresis example. In the discussion of that example, I indicated my reasons for suggesting that our deterministic hopes may simply be impossible to satisfy. In an attempt to undercut the hope generated in medical science that determinism would eventually triumph, I introduced the parallel example of the explanation of a radioactive decay in an alloy of lead and uranium, in which case there is strong reason to believe that the reference class of uranium 238 atoms is strictly homogeneous with respect to disintegration.

In order to avoid being victimized by the same deterministic hope in the present context, we could replace the coin-tossing example at the beginning of this section with another example from atomic physics. For instance, we could consider a mixture of uranium 238 atoms, whose half-life is 4.5×10^9 years, and polonium 214 atoms, whose half-life is 1.6×10^{-4} seconds.[19] The probability of disintegration of an unspecified atom in the mixture is between that for atoms of U^{238} and Po^{214}. Suppose that within some small specified time interval a decay occurs. There is a high probability of a polonium atom disintegrating within that interval, but a very low probability for a uranium atom. Nevertheless, a given disintegration may be of a uranium atom, so the transition from the reference class of a mixture of atoms of the two types to a reference class of atoms of U^{238} may result in a considerable lowering of the weight. Nevertheless, the latter reference class may be unqualifiedly homogeneous. When we ask why that particular atom disintegrated, the answer is that it was a U^{238} atom, and there is a small probability that such an atom will disintegrate in a short time interval.

If, in the light of modern developments in physics and philosophy, determinism no longer seems tenable as an a priori principle, we may try to salvage what we can by demanding that an explanation that does not necessitate its explanandum must at least make it highly probable. This is what Hempel's account requires. I have argued above, in the light of various examples, that even this demand is excessive and that we must accept explanations in which the explanandum event ends up with a low posterior weight. "Well, then," someone might say, "if the explanation does not show us that the event was to be expected, at least it ought to show us that the event was to be expected somewhat more than it otherwise would have been." But this attitude seems to derive in an unfortunate way from regarding explanations as arguments. At this juncture it is crucial to point out that the emphasis in the present account of explanation is upon achieving a relevant partition of an inhomogeneous reference class into homogeneous subclasses. On this conception an explanation is not an argument that is intended to produce conviction; instead, it is an attempt to assemble the factors that are relevant to the occurrence of an event. There is no more reason to suppose that such a process will increase the weight we attach to such an occurrence than there is to suppose that it will lower it. Whether the posterior weight is higher than or lower than the prior weight is really beside the point. I shall have more to say later about the function of explanations.

Before leaving this topic, I must consider one more tempting principle. It may seem evident from the examples thus far considered that an explanation must result in a change – an increase or a decrease – in the transition from the prior weight to the posterior weight. Even this need not occur. Suppose, for instance, that we change the coin-tossing game mentioned at the outset of this section by introducing a fair penny into the play. Now there are three pennies brought into play randomly: one with a probability of 0.9 for heads, one with a probability of 0.5 for heads, and one with a probability of 0.1 for heads. Overall, the probability of heads in the game is still one-half. Now, if we attempt to explain a given instance of a head coming up, we may partition the original reference class into three homogeneous subclasses, but in one of these three the probability of heads is precisely the same as it is in the entire original class. Suppose our particular head happens to belong to that subclass. Then its prior weight is exactly the same as its posterior weight, but I would claim that explanation has occurred simply by virtue of the relevant partition of the original nonhomogeneous reference class. This makes sense if one does not insist upon regarding explanations as arguments.

[...]

The Temporal Asymmetry of Explanation

It is also a virtue of the present account, I believe, that it seems to accommodate the rather puzzling temporal asymmetry of explanation already noted above – the fact that we seem to insist upon explaining events in terms of earlier rather than later initial conditions. Having noted that both the microstatistical approach and the macrostatistical approach yield a fundamental temporal asymmetry, let us now use that fact to deal with a familiar example. I shall attempt to show how Reichenbach's principle of the common cause, introduced in connection with the macrostatistical examples discussed above, helps us to establish the temporal asymmetry of explanation.

Consider Silvain Bromberger's flagpole example, which goes as follows.[20] On a sunny day a flagpole of a particular height casts a shadow of some particular length depending upon the elevation of the sun in the sky. We all agree that the position of the sun and the height of the flagpole explain the length of the shadow. Given the length of the shadow, however, and the position of the sun in the sky, we can equally infer the height of the flagpole, but we rebel at the notion that the length of the shadow explains the height of the flagpole. It seems to me that the temporal relations are crucial in this example. Although the sun, flagpole, and shadow are perhaps commonsensically regarded as simultaneous, a more sophisticated analysis shows that physical processes going on in time are involved. Photons are emitted by the sun, they travel to the vicinity of the flagpole where some are absorbed and some are not, and those which are not go on to illuminate the ground. A region of the ground is not illuminated, however, because the photons traveling toward that region were absorbed by the flagpole. Clearly the interaction between the photons and the flagpole temporally precedes the interaction between the neighboring photons and the ground. The reason that the explanation of the length of the shadow in terms of the height of the flagpole is acceptable, whereas the "explanation" of the height of the flagpole in terms of the length of the shadow is not acceptable, seems to me to hinge directly upon the fact that there are causal processes with earlier and later temporal stages. It takes only a very moderate extension of Reichenbach's terminology to conclude that the flagpole produces the shadow in a sense in which the shadow certainly does not produce the flagpole.

If we give up the notion that explanations are arguments, there is no need to be embarrassed by the fact that we can oftentimes infer earlier events from later ones by means of nomological relations, as in the cases of the eclipse and the flagpole. I have been arguing that relevance considerations are preeminent for explanations; let us see whether this approach provides a satisfactory account of the asymmetry of explanation in the flagpole case. The apparent source of difficulty is that, under the general conditions of the example, the flagpole is relevant to the shadow and the shadow is relevant to the flagpole. It does not follow, however, that the shadow explains the flagpole as well as the flagpole explains the shadow.

In order to analyze the relevance relations more carefully, I shall reexamine a simplified version of the example of the illness in the theatrical company and compare it with a simplified version of the flagpole example. In both cases the screening-off relation will be employed in an attempt to establish the temporal asymmetry of the explanation. The general strategy will be to show that a common cause screens off a common effect and, consequently, by using the screening-off rule, that the explanation must be given in terms of the common cause and not in terms of the common effect. In order to carry out this plan, I shall regard the flagpole as an orderly arrangement of parts that requires an explanation of the same sort as does the coincidental illnesses of the actors.

Consider, then, the simple case of a theatrical company consisting of only two people, the leading lady and the leading man. Let A be our general reference class of days, let M be the illness of the leading man, let L be the illness of the leading lady, let F be a meal of spoiled food that both eat, and let C be the cancellation of the performance. Our previous discussion of the example has shown that the simultaneous illness occurs more frequently than it would if the two were independent, that is,

$$P(A,L.M) > P(A,L) \times P(A,M),$$

and that the common meal is highly relevant to the concurrent illnesses, that is,

$$P(A.F,L.M) > P(A,L.M) > P(A.\bar{F},L.M).$$

It is not true, without further qualification, that the eating of spoiled food F screens off the cancellation of the performance C from the joint illness $L.M$. Although the eating of spoiled food does make it probable that the two actors will be ill, the fact that the performance has been cancelled supplies further evidence of the illness and makes the joint illness more probable. At this point experiment must be admitted. It is clearly possible to arrange things so that the play goes on, illness or no illness, by providing substitutes who never eat with the regular cast. Likewise, it is a simple matter to see to it that the performance is cancelled, whether or not anyone is ill. Under these experimental conditions we can see that the probability of the joint illness, given the common meal with spoiled food, is the same whether or not the play is performed, that is,

$$P(A.F.C,L.M) = P(A.F.\bar{C},L.M).$$

From this it follows that

$$P(A.F.C,L.M) = P(A.F,L.M).$$

If we approach the common meal of spoiled food in the same experimental fashion, we can easily establish that it is not screened off by the cancellation of the performance. We can arrange for the two leading actors to have no meals supplied from the same source and ascertain the probability $P(A.\bar{F}.C,L.M)$. This can be compared with the probability $P(A.F.C,L.M)$ which arises under the usual conditions of the two actors eating together in the same restaurants. Since they are not equal, the common spoiled food is statistically relevant to the joint illness, even in the presence of the cancellation of the performance, that is,

$$P(A.F.C,L.M) \neq P(A.C,L.M).$$

Thus, although the common cause F is relevant to the coincidence $L.M$ and the coincidence $L.M$ is relevant to C (from which it follows that C is relevant to $L.M$), it turns out that the common cause F screens off the common effect C from the coincidence to be explained. By the screening-off rule, F must be used and C must not be used to partition the reference class A. These considerations express somewhat formally the fact that tampering with the frequency of F without changing the frequency of C will affect the frequency of $L.M$, but tampering with the frequency of C without changing the frequency of F will have no effect on the frequency of $L.M$. This seems to capture the idea that we can influence events by influencing their causal antecedents, but not by influencing their causal consequents.

Let us apply the same analysis to the flagpole example. Again, let A be the general reference class, and let us for simplicity suppose that the flagpole is composed of two parts, a top and a bottom. The flagpole is in place when the two parts are in place; let T be the proper positioning of the top, and B the proper positioning of the bottom. Let M represent the flagpole makers' bringing the pieces together in the appropriate positions. Let S be the occurrence of the full shadow of the flagpole. Now, since the flagpole's existence consists in the two pieces being put together in place and since that hardly happens except when the flagpole makers bring them together and assemble them, it is clear that

$$P(A,T.B) > P(A,T) \times P(A,B).$$

Hence, the existence of the flagpole is something to be explained. Since we know that

$$P(A.M,T.B) > P(A,T.B) > P(A.\bar{M},T.B),$$

M effects a relevant partition in the general reference class A.

The question of whether the flagpole gets put together when the flagpole makers go about putting it together is unaffected by the existence or nonexistence of a shadow – for example, the ground might be illuminated by other sources of light besides the sun, or mirrors might deflect some of the sun's rays, with the result that there is no shadow – but the flagpole is there just the same; hence,

$$P(A.M.S,T.B) = P(A.M,T.B),$$

but

$$P(A.M.S,T.B) \neq P(A.S,T.B),$$

since the shadow can easily be obliterated at will or produced by other means without affecting the existence of the flagpole. We, therefore, conclude again that the common cause screens off the common effect and that, by virtue of the screening-off rule, the causal antecedent M – and not the causal consequent S – must be invoked to effect a relevant partition in the reference class A. In this highly schematic way, I hope I have shown that the approach to explanation via statistical relevance has allowed us to establish the temporal asymmetry of explanation. Once more, the intuitive idea is that manipulating the pieces of the flagpole, without otherwise tampering with the shadow, affects the shadow; contrariwise, tampering with the shadow, without otherwise manipulating the pieces of the flagpole, has no effect upon the flagpole. The analysis of the flagpole example may, of course, require the same sort of appeal to experiment as we invoked in the preceding example.

[. . .]

The general fact about the world that seems to be involved is that causal processes very frequently exhibit the following sort of structure: a process leading up to a given event E consists of a series of events earlier than E, but such that later ones screen off earlier ones. In other words, a given antecedent event A_1 will be relevant to E, but it will be screened off by a later antecedent A_2 that intervenes between A_1 and E. This situation obtains until we get to E, and then every subsequent event is screened off by some causal antecedent or other. Thus, in some deeply significant sense, the causal consequents of an event are made irrelevant to its occurrence in a way in which the causal antecedents are not. If, in Velikovsky-like cataclysm, a giant comet should disrupt the solar system, it would have enormous bearing upon subsequent eclipses but none whatever upon previous ones. The working out of the details of the eclipse example, along the lines indicated by the analysis of the flagpole example, is left as an exercise for the reader.

[. . .]

The Nature of Statistical Explanation

Let me now, at long last, offer a general characterization of explanations of particular events. As I have suggested earlier, we may think of an explanation as an answer to a question of the form, "Why does this x which is a member of A have the property B?" The answer to such a question consists of a partition of the reference class A into a number of subclasses, all of which are homogeneous with respect to B, along with the probabilities of B within each of these subclasses. In addition, we must say which of the members of the partition contains our particular x. More formally, an explanation of the fact that x, a member of A, is a member of B would go as follows:

$$\begin{aligned} P(A.C_1,B) &= p_1 \\ P(A.C_2,B) &= p_2 \\ &\vdots \\ P(A.C_n,B) &= p_n \end{aligned}$$

where

$A.C_1, A.C_2, \ldots, A.C_n$ is a homogeneous partition of A with respect to B,
$p_i = p_j$ only if $i = j$, and
$x \in A.C_k$.

With Hempel, I regard an explanation as a linguistic entity, namely, a set of statements, but unlike him, I do not regard it as an argument. On my view, an explanation is a set of probability statements, qualified by certain provisos, plus a statement specifying the compartment to which the explanandum event belongs.

[. . .]

One might ask on what grounds we can claim to have characterized explanation. The answer is this. When an explanation (as herein explicated) has been provided, we know exactly how to regard any *A* with respect to the property *B*. We know which ones to bet on, which to bet against, and at what odds. We know precisely what degree of expectation is rational. We know how to face uncertainty about an *A*'s being a *B* in the most reasonable, practical, and efficient way. We know every factor that is relevant to an *A* having property *B*. We know exactly the weight that should have been attached to the prediction that this *A* will be a *B*. We know all of the regularities (universal or statistical) that are relevant to our original question. What more could one ask of an explanation?

There are several general remarks that should be added to the foregoing theory of explanation:

a. It is evident that explanations as herein characterized are nomological. For the frequency interpretation probability statements are statistical generalizations, and every explanation must contain at least one such generalization. Since an explanation essentially consists of a set of statistical generalizations, I shall call these explanations "statistical" without qualification, meaning thereby to distinguish them from what Hempel has recently called "inductive-statistical."[21] His inductive-statistical explanations contain statistical generalizations, but they are inductive inferences as well.
b. From the standpoint of the present theory, deductive-nomological explanations are just a special case of statistical explanation. If one takes the frequency theory of probability as literally dealing with infinite classes of events, there is a difference between the universal generalization, "All *A* are *B*," and the statistical generalization, "$P(A,B) = 1$," for the former admits no *A*s that are not *B*s, whereas the latter admits of infinitely many *A*s that are not *B*s. For this reason, if the universal generalization holds, the reference class *A* is homogeneous with respect to *B*, whereas the statistical generalization may be true even if *A* is not homogeneous. Once this important difference is noted, it does not seem necessary to offer a special account of deductive-nomological explanations.
c. The problem of symmetry of explanation and prediction, which is one of the most hotly debated issues in discussions of explanation, is easily answered in the present theory. To explain an event is to provide the best possible grounds we could have had for making predictions concerning it. An explanation does not show that the event was to be expected; it shows what sorts of expectations would have been reasonable and under what circumstances it was to be expected. To explain an event is to show to what degree it was to be expected, and this degree may be translated into practical predictive behavior such as wagering on it. In some cases the explanation will show that the explanandum event was not to be expected, but that does not destroy the symmetry of explanation and prediction. The symmetry consists in the fact that the explanatory facts constitute the fullest possible basis for making a prediction of whether or not the event would occur. To explain an event is not to predict it ex post facto, but a complete explanation does provide complete grounds for rational prediction concerning that event. Thus, the present account of explanation does sustain a thoroughgoing symmetry thesis, and this symmetry is not refuted by explanations having low weights.
d. In characterizing statistical explanation, I have required that the partition of the reference class yield subclasses that are, in fact, homogeneous. I have not settled for practical or epistemic homogeneity. The question of whether actual homogeneity or epistemic homogeneity is demanded is, for my view, analogous to the question of whether the premises of the explanation must be true or highly confirmed for Hempel's view.[22] I have always felt that truth was the appropriate requirement, for I believe Carnap has shown that the concept of truth is harmless enough.[23] However, for those who feel too uncomfortable with the stricter requirement, it would be possible to characterize statistical explanation in terms of epistemic homogeneity instead of actual homogeneity. No fundamental problem about the nature of explanation seems to be involved.
e. This paper has been concerned with the explanation of single events, but from the standpoint of probability theory, there is no

significant distinction between a single event and any finite set of events. Thus, the kind of explanation appropriate to a single result of heads on a single toss of a coin would, in principle, be just like the kind of explanation that would be appropriate to a sequence of ten heads on ten consecutive tosses of a coin or to ten heads on ten different coins tossed simultaneously.

f. With Hempel, I believe that generalizations, both universal and statistical, are capable of being explained. Explanations invoke generalizations as parts of the explanans, but these generalizations themselves may need explanation. This does not mean that the explanation of the particular event that employed the generalization is incomplete; it only means that an additional explanation is possible and may be desirable. In some cases it may be possible to explain a statistical generalization by subsuming it under a higher level generalization; a probability may become an instance for a higher level probability. For example, Reichenbach offered an explanation for equiprobability in games of chance, by constructing, in effect, a sequence of probability sequences.[24] Each of the first level sequences is a single case with respect to the second level sequence. To explain generalizations in this manner is simply to repeat, at a higher level, the pattern of explanation we have been discussing. Whether this is or is not the only method of explaining generalizations is, of course, an entirely different question.

g. In the present account of statistical explanation, Hempel's problem of the "nonconjunctiveness of statistical systematization"[25] simply vanishes. This problem arises because in general, according to the multiplication theorem for probabilities, the probability of a conjunction is smaller than that of either conjunct taken alone. Thus, if we have chosen a value *r*, such that explanations are acceptable only if they confer upon the explanandum an inductive probability of at least *r*, it is quite possible that each of the two explananda will satisfy that condition, whereas their conjunction fails to do so. Since the characterization of explanation I am offering makes no demands whatever for high probabilities (weights), it has no problem of nonconjunctiveness.

Conclusion

Although I am hopeful that the foregoing analysis of statistical explanation of single events solely in terms of statistical relevance relations is of some help in understanding the nature of scientific explanation, I should like to cite, quite explicitly, several respects in which it seems to be incomplete.

First, and most obviously, whatever the merits of the present account, no reason has been offered for supposing the type of explanation under consideration to be the only legitimate kind of scientific explanation. If we make the usual distinction between empirical laws and scientific theories, we could say that the kind of explanation I have discussed is explanation by means of empirical laws. For all that has been said in this paper, theoretical explanation – explanation that makes use of scientific theories in the fullest sense of the term – may have a logical structure entirely different from that of statistical explanation. Although theoretical explanation is almost certainly the most important kind of scientific explanation, it does, nevertheless, seem useful to have a clear account of explanation by means of empirical laws, if only as a point of departure for a treatment of theoretical explanation.

Second, in remarking above that statistical explanation is nomological, I was tacitly admitting that the statistical or universal generalizations invoked in explanations should be lawlike. I have made no attempt to analyze lawlikeness, but it seems likely that an adequate analysis will involve a solution to Nelson Goodman's "grue-bleen" problem.[26]

Third, my account of statistical explanation obviously depends heavily upon the concept of *statistical relevance* and upon the *screening-off relation*, which is defined in terms of statistical relevance. In the course of the discussion, I have attempted to show how these tools enable us to capture much of the involvement of explanation with causality, but I have not attempted to provide an analysis of causation in terms of these statistical concepts alone. Reichenbach has attempted such an analysis,[27] but whether his – or any other – can succeed is a difficult question. I should be inclined to harbor serious misgivings about the adequacy of my view of statistical explanation if the statistical analysis of causation cannot be carried through successfully, for the relation between causation and explanation seems extremely intimate.

Finally, although I have presented my arguments in terms of the limiting frequency conception of probability, I do not believe that the fundamental correctness of the treatment of statistical explanation hinges upon the acceptability of that interpretation of probability. Proponents of other theories of probability, especially the personalist and the propensity interpretations, should be able to adapt this treatment of explanation to their views of probability with a minimum of effort. That, too, is left as an exercise for the reader.[28]

Notes

1 Salmon, "The Status of Prior Probabilities in Statistical Explanation," *Philosophy of Science*, XXXII, no. 2 (April, 1965).

2 See Salmon, *The Foundations of Scientific Inference* (Pittsburgh: University of Pittsburgh Press, 1967), pp. 83–96, for fuller explanations. Note that, contrary to frequent usage, the expression "$P(A,B)$" is read "the probability *from A to B*." This notation is Reichenbach's.

3 Reichenbach, *The Theory of Probability* (Berkeley and Los Angeles: University of California Press, 1949), sec. 72. John Venn, *The Logic of Chance*, 4th ed. (New York: Chelsea Publishing Co., 1962), chap. IX, sec. 12–32. Venn was the first systematic exponent of the frequency interpretation, and he was fully aware of the problem of the single case. He provides an illuminating account, and his discussion is an excellent supplement to Reichenbach's well-known later treatment.

4 Richard von Mises, *Probability, Statistics and Truth*, 2d rev. edn. (London: Allen and Unwin, 1957), p. 25.

5 Also, of course, there are cases in which it would be possible in principle to make a relevant partition, but we are playing a game in which the rules prevent it. Such is the case in roulette, where the croupier prohibits additional bets after a certain point in the spin of the wheel. In these cases also we shall speak of practical homogeneity.

6 See Michael J. Scriven, "Explanation and Prediction in Evolutionary Theory," *Science*, CXXX, no. 3374 (Aug. 28, 1959).

7 Ibid., p. 480.

8 See Adolf Grünbaum, *Philosophical Problems of Space and Time* (New York: Alfred A. Knopf, 1963), pp. 309–11.

9 Hans Reichenbach, *The Direction of Time* (Berkeley and Los Angeles: The University of California Press, 1956), p. 189.

10 Since $P(A.C.B) = P(A,B)$ entails $P(A.B,C) = P(A,C)$, provided $P(A.B,C) \neq 0$, the relevance relation is symmetrical. The screening-off relation is a three-place relation; it is nonsymmetrical in its first and second arguments, but it is symmetrical in the second and third arguments. If D screens off C from B, then D screens off B from C.

11 Reichenbach, *The Direction of Time*, sec. 22.

12 This example has received considerable attention in the recent literature on explanation. Introduced in Scriven, "Explanation and Prediction," it has been discussed by (among others) May Brodbeck, "Explanation, Prediction, and 'Imperfect Knowledge,'" in *Minnesota Studies in Philosophy of Science*, III, eds. Herbert Feigl and Grover Maxwell (Minneapolis: University of Minnesota Press, 1962); Adolf Grünbaum, *Philosophical Problems*, pp. 303–08; and Carl G. Hempel, "Explanation and Prediction by Covering Laws," *Philosophy of Science: The Delaware Seminar*, I, ed. Bernard H. Baumrin (New York: John Wiley and Sons, 1963).

13 "72 out of 100 untreated persons [with latent syphilis] go through life without the symptoms of late [tertiary] syphilis, but 28 out of 100 untreated persons were known to have developed serious outcomes [paresis and others] and there is no way to predict what will happen to an untreated infected person" (Edwin Gurney Clark, MD, and William D. Mortimer Harris, MD, "Venereal Diseases," *Encyclopedia Britannica*, XXIII [1961], p. 44).

14 Ralph E. Lapp and Howard L. Andrews, *Nuclear Radiation Physics*, 3rd edn. (Englewood Cliffs, N.J.: Prentice-Hall, 1963), p. 73.

15 Cf. Hempel's discussion of this example in *Aspects of Scientific Explanation* (New York: Free Press, 1965), pp. 369–74.

16 Here I am assuming the separate trials to be independent events.

17 George Gamow, *The Atom and Its Nucleus* (Englewood Cliffs, N.J.: Prentice-Hall, 1961), p. 114.

18 See Jeffrey, "Statistical Explanation vs. Statistical Inference", paper presented at the meeting of the American Association of Science, Section L, New York, 1967, for a lucid and eloquent discussion of this point. In this context I am, of course, assuming that the pertinent probabilities exist.

19 Lapp and Andrews, *Nuclear Radiation Physics*, p. 73.
20 See Hempel, "Deductive-Nomological vs. Statistical Explanation," in *Minnesota Studies in the Philosophy of Science*, III, eds. Herbert Feigl and Grover Maxwell (Minneapolis: University of Minnesota Press, 1962), pp. 109–10; also Grünbaum, *Philosophical Problems*, pp. 307–08.
21 See *Aspects of Scientific Explanation*, secs. 3.2–3.3. In the present essay I am not at all concerned with explanations of the type Hempel calls "deductive-statistical." For greater specificity, what I am calling "statistical explanation" might be called "statistical-relevance explanation," or "S-R explanation" as a handy abbreviation to distinguish it from Hempel's D-N, D-S, and I-S types.
22 Hempel, "Deductive-Nomological vs. Statistical Explanation," sec. 3.
23 Rudolf Carnap, "Truth and Confirmation," in *Readings in Philosophical Analysis*, eds. Herbert Feigl and Wilfrid Sellars (New York: Appleton-Century-Crofts, 1949), pp. 119–27.
24 Reichenbach, *Theory of Probability*, sec. 69.
25 Hempel, "Deductive-Nomological vs. Statistical Explanation," sec. 13, and *Aspects of Scientific Explanation*, sec. 3.6. Here, Hempel says, "Non-conjunctiveness presents itself as an inevitable aspect of [inductive-statistical explanation], and thus as one of the fundamental characteristics that set I-S explanation apart from its deductive counterparts."
26 See Nelson Goodman, *Fact, Fiction, and Forecast*, 2d edn. (Indianapolis: Bobbs-Merrill Co., 1965), chap. III. I have suggested a resolution in "On Vindicating Induction," *Philosophy of Science*, XXX (July 1963), pp. 252–61, reprinted in Henry E. Kyburg and Ernest Nagel, eds., *Induction: Some Current Issues* (Middletown, Conn.: Wesleyan University Press, 1963).
27 Reichenbach, *The Direction of Time*, chap. IV.
28 The hints are provided in sec. 3.

8.3

Why Ask, "Why"?

Wesley Salmon

Underlying his earlier S-R model was Salmon's view that explanations describe the causal history leading up to the explanandum. He assumed that the causal relations described in that history are captured by the statistical-relevance relations that are presented in an S-R explanation. But he eventually realized that this is not correct, and formulated a model of explanation that attempts to more directly describe the causal history. In the selection below Salmon presents his new *causal-mechanical* (C-M) model, which invokes causal processes and their interactions.

[. . .]

Developments in twentieth-century science should prepare us for the eventuality that some of our scientific explanations will have to be statistical – not merely because our knowledge is incomplete (as Laplace would have maintained), but rather, because nature itself is inherently statistical. Some of the laws used in explaining particular events will be statistical, and some of the regularities we wish to explain will also be statistical. I have been urging that causal considerations play a crucial role in explanation; indeed, I have just said that regularities – and this certainly includes statistical regularities – require causal explanation. I do not believe there is any conflict here. It seems to me that, by employing a statistical conception of causation along the lines developed by Patrick Suppes and Hans Reichenbach,[1] it is possible to fit together harmoniously the causal and statistical factors in explanatory contexts. Let me attempt to illustrate this point by discussing a concrete example.

A good deal of attention has recently been given in the press to cases of leukemia in military personnel who witnessed an atomic bomb test (code name "Smokey") at close range in 1957.[2] Statistical studies of the survivors of the bombings of Hiroshima and Nagasaki have established the fact that exposure to high levels of radiation,

From The Proceedings and Addresses of the American Philosophical Association, Vol. 51, No. 6 (August, 1978), extracts from pp. 688–96, 697, 699–700, plus notes 702–5.

such as occur in an atomic blast, is statistically relevant to the occurrence of leukemia – indeed, that the probability of leukemia is closely correlated with the distance from the explosion.[3] A clear pattern of statistical relevance relations is exhibited here. If a particular person contracts leukemia, this fact may be explained by citing the fact that he was, say, 2 kilometers from the hypocenter at the time of the explosion. This relationship is further explained by the fact that individuals located at specific distances from atomic blasts of specified magnitude receive certain high doses of radiation.

This tragic example has several features to which I should like to call special attention:

(1) The location of the individual at the time of the blast is statistically relevant to the occurrence of leukemia; the probability of leukemia for a person located 2 kilometers from the hypocenter of an atomic blast is radically different from the probability of the disease in the population at large. Notice that the probability of such an individual contracting leukemia is not high; it is much smaller than one-half – indeed, in the case of Smokey it is much less than 1/100. But it is markedly higher than for a random member of the entire human population. It is the *statistical relevance* of exposure to an atomic blast, not a high probability, which has explanatory force.[4] Such examples defy explanation according to an inferential view which requires high inductive probability for statistical explanation.[5] The case of leukemia is subsumed under a statistical regularity, but it does not "follow inductively" from the explanatory facts.

(2) There is a *causal process* which connects the occurrence of the bomb blast with the physiological harm done to people at some distance from the explosion. High energy radiation, released in the nuclear reactions, traverses the space between the blast and the individual. Although some of the details may not yet be known, it is a well-established fact that such radiation does interact with cells in a way which makes them susceptible to leukemia at some later time.

(3) At each end of the causal process – i.e., the transmission of radiation from the bomb to the person – there is a *causal interaction*. The radiation is emitted as a result of a nuclear interaction when the bomb explodes, and it is absorbed by cells in the body of the victim. Each of these interactions may well be irreducibly statistical and indeterministic, but that is no reason to deny that they are causal.

(4) The causal processes begin at a central place, and they travel outward at a finite velocity. A rather complex set of statistical relevance relations is explained by the propagation of a process, or set of processes, from a common central event.

In undertaking a general characterization of causal explanation, we must begin by carefully distinguishing between causal processes and causal interactions. The transmission of light from one place to another, and the motion of a material particle, are obvious examples of causal processes. The collision of two billiard balls, and the emission or absorption of a photon, are standard examples of causal interactions. Interactions are the sorts of things we are inclined to identify as events. Relative to a particular context, an event is comparatively small in its spatial and temporal dimensions; processes typically have much larger durations, and they may be more extended in space as well. A light ray, traveling to earth from a distant star, is a process which covers a large distance and lasts for a long time. What I am calling a "causal process" is similar to what Russell called a "causal line."[6]

When we attempt to identify causal processes, it is of crucial importance to distinguish them from such pseudo-processes as a shadow moving across the landscape. This can best be done, I believe, by invoking Reichenbach's *mark criterion*.[7] Causal processes are capable of propagating marks or modifications imposed upon them; pseudo-processes are not. An automobile traveling along a road is an example of a causal process. If a fender is scraped as a result of a collision with a stone wall, the mark of that collision will be carried on by the car long after the interaction with the wall occurred. The shadow of a car moving along the shoulder is a pseudo-process. If it is deformed as it encounters a stone wall, it will immediately resume its former shape as soon as it passes by the wall. It will not transmit a mark or modification. For this reason, we say that a causal process can transmit information or causal influence; a pseudo-process cannot.[8]

When I say that a causal process has the capability of transmitting a causal influence, it might be supposed that I am introducing precisely the sort of mysterious power Hume warned us against. It seems to me that this danger can be circumvented by employing an adaptation of the "at-at" theory of motion, which Russell used so effectively in dealing with Zeno's paradox of the flying arrow.[9] The flying arrow – which is, by the way, a causal process – gets from one place to another by being *at* the appropriate intermediate points of space *at* the appropriate instants of time. Nothing more is involved in getting *from* one point *to* another. A mark, analogously, can be said to be propagated from the point of interaction at which it is imposed to later stages in the process if it appears *at* the appropriate intermediate stages in the process *at* the appropriate times without additional interactions which regenerate the mark. The precise formulation of this condition is a bit tricky, but I believe the basic idea is simple, and that the details can be worked out.[10]

If this analysis of causal processes is satisfactory, we have an answer to the question, raised by Hume, concerning the connection between cause and effect. If we think of a cause as one event, and of an effect as a distinct event, then the connection between them is simply a spatio-temporally continuous causal process. This sort of answer did not occur to Hume because he did not distinguish between causal processes and causal interactions. When he tried to analyze the connections between distinct events, he treated them as if they were chains of events with discrete links, rather than processes analogous to continuous filaments. I am inclined to attribute considerable philosophical significance to the fact that each link in a chain has adjacent links, while the points in a continuum do not have next-door neighbors. This consideration played an important role in Russell's discussion of Zeno's paradoxes.[11]

After distinguishing between causal interactions and causal processes, and after introducing a criterion by means of which to discriminate the pseudo-processes from the genuine causal processes, we must consider certain configurations of processes which have special explanatory import. Russell noted that we often find similar structures grouped symmetrically about a center – for example, concentric waves moving across an otherwise smooth surface of a pond, or sound waves moving out from a central region, or perceptions of many people viewing a stage from different seats in a theatre. In such cases, Russell postulates the existence of a central event – a pebble dropped into the pond, a starter's gun going off at a race-track, or a play being performed upon the stage – from which the complex array emanates.[12] It is noteworthy that Russell never suggests that the central event is to be explained on the basis of convergence of influences from remote regions upon that locale.

Reichenbach articulated a closely-related idea in his *principle of the common cause*. If two or more events of certain types occur at different places, but occur at the same time more frequently than is to be expected if they occurred independently, then this apparent coincidence is to be explained in terms of a common causal antecedent.[13] If, for example, all of the electric lights in a particular area go out simultaneously, we do not believe that they just happened by chance to burn out at the same time. We attribute the coincidence to a common cause such as a blown fuse, a downed transmission line, or trouble at the generating station. If all of the students in a dormitory fall ill on the same night, it is attributed to spoiled food in the meal which all of them ate. Russell's similar structures arranged symmetrically about a center obviously qualify as the sorts of coincidences which require common causes for their explanations.[14]

In order to formulate his common cause principle more precisely, Reichenbach defined what he called a *conjunctive fork*. Suppose we have events of two types, A and B, which happen in conjunction more often than they would if they were statistically independent of one another. For example, let A and B stand for colorblindness in two brothers. There is a certain probability that a male, selected from the population at random, will have that affliction, but since it is hereditary, occurrences in male siblings are not independent. The probability that both will have it is greater than the product of the two respective probabilities. In cases of such statistical dependencies, we invoke a common cause C which accounts for them; in this case, it is a genetic factor carried by the mother. In order to satisfy the conditions for a conjunctive fork, events of the types *A* and *B* must occur independently in the absence of the common cause *C* – that is, for two unrelated males, the probability of both being colorblind is equal to the product of the two separate probabilities. Furthermore, the probabilities of *A* and *B* must each be increased above their overall values if *C* is present. Clearly the probability of colorblindness is greater in sons of mothers carrying the genetic factor than it is among all male children regardless of the genetic make-up of their mothers. Finally, Reichenbach stipulates, the dependency between *A* and *B* is absorbed into the occurrence of the common cause *C*, in the sense that the probability of *A* and *B* given *C* equals the product of the probability of *A* given *C* and the probability of *B* given *C*. This is true in the colorblindness case. Excluding pairs of identical twins, the question of whether a male child inherits colorblindness from the mother who carries the genetic trait depends only upon the genetic relationship between that child and his mother, not upon whether other sons happened to inherit the trait.[15] Note that screening-off occurs here.[16] While the colorblindness of a brother is statistically relevant to colorblindness in a boy, it becomes irrelevant if the genetic factor is known to be present in the mother.

Reichenbach obviously was not the first philosopher to notice that we explain coincidences in terms of common causal antecedents. Leibniz postulated a pre-established harmony for his windowless monads which mirror the same world, and the occasionalists postulated God as the coordinator of mind and body. Reichenbach was, to the best of my knowledge, the first to give a precise characterization of the conjunctive fork, and to formulate the general principle that conjunctive forks are open only to the future, not to the past.[17] The result is that we cannot explain coincidences on the basis of future effects, but only on the basis of antecedent causes. A widespread blackout is explained by a power failure, not by the looting which occurs as a consequence. (A common effect *E* may form a conjunctive fork with *A* and *B*, but only if there is also a common cause *C*.) The principle that conjunctive forks are not open to the past accounts for Russell's principle that symmetrical patterns emanate from a central source – they do not converge from afar upon the central point. It is also closely related to the operation of the second law of thermodynamics and the increase of entropy in the physical world.

The common cause principle has, I believe, deep explanatory significance. Bas van Fraassen has recently subjected it to careful scrutiny, and he has convinced me that Reichenbach's formulation in terms of the conjunctive fork, as he defined it, is faulty.[18] (We do not, however, agree about the nature of the flaw.) There are, it seems, certain sorts of causal *interactions* in which the resulting effects are more strongly correlated with one another than is allowed in Reichenbach's conjunctive forks. If, for example, an energetic photon collides with an electron in a Compton scattering experiment, there is a certain probability that a photon with a given smaller energy will emerge, and there is a certain probability that the electron will be kicked out with a given kinetic energy (see Figure 16). However, because of the law of conservation of energy, there is a strong correspondence between the two energies – their sum must be close to the energy of the incident photon. Thus, the probability of getting a photon with energy E_1 and an electron with energy E_2, where $E_1 + E_2$ is approximately equal to E (the energy of the incident photon), is much greater than the product of the probabilities of each energy occurring separately. Assume, for example, that there is a probability of 0.1 that a photon of energy E_1 will emerge if a photon of energy E impinges on a given target, and assume that there is a probability of 0.1 that an electron with kinetic energy E_2 will emerge under the same circumstances (where E, E_1 and E_2 are related as the law of conservation of energy demands).

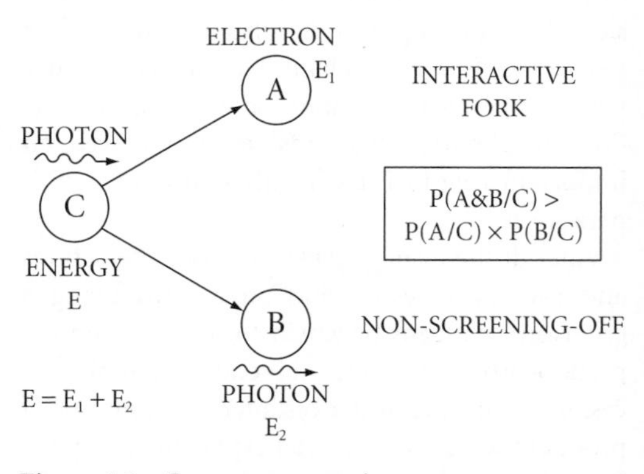

Figure 16 Compton scattering

In this case the probability of the joint result is not 0.01, the product of the separate probabilities, but 0.1, for each result will occur if and only if the other does.[19] The same relationships could be illustrated by such macroscopic events as collisions of billiard balls, but I have chosen Compton scattering because there is good reason to believe that events of that type are irreducibly statistical. Given a high energy photon impinging upon the electron in a given atom, there is no way, even in principle, of predicting with certainty the energies of the photon and electron which result from the interaction.

This sort of interaction stands in sharp contrast with the sort of statistical dependency we have in the leukemia example (see Figure 17, which also represents the relationships in the colorblindness case). In the absence of a strong source of radiation, such as the atomic blast, we may assume that the probability of next-door neighbors contracting the disease equals the product of the probabilities for each of them separately. If, however, we consider two next-door neighbors who lived at a distance of 2 kilometers from the hypocenter of the atomic explosion, the probability of both of them contracting leukemia is much greater than it would be for any two randomly selected members of the population at large. This apparent dependency between the two leukemia cases is not a direct physical dependency between them; it is merely a statistical result of the fact that the probability for each of them has been enhanced independently of the other by being located in close proximity to the atomic explosion. But the individual photons of radiation which impinge upon the two victims are emitted independently, travel independently, and damage living tissues independently.

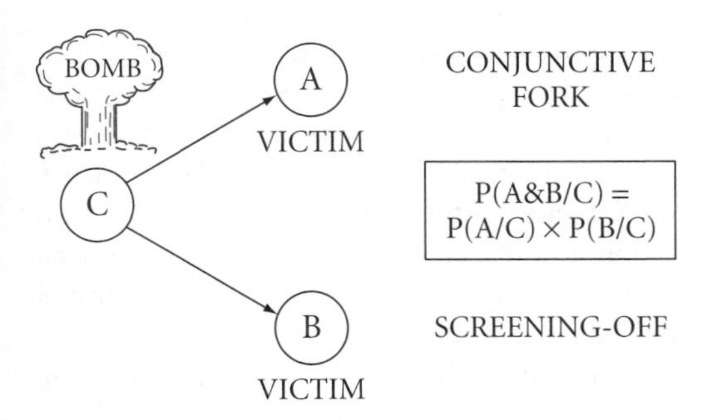

Figure 17 Leukemia

It thus appears that there are two kinds of causal forks: (1) Reichenbach's *conjunctive forks*, in which the common cause screens-off the one effect from the other, which are exemplified by the colorblindness and leukemia cases, and (2) *interactive forks*, exemplified by the Compton scattering of a photon and an electron. In forks of the interactive sort, the common cause does not screen-off the one effect from the other. The probability that the electron will be ejected with kinetic energy E_2 given an incident photon of energy E is *not equal to* the probability that the electron will emerge with energy E_2 given an incident photon of energy E and a scattered photon of energy E_1. In the conjunctive fork, the common cause C absorbs the dependency between the effects A and B, for the probability of A and B given C is *equal to* the product of the probability of A given C and the probability of B given C. In the interactive fork, the common cause C does not absorb the dependency between the effects A and B, for the probability of A and B given C is *greater than* the product of the two separate conditional probabilities.[20]

Recognition and characterization of the interactive fork enables us to fill a serious lacuna in the treatment up to this point. I have discussed causal processes, indicating roughly how they are to be characterized, and I have mentioned causal interactions, but have said nothing about their characterization. Indeed, the criterion by which we distinguished causal processes from pseudo-processes involved the use of marks, and marks are obviously results of causal interactions. Thus, our account stands in serious need of a characterization of causal interactions, and the interactive fork enables us, I believe, to furnish it.

There is a strong temptation to think of events as basic types of entities, and to construe processes – real or pseudo – as collections of events. This viewpoint may be due, at least in part, to the fact that the space-time interval between events is a fundamental invariant of the special theory of relativity, and that events thus enjoy an especially fundamental status. I suggest, nevertheless, that we reverse the approach. Let us begin with processes (which have not yet been sorted out into causal and pseudo) and look at their intersections. We can be reassured about the legitimacy of this new orientation by the fact that the basic spacetime structure of both special relativity and general relativity can be built upon processes without direct recourse to events.[21] An electron traveling through space is a process, and so is a photon; if they collide, that is an intersection. A light pulse traveling from a beacon to a screen is a process, and a piece of red glass standing in the path is another; the light passing through the glass is an intersection. Both of these intersections constitute interactions. If two light beams cross one another, we have an intersection without an interaction – except in the extremely unlikely event of a particle-like collision between photons. What we want to say, very roughly, is that when two processes intersect, and both are modified in such ways that the changes in one are correlated with changes in the other – in the manner of an interactive fork (see Figure 18) – we have a causal interaction. There are technical details to be worked out before we can claim to have a satisfactory account, but the general idea seems clear enough.[22]

[. . .]

Let me now summarize the picture of scientific explanation I have tried to outline. If we wish to explain a particular event, such as death by leukemia of GI Joe, we begin by assembling the factors statistically relevant to that occurrence – for example, his distance from the atomic explosion, the magnitude of the blast, and the type of shelter he was in. There will be many others, no doubt, but these will do for purposes of illustration. We must also obtain the probability values associated with the relevancy relations. The statistical relevance relations are statistical regularities, and we proceed to explain them. Although this differs substantially from things I have said previously, I no longer believe that the assemblage of relevant factors provides a complete explanation – or much of anything in the way of an explanation.[23] We do, I believe, have a bona fide explanation of an event if we have a complete set of statistically relevant factors, the pertinent probability values, *and* causal explanations of the relevance relations. Subsumption of a particular occurrence under statistical regularities – which, we recall, does not imply anything about the construction of deductive or inductive arguments – is a necessary part of any adequate explanation of its occurrence, but it is not the whole story. The causal explanation of the regularity is also needed. This claim, it should be noted, is in direct conflict with the received view, according to which the mere subsumption – deductive or inductive – of an event under a lawful regularity constitutes a complete explanation. One can, according to the received view, go on to ask for an explanation of any law used to explain a given event, but that is a different explanation. I am suggesting, on the contrary, that if the regularity invoked is not a causal regularity, then a causal explanation of that very regularity must be made part of the explanation of the event.

If we have events of two types, *A* and *B*, whose respective members are not spatio-temporally contiguous, but whose occurrences are correlated with one another, the causal explanation of this regularity may take either of two forms. Either there is a direct causal connection from *A* to *B* or from *B* to *A*, or there is a common cause *C* which accounts for the statistical dependency. In either case, those events which stand in the cause-effect relation to one another are joined by a causal process.[24] The distinct events *A*, *B*, and *C* which are thus related constitute interactions – as defined in terms of an interactive fork – at the appropriate places in the respective causal processes. The interactions *produce* modifications in the causal processes, and the causal processes *transmit* the modifications. Statistical dependency relations arise out of local interactions – there is no action-at-a-distance (as far as macro-phenomena are concerned, at least) – and they are propagated through the world by causal processes. In our leukemia example, a slow neutron, impinging upon a uranium atom, has a certain probability of inducing nuclear fission, and if fission occurs,

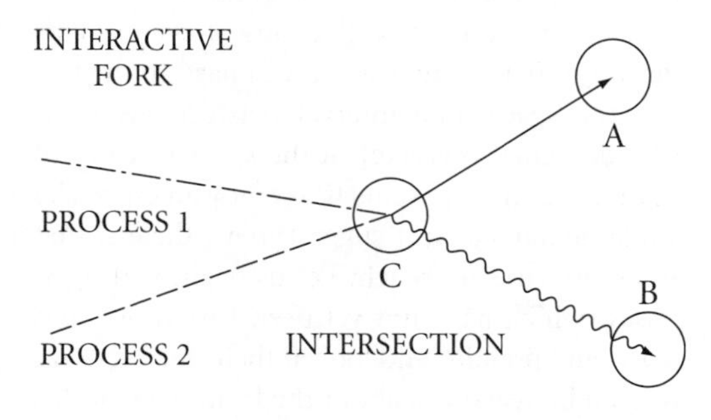

Figure 18 Causal interaction

gamma radiation is emitted. The gamma ray travels through space, and it may interact with a human cell, producing a modification which may leave the cell open to attack by the virus associated with leukemia. The fact that many such interactions of neutrons with fissionable nuclei are occurring in close spatio-temporal proximity, giving rise to processes which radiate in all directions, produces a pattern of statistical dependency relations. After initiation, these processes go on independently of one another, but they do produce relationships which can be described by means of the conjunctive fork.

Causal processes and causal interactions are, of course, governed by various laws – e.g., conservation of energy and momentum. In a causal process, such as the propagation of a light wave or the free motion of a material particle, energy is being transmitted. The distinction between causal processes and pseudo-processes lies in the distinction between the transmission of energy from one space-time locale to another and the mere appearance of energy at various space-time locations. When causal interactions occur – not merely intersections of processes – we have energy and/or momentum transfer. Such laws as conservation of energy and momentum are causal laws in the sense that they are regularities exhibited by causal processes and interactions.

Near the beginning, I suggested that deduction of a restricted law from a more general law constitutes a paradigm of a certain type of explanation. No theory of scientific explanation can hope to be successful unless it can handle cases of this sort. Lenz's law, for example, which governs the direction of flow of an electric current generated by a changing magnetic field, can be deduced from the law of conservation of energy. But this deductive relation shows that the more restricted regularity is simply part of a more comprehensive physical pattern expressed by the law of conservation of energy. Similarly, Kepler's laws of planetary motion describe a restricted subclass of the class of all motions governed by Newtonian mechanics. The deductive relations *exhibit* what amounts to a part-whole relationship, but it is, in my opinion, the physical relationship between the more comprehensive physical regularity and the less comprehensive physical regularity which has explanatory significance. I should like to put it this way. An explanation may sometimes provide the materials out of which an argument, deductive or inductive, can be constructed; an argument may sometimes exhibit explanatory relations. It does not follow, however, that explanations are arguments.

Notes

1 Patrick Suppes, *A Probabilistic Theory of Causation* (Amsterdam: North-Holland Publishing Co., 1970), Hans Reichenbach, *The Direction of Time* (Berkeley & Los Angeles: University of California Press, 1956), Chap. IV.
2 See *Nature*, Vol. 271 (2 Feb. 1978), p. 399.
3 Irving Copi, *Introduction to Logic*. 4th ed. (New York: Macmillan Publishing Co., 1972), pp. 396–7, cites this example from *No More War* by Linus Pauling.
4 According to the article in *Nature* (note 2), "the eight reported cases of leukaemia among 2235 [soldiers] was 'out of the normal range'." Dr Karl Z. Morgan "had 'no doubt whatever' that [the] radiation had caused the leukaemia now found in those who had taken part in the manoeuvers."
5 Hempel's inductive-statistical model, as formulated in "Aspects of Scientific Explanation" (1965) embodied such a high probability requirement, but in "Nachwort 1976" inserted into a German translation of this article (*Aspekte wissenschaftlicher Erklarung*, Walter de Gruyter, 1977) this requirement is retracted.
6 Bertrand Russell, *Human Knowledge, Its Scope and Limits* (New York: Simon and Schuster, 1948), p. 459.
7 Hans Reichenbach, *The Philosophy of Space and Time* (New York: Dover Publications, 1958), Sec. 21.
8 See my "Theoretical Explanation" Sec. 3, pp. 129–34, in Stephan Körner, ed., *Explanation* (Oxford: Basil Blackwell, 1975), for a more detailed discussion of this distinction. It is an unfortunate lacuna in Russell's discussion of causal lines – though one which can easily be repaired – that he does not notice the distinction between causal processes and pseudo-processes.
9 See Wesley C. Salmon, ed., *Zeno's Paradoxes* (Indianapolis: Bobbs-Merrill, 1970), p. 23, for a description of this "theory."
10 I have made an attempt to elaborate this idea in "a 'At-At' Theory of Causal Influence," *Philosophy of Science*, Vol. 44, No. 2 (June 1977), pp. 215–24. Because of a criticism due to Nancy Cartwright, I now realize that the formulation

given in this article is not entirely satisfactory, but I think the difficulty can be repaired.

11 Russell, *Our Knowledge of the External World*, Lecture VI, "The Problem of Infinity Considered Historically." The relevant portions are reprinted in my anthology, *Zeno's Paradoxes*.

12 Russell, *Human Knowledge*, pp. 460–475.

13 Reichenbach, *The Direction of Time*, Sec. 19.

14 In "Theoretical Explanation" I discuss the explanatory import of the common cause principle in greater detail.

15 Reichenbach offers the following formal definition of a conjunctive fork *ACB*

$$P(A\&B/C) = P(A/C) \times P(B/C)$$
$$P(A\&B/\bar{C}) = P(A/\bar{C}) \times P(B/\bar{C})$$
$$P(A/C) > P(A/\bar{C})$$
$$P(B/C) > P(B/\bar{C})$$

in *The Direction of Time*, p. 159. I have changed these formulas from Reichenbach's notation into a more standard one.

16 *C* screens-off *A* from *B* if

$$P(A/C\&B) = P(A/C) \neq P(A/B)$$

17 *The Direction of Time*, pp. 162–3.

18 Bas C. van Fraassen, "The Pragmatics of Explanation," *American Philosophical Quarterly*, Vol. 14, No. 2 (April 1977), pp. 143–50. This paper was presented at the 51st Annual Meeting of the American Philosophical Association, Pacific Division, March 1977.

19 The relation between $E_1 + E_2$ and E is an approximate rather than a precise equality because the ejected electron has some energy of its own before scattering, but this energy is so small compared with the energy of the incident X-ray or Y-ray photon that it can be neglected. When I refer to the probability that the scattered photon and electron will have energies E_1 and E_2 respectively, this should be taken to mean that these energies fall within some specified interval, not that they have exact values.

20 As the boxed formulas in Figures 1 and 2 indicate, the difference between a conjunctive fork and an interactive fork lies in the difference between

$$P(A\&B/C) = P(A/C) \times P(B/C)$$

and

$$P(A\&B/C) > P(A/C) \times P(B/C).$$

The remaining formulas given in Note 15 may be incorporated into the definitions of both kinds of forks.

One reason why Reichenbach may have failed to notice the interactive fork is that, in the special case in which

$$P(A/C) = P(B/C) = 1,$$

the conjunctive fork shares a fundamental property of the interactive fork, namely, a perfect correlation between *A* and *B* given *C*. Many of his illustrative examples are instances of this special case.

21 For the special theory of relativity, this has been shown by John Winnie in "The Causal Theory of Space-time" in John S. Earman, Clark N. Glymour, and John J. Stachel, eds., *Foundations of Space-Time Theories, Minnesota Studies in the Philosophy of Science*, Vol. VIII (Minneapolis University of Minnesota Press, 1977) pp. 134–205, which utilizes much earlier results of A. A. Robb. For general relativity, the approach is discussed under the heading "The Geodesic Method" in Adolf Grünbaum, *Philosophical Problems of Space and Time*, 2nd ed. (Dordrecht: D. Reidel Publishing Co., 1973), pp. 735–50.

22 The whole idea of characterizing causal interactions in terms of forks was suggested by Philip von Bretzel in "Concerning a Probabilistic Theory of Causation Adequate for the Causal Theory of Time," *Synthese*, Vol. 35, No. 2 (June 1977), pp. 173–90, especially Note 13.

23 Compare Wesley C. Salmon, et al., *Statistical Explanation and Statistical Relevance* (Pittsburgh: University of Pittsburgh Press, 1971), p. 78. There I ask, "What more could one ask of an explanation?" The present paper attempts to present at least part of the answer.

24 Reichenbach believed that various causal relations, including conjunctive forks, could be explicated entirely in terms of the statistical relations among the events involved. I do not believe this is pos-sible; it seems to me that we must also establish the appropriate connections via causal processes.

Figure acknowledgments

Figures 16, 17, and 18: from Salmon, "Why Ask 'Why'?" from *The Proceedings and Addresses of the American Philosophical Association* 51(6) (August, 1978), pp. 683–705; pp. 693, 695 and 697.

8.4

Explanatory Unification

Philip Kitcher

Philip Kitcher's (b. 1947) work has had a significant impact in a wide variety of fields, including the philosophy of mathematics, the philosophy of biology, and the relation between science and society. In the selection below, he develops a *unificationist* approach to explanation. According to this approach, explanations are arguments as Hempel suggested. But explanations cannot be distinguished from other arguments by criteria that apply to individual candidates one by one. Instead we must determine which collections of arguments do the best job of unifying our body of scientific knowledge. According to Kitcher, unification is (roughly) a matter of deriving the greatest number of conclusions from arguments that are instances of the smallest number of argument patterns.

[. . .]

2 Explanation: Some Pragmatic Issues

Our first task is to formulate the problem of scientific explanation clearly, filtering out a host of issues which need not concern us here. The most obvious way in which to categorize explanation is to view it as an activity. In this activity we answer the actual or anticipated questions of an actual or anticipated audience. We do so by presenting reasons. We draw on the beliefs we hold, frequently using or adapting arguments furnished to us by the sciences.

Recognizing the connection between explanations and arguments, proponents of the covering law model (and other writers on explanation) have identified explanations as special types of arguments. But although I shall follow the covering law model in employing the notion of argument to characterize that of explanation, I shall not adopt the ontological thesis that explanations are arguments. Following Peter Achinstein's thorough discussion of ontological issues concerning

From *Philosophy of Science*, 48 (1981), extracts from pp. 509–10, 512–26, 529–31.

explanation in his (1977), I shall suppose that an explanation is an ordered pair consisting of a proposition and an act type.[1] The relevance of arguments to explanation resides in the fact that what makes an ordered pair (p, explaining q) an explanation is that a sentence expressing p bears an appropriate relation to a particular argument. (Achinstein shows how the central idea of the covering law model can be viewed in this way.) So I am supposing that there are acts of explanation which draw on arguments supplied by science, reformulating the traditional problem of explanation as the question: What features should a scientific argument have if it is to serve as the basis for an act of explanation?[2]

The complex relation between scientific explanation and scientific argument may be illuminated by a simple example. Imagine a mythical Galileo confronted by a mythical fusilier who wants to know why his gun attains maximum range when it is mounted on a flat plain, if the barrel is elevated at 45° to the horizontal. Galileo reformulates this question as the question of why an ideal projectile, projected with fixed velocity from a perfectly smooth horizontal plane and subject only to gravitational acceleration, attains maximum range when the angle of elevation of the projection is 45°. He defends this reformulation by arguing that the effects of air resistance in the case of the actual projectile, the cannonball, are insignificant, and that the curvature of the earth and the unevenness of the ground can be neglected. He then selects a kinematical argument which shows that, for fixed velocity, an ideal projectile attains maximum range when the angle of elevation is 45°. He adapts this argument by explaining to the fusilier some unfamiliar terms ('uniform acceleration', let us say), motivating some problematic principles (such as the law of composition of velocities), and by omitting some obvious computational steps. Both Galileo and the fusilier depart satisfied.

[. . .]

. . . [W]e can use the example of Galileo and the fusilier to achieve a further refinement of our problem. Galileo selects and adapts an argument from his new kinematics – that is, he draws an argument from a set of arguments available for explanatory purposes, a set which I shall call the *explanatory store*. We may think of the sciences not as providing us with many unrelated individual arguments which can be used in individual acts of explanation, but as offering a reserve of explanatory arguments, which we may tap as need arises. Approaching the issue in this way, we shall be led to present our problem as that of specifying the conditions which must be met by the explanatory store.

The set of arguments which science supplies for adaptation in acts of explanation will change with our changing beliefs. Therefore the appropriate *analysandum* is the notion of the store of arguments relative to a set of accepted sentences. Suppose that, at the point in the history of inquiry which interests us, the set of accepted sentences is K. (I shall assume, for simplicity's sake, that K is consistent. Should our beliefs be inconsistent then it is more appropriate to regard K as some tidied version of our beliefs.) The general problem I have set is that of specifying $E(K)$, the *explanatory store over* K, which is the set of arguments acceptable as the basis for acts of explanation by those whose beliefs are exactly the members of K. (For the purposes of this paper I shall assume that, for each K there is exactly one $E(K)$.)

The unofficial view answers the problem: for each K, $E(K)$ is the set of arguments which best unifies K. My task is to articulate the answer. I begin by looking at two historical episodes in which the desire for unification played a crucial role. In both cases, we find three important features: (i) prior to the articulation of a theory with high predictive power, certain proposals for theory construction are favored on grounds of their explanatory promise; (ii) the explanatory power of embryonic theories is explicitly tied to the notion of unification; (iii) particular features of the theories are taken to support their claims to unification. Recognition of (i) and (ii) will illustrate points that have already been made, while (iii) will point towards an analysis of the concept of unification.

3 A Newtonian Program

Newton's achievements in dynamics, astronomy and optics inspired some of his successors to undertake an ambitious program which I shall call "dynamic corpuscularianism".[3] *Principia* had shown how to obtain the motions of bodies from a knowledge of the forces acting on them, and had also demonstrated the possibility of dealing with gravitational systems in a unified way. The next step would be to isolate a few basic force laws, akin to the law of universal gravitation, so that, applying the basic laws to specifications of the dispositions of the ultimate parts of bodies, all of the phenomena of nature could be derived. Chemical reactions, for example, might be understood in terms of the rearrangement of ultimate parts under the action of cohesive and repulsive forces. The phenomena of reflection, refraction and diffraction of light might be viewed as resulting from a special force of attraction between light corpuscles and ordinary matter. These speculations encouraged eighteenth century Newtonians to construct very general hypotheses about inter-atomic forces – even in the absence of any confirming evidence for the existence of such forces.

In the preface to *Principia*, Newton had already indicated that he took dynamic corpuscularianism to be a program deserving the attention of the scientific community:

> I wish we could derive the rest of the phenomena of Nature by the same kind of reasoning from mechanical principles, for I am induced by many reasons to suspect that they may all depend upon certain forces by which the particles of bodies, by some causes hitherto unknown, are either mutually impelled towards one another, and cohere in regular figures, or are repelled and recede from one another (Newton 1962, p. xviii. See also Newton 1952, pp. 401–2).

This, and other influential passages, inspired Newton's successors to try to complete the unification of science by finding further force laws analogous to the law of universal gravitation. Dynamic corpuscularianism remained popular so long as there was promise of significant unification. Its appeal began to fade only when repeated attempts to specify force laws were found to invoke so many different (apparently incompatible) attractive and repulsive forces that the goal of unification appeared unlikely. Yet that goal could still motivate renewed efforts to implement the program. In the second half of the eighteenth century Boscovich revived dynamic corpuscularian hopes by claiming that the whole of natural philosophy can be reduced to "one law of forces existing in nature."[4]

The passage I have quoted from Newton suggests the nature of the unification that was being sought. *Principia* had exhibited how one style of argument, one "kind of reasoning from mechanical principles", could be used in the derivation of descriptions of many, diverse, phenomena. The unifying power of Newton's work consisted in its demonstration that one *pattern* of argument could be used again and again in the derivation of a wide range of accepted sentences. (I shall give a representation of the Newtonian pattern in Section 5.) In searching for force laws analogous to the law of universal gravitation, Newton's successors were trying to generalize the pattern of argument presented in *Principia*, so that one "kind of reasoning" would suffice to derive all phenomena of motion. If, furthermore, the facts studied by chemistry, optics, physiology and so forth, could be related to facts about particle motion, then one general pattern of argument would be used in the derivation of all phenomena. I suggest that this is the ideal of unification at which Newton's immediate successors aimed, which came to seem less likely to be attained as the eighteenth century wore on, and which Boscovich's work endeavored, with some success, to reinstate.

4 The Reception of Darwin's Evolutionary Theory

The picture of unification which emerges from the last section may be summarized quite simply: a theory unifies our beliefs when it provides one (or more generally, a few) pattern(s) of argument which can be used in the derivation of a large number of sentences which we accept. I shall try to develop this idea more precisely in later sections. But first I want to show how a different example suggests the same view of unification.

In several places, Darwin claims that his conclusion that species evolve through natural selection should be accepted because of its explanatory power, that ". . . the doctrine must sink or swim according as it groups and explains phenomena" (F. Darwin 1887; Vol. 2, p. 155, quoted in Hull 1974, p. 292). Yet, as he often laments, he is unable to provide any complete derivation of any biological phenomenon – our ignorance of the appropriate facts and regularities is "profound". How, then, can he contend that the primary virtue of the new theory is its explanatory power?

The answer lies in the fact that Darwin's evolutionary theory promises to unify a host of biological phenomena (C. Darwin 1964, pp. 243–4). The eventual unification would consist in derivations of descriptions of these phenomena which would instantiate a common pattern. When Darwin expounds his doctrine what he offers us is the pattern. Instead of detailed explanations of the presence of some particular trait in some particular species, Darwin presents two "imaginary examples" (C. Darwin 1964, pp. 90–96) and a diagram, which shows, in a general way, the evolution of species *represented by schematic letters* (1964, pp. 116–26). In doing so, he exhibits a pattern of argument, which, he maintains, can be instantiated, *in principle*, by a complete and rigorous derivation of descriptions of the characteristics of any current species. The derivation would employ the principle of natural selection – as well as premises describing ancestral forms and the nature of their environment and the (unknown) laws of variation and inheritance. In place of detailed evolutionary stories, Darwin offers *explanation-sketches*. By showing how a particular characteristic would be advantageous to a particular species, he indicates an explanation of the emergence of that characteristic in the species, suggesting the outline of an argument instantiating the general pattern.

From this perspective, much of Darwin's argumentation in the *Origin* (and in other works) becomes readily comprehensible. Darwin attempts to show how his pattern can be applied to a host of biological phenomena. He claims that, by using arguments which instantiate the pattern, we can account for analogous variations in kindred species, for the greater variability of specific (as opposed to generic) characteristics, for the facts about geographical distribution, and so forth. But he is also required to resist challenges that the pattern cannot be applied in some cases, that premises for arguments instantiating the pattern will not be forthcoming. So, for example, Darwin must show how evolutionary stories, fashioned after his pattern, can be told to account for the emergence of complex organs. In both aspects of his argument, whether he is responding to those who would limit the application of his pattern or whether he is campaigning for its use within a realm of biological phenomena, Darwin has the same goal. He aims to show that his theory should be accepted because it unifies and explains.

5 Argument Patterns

Our two historical examples[5] have led us to the conclusion that the notion of an argument pattern is central to that of explanatory unification. Quite different considerations could easily have pointed us in the same direction. If someone were to distinguish between the explanatory worth of two arguments instantiating a common pattern, then we would regard that person as an explanatory deviant. To grasp the concept of explanation is to see that if one accepts an argument as explanatory, one is thereby committed to accepting as explanatory other arguments which instantiate the same pattern.

To say that members of a set of arguments instantiate a common pattern is to recognize that the arguments in the set are similar in some interesting way. With different interests, people may fasten on different similarities, and may arrive at different notions of argument pattern. Our enterprise is to characterize the concept of argument pattern which plays a role in the explanatory activity of scientists.

Formal logic, ancient and modern, is concerned in one obvious sense with patterns of argument. The logician proceeds by isolating a small set of expressions (the logical vocabulary), considers the schemata formed from sentences by replacing with dummy letters all expressions which do not belong to this set, and tries to specify which sequences of these schemata are valid patterns of argument. The pattern of argument which is taught to students of Newtonian dynamics is not a pattern of the kind which interests logicians. It has instantiations with different logical structures. (A rigorous derivation of the equations of motion of different dynamical systems would have a logical structure depending on the number of bodies involved and the mathematical details of the integration.) Moreover, an argument can only instantiate the Newtonian pattern if particular *non*logical terms, 'force', 'mass' and 'acceleration', occur in it in particular ways. However, the logician's approach can help us to isolate the notion of argument pattern which we require.

Let us say that a *schematic sentence* is an expression obtained by replacing some, but not necessarily all, the nonlogical expressions occurring in a sentence with dummy letters. A set of *filling instructions* for a schematic sentence is a set of directions for replacing the dummy letters of the schematic sentence, such that, for each dummy letter, there is a direction which tells us how it should be replaced. A *schematic argument* is a sequence of schematic sentences. A *classification* for a schematic argument is a set of sentences which describe the inferential characteristics of the schematic argument: its function is to tell us which terms in the sequence are to be regarded as premises, which are to be inferred from which, what rules of inference are to be used, and so forth.

We can use these ideas to define the concept of a *general argument pattern*. A general argument pattern is a triple consisting of a schematic argument, a set of sets of filling instructions containing one set of filling instructions for each term of the schematic argument, and a classification for the schematic argument. A sequence of sentences instantiates the general argument pattern just in case it meets the following conditions:

(i) The sequence has the same number of terms as the schematic argument of the general argument pattern.
(ii) Each sentence in the sequence is obtained from the corresponding schematic sentence in accordance with the appropriate set of filling instructions.
(iii) It is possible to construct a chain of reasoning which assigns to each sentence the status accorded to the corresponding schematic sentence by the classification.

We can make these definitions more intuitive by considering the way in which they apply to the Newtonian example. Restricting ourselves to the basic pattern used in treating systems which contain one body (such as the pendulum and the projectile) we may represent the schematic argument as follows:

(1) The force on α is β.
(2) The acceleration of α is γ.
(3) Force = mass × acceleration.
(4) (Mass of α) × (γ) = β
(5) $\delta = \theta$

The filling instructions tell us that all occurrences of 'α' are to be replaced by an expression referring to the body under investigation; occurrences of 'β' are to be replaced by an algebraic expression referring to a function of the variable coordinates and of time; 'γ' is to be replaced by an expression which gives the acceleration of the body as a function of its coordinates and their time-derivatives (thus, in the case of a one-dimensional motion along the *x*-axis of a Cartesian coordinate system, 'γ' would be replaced by the expression 'd^2x/dt^2'); 'δ' is to be replaced by an expression referring to the variable coordinates of the body, and 'θ' is to be replaced by an explicit function of time, (thus the sentences which instantiate (5) reveal the dependence of the variable coordinates on time, and so provide specifications of the positions of the body in question throughout the motion). The classification of the argument tells us that (1)–(3) have the status of premises, that (4) is obtained from them by substituting identicals, and that (5) follows from (4) using algebraic manipulation and the techniques of the calculus.

Although the argument patterns which interest logicians are general argument patterns in the sense just defined, our example exhibits clearly the features which distinguish the kinds of patterns which scientists are trained to use. Whereas logicians are concerned to display all the schematic premises which are employed and to specify exactly which rules of inference are used, our example allows for the use of premises (mathematical assumptions) which do not occur as terms of the schematic argument and it does not give a complete description of the way in which the route from (4) to (5) is to go. Moreover, our pattern does not replace all nonlogical expressions by dummy letters. Because some nonlogical expressions remain, the pattern imposes special demands on arguments which instantiate it. In a different way, restrictions are set by the instructions for replacing dummy letters. The patterns of logicians are very liberal in both these latter respects. The conditions for replacing dummy letters in Aristotelian syllogisms, or first-order schemata, require only that some letters be replaced with predicates, others with names.

Arguments may be similar either in terms of their logical structure or in terms of the non-logical vocabulary they employ at corresponding places. I think that the notion of similarity (and the corresponding notion of pattern) which is central to the explanatory activity of scientists results from a compromise in demanding these two kinds of similarity. I propose that scientists are interested in *stringent* patterns of argument, patterns which contain some nonlogical expressions and which are fairly similar in terms of logical structure. The Newtonian pattern cited above furnishes a good example. Although arguments instantiating this pattern do not have exactly the same logical structure, the classification imposes conditions which ensure that there will be similarities in logical structure among such arguments. Moreover, the presence of the nonlogical terms sets strict requirements on the instantiations and so ensures a different type of kinship among them. Thus, without trying to provide an exact analysis of the notion of stringency, we may suppose that the stringency of a pattern is determined by two different constraints: (1) the conditions on the substitution of expressions for dummy letters, jointly imposed by the presence of nonlogical expressions in the pattern and by the filling instructions; and, (2) the conditions on the logical structure, imposed by the classification. If both conditions are relaxed completely then the notion of pattern degenerates so as to admit *any* argument. If both conditions are simultaneously made as strict as possible, then we obtain another degenerate case, a "pattern" which is its own unique instantiation. If condition (2) is tightened at the total expense of (1), we produce the logician's notion of pattern. The use of condition (1) requires that arguments instantiating a common pattern draw on a common nonlogical vocabulary. We can glimpse here that ideal of unification through the use of a few theoretical concepts which the remarks of Hempel and Feigl suggest.

Ideally, we should develop a precise account of how these two kinds of similarity are weighted against one another. The best strategy for obtaining such an account is to see how claims about stringency occur in scientific discussions. But scientists do not make explicit assessments of the stringency of argument patterns. Instead they evaluate the ability of a theory to explain and to unify. The way to a refined account of stringency lies through the notions of explanation and unification.

6 Explanation as Unification

As I have posed it, the problem of explanation is to specify which set of arguments we ought to accept for explanatory purposes given that we hold certain sentences to be true. Obviously this formulation can encourage confusion: we must not think of a scientific community as *first* deciding what sentences it will accept and *then* adopting the appropriate set of arguments. The Newtonian and Darwinian examples should convince us that the promise of explanatory power enters into the modification of our beliefs. So, in proposing that $E(K)$ is a function of K, I do not mean to suggest that the acceptance of K must be temporally prior to the adoption of $E(K)$.

$E(K)$ is to be that set of arguments which best unifies K. There are, of course, usually many ways of deriving some sentences in K from others. Let us call a set of arguments which derives some members of K from other members of K a *systematization* of K. We may then think of $E(K)$ as the best systematization of K.

Let us begin by making explicit an idealization which I have just made tacitly. A set of arguments will be said to be *acceptable relative to K* just in case every argument in the set consists of a sequence of steps which accord with elementary valid rules of inference (deductive or inductive) and if every premise of every argument in the set belongs to K. When we are considering ways of systematizing K we restrict our attention to those sets of arguments which are acceptable relative to K. This is an idealization because we sometimes use as the basis of acts of explanation arguments furnished by theories whose principles we no longer believe. I shall not investigate this practice nor the considerations which justify us in engaging in it. The most obvious way to extend my idealized picture to accommodate it is to regard

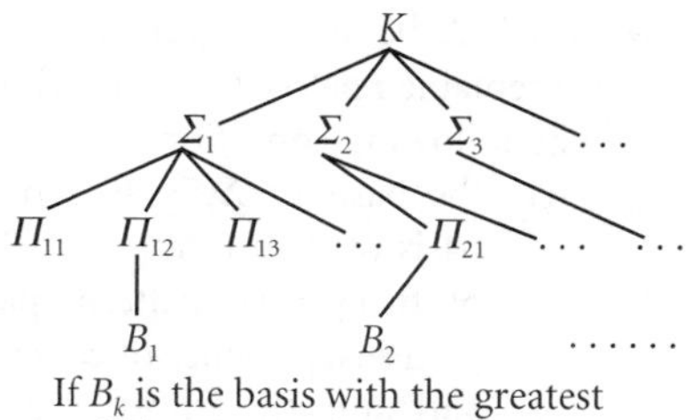

Figure 19 Diagram of generating a set

the explanatory store over K, as I characterize it here, as being supplemented with an extra class of arguments meeting the following conditions: (a) from the perspective of K, the premises of these arguments are approximately true; (b) these arguments can be viewed as approximating the structure of (parts of) arguments in $E(K)$; (c) the arguments are simpler than the corresponding arguments in $E(K)$. Plainly, to spell out these conditions precisely would lead into issues which are tangential to my main goal in this paper.

The moral of the Newtonian and Darwinian examples is that unification is achieved by using similar arguments in the derivation of many accepted sentences. When we confront the set of possible systematizations of K we should therefore attend to the *patterns* of argument which are employed in each systematization. Let us introduce the notion of a *generating set*: if Σ is a set of arguments then a generating set for Σ is a set of argument patterns Π such that each argument in Σ is an instantiation of some pattern in Π. A generating set for Σ will be said to be *complete with respect to* K if and only if every argument which is acceptable relative to K and which instantiates a pattern in Π belongs to Σ. In determining the explanatory store $E(K)$ we first narrow our choice to those sets of arguments which are acceptable relative to K, the systematizations of K. Then we consider, for each such set of arguments, the various generating sets of argument patterns which are complete with respect to K. (The importance of the requirement of completeness is to debar explanatory deviants who use patterns selectively.) Among these latter sets we select that set with the greatest unifying power (according to criteria shortly to be indicated) and we call the selected set the *basis* of the set of arguments in question. The explanatory store over K is that systematization whose basis does best by the criteria of unifying power.

This complicated picture can be made clearer, perhaps, with the help of a diagram [Fig. 19].

The task which confronts us is now formulated as that of specifying the factors which determine the unifying power of a set of argument patterns. Our Newtonian and Darwinian examples inspire an obvious suggestion: unifying power is achieved by generating a large number of accepted sentences as the conclusions of acceptable arguments which instantiate a few, stringent patterns. With this in mind, we define the *conclusion set* of a set of arguments Σ, $C(\Sigma)$, to be the set of sentences which occur as conclusions of some argument in Σ. So we might propose that the unifying power of a basis B_i with respect to K varies directly with the size of $C(\Sigma_i)$, varies directly with the stringency of the patterns which belong to B_i, and varies inversely with the number of members of B_i. This proposal is along the right lines, but it is, unfortunately, too simple.

The pattern of argument which derives a specification of the positions of bodies as explicit functions of time from a specification of the forces acting on those bodies is, indeed, central to Newtonian explanations. But not every argument used in Newtonian explanations instantiates this pattern. Some Newtonian derivations consist of an argument instantiating the pattern followed by further derivations from the conclusion. Thus, for example, when we explain why a pendulum has the period it does we may draw on an argument which *first* derives the equation of motion of the pendulum and *then* continues by deriving the period. Similarly, in explaining why projectiles projected with fixed velocity obtain maximum range when projected at 45° to the horizontal, we first show how the values of the horizontal and vertical coordinates can be found as functions of time and the angle of elevation, use our results to compute the horizontal distance travelled by the time the projectile returns to the horizontal, and then show how this distance is a maximum when the angle of elevation of projection is 45°. In both cases we take further steps beyond the computation of the explicit equations of motion – and the further steps in each case are different.

If we consider the entire range of arguments which Newtonian dynamics supplies for explanatory purposes, we find that these arguments instantiate a number of different patterns. Yet these patterns are not entirely distinct, for all of them proceed by using the computation of explicit equations of motion as a prelude to further derivation. It is natural to suggest that the pattern of computing equations of motion is the *core* pattern provided by Newtonian theory, and that the theory also shows how conclusions generated by arguments instantiating the core pattern can be used to derive further conclusions. In some Newtonian explanations, the core pattern is supplemented by a *problem-reducing pattern*, a pattern of argument which shows how to obtain a further type of conclusion from explicit equations of motion.

This suggests that our conditions on unifying power should be modified, so that, instead of merely counting the number of different patterns in a basis, we pay attention to similarities among them. All the patterns in the basis may contain a common core pattern, that is, each of them may contain some pattern as a subpattern. The unifying power of a basis is obviously increased if some (or all) of the patterns it contains share a common core pattern.

As I mentioned at the beginning of this paper, the account of explanation as unification is complicated. The explanatory store is determined on the basis of criteria which pull in different directions, and I shall make no attempt here to specify precisely the ways in which these criteria are to be balanced against one another. Instead, I shall show that some traditional problems of scientific explanation can be solved without more detailed specification of the conditions on unifying power. For the account I have indicated has two important corollaries.

(A) Let Σ, Σ' be sets of arguments which are acceptable relative to K and which meet the following conditions:
  (i) the basis of Σ' is as good as the basis of Σ in terms of the criteria of stringency of patterns, paucity of patterns, presence of core patterns, and so forth.
  (ii) $C(\Sigma)$ is a proper subset of $C(\Sigma')$.
  Then $\Sigma \neq E(K)$.

(B) Let Σ, Σ' be sets of arguments which are acceptable relative to K and which meet the following conditions:
  (i) $C(\Sigma) = C(\Sigma')$
  (ii) the basis of Σ' is a proper subset of the basis of Σ.
  Then $\Sigma \neq E(K)$

(A) and (B) tell us that sets of arguments which do equally well in terms of some of our conditions are to be ranked according to their relative ability to satisfy the rest. I shall try to show that (A) and (B) have interesting consequences.

7 Asymmetry, Irrelevance and Accidental Generalization

Some familiar difficulties beset the covering law model. The *asymmetry problem* arises because some scientific laws have the logical form of equivalences. Such laws can be used "in either direction". Thus a law asserting that the satisfaction of a condition C_1 is equivalent to the satisfaction of a condition C_2 can be used in two different kinds of argument. From a premise asserting that an object meets C_1, we can use the law to infer that it meets C_2; conversely, from a premise asserting that an object meets C_2, we can use the law to infer that it meets C_1. The asymmetry problem is generated by noting that in many such cases one of these derivations can be used in giving explanations while the other cannot.

Consider a hoary example. (For further examples, see Bromberger 1966.) We can explain why a simple pendulum has the period it does by deriving a specification of the period from a specification of the length and the law which relates length and period. But we cannot explain the length of the pendulum by deriving a specification of the length from a specification of the period and the same law. What accounts for our different assessment of these two arguments? Why does it seem that one is explanatory while the other "gets things backwards"? The covering law model fails to distinguish the two, and thus fails to provide answers.

The *irrelevance problem* is equally vexing. The problem arises because we can sometimes find a lawlike connection between an accidental and irrelevant occurrence and an event or state which would have come about independently of that occurrence. Imagine that Milo the magician waves his hands over a sample of table salt, thereby "hexing" it. It is true (and I shall suppose, lawlike) that all hexed samples of table salt dissolve when placed in water. Hence we can construct a derivation of the dissolving of Milo's hexed sample of salt by citing the circumstances of the hexing. Although this derivation fits the covering law model, it is, by our ordinary lights, nonexplanatory. (This example is given by Wesley Salmon in his (1970); Salmon attributes it to Henry Kyburg. For more examples, see Achinstein 1971.)

The covering law model explicitly debars a further type of derivation which any account of explanation ought to exclude. Arguments whose premises contain no laws, but which make essential use of accidental generalizations are intuitively nonexplanatory. Thus, if we derive the conclusion that Horace is bald from premises stating that Horace is a member of the Greenbury School Board and that all members of the Greenbury School Board are bald we do not thereby explain why Horace is bald. (See Hempel 1965, p. 339.) We shall have to show that our account does not admit as explanatory derivations of this kind.

I want to show that the account of explanation I have sketched contains sufficient resources to solve these problems.[6] In each case we shall pursue a common strategy. Faced with an argument we want to exclude from the explanatory store we endeavor to show that any set of arguments containing the unwanted argument could not provide the best unification of our beliefs. Specifically, we shall try to show either that any such set of arguments will be more limited than some other set with an equally satisfactory basis, or that the basis of the set must fare worse according to the criterion of using the smallest number of most stringent patterns. That is, we shall appeal to the corollaries (A) and (B) given above. In actual practice, this strategy for exclusion is less complicated than one might fear, and, as we shall see, its applications to the examples just discussed brings out what is intuitively wrong with the derivations we reject.

Consider first the irrelevance problem. Suppose that we were to accept as explanatory the argument which derives a description of the dissolving of the salt from a description of Milo's act of hexing. What will be our policy for explaining the dissolving of samples of salt which have not been hexed? If we offer the usual chemical arguments in these latter cases then we shall commit ourselves to an inflated basis for the set of arguments we accept as explanatory. For, unlike the person who explains *all* cases of dissolving of samples of salt by using the standard chemical pattern of argument, we shall be committed to the use of two different patterns of argument in covering such cases. Nor is the use of the extra pattern of argument offset by its applicability in explaining other phenomena. Our policy employs one extra pattern of argument without extending the range of things we can derive from our favored set of arguments. Conversely, if we eschew the standard chemical pattern of argument (just using the pattern which appeals to the hexing) we shall find ourselves unable to apply our favored pattern to cases in which the sample of salt dissolved has not been hexed. Moreover, the pattern we use will not fall under the more general patterns we employ to explain chemical phenomena such as solution, precipitation and so forth. Hence the unifying power of the basis for our preferred set of arguments will be less than that of the basis for the set of arguments we normally accept as explanatory.[7]

If we explain the dissolving of the sample of salt which Milo has hexed by appealing to the hexing then we are faced with the problems of explaining the dissolving of unhexed samples of salt. We have two options: (a) to adopt two patterns of argument corresponding to the two kinds of case; (b) to adopt one pattern of argument whose instantiations apply just to the cases of hexed salt. If we choose (a) then we shall be in conflict with [B], whereas choice of (b) will be ruled out by [A]. The general moral is that appeals to hexing fasten on a local and accidental feature of the cases of solution. By contrast our standard arguments instantiate a pattern which can be generally applied.[8]

A similar strategy succeeds with the asymmetry problem. We have general ways of explaining why bodies have the dimensions they do. Our practice is to describe the circumstances leading to the formation of the object in question and then to show how it has since been modified. Let us call explanations of this kind "origin and development derivations". (In some cases, the details of the original formation of the object are more important; with other objects, features of its subsequent modification are crucial.) Suppose now that we admit as explanatory a derivation of the length of a simple pendulum from a specification of the period. Then we shall either have to explain the lengths of *non*swinging bodies by employing quite a different style of explanation (an origin and development derivation) or we shall have to forego explaining the lengths of such bodies. The situation is exactly parallel to that of the irrelevance problem. Admitting the argument which is intuitively nonexplanatory saddles us with a set of arguments which is less good at unifying our beliefs than the set we normally choose for explanatory purposes.

Our approach also solves a more refined version of the pendulum problem (given by Paul Teller in his (1974)). Many bodies which are not currently executing pendulum motion *could* be making small oscillations, and, were they to do so, the period of their motion would be functionally related to their dimensions. For such bodies we can specify the *dispositional period* as the period which the body would have if it were to execute small oscillations. Someone may now suggest that we can construct derivations of the dimensions of bodies from specifications of their dispositional periods, thereby generating an argument pattern which can be applied as generally as that instantiated in origin and development explanations. This suggestion is mistaken. There are some objects – such as the Earth and the Crab Nebula – which *could not* be pendulums, and for which the notion of a dispositional period makes no sense. Hence, the argument pattern proposed cannot entirely supplant our origin and development derivations, and, in consequence, acceptance of it would fail to achieve the best unification of our beliefs.

The problem posed by accidental generalizations can be handled in parallel fashion. We have a general pattern of argument, using principles of physiology, which we apply to explain cases of baldness. This pattern is generally applicable, whereas that which derives ascriptions of baldness using the principle that all members of the Greenbury School Board are bald is not. Hence, as in the other cases, sets which contain the unwanted derivation will be ruled out by one of the conditions (A), (B).

Of course, this does not show that an account of explanation along the lines I have suggested would sanction only derivations which satisfy the conditions imposed by the covering law model. For I have not argued that an explanatory derivation need contain *any* sentence of universal form. What *does* seem to follow from the account of explanation as unification is that explanatory arguments must not use accidental generalization, and, in this respect, the new account appears to underscore and generalize an important insight of the covering law model. Moreover, our success with the problems of asymmetry and irrelevance indicates that, even in the absence of a detailed account of the notion of stringency and of the way in which generality of the consequence set is weighed against paucity and stringency of the patterns in the basis, the view of explanation as unification has the resources to solve some traditional difficulties for theories of explanation.

[. . .]

9 Conclusions

I have sketched an account of explanation as unification, attempting to show that such an account has the resources to provide insight into episodes in the history of science and to overcome some traditional problems for the covering law model. In conclusion, let me indicate very briefly how my view of explanation as unification suggests how scientific explanation yields understanding. By using a few patterns of argument in the derivation of many beliefs we minimize the number of *types* of premises we must take as underived. That is, we reduce, in so far as possible, the number of types of facts we must accept as brute. Hence we can endorse something close to Friedman's view of the merits of explanatory unification (Friedman 1974, pp. 18–19).

Quite evidently, I have only *sketched* an account of explanation. To provide precise analyses of the notions I have introduced, the basic approach to explanation offered here must be refined against concrete examples of scientific practice. What needs to be done is to look closely at the argument patterns favored by scientists and attempt to understand what characteristics they share. If I am right, the scientific search for explanation is governed by a maxim, once formulated succinctly by E. M. Forster. Only connect.

Notes

1 Strictly speaking, this is one of two views which emerge from Achinstein's discussion and which he regards as equally satisfactory. As Achinstein goes on to point out, either of these ontological theses can be developed to capture the central idea of the covering law model.

2 To pose the problem in this way we may still invite the charge that *arguments* should not be viewed as the bases for acts of explanation. Many of the criticisms levelled against the covering law model by Wesley Salmon in his seminal paper on statistical explanation (Salmon 1970) can be reformulated to support this charge. My discussion in section 7 will show how some of the difficulties raised by Salmon for the covering law model do not bedevil my account. However, I shall not respond directly to the points about statistical explanation and statistical inference advanced by Salmon and by Richard Jeffrey in his (1970). I believe that Peter Railton has shown how these specific difficulties concerning statistical explanation can be accommodated by an approach which takes explanations to be (or be based on) arguments (see Railton 1978), and that the account offered in section 4 of his paper can be adapted to complement my own.

3 For illuminating accounts of Newton's influence on eighteenth century research see Cohen (1956) and Schofield (1969). I have simplified the discussion by considering only *one* of the programs which eighteenth century scientists derived from Newton's work. A more extended treatment would reveal the existence of several different approaches aimed at unifying science, and I believe that the theory of explanation proposed in this paper may help in the historical task of understanding the diverse aspirations of different Newtonians. (For the problems involved in this enterprise, see Heimann and McGuire 1971.)

4 See Boscovich (1966) Part III, especially p. 134. For an introduction to Boscovich's work, see the essays by L. L. Whyte and Z. Markovic in Whyte (1961). For the influence of Boscovich on British science, see the essays of Pearce Williams and Schofield in the same volume, and Schofield (1969).

5 The examples could easily be multiplied. I think it is possible to understand the structure and explanatory power of such theories as modern evolutionary theory, transmission genetics, plate tectonics, and sociobiology in the terms I develop here.

6 More exactly, I shall try to show that my account can solve some of the principal versions of these difficulties which have been used to discredit the covering law model. I believe that it can also overcome more refined versions of the problems than I consider here, but to demonstrate that would require a more lengthy exposition.

7 There is an objection to this line of reasoning. Can't we view the arguments <(*x*)((*Sx* & *Hx*) → *Dx*), *Sa* & *Ha*, *Da*>, <(*x*)((*Sx* & ~*Hx*) → *Dx*), *Sb* & ~*Hb*, *Db*> as instantiating a common pattern? I reply that, insofar as we can view these arguments as instantiating a common pattern, the standard pair of comparable (low-level) derivations – <(*x*)(*Sx* → *Dx*), *Sa*, *Da*>, <(*x*)(*Sx* → *Dx*), *Sb*, *Db*> – share a more stringent common pattern. Hence, incorporating the deviant derivations in the explanatory store would give us an inferior basis. We can justify the claim that the pattern instantiated by the standard pair of derivations is more stringent than that shared by the deviant derivations, by noting that representation of the deviant pattern would compel us to broaden our conception of schematic sentence, and, even were we to do so, the deviant pattern would contain a "degree of freedom" which the standard pattern lacks. For a representation of the deviant "pattern" would take the form <(*x*)((*Sx* & α*Hx*) → *Dx*), *Sa* & α*Ha*, *Da*>, where 'α' is to be replaced uniformly either with the null symbol or with '~'. Even if we waive my requirement that, in schematic sentences, we substitute for *non*logical vocabulary, it is evident that this "pattern" is more accommodating than the standard pattern.

8 However, the strategy I have recommended will not avail with a different type of case. Suppose that a deviant wants to explain the dissolving of the salt by appealing to some property which holds universally. That is, the "explanatory" arguments are to begin from some premise such as "(*x*)((*x* is a sample of salt & *x* does not violate conservation of energy) → *x* dissolves in water)" or "(*x*)((*x* is a sample of salt & *x* = *x*) → *x* dissolves in water)." I would handle these cases somewhat differently. If

the deviant's explanatory store were to be as unified as our own, then it would contain arguments corresponding to ours in which a redundant conjunct systematically occurred, and I think it would be plausible to invoke a criterion of simplicity to advocate dropping that conjunct.

References

Achinstein, P. (1971), *Law and Explanation*. Oxford: Oxford University Press.

Achinstein, P. (1977), "What is an Explanation?", *American Philosophical Quarterly* 14: pp. 1–15.

Belnap, N. and Steel, T. B. (1976), *The Logic of Questions and Answers*. New Haven: Yale University Press.

Boscovich, R. J. (1966), *A Theory of Natural Philosophy* (trans. J. M. Child). Cambridge: M.I.T. Press.

Bromberger, S. (1962), "An Approach to Explanation", in R. J. Butler (ed.), *Analytical Philosophy* (First Series). Oxford: Blackwell.

Bromberger, S. (1966), "Why-Questions", in R. Colodny (ed.), *Mind and Cosmos*. Pittsburgh: University of Pittsburgh Press.

Cohen, I. B. (1956), *Franklin and Newton*. Philadelphia: American Philosophical Society.

Darwin, C. (1964), *On the Origin of Species*, Facsimile of the First Edition, edited by E. Mayr. Cambridge: Harvard University Press.

Darwin, F. (1887), *The Life and Letters of Charles Darwin*. London: John Murray.

Eberle, R., Kaplan, D., and Montague, R. (1961), "Hempel and Oppenheim on Explanation", *Philosophy of Science* 28: pp. 418–28.

Feigl, H. (1970), "The 'Orthodox' View of Theories: Remarks in Defense as well as Critique", in M. Radner and S. Winokur (eds.), *Minnesota Studies in the Philosophy of Science*, Volume IV. Minneapolis: University of Minnesota Press.

Friedman, M. (1974), "Explanation and Scientific Understanding", *Journal of Philosophy LXXI*: pp. 5–19.

Heimann, P., and McGuire, J. E. (1971), "Newtonian Forces and Lockean Powers", *Historical Studies in the Physical Sciences 3*: pp. 233–306.

Hempel, C. G. (1965), *Aspects of Scientific Explanation*. New York: The Free Press.

Hempel, C. G. (1962), "Deductive-Nonlogical vs. Statistical Explanation", in H. Feigl and G. Maxwell (eds.) *Minnesota Studies in the Philosophy of Science*, Volume III. Minneapolis: University of Minnesota Press.

Hempel, C. G. (1966), *Philosophy of Natural Science*. Englewood Cliffs: Prentice-Hall.

Hull, D. (ed.) (1974), *Darwin and his Critics*. Cambridge: Harvard University Press.

Jeffrey, R. (1970), "Statistical Explanation vs. Statistical Inference", in N. Rescher (ed.), *Essays in Honor of Carl G. Hempel*. Dordrecht: D. Reidel.

Kitcher, P. S. (1976), "Explanation, Conjunction and Unification", *Journal of Philosophy*, LXXIII: pp. 207–12.

Lavoisier, A. (1862), *Oeuvres*. Paris.

Newton, I. (1952), *Opticks*. New York: Dover.

Newton, I. (1962), *The Mathematical Principles of Natural Philosophy* (trans. A. Motte and F. Cajori). Berkeley: University of California Press.

Railton, P. (1978), "A Deductive-Nomological Model of Probabilistic Explanation", *Philosophy of Science 45:* pp. 206–26.

Salmon, W. (1970), "Statistical Explanation", in R. Colodny (ed.), *The Nature and Function of Scientific Theories*. Pittsburgh: University of Pittsburgh Press.

Schofield, R. E. (1969), *Mechanism and Materialism*. Princeton: Princeton University Press.

Teller, P. (1974), "On Why-Questions", *Noûs* VIII: pp. 371–80.

van Fraassen, B. (1977), "The Pragmatics of Explanation", *American Philosophical Quarterly 14:* pp. 143–50.

Whyte, L. L. (ed.) (1961), *Roger Joseph Boscovich*. London: Allen and Unwin.

Figure acknowledgments

Figure 19: Kitcher, "Explanatory Unification" from *Philosophy of Science* 48 (1981), pp. 507–31, p. 520.

Unit 9

After the Received View: The Realism Debate

1 Characterizing Scientific Realism: Modesty and Presumption

One area of philosophy is called *ontology*. Ontology concerns existence: what the different kinds of entities are that make up reality, and what they are like. The kinds with which the ontologist is concerned are very general. She is no better placed than anyone else to determine whether there are, say, brick houses on Elm Street. But she does concern herself with the question whether there are material objects in general and, if there are, whether they are among the basic constituents of reality or instead somehow manifestations of a more basic kind.[1] A particular dispute concerning the existence and character of such a putatively basic kind of entity is called a *realism* debate.

There are many realism debates: about theoretical entities in science (our present concern), numbers, ethical properties, mental states, and so on. These entities are associated with a region of discourse – science, mathematics, ethics, or psychology – within which assertions about that kind of entity are made (those entities constitute the discourse's *domain*). The discourse is *disciplined*: it includes procedures for determining what should be asserted about that domain.[2] In mathematics, for example, "there is an even prime" is correct and "three is the square root of eight" is not. The realism debate concerning a domain is not, however, responsive to that discipline. Primes are numbers; so since there is an even prime, there is a number. But that is not taken to decide the debate in favor of mathematical realism. That debate is about mathematics rather than within mathematics; it concerns whether the discipline integral to it reveals some aspect of reality.

So we cannot identify the realist by finding out who is willing to affirm sentences that (appear to) assert the existence of the entities in question. Virtually everyone is so willing when doing their taxes, for example, but that does not make them mathematical realists. A better approach is to begin with the recognition that realism always involves a mixture of *modesty* and *presumption*.[3] The modesty in question is the conviction that the world described by the discourse is independent of us, that we discover rather than create truths about it. This is sometimes called *mind-independence*, but that description does not allow for realism about pain, for example, which is real despite being mind-dependent. We will do better if we identify two requirements of modesty: *conceptual independence* and *epistemic independence*.

According to conceptual independence, how things are within the domain does not depend on how we conceptualize, represent, or describe how they are. This rules out the possibility that our "conceptual scheme" plays any role in determining how things are. Conceptual independence ensures that how we conceive of matters within the domain is answerable to how they in fact are, rather than the other way around.

According to epistemic independence, how things are within the domain does not depend on how our beliefs about it come to be justified. This rules out *epistemically constrained* theories of meaning, according to which the meaning of assertions about the domain is a function of how those assertions are evaluated. Verificationism is a paradigmatic example of such a theory since it ties the meaning of a statement to its method of verification (or confirmation). It also rules out epistemically constrained theories of truth, according to which what is true in the domain depends on what assertions about it are justified or would be affirmed in certain idealized circumstances.[4]

Presumption, the other component of realism, is the conviction that, notwithstanding the domain's independence, we are nevertheless in a position to know, or at least be justified in believing, claims about it. This is not merely endorsement of the discipline that the discourse imposes but, in addition, a view about that discipline: it really does put us in contact with the kind of reality required by the modesty constraint. This rules out the denial that such entities exist. It also rules out the denial that we know (or are justified in believing) that they exist.[5]

For the realist in the philosophy of science, presumption is embodied in two claims. The first is that established contemporary scientific theories are approximately true (and we are justified in taking them to be so).[6] The second is that the history of science has consisted in a progression of theories that constitute closer and closer approximations to the truth.

The realist standpoint in the philosophy of science also excludes a reductionist position according to which the entities do exist, but are in fact reducible to, or characterizable as, the entities in another domain. A scientific realist, for example, cannot say that electrons exist, but that "electron" is completely translatable into an assertion about observable physical objects. The truth that the theories increasingly approximate is in that sense unique.[7]

2 Realism Reborn

We saw in §9 of the Unit 5 commentary that Rudolf Carnap and the positivists are traditionally understood to be anti-realists with respect to theoretical entities. We discussed some problems with this understanding, noting that it is rather difficult to classify Carnap's position on the issue very precisely. But whether Carnap is best seen as an antirealist of either the reductionist or instrumentalist variety or instead as rejecting the debate itself as a meaningless metaphysical dispute, one thing seems clear: he was not a full-fledged realist.[8]

The fall of the received view led to a reconsideration of the realist position. That position was, with some irony, encouraged by Thomas Kuhn's influence. The complete rift that Kuhnian revolutions represent was widely taken to rule out a conception of those transitions as rational and progressive. It was also taken to rule out a conception of those transitions as successive approximations toward the truth as realism requires.[9] But, excepting those philosophers in the *constructivist* or *relativist* traditions (see §3 below), most philosophers of science were unwilling to swallow the irrationalist, non-progressivist consequences of this view.[10] So if Kuhnian pessimism about progress and rationality was untenable, so also, perhaps, was its skepticism concerning the possibility of successive approximation toward the truth.

Kuhn's account might even be viewed as a *reductio ad absurdum* of the received view.[11] Essential to both his account and the received view, after all, is semantic holism, according to which the meaning of a theoretical term is derived from its place in the theory and therefore forever tied to that theory. If that is correct, then there could indeed be no continuous improvement in truth-content across paradigm shifts such as the realist affirms since such changes imply incommensurable changes in meaning. There is then no content that the theories share that the later theory can improve upon. But then perhaps the problem is not with the idea that such improvement is possible, but instead with the theory of meaning presupposed by both Kuhn and the received view.

There was no denying, however, that Kuhn was right about one thing: the positivist's conception of progress – which we called *diachronic ontological reductionism* in §8 of the Unit 5 commentary – is not tenable. According to that view, earlier theories are essentially correct within the domains they were designed to accommodate, but were absorbed by successor theories that covered a broader domain. But it is just not tenable, for example, to view earlier geocentric astronomical theories as correct within a limited domain (which domain?) but incorrect within a larger domain covered by heliocentric theories.

The prospects for a realist account were dramatically improved by the development of the *causal theory of reference.*[12] On the older *descriptivist* account, the referent of a term is fixed by a description that is associated with that term; roughly, if the description changes, then so does the referent.[13] So if 'mass' is characterized within Newtonian mechanics as independent of inertial frame but as relative to inertial frame in Einstein's theory, then, since these are incompatible descriptions, they cannot refer to the same property. We cannot, therefore, view the transition from Newton's to Einstein's theory as one from a less accurate to a more accurate characterization of that same property.

On the causal theory, however, reference to an entity (or property) is secured independently of the description associated with it. The referent is determined instead by a sequence of causal relations (a "causal chain") between the referent and our use of the term. When the ancients looked up in the night sky and saw Mars, for example, they certainly did not take themselves to be viewing a heavy earthlike object orbiting the sun; their description of it was almost entirely incorrect. Their visual contact with it, however, allowed them to refer to the existing Mars nonetheless, no matter how inaccurate their description. And thanks to our intention to refer to the same object to which they referred, we also are able to refer to the very same body, no matter how dramatically our understanding of what Mars actually is differs from theirs.[14]

The causal theory has two beneficial consequences for the realist. First, it means that it is possible for a theory's description of a theoretical entity to be incorrect, even dramatically so, while the term itself still succeeds in referring. Second, it is possible for two theories that advocate incompatible descriptions to nevertheless both refer to the same theoretical entity. This latter consequence is very important for the realist, since it is difficult to make sense of progress toward the truth otherwise. How could the sequence of theories about atoms, for example, be coming closer to the truth about what atoms are like unless they were all referring to the same entities notwithstanding the changes in their characterization of those entities? It is hard to see how this could be possible without the independence between description and reference embodied in the causal theory.

These developments in the theory of reference of course only make scientific realism possible; they do not make it right. We will now turn to the arguments.

3 The Empiricist and Constructivist Challenges

Richard Boyd's "On the Current Status of Scientific Realism" (reading 9.1) is one of the best-known defenses of scientific realism. He begins by considering two challenges to realism, one from the empiricist, and the other from the constructivist.

However untenable the received view might be now, its being so does nothing to mitigate the force of the most influential empiricist argument against realism, which is the *underdetermination argument* (which Boyd simply calls "the empiricist argument"). The argument begins by pointing out that no theory is implied by the body of empirical evidence that supports it; that is, the theory is *underdetermined* by the evidence. There are, therefore, always alternative theories that are equally compatible with, or indeed that imply, the same body of empirical evidence.[15] It is then claimed that there is no evidence other than empirical evidence. Subjecting a theory to empirical test is, after all, the only way in which the world itself gets to have a say in how theories are evaluated. But if empirical evidence is indeed the only evidence, and does not favor any particular theory over its empirically equivalent alternatives, then we have no rational grounds for favoring one over the other, and therefore no grounds for believing one of those theories rather than any of its competitors.

Although the empiricists that Boyd has in mind here are pretty clearly the positivists, advocates of the underdetermination argument typically cite the Duhem-Quine thesis, according to which any particular hypothesis can be saved from refutation by rejecting, instead of the hypothesis, any of the auxiliary claims that were required in order to derive a prediction from that hypothesis. This suggests that there are any number of alternative ways of resolving the conflict between hypothesis and failed prediction, generated by placing the blame on a different auxiliary assumption. Quine, as we saw (§6 of Unit 6 commentary), invoked conservatism and simplicity as determining what response would be best. But if it is indeed correct that the only real evidence is empirical evidence, then the virtues of simplicity and conservatism – now sometimes called *super-empirical virtues*[16] – can only be pragmatic. They reflect only our practical interests in not changing things too much and keeping things relatively simple and manageable. But they do not provide any additional reason to believe that the conservative and simple theory is more likely to be true than its empirically equivalent competitors.[17]

The constructivists interpreted Kuhn's claim that scientists working within different paradigms "live in different worlds" literally, and claimed that it is correct: the reality represented by the paradigm is itself paradigm-dependent. The world is *constructed* by the paradigm, so that the only truth is truth relative to that paradigm.

However difficult this view might be to swallow – did the sun really revolve around the earth during the heyday of geocentric astronomy? – Boyd points out that constructivism does have one thing going for it. Kuhn was, Boyd suggests, correct to point out that scientific methodology is heavily theory-laden. That is, the evaluation of subsequent theories, the quality of the evidence for them, whether the problems they solve are significant ones, and the projectability of the predicates they employ (see §3 of the Unit 6 commentary) are all dependent on the theories that are currently accepted. If we insist on thinking that scientific theories get at the truth, then we are left with a puzzle: how can such a theory-dependent methodology be a reliable guide to the truth? The constructivist answer is that the truth to which it is a guide is itself theory-dependent. One might perhaps think that this is not very plausible. But at least it is an answer; if the realist is going to counter this challenge, she is going to need another answer.

4 The Explanationist Defense

Boyd's response to both the empiricist and constructivist challenges is essentially the same: neither empiricism nor constructivism has the resources to account for a striking feature of scientific inquiry, namely, its instrumental reliability, by which Boyd means its success in making predictions and in generating technological advances. Contemporary scientific theories, at least in the physical sciences, are very reliable in this sense; the successful predictions derived from quantum mechanics, for example, demonstrate a spectacular level of accuracy. There is, moreover, no question that this predictive success has been improving over time; no one would seriously suggest that Aristotelian physics, or even Newtonian physics, is as accurate as our current physical theories. The point can also be made by reference to the technological products of scientific inquiry. Airplanes, plastic, microwave ovens, hydrogen bombs, computers, MRIs, and much of contemporary life in the developed world is ultimately a product of scientific inquiry. One might applaud or lament this, but the dramatic control over our environment that this involves is surely impressive either way.

Suppose that contemporary science, which generates all of this predictive accuracy and technological control, is wildly off the mark. It would then be a tremendous coincidence, nothing short of miraculous, that these erroneous theories nevertheless display such incredible instrumental reliability. The only reasonable explanation, then, for that success is simply that the contemporary theories are *not* wildly off the mark, but are in fact close to the truth. The success that such theories demonstrate would then be no miracle at all. This argument – called the *no miracles argument* – is by far the most popular argument for realism, and is the core of Boyd's response to the empiricist and constructivist challenges.[18]

Boyd's version of the argument begins with the feature of science that the constructivist highlighted above, that its methodology is highly theory-laden. Boyd argues that the best way to account for *both* instrumental reliability and the theory-ladenness of methodology is to claim that past theories were approximately true and that the succession of theories constitutes a closer and closer approximation to the truth. Suppose that a theory is approximately true, but its known faults motivate the selection of a successor theory from among a range of options. We then evaluate the plausibility of those alternative theories from the standpoint of the current theory. Since the current theory is approximately true, this evaluation is likely to guide us to select as successor the theory that is the likeliest among those alternatives, and so that theory which is a closer approximation to the truth than the current theory. The result is a succession of increasingly approximately true theories.

This history is, Boyd suggests, the only one that will explain the instrumental reliability of previous theories and the increase in that reliability over successive changes in theory, arriving at the highly reliable theories of the present day. The constructivist can explain only the theory-ladenness of methodology; she cannot, however, explain either the instrumental reliability of present theories arrived at by that methodology, or the history of increasing instrumental reliability. And the empiricist has no explanation to offer at all. So only the dialectical conception of the progression of scientific theories offered by the realist can explain these dramatic features of scientific inquiry.

Assuming that history to be correct, we are now in a position to respond more directly to the empiricist's challenge. That challenge was, again, that if the only evidence is empirical evidence, then the other super-empirical factors that we appeal to in theory choice are merely pragmatic, and so of no evidential significance. But if the history sketched in the last paragraph is indeed correct, then this is false. Those factors played an essential role in producing a sequence of theories that are increasingly approximately true, and are evidential factors for that reason. This does not require, however, denying the empiricist's plausible claim that all factual knowledge ultimately rests upon observation. For the reason why we believe that that history *is* correct is because that history is the only explanation for the impressive, and increasing, instrumental reliability. But the instrumental reliability of science is a fact gleaned from experience. The super-empirical virtues are, in this way, *indirectly* justified by observation.

5 Fine's Challenge

As presented above, Boyd's argument has a missing premise, namely, that the best explanation for a phenomenon is the most likely to be correct, so that realism's being the best explanation of instrumental reliability makes it the most plausible. Such an argument is called an *inference to the best explanation* (IBE); an older name is *abduction* (see reading 4.10).

The status of IBE is hotly contested to this day. But for Boyd, the justification for its employment in the philosophy of science is simply that it is a ubiquitous rule of inference in science itself. Boyd offers his realism, and his philosophy of science generally, as a scientific theory that explains a phenomenon, namely the success of science itself. This marks the rise of naturalistic approaches in the philosophy of science, which treat philosophical issues, including philosophical issues concerning science, as continuous with the scientific endeavor itself. Naturalists therefore reject the conception of philosophy as prior to, and more fundamental than, the scientific endeavor (a view often called *first philosophy*).

In the first half of reading 9.4 (and elsewhere), Arthur Fine has presented a serious challenge to Boyd's use of IBE so understood.[19] Fine points out that the status of IBE is itself part of what is at issue between the realist and the anti-realist. The anti-realist does not view IBE as a guide to truth; she sees it instead as serving other purposes (most notably, getting at theories that are empirically successful rather than true; see §9 below). The anti-realist must view IBE that way, precisely because it is used ubiquitously in scientific practice. For if scientists' routine use of IBE were seen as inferences that track the truth, then that would decide the issue against the anti-realist. This is because the conclusions of those IBEs often posit unobservable entities (e.g., electrons); if those conclusions are correct, then the anti-realist is wrong to deny those entities' existence. But she does not take that practice to settle the debate in the realist's favor. And nor, presumably, does Boyd, since then there would be no need for his second-order use of IBE in his explanationist defense of realism. The use of IBE in science itself, rather than his use of it to explain the instrumental reliability of science, would suffice on its own to establish realism.

Recall the illustration of mathematical realism in §1 above. Affirming that there is an even prime when doing your math homework does not resolve the philosophical debate in the mathematical realist's favor. The question is *about* mathematics, and in particular about whether the discipline that it imposes tracks an independent reality. Similarly, the issue between the scientific realist and anti-realist concerns, in part, whether the discipline imposed upon assertions in science by the use of IBE also tracks an independent reality. The mere fact that it is used does not settle that issue. So to assume, in Boyd's explanationist defense of realism, that IBE is a guide to the truth – the missing premise in his argument – is to beg the question against the anti-realist.

Boyd's response is detailed, but there are two themes. The first is that the price for not treating IBE as a guide to the truth is very high indeed. Its use is ubiquitous, not just in science, but in human inquiry generally, and it is used even in determining what inductions are legitimate. So the result of rejecting it will not be a viable alternative conception of science, but instead will constitute wholesale skepticism. His second response, which works in conjunction with the first, is that we should evaluate, not this or that thesis, argument, response, etc., but instead the philosophical "packages" offered up by the different parties to the debate. If, for example, the anti-realist package amounts to wholesale skepticism, then the realist package is superior to it, however much it might appear to beg the question.

6 Laudan's Pessimistic Meta-Induction

Like Boyd, Larry Laudan is also a philosophical naturalist, and so also takes the philosophy of science to be legitimately informed by empirical facts drawn from the history of science. Far from supporting realism as Boyd suggests, however, Laudan argues that that history supports anti-realism instead.

The realist's claim might be put in terms of reference or in terms of truth. First, the realist might suggest that if a theory's central terms refer then it will be empirically successful; so if we discover that it is successful, that is evidence that those central terms refer. Second, the realist might suggest that if a theory is approximately true, then it will be successful. So, if we discover that it is successful, then that will be evidence that it is true. Laudan argues that there is substantial evidence from the history of science against both of these claims.

First, there are many examples of theories whose central terms refer but that were unsuccessful. The chemical atomic theory in the eighteenth century, the Proutian theory that the atoms of heavy elements are composed of hydrogen atoms, the Wegenerian continental plate theory, wave theories of light, and atomic theories of matter, were all unsuccessful, Laudan points out, at least for a considerable period of time. Nor should this be a surprise. It is quite possible – especially according to the realist (see §2 above) – for a theory to refer to an entity and nevertheless be deeply mistaken about it. And, of course, there is no reason to think that a deeply mistaken theory will be successful.[20] The idea, clearly, was supposed to be that a theory that successfully refers stands a good chance of being approximately true, and thereby successful. Laudan has argued against the first step. He now argues against the second: approximate truth is not tied to success.

In fact, truth *is* tied to success: a theory that is entirely correct will be successful, so long as you know how to apply it (if you really do know which path is the path out of the woods – and you know how to follow the path – then you know how to get out). But the realist is claiming that *approximate* truth is also likely to be successful. But a theory that is approximately true is still, strictly speaking, false; there is no guarantee that approximately true but strictly false theories will in fact be successful, and so no guarantee that success betokens approximate truth.

And the history of science provides many examples, Laudan insists, of successful theories that were not even approximately true. His argument presupposes that, whatever approximate truth of a theory is, it requires at least that the theory's central terms refer. (Approximate truth would otherwise not support realism concerning the entities to which those terms putatively refer.) He then presents a list of theories that were successful but non-referring, including: the crystalline spheres of ancient and medieval astronomy; the humoral theory of medicine; the theory of circular inertia; the phlogiston theory of chemistry; the caloric theory of heat; and the electromagnetic ether.

Laudan's overall argument is a *reductio ad absurdum*. Suppose that the success of our contemporary theories is evidence that they are true (or approximately true). Then the central terms of the theories on this list – spheres, phlogiston, caloric, ether – do not refer because, according to contemporary theories, there are no such things. So these theories were not approximately true. But they were all successful. So success is no indication of approximate truth. Therefore, the success of our present theories is no indication that they are true, which contradicts the first premise. The first premise is thus false; success of our contemporary theories is not evidence that they are true. This argument is called the *pessimistic meta-induction*.

There have been many responses to Laudan's argument, but two are prominent. First, some realists deny that Laudan's list of successful but non-referring theories is really as impressive as it looks. It is far from clear that the humoral theory of medicine was at all successful, for example; and the crystalline spheres – as opposed to the postulated path of the heavenly body itself – played little to no role in whatever accuracies were demonstrated by early astronomical theories (this is particularly clear, for example, in Ptolemy's account).

Other realists challenge the structure of the argument. For example, even if Laudan's list is left alone, the number of successful theories that did turn out to be approximately true is, one might argue, much larger. And it is the comparison between these classes that matters to the question whether success is evidence for truth, not just the existence of members in one or the other class.

7 Van Fraassen's Constructive Empiricism

In *The Scientific Image* (from which reading 9.3 is excerpted) Bas van Fraassen articulated and defended an anti-realist position called *constructive empiricism* (CE); much of the realism debate since its publication has centered on his view.

The realist, he says, claims both that science is aimed at the truth and that acceptance of a theory is acceptance of its truth. An odd feature of this definition of realism is that it does not require that the realist believe that science has actually achieved the truth (an endeavor can aim at something without succeeding). To some extent this is an advantage, since the realist will not want to commit to the complete accuracy of contemporary scientific theory (since this would suggest, implausibly, that there is no more scientific inquiry to conduct and no more corrections to be made of current theory (see note 6)). But realists do think that those theories are at least approximately true. The presumption that we are able to know that our theories are at least approximately true, which we identified in §1 as essential to realism, is absent from van Fraassen's definition since it is compatible with this definition that science has utterly failed to achieve its aim.

This is rectified by a presupposition shared by many participants in the debate, both realist and anti-realist, that science is in fact reasonably successful in achieving its aim. There is, however, no such accord concerning what that aim actually is. This might seem to get things backward; after all, do we not have to find out what science is aimed at before we can find out whether it is successful in achieving that aim? However, this is not how the debate has proceeded. Most philosophers of science are considerably more convinced that science is a paradigmatically rational and successful activity than they are that one or the other account of its aim is correct. Recall the normative constraint from §4 of Unit 7 commentary, according to which the rationality of science is an underlying presumption of much philosophy of science in general. Recall also Boyd's claim that the dispute between the realist and the anti-realist comes down to which offers a better philosophical package that makes sense of the scientific endeavor.

So with that presumption in the background, the debate concerns which account provides the most coherent interpretation of scientific practice. Van Fraassen cheerfully concedes that the received view is untenable, but he proceeds to formulate an empiricist interpretation that is immune to the criticisms that brought that earlier view down. There are, he says, two ways to reject realism. The first is to agree that science aims at the truth, but then to provide a non-literal interpretation of what scientific theories say, the result of which interpretation is that those theories only make claims concerning observable entities. This will include, for example, operational definitions, which define assertions that seem to posit unobservable (or theoretical) entities as in fact referring to observables. It is also supposed to rule out the partial interpretation approach since, according to that approach, theoretical terms derive the only content they have from their indirect relation to their empirical consequences.[21] This seems to many (including van Fraassen) to amount to eliminating reference to unobservables, which reference is in fact there in the scientific doctrine to start with.[22]

But there is a second way. We could concede that scientific doctrine speaks of entities that are truly unobservable (thereby interpreting it literally). But we do not have to view the scientific endeavor as aimed at telling the truth about such entities. We could instead view it as only aimed at telling the truth about observable entities. (When a theory is accurate in its description of observable entities, it is *empirically adequate.*) The reference to unobservable entities in scientific doctrine is then understood to play a merely pragmatic or instrumental role. It has long been noted that the development of empirically powerful theories – theories that generate a wide variety of accurate observable predictions – seems to require the formulation of theories that posit unobservable entities and processes "behind" the observable phenomena.[23] The realist takes the scientist to be fundamentally interested in accurately representing these unobservable entities and processes. Van Fraassen proposes instead that the scientist is fundamentally interested in generating empirically powerful theories. The positing is a means to this end, but the existence of the posits themselves is neither affirmed nor denied. (Our attitude toward them is in this way akin to the agnostic's attitude to the question whether God exists.)

This is the constructive empiricist's approach. According to it, science is aimed at empirical adequacy, and acceptance of a theory is acceptance of it as empirically adequate. To say that a theory is empirically adequate is to say that what it asserts about observable entities is correct, but no commitment is made one way or the other concerning the unobservables referred to in the theory.

8 Observability

Van Fraassen is primarily concerned to defend the legitimacy of CE rather than to attack realism. His first task is to defend the much-maligned observable/unobservable distinction.[24] In fact, he accepts the criticisms of both Hanson and Putnam: he grants that observation is theory-laden; and he agrees that the observable/unobservable entity distinction does not correlate with the theoretical/non-theoretical language distinction (if indeed the latter distinction is tenable at all, which he doubts).

But then the observable/unobservable entity distinction is unaffected by the charge of theory-ladenness of either experience or language. It does not matter if our perceptual experience, or our description of that experience, is thoroughly theory-infected. The observable/unobservable distinction does not concern perception or language; it concerns objects.

So the only criticisms that van Fraassen has to deal with are those of Maxwell, which are directed against the observable/unobservable distinction itself. Recall that Maxwell presented three arguments: first, the continuity argument from naked-eye vision to electron microscopy; second, the continuity argument from sub-visible molecules to polymeric lumps of plastic; and third, the argument from the possibility of humans evolving in such a way that they are able to observe what is unobservable now.[25]

Against the first argument, van Fraassen points out that all that it demonstrates is that the concept of the observable is vague. But so are a tremendous number of other entirely legitimate concepts. For example, 'red', 'automobile', 'adult', 'pornography', 'violence', and 'art' are all vague concepts, as indeed are many expressions in natural language. Vague concepts do present interesting logical problems, and philosophers have offered many attempted solutions.[26] But rarely do those solutions claim that vague concepts are illegitimate.

Against the second argument, van Fraassen cheerfully grants that some entities that are characterized as very large single molecules are observable. That is only a problem if the word 'molecule' could only refer to unobservables; van Fraassen, however, has already agreed that the observable/unobservable distinction is not a linguistic distinction. (Notice that the fact that we call a visible lump of plastic a molecule does not imply that we can *see* that it is a molecule. Van Fraassen is not, in permitting the use of that term, forced to grant the truth of molecular theory generally.)

Against the third argument, van Fraassen points out that the Empire State Building is not portable just because we could evolve into giants that could lift it, and a block of lead is not fragile just because we could evolve into superhumans that could break it. Such "x-able" (or "x-ile") terms are implicitly relative to our present capacities. If we – or the beings we count as within our "epistemic community" – change, then what will be observable then will be different. But that does not change what is observable now. The limits of observability, so understood, depend on the biological characteristics of human beings: there are certain objects that are detectable by the human senses and others that are not (e.g., those that are too small to see). And the biological characteristics that fix those limits is the subject matter of biological science; it is science itself, not philosophy, that determines what is observable and, therefore, which scientific claims should be accepted.[27]

9 Explanation and Realism

Realists have argued, in a number of different ways, that reflection on the concept of explanation and its role in science demonstrates that such a selectively skeptical position as CE is not tenable. One argument along these lines begins by pointing out that we use IBE routinely in everyday life.[28] We hear the patter of little footsteps, the cheese disappears, there are holes in the baseboard, and we infer that a mouse has come to live with us. Such inferences are part of our everyday cognitive equipment, and we certainly could not get along without them. But consistent interpretation of the use of IBE within both everyday life and in science requires that we accept its conclusions in science as well; and in science the conclusions of such inferences refer to unobservables. So we are committed to accepting the existence of unobservables.

Van Fraassen responds that we might instead be inferring to the empirical adequacy of the best explanation rather than to its truth. The fact that we do believe that there is a mouse does not show this to be wrong. For the mouse is observable; and inference to empirical adequacy *is* inference to the truth when the explanation concerns observables. So such examples from everyday life do not support the conclusion that IBE is to the truth rather than to empirical adequacy, and so do not force belief in unobservables on us.

Van Fraassen argues, further, that even if we accept IBE as an inference to the truth, this does not force us to accept realism. IBE requires that we select the most explanatory hypothesis from some number of candidates; it does not tell us how to ensure that the true hypothesis is among those candidates. But then the chosen hypothesis might be the *best of a bad lot*: it might be the best among those under consideration, but nowhere near the not-yet-conceived true hypothesis.

Another argument for realism appeals to the empirical regularities that scientific theories bring to our attention (that pressure, temperature, and volume of a gas are always related in accordance with the gas laws drawn from the kinetic theory, for example). Such regularities (or correlations, or coincidences) demand a common cause, which will inevitably invoke the activity of an underlying unobservable reality. To this van Fraassen points out that that unobservable reality itself manifests regularities, namely those enshrined in its laws. And if those regularities are accounted for by a more fundamental reality, then the laws of that more fundamental reality will not be explained. Eventually some regularities must go unaccounted for. So there can be no universal demand that every regularity must be explained; the realist then needs an additional reason for insisting that observable regularities are uniquely subject to this demand.

Finally, van Fraassen argues that even if we insist on an explanation for the success of science which must be regarded as true, this still will not deliver realism; for he has a different explanation to offer from that of the realist. The evolutionary explanation for why the mouse runs from the cat is that animals that did not do so were eaten before they had the opportunity to pass on their genes. The scientific environment, like the natural environment, is similarly "red in tooth and claw": the history of science consists in a sequence of increasingly successful theories simply because the unsuccessful ones were ruthlessly weeded out.

Van Fraassen does agree that the pursuit of explanation is a fundamental aspect of the scientific endeavor, and he encourages its pursuit. But he does so because the pursuit of explanation has paid off in generating empirically successful theories, not because it reveals unobservable reality. Like simplicity, conservatism, and the other super-empirical virtues, explanation is a pragmatic value, not an epistemic one. Pursuit of explanation, for the constructive empiricist, is a tactical strategy contributing to the empirical adequacy that is science's aim; it is not the inescapable path to the truth as the realist conceives of it.

10 Why Constructive Empiricism?

These arguments seem to leave us at something of an impasse. Realism is not forced on us by reflection on the role of explanation in science or in our intellectual lives. But neither is it ruled out by these arguments; van Fraassen has attempted to put CE on the table as an option, but has not argued that realism must be taken off. He has often been taken to claim that realism is irrational because belief in observables is supported by the evidence whereas belief in unobservables is not. And he does present arguments that suggest this. One such argument points out that the claim that a theory T is empirically adequate (call this TE) is itself empirically equivalent to the claim that T is true: they have, by definition, the same empirical consequences. Since the claim that T is true is less probable (because T asserts what TE asserts, and more), and there is (van Fraassen claims) no evidence other than empirical evidence, we cannot be warranted in believing T over TE.[29]

But critics have pointed out that TE itself goes beyond the empirical data, since it claims that the theory is correct about *all* observable phenomena, most of which will never be observed. So if van Fraassen presses this line, then he will end up with what Gideon Rosen calls *manifestationalism* – belief in nothing but what *is* observed – rather than CE.[30]

Van Fraassen's position is, however, more subtle. It is true that CE goes beyond the empirical data. But, he argues, it goes no further beyond that data than is required in order to make sense of scientific practice. (Manifestationalism cannot do this, since it provides no reason for the scientist to uncover *new* evidence that might, after all, conflict with the theory.) He does not deny that realism is rational, but he does deny that realism is forced on us.[31] And, for the empiricist at least, there is virtue in a view of science that strays no further from the empirical ground than is necessary to make sense of the scientific endeavor.[32]

The truth be told, there are not many constructive empiricists. There is, however, considerable disagreement concerning just what is wrong with the view; one gets the sense that realists find it deeply disturbing, notwithstanding the barrage of arguments that they have hurled against it. The charge that realism involves flights of intellectual fancy beyond what the empirical evidence and scientific practice can justify goes some way to explain that malaise.

11 Fine's Natural Ontological Attitude

Arthur Fine rejects both realism and anti-realism, and offers in their stead the *natural ontological attitude* (NOA). Both the realist and the anti-realist, he argues, endorse what he calls the *core position*. The core position requires treating both everyday and scientific investigations as continuous. However we understand confirmation to work, for example, or interpret IBE, we must treat them the same way in these two contexts. So, he says, we should accept the results of scientific investigations as true, on a par with more homely truths.

The anti-realist adds to this core position a pragmatist, instrumentalist or conventionalist account of truth. Or they add an idealist, constructivist or phenomenalist analysis of concepts. Or they add an anti-realist construal of particular inferences, explanations, or laws. The realist, on the other hand, just stamps her feet, shouting "it's all really true!" Or, when she has quieted down, she adds a correspondence theory of truth. When we recognize that realism and anti-realism constitute additions to the core position, another, minimal, alternative emerges: that we endorse the core position all by itself. Endorsing it, and nothing more, is NOA.

Making sense of NOA is challenging. Van Fraassen, for example, certainly would not agree that we accept claims about unobservables as true, either inside or outside science; we accept them as empirically adequate instead. In light of his insistence that claims within and outside science should be considered true, Fine has often been taken to have really sided with the realist.[33] But a more charitable interpretation can result if we recall the distinction in §1 above between the discipline that the discourse imposes (resulting in the affirmation that there is an even prime, for example) and the realist's investigation of that discipline (by asking whether there really are numbers). Even van Fraassen endorses "immersion" in the scientific "world-picture," within which all manner of assertions about electrons etc. are licensed (pp. 642 of reading 9.3). Perhaps Fine can be read as suggesting that the presumption that such an "external" point of view from which scientific inquiry can be examined in this way is mistaken. If so, then the discipline that the discourse imposes is all there is, and the parties to the realism debate no longer have something to disagree about.

We will not attempt to evaluate this suggestion here. But it is interesting to notice the similarity between NOA and Carnap's position in reading 5.3, which also involves rejecting such "external" questions concerning the domain's discipline (or as Carnap put it, the "rules of the framework"). Whether or not such dismissals of the realism debate are ultimately successful, both of these positions can be appreciated for bringing to our attention the fact that the debate does presuppose the availability of a meta-scientific standpoint whose presupposition can perhaps be questioned.[34]

Notes

1 Bishop Berkeley, for example, affirmed the existence of material objects, but then insisted that they are really constituted by 'ideas'. He is as a result counted as an anti-realist concerning such objects (George Berkeley, *Three Dialogues between Hylas and Philonous*, Kessinger Publishing, 2005).

2 Those procedures are not necessarily exhaustive; there may be many assertions that can be formed within it that are neither legitimately affirmed nor denied. The expression "disciplined discourse" is from Crispin Wright, 1992, *Truth and Objectivity*, Cambridge: Harvard University Press.

3 Wright 1992 (see above), pp. 1–2. Wright's own procedure for distinguishing realism from anti-realism is considerably more elaborate and nuanced than that presented in this section; we do not claim to have reproduced it here.

4 An example is Hilary Putnam's "internal realism" presented in Hilary Putnam, 1981, *Reason, Truth and History* (Cambridge: Cambridge University Press), and a number of subsequent essays and books.

5 Notice that these are not the same. We have no evidence one way or the other for the existence of extra-terrestrial life, and so we should not affirm its existence. But, for the same reason, we should not deny its existence either.

6 Scientific realism does not just say that current theories are true. That would imply that current theories are complete, that there will be no improvements upon them. But no serious realist doubts that there will be future theoretical improvements.

7 Interestingly, anti-reductionism is not a constraint on realist positions generally. Reductive physicalism in the philosophy of mind, for example, is usually thought to be a way to save realism about the mind rather than rejecting it.

8 The same cannot be said, however, for every positivist. Herbert Feigl, for example, was clearly a realist.

9 "[T]he notion of a match between the ontology of a theory and its 'real' counterpart in nature now seems to me illusive in principle . . . I can see [in the succession of theories] no coherent direction of ontological development" (reading 7.2, p. 512).

10 Including Kuhn himself; see discussion in §7 of Unit 7 commentary.

11 In a *reductio ad absurdum*, an assumption is proven false by showing that a contradiction can be validly derived from it. Since a contradiction must be false, and the argument is valid, this implies that one of the premises is false (which is then understood to be the assumption).

12 The causal theory is widely seen to have originated with Kripke's *Naming and Necessity*, and to have been extended to natural kinds by Hilary Putnam in his (1975/1985) "The Meaning of 'Meaning'." in *Philosophical Papers, Vol. 2: Mind, Language and Reality*, Cambridge University Press.

13 This is not quite right. Different descriptions – 'the wife of Bill Clinton' and 'the senator from New York' – can pick out the same objects. But not so if the descriptions are incompatible; nothing can satisfy both 'Clinton's mother' and 'Clinton's daughter'.

14 The fact that theoretical entities are unobservable, at least by the naked senses, means that the causal theory has to be adjusted to accommodate reference to them. One approach is that of Richard Boyd, as described on p. 597 in reading 9.1 (Boyd

has, however, since revised his views on the issue: see articles referred to in note 26 of reading 9.1.)

15 The claim that there are alternative theories that are compatible with the same evidence is unquestionably true (so long as one is liberal enough in what one is willing to count as a theory); the claim that there are alternative theories that *imply* the same evidence is, however, more contentious. It is, nevertheless, widely accepted. See §5 of Unit 6 commentary for more discussion of underdetermination.

16 Simplicity and conservatism are, however, not the only super-empirical virtues; explanatory power is another and a highly contentious one as we will see.

17 Quine rejected the distinction between pragmatic and evidential constraints on the web of belief. But few scientific realists are willing to endorse as much pragmatism as does Quine in order to defend their realism. (Whether Quine was himself a realist is a difficult, and interesting, question.)

18 The No-Miracles Argument was also advanced in Hilary Putnam 1972, "Explanation and Reference," in G. Pearce and P. Maynard (eds), *Conceptual Change*, Dordrecht: Reidel, and is often now referred to as the "Putnam–Boyd No-Miracles Argument". This is, incidentally, the same Hilary Putnam who advocated the anti-realist position called "internal realism" mentioned in note 4 above. Putnam occupies a unique position in philosophy as one who has not only advocated a wide variety of competing positions at different stages of his career, but who also presented some of the strongest arguments for those positions while endorsing them, as well as some of the strongest arguments against them when advocating their competitors.

19 We will be considering Fine's positive view, presented in the second half of reading 9.4, in §11 below.

20 For example, consider the following recipe: take any collection of sentences that are true and refer, like "London is in England", "the earth revolves around the sun," and "cats are mammals." Then negate each sentence to form a new collection of sentences. The latter collection successfully *refers* to the same objects as the former. But it is completely wrong, and bound to be very unsuccessful.

21 See §7 of Unit 5 commentary.

22 Carnap would hardly have conceded that this is a "non-literal" interpretation, insisting instead that it is the only coherent one available. But history has not been kind to that way of viewing the situation.

23 Carl Hempel noted this, for example; see reading 6.1 and §1 of Unit 6 commentary.

24 See readings 6.5–6.7 and §§7–9 of the Unit 6 commentary.

25 See reading 6.7 and §9 of the Unit 6 commentary.

26 The best-known such problem is the Sorites paradox, or "paradox of the heap." Begin with a heap of sand. Take one grain away; the result is obviously still a heap. Take another grain away; obviously it is still a heap. In general, no one grain will change it from a heap to a non-heap. But keep going, and you will remove every grain; no heap remains. But if it was a heap to start with, and the removal of no one grain ever changed it from being a heap to not being a heap, then it should still be a heap.

27 This appeal to scientific doctrine in order to determine what in scientific doctrine we should believe has struck many as circular; but van Fraassen insists that this "hermeneutic circle" is unobjectionable. See reading 9.3, pp. 638–640. For critique of van Fraassen's use of the distinction, see Churchland 1985, "The Ontological Status of Observables: In Praise of the Superempirical Virtues," in Paul M. Churchland and Clifford A. Hooker (eds.), *Images of Science: Essays on Realism and Empiricism*, University of Chicago Press, and Alspector-Kelly 2004,"Seeing the Unobservable," *Synthese* 140(3) (June 2004): 331–53.

28 Recall Boyd's argument in §7 above that IBE is too ubiquitous in our intellectual lives to be eliminated.

29 A nice feature of this version of the empirical equivalence argument is that it does not require that there are, for every theory, empirically equivalent rivals that posit different unobservables (which claim has been challenged). Suppose that there is no alternative T′ which implies TE, and therefore the same observations. TE itself is nevertheless always available, and always does imply those same observations.

30 Gideon Rosen 1994, "What is Constructive Empiricism?" *Philosophical Studies* 74: 143–78.

31 Van Fraassen's conception of rationality is in fact very liberal, counting as rational any position that does not violate either the laws of logic or the axioms of probability.

32 For discussion and critique of this position see Alspector-Kelly 2001, "Should the Empiricist be a Constructive Empiricist?" *Philosophy of Science* 68(4) (December): 413–31, and Alspector-Kelly 2006, "Constructive Empiricism and Epistemic Modesty: Reply to Van Fraassen and Monton," *Erkenntnis* 64(3) (October): 371–9.

33 See, for example, Musgrave 1989, "NOA's Ark – Fine for Realism," *The Philosophical Quarterly* 39: 382–98.

34 That presupposition, indeed, has a non-naturalist air about it that should perhaps make naturalistic

realists – Boyd, in particular – uncomfortable. See Alspector-Kelly 2003, "The NOAer's Dilemma: Constructive Empiricism and the Natural Ontological Attitude," *Canadian Journal of Philosophy* 33(3) (September): 307–22, for discussion.

Suggestions for Further Reading

The causal theory of reference is widely understood to originate in Kripke 1972[1980]. Hilary Putnam applied it to natural-kind terms in Putnam 1975. Boyd develops his version of scientific realism further in Boyd 1973, Boyd 1982, and Boyd 1990 (and other essays). Van Fraassen develops his constructive empiricist approach in van Fraassen 1980 (from which reading 9.3 is excerpted). An important collection of responses to his view is Churchland and Hooker 1985. Reading 9.4 by Fine is chapter 7 of Fine 1986a; chapter 8 of the same book continues his presentation of NOA and distinguishes it from anti-realism (Fine 1986b). He develops arguments against both realism and anti-realism in Fine 1986c and Fine 1991. An influential critique of NOA is Musgrave 1989. Another position in the realism debate is called *entity realism*, according to which belief in the existence of unobservables is a consequence of their experimental manipulation rather than the truth of the theories that refer to them. Entity realism was originally developed in Hacking 1983 and Cartwright 1983. Yet another approach is called *structural realism*, according to which we can legitimately believe in the (approximate) truth of the structural relations represented by contemporary theories, but we cannot claim to know what the entities are that realize those relations. See Worrall 1989.

Boyd, R., 1973, "Realism, Underdetermination and a Causal Theory of Evidence," *Nous* 7: 1–12.

Boyd, R., 1982, "Scientific Realism and Naturalistic Epistemology," *PSA* 80(2). East Lansing: Philosophy of Science Association.

Boyd, R., 1990, "Realism, Approximate Truth, and Philosophical Method" in W. Savage (ed.), *Minnesota Studies in the Philosophy of Science Vol. 14: Scientific Theories*. Minneapolis: University of Minnesota Press, pp. 355–91.

Cartwright, N., 1983, *How the Laws of Physics Lie*. Oxford: Oxford University Press.

Churchland, P. and Hooker, C. (eds.), 1985, *Images of Science*. Chicago: University of Chicago Press.

Fine, A., 1986a, *The Shaky Game: Einstein, Realism, and the Quantum Theory*. Chicago: University of Chicago Press.

Fine, A., 1986b, "And Not Anti-Realism Either," chapter 8 of Fine 1986a.

Fine, A., 1986c, "Unnatural Attitudes: Realist and Instrumentalist Attachments to Science," *Mind* 95(378): 149–79.

Fine, A., 1991, "Piecemeal Realism," *Philosophical Studies* 61: 79–96.

Hacking, I., 1983, *Representing and Intervening*. Cambridge: Cambridge University Press.

Kripke, S., 1972, "Naming and Necessity," in *Semantics of Natural Language*, D. Davidson and G. Harman (eds.), Dordrecht: Reidel. (Reprinted as *Naming and Necessity*, Cambridge: Harvard University Press, 1980).

Musgrave, A., 1989, "NOA's Ark: Fine for Realism," *The Philosophical Quarterly*, 39(157) (October): 383–98.

Putnam, H., 1975, "The Meaning of 'Meaning,'" in Putnam, H., *Philosophical Papers, Vol. 2: Mind, Language and Reality*. Cambridge: Cambridge University Press.

van Fraassen, B., 1980, *The Scientific Image*. Oxford: Oxford University Press.

Worrall, J., 1989, "Structural Realism: The Best of Both Worlds?" *Dialectica* 43: 99–124.

9.1

The Current Status of Scientific Realism

Richard N. Boyd

Richard Boyd (b. 1942) is one of the foremost advocates of scientific realism. He is perhaps best known for his presentation and elaboration of the no-miracles argument (NMA) for scientific realism, according to which the best explanation for the dramatic and increasing empirical success of science is that present scientific theories are approximately true and have been increasing in their truth-content. In the selection below, he considers objections to realism from empiricist and constructivist opponents and formulates his version of the NMA in response.

Introduction

The aim of this essay is to assess the strengths and weaknesses of the various traditional arguments for and against scientific realism. I conclude that the typical realist rebuttals to empiricist or constructivist arguments against realism are, in important ways, inadequate. I diagnose the source of the inadequacies in these arguments as a failure to appreciate the extent to which scientific realism requires the abandonment of central tenets of modern epistemology, and I offer an outline of a defense of scientific realism that avoids the inadequacies in question.

Scientific Realism Defined

By 'scientific realism' philosophers typically understand a doctrine which we may think of as embodying four central theses:

1. Theoretical terms in scientific theories (i.e., nonobservational terms) should be thought of as putatively referring expressions; that is, scientific theories should be interpreted "realistically."
2. Scientific theories, interpreted realistically, are confirmable and in fact are often confirmed as approximately true by ordinary scientific

From "The Current Status of the Issue of Scientific Realism," in *Erkenntnis*, Vol. 19, Nos. 1–3 (May 1983), selections from pp. 45–90.

evidence interpreted in accordance with ordinary methodological standards.

3. The historical progress of mature sciences is largely a matter of successively more accurate approximations to the truth about both observable and unobservable phenomena. Later theories typically build upon the (observational and theoretical) knowledge embodied in previous theories.
4. The reality which scientific theories describe is largely independent of our thoughts or theoretical commitments.

Critics of realism in the empiricist tradition typically deny theses 1 and 2, and qualify their acceptance of 3 so as to avoid commitment to the possibility of theoretical knowledge (however, van Fraassen accepts 1).[1] Antirealists in the constructivist tradition, such as Kuhn,[2] deny 4; however, they may well affirm 1 through 3 on the understanding that the "reality" which scientific theories describe is somehow a social and intellectual construct. As Kuhn and Hanson both argue,[3] a constructivist perspective limits, however, the scope of application of 3, since successive theories can be understood as approximating the truth more closely only when they are part of the same general constructive tradition or paradigm. J. C. C. Smart's version of scientific realism departs from the typical conception in that he rejects 2,[4] holding that distinctively philosophical considerations are required, over and above ordinary standards of scientific evidence, in order to justify our acceptance of the theoretical claims of scientific theories. Since Smart appears to hold that these philosophical considerations are nonevidential, it is perhaps appropriate to treat his position as intermediate between realism and constructivism.

In any event, the principal challenges to scientific realism arise from quite deep epistemological criticisms of 1 through 4. The key antirealist arguments, the standard rebuttals to them in the literature, as well as certain weaknesses in these rebuttals are summarized in the chart in the next section.

Antirealism in the Empiricist Tradition

There is a single, simple, and very powerful epistemological argument that represents the basis for the rejection of scientific realism by philosophers in the empiricist tradition. Suppose that *T* is a proposed theory of unobservable phenomena, which can be subjected to experimental testing. A theory is said to be empirically equivalent to *T* just in case it makes the same predictions about observable phenomena that *T* does. Now, it is always possible, given *T*, to construct arbitrarily many alternative theories that are empirically equivalent to *T* but which offer contradictory accounts of the nature of unobservable phenomena. Since scientific evidence for or against a theory consists in the confirmation or disconfirmation of one of its observational predictions, *T* and each of the theories empirically equivalent to it will be equally well confirmed or disconfirmed by any possible observational evidence. Therefore, no scientific evidence can bear on the question of which of these theories provides the correct account of unobservable phenomena; at best, it might be possible to confirm or disconfirm the claim that each of these theories is a reliable instrument for the prediction of observable phenomena. Since this construction is possible for any theory *T*, it follows that scientific evidence can never decide the question between theories of unobservable phenomena and, therefore, knowledge of unobservable phenomena is impossible.

This is the central argument of the verificationist tradition. If sound, it refutes scientific realism even if it is not associated with a version of the verifiability theory of meaning. Meaningful or not, theoretical claims are incapable of confirmation or disconfirmation. We may choose the "simplest model" for "pragmatic" reasons, but if evidence in science is experimental evidence, then pragmatic standards for theory choice have nothing to do with truth or knowledge. Scientific realism promises theoretical knowledge of the world, where, at best, it can deliver only formal elegance, or computational convenience.

As I have indicated in the chart, the empiricist argument we have been considering depends on the epistemological principle that empirically equivalent theories are evidentially indistinguishable. The evidential indistinguishability thesis (whether explicit or implicit) represents the key epistemological doctrine of contemporary empiricism and may be thought of as a precise formulation of the traditional empiricist doctrine

Table 1 Basic antirealist arguments, the standard rebuttals, and their weaknesses (Boyd, "The Current Status of Scientific Realism" from *Scientific Realism*, ed. Jarrett Leplin [Berkeley: University of California Press, 1984], pp. 41–81, p. 43)

Antirealist Argument	*Standard Rebuttal*	*Weakness*
1. The empiricist argument: empirically equivalent theories are evidentially indistinguishable; therefore, knowledge cannot extend to 'unobservables.'	1.*a.* There is no sharp distinction between "observables" and "unobservables."	1.*a.* (i) A sharp distinction can be drawn in a well-motivated way. (ii) In any event, distinction need not be sharp.
	1.*b.* The empiricist argument ignores the role of auxiliary hypotheses in assessing empirical equivalence.	1.*b.* The empiricist argument can be reformulated to apply to "total sciences."
	1.*c.* The "no miracles" argument: If scientific theories weren't (approximately) true, it would be miraculous that they yield such accurate observational predictions.	1.*c.* It does not address the crucial epistemological claim of the empiricist argument: that since factual knowledge is grounded in experience, it can extend only to observable phenomena.
2. Constructivist arguments:		
2.*a.* Scientific methodology is so theory-dependent that it is, at best, a construction procedure, not a discovery procedure.	2.*a.* Pair-wise theory-neutrality of method: for any two rival theories, there are experimental tests based on a method legitimized by both theories.	2.*a.* It does not address the epistemological point that theory-dependent methodology must be a construction procedure.
2.*b.* Consecutive "paradigms" in the history of science are not logically commensurable in the way they would be if they embodied theories about a paradigm-independent world.[5]	2.*b.* It is possible to give an account of continuity of reference for theoretical terms that allows for commensurability of paradigms.	2.*b.* If the antirealist *epistemological* argument (2.*a.*) is sound, then such continuity of reference is itself a construct, or at best a matter of continuity of reference to constructs, so that the realist's conception of scientific knowledge of theory-independent reality is still not vindicated.

("knowledge empiricism" in the phrase of Bennett)[6] that factual knowledge must always be grounded in experiences, and that there is no a priori factual knowledge. (As I shall argue later, the evidential indistinguishability thesis is the wrong formulation of the important epistemological truth in that doctrine; still, it represents the way in which empiricist philosophers of science – and most other empiricists, for that matter – have understood the fundamental doctrine of empiricist epistemology.)

Let us now turn to the standard rebuttals to the antirealist application of the indistinguishability thesis. Perhaps the most commonplace rebuttal to verificationist or empiricist arguments against realism is that the distinction between observable and unobservable phenomena is not a sharp one, and that the fundamental empiricist antirealist argument therefore rests upon an arbitrary distinction.[7] In assessing this rebuttal, it is important to distinguish between the question of the truth of the claim that the distinction between observable and theoretical entities is not sharp, and the question of the appropriateness of this claim as a rebuttal to empiricist antirealism. If scientific realism has somehow been established, then it may well be evident that the distinction in question is epistemologically arbitrary: if we are able to confirm theories of, say, electrons, then we may be able to employ such theories to design electron detecting instruments whose "readings" may have an epistemological status essentially like that of ordinary observations. If, however, it is scientific realism that is in dispute, then the considerations just presented would be inappropriately circular, even if their conclusion is ultimately sound. Only a non-question-begging demonstration that the distinction in question is arbitrary would constitute an adequate rebuttal to the empiricist's strong prima facie case that experimental knowledge cannot extend to the unobservable realm.

If we understand the rebuttal in question in this light, then several responses are available to the empiricist which indicate its weakness as a response to the central epistemological principle of empiricism. In the first place, it is by no means clear that the empiricist need hold that there is a *sharp* distinction between observable and unobservable phenomena in order to show that the distinction is epistemologically nonarbitrary. Suppose that there are entities that represent borderline cases of observability and suppose that there are cases in which it is not clear whether something is being observed or not. Then there will be some entities about which our knowledge will be limited by our capacity to observe them, and there will be cases in which the evidence is equivocal about whether there are entities of a certain sort at all. But the empiricist need hardly resist these conclusions: they are independently plausible, and – provided that there are some clear cases of putative unobservable entities (atoms, elementary particles, magnetic fields, etc.) – the antirealist claims of the empiricist are essentially unaffected.

Moreover, there are at least three ways in which the distinction in question can be made sharper in an epistemologically motivated way. In the first place, there is nothing obviously wrong with the traditional empiricist distinction between sense data and putative external objects. It is often claimed that the failure of logical positivists to construct a sense-datum language shows that the observation-theory dichotomy cannot be formulated in such terms, because it would be impossible to say of a theory that evidence for or against it consists in the confirmation or disconfirmation of observational (that is, sense-datum) predictions that are *deduced* from the theory. Quite so, but the fact remains that some experiences are of the sort we expect on the basis of the acceptance of a given theory, and others are of the sort we would not expect. Whatever the relation of expectation is between theories and sensory experiences, we may define empirical equivalence with respect to it, and affirm the empiricist thesis of the evidential indistinguishability of empirically equivalent theories. The result is *the* classical empiricist formulation of "knowledge empiricism." Insofar as it is plausible, this version of knowledge empiricism provides an argument against scientific realism, even though it also poses the philosophical problem of explicating the relevant expectation relation. In any event, that relation might well be taken to be given by empirical facts about human understanding, rather than by philosophical analysis.

It is true, of course, that the sense-datum formulation of the evidential indistinguishability thesis leads to phenomenalism (at best) about physical objects and other persons. As early logical positivists recognized, this consequence makes it difficult to account for the apparent social and intersubjective character of scientific knowledge. To be sure, this difficulty provides a reason to doubt the truth of the evidential indistinguishability thesis in its sense-datum formulation. But it does not constitute a satisfactory rebuttal to that thesis, nor a satisfactory rebuttal to the antirealist argument we are considering. The sense-datum version of the indistinguishability thesis is, after all, the obvious precise formulation of the doctrine that factual knowledge is always grounded in experience.The empiricist argument against realism is a straightforward application of that thesis. The fact that the thesis in question has inconvenient consequences shows neither that factual knowledge is not grounded in experience, nor that the (sense-datum version of) the indistinguishability thesis is not the appropriate explication of the doctrine that factual knowledge is grounded in this way. Considerations about the public character of science may provide us with reason to think that there must be *something* wrong with the phenomenalist's argument against scientific realism, but these considerations do not provide us with any plausible account of *what* is wrong with it. If I am right, the rebuttal to the sense-datum version of the evidential indistinguishability thesis displays a weakness common to all of the rebuttals to anti-empiricist arguments described in the chart. Each of the principal anti-empiricist arguments raises deep questions in epistemology or semantic theory against scientific realism. The standard rebuttals, insofar as they are effective at all, provide some reason to think that the antirealist arguments in question are unsound, or that realism is true, but they do not succeed in diagnosing the error in these arguments, nor do they point the way to alternative and genuinely realist conceptions of the central issues in epistemology or semantic theory.

It remains to examine the other two ways in which the dichotomy between observable and unobservable phenomena can be sharpened. On the one hand, phenomena might be classed as observable if they are quite plainly observable to persons with normal perceptual abilities. On the other hand, there is the proposal, which seems to be implicit in Maxwell, that entities that may not be directly observable to the unaided senses should count as observable for the purposes of the epistemology of science if they can be detected by the senses when the senses have been aided by devices whose reliability has been previously established by procures that do not beg the question between empiricists and scientific realists.

[. . .]

. . . [E]ach of these proposals reflects an important aspect of the intuitive conception that experimental knowledge is grounded in observation. What is important for our purposes is that *either* account of unobservability is sufficient to sustain a significant antirealist application of the evidential indistinguishability thesis.

[. . .]

We may apparently conclude the following about the rebuttal to empiricist antirealist arguments which turns on the claim that the distinction between observable entities and unobservable ones is not sharp and that the empiricist argument therefore rests upon an epistemologically arbitrary distinction: The distinction in question need not be sharp in order to be non-arbitrary. Moreover, there are at least three epistemologically motivated ways of making it sharper. An examination of each of these refinements of the distinction indicates features that might make it reasonable to suppose that there is something problematic about the basic empiricist argument against realism, but none of these considerations provides any diagnosis of the error, nor do any of them allow us to foresee any alternative to the doctrine of evidential indistinguishability of empirically equivalent theories upon which the empiricist argument depends. The standard rebuttals are inadequate in the face of the serious epistemological issues raised by the empiricist position.

[. . .]

There remains one rebuttal among the standard responses to empiricist antirealism, and it does seem to challenge directly the evidential indistinguishability thesis. The evidential indistinguishability thesis asserts that empirically equivalent theories are evidentially indistinguishable. But it has been widely recognized by philosophers of science that this is wrong. It might be right, they would argue, if the only predictions from a theory that are appropriate to test are those that can be deduced from the theory in isolation. But it is universally acknowledged that in theory testing we are permitted to use various well-confirmed theories as "auxiliary hypotheses" in the derivation of testable predictions. Thus, two different theories might be empirically equivalent – they might have the same consequences about observable phenomena – but it might be easy to design a crucial experiment for deciding between the theories if one could find a suitable set of auxiliary hypotheses such that when they were brought into play as additional premises, the theories (so expanded) were no longer empirically equivalent.

There is almost no doubt that considerations of this sort rebut any verificationist attempt to classify individual statements or theories as literally meaningful or literally meaningless by the criterion of verifiability in principle. But there is no reason to suppose that the rebuttal based on the role of auxiliary hypotheses is fatal to the basic claim of the evidential indistinguishability thesis, or to its antirealistic application. The reason is this: we may reformulate the evidential indistinguishability thesis so that it applies not to individual theories, but to "total sciences." The thesis, so understood, then asserts that empirically equivalent total sciences are evidentially indistinguishable. Since total sciences are self-contained with respect to auxiliary hypotheses, the rebuttal we have been considering does not apply, and the revised version of the evidential indistinguishability thesis entails that at no point in the history of science could we have knowledge that the theoretical claims of the existing total science are true or approximately true.[8]

[...]

There *is* a point regarding the use of auxiliary hypotheses that can be made the basis for a very strong defense of scientific realism. The use of auxiliary hypotheses, like other applications of what positivists called the "unity of science" principle, depends upon judgments of univocality regarding different occurrences of the same theoretical terms. It is possible to argue that only a realist conception of the semantics and epistemology of science can account for the role of such univocality judgments in contributing to the reliability of scientific methodology.[9] But this argument is not anticipated in the standard rebuttals to empiricist antirealism.

One must conclude that the standard rebuttals to the central empiricist argument against scientific realism are significantly flawed. Where they do provide reason to suspect that the empiricist argument is unsound (or, more directly, that realism is true), they do not provide any effective rebuttal to the main epistemological principle (the evidential indistinguishability thesis) upon which the empiricist argument depends, nor do they indicate respects in which the application of that principle to the question of realism is unwarranted.

Constructivist Antirealism

There is a single basic empiricist argument against realism, and it is an argument of striking simplicity and power. In the case of constructivist antirealism, the situation is much more complex. In part, at least, this is so because constructivist philosophers of science have typically been led to antirealist conclusions by reflections upon the results of *detailed* examinations of the history and actual methodological practices of science as well as by reflections on the psychology of scientific understanding. Different philosophers have focused on different aspects of the complex procedures of actual science as a basis for antirealist conclusions. Nevertheless, it is possible, I believe, to identify the common thread in all of these diverse arguments. Roughly, the constructivist antirealist reasons as follows: The actual methodology of science is profoundly theory-dependent. What scientists count as an acceptable theory, what they count as an observation, which experiments they take to be well designed, which measurement procedures they consider legitimate, what problems they seek to solve, and what sorts of evidence they require before accepting a theory – which are all features of scientific methodology – are in practice determined by the theroetical tradition within which scientists work. What sort of world must there be, the constructivist asks, for this sort of theory-dependent methodology to constitute a vehicle for gaining knowledge? The answer, according to the constructivist, is that the world that scientists study, in some robust sense must be defined or constituted by or "constructed" from the theoretical tradition in which the scientific community in question works. If the world that scientists study were not partly constituted by their theoretical tradition, then, so the argument goes, there would be no way of explaining why the theory-dependent methods that scientists use are a way of finding out what is true.

To this argument there is typically added another which addresses an apparent problem with constructivism. The problem is that scientists seem sometimes to be forced by new data to abandon important features of their current theories and to adopt radically new theories in their place. This phenomenon, it would seem, must be an example of scientific theories being brought into conformity with a theory-independent world, rather than an example of the construction of reality within a theoretical tradition. In response to this problem, constructivism often asserts that successive theories in science that represent the sort of radical "breaks" in tradition at issue are "incommensurable."[10] The idea here is that the standards of evidence, interpretation, and understanding dictated by the old theory, on the one hand, and by the new theory, on the other hand, are so different that the transition between them cannot be interpreted as having been dictated by any common standards of rationality. Since there are no significant theory-independent standards of rationality, it follows that the transition in question is not a matter of rationally adopting a new conception of (theory-independent) reality in the light of new evidence; instead, what is involved is the adoption of a wholly new conception of the world, complete with its own distinctive standards of rationality. In its most influential version, this argument incorporates the claim that the semantics of the two consecutive theories change to such an extent that those terms that they have in common should not be thought of as having the same referents in the two theories.[11] Thus, transitions of the sort we are discussing ("scientific revolutions" in Kuhn's terminology) involve a total change of theoretical subject matter.

There are two closely related standard rebuttals to these antirealist arguments. In the first place, against the claim that realism must be abandoned because scientific methodology is too theory-dependent to constitute a discovery (as opposed to a construction) procedure, it is often replied that for any two rival scientific theories it is always possible to find a methodology for testing them that is neutral with respect to the theories in question. Thus the choice between rival scientific theories on the basis of experimental evidence can be rational even though experimental methodology is theory-dependent. The outcome of a "crucial experiment" that pits one rival theory against another need not be biased, since such an experiment can be conducted on the basis of a methodology that – however theory-dependent – is not committed to either of the two contesting theories.

Against the incommensurability claim, it is often argued that an account of reference for theoretical expressions can be provided that makes it possible to describe scientific revolutions as involving continuity in reference for the theoretical terms common to the laws of the earlier and later theoretical traditions or "paradigms." With such referential continuity comes a kind of continuity of methodology as well, because (assuming continuity of reference) the actual cases of scientific revolutions typically result in the preservation of some of the theoretical machinery of the earlier paradigm in the structure of the new one, and this, in turn, guarantees a continuity of methodology.

Neither of these rebuttals is fully adequate as a response to constructivist antirealism. Consider first the claim that for any two rival theories there is a methodology for testing them that is neutral with respect to the issues on which they differ ("pair-wise theory-neutrality of method" in the chart). It is generally true that for theoretical rivalries that arise in actual science, a relevantly neutral testing methodology will exist. Indeed, the use of such "neutral" testing methodologies is a routine part of what Kuhn calls "normal science." And indeed, the existence of such methodologies helps to explain how scientists can appeal to common standards of rationality even when they have theoretical differences of the sort that influence methodological judgments. Nevertheless, pair-wise theory-neutrality of method does not provide a reason to reject the antirealist conclusions of the constructivist.

Remember that what the constructivist argues is that a general methodology that is predicated upon a particular theoretical tradition and is theory-determined to its core cannot be understood as a methodology for discovering features of a world that is not in some significant way defined by that tradition. All that the doctrine of pair-wise theory-neutral methods asserts is that within the theoretical and methodological tradition in question, there are available experimental procedures that are neutral with respect to quite particular disputes between alternative ways of modifying or extending that very tradition. There is no suggestion of a procedure by which scientific methodology can escape from the presuppositions of the tradition and examine objectively the structure of a theory-independent world. Insofar as the profound theory-dependence of method raises an epistemological problem for realism, the pair-wise theory-neutrality of methods does not provide an answer to it.

Perhaps surprisingly, it does not help either to demonstrate that successive paradigms are commensurable. Suppose that a satisfactory account of referential continuity for theoretical terms during scientific revolutions is available.[12] Suppose further (what is not implied by the former claim) that the theoretical continuity thus established during revolutionary periods is such that the transition between the prerevolutionary theory and the postrevolutionary one is governed by a continuously evolving standard of scientific rationality. If these suppositions are true, then much of what Kuhn, for example, has claimed about the history of science will be mistaken: postrevolutionary scientists will (contrary to Kuhn) be building on the theoretical achievements of their prerevolutionary predecessors; the adoption of new "paradigms" will be scientifically rational; and it will not involve a "Gestalt shift" in the scientific community's understanding of the world, whatever may be the case for some individual scientists. *But*, the basic constructivist epistemological objection to scientific realism will still be unrebutted. If the theory-dependence of methodology provides reason to doubt that scientific inquiry possesses the right sort of objectivity for the study of a theory-independent world, then the sort of historical continuity through scientific revolutions we are considering will not address that doubt. Only if the transitional methodology during revolutions were largely theory-neutral would the fact of methodological and semantic continuity between revolutions provide, by itself, a rebuttal to the constructivist antirealist; but there is no chance that such theory-independence could be demonstrated by the sort of rebuttal to incommensurability we are considering. Indeed, there is no reason of any sort to suppose that such a theory-neutral method ever prevails.

In the present case, as in the case of the standard rebuttals to empiricist antirealism, it is by no means true that the standard rebuttals to the constructivist arguments are irrelevant to the issue of scientific realism. If there were no such phenomenon as pair-wise theory-neutrality of method, then it would be hard to see how there could be any sort of scientific objectivity, realist or constructivist. If there is no way of defending the continuity of subject matter and methodology during most of the episodes which Kuhn calls scientific revolutions, then the realist conception of science is rendered most implausible. The point is that, even though these prorealist rebuttals to constructivist antirealism do provide some support for aspects of the realist position, they fail to offer any reason to reject the basic epistemological argument against realism which the constructivist offers.

Empiricism and Constructivism

Kuhn presents his constructivist account of science as an alternative to the tradition of logical empiricism and, indeed, there is much he says with which traditional positivists would disagree. There are, nevertheless, important similarities between the constructivist and the empiricist approach to the philosophy of science. Kuhn, for example, relies on the late positivist 'law-cluster' account of the meaning of theoretical terms in his famous argument against the semantic commensurability of successive paradigms.[13] Similarly, R. Carnap's mature positivism of the early 1950s has much in common with Kuhn's views. In particular, "Empiricism, Semantics, and Ontology"[14] offers an account of the criteria for the rational acceptance of a linguistic framework which is surprisingly like a formalized version of Kuhn's view.[15] We may say with some precision what the points of similarity between Kuhn and Carnap are. In the first place, they are agreed that the day-to-day business of the development and testing of scientific theories is governed by broader and more basic theoretical principles, including the most basic laws and definitions of the relevant sciences.

There is a far deeper point of agreement. Kuhn, and constructivists generally, cannot consistently accept the principle of the evidential indistinguishability of empirically equivalent total sciences; they hold, after all, that 'facts' – insofar as they are the subject matter of the sciences – are partly constituted or defined by the adoption of "paradigms" or theoretical traditions, so that there is a sort of a priori character to the scientist's knowledge of the fundamental laws in the relevant tradition or paradigm. But they agree with logical empiricists in holding that any rational constraint on theory acceptance that is not purely pragmatic and that does not accord with the evidential indistinguishability thesis must be essentially conventional. For Carnap and other positivists, the conventions are essentially linguistic: they amount to the conventional adoption of one set of "L-truths" rather than another. For Kuhn and other constructivists, the conventions go far deeper: they amount to the social construction of reality and of experimental "facts." What neither empiricists nor constructivists accept is the idea that the regulation of theory acceptance by features (linguistic or otherwise) of the existing theoretical tradition can be a reliable guide to the discovery of theory-independent matters of fact.

Of course, empiricists and constructivists differ, especially regarding the extent to which experimental observations can be divorced from theoretical considerations, and (if constructivists are "relativists" in the Kuhnian tradition) about the methodological commensurability of successive theoretical traditions or paradigms. It is interesting to note that Kuhn and the Carnap of the early 1950s do not disagree about the *semantic* incommensurability of the theoretical portions of alternative linguistic frameworks for science – neither accepts any doctrine of continuity of *reference* for theoretical terms in the transition to alternative frameworks. Indeed, for Carnap, questions of reference and ontology are meaningless when raised outside the scope of some particular linguistic framework. That Kuhn and Carnap should agree to this extent about the semantics of theoretical terms is less surprising when one realizes that Kuhn's account of the meaning of such terms is simply a more subtle and historically more accurate version of Carnap's.[16]

One further point of agreement between empiricists and constructivists is significant for our purposes. Empiricist philosophers of science deny that knowledge of theoretical entities is possible. But it is no part of contemporary empiricism to deny that the scientific method yields objective instrumental knowledge: knowledge of regularities in the behavior of observable phenomena. It is important to see that this point is not seriously contested by constructivist philosophers of science. It is true that constructivists insist that observation in science is significantly theory-determined, and that Kuhn, for example, emphasizes that experimental results that are anomalous in the light of the prevailing theoretical conceptions are typically ignored if they cannot readily be assimilated into the received theoretical framework. But no serious constructivist maintains that the predictive reliability of theories in mature science or the reliability of scientific methodology in identifying predictively reliable theories is largely an artifact of the tendency to ignore anomalous results. Such a view would be nonsensical in the light of the contributions of pure science to technological advance.

There is one point that, whether ultimately compatible with empiricism or not, is certainly emphasized by constructivists much more than by empiricists, and is especially relevant when one considers the role of scientific methodology in producing instrumental knowledge. It was recognized early on by logical empiricists that any account of the methodology of science requires some account of the way in which the "degree of confirmation" of a theory, given a body of observational evidence, is to be determined. More recently, N. Goodman has, following Locke, raised a question that is really a special case of the problem of determining degree of confirmation.[17] Any account of the methodology of science must account for judgments of 'projectability' of predicates or, to put the issue more broadly, it must provide an account of the standards by which scientists determine which general conclusions are even real candidates for acceptance given an (always finite) body of available data.[18] This question is interesting precisely because, given any finite body of data, there are infinitely many different general theories that are logically consistent with those data (indeed, there will be infinitely many such theories that are pair-wise empirically inequivalent, given the existing total science as a source of auxiliary hypotheses).

What Kuhn and other constructivists insist (correctly, I believe) is that judgments of projectability and of degrees of confirmation are quite profoundly dependent upon the theories that make up the existing theoretical tradition or paradigm. The theoretical tradition dictates the terms in which questions are posed and the terms in which possible answers are articulated. In a similar way, theoretical considerations dictate the standards for experimental design and for the assessment of the experimental evidence. Assuming this to be true, and assuming, as reasonable constructivists must, that the reliability of scientific methodology in producing instrumental knowledge is not to be explained largely by the tendency to ignore anomalous data, we can see that an important epistemological issue emerges regarding judgments of projectability and of degree of confirmation: why should so theory-dependent a methodology be reliable at producing knowledge about (largely theory-independent) observable phenomena?

A related question about what we might call the "instrumental reliability" of scientific method should prove challenging both to Kuhn, and to empiricists who share with Kuhn the law-cluster theory of the meaning of theoretical terms. Judgment of univocality for particular occurrences of the (lexicographically) same theoretical term play an important epistemological role in scientific methodology. This is evident since commonplaces such as the use of auxiliary hypotheses in theory testing, or applications of the principle of "unity of science" in the derivation of observational predictions from theories that have already been accepted, depend upon prior assessments of univocality. This means that scientific standards for the assessment of univocality for token occurrences of theoretical terms must play a crucial epistemological role, and it must be the business of an adequate account of the language of science to say what those standards are *and* why they are such as to render instrumentally reliable the methodological principles in actual science which depend upon univocality judgments.[19]

Unlike earlier positivist theories of meaning for theoretical terms (like operationalism, for example) the law-cluster theory does not say what it is for two tokens of orthographically the same theoretical term to occur with the same meaning or reference. The meaning of a theoretical term is given by the most basic laws in which it occurs; this may possibly tell us something about diachronic questions about univocality of theoretical terms. But suppose that t and t' are two tokens of orthographically the same theoretical term, used at the same time, and that neither t nor t' occurs in a law that is fundamental in the sense relevant to the law-cluster theory. This latter condition describes the circumstances of almost all tokens of theoretical terms in actual scientific usage. Under the circumstances in question, the law-cluster theory says nothing about the question of whether t and t' have the same meaning or reference. Only when the synchronic problem of univocality in such cases is presumed to have already been solved does the law-cluster theory have anything to say about univocality for theoretical terms. The law-cluster theory is thus entirely without the resources to address the important question of the contribution that judgments of univocality for theoretical terms make to the instrumental reliability of scientific methodology.

Thus we have identified two questions that pose especially sharp challenges to both empiricist and constructivist conceptions of science: why are theory-dependent standards for assessing projectability and degrees of confirmation instrumentally reliable? and, how do judgments of univocality for theoretical terms contribute to the instrumental reliability of scientific methodology? I shall argue in the next section that answers to these challenges provide the basis for a new and more effective defense of scientific realism.

Defending Scientific Realism

Elsewhere, I have offered a defense of scientific realism against empiricist antirealism which proceeds by proposing that a realistic account of scientific theories is a component in the only scientifically plausible explanation for the instrumental reliability of scientific methodology.[20] What I propose to do here is to summarize this defense very briefly and to indicate how it also constitutes a defense of scientific realism against constructivist criticisms, and how it avoids the weaknesses in the traditional rebuttals to antirealist arguments.

The proposal that scientific realism might be required in order to explain adequately the instrumental reliability of scientific methodology can be motivated by reexamining the principal constructivist argument against scientific realism (see 2.*a* in the chart). The constructivist asks, What must the world be like in order that a methodology so theory-dependent as ours could constitute a way of finding out what is true? She answers: The world would have to be largely defined or constituted by the theoretical tradition that defines that methodology. It is clear that another answer is at least possible: the world might be one in which the laws and theories embodied in our actual theoretical tradition are approximately true. In that case, the methodology of science might progress dialectically. Our methodology, based on approximately true theories, would be a reliable guide to the discovery of new results and the improvement of older theories. The resulting improvement in our knowledge of the world would result in a still more reliable methodology leading to still more accurate theories, and so on.[21]

What I have argued in the works cited above is that this conception of the enterprise of science provides the only scientifically plausible explanation for the instrumental reliability of the scientific method. In particular, I argue that the reliability of theory-dependent judgments of projectability and degrees of confirmation can only be satisfactorily explained on the assumption that the theoretical claims embodied in the background theories which determine those judgments are relevantly approximately true, and that scientific methodology acts dialectically so as to produce in the long run an increasingly accurate theoretical picture of the world. Since logical empiricists accept the instrumental reliability of actual scientific methodology, this defense of realism represents a cogent challenge to logical empiricist antirealism. It remains to be seen whether it has the weaknesses of more traditional responses to empiricist antirealism, but, first, let us examine its relevance to constructivism.

First, it should be observed that the argument for realism that I have indicated is a direct response to the central constructivist argument against realism. If the argument for realism is correct, then we can see *what* is wrong with the central constructivist argument: the constructivist's epistemological challenge to scientific realism rests upon the wrong explanation for the reliability of the scientific method as a guide to truth.

It is equally important to see that there is no answer within a purely constructivist framework to the question of why the methods of science are *instrumentally* reliable. The instrumental reliability of particular scientific theories cannot be an artifact of the social construction of reality. Even within "pure" science this is acknowledged, for example, by Kuhn. The anomalous observations that (sometimes) give rise to "scientific revolutions" cannot be reflections of a fully paradigm-dependent world: anomalies are defined as observations that are inexplicable within the relevant paradigm. It is even more evident that theory-dependent technological progress (the most striking example of the instrumental reliability of scientific *methods* as well as theories) cannot be explained by an appeal to social construction of reality. It cannot be that the explanation for the fact that airplanes, whose design rests upon enormously sophisticated theory, do not often crash is that the paradigm *defines* the concept of an airplane in terms of crash resistance. If the empiricist cannot offer a satisfactory account of the instrumental reliability of scientific method (as I have argued in the works cited), then the constructivist – who even more than the empiricist emphasizes the theory dependence of that method – cannot do so either. Thus, the epistemological thrust of constructivism is directly challenged by the argument for scientific realism under consideration.

It is clear, moreover, that if scientific realism is defended in this way, then the more traditional rebuttals to constructivist antirealism are rendered fully effective. If the fundamental epistemological thrust of constructivism is mistaken, then (as I indicated earlier) the pair-wise theory-neutrality of scientific methodology, and the continuity of reference of theoretical terms and methods across "revolutions" are crucial components in the defense of scientific realism.

Let us turn now to the question of whether the defense of realism we are considering has the weakness of the more traditional rebuttals to empiricist antirealism. Those rebuttals had the defect that, while they provided some reason to believe that scientific realism is true, they offered no insight into the question of what is wrong with the crucial empiricist argument against realism. Here, the argument under consideration succeeds where the more traditional arguments fail. What is wrong with the fundamental empiricist argument is that the principle that empirically equivalent total sciences are evidentially indistinguishable is false, and it represents the wrong reconstruction of the perfectly true doctrine that factual knowledge is grounded in observation.

The point here is that if the realist and dialectical conception of scientific methodology is right, then considerations of the theoretical plausibility of a proposed theory in the light of the *actual* (and approximately true) theoretical tradition are *evidential* considerations: results of such assessments of plausibility constitute evidence for or against proposed theories. Indeed, such considerations are a matter of theory-mediated empirical evidence, since the background theories, with respect to which assessments of plausibility are made, are themselves empirically tested (again, in a theory-mediated way). Theory-mediated evidence of this sort is no less empirical than more direct experimental evidence – largely because the evidential standards that apply to so-called direct experimental tests of theories are theory-determined in just the same way that judgments of plausibility are. In consequence, the *actual* theoretical tradition has an epistemically privileged position in the assessment of *empirical* evidence. Thus, a total science whose theoretical conception is significantly in conflict with the received theoretical tradition is, for that reason, subject to "indirect" but perfectly real prima facie disconfirmation relative to an empirically equivalent total science that reflects the existing tradition. The evidential indistinguishability thesis is therefore false, and the basic empiricist antirealist argument is fully rebutted.[22]

It might seem that the realist conception that theoretical considerations in science are evidential would reflect a weakening of ordinary standards of evidential rigor in science. After all, on the realist conception, a theory can get evidential support both from (direct) experimental evidence and from (indirect) theoretical considerations. Moreover, the realist proposal might seem to make it impossible to disconfirm traditional theories, treating them as a priori truths in much the same way that the constructivist conception does. Neither of these claims proves to be sound. In the first place, rigorous assessment of experimental evidence in science depends fundamentally upon just the principle that theoretical considerations are evidential: that is why a realist conception of theories is necessary to account for the instrumental reliability of our standards for assessing experimental evidence. Second, the realist conception of theory-mediated experimental evidence does not have the consequence that any traditional laws are immune from refutation. Instead, it provides the explanation of how rigorous testing of these and other laws is possible. The dialectical process of improvement in the theoretical tradition does not preclude, but instead requires, that particular laws or principles in the tradition may have to be abandoned in the light of new evidence.

Let us turn now to the second puzzle about the instrumental reliability of scientific method which was raised at the end of the preceding section: how to account for the epistemic reliability of judgments of univocality for theoretical terms. The realistic account of the instrumental reliability of judgments of projectability requires that the kinds or categories into which features of the world are sorted for the purpose of inductive inference be determined by theoretical considerations rather than being fixed by conventional definitions, however abstract.[23] In particular, the law-cluster theory of meaning, understood conventionally, is inadequate as an account of the "definitions" of theoretical terms in science. It has been widely recognized that if theoretical terms in science are to refer to entities or kinds whose "essences" are determined by empirical investigation rather than by stipulation, then the traditional conception of reference fixing by stipulatory conventions must be abandoned for such terms in favor of some "causal" or "naturalistic" theory of reference.[24]

Given the distinctly realistic conception of scientific knowledge described previously, it is possible to offer a naturalistic theory of reference which is especially appropriate to an understanding of the role of theoretical considerations in scientific reasoning. Such a theory defines reference in terms of relations of "epistemic access."[25] Roughly, a (type) term *t* refers to some entity *e* just in case complex causal interactions between features of the world and human social practices bring it about that what is said of *t* is, generally speaking and over time, reliably regulated by the real properties of *e*. Because such regulation of what we say by the real features of the world depends upon the approximate truth of background theories, the approximate reliability of measurement and detection procedures, and the like, the epistemic access account of reference can explain the grains of truth in such previous accounts of reference as the law-cluster theory, or operationalism.[26]

Consider now the question of univocality for two token occurrences of orthographically the same theoretical term. Such a pair of terms will be coreferential just in case the social history of each of their occurrences links them, by the relevant sort of causal relations, to a situation of reliable belief regulation by the actual properties of the same feature of the world. The relevant sorts of causal relations are to be determined by epistemology, construed as an empirical investigation into the mechanisms of reliable belief regulation. Thus it is an empirical question, not a conceptual one, whether two such tokens are univocal.

Because the epistemic access account of reference can explain the grains of truth in the other theories of reference for theoretical terms which have been advanced to explain the actual judgments of scientists and historians about issues of univocality, there is every reason to believe that the epistemic access account can explain why the ordinary standards for judging univocality that prevail in science are reliable indicators of actual coreferentiality. Together with the realist's conception that scientific methodology produces (typically and over time) approximately true beliefs about theoretical entities, the epistemic access account of reference provides an explanation of how univocality judgments contribute to the reliability of scientific methodology, an explanation that is fully in accord with the general realist conception of scientific methodology described here.

Finally, the epistemic access account provides a precise formulation of the crucial realist claim that typically (perhaps despite changes in law-clusters) there is continuity of reference across "scientific revolutions."[27] Indeed, it permits us to integrate cases of what H. Field calls "partial denotation"[28] into a general theory of reference and thus to treat cases of "denotational refinement" as establishing referential continuity in the relevant sense.

If the dialectical and realistic conception of scientific methodology described here and the related epistemic access conception of reference are approximately correct, then together they constitute a rebuttal to both empiricist and constructivist antirealism which suffers none of the shortcomings of the more traditional rebuttals, while at the same time accommodating the insights that the more traditional rebuttals provide.

Scientific Realism and Metaphilosophy

I have examined traditional rebuttals to antirealist arguments in the empiricist and constructivist traditions and have suggested that these rebuttals have the weakness that they do not provide a diagnosis of the epistemological errors that must – if realism is true – lie behind the standard argument against realism. I indicated how a distinctly realistic and dialectical conception of scientific methodology, together with a closely related naturalistic conception of reference, could provide the basis for a defense of realism that does diagnose the epistemological errors in antirealist arguments. If the conception of scientific knowledge and language described here is correct, then it has implications for philosophical methodology which are sufficiently startling that they may help to explain why the dialectical and realist account of the reliability of scientific methodology was not put forward earlier as the epistemological foundation for scientific realism.

I believe it is fair to say that scientific realists have had a conception of their dispute with empiricist and (more recently) with constructivist antirealists according to which they shared with their opponents a general conception of the logic and methods of science, and according to which the dispute between realists and antirealists was over whether that logic and those methods were adequate to secure theoretical knowledge of a theory-independent reality. It was not anticipated that a new and distinctly realist general account of the methods of science would be necessary in order to defend scientific realism. This conception of a shared account of the logic and methods of science was advanced explicitly by E. Nagel, in discussing the realism-empiricist dispute:

> It is difficult to escape the conclusion that when the two opposing views on the cognitive status of theories are stated with some circumspection, each can assimilate into its formulation not only the facts concerning the primary subject matter explored by experimental inquiry but also the relevant facts concerning the logic and procedures of science. In brief, the opposition between these views is a conflict over preferred mode of speech.[29]

It is evident that the argument for scientific realism described in the preceding section departs from this understanding. According to that argument, no empiricist or constructivist account of the methods of science can explain the phenomenon of instrumental knowledge in science, the very kind of scientific knowledge about which realists, empiricists, and constructivists largely agree. Only on a distinctly realist conception of the logic and methods of science – a conception that empiricists and constructivists cannot share – can instrumental knowledge be explained.

The distinctly realist conception of the methodology of science departs even farther from the normal conception of the epistemology of science. At least since Descartes, the characteristic conception of epistemology in general has been that the most basic epistemological principles – the basic canons of reasoning or justification – should be defensible a priori. Thus, for example, almost all empiricists have thought that "knowledge empiricism" represented an a priori truth about knowledge, and that the most basic principles of inductive reasoning, whatever they are, can be defended a priori. Similar conceptions are even more clearly seen in the rationalist and Kantian traditions. What is striking is that, if the distinctly realist account of scientific knowledge is sound, then the most basic principles of inductive inference lack any a priori justification. That this is so can be seen by reflecting on what the scientific realist must say about the history of the scientific method.

According to the distinctly realist account of scientific knowledge, the reliability of the scientific method as a guide to (approximate) truth is to be explained only on the assumption that the theoretical tradition that defines our actual methodological principles reflects an approximately true account of the natural world. On that assumption, scientific methods will lead to successively more accurate theories and to successively more reliable methodological practices.[30] If we now inquire how the theoretical tradition came to embody sufficiently accurate theories in the first place, the scientific realist cannot appeal to the scientific method as in explanation, because that method is epistemically reliable only on the assumption that the relevant theoretical tradition already embodies a sufficiently good approximation to the truth. The realist, as I have portrayed here, must hold that the reliability of the scientific method rests upon the logically, epistemically, and historically contingent emergence of suitably approximately true theories. Like the causal theorist of perception or other "naturalistic" epistemologists, the scientific realist must deny that the most basic principles of inductive inference or justification are defensible a priori. In a word, the scientific realist must see epistemology as an *empirical* science.[31]

Closely analogous consequences follow from the epistemic access account of reference when it is applied in the light of scientific realism. The question of whether two tokens of a theoretical term are coreferential is, for example, a purely empirical question that cannot be resolved by conceptual analysis. If we think of the "meaning" of a theoretical term as comprising those features of its use in virtue of which it has whatever referent it in fact has, then meanings of theoretical terms are not given by a priori stipulations or social conventions. It is logically, historically, and epistemically contingent matter which features of the use of a given term constitute its meaning, in the sense of meaning relevant to referential semantics. There simply are not going to be any important analytic or conceptual truths about any scientifically interesting subject matter.

If these controversial consequences of a thoroughgoing realist conception of scientific knowledge are sound, then it would be hard to escape a still more controversial conclusion: philosophy is itself a sort of empirical science. It may well be a normative science – epistemology, for example, may aim at understanding which belief-regulating mechanisms are reliable guides to the truth – but it will be no less an empirical science for being normative in this way.

Issues of Philosophical Method

In this section, I shall discuss two issues of philosophical methodology raised by the arguments for scientific realism described in the section "Defending Scientific Realism." First, I shall discuss, at some length, an important challenge raised by Arthur Fine against the basic strategy of those arguments. Then, I shall discuss, somewhat more briefly, certain questions about the ways in which evidence from the history of science bears upon the arguments in question.

The challenge to abduction

Fine raises a number of interesting objections to the arguments for scientific realism outlined earlier.[32] Of these objections, one is particularly striking because it challenges not the details of the argument for realism, but its basic philosophical strategy.

Fine's objection is extremely simple and elegant. The proposed defense of realism proceeds by an abductive argument: we are encouraged to accept realism because, realists maintain, realism provides the best explanation of the instrumental reliability of scientific methodology. Suppose for the sake of argument that this is true. We are still not justified in believing that realism is true. This is so because the issue between realists and empiricists is precisely over the question of whether or not abduction is an epistemologically justifiable inferential principle, especially when, as in the present case, the explanation postulated involves the operation of unobservable mechanisms. After all, if abductive inference is justifiable, then there is no epistemological problem about the theoretical postulation of unobservables in the first place. It is precisely abductive inference to unobservables that the standard empiricist arguments call into question. Thus, the abductive defense of realism we are considering is viciously circular.

It is reasonable to think of Fine's objection in the light of the previous discussion of the "no miracles" argument for realism (discussed in the section "Antirealism in the Empiricist Tradition"). Against the "no miracles" argument I argued that, even if realism provides the best explanation for the predictive reliability of scientific theories, there remains for the realist the problem that this fact does not constitute a rebuttal to the very powerful epistemological considerations that form the basis for empiricist antirealism. Fine, in effect, presents a generalized version of this response to the "no miracles" argument. In the first place, Fine's version of the response in question applies not only to the "no miracles" argument but to any argument for realism that adduces realism as (a component of) the best explanation for some natural phenomenon. In particular, Fine's objection applies to the argument for realism offered in the section on "Defending Scientific Realism." Suppose now that scientific realism provides the best explanation for the reliability (not just of individual theories but) of the methodology of science as a whole. This fact *by itself* does not constitute a rebuttal to the epistemological principles upon which the empiricist criticism of realism rests.

Moreover, Fine's objection diagnoses not only a weakness in such arguments for realism but a circularity as well. The issue of scientific realism is – at least insofar as the dispute between realists and empiricists is concerned – a debate over the legitimacy of inductive inference to the best explanation, at least in those cases in which the explanation in question postulates unobservable entities. Arguments for realism of the sort which Fine criticizes employ just this sort of inference, and, thus, simply beg the question between realists and empiricist antirealists.

Several things must be said in reply to Fine's subtle and elegant objection. In the first place, Fine's entirely correct insistence that the issue between empiricists and realists is over the legitimacy of abductive inference is a double-edged sword. While it facilitates the identification of a sort of circularity in arguments for realism, it also highlights the epistemological oddity of consistent empiricism. The rejection of abduction or inference to the best explanation would place quite remarkable strictures on intellectual inquiry. In particular, it is by no means clear that students of the sciences, whether philosophers or historians, would have any methodology left if abduction were abandoned. If the fact that a theory provides the best available explanation for some important phenomenon is not a justification for believing that the theory is at least approximately true, then it is hard to see how intellectual inquiry could proceed. Of course, the antirealist might accept abductive inferences whenever their conclusions do not postulate unobservables, while rejecting such inferences to "theoretical" conclusions. In this case, however, the burden of proof would no longer lie exclusively on the realist's side: the antirealist must justify the proposed limitation on an otherwise legitimate principle of inductive inference.

This difficulty for the antirealist is exacerbated when one considers the issue of inductive inference in science itself. It must be remembered that empiricist philosophers of science do not intend to be fully skeptical: it is no part of standard empiricist philosophy of science to reject all nondeductive inferences. Instead, a selective skepticism is intended: (some) inductive generalizations about observables are to be epistemologically legitimate, while inferences to conclusions about unobservables are to be rejected. As Hanson,

Kuhn, and others have shown, the actual methods of science are profoundly theory-dependent. I have emphasized in previously cited publications that this theory-dependence extends to the methods scientists employ in making inductive generalizations about observable phenomena. Both the choice of generalizations that are seriously advanced and the assessment of the evidence for or against them rest upon theoretical inferences that manifest, or depend upon, the sort of abductive inferences to which the empiricist objects. In the terminology of recent empiricism, both the assessment of "projectability" of predicates, and the assessment of "degree of confirmation" of generalizations about observables depend in practice, upon inferences about "theoretical entities." Of course, acknowledging these facts about scientific practice would not commit the empiricist to agreeing that realism provides the best explanation for the instrumental reliability of scientific methodology nor, as Fine insists, would agreeing to that proposition commit the empiricist to holding that there is any reason to believe that realism is true. Nevertheless it certainly seems that, unless – as is very unlikely – the apparent theory-dependence of inductive inference about observables is really only apparent, the empiricist who rejects abductive inferences regarding unobservables must hold that even the inductive inferences scientists make about observables are unjustified.

It might seem that there is an easy way out of this last difficulty for the empiricist. Suppose that inductive inferences about observables in science are genuinely theory-dependent and that, therefore, the (necessarily theoretical) justifications, which scientists would ordinarily offer in defense of their inductive inferences about observables, themselves rest on theoretical claims that are without justification. Still, a philosopher might propose a sort of inductive justification of theory-dependent scientific inductions. Let the inductive procedures of science be as theory-dependent as you like, and let the justifications offered for individual inferences by scientists be as faulty as the empiricist claims. The fact remains that the (theory-dependent) methodology of science gives evidence of being instrumentally reliable. Let *that* constitute the justification for the inferences which scientists make. The thesis that the methodology of science is instrumentally reliable is, after all, a thesis about observable phenomena. It is, moreover, well confirmed by the observational evidence presented by the recent history of science and technology. Since no abductive inference objectionable from an empiricist perspective is required to establish the generalization that scientific methodology is instrumentally reliable, we may accept this generalization and then apply it to justify the acceptance of the inductive generalizations scientists arrive at by employing the scientific method. Even though the theoretical reasoning that underlies inductive inferences about observables may not be justificatory, a second-order induction about the instrumental reliability of such reasoning might still afford a justification for that part of scientific practice that is supposed to be immune from the empiricist's selective skepticism.

It is very doubtful that this application of the inductive justification of induction can help the empiricist we are considering to avoid the conclusion that inductive generalizations in science about observables are unjustified. The hypothesis that scientific methodology is instrumentally reliable (henceforth, the "reliability hypothesis") is itself an inductive generalization about observable phenomena. If, as I have suggested earlier, the confirmation or disconfirmation of such generalizations typically presupposes theoretical considerations of the sort our empiricist cannot accept, then we should expect that this might be true of the confirmation of the reliability hypothesis itself. If this is so, then the effort to circumvent the empiricist's conclusion that inductive generalizations in science are unjustified because they are theory-dependent, by appealing to the confirmation of the reliability hypothesis, will have failed. The reliability hypothesis will itself be unjustified by the standards of the empiricist we are considering.

[. . .]

I conclude that the empiricist who rejects abductive inferences is probably unable to avoid, in any plausible way, the conclusion that the inductive inferences that scientists make about observables are unjustified. Nevertheless, even if this is so, Fine's criticism of abductive arguments for realism still has force. If what is at issue is the legitimacy of abductive inferences to theoretical explanations in general, then there is a kind of circularity in the appeal to a particular abduction of this sort in the defense of scientific realism. I suggested earlier in this paper that while standard rebuttals to empiricist antirealism provide some reason to believe that scientific realism is true, these rebuttals fail to respond to the strong epistemological challenge that empiricist antirealism offers. Should we take the circularity which Fine discerns to indicate that the same is true for the abductive argument for scientific realism as a component in the best explanation for the instrumental reliability of scientific method? I want to argue that the answer should be no.

If abduction were prima facie suspect, in the way that palm reading or horoscope casting now are, then surely it would be inappropriate to appeal to some particular abductive inference in defense of abductive inference in general. Abduction is, however, prima facie legitimate; it is seen as suspect only in the light of certain distinctly empiricist epistemological considerations. To assess the import of the circularity of appealing to abduction in replying to empiricist antirealism, we must examine more closely the relation between the particular abductive inferences in question, and the empiricist's arguments against realism.

I suggest that an assessment of the import of the circularity in question should focus not on the legitimacy of the realist's abductive inference considered in isolation, but rather on the relative merits of the overall accounts of scientific knowledge which the empiricist and the realist defend. Such an assessment strategy is familiar from many areas of intellectual inquiry, scientific and scholarly. Defenders of rival positions often reach their distinctive conclusions via forms of inference which their rivals think unjustified. The pairwise theory-neutral procedure for addressing such disputes typically consists in an assessment of the overall adequacy of the theories put forward, rather than in an assessment of the particular controversial inference forms considered in isolation.

If we consider the present dispute in this light, then there are two especially important considerations. First, the empiricist's objection to abductive inferences (at least to those that yield conclusions about unobservable phenomena) rests upon the powerful and sophisticated epistemological argument rehearsed in my discussion of empiricist antirealism. That argument depends upon the evidential indistinguishability thesis. Moreover, the evidential indistinguishability thesis itself is put forward by empiricists (tacitly or explicitly) on the understanding that it captures the truth reflected in the doctrine of knowledge empiricism: the doctrine that all factual knowledge must be grounded in observation. If either knowledge empiricism is basically false, or the indistinguishability thesis represents a seriously misleading interpretation of it, then the empiricist's argument against abduction to theoretical explanation fails.

Second, the empiricist aims at a selectively skeptical account of scientific knowledge: knowledge of unobservables is impossible, but inductive generalizations about observables are sometimes epistemologically legitimate. It turns out, however, that the empiricist's commitment to knowledge empiricism, together with her adoption of the evidential indistinguishability thesis as an interpretation of it, threaten to dictate the unwelcome and implausible conclusion that even inductive inferences regarding observables are always unjustified.

The rebuttals to empiricist antirealism discussed earlier strengthen the case for realism as an account of the structure of scientific knowledge, yet they provide no direct argument either against knowledge empiricism or against the evidential indistinguishability thesis as an interpretation of it. The situation of the abductive argument for scientific realism sketched previously is quite different. If we accept the abductive inference to a distinctly realistic account of scientific methodology, then we can see *why* the evidential indistinguishability thesis is false. Moreover, we can see that the distinctly realistic conception of scientific methodology retains the central core of the doctrine of knowledge empiricism: all factual knowledge *does* depend upon observation; there are no a priori factual statements immune from empirical refutation.

I think it is fair to say that, given the difficulties that plague empiricist antirealism in the philosophy of science, the only philosophically cogent reason for rejecting scientific realism in favor of instrumentalism, or some other variant of empiricism, lies in the conviction that only from an empiricist perspective can one be faithful to the basic idea that factual knowledge must be experimental knowledge, that is, to the grain of truth in knowledge empiricism. The abductive argument for scientific realism that we are considering is best thought of as a component of an alternative realistic conception of scientific knowledge which preserves the empiricist insight that factual knowledge rests on the senses without the cost of an inadequate and potentially wholly skeptical treatment of scientific inquiry.

[...]

Neither the empiricist nor the constructivist can explain the most striking feature of the recent history of science, that is, the instrumental reliability of its methods. Only scientific realism provides the resources for explaining this crucial historical phenomenon. It is for this reason that realism is to be preferred to rival accounts of scientific knowledge, and for this reason that the realist account of semantic and methodological continuity is to be preferred to the alternative account presented in various forms by empiricists and constructivists.

Notes

1 B. van Fraassen, *The Scientific Image* (Oxford: The Clarendon Press, 1980).
2 T. Kuhn, *The Structure of Scientific Revolutions* (Chicago: University of Chicago Press, 1970).
3 N. R. Hanson, *Patterns of Discovery* (Cambridge: Cambridge University Press, 1958).
4 J. J. C. Smart, *Philosophy and Scientific Realism* (London: Routledge and Kegan Paul, 1963).
5 Kuhn, *The Structure of Scientific Revolutions.*
6 J. Bennett, *Locke, Berkeley, Hume* (Oxford: Oxford University Press, 1971).
7 See, for example, G. Maxwell, "The Ontological Status of Theoretical Entities," in *Scientific Explanation, Space and Time*, ed. H. Feigl and G. Maxwell (Minneapolis: University of Minnesota Press, 1963).
8 R. Boyd, "Scientific Realism and Naturalistic Epistemology," *PSA* (1980), Vol. 2, ed. P. D. Asquith and R. N. Giere.
9 R. Boyd, "Metaphor and Theory Change," *Metaphor and Thought*, ed. Andrew Ortony (Cambridge: Cambridge University Press, 1979); "Scientific Realism and Naturalistic Epistemology"; *Realism and Scientific Epistemology.*
10 This is Kuhn's term; see Kuhn, *The Structure of Scientific Revolutions.*
11 Ibid.
12 Boyd, "Metaphor and Theory Change."
13 Kuhn, *The Structure of Scientific Revolutions*, 101–2; see Boyd, "Metaphor and Theory Change," for a discussion.
14 R. Carnap, *Meaning and Necessity* (Chicago: University of Chicago Press, 1950).
15 See M. Schlick, "Positivism and Realism," *Erkenntnis* 3 (1932–33), in *Logical Positivism*, ed. A. J. Ayer (New York: Free Press, 1959), for an anticipation of Carnap's later position.
16 Boyd, "Metaphor and Theory Change," especially 397–8.
17 N. Goodman, *Fact, Fiction and Forecast*, 3d ed. (Indianapolis and New York: Bobbs-Merrill Co., 1973).
18 For further discussion of this issue, see W. V. O. Quine, "Natural Kinds," in W. V. O. Quine, *Ontological, Relativity and Other Essays* (New York: Columbia University Press, 1969); and Boyd, "Metaphor and Theory Change"; R. Boyd, "Materialism without Reductionism: What Physicalism Does Not Entail," in *Readings in Philosophy of Psychology*, Vol. 1, ed. Ned Block (Cambridge: Harvard University Press, 1980); and Boyd, "Scientific Realism and Naturalistic Epistemology."
19 See Boyd, "Scientific Realism and Naturalistic Epistemology," and Boyd, *Realism and Scientific Epistemology*, for discussion.
20 R. Boyd, "Determinism, Laws and Predictability in Principle," *Philosophy of Science* 39 (1972); "Realism, Underdetermination and A Causal Theory of Evidence," *Nous* (March 1973); "Metaphor and Theory Change"; "Scientific Realism and Naturalistic Epistemology"; "Materialism without Reductionism: Non-Humean Causation and the Evidence for Physicalism," in *The Physical Basis of Mind*, ed. R. Boyd (Cambridge, Mass.: Harvard University Press, forthcoming); *Realism and Scientific Epistemology* (Cambridge: Cambridge University Press, forthcoming).
21 Boyd, "Scientific Realism and Naturalistic Epistemology."
22 See Boyd, "Metaphor and Theory Change"; "Materialism without Reductionism: What Physicalism Does Not Entail"; "Scientific Realism and Naturalistic Epistemology"; "Materialism without Reductionism: Non-Humean Causation"; and *Realism and Scientific Epistemology*, for discussion of these points.
23 Boyd, "Scientific Realism and naturalistic Epistemology"; see also Quine, "natural Kinds."
24 H. Feigl, "Some Major Issues and Developments in the Philosophy of Science of Logical Empiricism,"

in *Minnesota Studies in the Philosophy of Science*, Vol. 1, ed. H. Feigl and M. Scriven (Minneapolis: University of Minnesota Press, 1956); S. Kripke, "Naming and Necessity," in *The Semantics of Natural Language*, eds. G. Harman and D. Davidson (Dordrecht: D. Reidel, 1972); H. Putnam, "The Meaning of 'Meaning,'" in *Mind, Language and Reality*, ed. H. Putnam, Philosophical Papers, Vol. 2 (Cambridge: Cambridge University Press, 1975).

25 Boyd, "Metaphor and Theory Change"; "Scientific Realism and Naturalistic Epistemology"; *Realism and Scientific Epistemology*.

26 Boyd, "Metaphor and Theory Change"; "Scientific Realism and Naturalistic Epistemology." [Boyd's treatment of the theory of reference and of natural kinds and magnitudes has undergone considerable development since this paper was published. See, e.g. Boyd, 1999, "Homeostasis, Species, and Higher Taxa," in R. Wilson (ed.), *Species: New Interdisciplinary Essays*, Cambridge: MIT Press, and Boyd, 2001, "Truth Through Thick and Thin," in Richard Schantz (ed.), *What is Truth?* Berlin and New York: Walter de Gruyter.]

27 Boyd, "Metaphor and Theory Change."

28 H. Field, "Thoery Change and the Indeterminacy of Reference," *Journal of Philosophy* 70 (1973).

29 E. Nagel, *The Structure of Science* (New York: Harcourt Brace, 1961), 151–2.

30 For a discussion of limitations of this process of successive approximation see Boyd, "Scientific Realism and Naturalistic Epistemology."

31 See ibid., for a discussion of the relation between scientific realism and other recent naturalistic trends in epistemology.

32 See "The Natural Ontological Attitude" in this volume. I am extremely grateful to Professor Fine for the opportunity to read a prepublication copy of this paper.

9.2

A Confutation of Convergent Realism

Larry Laudan

Although a staunch critic of relativist and constructivist versions of anti-realism, Larry Laudan (b. 1941) is equally a critic of realism. In the selection below he presents his argument, now popularly referred to as the *pessimistic meta-induction*. According to this argument, the history of science provides many examples of theories that were both successful and that have been rejected as false by successor theories. Since we have no reason not to think that the same fate awaits our present theories, we should concede that it will also be counted as false by future theories. The empirical success of those present theories, therefore, does not provide the grounds for belief in their approximate truth as the realist suggests.

The positive argument for realism is that it is the only philosophy that doesn't make the success of science a miracle.

H. Putnam (1975)

The Problem

It is becoming increasingly common to suggest that epistemological realism is an empirical hypothesis, grounded in, and to be authenticated by, its ability to explain the workings of science. A growing number of philosophers (including Boyd, W. Newton-Smith, A. Shimony, Putnam, and I. Niiniluoto) have argued that the theses of epistemic realism are open to empirical test.[1] The suggestion that epistemological doctrines have much the same empirical status as the sciences is a welcome one; for, whether it stands up to detailed scrutiny or not, this suggestion marks a significant facing-up by the philosophical community to one of the most neglected (and most notorious) problems of philosophy: the status of epistemological claims.

There are, however, potential hazards as well as advantages associated with the "scientizing" of

From *Philosophy of Science* 48 (1981), pp. 19–49.

epistemology. Specifically, once one concedes that epistemic doctrines are to be tested in the court of experience, it is possible that one's favorite epistemic theories may be refuted rather than confirmed. It is the thesis of this paper that precisely such a fate afflicts a form of realism advocated by those who have been in the vanguard of the move to show that realism is supported by an empirical study of the development of science. Specifically, I will show that epistemic realism, at least in certain of its extant forms, is neither supported by, nor has it made sense of, much of the available historical evidence.

Convergent Realism

Like other philosophical "isms," the term 'realism' covers a variety of sins. Many of these will not be at issue here. For instance, 'semantic realism' (in brief, the claim that all theories have truth values and that some theories are true, although we know not which) is not in dispute. Nor shall I discuss what one might call 'intentional realism' (i.e., the view that theories are generally intended by their proponents to assert the existence of entities corresponding to the terms in those theories). What I will focus on instead are certain forms of epistemological realism. As Hilary Putnam has pointed out, although such realism has become increasingly fashionable, "very little is said about what realism *is*." The lack of specificity about what realism asserts makes it difficult to evaluate its claims, since many formulations are too vague and sketchy to get a grip on. At the same time, any efforts to formulate the realist position with greater precision lay the critic open to charges of attacking a straw man. In the course of this paper, I shall attribute several theses to the realists. Although there is probably no realist who subscribes to all of them, most of them have been defended by some self-avowed realist or other; taken together, they are perhaps closest to that version of realism advocated by Putnam, Boyd, and Newton-Smith. Although I believe the views I shall be discussing can be legitimately attributed to certain contemporary philosophers (and I will cite the textual evidence for such attributions), it is not crucial to my case that such attributions can be made. Nor will I claim to do justice to the complex epistemologies of those whose work I will criticize. Rather, my aim is to explore certain epistemic claims which those who are realists might be tempted (and in some cases have been tempted) to embrace. If my arguments are sound, we will discover that some of the most intuitively tempting versions of realism prove to be chimeras.

The form of realism I shall discuss involves variants of the following claims:

(Rl) Scientific theories (at least in the 'mature' sciences) are typically approximately true, and more recent theories are closer to the truth than older theories in the same domain.

(R2) The observational and theoretical terms within the theories of a mature science genuinely refer (roughly, there are substances in the world that correspond to the ontologies presumed by our best theories).

(R3) Successive theories in any mature science will be such that they preserve the theoretical relations and the apparent referents of earlier theories, that is, earlier theories will be limiting cases of later theories.[2]

(R4) Acceptable new theories do and should explain why their predecessors were successful insofar as they were successful.

To these semantic, methodological, and epistemic theses is conjoined an important metaphilosophical claim about how realism is to be evaluated and assessed. Specifically, it is maintained that:

(R5) Theses (Rl) to (R4) entail that ('mature') scientific theories should be successful; indeed, these theses constitute the best, if not the only, explanation for the success of science. The empirical success of science (in the sense of giving detailed explanations and accurate predictions) accordingly provides striking empirical confirmation for realism.

I shall call the position delineated by (Rl) to (R5) *convergent epistemological realism*, or CER for short. Many recent proponents of CER maintain that (Rl), (R2), (R3), and (R4) are empirical hypotheses that, via the linkages postulated in (R5), can be tested by an investigation of science itself. They propose two elaborate abductive arguments. The structure of the first (argument 1) which is germane to (Rl), is something like this:

1 If scientific theories are approximately true, then they typically will be empirically successful.

2 If the central terms in scientific theories genuinely refer, then those theories generally will be empirically successful.
3 Scientific theories are empirically successful.
4 (Probably) theories are approximately true and their terms genuinely refer.

The structure of the second abductive argument (argument 2), which is relevant to (R3), is of slightly different form, specifically:

1 If the earlier theories in a "mature" science are approximately true, and if the central terms of those theories genuinely refer, then later, more successful theories in the same science will preserve the earlier theories as limiting cases.
2 Scientists seek to preserve earlier theories as limiting cases and generally succeed in doing so.
3 (Probably) earlier theories in a "mature" science are approximately true and genuinely referential.

Taking the success of present and past theories as givens, proponents of CER claim that *if* CER were true, it would follow, as a matter of course, that science would be successful and progressive. Equally, they allege that if CER were false, the success of science would be "miraculous" and without explanation.[3] Because (on their view) CER explains the fact that science is successful, the theses of CER are thereby confirmed by the success of science, and nonrealist epistemologies are discredited by the latter's alleged inability to explain both the success of current theories and the progress which science historically exhibits.

As Putnam and certain others (e.g., Newton-Smith) see it, the fact that statements about reference (R2, R3) or about approximate truth (Rl, R3) function in the explanation of a contingent state of affairs, establishes that "the notions of 'truth' and 'reference' have a causal explanatory role in epistemology."[4] In one fell swoop, both epistemology and semantics are 'naturalized' and, to top it all off, we get an explanation of the success of science thrown into the bargain!

The central question before us is whether the realist's assertions about the interrelations between truth, reference, and success are sound. It will be the burden of this paper to raise doubts about both arguments 1 and 2. Specifically, I will argue that four of the five premises of those abductions are either false or too ambiguous to be acceptable. I will also seek to show that, even if the premises were true, they would not warrant the conclusions that realists draw from them. The next three sections of this essay deal with the first abductive argument. Then I turn to the second.

Reference and Success

The specifically referential side of the empirical argument for realism has been developed chiefly by Putnam, who talks explicitly of reference rather more than most realists. However, reference is usually implicitly smuggled in, since most realists subscribe to the (ultimately referential) thesis that "the world probably contains entities very like those postulated by our most successful theories."

If (R2) is to fulfill Putnam's ambition that reference can explain the success of science, and that the success of science establishes the presumptive truth of (R2), it seems he must subscribe to claims similar to these:

(S1) The theories in the advanced or mature sciences are successful.

(S2) A theory whose central terms genuinely refer will be a successful theory.

(S3) If a theory is successful, we can reasonably infer that its central terms genuinely refer.

(S4) All the central terms in theories in the mature sciences do refer.

There are complex interconnections here. (S2) and (S4) explain (S1), while (S1) and (S3) provide the warrant for (S4). Reference explains success, and success warrants a presumption of reference. The arguments are plausible, given the premises. But there is the rub for with the possible exception of (S1), none of the premises is acceptable.

The first and toughest problem involves getting clearer about the nature of that "success" which realists are concerned to explain. Although Putnam, W. Sellars, and Boyd all take the success of certain sciences as a given, they say little about what this success amounts to. So far as I can see, they are working with a largely *pragmatic* notion to be couched in terms of a theory's workability or applicability. On this account, we would say that a theory is successful if it makes substantially correct predictions, if it leads to efficacious interventions in the natural order, and if it passes a battery of standard tests. One would like to be able to be more specific about what success amounts to, but the lack of a coherent theory of confirmation makes further specificity very difficult.

Moreover, the realist must be wary, at least for these purposes, of adopting too strict a notion of success, for a highly robust and stringent construal of "success" would defeat the realist's purposes. What he wants to explain, after all, is why science in general has worked so well. If he were to adopt a very demanding characterization of success (such as those advocated by inductive logicians or Popperians), then it would probably turn out that science has been largely "unsuccessful" (because it does not have high confirmation), and the realist's avowed explanandum would thus be a nonproblem. Accordingly, I will assume that a theory is successful so long as it has worked well that is, so long as it has functioned in a variety of explanatory contexts, has led to confirmed and has been of broad explanatory scope. As I understand the realist's position, his concern is to explain why certain theories have enjoyed this kind of success.

If we construe 'success' in this way, (S1) can be conceded. Whether one's criterion of success is broad explanatory scope, possession of a large number of confirming instances, or conferring manipulative or predictive control, it is clear that science, by and large, is a successful activity.

What about (S2)? I am not certain that any realist would or should endorse it, although it is a perfectly natural construal of the realist's claim that "reference explains success." The notion of reference that is involved here is highly complex and unsatisfactory in significant respects. Without endorsing it, I shall use it frequently in the ensuing discussion. The realist sense of reference is a rather liberal one according to which the terms in a theory may be genuinely referring even if many of the claims the theory makes about the entities to which it refers are false. Provided that there are entities that "approximately fit" a theory's description of them, Putnam's charitable account of reference allows us to say that the terms of a theory genuinely refer.[5] On this account (and these are Putnam's examples), Bohr's 'electron', Newton's 'mass', Mendel's 'gene', and Dalton's 'atom' are all referring terms, while 'phlogiston' and 'ether' are not.[6]

Are genuinely referential theories (i.e., theories whose central terms genuinely refer) invariably or even generally successful at the empirical level, as (S2) states? There is ample evidence that they are not. The chemical atomic theory in the eighteenth century was so remarkably unsuccessful that most chemists abandoned it in favor of a more phenomenological, elective affinity chemistry. The Proutian theory that the atoms of heavy elements are composed of hydrogen atoms had, through most of the nineteenth century, a strikingly unsuccessful career, confronted by a long string of apparent refutations. The Wegenerian theory that the continents are carried by large subterranean objects moving laterally across the earth's surface was, for some thirty years in the recent history of geology, a strikingly unsuccessful theory until, after major modifications, it became the geological orthodoxy of the 1960s and 1970s. Yet all of these theories postulated basic entities which (according to Putnam's "principle of charity") genuinely exist.

The realist's claim that we should expect referring theories to be empirically successful is simply false. And, with a little reflection, we can see good reasons why it should be. To have a genuinely referring theory is to have a theory that "cuts the world at its joints, a theory that postulates entities of a kind that really exist. But a genuinely referring theory need not be such that all – or even most – of the specific claims it makes about the properties of those entities and their modes of interaction are true. Thus, Dalton's theory makes many false claims about atoms; Bohr's early theory of the electron was similarly flawed in important respects. Contra-(S2), genuinely referential theories need not be strikingly successful, since such theories may be 'massively false' (i.e., have far greater falsity content than truth content).

(S2) is so patently false that it is difficult to imagine that the realist need be committed to it. But what else will do? The (Putnamian) realist wants attributions of reference to a theory's terms to function in an explanation of that theory's success. The simplest and crudest way of doing that involves a claim like (S2). A less outrageous way of achieving the same end would involve the weaker:

> (S2′) A theory whose terms refer will usually (but not always) be successful.

Isolated instances of referring but unsuccessful theories, sufficient to refute (S2), leave (S2′) unscathed. But, if we were to find a broad range of referring but unsuccessful theories, that would be evidence against (S2′). Such theories can be generated at will. For instance, take any set of terms which one believes to be genuinely referring. In any language rich enough to contain negation, it will be possible to construct indefinitely many unsuccessful theories, all of whose substantive terms are genuinely referring. Now, it is always open to the realist to claim that such "theories" are not really theories at all, but mere conjunctions of isolated statements – lacking that sort of conceptual integration we associate with "real" theories. Sadly, a parallel argument can be made for genuine theories. Consider, for instance, how many inadequate versions of the atomic theory there were in the 2,000 years of atomic speculating, before a genuinely successful theory emerged. Consider how many unsuccessful versions there were of the wave theory of light before the 1820s, when a successful wave theory first emerged. Kinetic theories of heat in the seventeenth and eighteenth century, and developmental theories of embryology before the late nineteenth century, sustain a similar story. (S2′), every bit as much as (S2), seems hard to reconcile with the historical record.

As Richard Burian has pointed out to me (personal communication), a realist might attempt to dispense with both of those theses and simply rest content with (S3) alone. Unlike (S2) and (S2′), (S3) is not open to the objection that referring theories are often unsuccessful, for it makes no claim that referring theories are always or generally successful. But (S3) has difficulties of its own. In the first place it seems hard to square with the fact that the central terms of many relatively successful theories (e.g., ether theories or phlogistic theories) are evidently nonreferring. I shall discuss this tension in detail below. More crucial for our purposes here is that (S3) is *not strong enough* to permit the realist to utilize reference to explain success. Unless genuineness of reference entails that all or most referring theories will be successful, then the fact that a theory's terms refer scarcely provides a convincing explanation of that theory's success. If, as (S3) allows, many (or even most) referring theories can be unsuccessful, how can the fact that a successful theory's terms refer be taken to explain why it is successful? (S3) may or may not be true but in either case it arguably gives the realist no explanatory access to scientific success.

A more plausible construal of Putnam's claim that reference plays a role in explaining the success of science involves a rather more indirect argument. It might be said (and Putnam does say this much) that we can explain why a theory is successful by assuming that the theory is true or approximately true. Since a theory can only be true or nearly true (in any sense of those terms open to the realist) if its terms genuinely refer, it might be argued that reference gets into the act willy-nilly when we explain a theory's success in terms of its truthlike status. On this account, reference is piggybacked on approximate truth. The viability of this indirect approach is treated at length in the next section, so I will not discuss it here except to observe that if the only contact point between reference and success is provided through the medium of approximate truth, then the link between reference and success is extremely tenuous.

What about (S3), the realist's claim that success creates a rational presumption of reference? We have already seen that (S3) provides no explanation of the success of science, but does it have independent merits? The question specifically is whether the success of a theory provides a warrant for concluding that its central terms refer. Insofar as this is, as certain realists suggest, an empirical question, it requires us to inquire whether past theories which have been successful are ones whose central terms genuinely referred (according to the realist's own account of reference).

A proper empirical test of this hypothesis would require an extensive sifting of the historical record that is not possible to perform here. What I can do is to mention a range of once successful, but (by present lights) nonreferring, theories. A fuller list will come later, but for now we will focus on a whole family of related theories, namely, the subtle fluids and ethers of eighteenth- and nineteenth-century physics and chemistry.

Consider specifically the state of etherial theories in the 1830s and 1840s. The electrical fluid, a substance that was generally assumed to accumulate on the surface rather than to permeate the interstices of bodies, had been utilized to explain inter alia the attraction of oppositely charged bodies, the behavior of the Leyden jar, the similarities between atmospheric and static electricity, and many phenomena of current electricity. Within chemistry and heat theory, the caloric ether had been widely utilized since H. Boerhaave (by, among others, A. L. Lavoisier, P. S. Laplace, J. Black, Count Rumford, J. Hutton, and H. Cavendish) to explain everything from the role of heat in chemical reactions to the conduction and radiation of heat and several standard problems of thermometry. Within the theory of light, the optical ether functioned centrally in explanations of reflection, refraction, interference, double refraction, diffraction, and polarization. (Of more than passing interest, optical ether theories had also made some very startling predictions, e.g., A. Fresnel's prediction of a bright spot at the center of the shadow of a circular disc; a surprising prediction which, when tested, proved correct. If that does not count as empirical success, nothing does!) There were also gravitational (e.g., G. LeSage's) and physiological (e.g., D. Hartley's) ethers which enjoyed some measure of empirical success. It would be difficult to find a family of theories in this period as successful as ether theories; compared with them, nineteenth-century atomism (for instance), a genuinely referring theory (on realist accounts), was a dismal failure. Indeed, on any account of empirical success which I can conceive of, nonreferring nineteenth-century ether theories were more successful than contemporary, referring atomic theories. In this connection, it is worth recalling the remark of the great theoretical physicist J. C. Maxwell to the effect that the ether was better confirmed than any other theoretical entity in natural philosophy.

What we are confronted with in nineteenth-century ether theories, then, is a wide variety of once successful theories, whose central explanatory concept Putnam singles out as a prime example of a nonreferring one.[7] What are (referential) realists to make of this historical case? On the face of it, this case poses two rather different kinds of challenges to realism: first, it suggests that (S3) is a dubious piece of advice in that *there can be* (and have been) *highly successful theories some central terms of which are nonreferring*, and second, it suggests that *the realist's claim that he can explain why science is successful is false at least insofar as a part of the historical success of science has been success exhibited by theories whose central terms did not refer*.

But perhaps I am being less than fair when I suggest that the realist is committed to the claim that *all* the central terms in a successful theory refer. It is possible than when Putnam, for instance, says that "terms in a mature [or successful] science typically refer,"[8] he only means to suggest that *some* terms in a successful theory or science genuinely refer. Such a claim is fully consistent with the fact that certain other terms (e.g., 'ether') in certain successful, mature sciences (e.g., nineteenth-century physics) are nonetheless nonreferring. Put differently, the realist might argue that the success of a theory warrants the claim that at least some (but not necessarily all) of its central concepts refer.

Unfortunately, such a weakening of (S3) entails a theory of evidential support which can scarcely give comfort to the realist. After all, part of what separates the realist from the positivist is the former's belief that the evidence for a theory is evidence for *everything* the theory asserts. Where the stereotypical positivist argues that the evidence selectively confirms only the more 'observable' parts of a theory, the realist generally asserts (in the language of Boyd) that:

> the sort of evidence which ordinarily counts in favor of the acceptance of a scientific law or theory is, ordinarily, evidence for the (at least approximate) truth of the law or theory as an account of the causal relations obtaining between the entities ["observational or theoretical"] quantified over in the law or theory in question.[9]

For realists such as Boyd, either all parts of a theory (both observational and nonobservational) are confirmed by successful tests or none are. In general, realists have been able to utilize various holistic arguments to insist that it is not merely the lower-level claims of a well-tested theory that are confirmed but its deep-structural assumptions as well. This tactic has been used to good effect by realists in establishing that inductive support 'flows upward' so as to authenticate the most 'theoretical' parts of our theories. Certain latter-day realists (e.g., Glymour) want to break out of this holist web and argue that certain components of theories can be 'directly' tested. This approach runs the very grave risk of undercutting what the realist desires most: a rationale for taking our deepest-structure theories seriously, and a justification for linking reference and success. After all, if the tests to which we subject our theories only test *portions* of those theories, then even highly successful theories may well have central terms that are nonreferring and central tenets that, because untested, we have no grounds for believing to be approximately true. Under those circumstances, a theory might be highly successful and yet contain important constituents that were patently false. Such a state of affairs would wreak havoc with the realist's presumption (thesis R1) that success betokens approximate truth. In short, to be less than a holist about theory testing is to put at risk precisely that predilection for deep-structure claims which motivates much of the realist enterprise.

There is, however, a rather more serious obstacle to this weakening of referential realism. It is true that by weakening (S3) to only certain terms in a theory, one would immunize it from certain obvious counterexamples. But such a maneuver has debilitating consequences for other central realist theses. Consider the realist's thesis (R3) about the retentive character of intertheory relations (discussed below in detail). The realist both recommends as a matter of policy and claims as a matter of fact that successful theories are (and should be) rationally replaced only by theories that preserve reference for the central terms of their successful predecessors. The rationale for the normative version of this retentionist doctrine is that the terms in the earlier theory, *because it was successful, must have been referential*, and thus a constraint on any successor to that theory is that reference should be retained for such terms. This makes sense just in case success provides a blanket warrant for presumption of reference. But if (S3) were weakened so as to say merely that it is reasonable to assume that *some* of the terms in a successful theory genuinely refer, then the realist would have no rationale for his retentive theses (variants of R3), which have been a central pillar of realism for several decades.[10]

Something apparently has to give. A version of (S3) strong enough to license (R3) seems incompatible with the fact that many successful theories contain nonreferring central terms. But any weakening of (S3) dilutes the force of, and removes the rationale for, the realist's claims about convergence, retention, and correspondence in intertheory relations.[11] If the realist once concedes that some unspecified set of the terms of a successful theory may well not refer, then his proposals for restricting "the class of candidate theories" to those that retain reference for the prima facie referring terms in earlier theories is without foundation.[12]

More generally, we seem forced to say that such linkages as there are between reference and success are rather murkier than Putnam's and Boyd's discussions would lead us to believe. If the realist is going to make his case for CER, it seems that it will have to hinge on approximate truth, (R1), rather than reference, (R2).

Approximate Truth and Success: The Downward Path

Ignoring the referential turn among certain recent realists, most realists continue to argue that, at bottom, epistemic realism is committed to the view that successful scientific theories, even if strictly false, are nonetheless 'approximately true' or 'close to the truth' or 'verisimilar'.[13] The claim generally amounts to this pair:

(T1) If a theory is approximately true, then it will be explanatorily successful.

(T2) If a theory is explanatorily successful, then it is probably approximately true.

What the realist would *like* to be able to say, of course, is:

(T1′) If a theory is true, then it will be successful.

(T1′) is attractive because self-evident. Most realists, however, balk at invoking (T1′) because they are (rightly) reluctant to believe that we can reasonably presume of any given scientific theory that it is true. If all the realist could explain was the success of theories that were true *simpliciter*, his explanatory repertoire would be acutely limited. As an attractive move in the direction of broader explanatory scope, (T1) is rather more appealing. After all, presumably many theories which we believe to be false (e.g., Newtonian mechanics, thermodynamics, wave optics) were – and still are – highly successful across a broad range of applications.

Perhaps, the realist evidently conjectures, we can find an *epistemic* account of that pragmatic success by assuming such theories to be 'approximately true'. But we must be wary of this potential sleight of hand. It may be that there is a connection between success and approximate truth; *but if there is such a connection it must be independently argued for*. The acknowledgedly uncontroversial character of (T1′) must not be surreptitiously invoked – as it sometimes seems to be – in order to establish T1. When the antecedent of (T1′) is appropriately weakened by speaking of approximate truth, it is by no means clear that (T1) is sound.

Virtually all the proponents of epistemic realism take it as unproblematic that if a theory were approximately true, it would deductively follow that the theory would be a relatively successful predictor and explainer of observable phenomena. Unfortunately, few of the writers of whom I am aware have defined what it means for a statement or theory to be 'approximately true'. Accordingly, it is impossible to say whether the alleged entailment is genuine. This reservation is more than perfunctory. Indeed, on the best known account of what it means for a theory to be approximately true, it does *not* follow that an approximately true theory will be explanatorily successful.

Suppose, for instance, that we were to say in a Popperian vein that a theory, T_1, is approximately true if its truth content is greater than its falsity content, that is,

$$Ct_T(T_1) \gg Ct_F(T_1),$$[14]

where $Ct_T(T_1)$ is the cardinality of the set of true sentences entailed by T_1, and $Ct_F(T_1)$ is the cardinality of the set of false sentences entailed by T_1. When approximate truth is so construed, it does *not* logically follow that an arbitrarily selected class of a theory's entailments (namely, some of its observable consequences) will be true. Indeed, it is entirely conceivable that a theory might be approximately true in the indicated sense and yet be such that *all* of its consequences tested thus far are *false*.[15]

Some realists concede their failure to articulate a coherent notion of approximate truth or verisimilitude but insist that this failure in no way compromises the viability of (T1). Newton-Smith, for instance, grants that "no one has given a satisfactory analysis of the notion of verisimilitude,"[16] but insists that the concept can be legitimately invoked "even if one cannot at the time give a philosophically satisfactory analysis of it." He quite rightly points out that many scientific concepts were explanatorily useful long before a philosophically coherent analysis was given for them. But the analogy is unseemly, for what is being challenged is not whether the concept of approximate truth is philosophically rigorous, but rather whether it is even clear enough for us to ascertain whether it entails what it purportedly explains. Until someone provides a clearer analysis of approximate truth than is now available, it is not even clear whether truthlikeness would explain success, let alone whether, as Newton-Smith insists,[17] "the concept of verisimilitude is *required* in order to give a satisfactory theoretical explanation of an aspect of the scientific enterprise." If the realist would demystify the "miraculousness" (Putnam) or the "mysteriousness" (Newton-Smith) of the success of science, he needs more than a promissory note that somehow, someday, someone will show that approximately true theories must be successful theories.[18]

It is not clear whether there is some definition of approximate truth that does indeed entail that approximately true theories will be predictively successful (and yet still probably false).[19] What can be said is that, promises to the contrary notwithstanding, none of the proponents of realism has yet articulated a coherent account of approximate truth which entails that approximately true theories will, across the range where we can test them, be successful predictors. Further difficulties abound. Even if the realist had a semantically adequate characterization of approximate or partial truth, and even if that semantics entailed that most of the consequences of an approximately true theory would be true, he would still be without any criterion that would *epistemically* warrant the ascription of approximate truth to a theory. As it is, the realist seems to be long on intuitions and short on either a semantics or an epistemology of approximate truth.

These should be urgent items on the realists' agenda since, until we have a coherent account of what approximate truth is, central realist theses such as (R1), (T1), and (T2) are just so much mumbo jumbo.

Approximate Truth and Success: The Upward Path

Despite the doubts voiced in the previous section, let us grant for the sake of argument that if a theory is approximately true, then it will be successful. Even granting (T1), is there any plausibility to the suggestion of (T2) that explanatory success can be taken as a rational warrant for a judgment of approximate truth? The answer seems to be "no."

To see why, we need to explore briefly one of the connections between "genuinely referring" and being "approximately true." However the latter is understood, I take it that *a realist would never want to say that a theory was approximately true if its central terms failed to refer*. If there were nothing like genes, then a genetic theory, no matter how well confirmed it was, would not be approximately true. If there were no entities similar to atoms, no atomic theory could be approximately true; if there were no subatomic particles, then no quantum theory of chemistry could be approximately true. In short, a necessary condition, especially for a scientific realist, for a theory being close to the truth is that its central explanatory terms genuinely refer. (An *instrumentalist*, of course, could countenance the weaker claim that a theory was approximately true so long as its directly testable consequences were close to the observable values. But as I argued above, the realist must take claims about approximate truth to refer alike to the observable and the deep-structural dimensions of a theory.)

Now, what the history of science offers us is a plethora of theories that were both successful and (so far as we can judge) nonreferential with respect to many of their central explanatory concepts. I discussed earlier one specific family of theories that fits this description. Let me add a few more prominent examples to the list:

- the crystalline spheres of ancient and medieval astronomy;
- the humoral theory of medicine;
- the effluvial theory of static electricity;
- "catastrophist" geology, with its commitment to a universal (Noachian) deluge;
- the phlogiston theory of chemistry;
- the caloric theory of heat;
- the vibratory theory of heat;
- the vital force theories of physiology;
- the electromagnetic ether;
- the optical ether;
- the theory of circular inertia; and
- theories of spontaneous generation.

This list, which could be extended ad nauseam, involves in every case a theory that was once successful and well confirmed, but which contained central terms that (we now believe) were nonreferring. Anyone who imagines that the theories that have been successful in the history of science have also been, with respect to their central concepts, genuinely referring theories has studied only the more whiggish versions of the history of science (i.e., the ones which recount only those past theories that are referentially similar to currently prevailing ones).

It is true that proponents of CER sometimes hedge their bets by suggesting that their analysis applies exclusively to "the mature sciences" (e.g., Putnam and W. Krajewski). This distinction between mature and immature sciences proves convenient to the realist since he can use it to dismiss any prima facie counterexample to the empirical claims of CER on the grounds that the example is drawn from a so-called immature science. But this insulating maneuver is unsatisfactory in two respects. In the first place, it runs the risk of making CER vacuous since these authors generally define a mature science as one in which correspondence or limiting-case relations obtain invariably between any successive theories in the science once it has passed "the threshold of maturity." Krajewski grants the tautological character of this view when he notes that "the thesis that there is [correspondence] among successive theories becomes, indeed, analytical."[20]

Nonetheless, he believes that there is a version of the maturity thesis which "may be and must be tested by the history of science." That version is that "every branch of science crosses at some period the threshold of maturity." But the testability of this hypothesis is dubious at best. There is no historical observation that could conceivably *refute* it since, even if we discovered that no sciences yet possessed "corresponding" theories, it could be maintained that eventually every science will become corresponding. It is equally difficult to *confirm* it since, even if we found a science in which corresponding relations existed between the latest theory and its predecessor, we would have no way of knowing whether that relation will continue to apply to subsequent changes of theory in that science. In other words, the much-vaunted empirical testability of realism is seriously compromised by limiting it to the mature sciences.

But there is a second unsavory dimension to the restriction of CER to the mature sciences. The realists' avowed aim, after all, is to explain why science is successful: that is the "miracle" they allege the nonrealists leave unaccounted for. The fact of the matter is that parts of science, including many immature sciences, have been successful for a very long time; indeed, many of the theories I alluded to above were empirically successful by any criterion I can conceive of (including fertility, intuitively high confirmation, successful prediction, etc.). If the realist restricts himself to explaining only how the mature sciences work (and recall that very few sciences indeed are yet mature as the realist sees it), then he will have completely failed in his ambition to explain why science in general is successful. Moreover, several of the examples I have cited above come from the history of mathematical physics in the last century (e.g., electromagnetic and optical ethers) and, as Putnam himself concedes, "*physics* surely counts as a 'mature' science if any science does."[21] Since realists would presumably insist that many of the central terms of the theories enumerated above do not genuinely refer, it follows that none of those theories could be approximately true (recalling that the former is a necessary condition for the latter). Accordingly, cases of this kind cast very grave doubts on the plausibility of (T2), that is, the claim that nothing succeeds like approximate truth.

I daresay that for every highly successful theory in the history of science that we now believe to be a genuinely referring theory, one could find half a dozen once successful theories that we now regard as substantially nonreferring. If the proponents of CER are the empiricists they profess to be about matters epistemological, cases of this kind and this frequency should give them pause about the well-foundedness of (T2).

But we need not limit our counterexamples to nonreferring theories. There were many theories in the past that (so far as we can tell) were both genuinely referring and empirically successful which we are nonetheless loathe to regard as approximately true. Consider, for instance, virtually all those geological theories prior to the 1960s which denied any lateral motion to the continents. Such theories were, by any standard, highly successful (and apparently referential); but would anyone today be prepared to say that their constituent theoretical claims – committed as they were to laterally stable continents – are almost true? Is it not the fact of the matter that structural geology was a successful science between (say) 1920 and 1960, even though geologists were fundamentally mistaken about many (perhaps even most) of the basic mechanisms of tectonic construction? Or what about the chemical theories of the 1920s which assumed that the atomic nucleus was structurally homogenous? Or those chemical and physical theories of the late nineteenth century which explicitly assumed that matter was neither created nor destroyed? I am aware of no sense of approximate truth (available to the realist) according to which such highly successful, but evidently false, theoretical assumptions could be regarded as "truthlike."

More generally, the realist needs a riposte to the prima facie plausible claim that there is no necessary connection between increasing the accuracy of our deep-structural characterizations of nature and improvements at the level of phenomenological explanations, predictions, and manipulations. It *seems* entirely conceivable intuitively that the theoretical mechanisms of a new theory, T_2, might be closer to the mark than those of a rival T_1, and yet T_1 might be more accurate at the level of testable predictions. In the absence of an argument that greater correspondence at the level of unobservable claims is more likely than not to reveal itself in greater accuracy at the experimental level, one is obliged to say that the realist's hunch that increasing deep-structural fidelity must manifest itself pragmatically in the form of heightened experimental accuracy has yet to be made cogent. (Equally problematic, of course, is the inverse argument to the effect that increasing experimental accuracy betokens greater truthlikeness at the level of theoretical, i.e., deep-structural, commitments.)

Confusions About Convergence and Retention

Thus far, I have discussed only the static or synchronic versions of CER, versions that make absolute rather than relative judgments about truth-likeness. Of equal appeal have been those variants of CER that invoke a notion of what is variously called "convergence," "correspondence," or "cumulation." Proponents of the diachronic version of CER supplement the arguments discussed above [(S1)-(S4) and (T1)-(T2)] with an additional set. They tend to be of this form:

(C1) If earlier theories in a scientific domain are successful and thereby, according to realist principles [e.g., (S3) above], approximately true, then scientists should only accept later theories that retain appropriate portions of earlier theories.

(C2) As a matter of fact, scientists do adopt the strategy of (C1) and manage to produce new, more successful theories in the process.

(C3) The "fact" that scientists succeed at retaining appropriate parts of earlier theories in more successful successors shows that the earlier theories did genuinely refer and that they are approximately true. And thus, the strategy propounded in (C1) is sound.[22]

Perhaps the prevailing view here is Putnam's and (implicitly) Popper's, according to which rationally warranted successor theories in a mature science must contain reference to the entities apparently referred to in the predecessor theory (since, by hypothesis, the terms in the earlier theory refer), and also contain the theoretical laws and mechanisms of the predecessor theory as limiting cases. As Putnam tells us, a realist should insist that *any* viable successor to an old theory T_1 must "contain the laws of T as a limiting case."[23] John Watkins, a like-minded convergentist, puts the point this way:

> It typically happens in the history of science that when some hitherto dominant theory T is superceded by T^1, T^1 is in the relation of correspondence to T [i.e., T is a 'limiting case' of T^1].[24]

Numerous recent philosophers of science have subscribed to a similar view, including Popper, H. R. Post, Krajewski, and N. Koertge.[25]

This form of retention is not the only one to have been widely discussed. Indeed, realists have espoused a wide variety of claims about what is or should be retained in the transition from a once successful predecessor (T_1) to a successor theory (T_2). Among the more important forms of realist retention are the following cases: (1) T_2 entails T_1 (W. Whewell); (2) T_2 retains the true consequences or truth content of T_1 (Popper); (3) T_2 retains the "confirmed" portions of T_2 (Post, Koertge); (4) T_2 preserves the theoretical laws and mechanisms of T_1 (Boyd, McMullin, Putnam); (5) T_2 preserves T_1 as a limiting case (J. Watkins, Putnam, Krajewski); (6) T_2 explains why T_1 succeeded insofar as T_1 succeeded (W. Sellars); and (7) T_2 retains reference for the central terms of T_1 (Putnam, Boyd).

The question before us is whether, when retention is understood in *any* of these senses, the realist's theses about convergence and retention are correct.

Do scientists adopt the retentionist strategy of CER?

One part of the convergent realist's argument is a claim to the effect that scientists generally adopt the strategy of seeking to preserve earlier theories in later ones. As Putnam puts it:

> preserving the *mechanisms* of the earlier theory as often as possible, which is what scientists try to do. . . . That scientists try to do this . . . is a fact, and that this strategy has led to important discoveries . . . is also a fact.[26]

In a similar vein, I. Szumilewicz (although not stressing realism) insists that many eminent scientists made it a main heuristic requirement of their research programs that a new theory stand in a relation of 'correspondence' with the theory it supersedes.[27] If Putnam and the other retentionists are right about the strategy that most scientists have adopted, we should expect to find the historical literature of science abundantly provided with proofs that later theories do indeed contain earlier theories as limiting cases, or outright rejections of later theories that fail to contain earlier theories. Except on rare occasions (coming primarily from the history of mechanics), one finds neither of these concerns prominent in the literature of science. For instance, to the best of my knowledge, literally no one criticized the wave theory of light because it did not preserve the theoretical mechanisms of the earlier corpuscular theory; no one faulted C. Lyell's uniformitarian geology on the grounds that it dispensed with several causal processes prominent in catastrophist geology; Darwin's theory was not criticized by most geologists for its failure to retain many of the mechanisms of Lamarckian evolutionary theory.

For all the realist's confident claims about the prevalence of a retentionist strategy in the sciences, I am aware of *no* historical studies that would sustain as a *general* claim this hypothesis about the evaluative strategies utilized in science. Moreover, insofar as Putnam and Boyd claim to be offering "an explanation of the retentionist behavior of scientists,"[28] they have the wrong explanandum, for if there is any widespread strategy in science, it is one that says, "accept an empirically successful theory, regardless of whether it contains the theoretical laws and mechanisms of its predecessors."[29] Indeed, one could take a leaf from the realist's (C2) and claim that the success of the strategy of assuming that earlier theories do not generally refer shows that it is true that earlier theories generally do not!

(One might note in passing how often, and on what evidence, realists imagine that they are speaking for the scientific majority. Putnam, for instance, claims that "realism is, so to speak, 'science's philosophy of science'" and that "science taken at 'face value' *implies* realism."[30] C. A. Hooker insists that to be a realist is to take science "seriously,"[31] as if to suggest that conventionalists, instrumentalists, and positivists such as Duhem, Poincaré, and Mach did not take science seriously. The willingness of some realists to attribute realist strategies to working scientists – on the strength of virtually no empirical research into the principles which *in fact* have governed scientific practice – raises doubts about the seriousness of their avowed commitment to the empirical character of epistemic claims.)

Do later theories preserve the mechanisms, models, and laws of earlier theories?

Regardless of the explicit strategies to which scientists have subscribed, are Putnam and several other retentionists right that later theories "typically" entail earlier theories, and that "earlier theories are, very often, limiting cases of later theories?"[32] Unfortunately, answering this question is difficult, since 'typically' is one of those weasel words that allows for much hedging. I shall assume that Putnam and Watkins mean that "most of the time (or perhaps in most of the important cases) successor theories contain predecessor theories as limiting cases." So construed, the claim is patently false. Copernican astronomy did not retain all the key mechanisms of Ptolemaic astronomy (e.g., motion along an equant); Newton's physics did not retain all (or even most of) the theoretical laws of Cartesian mechanics, astronomy, and optics; Franklin's electrical theory did not contain its predecessor (J. A. Nollet's) as a limiting case. Relativistic physics did not retain the ether, nor the mechanisms associated with it; statistical mechanics does not incorporate all the mechanisms of thermodynamics; modern genetics does not have Darwinian pangenesis as a limiting case; the wave theory of light did not appropriate the mechanisms of corpuscular optics; modern embryology incorporates few of the mechanisms prominent in classical embryological theory. As I have shown elsewhere,[33] loss occurs at virtually every level: the confirmed predictions of earlier theories are sometimes not explained by later ones; even the 'observable' laws explained by earlier theories are not always retained, not even as limiting cases; theoretical processes and mechanisms of earlier theories are, as frequently as not, treated as flotsam.

The point is that some of the most important theoretical innovations have been due to a willingness of scientists to violate the cumulationist or retentionist constraint which realists enjoin 'mature' scientists to follow.

There is a deep reason why the convergent realist is wrong about these matters. It has to do, in part, with the role of ontological frameworks in science and with the nature of limiting case relations. As scientists use the term 'limiting case', T_1 can be a limiting case of T_2 only if *all* the variables (observable and theoretical) assigned a value in T_1 are assigned a value by T_2, and if the values assigned to every variable of T_1 are the same as, or very close to, the values T_2 assigns to the corresponding variable when certain initial and boundary conditions – consistent with T_2[34] – are specified. This seems to require that T_1 can be a limiting case of T_2 only if *all* the entities postulated by T_1 occur in the ontology of T_2. Whenever there is a change of ontology accompanying a theory transition such that T_2 (when conjoined with suitable initial and boundary conditions) fails to capture the ontology of T_1, then T_1 cannot be a limiting case of T_2. Even where the ontologies of T_1 and T_2 overlap appropriately (i.e., where T_2's ontology embraces all of T_1's), T_1 is a limiting case of T_2 only if *all* the laws of T_1 can be derived from T_2, given appropriate limiting conditions. It is important to stress that *both* these conditions (among others) must be satisfied before one theory can be a limiting case of another. Where "closet" positivists might be content with capturing only the formal mathematical relations or only the observable consequences of T_1 within a successor T_2, any genuine realist must insist that T_1's underlying ontology is preserved in T_2's, *for it is that ontology above all which he alleges to be approximately true.*

Too often, philosophers (and physicists) infer the existence of a limiting-case relation between T_1 and T_2 on substantially less than this. For instance, many writers have claimed one theory to be a limiting case of another when only some, but not all, of the laws of the former are derivable from the latter. In other cases, one theory has been said to be a limiting case of a successor when the mathematical laws of the former find homologies in the latter but where the former's ontology is not fully extractable from the latter's.

Consider one prominent example which has often been misdescribed, namely, the transition from the classical ether theory to relativistic and quantum mechanics. It can, of course, be shown that *some* laws of classical mechanics are limiting cases of relativistic mechanics. But there are other laws and general assertions made by the classical theory (e.g., claims about the density and fine structure of the ether, general laws about the character of the interaction between ether and matter, models and mechanisms detailing the compressibility of the ether) which could not conceivably be limiting cases of modern mechanics. The reason is a simple one: a theory cannot assign values to a variable that does not occur in that theory's language (or, more colloquially, it cannot assign properties to entities whose existence it does not countenance). Classical ether physics contained a number of postulated mechanisms for dealing inter alia with the transmission of light through the ether. Such mechnisms could not possibly appear in a successor theory like the special theory of relativity which denies the very existence of an etherial medium and which accomplishes the explanatory tasks performed by the ether via very different mechanisms.

Nineteenth-century mathematical physics is replete with similar examples of evidently successful mathematical theories which, because some of their variables refer to entities whose existence we now deny, cannot be shown to be limiting cases of our physics. As Adolf Grünbaum has cogently argued, when we are confronted with two incompatible theories, T_1 and T_2, such that T_2 does not "contain" all of T_1's ontology, then not all the mechanisms and theoretical laws of T_1 that involve those entities of T_1 not postulated by T_2 can possibly be retained – not even as limiting cases – in T_2.[35] This result is of some significance. What little plausibility convergent or retentive realism has enjoyed derives from the presumption that it correctly describes the relationship between classical and postclassical mechanics and gravitational theory. Once we see that even in this prima facie most favorable case for the realist (where *some* of the laws of the predecessor theory are genuinely limiting cases of the successor), changing ontologies or conceptual frameworks make it impossible to capture many of the central theoretical laws and mechanisms postulated by the earlier theory, then we can see how misleading Putnam's claim, is that "what scientists try to do [is to preserve] the *mechanisms* of the earlier theory as often as possible – or to show that they are 'limiting cases' of new mechanisms."[36] Where the mechanisms of the earlier theory involve entities whose existence the later theory denies, no scientist does (or should) feel any compunction about wholesale repudiation of the earlier mechanisms.

But even where there is no change in basic ontology, many theories (even in mature sciences such as physics) fail to retain all the explanatory successes of their predecessors. It is well known that statistical mechanics has yet to capture the irreversibility of macrothermodynamics as a genuine limiting case. Classical continuum mechanics has not yet been reduced to quantum mechanics or relativity. Contemporary field theory has yet to replicate the classical thesis that physical laws are invariant under reflection in space. If scientists had accepted the realist's constraint (namely, that new theories must have old theories as limiting cases), neither relativity nor statistical mechanics would have been viewed as viable theories. It has been said before, but it needs to be reiterated over and again: *a proof of the existence of limiting relations between selected components of two theories is a far cry from a systematic proof that one theory is a limiting case of the other.* Even if classical and modern physics stood to one another in the manner in which the convergent realist erroneously imagines they do, his hasty generalization that theory successions in all the advanced sciences show limiting-case relations is patently false.[37] But, as this discussion shows, not even the realist's paradigm case will sustain the claims he is apt to make about it.

What this analysis underscores is just how reactionary many forms of convergent epistemological realism are. If one took seriously CER's advice to reject any new theory that did not capture existing mature theories as referential and existing laws and mechanisms as approximately authentic, then any prospect for deep-structure, ontological changes in our theories would be foreclosed. Equally outlawed would be any significant repudiation of our theoretical models. In spite of his commitment to the growth of knowledge, the realist would unwittingly freeze science in its present state by forcing all future theories to accommodate the ontology of contemporary (mature) science and by foreclosing the possibility that some future generation may come to the conclusion that some (or even most) of the central terms in our best theories are no more referential than was 'natural place', 'phlogiston', 'ether', or 'caloric'.

Could theories converge in ways required by the realist?

These violations, within genuine science, of the sorts of continuity usually required by realists are by themselves sufficient to show that the form of scientific growth which the convergent realist takes as his explicandum is often absent, even in the mature sciences. But we can move beyond these specific cases to show in principle that the kind of cumulation demanded by the realist is unattainable. Specifically, by drawing on some results established by David Miller and others, the following can be shown:

1 The familiar requirement that a successor theory, T_2, must both preserve as true the true consequences of its predecessor, T_1, and explain T_1's anomalies is contradictory.
2 If a new theory, T_2, involves a change in the ontology or conceptual framework of a predecessor, T_1, then T_1 will have true and determinate consequences not possessed by T_2.
3 If two theories, T_1 and T_2, disagree, then each will have true and determinate consequences not exhibited by the other.

To establish these conclusions, one needs to utilize a "syntactic" view of theories according to which a theory is a conjunction of statements and its consequences are defined à la Tarski in terms of content classes. Needless to say, this is neither the only, nor necessarily the best, way of thinking about theories; but it happens to be the way in which most philosophers who argue for convergence and retention (e.g., Popper, Watkins, Post, Krajewski, and I. Niiniluoto) tend to conceive of theories. What can be said is that if one utilizes the Tarskian conception of a theory's content and its consequences as they do, then the familiar convergentist theses alluded to in conclusions 1 through 3 make no sense.

The elementary but devastating consequences of Miller's analysis establish that virtually any effort to link scientific progress or growth to the wholesale retention of a predecessor theory's Tarskian content *or* logical consequences *or* true consequences *or* observed consequences *or* confirmed consequences, is evidently doomed. Realists have not only got their history wrong insofar as they imagine that cumulative retention has prevailed in science, but we can also see that, given their views on what should be retained through theory change, history could not possibly have been the way their models require it to be. The realist's strictures on cumulativity are as ill advised normatively as they are false historically.

Along with many other realists, Putnam has claimed that "the mature sciences do converge ... and that that convergence has great explanatory value for the theory of science."[38] As this section should show, Putnam and his fellow realists are arguably wrong on *both* counts. Popper once remarked that "no theory of knowledge should attempt to explain why we are successful in our attempts to explain things."[39] Such a dogma is too strong. But what the foregoing analysis shows is that an occupational hazard of recent epistemology is imagining that convincing explanations of our success come easily or cheaply.

Should new theories explain why their predecessors were successful?

An apparently more modest realism than that outlined above is familiar in the form of the requirement (R4) often attributed to Sellars – that every satisfactory new theory must be able to explain why its predecessor was successful insofar as it was successful. On this view, viable new theories need not preserve all the content of their predecessors, nor capture those predecessors as limiting cases. Rather, it is simply insisted that a viable new theory, T_N, must explain why, when we conceive of the world according to the old theory T_O, there is a range of cases where our T_O-guided expectations are correct or approximately correct.

What are we to make of this requirement? In the first place, it is clearly *gratuitous*. If T_N has more confirmed consequences (and greater conceptual simplicity) than T_O, then T_N is preferable to T_O even if T_N cannot explain why T_O is successful. Contrariwise, if T_N has fewer confirmed consequences than T_O, then T_N cannot be rationally preferred to T_O even if T_N explains why T_O is successful. In short, a theory's ability to explain why a rival is successful is neither a necessary nor a sufficient condition for saying that it is better than its rival.

Other difficulties likewise confront the claim that new theories should explain why their predecessors were successful. Chief among them is the ambiguity of the notion itself. One way to show that an older theory, T_O, was successful is to show that it shares many confirmed consequences with a newer theory, T_N, which is highly successful. But this is not an "explanation" that a scientific realist could accept, since it makes no reference to, and thus does not depend upon, an epistemic assessment of either T_O or T_N. (After all, an instrumentalist could quite happily grant that if T_N "saves the phenomena" then T_O – insofar as some of its observable consequences overlap with or are experimentally indistinguishable from those of T_N – should also succeed at saving the phenomena.)

The intuition being traded on in this persuasive account is that the pragmatic success of a new theory, combined with a partial comparison of the respective consequences of the new theory and its predecessor, will sometimes put us in a position to say when the older theory worked and when it failed. But such comparisons as can be made in this manner do not involve *epistemic* appraisals of either the new or the old theory qua theories. Accordingly, the possibility of such comparisons provides no argument for epistemic realism.

What the realist apparently needs is an epistemically robust sense of "explaining the success of a predecessor." Such an epistemic characterization would presumably begin with the claim that T_N, the new theory, was approximately true and would proceed to show that the 'observable' claims of its predecessor, T_O, deviated only slightly from (some of) the 'observable' consequences of T_N. It would then be alleged that the (presumed) approximate truth of T_N and the partially overlapping consequences of T_O and T_N jointly explained why T_O was successful insofar as it was successful. But this is a non-sequitur. As I have shown above, the fact that a T_N is approximately true does not even explain why it is successful; how, under those circumstances, can the approximate truth of T_N explain why some theory different from T_N is successful? Whatever the nature of the relations between T_N and T_O (entailment, limiting case, etc.), the epistemic ascription of approximate truth to either T_O or T_N (or both) apparently leaves untouched questions of how successful T_O or T_N are.

The idea that new theories should explain why older theories were successful (insofar as they were) originally arose as a rival to the "levels" picture of explanation according to which new theories fully explained, because they entailed, their predecessors. It is clearly an improvement over the levels picture (for it does recognize that later theories generally do not entail their predecessors). But when it is formulated as a general thesis about intertheory relations, designed to buttress a realist epistemology, it is difficult to see how this position avoids difficulties similar to those discussed in earlier sections.

The Realists' Ultimate *Petitio Principii*

It is time to step back a moment from the details of the realists' argument to look at its general strategy. Fundamentally, the realist is utilizing, as we have seen, an abductive inference which proceeds from the success of science to the conclusion that science is approximately true, verisimilar, or referential (or any combination of these). This argument is meant to show the skeptic that theories are not ill gotten, the positivist that theories are not reducible to their observational consequences, and the pragmatist that classical epistemic categories (e.g., "truth" and "falsehood") are a relevant part of metascientific discourse.

It is little short of remarkable that realists would imagine that their critics would find the argument compelling. As I have shown elsewhere,[40] ever since antiquity critics of epistemic realism have based their skepticism upon a deep-rooted conviction that the fallacy of affirming the consequent is indeed fallacious. When E. Sextus or R. Bellarmine or Hume doubted that certain theories which saved the phenomena were warrantable as true, their doubts were based on a belief that the exhibition that a theory had some true consequences left entirely open the truth-status of the theory. Indeed, many nonrealists have been nonrealists precisely because they believed that false theories, as well as true ones, could have true consequences.

Now enters the new breed of realist (e.g., Putnam, Boyd, Newton-Smith) who wants to argue that epistemic realism can reasonably be presumed true by virtue of the fact that it has true consequences. But this is a monumental case of begging the question. The nonrealist refuses to admit that a *scientific* theory can be warrantedly judged to be true simply because it has some true consequences. Such nonrealists are not likely to be impressed by the claim that a philosophical theory such as realism can be warranted as true because it arguably has some true consequences. If nonrealists are chary about first-order abductions to avowedly true conclusions, they are not likely to be impressed by second-order abductions, particularly when, as I have tried to show above, the premises and conclusions are so indeterminate.

But, it might be argued, the realist is not out to convert the intransigent skeptic or the determined instrumentalist.[41] Perhaps, he is seeking to show that realism can be tested like any other scientific hypothesis, and that realism is at least as well confirmed as some of our best scientific theories. Such an analysis, however plausible initially, will not stand up to scrutiny. I am aware of no realist who is willing to say that a scientific theory can be reasonably presumed to be true or even regarded as well confirmed just on the strength of the fact that its thus-far-tested consequences are true. Realists have long been in the forefront of those opposed to ad hoc and post hoc theories. Before a realist accepts a scientific hypothesis, he generally wants to know whether it has explained or predicted more than it was devised to explain, whether it has been subjected to a battery of controlled tests, whether it has successfully made novel predictions, and whether there is independent evidence for it.

What, then, of realism itself as a "scientific" hypothesis?[42] Even if we grant (contrary to what I argued in the section on "Approximate Truth and Success") that realism entails and thus explains the success of science, ought that (hypothetical) success warrant, by the realist's own construal of scientific acceptability, the acceptance of realism? Since realism was devised to explain the success of science, it remains purely ad hoc with respect to that success. If realism has made some novel predictions or has been subjected to carefully controlled tests, one does not learn about it from the literature of contemporary realism. At the risk of apparent inconsistency, the realist repudiates the instrumentalist's view that saving the phenomena is a significant form of evidential support while endorsing realism itself on the transparently instrumentalist grounds that it is confirmed by those very facts it was invented to explain. No proponent of realism has sought to show that realism satisfies those stringent empirical demands which the realist himself minimally insists on when appraising scientific theories. The latter-day realist often calls realism a "scientific" or "well-tested" hypothesis but seems curiously reluctant to subject it to those controls he otherwise takes to be a sine qua non for empirical well-foundedness.

Conclusion

The arguments and cases discussed above seem to warrant the following conclusions:

1 The fact that a theory's central terms refer does not entail that it will be successful, and a theory's success is no warrant for the claim that all or most of its central terms refer.
2 The notion of approximate truth is presently too vague to permit one to judge whether a theory consisting entirely of approximately true laws would be empirically successful. What is clear is that a theory may be empirically successful even if it is not approximately true.
3 Realists have no explanation whatever for the fact that many theories which are not approximately true and whose "theoretical" terms seemingly do not refer are, nonetheless, often successful.
4 The convergentist's assertion that scientists in a "mature" discipline usually preserve, or seek to preserve, the laws and mechanisms of earlier theories in later ones is probably false. His assertion that when such laws are preserved in a successful successor, we can explain the success of the latter by virtue of the truthlikeness of the preserved laws and mechanisms, suffers from all the defects noted above confronting approximate truth.
5 Even if it could be shown that referring theories and approximately true theories would be successful, the realist's argument that successful theories are approximately true and genuinely referential takes for granted precisely what the nonrealist denies, namely, that explanatory success betokens truth.
6 It is not clear that acceptable theories either *do* or *should* explain why their predecessors succeeded or failed. If a theory is better supported than its rivals and predecessors, then it is not epistemically decisive whether it explains why its rivals worked.
7 If a theory has once been falsified, it is unreasonable to expect that a successor should retain either all of its content *or* its confirmed consequences *or* its theoretical mechanisms.
8 Nowhere has the realist established, except by fiat, that nonrealist epistemologists lack the resources to explain the success of science.

With these specific conclusions in mind, we can proceed to a more global one: it is not yet established – Putnam, Newton-Smith, and Boyd notwithstanding – that realism can explain *any* part of the successes of science. What is very clear is that realism *cannot*, even by its own lights, explain the success of those many theories whose central terms have evidently not referred and whose theoretical laws and mechanisms were not approximately true. The inescapable conclusion is that insofar as many realists are concerned with explaining how science works and with assessing the adequacy of their epistemology by that standard, they have, thus far, failed to explain very much. Their epistemology is confronted by anomalies that seem beyond its resources to grapple with.

It is important to guard against a possible misinterpretation of this essay. Nothing I have said here refutes the possibility, in principle, of a realistic epistemology of science. To conclude as much would be to fall prey to the same inferential prematurity with which many realists have rejected in principle the possibility of explaining science in a nonrealist way. My task here is, rather, that of reminding ourselves that there *is* a difference between wanting to believe something and having good reasons for believing it. All of us would like realism to be true; we would like to think that science works because it has got a grip on how things really are. But such claims have yet to be made out. Given the present state of the art, it can only be wish fulfillment that gives rise to the claim that realism, and realism alone, explains why science works.

Notes

1 R. Boyd, "Realism, Underdetermination, and a Causal Theory of Evidence," *Nous* 7 (1973): 1–12; W. Newton-Smith, "The Underdetermination of Theories by Data," *Proceedings of the Aristotelian Society* (1978), 71–91; H. Putnam, *Mathematics, Matter, and Method*, Vol. 1 (Cambridge: Cambridge University Press, 1975); Ilkka Niiniluoto, "On the Truthlikeness of Generalizations," in *Basic Problems in Methodology and Linguistics*, ed. R. Butts and J. Hintikka (Dordrecht: D. Reidel, 1977), 121–47.

2 Putnam, evidently following Boyd, sums up theses R1 and R3 in these words:

> "1) Terms in a mature science typically *refer*.
> 2) The laws of a theory belonging to a mature science are typically approximately true.
> . . . I will only consider [new] theories . . . which have this property – [they] contain the [theoretical] laws of [their predecessors] as a limiting case."

H. Putnam, *Meaning and the Moral Sciences* (London: Routledge and Kegan Paul, 1978), pp. 20–1.

3 Putnam insists, for instance, that if the realist is wrong about theories being referential, then "the success of science is a miracle" (Putnam, *Mathematics, Matter, and Method*, 1: 69).

4 Boyd remarks: "Scientific realism offers an *explanation* for the legitimacy of ontological commitment to theoretical entities" (Putnam, *Meaning and the Moral Sciences*, p. 2). It allegedly does so by explaining why theories containing theoretical entities work so well: because such entities genuinely exist.

5 Whether one utilizes Putnam's earlier or later versions of realism is irrelevant for the central arguments of this essay.

6 Putnam, *Meaning and the Moral Sciences*, pp. 20–2.

7 Ibid., p. 22.

8 Ibid., p. 20.

9 "Realism, Underdeterminism, and a Causal Theory of Evidence," 1. See also p. 3: "experimental evidence for a theory is evidence for the truth of even its non-observational laws." See also W. Sellars, *Science, Perception and Reality* (New York: The Humanities Press, 1963), p. 97.

10 A caveat is in order here. *Even* if all the central terms in some theory refer, it is not obvious that every rational successor to that theory must preserve all the referring terms of its predecessor. One can easily imagine circumstances when the new theory is preferable to the old one even though the range of application of the new theory is less broad than the old. When the range is so restricted, it may well be entirely appropriate to drop reference to some of the entities that figured in the earlier theory.

11 For Putnam and Boyd both, "it will be a constraint on T_2 [i.e., any new theory in a domain] . . . that T_2 must have this property, the property that *from its standpoint* one can assign referents to the terms of T_1 [i.e., an earlier theory in the same domain]". For Boyd, see "Realism, Underdeterminism, and a Causal Theory of Evidence," p. 8: "new theories should, *prima facie*, resemble current theories with respect to their accounts of causal relations among theoretical entities."

12 Putnam, *Mathematics, Matter, and Method*, p. 22.

13 For just a small sampling of this view, consider the following: "the claim of a realist ontology of science is that the only way of explaining why the models of science function so successfully . . . is that they approximate in some way the structure of the object" (Ernan McMullin, "The History and Philosophy of Science: A Taxonomy," in *Minnesota Studies in the Philosophy of Science*, ed. R. Stuewer, 5 [1970]: 63–4); "the continued success of confirmed theories can be *explained* by the hypothesis that they are in fact close to the truth" (Niiniluoto); and the claim that "the laws of a theory belonging to a mature science are typically approximately *true* . . . [provides] an explanation of the behavior of scientists and the success of science" (Putnam, *Meaning and the Moral Sciences*, pp. 20–1). J. J. Smart, W. Sellars, and Newton-Smith, among others, share a similar view.

14 Although Popper is generally careful not to assert that actual historical theories exhibit ever-increasing truth content (for an exception, see his *Conjectures and Refutations* [London: Routledge and Kegan Paul, 1963], p. 220), other writers have been more bold. Thus, Newton-Smith writes that "the historically generated sequence of theories of a mature science is a sequence in which succeeding theories are increasing in truth content without increasing in falsity content" (W. Newton-Smith, "In Defense of Truth," in *The Philosophy of Evolution*, ed. Uffe J. Jensen and Rom Harre, 1981, New York: St Martin's Press, pp. 269–89.).

15 On the more technical side, Niiniluoto has shown that a theory's degree of corroboration covaries with its "estimated verisimilitude" (Niiniluoto, "On the Truthlikeness of Generalizations," pp. 121–47). Roughly speaking, "estimated truth-likeness" is a measure of how closely (the content of) a

theory corresponds to *what we take to be* the best conceptual systems that, so far, we have been able to find (Ilkka Niiniluoto, "Scientific Progress," *Synthese* 45 (1980): 443 ff.). If Niiniluoto's measures work, it follows from the above-mentioned covariance that an empirically successful theory will have a high degree of estimated truthlikeness. But because estimated truthlikeness and genuine verisimilitude are not necessarily related (the former being parasitic on existing evidence and available conceptual systems), it is an open question whether, as Niiniluoto asserts, the continued success of highly confirmed theories can be *explained* by the hypothesis that they in fact are so close to the truth, at least in the relevant respects. Unless I am mistaken, this remark of his betrays a confusion between "true verisimilitude" (to which we have no epistemic access) and "estimated verisimilitude" (which is accessible but nonepistemic).

16 Newton-Smith, "In Defense of Truth," p. 16.

17 Newton-Smith claims that the increasing predictive success of science through time "would be totally mystifying ... if it were not for the fact that theories are capturing more and more truth about the world" (Newton-Smith, "In Defense of Truth," p. 15).

18 I must stress again that I am *not* denying that there *may* be a connection between approximate truth and predictive success. I am only observing that until the realists show us what that connection is, they should be more reticent than they are about claiming that realism can explain the success of science.

19 A *nonrealist* might argue that a theory is approximately true just in case all its *observable* consequences are true or within a specified interval from the true value. Theories that were "approximately true" in this sense would indeed be demonstrably successful. But, the realist's (otherwise commendable) commitment to taking seriously the theoretical claims of a theory precludes him from utilizing any such construal of approximate truth, since he wants to say that the theoretical as well as the observational consequences are approximately true.

20 W. Krajewski, *Correspondence Principle and Growth of Science* (Dordrecht: D. Reidel, 1977), p. 91.

21 Putnam, *Meaning and the Moral Sciences*, p. 21.

22 If this argument, which I attribute to the realists, seems a bit murky, I challenge any reader to find a more clear-cut one in the literature! Overt formulations of this position can be found in Putnam, Boyd, and Newton-Smith.

23 Putnam, *Meaning and the Moral Sciences*, p. 21.

24 John Watkins, "Corroboration and the Problem of Content-Comparison," in *Progress and Rationality in Science*, ed. G. Radnitzky and G. Anderson (Dordrecht: D. Reidel, 1978), pp. 376–7.

25 Popper: "A theory which has been well-corroborated can only be superseded by one ... [which] *contains* the old well-corroborated theory – or at least a good approximation to it." K. Popper, *Logic of Scientific Discovery* (New York: Basic Books, 1959), p. 276.

Post: "I shall even claim that, as a matter of empirical historical fact, [successor] theories [have] always explained the *whole* of [the well-confirmed part of their predecessors]." H. R. Post, "Correspondence, Invariance and Heuristics: In Praise of Conservative Induction," *Studies in the History and Philosophy of Science* 2 (1971): 229.

Koertge: "Nearly all parts of successive theories in the history of science stand in a correspondence relation and ... where there is no correspondence to begin with, the new theory will be developed in such a way that it comes more nearly into correspondence with the old." N. Koertge, "Theory Change in Science," in *Conceptual Change*, ed. G. Pearce and P. Maynard (Dordrecht: D. Reidel, 1973), pp. 176–7.

Among other authors who have defended a similar view, one should mention A. Fine, "Consistency, Derivability and Scientific Change," *Journal of Philosophy* 64 (1967): 231 ff.; C. Kordig, "Scientific Transitions, Meaning Invariance, and Derivability," *Southern Journal of Philosophy* (1971): 119–25; H. Margenau, *The Nature of Physical Reality* (New York: McGraw-Hill, 1950); and L. Sklar, "Types of Inter-Theoretic Reductions," *British Journal for Philosophy of Science* 18 (1967): 190–224.

26 Putnam fails to point out that it is also a fact that many scientists do *not* seek to preserve earlier theoretical mechanisms and that theories which have not preserved earlier theoretical mechanisms (whether the germ theory of disease, plate tectonics, or wave optics) have led to important discoveries is also a fact.

27 I. Szumilewicz, "Incommensurability and the Rationality of the Development of Science," *British Journal for the Philosophy of Science* 28 (1977): 348.

28 Putnam, *Meaning and the Moral Sciences*, p. 21.

29 I have written a book about this strategy. See Larry Laudan, *Progress and Its Problems* (Berkeley, Los Angeles, London: University of California Press, 1977).

30 After the epistemological and methodological battles about science during the last three hundred

years, it should be fairly clear that science, taken at its face value, implies no particular epistemology.

31 Clifford Hooker, "Systematic Realism," *Synthese* 26 (1974): 467–72.

32 Putnam, *Meaning and the Moral Sciences*, pp. 20, 123.

33 Larry Laudan, "Two Dogmas of Methodology," *Philosophy of Science* 43 (1976): 467–72.

34 This matter of limiting conditions consistent with the "reducing" theory is curious. Some of the best-known expositions of limiting-case relations depend (as Krajewski has observed) upon showing an earlier theory to be a limiting case of a later theory only by adopting limiting assumptions *explicitly denied by the later theory*. For instance, several standard textbook discussions present (a portion of) classical mechanics as a limiting case of special relativity, provided *c* approaches infinity. But special relativity is committed to the claim that *c* is a constant. Is there not something suspicious about a "derivation" of T_1 from a T_2 which essentially involves an assumption inconsistent with T_2? If T_2 is correct, then it forbids the adoption of a premise commonly used to derive T_1 as a limiting case. (It should be noted that most such proofs can be reformulated unobjectionably, e.g., in the relativity case, by letting $\upsilon \rightarrow 0$ rather than $c \rightarrow \infty$.)

35 Adolf Grünbaum, "Can a Theory Answer More Questions than One of Its Rivals?" *British Journal for Philosophy of Science* 27 (1976): 1–23.

36 Putnam, *Meaning and the Moral Sciences*, p. 20.

37 As Mario Bunge has cogently put it: "The popular view on inter-theory relations . . . that every new theory includes (as regards its extension) its predecessors . . . is philosophically superficial . . . and it is false as a historical hypothesis concerning the advancement of science." M. Bunge, "Problems Concerning Intertheory Relations," in *Induction, Physics and Ethics*, ed. P. Weingartner and G. Zecha (Dordrecht: D. Reidel, 1970), pp. 309–10.

38 Putnam, *Meaning and the Moral Sciences*, p. 37.

39 K. Popper, *Conjectures and Refutations* (London: Routledge and Kegan Paul, 1963), Introduction.

40 Larry Laudan, "Ex-Huming Hacking," *Erkenntis* 13 (1978).

41 I owe the suggestion of this realist response to Andrew Lugg.

42 I find Putnam's views on the "empirical" or "scientific" character of realism rather perplexing. At some points, he seems to suggest that realism is both empirical and scientific. Thus, he writes: "If realism is an explanation of this fact [namely, that science is successful], realism must itself be an over-arching scientific *hypothesis*". Since Putnam clearly maintains the antecedent, he seems committed to the consequent. Elsewhere he refers to certain realist tenets as being "our highest level empirical generalizations about knowledge" (*Meaning and the Moral Sciences*, p. 37). He says, moreover, that realism "could be false," and that "facts are relevant to its support (or to criticize it)" (pp. 78–9). Nonetheless, for reasons he has not made clear, Putnam wants to deny that realism is either scientific or a hypothesis (p. 79). How realism can consist of doctrines which explain facts about the world, are empirical generalizations about knowledge, and can be confirmed or falsified by evidence, and *yet* be neither scientific nor hypothetical, is left opaque.

9.3

Constructive Empiricism

Bas van Fraassen

Bas van Fraassen's (b. 1941) book *The Scientific Image* (from which the selection below has been excerpted) has done much to keep the realism debate vigorous by presenting and defending a version of anti-realist empiricism that is immune to many of the criticisms of the received view. According to van Fraassen's position, called "constructive empiricism," contemporary scientific theories are to be interpreted literally as referring to entities that are incapable of being observed by the naked human senses. But the constructive empiricist limits her commitment to only what those theories claim concerning what is observable. Claims concerning the unobservable are understood to play only a pragmatic role in developing empirically successful theories; the existence of the unobservables themselves is neither affirmed nor denied.

Introduction

[I]t is easy to indulge the commonplace metaphysical instinct. But a taste for metaphysics may be one of those things which we must renounce, if we mean to mould our lives to artistic perfection. Philosophy serves culture, not by the fancied gift of absolute or transcendental knowledge, but by suggesting questions . . .

Walter Pater, *The Renaissance*

The opposition between empiricism and realism is old, and can be introduced by illustrations from many episodes in the history of philosophy. The most graphic of these is perhaps provided by the sense of philosophical superiority the participants in the early development of modern science felt toward the Aristotelian tradition. In that tradition, the realists held that regularities in the natural phenomena must have a reason (cause, explanation), and they sought this reason

From Bas C. van Fraassen, *The Scientific Image* (Oxford: Clarendon Press, 1980), pp. 1–21, 23–5, 38–40, 56–9, 64, 67–9, 80–3.

in the causal properties, constituting what they called the substantial forms or natures, of the substances involved in natural processes. The nominalists, who denied the reality of these properties, were in the position of having to reject such requests for explanation.[1]

The philosophers engaged in developing the philosophical foundations of modern science had apparently escaped this dilemma. Without postulating such causal properties, forms, or 'occult qualities', they could still explain the regularities that are observed in nature. Thus Robert Boyle writes,

> That which I chiefly aim at, is to make it probable to you by experiments, that almost all sorts of qualities, most of which have been by the schools either left unexplicated, or generally referred to I know not what incomprehensible substantial forms, may be produced mechanically; I mean by such corporeal agents as do not appear either to work otherwise than by virtue of the motion, size, figure, and contrivance of their own parts (which attributes I call the mechanical affections of matter).[2]

To give an account of such phenomena as heat or chemical reactions in terms only of mechanical attributes, they realized quite well, required at least an atomic theory of matter. But I suppose it is clear that they will face that same dilemma again for the regularities they postulate in the behaviour of the atomic parts. No mechanical explanations are possible there, since the atoms have no further parts. So either they must attribute specific powers, qualities, and causal properties to those atoms to explain why they act and react in the way they actually do; or else they must, like the nominalists before them, reject the request for explanation.

In addition, they have gained a problem. Part of the motivation for the nominalist rejection of the Aristotelian realists' world of powers, properties, dispositions (made famous by Molière's *virtus dormitiva*) was epistemological. The observation of the phenomena did not point unambiguously to the supposed causal connections behind them. This problem exists similarly for the atomic hypotheses: the phenomena do not decide their truth or falsity, though they are perhaps better explained by one hypothesis than by another. Subsequent scientists intent on clarifying the philosophical basis of their discipline found it ever more difficult to reconcile their professed empiricism and antipathy to metaphysics with an unqualified belief in hypotheses that describe a supposed world behind the phenomena.

This led in the nineteenth century to the phenomenalism of Ernst Mach, the conventionalism of Henri Poincaré, and the fictionalism of Pierre Duhem. In the twentieth, the logical empiricism of Hans Reichenbach and logical positivism of Rudolph Carnap were further developments in this radical turn to empiricism.

Today, however, no one can adhere to any of these philosophical positions to any large extent. Logical positivism, especially, even if one is quite charitable about what counts as a development rather than a change of position, had a rather spectacular crash. So let us forget these labels which never do more than impose a momentary order on the shifting sands of philosophical fortune, and let us see what problems are faced by an *aspirant* empiricist today. What sort of philosophical account is possible of the aim and structure of science?

Studies in philosophy of science divide roughly into two sorts. The first, which may be called foundational, concerns the content and structure of theories. The other sort of study deals with the relations of a theory on the one hand, to the world and to the theory-user on the other.

There are deep-going philosophical disagreements about the general structure of scientific theories, and the general characterization of their content. A current view, not altogether uncontroversial but still generally accepted, is that theories account for the phenomena (which means, the observable processes and structures) by postulating other processes and structures not directly accessible to observation; and that a system of any sort is described by a theory in terms of its possible states. This is a view about the structure of theories shared by many philosophers who nevertheless disagree on the issues concerning a theory's relation to the world and to its users. Opponents of that view will at least say, I think, that this account of what science is like is true 'on the face of it', or correct as a first approximation.

One relation a theory may have to the world is that of being true, of giving a true account of the facts. It may at first seem trivial to assert that science aims to find true theories. But coupled with the preceding view of what theories are like, the triviality disappears. Together they imply that science aims to find a true description of unobservable processes that explain the observable ones, and also of what are possible states of affairs, not just of what is actual. Empiricism has always been a main philosophical guide in the study of nature. But empiricism requires theories only to give a true account *of what is observable*, counting further postulated structure as a means to that end. In addition empiricists have always eschewed the reification of possibility (or its dual, necessity). Possibility and necessity they relegate to relations among ideas, or among words, as devices to facilitate the description of what is actual. So from an empiricist point of view, to serve the aims of science, the postulates need not be true, except in what they say about what is actual and empirically attestable.

When this empiricist point of view was represented by logical positivism, it had added to it a theory of meaning and language, and generally a linguistic orientation. Today that form of empiricism is opposed by scientific realism, which rejects not only the views on meaning of the positivists, but also those empiricist tenets which I outlined in the preceding paragraph. My own view is that empiricism is correct, but could not live in the linguistic form the positivists gave it. They were right to think in some cases that various philosophical perplexities, misconceived as problems in ontology and epistemology, were really at bottom problems about language. This opinion is correct especially, I think, about problems concerning possibility and necessity. The language of science, being a proper part of natural language, is clearly part of the subject of general philosophy of logic and language. But this only means that *certain* problems can be set aside when we are doing philosophy of science, and emphatically does *not* mean that philosophical concepts must be one and all linguistically explicated. The logical positivists, and their heirs, went much too far in this attempt to turn philosophical problems into problems about language. In some cases their linguistic orientation had disastrous effects in philosophy of science. Scientific realism, however, pursues the antithetical error of reifying whatever cannot be defined away.

Correlative to discussions of the relation between a theory and the world, is the question what it is to accept a scientific theory. This question has an epistemic dimension (how much belief is involved in theory acceptance?) and also a pragmatic one (what else is involved besides belief?). On the view I shall develop, the belief involved in accepting a scientific theory is only that it 'saves the phenomena', that is, correctly describes what is observable. But acceptance is not merely belief. We never have the option of accepting an all-encompassing theory, complete in every detail. So to accept one theory rather than another one involves also a commitment to a research programme, to continuing the dialogue with nature in the framework of one conceptual scheme rather than another. Even if two theories are empirically equivalent, and acceptance of a theory involves as belief only that it is empirically adequate, it may still make a great difference which one is accepted. The difference is pragmatic, and I shall argue that pragmatic virtues do not give us any reason over and above the evidence of the empirical data, for thinking that a theory is true.

So I shall argue for an empiricist position, and against scientific realism. In some ways, philosophy is a subject of fashions – not more so than other intellectual disciplines, I suppose, but at least to the extent that almost any philosopher will begin by explaining that he opposes the 'dominant' or 'received' view, and present his own as revolutionary. It would be quite suspicious therefore if I were to say at this point that scientific realism has become dominant in philosophy of science. Others have certainly characterized it as the emerging victor: Isaac Levi recently wrote, 'My own view is that the coffin of empiricism is already sealed tight.'[3] And Arthur Fine, in a reply to Richard Healey:

> The objections that he raises to a realist understanding of [quantum mechanics] are . . . supposed to move my philosophical colleagues to the same anti-realist convictions that Mr. Healey thinks are held by many physicists. I am not sure how many physicists do hold such anti-realist convictions these days . . . I suspect . . . that most physicists who do shy away from realism are influenced more by the tradition in which they are schooled than they are by these

> rather recent and sophisticated arguments. That tradition is the deeply positivist legacy of Bohr and Heisenberg . . . I am not much worried that my philosophical colleagues will be seduced by positivist considerations coupled with insubstantial reasons, for we are differently schooled.[4]

There is therefore at least already considerable sentiment on the side of realists that they have replaced the ametaphysical empiricism of the positivists. The empiricist position I mean to advocate will be strongly dissociated from both.

In part my argument will be destructive, countering the arguments brought forward by scientific realists against the empiricist point of view. I shall give a momentary name, 'constructive empiricism', to the specific philosophical position I shall advocate. The main part of that advocacy will be the development of a constructive alternative to scientific realism, on the main issues that divide us: the relation of theory to world, the analysis of scientific explanation, and the meaning of probability statements when they are part of a physical theory. I use the adjective 'constructive' to indicate my view that scientific activity is one of construction rather than discovery: construction of models that must be adequate to the phenomena, and not discovery of truth concerning the unobservable. The baptism of this philosophical position as a specific 'ism' is not meant to imply the desire for a school of thought; only to reflect that scientific realists have appropriated a most persuasive name for themselves (aren't we all scientific, and realists, nowadays?), and that there is after all something in a name.

Arguments Concerning Scientific Realism

> *The rigour of science requires that we distinguish well the undraped figure of nature itself from the gay-coloured vesture with which we clothe it at our pleasure.*
>
> Heinrich Hertz, quoted by Ludwig Boltzmann, letter to *Nature*, 28 February 1895

In our century, the first dominant philosophy of science was developed as part of logical positivism. Even today, such an expression as 'the received view of theories' refers to the views developed by the logical positivists, although their heyday preceded the Second World War.

In this chapter I shall examine, and criticize, the main arguments that have been offered for scientific realism. These arguments occurred frequently as part of a critique of logical positivism. But it is surely fair to discuss them in isolation, for even if scientific realism is most easily understood as a reaction against positivism, it should be able to stand alone. The alternative view which I advocate – for lack of a traditional name I shall call it *constructive empiricism* – is equally at odds with positivist doctrine.

§1 Scientific realism and constructive empiricism

In philosophy of science, the term 'scientific realism' denotes a precise position on the question of how a scientific theory is to be understood, and what scientific activity really is. I shall attempt to define this position, and to canvass its possible alternatives. Then I shall indicate, roughly and briefly, the specific alternative which I shall advocate and develop in later chapters.

§1.1 Statement of scientific realism

What exactly is scientific realism? A naive statement of the position would be this: the picture which science gives us of the world is a true one, faithful in its details, and the entities postulated in science really exist: the advances of science are discoveries, not inventions. That statement is too naive; it attributes to the scientific realist the belief that today's theories are correct. It would mean that the philosophical position of an earlier scientific realist such as C. S. Peirce had been refuted by empirical findings. I do not suppose that scientific realists wish to be committed, as such, even to the claim that science will arrive in due time at theories true in all respects – for the growth of science might be an endless self-correction; or worse, Armageddon might occur too soon.

But the naive statement has the right flavour. It answers two main questions: it characterizes a scientific theory as a story about what there really is, and scientific activity as an enterprise of discovery, as opposed to invention. The two questions of what a scientific theory is, and what a scientific theory does, must be answered by any philosophy of science. The task we have at this point is to find a statement of scientific realism that shares these features with the naïve statement, but does not saddle the realists with unacceptably strong consequences. It is especially important to make the statement as weak as possible if we wish to argue against it, so as not to charge at windmills.

As clues I shall cite some passages most of which will also be examined below in the contexts of the authors' arguments. A statement of Wilfrid Sellars is this:

> to have good reason for holding a theory is *ipso facto* to have good reason for holding that the entities postulated by the theory exist.[5]

This addresses a question of epistemology, but also throws some indirect light on what it is, in Sellars's opinion, to hold a theory. Brian Ellis, who calls himself a scientific entity realist rather than a scientific realist, appears to agree with that statement of Sellars, but gives the following formulation of a stronger view:

> I understand scientific realism to be the view that the theoretical statements of science are, or purport to be, true generalized descriptions of reality.[6]

This formulation has two advantages: It focuses on the understanding of the theories without reference to reasons for belief, and it avoids the suggestion that to be a realist you must believe current scientific theories to be true. But it gains the latter advantage by use of the word 'purport', which may generate its own puzzles.

Hilary Putnam, in a passage which I shall cite again in Section 7, gives a formulation which he says he learned from Michael Dummett:

> A realist (with respect to a given theory or discourse) holds that (1) the sentences of that theory are true or false; and (2) that what makes them true or false is something external – that is to say, it is not (in general) our sense data, actual or potential, or the structure of our minds, or our language, etc.[7]

He follows this soon afterwards with a further formulation which he credits to Richard Boyd:

> That terms in mature scientific theories typically refer (this formulation is due to Richard Boyd), that the theories accepted in a mature science are typically approximately true, that the same term can refer to the same thing even when it occurs in different theories – these statements are viewed by the scientific realist . . . as part of any adequate scientific description of science and its relations to its objects.[8]

None of these were intended as definitions. But they show I think that truth must play an important role in the formulation of the basic realist position. They also show that the formulation must incorporate an answer to the question what it is to *accept* or *hold* a theory. I shall now propose such a formulation, which seems to me to make sense of the above remarks, and also renders intelligible the reasoning by realists which I shall examine below – without burdening them with more than the minimum required for this.

Science aims to give us, in its theories, a literally true story of what the world is like; and acceptance of a scientific theory involves the belief that it is true. This is the correct statement of scientific realism.

Let me defend this formulation by showing that it is quite minimal, and can be agreed to by anyone who considers himself a scientific realist. The naive statement said that science tells a true story; the correct statement says only that it is the aim of science to do so. The aim of science is of course not to be identified with individual scientists' motives. The aim of the game of chess is to checkmate your opponent; but the motive for playing may be fame, gold, and glory. What the aim is determines what counts as success in the enterprise as such; and this aim may be pursued for any number of reasons. Also, in calling something *the* aim, I do not deny that there are other subsidiary aims which may or may not be means to that end: everyone will readily agree that simplicity, informativeness, predictive power, explanation are (also) virtues. Perhaps my formulation can even be accepted by any philosopher who considers the most important aim of science to be something which only *requires* the finding of true theories – given that I wish to give the weakest formulation of the doctrine that is generally acceptable.

I have added 'literally' to rule out as realist such positions as imply that science is true if 'properly understood' but literally false or meaningless. For that would be consistent with conventionalism, logical positivism, and instrumentalism. I will say more about this below.

The second part of the statement touches on epistemology. But it only equates acceptance of a theory with belief in its truth.[9] It does not imply that anyone is ever rationally warranted in forming such a belief. We have to make room for the epistemological position, today the subject of considerable debate, that a rational person never assigns personal probability 1 to any proposition except a tautology. It would, I think, be rare for a scientific realist to take this stand in epistemology, but it is certainly possible.[10]

To understand qualified acceptance we must first understand acceptance *tout court*. If acceptance of a theory involves the belief that it is true, then tentative acceptance involves the tentative adoption of the belief that it is true. If belief comes in degrees, so does acceptance, and we may then speak of a degree of acceptance involving a certain degree of belief that the theory is true. This must of course be distinguished from belief that the theory is approximately true, which seems to mean belief that some member of a class centring on the mentioned theory is (exactly) true. In this way the proposed formulation of realism can be used regardless of one's epistemological persuasion.

§1.2 Alternatives to realism

Scientific realism is the position that scientific theory construction aims to give us a literally true story of what the world is like, and that acceptance of a scientific theory involves the belief that it is true. Accordingly, anti-realism is a position according to which the aim of science can well be served without giving such a literally true story, and acceptance of a theory may properly involve something less (or other) than belief that it is true.

What does a scientist do then, according to these different positions? According to the realist, when someone proposes a theory, he is asserting it to be true. But according to the anti-realist, the proposer does not assert the theory to be true; *he displays it*, and claims certain virtues for it. These virtues may fall short of truth: empirical adequacy, perhaps; comprehensiveness, acceptability for various purposes. This will have to be spelt out, for the details here are not determined by the denial of realism. For now we must concentrate on the key notions that allow the generic division.

The idea of a literally true account has two aspects: the language is to be literally construed; and so construed, the account is true. This divides the anti-realists into two sorts. The first sort holds that science is or aims to be true, properly (but not literally) construed. The second holds that the language of science should be literally construed, but its theories need not be true to be good. The anti-realism I shall advocate belongs to the second sort.

It is not so easy to say what is meant by a literal construal. The idea comes perhaps from theology, where fundamentalists construe the Bible literally, and liberals have a variety of allegorical, metaphorical, and analogical interpretations, which 'demythologize'. The problem of explicating 'literal construal' belongs to the philosophy of language. In Section 7, where I briefly examine some of Michael Dummett's views, I shall emphasize that 'literal' does not mean 'truth-valued'. The term 'literal' is well enough understood for general philosophical use, but if we try to explicate it we find ourselves in the midst of the problem of giving an adequate account of natural language. It would be bad tactics to link an inquiry into science to a commitment to some solution to that problem. The following remarks, and those in Section 7, should fix the usage of 'literal' sufficiently for present purposes.

The decision to rule out all but literal construals of the language of science, rules out those forms of anti-realism known as *positivism* and *instrumentalism*. First, on a literal construal, the apparent statements of science really are statements, *capable of* being true or false. Secondly, although a literal construal can elaborate, it cannot change logical relationships. (It is possible to elaborate, for instance, by identifying what the terms designate. The 'reduction' of the language of phenomenological thermodynamics to that of statistical mechanics is like that: bodies of gas are identified as aggregates of molecules, temperature as mean kinetic energy, and so on.) On the positivists' interpretation of science, theoretical terms have meaning only through their connection with the observable. Hence they hold that two theories may in fact *say the same thing* although in form they contradict each other. (Perhaps the one says that all matter consists of atoms, while the other postulates instead a universal continuous medium; they will say the same thing nevertheless if they agree in their observable consequences, according to the positivists.) But two theories which contradict each other in such a way can 'really' be saying the same thing only if they are not literally construed. Most specifically, if a theory says that something exists, then a literal construal may elaborate on what that something is, but will not remove the implication of existence.

There have been many critiques of positivist interpretations of science, and there is no need to repeat them. I shall add some specific criticisms of the positivist approach in the next chapter.

§1.3 Constructive empiricism

To insist on a literal construal of the language of science is to rule out the construal of a theory as a metaphor or simile, or as intelligible only after it is 'demythologized' or subjected to some other sort of 'translation' that does not preserve logical form. If the theory's statements include 'There are electrons', then the theory says that there are electrons. If in addition they include 'Electrons are not planets', then the theory says, in part, that there are entities other than planets.

But this does not settle very much. It is often not at all obvious whether a theoretical term refers to a concrete entity or a mathematical entity. Perhaps one tenable interpretation of classical physics is that there are no concrete entities which are forces – that 'there are forces such that . . .' can always be understood as a mathematical statement asserting the existence of certain functions. That is debatable.

Not every philosophical position concerning science which insists on a literal construal of the language of science is a realist position. For this insistence relates not at all to our epistemic attitudes toward theories, nor to the aim we pursue in constructing theories, but only to the correct understanding of *what a theory says*. (The fundamentalist theist, the agnostic, and the atheist presumably agree with each other (though not with liberal theologians) in their understanding of the statement that God, or gods, or angels exist.) After deciding that the language of science must be literally understood, we can still say that there is no need to believe good theories to be true, nor to believe *ipso facto* that the entities they postulate are real.

Science aims to give us theories which are empirically adequate; and acceptance of a theory involves as belief only that it is empirically adequate. This is the statement of the anti-realist position I advocate; I shall call it *constructive empiricism*.

This formulation is subject to the same qualifying remarks as that of scientific realism in Section 1.1 above. In addition it requires an explication of 'empirically adequate'. For now, I shall leave that with the preliminary explication that a theory is empirically adequate exactly if what it says about the observable things and events in this world, is true – exactly if it 'saves the phenomena'. A little more precisely: such a theory has at least one model that all the actual phenomena fit inside. I must emphasize that this refers to *all* the phenomena; these are not exhausted by those actually observed, nor even by those observed at some time, whether past, present, or future. The whole of the next chapter will be devoted to the explication of this term, which is intimately bound up with our conception of the structure of a scientific theory.

The distinction I have drawn between realism and anti-realism, in so far as it pertains to acceptance, concerns only how much belief is involved therein. Acceptance of theories (whether full, tentative, to a degree, etc.) is a phenomenon of scientific activity which clearly involves more than belief. One main reason for this is that we are never confronted with a complete theory. So if a scientist accepts a theory, he thereby involves himself in a certain sort of research programme. That programme could well be different from the one acceptance of another theory would have given him, even if those two (very incomplete) theories are equivalent to each other with respect to everything that is observable – in so far as they go.

Thus acceptance involves not only belief but a certain commitment. Even for those of us who are not working scientists, the acceptance involves a commitment to confront any future phenomena by means of the conceptual resources of this theory. It determines the terms in which we shall seek explanations. If the acceptance is at all strong, it is exhibited in the person's assumption of the role of explainer, in his willingness to answer questions *ex cathedra*. Even if you do not accept a theory, you can engage in discourse in a context in which language use is guided by that theory – but acceptance produces such contexts. There are similarities in all of this to ideological commitment. A commitment is of course not true or false: The confidence exhibited is that it will be *vindicated*.

This is a preliminary sketch of the *pragmatic* dimension of theory acceptance. Unlike the epistemic dimension, it does not figure overtly in the disagreement between realist and anti-realist. But because the amount of belief involved in acceptance is typically less according to anti-realists, they will tend to make more of the pragmatic aspects. It is as well to note here the important difference. Belief that a theory is true, or that it is empirically adequate, does not imply, and is not implied by, belief that full acceptance of the theory will be vindicated. To see this, you need only consider here a person who has quite definite beliefs about the future of the human race, or about the scientific community and the influences thereon and practical limitations we have. It might well be, for instance, that a theory which is empirically adequate will not combine easily with some other theories which we have accepted in fact, or that Armageddon will occur before we succeed. Whether belief that a theory is true, or that it is empirically adequate, can be equated with belief that acceptance of it would, under ideal research conditions, be vindicated in the long run, is another question. It seems to me an irrelevant question within philosophy of science, because an affirmative answer would not obliterate the distinction we have already established by the preceding remarks. (The question may also assume that counterfactual statements are objectively true or false, which I would deny.)

Although it seems to me that realists and anti-realists need not disagree about the pragmatic aspects of theory acceptance, I have mentioned it here because I think that typically they do. We shall find ourselves returning time and again, for example, to requests for explanation to which realists typically attach an objective validity which anti-realists cannot grant.

§2 The theory/observation 'dichotomy'

For good reasons, logical positivism dominated the philosophy of science for thirty years. In 1960, the first volume of *Minnesota Studies in the Philosophy of Science* published Rudolf Carnap's 'The Methodological Status of Theoretical Concepts', which is, in many ways, the culmination of the positivist programme. It interprets science by relating it to an observation language (a postulated part of natural language which is devoid of theoretical terms). Two years later this article was followed in the same series by Grover Maxwell's 'The Ontological Status of Theoretical Entities', in title and theme a direct counter to Carnap's. This is the *locus classicus* for the new realists' contention that the theory/observation distinction cannot be drawn.

I shall examine some of Maxwell's points directly, but first a general remark about the issue. Such expressions as 'theoretical entity' and 'observable–theoretical dichotomy' are, on the face of it, examples of category mistakes. Terms or concepts are theoretical (introduced or adapted for the purposes of theory construction); entities are observable or unobservable. This may seem a little point, but it separates the discussion into two issues. Can we divide our language into a theoretical and non-theoretical part? On the other hand, can we classify objects and events into observable and unobservable ones?

Maxwell answers both questions in the negative, while not distinguishing them too carefully. On the first, where he can draw on well-known supportive essays by Wilfrid Sellars and Paul Feyerabend, I am in total agreement. All our language is thoroughly theory-infected. If we could cleanse our language of theory-laden terms, beginning with the recently introduced ones like 'VHF receiver', continuing through 'mass' and 'impulse' to 'element' and so on into the prehistory of language formation, we would end up with nothing useful. The way we talk, and scientists talk, is guided by the pictures provided by previously accepted theories. This is true also, as Duhem already emphasized, of experimental reports. Hygienic reconstructions of language such as the positivists envisaged are simply not on. I shall return to this criticism of positivism in the next chapter.

But does this mean that we must be scientific realists? We surely have more tolerance of ambiguity than that. The fact that we let our language be guided by a given picture, at some point, does not show how much we believe about that picture. When we speak of the sun coming up in the morning and setting at night, we are guided by a picture now explicitly disavowed. When Milton wrote *Paradise Lost* he deliberately let the old geocentric astronomy guide his poem, although various remarks in passing clearly reveal his interest in the new astronomical discoveries and speculations of his time. These are extreme examples, but show that no immediate conclusions can be drawn from the theory-ladenness of our language.

However, Maxwell's main arguments are directed against the observable–unobservable distinction. Let us first be clear on what this distinction was supposed to be. The term 'observable' classifies putative entities (entities which may or may not exist). A flying horse is observable – that is why we are so sure that there aren't any – and the number seventeen is not. There is supposed to be a correlate classification of human acts: an unaided act of perception, for instance, is an observation. A calculation of the mass of a particle from the deflection of its trajectory in a known force field, is not an observation of that mass.

It is also important here not to confuse *observing* (an entity, such as a thing, event, or process) and *observing that* (something or other is the case). Suppose one of the Stone Age people recently found in the Philippines is shown a tennis ball or a car crash. From his behaviour, we see that he has noticed them; for example, he picks up the ball and throws it. But he has not seen *that* it is a tennis ball, or *that* some event is a car crash, for he does not even have those concepts. He cannot get that information through perception; he would first have to learn a great deal. To say that he does not see the same things and events as we do, however, is just silly; it is a pun which trades on the ambiguity between seeing and seeing that. (The truth-conditions for our statement 'x observes *that A*' must be such that what concepts x has, presumably related to the language x speaks if he is human, enter as a variable into the correct truth definition, in some way. To say that x observed the tennis ball, therefore, does not imply at all that x observed that it was a tennis ball; that would require some conceptual awareness of the game of tennis.)

The arguments Maxwell gives about observability are of two sorts: one directed against the possibility of drawing such distinctions, the other against the importance that could attach to distinctions that can be drawn.

The first argument is from the continuum of cases that lie between direct observation and inference:

> there is, in principle, a continuous series beginning with looking through a vacuum and containing these as members: looking through a windowpane, looking through glasses, looking through binoculars, looking through a low-power microscope, looking through a high-power microscope, etc., in the order given. The important consequence is that, so far, we are left without criteria which would enable us to draw a non-arbitrary line between 'observation' and 'theory'.[11]

This continuous series of supposed acts of observation does not correspond directly to a continuum in what is supposed observable. For if something can be seen through a window, it can also be seen with the window raised. Similarly, the moons of Jupiter can be seen through a telescope; but they can also be seen without a telescope if you are close enough. That something is observable does not automatically imply that the conditions are right for observing it now. The principle is:

> X is observable if there are circumstances which are such that, if X is present to us under those circumstances, then we observe it.

This is not meant as a definition, but only as a rough guide to the avoidance of fallacies.

We may still be able to find a continuum in what is supposed detectable: perhaps some things can only be detected with the aid of an optical microscope, at least; perhaps some require an electron microscope, and so on. Maxwell's problem is: where shall we draw the line between what is observable and what is only detectable in some more roundabout way?

Granted that we cannot answer this question without arbitrariness, what follows? That 'observable' is a *vague predicate*. There are many puzzles about vague predicates, and many sophisms designed to show that, in the presence of vagueness, no distinction can be drawn at all. In Sextus Empiricus, we find the argument that incest is not immoral, for touching your mother's big toe with your little finger is not immoral, and all the rest differs only by degree. But predicates in natural language are almost all vague, and there is no problem in their use; only in formulating the logic that governs them.[12] A vague predicate is usable provided it has clear cases and clear counter-cases. Seeing with the unaided eye is a clear case of observation. Is Maxwell then perhaps challenging us to present a clear counter-case? Perhaps so, for he says 'I have been trying to support the thesis that any (non-logical) term is a *possible* candidate for an observation term.'

A look through a telescope at the moons of Jupiter seems to me a clear case of observation, since astronauts will no doubt be able to see them as well from close up. But the purported observation of micro-particles in a cloud chamber seems to me a clearly different case – if our theory about what happens there is right. The theory says that if a charged particle traverses a chamber filled with saturated vapour, some atoms in the neighbourhood of its path are ionized. If this vapour is decompressed, and hence becomes supersaturated, it condenses in droplets on the ions, thus marking the path of the particle. The resulting silver-grey line is similar (physically as well as in appearance) to the vapour trail left in the sky when a jet passes. Suppose I point to such a trail and say: 'Look, there is a jet!'; might you not say: 'I see the vapour trail, but where is the jet?' Then I would answer: 'Look just a bit ahead of the trail . . . there! Do you see it?' Now, in the case of the cloud chamber this response is not possible. So while the particle is detected by means of the cloud chamber, and the detection is based on observation, it is clearly not a case of the particle's being observed.

As a second argument, Maxwell directs our attention to the 'can' in 'what is observable is what can be observed.' An object might of course be temporarily unobservable – in a rather different sense: it cannot be observed in the circumstances in which it actually is at the moment, but could be observed if the circumstances were more favourable. In just the same way, I might be temporarily invulnerable or invisible. So we should concentrate on 'observable' *tout court*, or on (as he prefers to say) 'unobservable in principle'. This Maxwell explains as meaning that the relevant scientific theory *entails* that the entities cannot be observed in any circumstances. But this never happens, he says, because the different circumstances could be ones in which we have different sense organs – electron-microscope eyes, for instance.

This strikes me as a trick, a change in the subject of discussion. I have a mortar and pestle made of copper and weighing about a kilo. Should I call it breakable because a giant could break it? Should I call the Empire State Building portable? Is there no distinction between a portable and a console record player? The human organism is, from the point of view of physics, a certain kind of measuring apparatus. As such it has certain inherent limitations – which will be described in detail in the final physics and biology. It is these limitations to which the 'able' in 'observable' refers – our limitations, *qua* human beings.

As I mentioned, however, Maxwell's article also contains a different sort of argument: even if there is a feasible observable/unobservable distinction, this distinction has no importance. The point at issue for the realist is, after all, the reality of the entities postulated in science. Suppose that these entities could be classified into observables and others; what relevance should that have to the question of their existence?

Logically, none. For the term 'observable' classifies putative entities, and has logically nothing to do with existence. But Maxwell must have more in mind when he says: 'I conclude that the drawing of the observational–theoretical line at any given point is an accident and a function of our physiological make-up, . . . and, therefore, that it has no ontological significance whatever.'[13] No ontological significance if the question is only whether 'observable' and 'exists' imply each other – for they do not; but significance for the question of scientific realism?

Recall that I defined scientific realism in terms of the aim of science, and epistemic attitudes. The question is what aim scientific activity has, and how much we shall believe when we accept a scientific theory. What is the proper form of acceptance: belief that the theory, as a whole, is true; or something else? To this question, what is observable by us seems eminently relevant. Indeed, we may attempt an answer at this point: to accept a theory is (for us) to believe that it is empirically adequate – that what the theory says *about what is observable* (by us) is true.

It will be objected at once that, on this proposal, what the anti-realist decides to believe about the world will depend in part on what he believes to be his, or rather the epistemic community's, accessible range of evidence. At present, we count the human race as the epistemic community to which we belong; but this race may mutate, or that community may be increased by adding other animals (terrestrial or extra-terrestrial) through relevant ideological or moral decisions ('to count them as persons'). Hence the anti-realist would, on my proposal, have to accept conditions of the form

> If the epistemic community changes in fashion *Y*, then my beliefs about the world will change in manner *Z*.

To see this as an objection to anti-realism is to voice the requirement that our epistemic policies should give the same results independent of our beliefs about the range of evidence accessible to us. That requirement seems to me in no way rationally compelling; it could be honoured, I should think, only through a thoroughgoing scepticism or through a commitment to wholesale leaps of faith. But we cannot settle the major questions of epistemology *en passant* in philosophy of science; so I shall just conclude that it is, on the face of it, not irrational to commit oneself only to a search for theories that are empirically adequate, ones whose models fit the observable phenomena, while recognizing that what counts as an observable phenomenon is a function of what the epistemic community is (that *observable* is *observable-to-us*).

The notion of empirical adequacy in this answer will have to be spelt out very carefully if it is not to bite the dust among hackneyed objections. I shall try to do so in the next chapter. But the point stands: even if observability has nothing to do with existence (is, indeed, too anthropocentric for that), it may still have much to do with the proper epistemic attitude to science.

§3 *Inference to the best explanation*

A view advanced in different ways by Wilfrid Sellars, J. J. C. Smart, and Gilbert Harman is that the canons of rational inference require scientific realism. If we are to follow the same patterns of inference with respect to this issue as we do in science itself, we shall find ourselves irrational unless we assert the truth of the scientific theories we accept. Thus Sellars says: 'As I see it, to have good reason for holding a theory is *ipso facto* to have good reason for holding that the entities postulated by the theory exist.[14]

The main rule of inference invoked in arguments of this sort is the rule of *inference to the best explanation*. The idea is perhaps to be credited to C. S. Peirce, but the main recent attempts to explain this rule and its uses have been made by Gilbert Harman.[15] I shall only present a simplified version. Let us suppose that we have evidence *E*, and are considering several hypotheses, say *H* and *H′*. The rule then says that we should infer *H* rather than *H′* exactly if *H* is a better explanation of *E* than *H′* is. (Various qualifications are necessary to avoid inconsistency: we should always try to move to the best over-all explanation of all available evidence.)

It is argued that we follow this rule in all 'ordinary' cases; and that if we follow it consistently everywhere, we shall be led to scientific realism, in the way Sellars's dictum suggests. And surely there are many telling 'ordinary' cases: I hear scratching in the wall, the patter of little feet at midnight, my cheese disappears – and I infer that a mouse has come to live with me. Not merely that these apparent signs of mousely presence will continue, not merely that all the observable phenomena will be as if there is a mouse; but that there really is a mouse.

Will this pattern of inference also lead us to belief in unobservable entities? Is the scientific realist simply someone who consistently follows the rules of inference that we all follow in more mundane contexts? I have two objections to the idea that this is so.

First of all, what is meant by saying that we all *follow* a certain rule of inference? One meaning might be that we deliberately and consciously 'apply' the rule, like a student doing a logic exercise. That meaning is much too literalistic and restrictive; surely all of mankind follows the rules of logic much of the time, while only a fraction can even formulate them. A second meaning is that we act in accordance with the rules in a sense that does not require conscious deliberation. That is not so easy to make precise, since each logical rule is a rule of permission (*modus ponens* allows you to infer *B* from *A* and (if *A then B*), but does not forbid you to infer (*B or A*) instead). However, we might say that a person behaved in accordance with a set of rules in that sense if every conclusion he drew could be reached from his premisses via those rules. But this meaning is much too loose; in this sense we always behave in accordance with the rule that any conclusion may be inferred from any premiss. So it seems that to be following a rule, I must be willing to believe all conclusions it allows, while definitely unwilling to believe conclusions at variance with the ones it allows – or else, change my willingness to believe the premisses in question.

Therefore the statement that we all follow a certain rule in certain cases, is a *psychological hypothesis* about what we are willing and unwilling to do. It is an empirical hypothesis, to be confronted with data, and with rival hypotheses. Here is a rival hypothesis: we are always willing to believe that the theory which best explains the evidence, is empirically adequate (that all the observable phenomena are as the theory says they are).

In this way I can certainly account for the many instances in which a scientist appears to argue for the acceptance of a theory or hypothesis, on the basis of its explanatory success. (A number of such instances are related by Thagard.[16]) For, remember: I equate the acceptance of a scientific theory with the belief that it is empirically adequate. We have therefore two rival hypotheses concerning these instances of scientific inference, and the one is apt in a realist account, the other in an anti-realist account.

Cases like the mouse in the wainscoting cannot provide telling evidence between those rival hypotheses. For the mouse *is* an observable thing; therefore 'there is a mouse in the wainscoting' and 'All observable phenomena are as if there is a mouse in the wainscoting' are totally equivalent; each implies the other (given what we know about mice).

It will be countered that it is less interesting to know whether people do follow a rule of inference than whether they ought to follow it. Granted; but the premiss that we all follow the rule of inference to the best explanation when it comes to mice and other mundane matters – that premiss is shown wanting. It is not warranted by the evidence, because that evidence is not telling *for* the premiss *as against* the alternative hypothesis I proposed, which is a relevant one in this context.

My second objection is that even if we were to grant the correctness (or worthiness) of the rule of inference to the best explanation, the realist needs some further premiss for his argument. For this rule is only one that dictates a choice when given a set of rival hypotheses. In other words, we need to be committed to belief in one of a range of hypotheses before the rule can be applied. Then, under favourable circumstances, it will tell us which of the hypotheses in that range to choose. The realist asks us to choose between different hypotheses that explain the regularities in certain ways; but his opponent always wishes to choose among hypotheses of the form 'theory T_i is empirically adequate'. So the realist will need his special extra premiss that every universal regularity in nature needs an explanation, before the rule will make realists of us all. And that is just the premiss that distinguishes the realist from his opponents (and which I shall examine in more detail in Section 4 below).

[. . .]

§4 Limits of the demand for explanation

In this section and the next two, I shall examine arguments for realism that point to explanatory power as a criterion for theory choice. That this is indeed a criterion I do not deny. But these arguments for realism succeed only if the demand for explanation is supreme – if the task of science is unfinished, *ipso facto*, as long as any pervasive regularity is left unexplained. I shall object to this line of argument, as found in the writings of Smart, Reichenbach, Salmon, and Sellars, by arguing that such an unlimited demand for explanation leads to a demand for hidden variables, which runs contrary to at least one major school of thought in twentieth-century physics. I do not think that even these philosophers themselves wish to saddle realism with logical links to such consequences:

but realist yearnings were born among the mistaken ideals of traditional metaphysics.

In his book *Between Science and Philosophy*, Smart gives two main arguments for realism. One is that only realism can respect the important distinction between *correct* and *merely useful* theories. He calls 'instrumentalist' any view that locates the importance of theories in their use, which requires only empirical adequacy, and not truth. But how can the instrumentalist explain the usefulness of his theories?

> Consider a man (in the sixteenth century) who is a realist about the Copernican hypothesis but instrumentalist about the Ptolemaic one. He can explain the instrumental usefulness of the Ptolemaic system of epicycles because he can prove that the Ptolemaic system can produce almost the same predictions about the apparent motions of the planets as does the Copernican hypothesis. Hence the assumption of the realist truth of the Copernican hypothesis explains the instrumental usefulness of the Ptolemaic one. Such an explanation of the instrumental usefulness of certain theories would not be possible if *all* theories were regarded as merely instrumental.[17]

What exactly is meant by 'such an explanation' in the last sentence? If no theory is assumed to be true, then no theory has its usefulness explained as following from the truth of another one – granted. But would we have less of an explanation of the usefulness of the Ptolemaic hypothesis if we began instead with the premiss that the Copernican gives implicitly a very accurate description of the motions of the planets as observed from earth? This would not assume the truth of Copernicus's heliocentric hypothesis, but would still entail that Ptolemy's simpler description was also a close approximation of those motions.

However, Smart would no doubt retort that such a response pushes the question only one step back: what explains the accuracy of predictions based on Copernicus's theory? If I say, the empirical adequacy of that theory, I have merely given a verbal explanation. For of course Smart does not mean to limit his question to actual predictions – it really concerns all actual and possible predictions and retrodictions. To put it quite concretely: what explains the fact that all observable planetary phenomena fit Copernicus's theory (if they do)? From the medieval debates, we recall the nominalist response that the basic regularities are merely brute regularities, and have no explanation. So here the anti-realist must similarly say: that the observable phenomena exhibit these regularities, because of which they fit the theory, is merely a brute fact, and may or may not have an explanation in terms of unobservable facts 'behind the phenomena' – it really does not matter to the goodness of the theory, nor to our understanding of the world.

Smart's main line of argument is addressed to exactly this point. In the same chapter he argues as follows. Suppose that we have a theory T which postulates micro-structure directly, and macro-structure indirectly. The statistical and approximate laws about macroscopic phenomena are only partially spelt out perhaps, and in any case derive from the precise (deterministic or statistical) laws about the basic entities. We now consider theory T', which is part of T, and says only what T says about the macroscopic phenomena. (How T' should be characterized I shall leave open, for that does not affect the argument here.) Then he continues:

> I would suggest that the realist could (say) . . . that the success of T' is explained by the fact that the original theory T is true of the things that it is ostensibly about; in other words by the fact that there really are electrons or whatever is postulated by the theory T. If there were no such things, and if T were not true in a realist way, would not the success of T' be quite inexplicable? One would have to suppose that there were innumerable lucky accidents about the behaviour mentioned in the observational vocabulary, so that they behaved miraculously *as if* they were brought about by the non-existent things ostensibly talked about in the theoretical vocabulary.[18]

In other passages, Smart speaks similarly of 'cosmic coincidences'. The regularities in the observable phenomena must be explained in terms of deeper structure, for otherwise we are left with a belief in lucky accidents and coincidences on a cosmic scale.

I submit that if the demand for explanation implicit in these passages were precisely formulated, it would at once lead to absurdity. For if the mere fact of postulating regularities, without explanation, makes T' a poor theory, T will do no better. If, on the other hand, there is some precise limitation on what sorts of regularities can be postulated as basic, the context of the argument provides no reason to think that T' must automatically fare worse than T.

In any case, it seems to me that it is illegitimate to equate being a lucky accident, or a coincidence, with having no explanation. It was by coincidence that I met my friend in the market – but I can explain why I was there, and he can explain why he came, so together we can explain how this meeting happened. We call it a coincidence, not because the occurrence was inexplicable, but because we did not severally go to the market in order to meet.[19] There cannot be a requirement upon science to provide a theoretical elimination of coincidences, or accidental correlations in general, for that does not even make sense. There is nothing here to motivate the demand for explanation, only a restatement in persuasive terms.

[. . .]

In any case Hilary Putnam . . . directs himself to scientific realism per se, and formulates it in terms borrowed, he says, from Richard Boyd. The new formulation comes in the course of a new argument for scientific realism, which I shall call the Ultimate Argument:

> the positive argument for realism is that it is the only philosophy that doesn't make the success of science a miracle. That terms in mature scientific theories typically refer (this formulation is due to Richard Boyd), that the theories accepted in a mature science are typically approximately true, that the same term can refer to the same thing even when it occurs in different theories – these statements are viewed by the scientific realist not as necessary truths but as part of the only scientific explanation of the success of science, and hence as part of any adequate scientific description of science and its relations to its objects.[20]

Science, apparently, is required to explain its own success. There is this regularity in the world, that scientific predictions are regularly fulfilled; and this regularity, too, needs an explanation. Once *that* is supplied we may perhaps hope to have reached the *terminus de jure*?

The explanation provided is a very traditional one – *adequatio ad rem*, the 'adequacy' of the theory to its objects, a kind of mirroring of the structure of things by the structure of ideas – Aquinas would have felt quite at home with it.

Well, let us accept for now this demand for a scientific explanation of the success of science. Let us also resist construing it as merely a restatement of Smart's 'cosmic coincidence' argument, and view it instead as the question why we have successful scientific theories at all. Will this realist explanation with the Scholastic look be a scientifically acceptable answer? I would like to point out that science is a biological phenomenon, an activity by one kind of organism which facilitates its interaction with the environment. And this makes me think that a very different kind of scientific explanation is required.

I can best make the point by contrasting two accounts of the mouse who runs from its enemy, the cat. St Augustine already remarked on this phenomenon, and provided an intentional explanation: the mouse *perceives that* the cat is its enemy, hence the mouse runs. What is postulated here is the 'adequacy' of the mouse's thought to the order of nature: the relation of enmity is correctly reflected in his mind. But the Darwinist says: Do not ask why the *mouse* runs from its enemy. Species which did not cope with their natural enemies no longer exist. That is why there are only ones who do.

In just the same way, I claim that the success of current scientific theories is no miracle. It is not even surprising to the scientific (Darwinist) mind. For any scientific theory is born into a life of fierce competition, a jungle red in tooth and claw. Only the successful theories survive – the ones which *in fact* latched on to actual regularities in nature.[21]

To Save the Phenomena

[. . .]

§7 *The hermeneutic circle*

We have seen that we cannot interpret science, and isolate its empirical content, by saying that our language is divided into two parts. Nor should that conclusion surprise us. The phenomena are saved when they are exhibited as fragments of a larger unity. For that very reason it would be strange if scientific theories described the phenomena, the observable part, in different terms from the rest of the world they describe. And so an attempt to draw the conceptual line between phenomena and the trans-phenomenal by means of a distinction of vocabulary, must always have looked too simple to be good.

Not all philosophers who have discussed the observable/unobservable distinction, by any means, have done so in terms of vocabulary. But there has been a further assumption common also to critics of that distinction: that the distinction is a philosophical one. To draw it, they seem to assume, is in principle anyway the task of the philosophy of perception. To draw it, in principle anyway, philosophy must mobilize theories of sensing and perceiving, sense data and experiences, *Erlebnisse* and *Protokolsaetze*. If the distinction is a philosophical one, then it is to be drawn, if at all, by philosophical analysis, and to be attacked, if at all, by philosophical arguments.

This attitude needs a Grand Reversal. If there are limits to observation, these are a subject for empirical science, and not for philosophical analysis. Nor can the limits be described once and for all, just as measurement cannot be described once and for all. What goes on in a measurement process is differently described by classical physics and by quantum theory. To find the limits of what is observable in the world described by theory *T* we must inquire into *T* itself, and the theories used as auxiliaries in the testing and application of *T*.

We have now come to the 'hermeneutic circle' in the interpretation of science. I want to spell this out in detail, because one might too easily get a feeling of vicious circularity. And I want to give specific details on how science exhibits clear limits on observability.

Recall the main difference between the realist and anti-realist pictures of scientific activity. When a scientist advances a new theory, the realist sees him as asserting the (truth of the) postulates. But the anti-realist sees him as displaying this theory, holding it up to view, as it were, and claiming certain virtues for it.

This theory draws a picture of the world. But science itself designates certain areas in this picture as observable. The scientist, in accepting the theory, is asserting the picture to be accurate in those areas. This is, according to the anti-realist, the only virtue claimed which concerns the relation of theory to world alone. Any other virtues to be claimed will either concern the internal structure of the theory (such as logical consistency) or be pragmatic, that is, relate specifically to human concerns.

To accept the theory involves no more belief, therefore, than that what it says about observable phenomena is correct. To delineate what is observable, however, we must look to science – and possibly to that same theory – for that is also an empirical question. This might produce a vicious circle if what is observable were itself not simply a fact disclosed by theory, but rather theory-relative or theory-dependent. It will already be quite clear that I deny this; I regard what is observable as a theory-independent question. It is a function of facts about us *qua* organisms in the world, and these facts may include facts about the psychological states that involve contemplation of theories – but there is not the sort of theory-dependence or relativity that could cause a logical catastrophe here.

Let us consider two concrete examples which have been found puzzling. The first, already mentioned by Grover Maxwell, concerns molecules. Certain crystals, modern science tells us, are single molecules; these crystals are large enough to be seen – so, some molecules are observable. The second was mentioned to me by David Lewis: astronauts reported seeing flashes, and NASA scientists came to the conclusion that what they saw were high-energy electrons.

Is there anything puzzling about these examples? Only to those who think there is an intimate link between theoretical terms and unobservable entities or events. Compare the examples with Eddington's famous table: that table is an aggregate of interacting electrons, protons, and neutrons, he said; but that table is easily seen. If a crystal or table is classified by a theory as a theoretically described entity, does the presence of this observable object become evidence for the reality of other, different but similarly classified entities? Everything in the world has a proper classification within the conceptual framework of modern science. And it is this conceptual framework which we bring to bear when we describe any event, including an observation. This does not obliterate the distinction between what is observable and what is not – for that is an empirical distinction – and it does *not* mean that a theory could not be right about the observable without being right about everything.

We should also note here the intertranslatability of statements about objects, events, and quantities. There is a molecule in this place; the event of there-being-a-molecule occurs in this place (this is, roughly, Reichenbach's event language); a certain quantity, which takes value *one* if there is a molecule here and value *zero* if there is not, has value *one*. There is little difference between saying that a human being is a good detector of molecules and saying that he is a good detector of the presence of molecules. Any such classification of what happens may be correct, relative to a given, accepted theory. If we follow the principles of the general theory of measurement used in discussions of the foundations of quantum mechanics, we call system Y a measurement apparatus for quantity A exactly if Y has a certain possible state (the ground-state) such that if Y is in that state and coupled with another system X in *any* of its possible states, the evolution of the combined system (X *plus* Y) is subject to a law of interaction which has the effect of correlating the values of A in X with distinct values of a certain quantity B (often called the 'pointer reading observable') in system Y. Since observation is a special subspecies of measurement, this is a good picture to keep in mind as a partial guide.

Science presents a picture of the world which is much richer in content than what the unaided eye discerns. But science itself teaches us also that it is richer than the unaided eye *can* discern. For science itself delineates, at least to some extent, the observable parts of the world it describes. Measurement interactions are a special subclass of physical interactions in general. The structures definable from measurement data are a subclass of the physical structures described. It is in this way that science itself distinguishes the observable which it postulates from the whole it postulates. The distinction, being in part a function of the limits science discloses on human observation, is an anthropocentric one. But since science places human observers among the physical systems it means to describe, it also gives itself the task of describing anthropocentric distinctions. It is in this way that even the scientific realist must observe a distinction between the phenomena and the trans-phenomenal in the scientific world-picture.

[. . .]

. . . To present a theory is to specify a family of structures, its *models*; and secondly, to specify certain parts of those models (the *empirical substructures*) as candidates for the direct representation of observable phenomena. The structures which can be described in experimental and measurement reports we can call *appearances*: the theory is empirically adequate if it has some model such that all appearances are isomorphic to empirical substructures of that model . . .

[. . .]

Logical strength is determined by the class of models (inversely: the fewer the models the (logically) stronger the theory!) and empirical strength is similarly determined by the classes of empirical substructures. If $T >_e T'$ and $T' >_e T$, then they are *empirically equivalent*. We may call a theory *empirically minimal* if it is empirically non-equivalent to all logically stronger theories – that is, exactly if we cannot keep its empirical strength the same while discarding some of the models of this theory.

The notions of empirical adequacy and empirical strength, added to those of truth and logical strength, constitute the basic concepts for the semantics of physical theories. Of course, this addition makes the semantics only one degree less shallow than the one we had before. The semantic analysis of physical theory needs to be elaborated further, preferably in response to specific, concrete problems in the foundations of the special sciences. Especially pressing is the need for more finely delineated concepts pertaining to probability for theories in which that is a basic item. . . .

Empirical minimality is emphatically *not* to be advocated as a virtue, it seems to me. The reasons for this point are pragmatic. Theories with some degree of sophistication always carry some 'metaphysical baggage'. Sophistication lies in the introduction of detours via theoretical variables to arrive at useful, adequate, manageable descriptions of the phenomena. The term 'metaphysical baggage' will, of course, not be used when the detour pays off; it is reserved for those detours which yield no practical gain. Even the useless metaphysical baggage may be intriguing, however, because of its potentialities for future use. An example may yet be offered by hidden variable theories in quantum mechanics.[22] The 'no hidden variables' proofs, as I have already mentioned, rest on various assumptions which may be denied. Mathematically speaking there exist hidden variable theories equivalent to orthodox quantum theory in the following sense: the algebra of observables, reduced *modulo* statistical equivalence, in a model of the one is isomorphic to that in a model of the other. It appears to be generally agreed that such theories confront the phenomena exactly by way of these algebras of statistical quantities. On that assumption, theories equivalent in this sense are therefore empirically equivalent. Such hidden variable models have much extra structure, now looked upon as 'metaphysical baggage', but capable of being mobilized should radically new phenomena come to light.

With this new picture of theories in mind, we can distinguish between two epistemic attitudes we can take up toward a theory. We can assert it to be true (i.e. to have a model which is a faithful replica, in all detail, of our world), and call for belief; or we can simply assert its empirical adequacy, calling for acceptance as such. In either case we stick our necks out: empirical adequacy goes far beyond what we can know at any given time. (All the results of measurement are not in; they will never all be in; and in any case, we won't measure everything that can be measured.) Nevertheless there is a difference: the assertion of empirical adequacy is a great deal weaker than the assertion of truth, and the restraint to acceptance delivers us from metaphysics.

[...]

Empiricism and Scientific Methodology

[...]

§2.4 *Phenomenology of scientific activity*

The working scientist is totally immersed in the scientific world-picture. And not only he – to varying degrees, so are we all. If I call a certain box a VHF receiver, if I call a fork electro-plated, if I so much as decide to turn on the microwave oven to heat my sandwich in the cafeteria, I am immersed in a language which is thoroughly theory-infected, living in a world my ancestors of two centuries ago could not enter.

In the language-oriented philosophy of science developed by the logical positivists, one could not say this while remaining empiricist. For the empirical import of a theory was defined via a division of its (*sic*) language into a theoretical and a non-theoretical part. This division was a philosophical one, that is, imposed from outside. And you could not limit your endorsement to the empirical import of the theory unless your language remained in principle limited to the non-theoretical part of the theory's language. To immerse yourself fully in the theoretical world-picture, hence to use the full theoretical language without qualm, branded one (once the irreducibility of theoretical terms was realized) with complete commitment to the veracity of that picture.

In the constructive empiricist alternative I have been developing, nothing is more natural, or more to be recommended than this total immersion. For the empirical import of the theory is now defined from within science, by means of a distinction between what is observable and what is not observable drawn by science itself. The epistemic commitment to the empirical import of the theory above (its empirical adequacy) can be stated using the language of science – and indeed, in no other way. It may be the case that I have no adequate way to describe this box, and the role it plays in my world, except as a VHF receiver. From this it does not follow that I believe that the concept of very high frequency electromagnetic waves corresponds to an individually identifiable element of reality. Concepts involve theories and are inconceivable without them, to paraphrase Sellars. But immersion in the theoretical world-picture does not preclude 'bracketing' its ontological implications.

After all, what is this world in which I live, breathe and have my being, and which my ancestors of two centuries ago could not enter? It is the intentional correlate of the conceptual framework through which I perceive and conceive the world. But our conceptual framework changes, hence the intentional correlate of our conceptual framework changes – but the real world is the same world.

What I have just said, denies what is called conceptual relativism. To be more precise, it denies it on this level, that is, on the level on which we interpret science and describe its role in our intellectual and practical life. Philosophy of science is not metaphysics – there may or may not be a deeper level of analysis on which that concept of the real world is subjected to scrutiny and found itself to be . . . what? I leave to others the question whether we can consistently and coherently go further with such a line of thought. Philosophy of science can surely stay nearer the ground.

Let us discuss then the notion of *objectivity* as it appears in science. To someone immersed in that world-picture, the distinction between *electron* and *flying horse* is as clear as between *racehorse* and *flying horse*: the first corresponds to something in the actual world, and the other does not. While immersed in the theory, *and* addressing oneself solely to the problems in the domain of the theory, this objectivity of *electron* is not and cannot be qualified. *But this is so whether or not one is committed to the truth of the theory*. It is so not only for someone who believes, full stop, that the theory is true, but also for a Bayesian who grants a degree of belief equal to 1 to tautologies alone, and also to someone who is not a Bayesian but holds commitment to the truth of the theory in abeyance. For to say that someone is immersed in theory, 'living' in the theory's world, is not to describe his epistemic commitment. And if he describes his own epistemic commitment, he is stepping back for a moment, and saying something like: the theory entails that electrons exist, *and* not all theories do, *and* my epistemic attitude towards this theory is *X*.

We cannot revert to an earlier world-picture, because so many experimental findings cannot be accommodated in the science of an earlier time. This is not an argument for the truth of the present world-picture, but for its empirical adequacy. It is, you may wish to say, indirect or partial evidence for its truth – but only by being evidence for its empirical adequacy. It is also an argument for learning to find our way around in the world depicted by contemporary science, of speaking its language like a native. Someone who learns a second language comes to an all-important transition at a certain point: when he stops speaking by translating from his first language, and begins to speak 'directly'. It is only then that he begins to have access to the nuances and intangible differences that distinguish the two languages. The transition is a leap of sorts, into an unknown of sorts.

Not only objectivity, however, but also observability, is an intra-scientific distinction, if the science is taken wide enough. For that reason, it is possible even after total immersion in the world of science to distinguish possible epistemic attitudes to science, and to state them, and to limit one's epistemic commitment while remaining a functioning member of the scientific community – one who is reflective, and philosophically autonomous as well.

Notes

1 See my 'A Re-examination of Aristotle's Philosophy of Science' (*Dialogue*, 1980) and 'Essence and Existence', pp. 1–25, in N. Rescher (ed.), *Studies in Ontology*, American Philosophical Quarterly, Monograph No. 12 (Oxford: Blackwell, 1978), for discussion of some philosophical issues relating to that tradition.

2 *The Works of the Honourable Robert Boyle*, ed. Birch (London, 1672), Vol. III, p. 13; I take the quotation from R. S. Woolhouse, *Locke's Philosophy of Science and of Language* (Oxford: Blackwell, 1971), which has an excellent discussion of the philosophical issues of that period and of Boyle's role.

3 I. Levi, 'Confirmational Conditionalization', *Journal of Philosophy*, 75 (1978), 730–7; p. 737.

4 A. Fine, 'How to Count Frequencies: A Primer for Quantum Realists', *Synthese*, 42 (1979); quotation from pp. 151–2.

5 *Science, Perception and Reality* (New York: Humanities Press, 1962); cf. the footnote on p. 97. See also my review of his *Studies in Philosophy and its History*, in *Annals of Science*, January 1977.

6 Brian Ellis, *Rational Belief Systems* (Oxford: Blackwell, 1979), p. 28.

7 Hilary Putnam, *Mathematics, Matter and Method* (Cambridge: Cambridge University Press, 1975), Vol. 1, pp. 69 ff.

8 Putnam, op. cit. (n. 7 above). The argument is reportedly developed at greater length in Boyd's forthcoming book *Realism and Scientific Epistemology* (Cambridge University Press).

9 Hartry Field has suggested that 'acceptance of a scientific theory involves the belief that it is true' be replaced by 'any reason to think that any part

of a theory is not, or might not be true, is reason not to accept it.' The drawback of this alternative is that it leaves open what epistemic attitude acceptance of a theory does involve. This question must also be answered, and as long as we are talking about full acceptance – as opposed to tentative or partial or otherwise qualified acceptance – I cannot see how a realist could do other than equate that attitude with full belief. (That theories believed to be false are used for practical problems, for example, classical mechanics for orbiting satellites, is of course a commonplace.) For if the aim is truth, and acceptance requires belief that the aim is served . . . I should also mention the statement of realism at the beginning of Richard Boyd. 'Realism, Underdetermination, and a Causal Theory of Evidence', *Noûs*, 7 (1973), 1–12. Except for some doubts about his use of the terms 'explanation' and 'causal relation' I intend my statement of realism to be entirely in accordance with his. Finally, see C. A. Hooker, 'Systematic Realism', *Synthese*, 26 (1974), 409–97; esp. pp. 409 and 426.

10 More typical of realism, it seems to me, is the sort of epistemology found in Clark Glymour's forthcoming book, *Theory and Evidence* (Princeton: Princeton University Press, 1980), except of course that there it is fully and carefully developed in one specific fashion. (See esp. his chapter 'Why I am not a Bayesian' for the present issue.) But I see no reason why a realist, as such, could not be a Bayesian of the type of Richard Jeffrey, even if the Bayesian position has in the past been linked with anti-realist and even instrumentalist views in philosophy of science.

11 G. Maxwell, 'The Ontological Status of Theoretical Entities', *Minnesota Studies in Philosophy of Science*, III (1962), p. 7.

12 There is a great deal of recent work on the logic of vague predicates; especially important, to my mind, is that of Kit Fine ('Vagueness, Truth, and Logic', *Synthese*, 30 (1975), 265–300) and Hans Kamp. The latter is currently working on a new theory of vagueness that does justice to the 'vagueness of vagueness' and the context-dependence of standards of applicability for predicates.

13 Op. cit., n. 11. In the next chapter I shall discuss further how observability should be understood. At this point, however, I may be suspected of relying on modal distinctions which I criticize elsewhere. After all, I am making a distinction between human limitations, and accidental factors. A certain apple was dropped into the sea in a bag of refuse, which sank; relative to that information it is necessary that no one ever observed the apple's core. That information, however, concerns an accident of history, and so it is not human limitations that rule out observation of the apple core. But unless I assert that some facts about humans are essential, or physically necessary, and others accidental, how can I make sense of this distinction? This question raises the difficulty of a philosophical retrenchment for modal language. This I believe to be possible through an ascent to pragmatics. In the present case, the answer would be, to speak very roughly, that the scientific theories we accept are a determining factor for the set of features of the human organism counted among the limitations to which we refer in using the term 'observable'. The issue of modality will occur explicitly again in the chapter on probability.

14 Op. cit., See n. 5.

15 'The Inference to the Best Explanation', *Philosophical Review*, 74 (1965), 88–95 and 'Knowledge, Inference, and Explanation', *American Philosophical Quarterly*, 5 (1968), 164–73. Harman's views were further developed in subsequent publications (*Noûs*, 1967; *Journal of Philosophy*, 1968; in M. Swain (ed.), *Induction*, 1970; in H.-N. Castañeda (ed.), *Action, Thought, and Reality*, 1975; and in his book *Thought*, Ch. 10). I shall not consider these further developments here.

16 Cf. P. Thagard, doctoral dissertation, University of Toronto, 1977, and 'The Best Explanation: Criteria for Theory Choice', *Journal of Philosophy*, 75 (1978), 76–92.

17 J. J. C. Smart, *Between Science and Philosophy* (New York: Random House, 1968), p. 151.

18 Ibid., pp. 150ff.

19 This point is clearly made by Aristotle, *Physics*, II, Chs. 4–6 (see esp. 196^a 1–20; 196^b 20–197^a 12).

20 Putnam, op. cit., n. 8.

21 Of course, we can ask specifically why the *mouse* is one of the surviving species, how *it* survives, and answer this, on the basis of whatever scientific theory we accept, in terms of its brain and environment. The analogous question for theories would be why, say, Balmer's formula for the line spectrum of hydrogen survives as a successful hypothesis. In that case too we explain, on the basis of the physics we accept now, why the spacing of those lines satisfies the formula. Both the question and the answer are very different from the global question of the success of science, and the global answer of realism. The realist may now make the *further* objection that the anti-realist cannot answer the question about the mouse specifically, nor the one about Balmer's formula, in this fashion, since the answer is in part an assertion that the scientific theory, used as basis of the explanation, is true. This is a quite different argument . . .

In his most recent publications and lectures Hilary Putnam has drawn a distinction between two doctrines, metaphysical realism and internal realism. He denies the former, and identifies his preceding scientific realism as the latter. While I have at present no commitment to either side of the metaphysical dispute, I am very much in sympathy with the critique of Platonism in philosophy of mathematics which forms part of Putnam's arguments. Our disagreement about scientific (internal) realism would remain of course, whenever we came down to earth after deciding to agree or disagree about metaphysical realism, or even about whether this distinction makes sense at all.

22 See Stanley Gudder, 'Hidden Variables in Quantum Mechanics Reconsidered', *Review of Modern Physics*, 40 (1968), 229–31; and Sect. III of my 'Semantic Analysis of Quantum Logic', pp. 80–113 in C. A. Hooker (ed.), *Contemporary Research in the Foundations and Philosophy of Quantum Theory* (Dordrecht: Reidel, 1973); and F. J. Belinfante, *A Survey of Hidden–Variable Theories* (New York: Pergamon Press, 1973).

9.4

The Natural Ontological Attitude

Arthur Fine

In the first half of the selection below, Arthur Fine (b. 1937) argues that the no-miracles argument advocated by Boyd and others is viciously circular, presupposing the very principle – that inferences to the best explanation reliably generate true conclusions – which is at issue between the realist and the anti-realist. In the second half Fine argues that there is a position – the "core position" – concerning which both realists and anti-realists are in agreement, but to which they both add further philosophical baggage. Fine argues that we can make do with the core position by itself and that the extra burdens imposed upon it by the realist and anti-realist should simply be shrugged off. He calls advocacy of the core position (and nothing else) the "natural ontological attitude."

Let us fix our attention out of ourselves as much as possible; let us chace our imagination to the heavens, or to the utmost limits of the universe; we never really advance a step beyond ourselves, nor can conceive any kind of existence, but those perceptions, which have appear'd in that narrow compass. This is the universe of the imagination, nor have we any idea but what is there produced.

Hume, *Treatise*, book 1, part II, section VI

Realism is dead. Its death was announced by the neopositivists who realized that they could accept all the results of science, including all the members of the scientific zoo, and still declare that the questions raised by the existence claims of realism were mere pseudo-questions. Its death was hastened by the debates over the interpretation of quantum theory, where Bohr's nonrealist philosophy was seen to win out over Einstein's passionate realism. Its death was certified, finally, as the last two generations of physical scientists

From A. Fine, *The Shaky Game: Einstein, Realism, and the Quantum Theory* (Chicago: University of Chicago Press, 1986), pp. 112–35.

turned their backs on realism and have managed, nevertheless, to do science successfully without it. To be sure, some recent philosophical literature has appeared to pump up the ghostly shell and to give it new life. I think these efforts will eventually be seen and understood as the first stage in the process of mourning, the stage of denial. But I think we shall pass through this first stage and into that of acceptance, for realism is well and truly dead, and we have work to get on with, in identifying a suitable successor. To aid that work I want to do three things in this essay. First, I want to show that the arguments in favor of realism are not sound, and that they provide no rational support for belief in realism. Then, I want to recount the essential role of nonrealist attitudes for the development of science in this century, and thereby (I hope) to loosen the grip of the idea that only realism provides a progressive philosophy of science. Finally, I want to sketch out what seems to me a viable nonrealist position, one that is slowly gathering support and that seems a decent philosophy for postrealist times.[1]

1 Arguments for Realism

Recent philosophical argument in support of realism tries to move from the success of the scientific enterprise to the necessity for a realist account of its practice. As I see it, the arguments here fall on two distinct levels. On the ground level, as it were, one attends to particular successes; such as novel, confirmed predictions, striking unifications of disparate-seeming phenomena (or fields), successful piggybacking from one theoretical model to another, and the like. Then, we are challenged to account for such success, and told that the best and, it is slyly suggested, perhaps, the *only* way of doing so is on a realist basis. I do not find the details of these ground-level arguments at all convincing. Neither does Larry Laudan (1984) and, fortunately, he has provided a forceful and detailed analysis which shows that not even with a lot of handwaving (to shield the gaps in the argument) and charity (to excuse them) can realism itself be used to explain the very successes to which it invites our attention. But there is a second level of realist argument, the methodological level, that derives from Popper's (1972) attack on instrumentalism, which he attacks as being inadequate to account for the details of his own, falsificationist methodology. Arguments on this methodological level have been skillfully developed by Richard Boyd (1981, 1984), and by one of the earlier Hilary Putnams (1975). These arguments focus on the methods embedded in scientific practice, methods teased out in ways that seem to me accurate and perceptive about ongoing science. We are then challenged to account for why these methods lead to scientific success and told that the best and, (again) perhaps, the only truly adequate way of explaining the matter is on the basis of realism.

I want to examine some of these methodological arguments in detail to display the flaws that seem to be inherent in them. But first I want to point out a deep and, I think, insurmountable problem with this entire strategy of defending realism, as I have laid it out above. To set up the problem, let me review the debates in the early part of this century over the foundations of mathematics, the debates that followed Cantor's introduction of set theory. There were two central worries here, one over the meaningfulness of Cantor's hierarchy of sets insofar as it outstripped the number-theoretic content required by Kronecker (and others); the second worry, certainly deriving in good part from the first, was for the consistency (or not) of the whole business. In this context, Hilbert devised a quite brilliant program to try to show the consistency of a mathematical theory by using only the most stringent and secure means. In particular, if one were concerned over the consistency of set theory, then clearly a set-theoretic proof of consistency would be of no avail. For if set theory were inconsistent, then such a consistency proof would be both possible and of no significance. Thus, Hilbert suggested that finite constructivist means, satisfactory even to Kronecker (or Brouwer) ought to be employed in meta-mathematics. Of course, Hilbert's program was brought to an end in 1931, when Gödel showed the impossibility of such a stringent consistency proof. But Hilbert's idea was, I think, correct even though it proved to be unworkable. Metatheoretic arguments must satisfy more stringent requirements than those placed on the arguments used by the theory in question, for otherwise the significance of reasoning about the theory is simply moot. I think this maxim applies with particular force to the discussion of realism.

Those suspicious of realism, from Osiander to Poincaré and Duhem to the "constructive empiricism" of van Fraassen,[2] have been worried about the significance of the explanatory apparatus in scientific investigations. While they appreciate the systematization and coherence brought about by scientific explanation, they question whether acceptable explanations need to be true and, hence, whether the entities mentioned in explanatory principles need to exist.[3] Suppose they are right. Suppose, that is, that the usual explanation-inferring devices in scientific practice do not lead to principles that are reliably true (or nearly so), nor to entities whose existence (or near existence) is reliable. In that case, the usual abductive methods that lead us to good explanations (even to "the best explanation") cannot be counted on to yield results even approximately true. But the strategy that leads to realism, as I have indicated, is just such an ordinary sort of abductive inference. Hence, if the nonrealist were correct in his doubts, then such an inference to realism as the best explanation (or the like), while possible, would be of no significance – exactly as in the case of a consistency proof using the methods of an inconsistent system. It seems, then, that Hilbert's maxim applies to the debate over realism: to argue for realism one must employ methods more stringent than those in ordinary scientific practice. In particular, one must not beg the question as to the significance of explanatory hypotheses by assuming that they carry truth as well as explanatory efficacy.

There is a second way of seeing the same result. Notice that the issue over realism is precisely the issue of whether we should believe in the reality of those individuals, properties, relations, processes, and so forth, used in well-supported explanatory hypotheses. Now what *is* the hypothesis of realism, as it arises as an explanation of scientific practice? It is just the hypothesis that our accepted scientific theories are approximately true, where "being approximately true" is taken to denote an extratheoretical relation between theories and the world. Thus, to address doubts over the reality of relations posited by explanatory hypotheses, the realist proceeds to introduce a further explanatory hypothesis (realism), itself positing such a relation (approximate truth). Surely anyone serious about the issue of realism, and with an open mind about it, would have to behave inconsistently if he were to accept the realist move as satisfactory.

Thus, both at the ground level and at the level of methodology, no support accrues to realism by showing that realism is a good hypothesis for explaining scientific practice. If we are open-minded about realism to begin with, then such a demonstration (even if successful) merely begs the question that we have left open ("need we take good explanatory hypotheses as true?"). Thus, Hilbert's maxim applies, and we must employ patterns of argument more stringent than the usual abductive ones. What might they be? Well, the obvious candidates are patterns of induction leading to empirical generalizations. But, to frame empirical generalizations, we must first have some observable connections between observables. For realism, this must connect theories with the world by way of approximate truth. But no such connections are observable and, hence, suitable as the basis for an inductive inference. I do not want to labor the points at issue here. They amount to the well-known idea that realism commits one to an unverifiable correspondence with the world. So far as I am aware, no recent defender of realism has tried to make a case based on a Hilbert strategy of using suitably stringent grounds and, given the problems over correspondence, it is probably just as well.

The strategy of arguments for realism as a good explanatory hypothesis, then, *cannot* (logically speaking) be effective for an open-minded nonbeliever. But what of the believer? Might he not, at least, show a kind of internal coherence about realism as an overriding philosophy of science, and should that not be of some solace, at least for the realist?[4] Recall, however, the analogue with consistency proofs for inconsistent systems. That sort of harmony should be of no solace to anyone. But for realism, I fear, the verdict is even harsher. For, so far as I can see, the arguments in question just do not work, and the reason for that has to do with the same question-begging procedures that I have already identified. Let me look closely at some methodological arguments in order to display the problems.

A typical realist argument on the methodological level deals with what I shall call the problem of the "small handful." It goes like this. At any time, in a given scientific area, only a small handful of alternative theories (or hypotheses) are in the field. Only such a small handful are seriously considered as competitors, or as possible successors to some theory requiring revision. Moreover, in general, this handful displays a sort of family resemblance in that none of these live options will be too far from the previously accepted theories in the field, each preserving the well-confirmed features of the earlier theories and deviating only in those aspects less confirmed. Why? Why does this narrowing down of our choices to such a small handful of cousins of our previously accepted theories work to produce good successor theories?

The realist answers this as follows. Suppose that the already existing theories are themselves approximately true descriptions of the domain under consideration. Then surely it is reasonable to restrict one's search for successor theories to those whose ontologies and laws resemble what we already have, especially where what we already have is well confirmed. And if these earlier theories were approximately true, then so will be such conservative successors. Hence, such successors will be good predictive instruments; that is, they will be successful in their own right.

The small-handful problem raises three distinct questions: (1) why only a small handful out of the (theoretically) infinite number of possibilities? (2) why the conservative family resemblance between members of the handful? and (3) why does the strategy of narrowing the choices in this way work so well? The realist response does not seem to address the first issue at all, for even if we restrict ourselves just to successor theories resembling their progenitors, as suggested, there would still, theoretically, always be more than a small handful of these. To answer the second question, as to why conserve the well-confirmed features of ontology and laws, the realist must suppose that such confirmation is a mark of an approximately correct ontology and approximately true laws. But how could the realist possibly justify such an assumption? Surely, there is no valid inference of the form, "*T* is well-confirmed; therefore, there exist objects pretty much of the sort required by *T* and satisfying laws approximating those of *T*." Any of the dramatic shifts of ontology in science show the invalidity of this schema. For example, the loss of the ether from the turn-of-the-century electrodynamic theories demonstrates this at the level of ontology, and the dynamics of the Rutherford–Bohr atom vis-à-vis the classical energy principles for rotating systems demonstrates it at the level of laws. Of course, the realist might respond that there is no question of a strict inference between being well confirmed and being approximately true (in the relevant respects), but there is a probable inference of some sort. But of what sort? Certainly there is no probability relation that rests on inductive evidence here. For there is no independent evidence for the relation of approximate truth itself; at least, the realist has yet to produce any evidence that is independent of the argument under examination. But if the probabilities are not grounded inductively, then how else? Here, I think the realist may well try to fall back on his original strategy and suggest that being approximately true provides the best explanation for being well confirmed. This move throws us back to the ground-level realist argument, the argument from specific success to an approximately true description of reality, which Laudan (1984) has criticized. I should point out, before looking at the third question, that if this last move is the one the realist wants to make, then his success at the methodological level can be no better than his success at the ground level. If he fails there, he fails across the board.

The third question, and the one I think the realist puts most weight on, is why does the small-handful strategy work so well. The instrumentalist, for example, is thought to have no answer here. He must just note that it does work well and be content with that. The realist, however, can explain why it works by citing the transfer of approximate truth from predecessor theories to the successor theories. But what does this explain? At best, it explains why the successor theories cover the same ground as well as their predecessors, for the conservative strategy under consideration assures that. But note that here the instrumentalist can offer the same account: if we insist on preserving the well-confirmed components of earlier theories in later theories, then, of course, the later ones will do well over the well-confirmed ground. The difficulty, however, is not here at all but rather in how to account for the successes of the later theories in new ground, or with respect to novel predictions, or in overcoming the anomalies of the earlier theories. And what can the realist possibly say in this area except that the theorist, in proposing a new theory, has happened to make a good guess? For nothing in the approximate truth of the old theory can guarantee (or even make it likely) that modifying the theory in its less-confirmed parts will produce a progressive shift. The history of science shows well enough how such tinkering succeeds only now and again, and fails for the most part. This history of failures can scarcely be adduced to explain the occasional success. The idea that by extending what is approximately true one is likely to bring new approximate truth is a chimera. It finds support neither in the logic of approximate truth nor in the history of science. The problem for the realist is how to explain the *occasional success* of a strategy that *usually fails*.[5] I think he has no special resources with which to do this. In particular, his usual fallback onto approximate truth provides nothing more than a gentle pillow. He may rest on it comfortably, but it does not really help to move his cause forward.

The problem of the small handful raises three challenges: why small? why narrowly related? and why does it work? The realist has no answer for the first of these, begs the question as to the truth of explanatory hypotheses on the second, and has no resources for addressing the third. For comparison, it may be useful to see how well his archenemy, the instrumentalist, fares on the same turf. The instrumentalist, I think, has a substantial basis for addressing the questions of smallness and narrowness, for he can point out that it is extremely difficult to come up with alternative theories that satisfy the many empirical constraints posed by the instrumental success of theories already in the field. Often it is hard enough to come up with even one such alternative. Moreover, the common apprenticeship of scientists working in the same area certainly has the effect of narrowing down the range of options by channeling thought into the commonly accepted categories. If we add to this the instrumentally justified rule, "If it has worked well in the past, try it again," then we get a rather good account, I think, of why there is usually only a small and narrow handful. As to why this strategy works to produce instrumentally successful science, we have already noted that for the most part it does not. Most of what this strategy produces are failures. It is a quirk of scientific memory that this fact gets obscured, much as do the memories of bad times during a holiday vacation when we recount all our "wonderful" vacation adventures to a friend. Those instrumentalists who incline to a general account of knowledge as a social construction can go further at this juncture and lean on the sociology of science to explain how the scientific community "creates" its knowledge. I am content just to back off here and note that over the problem of the small handful, the instrumentalist scores at least two out of three, whereas the realist, left to his own devices, has struck out.[6]

I think the source of the realist's failure here is endemic to the methodological level, infecting all of his arguments in this domain. It resides, in the first instance, in his repeating the question-begging move from explanatory efficacy to the truth of the explanatory hypothesis. And in the second instance, it resides in his twofold mishandling of the concept of approximate truth: first, in his trying to project from some body of assumed approximate truths *to* some further and novel such truths, and second, in his needing genuine access to the relation of correspondence. There are no general connections of this first sort, however, sanctioned by the logic of approximate truth, nor secondly, any such warranted access. However, the realist must pretend that there are in order to claim explanatory power for his realism. We have seen those two agents infecting the realist way with the problem of the small handful. Let me show them at work in another methodological favorite of the realist, the "problem of conjunctions."

The problem of conjunctions is this. If T and T' are independently well-confirmed, explanatory theories, and if no shared term is ambiguous between the two, then we expect the conjunction of T and T' to be a reliable predictive instrument (provided, of course, that the theories are not mutually inconsistent). Why? challenges the realist, and he answers as follows. If we make the realist assumption that T and T', being well confirmed, are approximately true of the entities (etc.) to which they refer, and if the unambiguity requirement is taken realistically as requiring a domain of common reference, then the conjunction of the two theories will also be approximately true and, hence, it will produce reliable observational predictions. Q.E.D.

But notice our agents at work. First, the realist makes the question-begging move from explanations to their approximate truth, and then he mistreats approximate truth. For nothing in the logic of approximate truth sanctions the inference from "T is approximately true" and "T' is approximately true" to the conclusion that the conjunction "$T \cdot T'$" is approximately true. Rather, in general, the tightness of an approximation dissipates as we pile on further approximations. If T is within ε, in its estimation of some parameter, and T' is also within ε, then the only general thing we can say is that the conjunction will be within 2ε of the parameter. Thus, the logic of approximate truth should lead us to the opposite conclusion here; that is, that the conjunction of two theories is, in general, *less* reliable than either (over their common domain). But this is neither what we expect nor what we find. Thus, it seems quite implausible that our actual expectations about the reliability of conjunctions rest on the realist's stock of approximate truths.

Of course, the realist could try to retrench here and pose an additional requirement of some sort of uniformity on the character of the approximations, as between T and T'.[7] It is difficult to see how the realist could do this successfully without making reference to the distance between the approximations and "the truth." For what kind of internalist requirement could possibly insure the narrowing of this distance? But the realist is in no position to impose such requirements, since neither he nor anyone else has the requisite access to "the truth." Thus, whatever uniformity-of-approximation condition the realist might impose, we could still demand to be shown that this leads closer to the truth, not farther away. The realist will have no demonstration, except to point out to us that it all works (sometimes!). But that was the original puzzle.[8] Actually, I think the puzzle is not very difficult. For surely, if we do not entangle ourselves with issues over approximation, there is no deep mystery as to why two compatible and successful theories lead us to expect their conjunction to be successful. For in forming the conjunction, we just add the reliable predictions of one onto the reliable predictions of the other, having antecedently ruled out the possibility of conflict.

There is more to be said about this topic. In particular, we need to address the question of why we expect the logical gears of the two theories to mesh. However, I think that a discussion of the realist position here would only bring up the same methodological and logical problems that we have already uncovered at the center of the realist argument.

Indeed, this schema of knots in the realist argument applies across the board and vitiates every single argument at the methodological level. Thus my conclusion here is harsh, indeed. The methodological arguments for realism fail, even though, were they successful, they would still not support the case. For the general strategy they are supposed to implement is just not stringent enough to provide rational support for realism. In the next two sections, I will try to show that this situation is just as well, for realism has not always been a progressive factor in the development of science and, anyway, there is a position other than realism that is more attractive.

2 Realism and Progress

If we examine the two twentieth-century giants among physical theories, relativity and the quantum theory, we find a living refutation of the realist's claim that only his view of science explains its progress, and we find some curious twists and contrasts over realism as well. The theories of relativity are almost single-handedly the work of Albert Einstein. Einstein's early positivism and his methodological debt to Mach (and Hume) leap right out of the pages of the 1905 paper on special relativity.[9] The same positivist strain is evident in the 1916 general relativity paper as well, where Einstein (in section 3 of that paper) tries to justify his requirement of general covariance by means of a suspicious-looking verificationist argument which, he says, "takes away from space and time the last remnants of physical objectivity" (Einstein et al. 1952, p. 117). A study of his tortured path to general relativity[10] shows the repeated use of this Machist line, always used to deny that some concept has a real referent. Whatever other, competing strains there were in Einstein's philosophical orientation (and there certainly were others), it would be hard to deny the importance of this instrumentalist/positivist attitude in liberating Einstein from various realist commitments. Indeed, on another occasion, I would argue in detail that without the "freedom from reality" provided by his early reverence for Mach, a central tumbler necessary to unlock the secret of special relativity would never have fallen into place.[11] A few years after his work on general relativity, however, roughly around 1920, Einstein underwent a philosophical conversion, turning away from his positivist youth (he was forty-one in 1920) and becoming deeply committed to realism (see chapter 6). In particular, following his conversion, Einstein wanted to claim genuine reality for the central theoretical entities of the general theory, the four-dimensional space-time manifold and associated tensor fields. This is a serious business, for if we grant his claim, then not only do space and time cease to be real but so do virtually all of the usual dynamical quantities.[12] Thus motion, as we understand it, itself ceases to be real. The current generation of philosophers of space and time (led by Howard Stein and John Earman) have followed Einstein's lead here. But, interestingly, not only do these ideas boggle the mind of the average man in the street (like you and me), they boggle most contemporary scientific minds as well.[13] That is, I believe the majority opinion among working, knowledgeable scientists is that general relativity provides a magnificent organizing tool for treating certain gravitational problems in astrophysics and cosmology. But few, I believe, give credence to the kind of realist existence and nonexistence claims that I have been mentioning. For relativistic physics, then, it appears that a nonrealist attitude was important in its development, that the founder nevertheless espoused a realist attitude to the finished product, but that most who actually use it think of the theory as a powerful instrument, rather than as expressing a "big truth."

With quantum theory, this sequence gets a twist. Heisenberg's seminal paper of 1925 is prefaced by the following abstract, announcing, in effect, his philosophical stance: "In this paper an attempt will be made to obtain bases for a quantum-theoretical mechanics based exclusively on relations between quantities observable in principle" (Heisenberg 1925, p. 879). In the body of the paper, Heisenberg not only rejects any reference to unobservables, he also moves away from the very idea that one should try to form any picture of a reality underlying his mechanics. To be sure, Schrödinger, the second father of quantum theory, seems originally to have had a vague picture of an underlying wavelike reality for his own equation. But he was quick to see the difficulties here and, just as quickly, although reluctantly, abandoned the attempt to interpolate any reference to reality.[14] These instrumentalist moves away from a realist construal of the emerging quantum theory were given particular force by Bohr's so-called philosophy of complementarity. This nonrealist position was consolidated at the time of the famous Solvay Conference, in October 1927, and is firmly in place today. Such quantum nonrealism is part of what every graduate physicist learns and practices. It is the conceptual backdrop to all the brilliant successes in atomic, nuclear, and particle physics over the past fifty years. Physicists have learned to think about their theory in a highly nonrealist way, and doing just that has brought about the most marvelous predictive success in the history of science.

The war between Einstein, the realist, and Bohr, the nonrealist, over the interpretation of quantum theory was not, I believe, just a sideshow in physics, nor an idle intellectual exercise. It was an important endeavor undertaken by Bohr on behalf of the enterprise of physics as a progressive science. For Bohr believed (and this fear was shared by Heisenberg, Sommerfeld, Pauli, and Born – and all the big guys) that Einstein's realism, if taken seriously, would block the consolidation and articulation of the new physics and, thereby, stop the progress of science. They were afraid, in particular, that Einstein's realism would lead the next generation of the brightest and best students into scientific dead ends. Alfred Landé, for example, as a graduate student, was interested in spending some time in Berlin to sound out Einstein's ideas. His supervisor was Sommerfeld, and recalling this period, Landé (1974, p. 460) writes, "The more pragmatic Sommerfeld . . . warned his students, one of them this writer, not to spend too much time on the hopeless task of "explaining" the quantum but rather to accept it as fundamental and help work out its consequences."

The task of "explaining" the quantum, of course, is the realist program for identifying a reality underlying the formulas of the theory and thereby explaining the predictive success of the formulas as approximately true descriptions of this reality. It is this program that I have criticized in the first part of this chapter, and this same program that the builders of quantum theory saw as a scientific dead end. Einstein knew perfectly well that the issue was joined right here. In the summer of 1935, he wrote to Schrödinger, "The real problem is that physics is a kind of metaphysics; physics describes 'reality.' But we do not know what 'reality' is. We know it only through physical description. . . . But the Talmudic philosopher sniffs at 'reality,' as at a frightening creature of the naive mind."[15]

By avoiding the bogey of an underlying reality, the "Talmudic" originators of quantum theory seem to have set subsequent generations on precisely the right path. Those inspired by realist ambitions have produced no predictively successful physics. Neither Einstein's conception of a unified field, nor the ideas of the de Broglie group about pilot waves, nor the Bohm-inspired interest in hidden variables has made for scientific progress. To be sure, several philosophers of physics, including another Hilary Putnam and myself, have fought a battle over the last decade to show that the quantum theory is at least consistent with some kind of underlying reality. I believe that Hilary has abandoned the cause, perhaps in part on account of the recent Bell-inequality problem over correlation experiments, a problem that van Fraassen (1982) calls "the Charybdis of realism." My own recent work in the area suggests that we may still be able to keep realism afloat in this whirlpool.[16] But the possibility (as I still see it) for a realist account of the quantum domain should not lead us away from appreciating the historical facts of the matter.

One can hardly doubt the importance of a nonrealist attitude for the development and practically infinite success of the quantum theory. Historical counterfactuals are always tricky, but the sterility of actual realist programs in this area at least suggests that Bohr and company were right in believing that the road to scientific progress here would have been blocked by realism. The founders of quantum theory never turned on the nonrealist attitude that served them so well. Perhaps that is because the central underlying theoretical device of quantum theory, the densities of a complex-valued and infinite-dimensional wave function, are even harder to take seriously than is the four-dimensional manifold of relativity. But now there comes a most curious twist. For just as the practitioners of relativity, I have suggested, ignore the *realist* interpretation in favor of a more pragmatic attitude toward the space/time structure, the quantum physicists would appear to make a similar reversal and to forget their nonrealist history and allegiance when it comes time to talk about new discoveries.

Thus, anyone in the business will tell you about the exciting period, in the fall of 1974, when the particle group at Brookhaven, led by Samuel Ting, discovered the J particle, just as a Stanford team at the Stanford Linear Accelerator Center, under Burton Richter, independently found a new particle they called ψ. These turned out to be one and the same, the so-called ψ/J particle (Mass 3,098 MeV, Spin 1, Resonance 67 keV, Strangeness 0). To explain this new entity, the theoreticians were led to introduce a new kind of quark, the so-called charmed quark. The ψ/J particle is then thought to be made up out of a charmed quark and an anticharmed quark, with their respective spins aligned. But if this is correct, then there ought to be other such pairs antialigned, or with variable spin alignments, and these ought to make up quite new observable particles. Such predictions from the charmed-quark model have turned out to be confirmed in various experiments.

I have gone on a bit in this story in order to convey the realist feel to the way scientists speak in this area. For I want to ask whether this is a return to realism or whether, instead, it can somehow be reconciled with a fundamentally nonrealist attitude.[17] I believe that the nonrealist option is correct.

3 Nonrealism

Even if the realist happens to be a talented philosopher, I do not believe that, in his heart, he relies for his realism on the rather sophisticated form of abductive argument that I have examined and rejected in the first section of this chapter, and which the history of twentieth-century physics shows to be fallacious. Rather, if his heart is like mine, then I suggest that a more simple and homely sort of argument is what grips him. It is this, and I will put it in the first person. I certainly trust the evidence of my senses, on the whole, with regard to the existence and features of everyday objects. And I have similar confidence in the system of "check, double-check, check, triple-check" of scientific investigation, as well as the other safeguards built into the institutions of science. So, if the scientists tell me that there really are molecules, and atoms, and ψ/J particles, and, who knows, maybe even quarks, then so be it. I trust them and, thus, must accept that there really are such things with their attendant properties and relations. Moreover, if the instrumentalist (or some other member of the species "nonrealistica") comes along to say that these entities and their attendants are just fictions (or the like), then I see no more reason to believe him than to believe that *he is* a fiction, made up (somehow) to do a job on me; which I do not believe. It seems, then, that I had better be a realist. One can summarize this homely and compelling line as follows: it is possible to accept the evidence of one's senses and to accept, *in the same way*, the confirmed results of science only for a realist; hence, I should be one (and so should you!).

What is it to accept the evidence of one's senses and, *in the same way*, to accept confirmed scientific theories? It is to take them into one's life as true, with all that implies concerning adjusting one's behavior, practical and theoretical, to accommodate these truths. Now, of course, there are truths, and truths. Some are more central to us and our lives, some less so. I might be mistaken about anything, but were I mistaken about where I am right now, that might affect me more than would my perhaps mistaken belief in charmed quarks. Thus, it is compatible with the homely line of argument that some of the scientific beliefs that I hold are less central than some, for example, perceptual beliefs. Of course, were I deeply in the charmed-quark business, giving up that belief might be more difficult than giving up some at the perceptual level. (Thus we get the phenomenon of "seeing what you believe," well known to all thoughtful people.) When the homely line asks us, then, to accept the scientific results "in the same way" in which we accept the evidence of our senses, I take it that we are to accept them both as true. I take it that we are being asked not to distinguish between kinds of truth or modes of existence or the like, but only among truths themselves in terms of centrality, degrees of belief, or such.

Let us suppose this understood. Now, do you think that Bohr, the archenemy of realism, could toe the homely line? Could Bohr, fighting for the sake of science (against Einstein's realism) have felt compelled either to give up the results of science, or else to assign its "truths" to some category different from the truths of everyday life? It seems unlikely. And thus, unless we uncharitably think Bohr inconsistent on this basic issue, we might well come to question whether there is any necessary connection moving us from accepting the results of science as true to being a realist.[18]

Let me use the term "antirealist" to refer to any of the many different specific enemies of realism: the idealist, the instrumentalist, the phenomenalist, the empiricist (constructive or not), the conventionalist, the constructivist, the pragmatist, and so forth. Then, it seems to me that both the realist and the antirealist must toe what I have been calling "the homely line." That is, they must both accept the certified results of science as on par with more homely and familiarly supported claims. That is not to say that one party (or the other) cannot distinguish more from less well-confirmed claims at home or in science; nor that one cannot single out some particular mode of inference (such as inference to the best explanation) and worry over its reliability, both at home and away. It is just that one must maintain parity. Let us say, then, that both realist and antirealist accept the results of scientific investigations as "true," on par with more homely truths. (I realize that some antirealists would rather use a different word, but no matter.) And call this acceptance of scientific truths the "core position."[19] What distinguishes realists from antirealists, then, is what they add onto this core position.

The antirealist may add onto the core position a particular analysis of the concept of truth, as in the pragmatic and instrumentalist and conventionalist conceptions of truth. Or the antirealist may add on a special analysis of concepts, as in idealism, constructivism, phenomenalism, and in some varieties of empiricism. These addenda will then issue in a special meaning, say, for existence statements. Or the antirealist may add on certain methodological strictures, pointing a wary finger at some particular inferential tool, or constructing his own account for some particular aspects of science (e.g., explanations or laws). Typically, the antirealist will make several such additions to the core.

What then of the realist, what does he add to his core acceptance of the results of science as really true? My colleague, Charles Chastain, suggested what I think is the most graphic way of stating the answer – namely, that what the realist adds on is a desk-thumping, foot-stamping shout of "Really!" So, when the realist and antirealist agree, say, that there really are electrons and that they really carry a unit negative charge and really do have a small mass (of about 9.1×10^{-28} grams), what the realist wants to add is the emphasis that all this is really so. "There really are electrons, really!" This typical realist emphasis serves both a negative and a positive function. Negatively, it is meant to deny the additions that the antirealist would make to that core acceptance which both parties share. The realist wants to deny, for example, the phenomenalistic reduction of concepts or the pragmatic conception of truth. The realist thinks that these addenda take away from the substantiality of the accepted claims to truth or existence. "No," says he, "they *really* exist, and not in just your diminished antirealist sense." Positively, the realist wants to explain the robust sense in which *he* takes these claims to truth or existence; namely, as claims about reality – what is really, really the case. The full-blown version of this involves the conception of truth as correspondence with the world, and the surrogate use of approximate truth as near-correspondence. We have already seen how these ideas of correspondence and approximate truth are supposed to explain what *makes* the truth *true* whereas, in fact, they function as mere trappings, that is, as superficial decorations that may well attract our attention but do not compel rational belief. Like the extra "really," they are an arresting foot thump and, logically speaking, of no more force.

It seems to me that when we contrast the realist and the antirealist in terms of what they each want to add to the core position, a third alternative emerges – and an attractive one at that. It is the core position itself, *and all by itself.* If I am correct in thinking that, at heart, the grip of realism only extends to the homely connection of everyday truths with scientific truths, and that good sense dictates our acceptance of the one on the same basis as our acceptance of the other, then the homely line makes the core position, all by itself, a compelling one, one that we ought to take to heart. Let us try to do so and see whether it constitutes a philosophy, and an attitude toward science, that we can live by.

The core position is neither realist nor antirealist; it mediates between the two. It would be nice to have a name for this position, but it would be a shame to appropriate another "ism" on its behalf, for then it would appear to be just one of the many contenders for ontological allegiance. I think it is not just one of that crowd but rather, as the homely line behind it suggests, it is for commonsense epistemology – the natural ontological attitude. Thus, let me introduce the acronym *NOA* (pronounced as in "Noah"), for *natural ontological attitude*, and, henceforth, refer to the core position under that designation.

To begin showing how NOA makes for an adequate philosophical stance toward science, let us see what it has to say about ontology. When NOA counsels us to accept the results of science as true, I take it that we are to treat truth in the usual referential way, so that a sentence (or statement) is true just in case the entities referred to stand in the referred-to relations. Thus, NOA sanctions ordinary referential semantics and commits us, via truth, to the existence of the individuals, properties, relations, processes, and so forth referred to by the scientific statements that we accept as true. Our belief in their existence will be just as strong (or weak) as our belief in the truth of the bit of science involved, and degrees of belief here, presumably, will be tutored by ordinary relations of confirmation and evidential support, subject to the usual scientific canons. In taking this referential stance, NOA is not committed to the progressivism that seems inherent in realism. For the realist, as an article of faith, sees scientific success, over the long run, as bringing us closer to the truth. His whole explanatory enterprise, using approximate truth, forces his hand in this way. But, a "NOAer" (pronounced as "knower") is not so committed. As a scientist, say, within the context of the tradition in which he works, the NOAer, of course, will believe in the existence of those entities to which his theories refer. But should the tradition change, say, in the manner of the conceptual revolutions that Kuhn dubs "paradigm shifts," then nothing in NOA dictates that the change be assimilated as being progressive, that is, as a change where we learn more accurately about *the same things*. NOA is perfectly consistent with the Kuhnian alternative, which counts such changes as wholesale changes of reference. Unlike the realist, adherents to NOA are free to examine the facts in cases of paradigm shift, and to see whether or not a convincing case for stability of reference across paradigms can be made without superimposing on these facts a realist-progressivist superstructure. I have argued elsewhere (Fine 1975) that if one makes oneself free, as NOA enables one to do, then the facts of the matter will not usually settle the case; and that this is a good reason for thinking that cases of so-called incommensurability are, in fact, genuine cases where the question of stability of reference is indeterminate. NOA, I think, is the right philosophical position for such conclusions. It sanctions reference and existence claims, but it does not force the history of science into prefit molds.

So far I have managed to avoid what, for the realist, is the essential point: what of the "external world"? How can I talk of reference and of existence claims unless I am talking about referring to things right out there in the world? And here, of course, the realist, again, wants to stamp his feet.[20] I think the problem that makes the realist want to stamp his feet, shouting "Really!" (and invoking the external world) has to do with the stance the realist tries to take vis-à-vis the game of science. The realist, as it were, tries to stand outside the arena watching the ongoing game and then tries to judge (from this external point of view) what the point is. It is, he says, *about* some area external to the game. The realist, I think, is fooling himself. For he cannot (really!) stand outside the arena, nor can he survey some area off the playing field and mark it out as what the game is about.

Let me try to address these two points. How are we to arrive at the judgment that, in addition to, say, having a rather small mass, electrons are objects "out there in the external world"? Certainly, we can stand off from the electron game and survey its claims, methods, predictive success, and so forth. But what stance could we take that would enable us to judge what the theory of electrons is *about*, other than agreeing that it is about electrons? It is not like matching a blueprint to a house being built, or a map route to a country road. For we are *in* the world, both physically and conceptually.[21] That is, *we* are among the objects of science, and the concepts and procedures that we use to make judgments of subject matter and correct application are themselves part of that same scientific world. Epistemologically, the situation is very much like the situation with regard to the justification of induction. For the problem of the external world (so-called) is how to satisfy the realist's demand that we justify the existence claims sanctioned by science (and, therefore, by NOA) as claims to the existence of entities "out there." In the case of induction, it is clear that only an inductive justification will do, and it is equally clear that no inductive justification will do at all. So too with the external world, for only ordinary scientific inferences to existence will do, and yet none of them satisfies the demand for showing that the existent is really "out there." I think we ought to follow Hume's prescription on induction with regard to the external world. There is no possibility for justifying the kind of externality that realism requires, yet it may well be that, in fact, we cannot help yearning for just such a comforting grip on reality.

If I am right, then the realist is chasing a phantom, and we cannot actually do more, with regard to existence claims, than follow scientific practice, just as NOA suggests. What then of the other challenges raised by realism? Can we find in NOA the resources for understanding scientific practice? In particular (since it was the topic of the first part of this chapter), does NOA help us to understand the scientific method, say, the problems of the small handful or of conjunctions? The sticking point with the small handful was to account for why the few and narrow alternatives that we can come up with, result in successful novel predictions, and the like. The background was to keep in mind that most such narrow alternatives are not successful. I think that NOA has only this to say. If you believe that guessing based on some truths is more likely to succeed than guessing pure and simple, then if our earlier theories were in large part true and if our refinements of them conserve the true parts, then guessing on this basis has some relative likelihood of success. I think this is a weak account, but then I think the phenomenon here does not allow for anything much stronger since, for the most part, such guesswork fails. In the same way, NOA can help with the problem of conjunctions (and, more generally, with problems of logical combinations). For if two consistent theories in fact have overlapping domains (a fact, as I have just suggested, that is not so often decidable), and if the theories also have true things to say about members in the overlap, then conjoining the theories just adds to the truths of each and, thus, *may*, in conjunction, yield new truths. Where one finds other successful methodological rules, I think we will find NOA's grip on the truth sufficient to account for the utility of the rules.

Unlike the realist, however, I would not tout NOA's success at making science fairly intelligible as an argument in its favor, vis-à-vis realism or various antirealisms. For NOA's accounts are available to the realist and the antirealist, too, provided what they add to NOA does not negate its appeal to the truth, as does a verificationist account of truth or the realist's longing for approximate truth. Moreover, as I made plain enough in the first section of this chapter, I am sensitive to the possibility that explanatory efficacy can be achieved without the explanatory hypothesis being true. NOA may well make science seem fairly intelligible and even rational, but NOA could be quite the wrong view of science for all that. If we posit as a constraint on philosophizing about science that the scientific enterprise should come out in our philosophy as not too unintelligible or irrational, then, perhaps, we can say that NOA passes a minimal standard for a philosophy of science.

Indeed, perhaps the greatest virtue of NOA is to call attention to just how minimal an adequate philosophy of science can be. (In this respect, NOA might be compared to the minimalist movement in art.) For example, NOA helps us to see that realism differs from various antirealisms in this way: realism adds an outer direction to NOA, that is, the external world and the correspondence relation of approximate truth; antirealisms (typically) add an inner direction, that is, human-oriented reductions of truth, or concepts, or explanations (as in my opening citation from Hume). NOA suggests that the legitimate features of these additions are already contained in the presumed equal status of everyday truths with scientific ones, and in our accepting them both as *truths*. No other additions are legitimate, and none are required.

It will be apparent by now that a distinctive feature of NOA, one that separates it from similar views currently in the air, is NOA's stubborn refusal to amplify the concept of truth by providing a theory or analysis (or even a metaphorical picture). Rather, NOA recognizes in "truth" a concept already in use and agrees to abide by the standard rules of usage. These rules involve a Davidsonian–Tarskian referential semantics, and they support a thoroughly classical logic of inference. Thus NOA respects the customary "grammar" of "truth" (and its cognates). Likewise, NOA respects the customary epistemology, which grounds judgments of truth in perceptual judgments and various confirmation relations. As with the use of other concepts, disagreements are bound to arise over what is true (for instance, as to whether inference to the best explanation is always truth-conferring). NOA pretends to no resources for settling these disputes, for NOA takes to heart the great lesson of twentieth-century analytic and continental philosophy, namely, that there *are* no general methodological or philosophical resources for deciding such things. The mistake common to realism and all the antirealisms alike is their commitment to the existence of such nonexistent resources. If pressed to answer the question of what, then, does it *mean* to say that something is true (or to what does the truth of so-and-so commit one), NOA will reply by pointing out the logical relations engendered by the specific claim and by focusing, then, on the concrete historical circumstances that ground that particular judgment of truth. For, after all, there *is* nothing more to say.[22]

Because of its parsimony, I think the minimalist stance represented by NOA marks a revolutionary approach to understanding science. It is, I would suggest, as profound in its own way as was the revolution in our conception of morality, when we came to see that founding morality on God and his order was *also* neither legitimate nor necessary. Just as the typical theological moralist of the eighteenth century would feel bereft to read, say, the pages of *Ethics*, so I think the realist must feel similarly when NOA removes that "correspondence to the external world" for which he so longs. I too have regret for that lost paradise, and too often slip into the realist fantasy. I use my understanding of twentieth-century physics to help me firm up my convictions about NOA, and I recall some words of Mach, which I offer as a comfort and as a closing. With reference to realism, Mach writes,

> It has arisen in the process of immeasurable time without the intentional assistance of man. It is a product of nature, and preserved by nature. Everything that philosophy has accomplished . . . is, as compared with it, but an insignificant and ephemeral product of art. The fact is, every thinker, every philosopher, the moment he is forced to abandon his one-sided intellectual occupation . . . , immediately returns [to realism].
>
> Nor is it the purpose of these "introductory remarks" to discredit the standpoint [of realism]. The task which we have set ourselves is simply to show why and for what purpose we hold that standpoint during most of our lives, and why and for what purpose we are . . . obliged to abandon it.

These lines are taken from Mach's *The Analysis of Sensations* (sec. 14). I recommend that book as effective realism-therapy, a therapy that works best (as Mach suggests) when accompanied by historicophysical investigations (real versions of the breakneck history of my second section). For a better philosophy, however, I recommend NOA.

Notes

1 In the final section, I call this postrealism "NOA." Among recent views that relate to NOA, I would include Hilary Putnam's "internal realism," Richard Rorty's "epistemological behaviorism," the "semantic realism" espoused by Paul Horwich, parts of the "Mother Nature" story told by William Lycan, and the defense of common sense worked out by Joseph Pitt (as a way of reconciling W. Sellars's manifest and scientific images). For references, see Putnam (1981a), Rorty (1979), Horwich (1982), Lycan (1982), and Pitt (1981).

2 Van Fraassen (1980). See especially pp. 97–101 for a discussion of the truth of explanatory theories. To see that the recent discussion of realism is joined right here, one should contrast van Fraassen with Newton-Smith (1981), especially chap. 8.

3 Cartwright (1983) includes some marvelous essays on these issues.

4 Some realists may look for genuine support, and not just solace, in such a coherentist line. They may see in their realism a basis for general epistemology, philosophy of language, and so forth (as does Boyd 1981, 1984). If they find in all this a coherent and comprehensive worldview, then they might want to argue for their philosophy as Wilhelm Wien argued (in 1909) for special relativity, "What speaks for it most of all is the inner consistency which makes it possible to lay a foundation having no self-contradictions, one that applies to the totality of physical appearances." (Quoted by Gerald Holton, "Einstein's Scientific Program: Formative Years," in H. Woolf (1980), p. 58.) Insofar as the realist moves away from the abductive defense of realism to seek support, instead, from the merits of a comprehensive philosophical system with a realist core, he marks as a failure the bulk of recent defenses of realism. Even so, he will not avoid the critique pursued in the text. For although my argument above has been directed, in particular, against the abductive strategy, it is itself based on a more general maxim; namely, that the form of argument used to support realism must be more stringent than the form of argument embedded in the very scientific practice that realism itself is supposed to ground – on pain of begging the question. Just as the abductive strategy fails because it violates this maxim, so too would the coherentist strategy, should the realist turn from one to the other. For, as we see from the words of Wien, the same coherentist line that the realist would appropriate for his own support is part of ordinary scientific practice in framing judgments about competing theories. It is, therefore, not a line of defense available to the realist. Moreover, just as the truth-bearing status of abduction is an issue dividing realists from various nonrealists, so too is the status of coherence-based inference. Turning from abduction to coherence, therefore, still leaves the realist begging the question. Thus, when we bring out into the open the character of arguments *for* realism, we see quite plainly that they do not work. See Fine (1986) for a more detailed discussion.

In support of realism there seem to be only those "reasons of the heart" which, as Pascal says, reason does not know. Indeed, I have long felt that belief in realism involves a profound leap of faith, not at all dissimilar from the faith that animates deep religious convictions. I would welcome engagement with realists on this understanding, just as I enjoy conversation on a similar basis with my religious friends. The dialogue will proceed more fruitfully, I think, when the realists finally stop pretending to a rational support for their faith, which they do not have. Then we can all enjoy their intricate and sometimes beautiful philosophical constructions (of, e.g., knowledge, or reference, etc.), even though to us, as nonbelievers, they may seem only wonder-full castles in the air.

5 I hope all readers of this essay will take this idea to heart. For in formulating the question as how to explain why the methods of science lead to instrumental success, the realist has seriously misstated the explanandum. Overwhelmingly, the results of the conscientious pursuit of scientific inquiry are failures: failed theories, failed hypotheses, failed conjectures, inaccurate measurements, incorrect estimations of parameters, fallacious causal inferences, and so forth. If explanations are appropriate here, then what requires explaining is why the very same methods produce an overwhelming background of failures and, occasionally, also a pattern of successes. The realist literature has not yet begun to address this question, much less to offer even a hint of how to answer it.

6 Of course, the realist can appropriate the devices and answers of the instrumentalist, but that would be cheating, and anyway, it would not provide the desired support of realism per se.

7 Paul Teller has made this suggestion to me in conversation.

8 Niiniluoto (1982) contains interesting formal constructions for "degree of truthlikeness," and related versimilia. As conjectured above, they rely on an unspecified correspondence relation to the truth and on measures of the "distance" from the truth. Moreover, they fail to sanction that projection, from some approximate truths to other, novel truths, which lies at the core of realist rationalizations.

9 See Gerald Holton, "Mach, Einstein, and the Search for Reality," in Holton (1973), pp. 219–59. I have tried to work out the precise role of this positivist methodology in chapter 2. See also Fine (1981a).

10 Earman and Glymour (1978). The tortuous path detailed by Earman is sketched by B. Hoffmann (1972), pp. 116–28. A nontechnical and illuminating account is given by John Stachel (1979).

11 I have in mind the role played by the analysis of simultaneity in Einstein's path to special relativity. Despite the important study by Arthur Miller (1981) and an imaginative pioneering work by John Earman et al. (1983), I think the role of positivist analysis in the 1905 paper has yet to be properly understood.

12 Roger Jones in "Realism about What?" (1991) explains very nicely some of the difficulties here.

13 I think the ordinary, deflationist attitude of working scientists is much like that of Steven Weinberg (1972).

14 See Wessels (1979) and A. Fine (1986) *The Shaky Game: Einstein, Realism and the Quantum Theory* (Chicago: University of Chicago Press), ch. 5.

15 Letter to Schrödinger, June 19, 1935.

16 See my (1982d) for part of the discussion and also Fine (1986), ch. 9.

17 The nonrealism that I attribute to students and practitioners of the quantum theory requires more discussion and distinguishing of cases and kinds than I have room for here. It is certainly not the all-or-nothing affair I make it appear in the text. I carry out some of the required discussion in Fine (1986), ch. 9. My thanks to Paul Teller and James Cushing, each of whom saw the need for more discussion.

18 I should be a little more careful about the historical Bohr than I am in the text. For Bohr himself would seem to have wanted to truncate the homely line somewhere between the domain of chairs and tables and atoms, whose existence he plainly accepted, and that of electrons, where he seems to have thought the question of existence (and of realism, more generally) was no longer well defined. An illuminating and provocative discussion of Bohr's attitude toward realism is given by Paul Teller (1981). Thanks, again, to Paul for helping to keep me honest.

19 In this context, for example, van Fraassen's "constructive empiricism" would prefer the concept of empirical adequacy, reserving "truth" for an (unspecified) literal interpretation and believing in that truth only among observables. I might mention here that in this classification Putnam's internal realism comes out as antirealist. For Putnam accepts the core position, but he would add to it a Peircean construal of truth as ideal rational acceptance. This is a mistake, which I expect that Putnam will realize and correct in future writings. He is criticized for it by Horwich (1982) whose own "semantic realism" turns out, in my classification, to be neither realist nor antirealist. Indeed, Horwich's views are quite similar to what is called "NOA" below, and could easily be read as sketching a philosophy of language compatible with NOA. Finally, the "epistemological behaviorism" espoused by Rorty (1979) is a form of antirealism that seems to me very similar to Putnam's position, but achieving the core parity between science and common sense by means of an acceptance that is neither ideal nor especially rational, at least in the normative sense. (I beg the

reader's indulgence over this summary treatment of complex and important positions. I have been responding to Nancy Cartwright's request to differentiate these recent views from NOA.)

20 In his remarks at the Greensboro conference, my commentator, John King, suggested a compelling reason to prefer NOA over realism; namely, because NOA is less percussive! My thanks to John for this nifty idea, as well as for other comments.

21 "There is, I think, no theory-independent way to reconstruct phrases like 'really there'; the notion of a match between the ontology of a theory and its 'real' counterpart in nature now seems to me illusive in principle." T. S. Kuhn (1970), p. 206. The same passage is cited for rebuttal by Newton-Smith (1981). But the "rebuttal" sketched there in chapter 8, sections 4 and 5, not only runs afoul of the objections stated here in my first section, it also fails to provide for the required theory-independence. For Newton-Smith's explication of verisimilitude (p. 204) makes explicit reference to some unspecified background theory. (He offers either current science or the Peircean limit as candidates.) But this is not to rebut Kuhn's challenge (and mine); it is to concede its force.

22 No doubt I am optimistic, for one can always think of more to say. In particular, one could try to fashion a general, descriptive framework for codifying and classifying such answers. Perhaps there would be something to be learned from such a descriptive, semantical framework. But what I am afraid of is that this enterprise, once launched, would lead to a proliferation of frameworks not so carefully descriptive. These would take on a life of their own, each pretending to ways (better than its rivals) to settle disputes over truth claims, or their import. What we need, however, is less bad philosophy, not more. So here, I believe, silence is indeed golden.

Bibliography

Boyd, R. 1981. "Scientific Realism and Naturalistic Epistemology," in *PSA: 1980*, ed. P. Asquith and R. Giere, Vol. 2. East Lansing, MI: Philosophy of Science Association, pp. 613–62.

Boyd, R. 1984. "The Current Status of Scientific Realism," in J. Leplin (ed.), *Scientific Realism*, Berkeley: University of California Press, pp. 41–82.

Cartwright, N. 1983. *How the Laws of Physics Lie*. New York: Clarendon Press.

Earman, J., and C. Glymour. 1978. "Lost in the Tensors." *Studies in History and Philosophy of Science* 9: 251–78.

Earman, J., et al. 1983. "On Writing the History of Special Relativity. In *PSA: 1982*, Vol. 2, edited by P. Asquith and T. Nichols, 403–16. East Lansing, MI: Philosophy of Science Association.

Einstein, A., et al. 1952. *The Principle of Relativity*. Translated by W. Perrett and G. B. Jeffrey. New York: Dover.

Fine, A. 1975. "How to Compare Theories: Reference and Change." *Nous* 9: 17–32.

Fine A. 1981a. "Conceptual Change in Mathematics and Science: Lakatos' Stretching Refined." In *PSA: 1978, Vol. 2*, edited by P. Asquith and I. Hacking, 328–41. East Lansing, Mich.: Philosophy of Science Association.

Fine, A. 1982d. "Antinomies of Entanglement: The Puzzling Case of the Tangled Statistics." *Journal of Philosophy* 79: 733–47.

Fine, A. 1986. "Unnatural Attitudes: Realist and Instrumentalist Attachments to Science." *Mind*.

Heisenberg, W. 1925. "Über quantentheoretische Umdeutung kinematischer und mechanischer Beziehungen." *Zeitschrift für Physik* 33: 879–93.

Hoffmann, B. 1972. *Albert Einstein, Creator and Rebel*. New York: Viking Press.

Holton, G. 1973. *Thematic Origins of Scientific Thought*. Cambridge: Harvard University Press.

Horwich, P. 1982. "Three Forms of Realism." *Synthese* 51: 181–201.

Jones, R. 1991. "Realism about what?" *Philosophy of Science* 58(2): 185–202.

Kuhn, T. S. 1970. *The Structure of Scientific Revolutions*. 2nd edn. Chicago: University of Chicago Press.

Landé, A. 1974. "Albert Einstein and the Quantum Riddle." *American Journal of Physics* 42: 459–64.

Laudan, L. 1984. "A Confutation of Convergent Realism," in J. Leplin (ed.), *Scientific Realism*, Berkeley: University of California Press.

Lycan, W. 1982. "Epistemic value." Preprint.

Miller, A. 1981. *Albert Einstein's Special Theory of Relativity*. Reading, MA: Addison-Wesley.

Newton-Smith, W. H. 1981. *The Rationality of Science*. London: Routledge and Kegan Paul.

Niiniluoto, I. 1982. "What shall we do with verisimilitude?" *Philosophy of Science* 49: 181–97.

Pitt, J. 1981. *Pictures, Images and Conceptual Change*. Dordrecht: Reidel.

Popper, K. 1972. *Conjectures and Refutations*. London: Routledge and Kegan Paul.

Putnam, H. 1981a. *Reason, Truth and History*. Cambridge: Cambridge University Press.

Rorty, R. 1979. *Philosophy and the Mirror of Nature*. Princeton: Princeton University Press.

Stachel, J. 1979. "The Genesis of General Relativity," in *Einstein Symposium Berlin*, ed. H. Nelkowski, pp. 428–42. Berlin: Springer-Verlag.

Teller, P. 1981. "The Projection Postulate and Bohr's Interpretation of Quantum Mechanics," in P. Asquith and R. Giere (eds.), *PSA: 1980*. Vol. 2. East Lansing, MI: Philosophy of Science Association, pp. 201–23.

van Fraassen, B. 1980. *The Scientific Image*. Oxford: Clarendon Press.

van Fraassen, B. 1982. "The Charybdis of Realism: Epistemological Implications of Bell's Inequality." *Synthese* 52: 25–38.

Weinberg, S. 1972. *Gravitation and Cosmology: Principles and Applications of the General Theory of Relativity*. New York: Wiley.

Wessels, L. 1979. "Schrödinger's Route to Wave Mechanics." *Studies in History and Philosophy of Science* 10: 311–40.

Woolf, H., ed. 1980. *Some Strangeness in the Proportion*. Reading: Addison-Wesley.